ersity

AF606187

UNIVERSITY OF NORTHUMBRIA
AT NEWCASTLE LIBRARY
WITHDRAWN

Analytical Toxicology for Clinical, Forensic and Pharmaceutical Chemists

CLINICAL BIOCHEMISTRY

PRINCIPLES · METHODS APPLICATIONS

5

Series Editors
H.Ch.Curtius
M.Roth

W DE G

Walter de Gruyter
Berlin · New York
1997

Analytical Toxicology
for Clinical, Forensic and Pharmaceutical Chemists

Editors
H. Brandenberger
R. A. A. Maes

Walter de Gruyter
Berlin · New York
1997

Editors

Professor Dr. Hans Brandenberger
Former Head of the Department
of Forensic Chemistry and
Professor of Chemical Toxicology
at the University of Zürich
Chalet Biene
CH-6356 Rigi Kaltbad
Switzerland

Dr. Robert A.A. Maes
Professor of Human Toxicology
Netherlands Institute for
Drugs and Doping Research
University of Utrecht
Sorbonnelaan 16
NL-3584 CA Utrecht
The Netherlands

Series Editors

Professor Dr. Hans-Christoph Curtius
Former Head of the Division of Clinical
Chemistry
Department of Pediatrics
University of Zürich/Switzerland
Gut Weiherhof
D-78315 Radolfzell
Germany

Professor Dr. Marc Roth
Former Head of the Central Laboratory for
Clinical Chemistry, now at the
Department of Physical Medicine and
Reeducation, University Hospital Geneva
Beau Séjour
CH-1211 Geneva 14
Switzerland

Publishers

Walter de Gruyter & Co.
Genthiner Strasse 13
D-10785 Berlin
Germany

Walter de Gruyter, Inc.
200 Saw Mill River Road
Hawthorne, N.Y. 10532
USA

With 164 figures, tables and formulas.

Library of Congress Cataloging-in-Publication Data

Analytical toxicology : for clinical, forensic, and pharmaceutical
chemists / editors, H. Brandenberger, R.A.A. Maes,
XXVII; 735 p. 24 × 17 cm. – (Clinical biochemistry ; 5)
Includes index.
ISBN 3-11-010731-7 (alk. paper)
1. Analytical toxicology. I. Brandenberger, H. (Hans) II. Maes,
R.A.A. (Robert A.A.), 1937 III. Series.
[DNLM; 1. Toxicology–methods. QV 602 A5327 1997]
RA1221.A49 1997
615.9'07–dc21
DNLM/DLC
for Library of Congress

Die Deutsche Bibliothek – Cataloging-in-Publication Data

Analytical toxicology for clinical, forensic and pharmaceutical chemists / ed. Hans Brandenberger ; Robert A.A. Maes. – Berlin ;
New York : de Gruyter, 1997
(Clinical biochemistry ; 5)
ISBN 3-11-010731-7

∞ Printed on acid-free paper which falls within the guidelines
of the ANSI to ensure permanence and durability.

Typesetting/Printing: TUTTE Druckerei GmbH, Salzweg–Passau.
Binding: Lüderitz & Bauer GmbH, Berlin – Cover Design: Hansbernd Lindemann, Berlin.

Editorial

Analytical toxicology is a rather complex science. Chemists venturing into this field must possess a high degree of flexibility. They should not only have an excellent theoretical and technical knowledge of analytical chemistry, but also be prepared to look beyond the doors of their laboratories, to ensure that the submitted requests agree with the problematic, and that the results of their investigations will be interpreted correctly.

The analytical part of the work of a chemical toxicologist, which takes up his main time and effort, is complicated by several circumstances:

- Most specimens submitted for analysis are of biological nature (body fluids, tissue samples) and possess a complex matrix.
- The poisons or drugs which must be detected and quantified may be present only in minute amounts, so that microanalytical techniques have to be used.
- Often, goal and direction of a search are not clearly determined before the start of the investigation (searches for "the general unknown"). They may only become obvious during the actual work, so that strategy and pathways must progressively be adapted to the new situation.
- Toxic effects can be caused by an enormous variety of substances: metal ions, anions, gases, solvents, chemical intermediates, many classes of differently structured pesticides and drugs, as well as substances of natural origin. An analyst specialized in only a few of these fields is not the ideal toxicologist. All-rounders are needed, capable of tackling inorganic and organic problems, coping with gases and solvents, as well as with compounds possessing only minute volatility.

But even the best analytical all-rounder is not an ideal toxicologist, if he does not place his laboratory studies into a larger context. He can obtain his commission from a physician or a hospital (in clinical toxicology) or from a judge or attorney (in forensic toxicology). In both cases, the commissioner may not always be fully aware of the possibilities and difficulties inherent in the requested laboratory investigation. Furthermore, his knowledge of the dissemination and action of poisons may also be insufficient to judge all intoxication possibilities. The toxicologist must therefore possess the initiative to call for the case documentation (anamnesis, clinical or pathological data), to study this information and discuss it with the mandator, should a rectification of the analytical request seem desirable. But he can only do this, if he is well-informed about the distribution of poisons in the environment, their uses in our society, as well as their action on the body. This knowledge is likewise needed for the interpretation of the analytical data and for a correct and useful reporting of the results. In this context, a solid understanding of the resorption pathways of poisons, their metabolic fate and their excretion is equally indispensable.

Our book is intended to help a toxicologist with both aspects of his work, with the actual analytical investigation, as well as with the "side-lines" of his job, the "take-over" of an investigation from a medical or forensic authority, the interpretation of the analytical data and communication of results to the mandator.

The bulk of the book is divided into 3 main parts. Part 1 contains chapters of general nature. Various branches of toxicology are described, the history of toxicological analysis is briefly reviewed, and outlooks into new methodological possibilities made. We have abstained from including chapters describing different analytical techniques, since this information can be found in many chemical textbooks. Methods which are of special importance to toxicologists and may not be adequately treated in general analytical texts are discussed in some of the chapters in Parts 2 and 3 of the book. So contain for example:

- chapter 2.3 on ethyl alcohol and related solvents a discussion of different gas chromatographic procedures, as well as a computer program for data processing in alcohol analysis,
- chapter 2.10 on toxic metals considerable information on the role of atomic absorption analysis in inorganic toxicology,
- chapter 3.1 on barbiturates a detailed description of liquid-liquid extraction procedures in drug analysis, controlled by UV-screening of extracts,
- chapter 3.9 and 3.10 many examples for the use of mass spectrometry in the identification of drugs and metabolites,
- chapters 3.11, 3.15 and 3.16 a wealth of information on the different possibilities for immunochemical analysis.

In Part 2 of the book, the chapters dealing with the most important types of poisons have been assembled, while the main classes of drugs with toxicological importance are discussed in Part 3. This subdivision, however, has not always been observed. Chapter 2.5 on volatile chlorinated hydrocarbons presents mainly toxicological information, with emphasis on compounds of special importance in the field of medicine, while chapter 3.10 deals also with the environmental and chemical aspects of volatile halogenated hydrocarbons and includes a large number of solvents not used as anesthetics. The 2 chapters complement each other and should be studied together.

Part 4 of the book contains an up-dated list of clinical and toxicological blood drug levels, and will certainly be a welcome help for the interpretation of analytical data.

We have tried to avoid overlaps between the different chapters, unfortunately not always with success, as may be expected from a book written by a large number of authors.

We hope that the book will be of use to clinical and forensic toxicologists as well as to pharmacists involved in analytical work. The critical evaluations of the analytical possibilities may help with the choice of methods, the additional information assist in determining the goals of an investigation and in interpreting its results. – Our book is not a "cookbook" (which could anyway only have very restricted importance in the vast field of toxicology). It also does not try to list all substances of toxicological importance. Such volumes (i.e. the Merck Index and Clarke's Isolation and Identification of Drugs) already have their established place on the desk of every toxicologist and could not be replaced. Our new "Analytical Toxicology" is intended as a textbook for younger toxicologists, and as a critical review and stimulus for more experienced colleagues in the field. In addition to all analytical and toxicological information presented, it tries to teach the analytical toxicologists to see and place their technical obligations in a larger context.

The editors have to thank the authors for their manuscripts. The senior editor is indebted to Roberta Hanson-Brandenberger for her help with the literature search, for correcting all his manuscripts, and for her useful and welcome criticism.

The Editors

Contributors

Douwe de Boer, Dr.
Netherlands Institute for Drugs
and Doping Research
Utrecht University
Sorbonnelaan 16
NL-3584 CA Utrecht
The Netherlands

Robin A. Braithwaite, PhD
Laboratory for Toxicology
City Hospital, NHS Trust
Dudley Road
GB-Birmingham B18 7QH
Great Britain

Hans Brandenberger, Professor Dr.
formerly at the
Department of Forensic Chemistry
University of Zürich,
private address:
Chalet Biene
CH-6356 Rigi Kaltbad
Switzerland

Robert Brandenberger, Professor Dr.
Department of Physics
Brown University
Providence, RI 02912
USA

Robert J. Flanagan, PhD
National Poison's Unit
Guy's Hospital, NHS Trust
Avonley Road
GB-London SE14 5ER
Great Britain

Albert D. Fraser, Professor, PhD
Clinical & Forensic Toxicology
Dalhousie University and
Victoria General Hospital
1278 Tower Road
Halifax, Nova Scotia B3H 2Y9,
Canada

Marika Geldmacher-v. Mallinckrodt
Professor Dr. Dr.
formerly at the
Institute for Legal Medicine
University of Erlangen-Nürnberg
private address:
Schlehenstraße 20,
D-91056 Erlangen
Germany

Käthy Halder, Dr.
formerly at the
Department of Forensic Chemistry
University of Zürich
private address:
Englischgruss-Str. 5
CH-3902 Brig-Glis,
Switzerland

Jürgen Hallbach, Dr.
Institute for Clinical Chemistry
City Hospital München-Bogenhausen
Englschalkinger Str. 77
D-81925 München
Germany

Gary L. Henderson, PhD
Department of Pharmacology
and Toxicology
School of Medicine
University of California
Davis, CA 95616
USA

David W. Holt, PhD
The Analytical Unit
Department of Cardiological Sciences
St. George's Hospital, Medical School
GB-London SW17 ORE
Great Britain

Kojiro Kimura, Professor Dr.
Department of Legal Medicine
School of Medicine
Shimane University
Enya Chour 89-1, Izumo
Shimane 693
Japan

Tohru Kojima, Professor Dr.
Department of Legal Medicine
School of Medicine
Hiroshima University
1-2-3 Kasumi-cho
Minami-ku
Hiroshima 734
Japan

Wolf-Rüdiger Külpmann,
Professor Dr.
Institute for Clinical Chemistry I
Medical University of Hannover
Konstanty-Gutschow-Strasse 8
D-30625 Hannover
Germany

Yukio Kuroiwa, Professor Dr.
School of Pharmaceutical Sciences
Showa University
1-5-4 Hatanodai
Shinagawa
Tokyo 142
Japan

Miriam S. Leloux, Dr.
Central Research
AKZO-Nobel Laboratories
Postbus 9300
NL-6800 SB Arnhem
The Netherlands

Günter Machbert
Professor Dr.
Institute for Legal Medicine
University of Erlangen-Nürnberg
Universitätsstr. 22
D-91054 Erlangen
Germany

Robert A. A. Maes, Professor Dr.
Netherlands Institute for Drugs
and Doping Research
Utrecht University
Sorbonnelaan 16
NL-3584 CA Utrecht
The Netherlands

Masataka Nagao, Professor Dr.
Department of Forensic Medicine
University of Tokyo
Bunkyo-ku
Tokyo 113
Japan

Takeaki Nagata, Professor Dr.
formerly at the
Department of Forensic Medicine
Faculty of Medicine
Kyushu University
private address:
3-15-15 Kashiwara, Minami
Fukuoka 815
Japan

Michel Pelletier, Dr.
Department of Mineral, Analytical and
Applied Chemistry
University of Geneva
Quai Ernest-Ansermet 30
CH-1211 Geneva 4
Switzerland

Marc Roth, Professor Dr.
former Head of the Central Laboratory
for Clinical Chemistry
now at the Department of Physical
Medicine and Reeducation
University Hospital Geneva
Beau Séjour
CH-1211 Geneva 14
Switzerland

Maurice Schmidt, Dr.
City Hospital La Chaux-de-Fonds
CH-2300 La Chaux-de Fonds
Switzerland

Harald Schütz, Professor Dr.
Institute for Legal Medicine
Justus-Liebig-University Giessen
Frankfurter Str. 58
D-35392 Gießen
Germany

Thérèse J.A.
Seppenwoolde-Waasdorp
Netherlands Institute for Drugs
and Doping Research
Utrecht University
Sorbonnelaan 16
NL-3584 CA Utrecht
The Netherlands

Takehiko Takatori, Professor Dr.
Department for Forensic Medicine
University of Tokyo
Bunkyo-ku
Tokyo 113
Japan

Donald R. A. Uges, Dr.
Laboratory of the Department
of Pharmacy
University Hospital of Groningen
Postbus 30.001
Oostersingel 59
NL-9700 RB Groningen
The Netherlands

Hermann Vogel, Dr.
Institute for Clinical Chemistry
City Hospital München-Bogenhausen
Englschalkinger Str. 77
D-81925 München
Germany

Tom B. Vree, Dr.
Departments of Clinical Pharmacy
and Anesthesiology
Hospital Nijmegen
Geert Grooteplein, Zuid 8,
NL-6525 GA Nijmegen
The Netherlands

Robin Whelpton, PhD
Department of Pharmacology
Queen Mary & Westerfield College
Mile End Road
GB-London E1 4NS
Great Britain

Bryan Widdop, PhD
National Poisons Unit
Guy's and St. Thomas's NHS Trust
Avonley Road
GB-London SE14 5ER
Great Britain

Takemi Yoshida, Professor Dr.
School of Pharmaceutical Sciences
Showa University
1-5-4 Hatanodai, Shinagawa
Tokyo 142
Japan

Contents

Part 1
Chapters of General Nature

1.1 Emergency Toxicology

M. Schmidt and M. Roth

1.1.1 Introduction

Intoxications have existed ever since the origin of humanity. An empirical knowledge of poisons of natural origin has enabled earlier populations to avoid to some extent the troublesome consequences of poisoning. Conversely, this knowledge has been used to kill enemies. Actually, the first important developments in the field of the identification of poisons occurred during the 16th century, when it became possible to provide analytical expertise against the authors of criminal poisoning.

On the other hand, clinical toxicology, the purpose of which is to establish an appropriate treatment of the intoxicated patient, is a field of recent origin. Its development has been fostered by the fact that a great number of compounds successfully used in industry and medicine are definite poisons if ingested by mistake, overdose or suicidal attempt.

1.1.2 Role and Location of the Laboratory

The role of the toxicology laboratory is not always well understood by the clinical staff, since in many situations a trained physician is able to define the diagnostics on the basis of the clinical status of the patient and the events observed. In-as-much as the treatment is then done with a reasonable chance of success by a standard methodology, the analytical identification and quantification of the absorbed toxin in biological fluids is considered to have but a small impact on the clinical decisions. A still widespread opinion is that the effect of the analytical results is merely to confirm the opinion of the medical staff and to document the pertinence of the diagnosis.

This point of view is progressively being challenged by the progress of analytical toxicology. In certain important medical centres, the laboratory is now able to identify rapidly the absorbed poisons and to provide data on their concentrations. It then becomes possible to adjust the treatment much closer to what is actually needed, and to properly decide whether or not the patient has to be submitted to such additional measures as hemodialysis, hemoperfusion, alkalinization, etc. If an antidote exists, fast laboratory results also help to decide its administration and to determine the dosage.

Ideally, the toxicological analysis must contribute rapidly to establish the diagnosis and eventually helps to manage the therapy.

Even the knowledge that toxic substances are absent is an important information which may lead the clinician to investigate other causes of a physical or mental alteration of a patient.

The above facts show that the clinical laboratory must provide a fast response about the identity and concentration of toxic substances. This is an emergency service which has to operate night and day.

Historiocally, forensic chemists were the first to provide this type of service to hospitals. Forensic chemistry, however, differs from clinical chemistry as its results must have legal strength. Confirmation of the results by an independent method is much more important than speed. Today, if a forensic laboratory wants to provide service to a hospital, it should have a special section working on a different basis than the forensic section.

In large medical centres, it is preferable to install a toxicology laboratory within the hospital itself. This will usually be a section of the central laboratory of clinical chemistry. This location has the advantage of ensuring a better contact with the rest of clinical chemistry, the results of which are also needed in cases of intoxication, and with the clinical staff.

But it may also be worthwile to compare clinical chemistry with analytical toxicology, just to identify the difficulties in analytical toxicology:

Clinical chemistry:
- 20 to 50 compounds tested,
- many routine kits available,
- automation usual, if size of laboratory is large enough.

Analytical toxicology:
- 8000 pharmaceuticals are currently used,
- 40000 other potentially toxic substances exist,
- routine kits available only for screening for some common drugs,
- automation is rarely possible.

Once a clinical toxicology laboratory is capable of working with high efficiency, its reputation will increase beyond the hospital. Its scope may then extend to a whole region, making the service available to smaller hospitals. With a good transportation system, this may be more rational than several smaller laboratories working just for themselves.

1.1.3 Types of Intoxication

The strategy to be used in the analytical process depends on the type of intoxication. Two main groups exists, which are:

1.1.3.1 Intoxications of known origin

Inhalation of carbon monoxide and cyanide during a fire, absorption of industrial products by accident, administration of an excess of a drug during a treatment (for example within the preliminary phase when the proper posology is investigated) are examples of situations where a specific method of assay is applicable. Any good laboratory of clinical chemistry possessing a relatively simple instrumentation is able to perform qualitative analysis with direct methods (without isolation and purification): spot tests, immunoassays (enzyme immunoassays, fluorescence polarisation immunoassays, immunoturbidimetry, etc.), photometry (CO-oximeter), flame photometry (lithium), etc.

1.1.3.2 Intoxications of unknown origin

These are essentially intoxications by drugs, mainly in a suicidal attempt. Occasionally, intoxications by accident, when the patient is in a coma or unable to speak, also belong to this group.

These have to be systematically treated as intoxications of unknown origin, even in the presence of indices such as empty vials or declarations from the patient or accompanying persons. This attitude is dictated by the fact that the information is not always available (unconscious patient) and that any information received, which can hardly be verified, may be wrong or incomplete. According to Bailey (1, 2), in a quarter of the cases a substance that could not be suspected from the available information is likely to be found, and in every sixth case only substances that had not been suspected are actually detected.

This group of intoxications belongs to the domain of a toxicology laboratory equipped with sophisticated instruments allowing for the systematic investigations of unknown compounds, by means of more complicated methods, including: extraction, isolation, clean-up and concentration steps followed by: thin layer chromatography (TLC), gas chromatography (GC), high performance liquid chromatography (HPLC), gas chromatography coupled with mass spectrometry (GC-MS), this one in single ion monitoring (SIM) mode for quantitative analysis.

1.1.4 Causes of Intoxications

1.1.4.1 Legal drugs

In view of the type of most intoxications, it must be kept in mind that many legal substances, such as ethanol and therapeutic drugs (mainly benzodiazepines, antidepressants and mixed analgesics), are commonly abused. They are by far the main cause of intoxication observed in a hospital emergency room. Overdose of therapeutic drugs e.g. antiarrhythmics (digitalis) can induce fatal intoxications. Table 1 shows the distribution of positive results in the intoxication searching (Hôpital cantonal universitaire, Geneva, Switzerland, 1991–1992).
Suitable techniques of assay should be available both for identification and quantitative assay of those compounds. Table 2 shows a list of tests which are essential for a primary screening.

Table 1. Relative frequency of the causes of intoxication observed in Geneva. The total exceeds 100% as certain patients ingested more than one compound

Compound or class of compounds	Relative frequency (± S.D.)*
Ethanol	32% (± 1)
Benzodiazepines	60% (± 7)
Tricyclic antidepressants and phenothiazines	13% (± 7)
Barbiturates and related compounds	6% (± 2)
Acetaminophen (paracetamol)	1% (± 1)
Salicylates	2% (± 1)
Miscellaneous	6% (± 3)

* Standard deviation was evaluated on 1200 requests

Table 2. Tests to be available for primary screen in emergency toxicology

Compound	Preferred sample	Identification and quantification preferred methods	Remarks
Blood screen			
Volatiles (ethanol, methanol, isopropanol, acetone etc.)	Whole blood preserved with sodium fluoride	GC (head-space) Enzymatic test (ethanol only)	
Pharmacopea drugs (most CNS-stimulants, psychomimetics, tranquillisers, analgesics, antidepressants, etc.)	Serum	GC-MS, GC-MS (SIM mode), GC, HPLC, some immunoassays	Some compounds (barbiturates) need derivatization for better determination.
Digoxin	Serum	Immunoassay	
Lithium	Serum	Flame photometry	
Carboxyhaemoglobin	Whole blood (heparin)	Spectroscopy (COHb), micro diffusion, head-space GC (CO)	
Urine screen			
Illegal drugs	Random urine	Immunoassays, GC-MS for confirmation, GC, quantitative results not significant in most cases	
Other drugs	Random urine	Spot tests, GC-MS, GC, quantitative results not significant in most cases	
Gastric fluid screen			
All compounds	Emesis, lavage	Spot tests, GC-MS, GC, quantitative results not significant in most cases	Metabolites are absent. Spot test for salicylates has to be done systematically (delayed absorption).

1.1.4.2 Illegal drugs

The problem of drug abuse is not new, but the past 20 years have seen the number of drug addicts increase so dramatically that we can no longer consider this as a marginal side effect of our society.

In Europe (3), the drugs most often encountered are: amphetamines, barbiturates, benzodiazepines, cocaine, cannabinoids and opiates. Methadone is encountered when the patient is under treatment. Other drugs, like hallucinogens, "designer drugs", "poppers", solvent (glue-sniffing) and other volatile substances are sometimes found. Phencyclidine is very rarely found.

1.1.4.3 Other substances

If a substance other than a drug is supposed to be the cause of an intoxication, one has to take into consideration the identification of products used in industry (metals, solvents, various alcohols, antifreeze, etc.) or agriculture (pesticides, herbicides) as well as vegetable toxins according to the alimentary habits of certain regions (e.g. intoxication through the ingestion of amanitine contained in the mushroom Amanita phalloides).

Table 3 shows the tests that should be available for secondary oriented toxicology screening. They are not systematically performed, but are only done on specific grounds.

Table 3. Tests to be available for oriented screen in emergency toxicology

Compound	Preferred sample	Intoxication origin	Preferred technique
Cyanide	Whole blood (heparinized)	Fire, suicide, potassium ferrocyanide treatment	microdiffusion + colorimetry, head-space GC
Ethylene glycol (antifreeze)	Serum	Abuse (with urinary excretion of oxalate)	GC
Cholinesterase activity	Whole blood (heparinized)	Organophosphates (pesticides)	Photometry
Bromide, fluoride	Serum	Industry	Selective electrode, ion chromatography
Pesticides (paraquat, diquat)	Whole blood (heparinized), random urine	Agriculture	Colour tests, GC, HPLC
Halogenated hydrocarbons (solvents, chloral hydrate)	Whole blood, random urine, gastric fluid	Industry, abuse	Colour test, GC
Heavy-metals screen	Whole blood (heparinized), random urine	Industry, accidental (children)	Atomic absorption, flame emission
Methaemoglobin	Whole blood (heparinized)	Nitrate and nitrite exposure, abuse of certain vasodilators ("poppers")	Spectroscopy, colour test, GC

1.1.5 Request

Since rapidity of the response is an essential goal for the toxicology laboratory, it is important that the clinician tells what problem he is faced with, namely whether the intoxication is of known or unknown origin.

1.1.5.1 Data and indices

Three types of data can be distinguished:

1. Substance oriented data, such as chemical, physical and biological properties (substance data), and behavior in screening and identification procedures (analytical data).
2. Clinical data, such as toxicokinetic, therapeutic and toxic blood levels, clinical course with and without therapy.
3. Clinical status of the patient: coma, shock, neurological disorders, cardiac disorders, etc., possible treatment before sampling, kind of sample, usual therapy if it exists, suspected substances, etc. A good transfer of information between the clinic and the laboratory, and vice-versa, is quite helpful in cases of severe intoxication.

Successful use of the toxicology laboratory requires the input of certain basic informations to direct the toxicologist in his search for toxic agents. When ordering a toxicology screen, the clinician should provide at least the following information to the laboratory in written form:

1.1.5.2 Clinical status of the patient

The knowledge of the level of consciousness of the patient may help the laboratory to decide which screens are most urgent, if multiple requests are simultaneously received. This may be summarised in a few words: coma, shock, confuse or waking.

Other informations like neurological disorders, cardiac or respiratory perturbations are sometimes useful. Any treatment undertaken before the sampling (gastric washing, antidote or drug administration) should be specified.

1.1.5.3 Pathological indices

This is of special interest to the laboratory. The presence of a metabolic acidosis, of an anion gap, of an osmolality gap or hyperglycaemia (for example in the case of acetone positive finding suggesting a diabetic coma) may help the toxicologist to grasp the problem.

Existing pathology e. g. renal diseases or liver injury or paratoxic situations like heat or cold, can influence the toxicity of the absorbed drugs and must be taken in account if known.

1.1.5.4 Other indices

Some informations about the case history, like data on previous treatments, usual drug therapy or addiction if known, and the suspected agent(s) with dose, time of ingestion and time of sampling, are quite useful.

Additional informations such as finding of empty drug vials after a suicidal attempt and indications from the family or the neighbours could be useful but must be always interpreted with caution. All data concerning the suspected cause of the intoxication, if available, are useful as a guide and a help for interpretation, but are in no case a substitute for laboratory results.

1.1.6 Sampling

Many toxic compounds are conveniently assayed in blood serum, with the advantage that a quantitative determination may give an indication on the severity of the situation. However, some substances ("street drugs" for example) rapidly leave the blood circulation to reach adipose tissues or urine, so that their concentration in blood is too low to be detected by standard procedures, whereas an identification in urine is quite possible. The most abundant compound in urine may be a metabolite rather than the native substance. However, requests for toxicology screening should be accompanied by four basic sample types whenever possible:

1.1.6.1 Blood

Blood (as whole blood, plasma or serum) is the only sample type which allows quantitative results from which the degree of the intoxication can be estimated. Whole blood anti-coagulated with potassium oxalate and preserved with sodium fluoride is a convenient material for identifying and determining alcohol and other volatile compounds. Heparinized blood is suitable for the assay of carbon monoxide in the hemolysate. Most assays, as for drugs and medicines, are usually performed on serum.

1.1.6.2 Random urine

Spot urine samples are used most often. If a relatively large volume of urine may be sampled, this has the advantage of increasing the sensitivity of the procedures, although the presence of metabolites often complicates the problem. Quantitative urine levels cannot be used to determine an influence of drugs on behaviour. The presence of a drug in a specimen of urine just means that it has been ingested or administered.

1.1.6.3 Gastric content

This is obtained through vomiting or aspiration. The sample may contain large quantities of the native(s) compound(s), allowing for straight identification. The analysis usefully complements the results obtained from blood or urine, especially in cases where absorption is still low. Stomach washing may in some instances favour absorption, as for example in the case of salicylates, which must be systematically tested in this sample by a spot test.

1.1.6.4 Other specimens

Any materials (tablets, powders, liquids, solvent containers, tablet bottles, syringes etc.) found with the patient should be retained for possible toxicological examination. Pills, tablets and other pharmaceutical preparations may be identified by computerised retrieval of data such as colour, shape, size, weight, etc. Although this is not the primary task of a toxicological laboratory, it may in some instances help to rapidly identify or confirm the cause of an intoxication.

1.1.6.5 Specimen collection

For an adult patient in whom general toxicological investigations are required, table 4 lists the types of samples to be sent to the toxicology laboratory. Whenever possible, the specimens should be obtained before further drug administration takes place to avoid any interference in the analyses.

Table 4. Samples needed for toxicology screening

N°	Sample	Volume	Analysis	Action
Sample 1	Whole blood (anticoagulated) preserved with fluoride	2 ml	Alcohols	Head-space GC
Sample 2	Serum (from native blood)	5–10 ml	Drug screening, and if suspected: digoxin, lithium	Extraction, GC-MS screening, or immunoassay (digoxin) or flame photometry (lithium)
Sample 3	Whole blood (heparinized)	2 ml	Carboxyhaemoglobin if suspected	Photometry
Sample 4	Random urine	20–50 ml	Drug screening	Spot tests (salicylates, etc.), immunoassays (drugs of abuse), extraction and GC-MS (confirmation)
Sample 5	Gastric fluid	20–50 ml	Drug screening	Spot tests (salicylates, etc.), visual examination (if solid matter as pills, capsules, etc. is present), extraction, GC-MS

1.1.7 Transport and Preservation of Samples

Since the analysis has to be done rapidly, no preservative is normally added to the samples. If the laboratory works on a regional basis, transport from remote locations may have to be done by a special messenger.

1.1.8 Analytical Techniques Used in Emergency Toxicology

In emergency toxicology, the primary question asked by the clinician is "what was ingested?" with secondary emphasis on "how much is present?".

The systematic search for the cause of intoxications should primarily include alcohol, the pharmacopoeia drugs most often used, and illegal drugs.

1.1.8.1 Spot tests

Some presumptive colour tests are quite useful to confirm the presence or the absence of certain groups of compounds, particularly in samples such as urine or gastric fluid (4, 5):

Table 5. Qualitative spot tests for drugs and other poisons

Compound	Sample	Analytical principle	Approximate limit of sensitivity
Paracetamol	Urine, gastric aspirate	Hydrolysis to p-aminophenol to form blue indophenol dye with o-cresol-ammonia	30 µmol/l (5 mg/l)
Paraquat	Urine, gastric aspirate	Forms blue compound with alkaline dithionite (diquat forms yellow-green compound)	5 µmol/l (1 mg/l)
Salicylates	Plasma, serum, urine or gastric aspirate	Form purple colour with ferric salts	360 µmol/l (50 mg/l)
Trichloro-compounds (e.g. chloral hydrate)	Urine, gastric aspirate	Fujiwara reaction[5]	5 µmol/l (1 mg/l)

1.1.8.2 Immunoassays

The development of fast and easy immunological techniques for assaying drugs had led many clinical chemical laboratories to offer a list of drugs they accept to assay, either quantitatively or not. This has resulted in costly requests in which the different drugs are asked for assay in a screening manner. The importance of being aware of the limited usefulness of immunoassays cannot be overemphasized:

- not all drugs in the same drug family are covered by cross reactivities in the assay,
- detection limits are not all the same within one drug family,
- difficulties of interpretation may occur with some unexpected cross reactivities.

Results from such investigations have to be interpreted with some caution. And since the sensitivity of such tests is sometime insufficient, a "negative" result does not mean that the compound under consideration is actually absent from the sample.

1.1.8.3 Gas chromatography (GC)

The analytical toxicologist faces a formidable task: to detect and identify compounds, to differentiate between closely resembling ones, and to determine the concentration of a compound among thousands of substances.

Fortunately, there are powerful analytical methods available to approach these problems. One of the most useful and frequently applied techniques is gas chromatography, either alone or in combination with other methods. Gas chromatography provides the retention time or the retention index of a substance, which can be used for its identification, and the size of the signal or peak which serves to determine the amount of the substance. A certain specificity of detection can be achieved by the use of specific detectors (flame ionisation detector, electron capture detector for halogen-containing substances, nitrogen-phosphorus detector). Tables of retention indices are available (6, 7).

However, the consultation of the tables is not sufficient to identify a compound with certainty. The confirmation must be done by the chromatography of the suspected compound in the same way. The limitation of this method is therefore that only the substance suitable as reference in the system can be identified.

Unfortunately, GC has certain limitations in that it is not well suited to handle compounds that are of low volatility and/or thermally labile. Although in many cases these problems can be overcome by derivatization (8, 9), it is sometimes difficult to obtain a suitable derivative.

1.1.8.4 Gas chromatography-mass spectrometry (GC-MS)

The development of low-cost bench top mass-selective detectors made the use of mass-spectrometry more popular among the analytical toxicology laboratories. The easy-to-learn software and efficient data processing now available have made this technique quite accessible for non-academic technicians. Combined GC-MS using electron impact (EI) or positive and negative chemical ionisation (Cl^+/Cl^-) has become an invaluable technique for screening toxicological samples (10). Identification of spectra is facilitated by the availability of many reference collections which may be accessed by computer library matching programmes. The most useful for the toxicologist are computerised full-scan mass spectral data of drugs and their metabolites (11), and other substances (12).

The quantitative determinations can be performed by GC-MS in selective ion monitoring mode (SIM), where the mass spectrometer continuously monitors a few preselected ion masses characteristic of the analyte of interest. The scanning time for any single ion is long, and a higher signal than in full-scan conditions (where the recording time for each ion is short), is obtained. The ion current recorded by the detector is then proportional to the amount of the selected mass. A calibration curve with known quantities of the compound can be done and a quantitative determination can be performed. The sensitivity of the mass spectrometer in SIM mode is increased about 10 to 100 fold in comparison with the full-scan mode.

The use of a mass spectrometer as a detector for HPLC provides a very specific method for the characterisation of compounds which are not amenable to GC. However, adequate interfaces are necessary, and too complicated mobile phases are to be avoided, limiting the application of the LC-MS to very "clean" samples.

1.1.8.5 Liquid chromatography (LC and HPLC)

For most routine drug assays, immunoassay is the predominant methodology. However, liquid chromatography is used for measurement of drugs not readily measured by immunoassay or as reference method (13).

With HPLC, the main problems associated with the use of gas chromatography, as thermal stability, molecular mass, polarity of the compounds, are avoided. In addition, several drugs and many metabolites can be detected simultaneously.

The use of HPLC in toxicology screening suffers from some limitations. The chromatographic conditions (stationary phase type, mobile phase composition, detector type, temperature) are generally well-defined for the analysis of each class of compounds (e. g. barbiturates vs. antidepressants), and are not easily transposable. In addition, the main mode of identification is the retention index, which has the same limitations as in gas chromatography.

The use of a more specific detector, such as the diode-array absorbance detector, combined to the retention data (14), can improve the quality of the identification. But the discriminating properties of an UV scan are not as effective as a mass spectrum, and the number of compounds which are actually identified is low (a few hundred) compared to the mass spectra libraries (more than 150000).

1.1.8.6 Thin layer chromatography (TLC)

Some papers have discussed the use of standardised TLC systems in toxicological analysis (15, 16). This technique is very efficient, particularly because it gives an immediate vision of the presence or absence of drugs, but it suffers from lack of sensitivity for the detection of many modern drugs which are active at low concentrations.

1.1.8.7 Preparation of the sample

Biological samples are very complex multi-component mixtures. Drugs and other toxic substances are usually present in these samples only as minor components. Some assays (the most part of the immunoassays, carboxyhaemoglobin, lithium, etc.) do not need any special treatment. On the other hand the sample preparation for all the types of chromatographic analysis is a critical step.

Extraction, clean-up or selective isolation of the relevant compounds from the samples and preconcentration procedures are necessary prior to separation and measurement. Solvent extraction, solid-phase extraction and related techniques are used (17, 18).

1.1.9 Expression of Results and Interpretation

1.1.9.1 Identification of substances

A specialised laboratory working with chromatography and mass spectrometry can communicate to the clinician the chemical identity of a compound detected in a biological sample. A negative result allows one to conclude with good certainty that any xenobiotic is absent. The detection limits must be defined in each case in order to distinguish "detected" from "non-detected" substances.

It is advisable to use generic names in the reporting, even if the physician has mentioned a commercial name in his request.

1.1.9.2 Quantitative determination

The use of SI units in reporting is recommended by the International Federation of Clinical Chemistry. The use of moles rather than grams is particularly useful if the method measures metabolites rather than the native compound, as in the example of acetylsalicylate which yields equimolar amounts of salicylate in vivo and contributes to the anion gap on a molar basis. On the other hand it has been advocated that drugs are dosed in grams and not moles, but indeed the relation between the absorbed amount and the serum concentration is generally established on an empirical manner.

1.1.9.3 Follow-up of the patient

Successive determination of a toxic compound serves to provide an idea of the intoxication kinetics. In some cases (e.g. salicylates absorption), it is mandatory to follow the evolution of the blood concentration of the substance, because the absorption can be delayed for several hours. The interval at which such estimations have to be made depends on the half-life of the compound. The information on this point must be available from the laboratory upon request.

1.1.10 Organisation of the Laboratory

1.1.10.1 Emergencies

Whereas in the non-specialised laboratory a certain information on intoxications can be provided by the personnel on duty for emergencies, the establishment of a laboratory capable of identifying with a high degree of certainty an unknown compound at any time represents an effort which is only justified in a large hospital, or even on a regional basis for a population of, say, above half a million.

1.1.10.2 Techniques and equipment

The basic techniques used in an emergency toxicology laboratory should be gas chromatography-mass spectrometry, which answers in most cases (more than 95%) the question "what?". Other techniques such as gas chromatography and liquid chromatography answer the question "how much?". Thin-layer chromatography does not always provide the required sensitivity, but may be useful in certain cases.

In addition, the standard equipment should include a CO-oximeter, a spectrophotometer, a flame photometer, an immunoassay system and material for solid-phase extraction or liquid-liquid extraction.

1.1.10.3 Personnel

For working on a 24 hours basis, a staff of about six technicians headed by a specialised clinical chemist represents a reasonable solution.

1.1.11 Conclusion

The contribution of the laboratory to the diagnosis of acute intoxication is becoming increasingly important. If done properly and rapidly, it may have life-saving consequences. Modern life has increased the number of substances identifiable as causes of intoxications, and this has made their correct identification more complex. The existence of specialised laboratories for solving these problems is therefore amply justified. But in the final analysis, the emergency toxicology laboratory must work in close concert with the emergency-room physician throughout the treatment of the patient. Failure on the part of the clinician to keep the laboratory informed of the clinical evolution may result in delayed and inappropriate screening. Failure of the laboratory to offer a screening compatible with the patient's needs as perceived by the clinician will result in waste of resources and inadequate support of the patient.

References

1. Bailey, D.N. & Guba, J.J. (1979) Survey of emergency toxicology screening in a university medical centre. J. Anal. Toxicol. *3*, 133–136.
2. Bailey, D.N. & Manoguerra, A.S. (1980) Survey of a drug abuse patterns and toxicology analysis in an emergency-room population. J. Anal. Toxicol. *4*, 199–203.
3. Wennig, R. (1992) Practical compendium for health professionals: Drugs of abuse currently used in Europe, Commission of the European Communities Publication, Luxembourg, ref. CEC/V/E/LUX/61/92.
4. Meade, B.W., Widdop, B., Blackmore, D.J., Brown, S.S., Curry, A.S., Goulding, R., Higgins, G., Matthew, H.J.S., Rinsler, M.G. (1972) Simple tests to detect poisons. Ann. Clin. Biochem. *9*, 35–46.
5. Maehly, A.C. (1967) Volatile toxic compounds, Stolman, A. (ed.), Progress in chemical toxicology, Academic Press, London, pp. 63–98.
6. Moffat, A.C. (1975) Use of SE-30 stationary phase for the gas-liquid chromatography of drugs. J. Chromatog. *113*, 69–95.
7. Gas-chromatographic retention indices of toxicologically relevant substances on SE-30 or OV-1, Second revised and enlarged edition, Report II of the DFG (Deutsche Forschungsgemeinschaft) commission for clinical-toxicological analysis, special issue of the TIAFT (The International Association of Forensic Toxicologists) bulletin (1985), CH Verlagsgesellschaft, Weinheim FRG.
8. Drozd, J. (1975) Chemical derivatization in gas chromatography. J. Chromatog. *113*, 303–356.
9. Quilliam, M.A. & Westmore, J.B. (1978) Sterically crowded trialkylsilyl derivatives for chromatography and mass spectrometry of biologically-important compounds. Anal. Chem. *50*, 59–68.
10. Lawson, A.M. (1989) Mass spectrometry, The fundamental principles. In: Mass Spectrometry, A.M. Lawson (ed.) Clinical Biochemistry, principles, method, applications, H.C. Curtius & M.Roth, ed. Walter de Gruyter, Berlin. New York.
11. Pfleger, K., Maurer, H. & Weber, A. (1985) Mass spectral and GC Data of drugs, poisons and their metabolites, Parts I & II, VCH Verlagsgesellschaft, Weinheim, FRG.
12. The Wiley/NBS registry of mass spectral data (1989), F.W. McLafferty & D.B. Stauffer, John Wiley & Sons ltd, Chichester West Sussex, Great Britain.
13. Wong, S.H.Y. (ed.) (1985) Therapeutic drug monitoring and toxicology by liquid chromatography, Marcel Dekker, New York.
14. Binder, S.R., Regalia, M., Biaggi-McEachern, M. & Mazhar, M. (1989) A drug profiling system for emergency toxicology, J. Chromatog. *473*, 325–341.
15. Franke, J.P., Schepers, P., Bosman, J. & de Zeeuw, R.A. (1982) Optimization of thin-layer chromatography for toxicological screening: Applicability of shorter development distances. J. Analyt. Toxicol. *6*, 131–134.
16. Stead, A.H., Gill, R., Wright, T., Gibbs, J.P., Moffat, A.C. (1982) Standardised thin-layer chromatographic systems for identification of drugs and poisons. Analyst (Lond.) *107*, 1106–1168.
17. Lim, C.K. (1988) Trends Anal. Chem. *7*, 340.
18. Gupta, R.M. (1992) Drug level monitoring: antidepressants, J. Chromatog. (Biomed. Appl.) 576, 183–211.

1.2 Therapeutic Drug Monitoring

A. D. Fraser

1.2.1 Introduction

Therapeutic drug monitoring (TDM), also referred to as applied clinical pharmacokinetics, has developed as an integral component of clinical medicine over the past 25 years. The concept is based on the principle that the dosing regimen for a patient can be adjusted based on single or repeated drug measurements in serum, plasma or whole blood. The rationale for TDM is based on the documented presence of a relationship between either a beneficial outcome or toxic event and the concentration of the drug and/or its metabolite in the biological fluid analysed. This relationship between drug measurements and effect has led to the concept of a "desirable target concentration" or "therapeutic range". Values measured below this range would be considered "sub-therapeutic" and concentrations greater than this range would be considered "potentially toxic or toxic".

Many previous reviews on TDM have presented the pharmacokinetic principles upon which TDM is based (1–6).

It is the objective of this chapter to discuss the concept of a therapeutic range and to use certain anticonvulsant drugs as specific examples. The need for review or auditing of TDM programmes will also be presented.

For the majority of therapeutic agents administered to patients, there are definite clinical indices of efficacy which preclude the analysis of a drug concentration in a biological fluid. The following assumptions must be considered prior to considering the usefulness of TDM for an individual drug being used to treat a specific disease or disorder (4, 7–9).

1. The drug should have a pharmacological effect which is difficult or impossible to monitor by direct clinical observation.
2. Drug concentration(s) should correlate with and reflect the intensity and duration of pharmacological effect at its site of action.
3. The drug concentration should not be predictable in an individual patient based on the administered dose.
4. Variability in the concentration-effect relationship should be less than the variability in the dose-concentration relationship.
5. There is a well-defined "target range" where the patient experiences maximal therapeutic benefit with minimal risk of toxicity.
6. Attainment and maintenance of biological fluid drug concentrations in the optimal target range by adjustment of dose strength and frequency leads to greater benefit than can be achieved by the exercise of sound clinical judgement.
7. There must be an accurate, precise, specific and readily available assay for the drug and/or its metabolites, in the appropriate biological fluid.

1.2.2 Concept of a Therapeutic Range

The idea of having a defined serum/plasma concentration range for a drug used in the treatment of a disease such as epilepsy has often been accepted enthusiastically but very difficult to establish in a patient population. A defined population based therapeutic range may or may not be appropriate for an individual patient.

Woo (10) evaluated the concept of a "lower limit" for the therapeutic range in 1988. In a randomized controlled trial, it was demonstrated that there is no advantage to increasing the phenytoin or phenobarbital dose in patients with "sub-therapeutic" serum drug concentrations who have been seizure free for 90 days or longer. The study concluded that no universal lower concentration limit exists for phenytoin and phenobarbital in the treatment of epilepsy. The investigators, however, did not address whether these longterm seizure free patients actually required any medication for seizure control.

Other studies have concluded that there is no value in attempting to define a lower limit of the therapeutic range for any drug (3, 5, 11).

From a toxicity perspective, one is generally more concerned about the upper limit of the range being near a concentration where toxicity could occur. Phenytoin, for example, often has an upper limit of the therapeutic range quoted as 80 μmol/l for seizure control. Individual patients who are not adequately controlled may be given higher doses until their serum concentrations reach 100 or 120 μmol/l. In other words, the individual patient may have an upper limit of the therapeutic range greater than the population based range of 80 μmol/l (2–3). There is increasing concern, however, about the potential adverse effects of long term exposure to high serum concentrations of several drugs, especially in children (12–13).

Most of the clinical studies on TDM involve measurement of total drug concentrations in the biological fluid analysed. This includes the portion bound to plasma proteins such as albumin and "free" drug which is not protein bound. The majority of the studies measure only total drug concentrations and assume that the per cent free fraction does not vary. If the binding of the drug to a protein does vary, however, then the total drug values may not necessarily correlate with efficacy and/or toxicity (14–15).

Certain basic drugs such as lidocaine, disopyramide, imipramine and quinidine bind to the acute phase reactant protein alpha 1-acid glycoprotein (16–18). The synthesis of this protein is induced in many conditions such as inflammation, infections, surgery, myocardial infarction, etc. The concentration of alpha 1-acid glycoprotein may be elevated for 21–60 days (or longer) after a myocardial infarction. Quinidine is normally 90 ± 3 % bound to serum proteins (such as alpha 1-acid glycoprotein) and often has a therapeutic range quoted as 6–18 μmol/l or 0.6–1.9 μmol/l as "free" quinidine. In one report (19) a patient with a total plasma quinidine concentration of 34 μmol/l had a free quinidine value of 1.1 μmol/l (3.2 % free fraction, not the normal 10 % free) following a myocardial infarction. Therefore, one must be aware of pathologic conditions which can alter the extent of drug binding to plasma proteins.

Another important consideration is whether a therapeutic range established for one disorder can be applied in the treatment of another condition. For example, one cannot necessarily assume that the therapeutic range established for the treatment of focal epileptic seizures is appropriate for generalized or other types of seizures (20–21).

Development of new applications of TDM may be limited by the non-availability of routine assays for all the significant metabolites of a drug which are known to be pharmacologically active. For example, investigations involving procainamide and the traditional antidepressant drugs generally include analysis of the major active metabolite. Amitriptyline, for example, has four known active metabolites in addition to nortriptyline (E and Z 10 OH-amitriptyline and E and Z 10 OH-nortriptyline). The only metabolite routinely measured today is nortriptyline except in specialized academic centres (22–24).

The commercial availability of immunoassay kits often determines whether or not an active metabolite is routinely monitored. Diagnostic kits are available to monitor procainamide and its active metabolite N-acetyl procainamide. For other drugs, however, the immunoassay companies have not marketed kits to monitor active metabolites. Many investigators advocate the routine monitoring of carbamazepine 10,11-epoxide which can readily be monitored by HPLC (6, 25) but not by any currently available immunoassay products.

A recent survey (26) was performed in Canada to evaluate the use of TDM for antiepileptic drugs by paediatric neurologists. One question asked in the survey was the therapeutic range given by the laboratory for the antiepileptic drugs being monitored. For phenytoin, there was widespread agreement on the lower and upper limits used (tab. 1). For carbamazepine, however, there were significant differences in the therapeutic ranges used (tab. 2). An overall summary of the therapeutic ranges used by 56 paediatric neurologists in Canada is found in table 3.

Table 1. Therapeutic range for phenytoin used by 43 pediatric neurologists in Canada.

Lower Limit	40 μmol/l	93 %
Range	20-60 μmol/l	
Upper Limit	80 μmol/l	98 %
Range	40-80 μmol/l	

Table 2. Therapeutic range for carbamazepine used by 45 pediatric neurologists in Canada.

Lower Limit	<18 μmol/l	44 %
	>30 μmol/l	33 %
Range	13-40 μmol/l	
Upper Limit	<35 μmol/l	9 %
	>50 μmol/l	60 %
Range	34-51 μmol/l	

Table 3. Summary of anticonvulsant drug therapeutic ranges used by pediatric neurologists in Canada. All units are μmol/l.

AED	Lower Range	Upper Range
Carbamazepine	13 - 40	34 - 51
Phenytoin	20 - 60	40 - 120
Phenobarbital	42 - 82	90 - 215
Valproate	276 -380	500 -1035
Ethosuximide	200 -500	400 -1125

Therefore, the appropriate application of TDM requires a knowledge of and accounting for a number of variables for effective patient care.

1.2.3 Individual Drugs

1.2.3.1 Phenytoin

The initial studies investigating the relationship between serum phenytoin (diphenylhydantoin) concentrations and therapeutic effects were reported over 30 years ago (27, 28). In the Buchtal study (28), involving 29 in-patients and 51 out-patients, the authors concluded that the minimum serum concentration of phenytoin required to achieve seizure control was 40 μmol/l. Subsequent studies (29–31) have demonstrated that there is a reduction in mean seizure frequency as serum phenytoin concentrations increase above 40 μmol/l. Other investigators, however, have observed good to excellent seizure control at steady state serum phenytoin concentrations of less than 40 μmol/l (32–34). In a long term study of 31 previously untreated patients, Reynolds (35) found that 80% of the patients experienced complete seizure control on monotherapy with phenytoin. The mean seizure frequency decreased by 90% and 8/25 patients had complete seizure control with serum phenytoin concentrations less than 40 μmol/l at steady state.

It is generally believed that partial seizures are more difficult to control than generalised seizures. The relationship between seizure type and control of these seizures with serum drug concentrations has not been thoroughly investigated (22).

With respect to the upper limit of the therapeutic range, there is some data indicating that certain patients fail to achieve maximal therapeutic benefit until phenytoin concentrations are > 80 μmol/l (2,3). Cobos (36) reported that 17/30 patients who required higher steady state phenytoin concentrations for seizure control had mean serum values of 124 μmol/l. These high values were apparently well-tolerated in this group.

It has been known that there is significant interindividual variation in serum phenytoin concentrations following standard doses (37–41). The accumulation rate of phenytoin and steady state serum concentrations is predicted by a Michaelis-Menten model:

$$\text{Elimination rate} = V_{max}C/K_m + C$$

where C = serum concentration, V_{max} = maximum rate of drug elimination and K_m = Michaelis-Menten constant. Phenytoin exhibits non-linear pharmacokinetics because the K_m (equivalent to the serum concentration at which the elimination ratio is 50% of the maximum) is in the same range as clinically effective concentrations. The V_{max} is not much greater than the usual daily dose required to achieve a therapeutic steady-state concentration. The clinical importance of this is that:

1. Steady-state phenytoin concentrations can change disproportionately with small changes in the dose.
2. The time required to achieve any proportion of the steady-state concentration varies with the dose rate.

In addition, the factors V_{max} and K_m are also affected by other drugs which can induce or inhibit the hepatic cytochrome P_{450} enzyme systems responsible for phenytoin metabolism (42, 43).

Advocates of routine TDM often point to phenytoin as one drug where indications are strongest for monitoring due to the problem of saturation pharmacokinetics as just described.

Two recent studies (44, 45) indicate that appropriate application of TDM can improve clinical outcome (seizure frequency, number of hospital admissions) for patients receiving phenytoin.

In summary, there is insufficient evidence to state unequivocally a lower or upper limit of the therapeutic range for phenytoin. There is ample evidence, however, to support the role of TDM for phenytoin in seizure control management.

1.2.3.2 Carbamazepine

Carbamazepine (CBZ) is indicated for the treatment of complex partial seizures, tonic clonic seizures and mixed seizures (complex partial, tonic clonic, and other generalized seizures). It is also prescribed for the treatment of panic disorders, trigeminal neuralgia and has been used experimentally in the treatment of cocaine addiction (46, 47).

The relationship between dose and serum CBZ concentrations is unpredictable and varies depending the length of time of drug administration, formulation, metabolism and other factors (48—52). In studies involving all age groups, it was found that the CBZ serum concentration-dose ratio decreased over time with continued drug administration. This finding is probably caused by the induction by CBZ of its own metabolism (auto-induction). The elimination half-life of CBZ measured after a single dose (approximately 30 hours) is much longer than the half-life at steady-state (generally less than 20 hours). During chronic monotherapy, the half-life is approximately 12 hours (monotherapy) and less than 10 hours (polytherapy).

Interpretation of CBZ serum concentrations relative to therapeutic efficacy and/or toxicity is complicated by the accumulation in serum of CBZ-10,11-epoxide (CBZ-

E), a pharmacologically active metabolite. This metabolite is less protein bound in serum than CBZ. The relative amount of CBZ-E is 20–25% that of CBZ during monotherapy and increases to 50% or greater with polydrug therapy. Valproic acid, valpromide, and progabide are all inhibitors of CBZ-epoxide hydrolase at therapeutically relevant concentrations of CBZ (53, 54).

The pharmacological activity of CBZ-E has been documented in rodents and the effects of CBZ-E have been evaluated clinically by administration of CBZ-E to out-patients with frequent seizures (56).

For optimal TDM of CBZ, many investigators have recommended the routine analysis of CBZ-E along with CBZ. Many of the earlier studies on CBZ serum concentrations versus efficacy/toxicity ignored the contribution of CBZ-E which can reach high concentrations therapeutically and after an overdose of CBZ (6, 57, 58).

The most significant drug interaction involving CBZ is with the two macrolide antimicrobial agents triacetyloleandomycin and erythromycin. The later drug inhibits the metabolism of CBZ to CBZ-E. The unpredictable nature of this serious interaction means that this combination should not be used except under unusual circumstances warranting this drug combination (59–61).

Institutions which routinely monitor CBZ-E add the CBZ and CBZ-E values together when evaluating if the total is within the therapeutic range (6, 26). With polydrug therapy, one usually observes a larger percentage of the total CBZ effect due to the action of CBZ-E.

1.2.3.3 Sodium valproate (valproic acid)

Sodium valproate is another major anticonvulsant drug. It is used in the treatment of simple or complex absence seizures and for primary generalized seizures with tonic-clonic manifestations. The minimum effective serum concentration has been reported as 140 μmol/l (62) and as 210 μmol/l (63, 64). All these authors noted that there was major interindividual variation in the serum concentration of valproate required for seizure control.

Several investigators have found no correlation between toxicity and serum valproate concentrations (65–67). Toxicity was not correlated with serum valproate concentrations in a further study involving 53 children receiving valproate (68). There are concerns about the hepatotoxicity of valproate thought to be due to the formation of unsaturated metabolites (69, 70). These metabolites, however, are not routinely measured and there is no correlation between hepatotoxicity and serum concentrations of valproate.

In summary, since the correlation between serum concentration and efficacy or toxicity is so poor, the major indication for monitoring valproate in serum is to assess patient compliance rather than for dose adjustment (71). There is no evidence to support TDM for valproate since a therapeutic range does not exist (72–74).

1.2.4 Audit of Therapeutic Drug Monitoring Programmes

The ability of a laboratory to provide reliable analysis of drugs/metabolites in biological fluids is just one component of an effective therapeutic drug monitoring programme.

Systems organized to bring all the key individuals together and establish an effective on-going monitoring system have been established only recently (75–79).

Several studies (80–89), however, have concluded that up to 69% of serum drug measurements for TDM are uninterpretable. The major reasons why TDM has been considered to be of no or limited value are:

1. blood specimens are not drawn at the appropriate time (within the dosing interval and/or prior to a steady state concentration being achieved);
2. TDM requested without a good clinical indication for analysis;
3. drug administration time has not been recorded;
4. blood specimen obtained after drug administration has been discontinued;
5. inappropriate specimen handling.

Wing and Duff (45) performed a study to evaluate the cost savings and need for a continuous TDM programme for digoxin. Their study was a prospective, randomized, crossover design of 4 phases lasting 6 weeks/phase. There were 207 patients who met the entrance criteria to enrol in the study. These investigators also found a large number of requested assays which were not indicated, specimens drawn at inappropriate times and/or misinterpreted. In the study, 57% and 74% of assays requested in phases I and III had inappropriate indications (based on criteria established by a clinical pharmacologist and a cardiologist). The authors concluded that a successful, effective TDM programme must have on-going education as the key to efficient utilization of the TDM service.

In another prospective study of digoxin requests over a 6 month period (90), similar findings were obtained. For 111 specimens with serum digoxin concentrations greater than the upper limit of the therapeutic range (2.6 nmol/l), 35 specimens were collected at an inappropriate time. In the 64 specimens collected at the proper time, appropriate action was taken in only 36/64 cases.

Other reviews have focused on other drugs such as phenytoin (91). In this retrospective review of 58 patients receiving phenytoin over a 7 week period, 8% of the TDM requests were justified on a pharmacokinetic basis (i.e. for evaluating a loading dose or a steady state serum concentration). Based on clinical criteria, 30% of the requests were justified. The majority of the requests (56%) were not justified because measurements were made within 3 days of starting the drug or changing the dose. Several patients actually developed phenytoin toxicity due to unwarranted dose changes based on pre steady-state serum concentrations which were assumed to be at steady state.

The Italian Collaborative Group on Utilization of TDM in Hospital Departments (92) reviewed data on TDM services in 28 general hospital wards. The study found that use of a TDM service was indicated for 5% of the patient population but only ordered in 1% of the population. The authors concluded that many patients who could benefit from a TDM service were not being tested for serum drug concentrations. In a further study (93), TDM utilization was studied in a nursing home for senior citizens. Fifty per cent of the serum drug measurements in the study were at sub-therapeutic concentrations. The time since the last serum drug measurement averaged 28 months. The authors concluded that the application of TDM in their patient population was very much underutilized.

Matzuk and colleagues (94) recognized that digoxin serum specimens were often collected at the wrong time after dosing at a teaching hospital. To correct this chronic problem in TDM, the parties involved (pharmacy, medicine, nursing, laboratory) devised a new digoxin dosing regimen. All digoxin dosing (oral and i.v.) would be given at 1700 h and blood collected at 0700 h for drug monitoring. The authors were able to reduce inappropriate digoxin measurements by implementation of this policy. This policy, however, did not prevent the ordering of daily or 3 times/week routine drug measurements or prevent serum analysis prior to the attainment of a steady state concentration.

1.2.5 Discontinuation of Antiepileptic Drug Therapy

The administration of drugs for the treatment of many disorders or diseases such as epilepsy presumes that the patient will require this medication permanently. What does one do for an individual who remains seizure free for 1, 2 or 5 years? One powerful argument for discontinuation of anticonvulsant drugs is concern about long term toxicity. Long term administration of these agents may have subtle adverse effects on systemic and neurologic function (95). Adverse effects of phenytoin and phenobarbital can include effects on the central and peripheral nervous system, bone marrow, immune system and skin and connective tissue. There is also concern about the effects of these drugs on the developing nervous system of children (96).

Several groups have performed studies assessing risk of relapse after the withdrawal of anticonvulsant drugs (97—99). Callaghan and co-workers (100) studied 92 patients who had been seizure free during treatment with a single drug (phenytoin, carbamazepine or sodium valproate). Upon withdrawal of drug therapy, 31 patients relapsed and 61 remained seizure free. There was no significant difference between the relapse rate among adults (35 %) and children (31 %). This study suggested that the number of seizures a patient had before control, the number of drugs tried as monotherapy and the type of seizures all influenced the outcome when drug therapy was withdrawn.

As stated in an accompanying editorial (101), questions still requiring answers include: is there justification for partial or relative drug withdrawal in some patients, can one remain seizure free while a "sub-therapeutic" dose is being used in an attempt to minimize toxicity and do the seizure threshold and minimal therapeutic drug level decline with an increased duration of effective seizure control?

Discontinuation of drug treatment holds promise in many respects for the patients involved but does complicate the interpretation of the role (if any) of TDM in this population.

1.2.6 Summary

The role of TDM in clinical medicine has expanded tremendously in the past 25 years. The initial enthusiasm based on the ability to monitor many drugs/metabolites has been tempered by the finding that basic pharmacokinetic principles fundamental to TDM are often ignored.

Many investigators, however, have shown convincingly that a well-organized TDM consultative service can lead to improved patient care in a cost-effective and efficient manner.

The future of TDM will involve broadening the scope of more traditional applications. Examples include using analysis of a drug-metabolite pair as a measure of liver function in patients with cirrhosis (102) or in liver transplantation (103). There are also significant pharmacokinetic differences between drug enantiomers in man after administration of a racemic drug mixture. The area of drug enantiomers and TDM requires further development (104).

In conclusion, major advances have been made in the application of TDM for improved patient care; with recent advances in new drug development and molecular biology, the role of TDM will have a vital but new direction in the future.

1.2.7 Additional Remarks (added in proof)

Therapeutic drug monitoring for anticonvulsant, antiarrhythmic and certain antibiotic drugs remain an important component of clinical care today. Over the past decade, the optimal use of immunosuppressive drugs used in patients undergoing organ transplantation has been enhanced by the therapeutic drug monitoring of several drugs. Cyclosporine A was the first immunosuppressant agent where therapeutic drug monitoring was used to monitor efficacy and to avoid toxic concentrations in biological fluids. This area remains active because new drugs are being introduced and the guidelines for drug monitoring vary dependent upon the organ system, time since transplantation, mode of administration, concomitant drugs administered, etc. Several conferences have been held with the objective being to establish a consensus on monitoring guidelines (105–108).

The analysis of immunosuppressive drugs is complicated by the presence of toxicologically active metabolites which may or may not be readily measured dependent on the methodology used (109). Chromatographic techniques such as HPLC are the most specific methods and may allow one to monitor drug metabolites. In a routine service laboratory setting however, immunoassay techniques remain the most widely used technique (110). This area promises to be active field for drug monitoring investigation.

Over the past fifteen years, theophylline has been often described as one drug with well-accepted indications for drug monitoring. Today, however, theophylline is used much less since other drugs have replaced it as a front-line therapeutic agent.

In summary, therapeutic drug monitoring remains an important component of clinical medicine where there are no direct or indirect clinical indices of efficacy and/or toxicity.

References

1. Friedman, H. & Greenblatt, D.J. (1986) Rational therapeutic drug monitoring. JAMA *256*, 2227–2233.
2. Brodie, M.J. & Feely, J. (1988) Practical clinical pharmacology. Therapeutic drug monitoring and clinical trials. Br. Med. J. *296*, 1110–1114.
3. Spector, R., Park, G.D., Johnson, G.F. & Vesell, E.S. (1988) Therapeutic drug monitoring. Clin. Pharmacol. Ther. *43*, 345–353.
4. Richens, A. & Marks, V. (1981) Therapeutic drug monitoring, Churchill Livingston, Edinburgh.
5. Gal, P. (1988) Therapeutic drug monitoring in neonates: problems and issues. Drug Intell. Clin. Pharm. *22*, 317–323.
6. Meijer, J.W.A. (1991) Knowledge, attitude and practice in antiepileptic drug monitoring. Acta Neurol. Scand. Supplementum 134, *83*, pp. 69–96, 97–122.
7. Kaufmann, R.E. (1981) The clinical interpretation and application of drug concentration data. Pediatr. Clin. North Am. *28*, 35–45.
8. Brass, E.P. (1985) Rational approach to therapeutic drug monitoring. Hosp. Formul. *20*, 882–889.
9. Kisor, D.F., Mehta, V., Viera-Fattahi, S. & Sterchele, J.A. (1990) Therapeutic drug monitoring. Practical considerations for primary care physicians. Postgrad. Med. J. *87*, 239–247.
10. Woo, E., Chan, Y.M., Yu, Y.L., Chan, Y.W. & Huang, C.Y. (1988) If a well stabilized epileptic patient has a subtherapeutic antiepileptic drug level, should the dose be increased? A randomized prospective study. Epilepsia *29*, 129–139.
11. Vajda, F.J.E. & Aicardi, J. (1983) Reassessment of the concept of a therapeutic range of anticonvulsant plasma levels. Develop. Med. Child. Neurol. *25*, 660–671.
12. Herranz, J.L., Armijo, J.A. & Arteaga, R. (1988) Clinical side effects of phenobarbital, primidone, phenytoin, carbamazepine, and valproate during monotherapy in children. Epilepsia *29*, 794–804.
13. Park, B.K. & Kitteringham, N.R. (1987) Adverse reactions and drug metabolism. Adverse Drug Reaction Bulletin *122*, 456–459.
14. Barre, J., Didey, F., Delion, F. & Tillemont, J.-P. (1988) Problems in therapeutic drug monitoring. Free drug level monitoring. Ther. Drug. Monit. *10*, 133–143.
15. Mehta, A. (1989) Therapeutic monitoring of free (unbound) drug levels: Analytical aspects. Trends in Anal. Chem. *8*, 107–112.
16. Kremer, J.M.H., Wilting, J. & Janssen, L.H.M. (1988) Drug binding to human alpha-1-acid glycoprotein in health and disease. Pharmacological Reviews *40*, 1–47.
17. Drayer, D. (1983) Clinical consequences of the lipophilicity and plasma protein binding of antiarrhythmic drugs and active metabolites in man. Ann. N.Y. Acad. Sci. *432*, 45—56.
18. Levy, R. & Shand, D. (1984) Clinical implications of drug-protein binding. Clin. Pharmacokinetics, Supplement 1, *9*, 1–104.
19. Garfinkel, D., Mamelok, R.D. & Blaschke, T.F. (1987) Altered therapeutic range for quinidine after myocardial infarction and cardiac surgery. Ann. Intern Med. *107*, 48–50.
20. Dam, M., Jensen, A. & Christiansen, J. (1975) Plasma level and effect in grand mal and psychomotor epilepsy. Acta Neurol. Scand. Supplementum 60, *75*, 33–38.
21. Schmidt, D., Einicke, I. & Haenel, F. (1986) The influence of seizure type on the efficacy of plasma concentration of phenytoin, phenobarbital and carbamazepine. Arch. Neurol. *43*, 263–265.
22. Lukey, B.J., Jones, D.R., Wright, J.H. & Hurst, H.E. (1989) Relationships among nortriptyline, 10-OH(E) nortriptyline, and 10-OH(Z) nortriptyline steady state plasma levels and nortriptyline dosage. Ther. Drug. Monit. *11*, 221–227.
23. Suckow, R.F. & Cooper, T.B. (1982) Simultaneous determination of amitriptyline, nortriptyline and their respective isomeric 10-hydroxy metabolites in plasma by liquid chromatography. J. Chromatogr. *230*, 391–400.
24. Bock, J.L., Giller, E., Gray, S. & Jatlow, P. (1982) Steady state plasma concentrations of cis- and trans-10-OH amitriptyline metabolities. Clin. Pharmacol. Ther. *31*, 609–616.
25. Fraser, A.D. & Isner, A.F. (1988) Optimal drug monitoring of carbamazepine by routine analysis of carbamazepine and carbamazepine 10,11-epoxide. Clin. Biochem. *21*, 391.
26. Dooley, J., Camfield, P., Camfield, C., Gordon, K. & Fraser, A. (1991) The use of antiepileptic drug levels in children: a survey of Canadian pediatric neurologists. Ameri-

can Epilepsy Society Meeting, Philadelphia, U.S.A.

27. Plaa, G.L. & Hine, C.H. (1956) A method for the simultaneous determination of phenobarbital and diphenylhydantoin in blood. J. Lab. Clin. Med. *47*, 649–652.
28. Buchtal, F., Svensmark, O.P., Schiller, P.J. (1960) Clinical and electroencephalographic correlations with serum levels of diphenylhydantoin. Arch. Neurol. *2*, 624–630.
29. Lambie, D.G., Nanda, R.N., Johnson, R.H. & Shakir, R.A. (1976) Therapeutic and pharmacokinetic effects of increasing phenytoin in chronic epileptics on multiple drug therapy. Lancet *2*, 386–389.
30. Lund, L. (1974) Anticonvulsant effect of diphenylhydantoin relative to plasma levels: a prospective three year study in ambulant patients with generalized epileptic seizures. Arch. Neurol. *31*, 289–294.
31. Reynolds, E.H., Chadwick, D. & Galbraith, A.W. (1976) One drug (phenytoin) in the treatment of epilepsy. Lancet *1*, 923–926.
32. Borofsky, L.G., Louis, S., Kutt, H. & Roginsky, M. (1972) Diphenylhydantoin: efficacy, toxicity, and dose – serum level relationships in children. J. Pediatr. *81*, 995–1002.
33. Feely, J., Duggan, B., O'Callaghan, M. & Callagan, N. (1979) The therapeutic range for phenytoin – a reappraisal. Irish J. of Med. Sci. *2*, 44–49.
34. Feldman, R.G. & Pippenger, C.E. (1976) The relation of anticonvulsant drug levels to complete seizure control. J. Clin. Pharmacol. *16*, 51–59.
35. Reynolds, E.H., Shorvon, S.D., Galbraith, A.W., Chadwick, D., Dellaportas, C.I. et. al. (1981) Phenytoin monotherapy for epilepsy: a long term prospective study, assisted by serum level monitoring, in previously untreated patients. Epilepsia *22*, 475–488.
36. Cobos, J.E. (1987) High dose phenytoin in the treatment of refractory epilepsy. Epilepsia *28*, 111–114.
37. Kutt, H., Winters, W., Kokenge, R. & McDowell, F. (1964) Diphenylhydantoin metabolism, blood levels and toxicity. Arch. Neurol. *11*, 642–648.
38. Levine, M., Orr, J. & Chang, T. (1987) Interindividual variation in the extent and rate of phenytoin accumulation. Ther. Drug Monit. *9*, 171–176.
39. Perucca, E. & Richens, A. (1981) In Therapeutic Drug Monitoring (Richens, A. & Marks, V., eds.), pp. 320–348, Churchill Livingstone, Edinburgh.
40. Levine, M. & Chang, T. (1990) Therapeutic drug monitoring of phenytoin. Clin. Pharmacokinet. *19*, 341–358.
41. Ludden, T.M., Allen, J.P. Valutsky, W.A., Vicuna, A.V., Nappi, J.M. et. al. (1978) Rate of phenytoin accumulation in man: a simulation study. J. of Pharmacokinet. and Biopharm. *6*, 399–414.
42. Winter, M. & Tozer, T.N. (1986) In: Applied Pharmacokinetics: principles of TDM, 2nd ed. (Evans, A. ed.) 493–539, Applied Therapeutics, Inc., Spokane.
43. Nation, R.L., Evans, A.M. & Milne, R.W. (1990) Pharmacokinetic drug interactions with phenytoin (part I). Clin. Pharmacokinet. *18*, 37–60.
44. Ioannides-Demos, L.L., Horne, M.K., Tong, N., Wodak, J., Harrison, P.M. et. al. (1988) Impact of a pharmacokinetics consultation service on clinical outcomes in an ambulatory care epilepsy clinic. Am. J. Hosp. Pharm. *45*, 1549–1551.
45. Wing, D.S. & Duff, H.J. (1989) The impact of a therapeutic drug monitoring program for phenytoin. Ther. Drug. Monit. *11*, 32–37.
46. Post, Robert M. (1988) Time course of clinical effects of carbamazepine: implications for mechanisms of action. J. Clin. Psychiat. *49*, 35–48 Supplement.
47. Tomson, T. & Bertilsson, L. (1984) Potent therapeutic effect of carbamazepine-10,11-epoxide in trigeminal neuralgia. Arch. Neurol. *41*, 598–601.
48. Choonara, I.A. & Rane, A. (1990) Therapeutic drug monitoring of anticonvulsants: state of the art. Clin. Pharmacokinet. *18*, 318–328.
49. Weaver, D.F. & Purdy, R.A. (1987) Clinical pharmacology of phenytoin and carbam-azepine. Drugs and Ther. *10*, 16–20.
50. Ferrendelli, J.A. (1987) Pharmacology of antiepileptic drugs. Epilepsia *28*, 14–16, Supplement 3.
51. Pynnonen, S. (1979) Pharmacokinetics of carbamazepine in man: a review. Ther. Drug Monit. *1*, 409–431.
52. Levy, R.H. & Kerr, B.M. (1988) Clinical pharmacokinetics of carbamazepine. J. Clin. Psychiat. *49*, 58–61.
53. Ramsay, R.E., McManus, D.Q., Guterman, A., Briggle, T.V., Vazquez, D., Perchalski, R., Yost, R.A. & Wong, P. (1990) Carbamazepine metabolism in humans: effect of concurrent anticonvulsant therapy. Ther. Drug. Monit. *12*, 235–241.

54. Rambeck, B., May, T. & Juergens, U. (1987) Serum concentrations of carbamazepine and its epoxide and diol metabolites in epileptic patients: the influence of dose and comedication. Ther. Drug Monit. *9*, 298–303.
55. Hundt, H.K.L., Aucamp, A.K., Muller, F.O. & Potgieter, M.A. (1983) Carbamazepine and its major metabolites in plasma: a summary of eight years of therapeutic drug monitoring. Ther. Drug Monit. *5*, 427–435.
56. Tomson, T., Almkrist, O., Nilsson, B.Y., Svensson, J.O. & Bertilsson, L. (1990) Carbamazepine 10,11 epoxide in epilepsy, a pilot study. Arch. Neurol. *47*, 888–892.
57. Vree, T.B., Janssen, T.J, Hekster, Y.A. Termond, E.F.S., Van der Dries, A.C.P. & Wignands, W.J.A. (1986) Clinical pharmacokinetics of carbamazepine and its epoxy and hydroxy metabolites in humans after an overdose. Ther. Drug Monit. *8*, 297–304.
58. Weaver, D., Camfield, P. & Fraser, A.D. (1988) Massive carbamazepine overdose: clinical and pharmacological observations in five episodes. Neurology *38*, 755–759.
59. Hedrick, R., Williams, F., Morin, R., Lamb, W.A. & Cate, J.C. (1983) Carbamazepine-erythromycin interaction leading to carbamazepine toxicity in four epileptic children. Ther. Drug Monit. *5*, 405–407.
60. Goulden, K.-J, Camfield, P., Dooley, J.M. Fraser, A., Meek, D.C., Renton, K.W. & Tibbles, J.A.R. (1986) Severe carbamazepine intoxication after coadministration of erythromycin. J. Pediatr. *109*, 135–138.
61. Miles, M.V. & Tennison, M.B. (1989) Erythromycin effects on multiple dose carbamazepine kinetics. Ther. Drug Monit. *11*, 47–52.
62. Turnbull, D.W., Rawlins, M.D., Weightman, D. & Chadwick, D.W. (1983) Plasma concentrations of sodium valproate: their clinical value. Ann. Neurol. *14*, 38–42.
63. Klotz, U. & Schweizer, C. (1980) Valproic acid in childhood epilepsy: anticonvulsant efficacy in relation to its plasma levels. Int. J. of Clin. Pharmacol. Ther. & Toxicol. *18*, 461–465.
64. Minns, R.A., Brown, J.K., Blackwood, D.H.R. & McQueen, J.K. (1982) Valproate levels in children with epilepsy. Lancet *1*, 677–678.
65. Gram, L., Flachs, H., Wurtz-Jorgensen, A., Parnas, J. & Andersen, B. (1979) Sodium valproate serum levels and clinical effect in epilepsy: a controlled study. Epilepsia 20: 303–312 (1979).
66. Ruuskanen, I., Kilpelainer, H.O. & Riekkinen, P.J. (1979) Side-effects of sodium valproate during long term treatment in epilepsy. Acta Neurol. Scand. *60*, 125–128.
67. Herranz, J.L., Arteag, R., Armijo, J.A. (1982) Side-effects of sodium valproate in monotherapy controlled by plasma levels: a study of 88 pediatric patients. Epilepsia *23*, 203–214.
68. Choonara, I.A. (1988) Anticonvulsant toxicity in paediatric out-patients. Br. J. Clin. Pract. *42*, 21–23.
69. Scheffner, D., Konig, S., Rauterberg-Ruland, I. Kochen, W., Hoffmann, W.J. et. al. (1988) Fatal liver failure in 16 children with valproate therapy, Epilepsia 29: 530–542.
70. Binkerd, P.E., Rowland, J.M., Nau, H. & Hendrickx, A.G. (1988) Evaluation of valproic acid (VPA) developmental toxicity and pharmacokinetics in Sprague-Dawley rats. Fund. Appl. Toxicology *11*, 485–493.
71. Knott, C. & Reynolds, F. (1990) Therapeutic drug monitoring in pregnancy: rationale and current status. Clin. Pharmacokinet. *19*, 425–433.
72. Haigh, D. & Forsythe, W.I. (1975) The treatment of childhood epilepsy with sodium valproate. Devel. Med. Child Neurol. *17*, 743–748.
73. Schobben, F., van der Kleijn, E. & Vree, T.B. (1980) Therapeutic monitoring of valproic acid. Ther. Drug Monit. *2*, 61–71.
74. Kilpatrick, C.J., Bury, R.W., Fullinfaxe, R. & Moulds, R.F. (1987) Plasma concentrations of unbound valproate and the management of epilepsy. Aust. N. Z. J. Med. *17*, 574–579.
75. Destache, C.J., Meyer, S.K., Bittner, M.J. & Hermann, K.G. (1990) Impact of a clinical pharmacokinetic service on patients treated with aminoglycosides: a cost-benefit analysis. Ther. Drug Monit. *12*, 419–426.
76. Crist, K.D., Nahata, M.C. & Ety, J. (1987) Positive impact of a therapeutic drug monitoring program on total aminoglycoside dose and cost of hospitalization. Ther. Drug Monit. *9*, 306–310.
77. Ried, L.D., McKenna, D.A. & Horn, J.R. (1989) Effect of therapeutic drug monitoring services on the number of serum drug assays ordered for patients: A meta analysis. Ther. Drug Monit. *11*, 253–263.
78. Koren, G., Soldin, S.J. & MacLeod, S.M. (1985) Organization and efficacy of a therapeutic drug monitoring consultation serv-

ice in a pediatric hospital. Ther. Drug Monit. *7*, 295–298.

79. Fraser, A. D. (1985) Experience with a specific request form for antidepressant drug monitoring. Ther. Drug Monit. *7*, 299–302.
80. Feinberg, J. L. (1990) Interdisciplinary success: therapeutic drug monitoring programmes. Clin. Lab. Sci. *3*, 75–76.
81. Cox, S. & Walson, P. D. (1989) Providing effective therapeutic drug monitoring services. Ther. Drug Monit. *11*, 310–322.
82. Levin, S., Cohen, S. S. & Birmingham, P. H. (1981) Effect of pharmacist intervention on the use of serum drug assays. Am. J. Hosp. Pharm. *38*, 845–851.
83. Clague, H. W., Twum-Barima, Y., Carruthers, S. G. (1983) An audit of requests for therapeutic drug monitoring for digoxin: problems and pitfalls. Ther. Drug Monit. *5*, 249–254.
84. Greenlaw, C. W., Geiser, G., S., Haug, M. T. (1980) Use of digoxin serum assays in a nonteaching hospital. Am. J. Hosp. Pharm. *37*, 1466–1470.
85. Horn, J. R., Christensen, D. B. & de Blaquiere, P. A. (1985) Evaluation of a digoxin pharmacokinetic monitoring service in a community hospital. Drug Intell. Clin. Pharm. *19*, 45–52.
86. Vlasses, P. H., DiPiro, C. A. & Chalupa, D. (1982) Appropriateness of sampling time for selected serum drug assays. Hosp. Pharm. *17*, 371–373.
87. Ives, T. J. & Guyther, R. E. (1984) Serum drug level utilization review in a family medicine residency program. J. Fam. Pract. *19*, 507–512.
88. Snidero, M., Traina, G. L., Bonati, M. (1985) Is ambulatory therapeutic digoxin monitoring useful? Drug Intell. Clin. Pharm. *19*, 660–661.
89. Giriner, P. D. (1979) Medical house staff, Strong Memorial Hospital: use of laboratory tests in a teaching Hospital: long term trends. Ann. Intern. Med. *90*, 243–248.
90. O'Donnell, J. G., Paterson, N. C., Pledger, D. R. & Stewart, M. J. (1987) Serum digoxin assays: investigation of use and abuse. Ann. Clin. Biochem. *24*, 408–410.
91. Wong, P., Brown, D. L., Bachman, K. A. et. al. (1989) Optimal dosing of phenytoin: an evaluation of the timing and appropriateness of serum level monitoring. Hosp. Formul. *24*, 219–223.
92. Italian Collaborative Study on the Utilization of TDM in Hospital Departments (1988) Ther. Drug Monit. *10*, 275–279.
93. Pucino, F., Baumgart, P. J., Strommen, G. L., Silbergleit, I.-L. et. al. (1988) Evaluation of TDM in a long term care facility: a pilot project. Drug Intell. Clin. Pharm. *22*, 594–596.
94. Matzuk, M. M., Shlomchick, M. & Shaw, L. M. (1991) Making digoxin TDM more effective. Ther. Drug Monit. *13*, 215–219.
95. Schmidt, D. (1986) In: Recent Advances in Epilepsy No. 3, (Pedley, T. A. & Meldrum, B. S., eds.) 211–232, Churchill Livingston, Edinburgh.
96. Reynolds, E. H. (1975) Chronic antiepileptic toxicity: a review. Epilepsia *16*, 319–352.
97. Reynolds, E. H. (1983) Chronic Toxicity of Antiepileptic Drugs, (Oxley, J., Janz, D. & Meinardi, H., eds.) 91–99, Raven Press, New York
98. Jeavons, P. (1983) Chronic Toxicity of Antiepileptic Drugs, (Oxley, J., Janz, D. & Meinardi, H., eds.) pp. 1–45, Raven Press, New York.
99. Logan, W. J. & Freeman, J. M. (1969) Pseudodegenerative disease due to diphenylhydantoin intoxication. Arch. Neurol. *21*, 631–637.
100. Callaghan, N., Garrett, A. & Goggin, T. (1988) Withdrawal of anticonvulsant drugs in patients free of seizures for two years. New Eng. J. Med. *318*, 942–946.
101. Pedley, T. A. (1988) Discontinuing antiepileptic drugs. New Eng. J. Med. *318*, 982–984.
102. Oellerich, M., Burdelski, M., Lantz, H.-U., Schulz, M., Schmidt, F.-W. & Herrmann, H. (1990) Lidocaine metabolite formation as a measure of liver function in patients with cirrhosis. Ther. Drug Monit. *12*, 219–226.
103. Oellerich, M., Burdelski, M., Lantz, H.-U., et al. (1988) Efficiency of the MEGX formation test in liver transplantation. Hepatology, *8*, 1220.
104. Drayer, D. E. (1988) Problems in therapeutic drug monitoring: the dilemma of enantiomeric drugs in man. Ther. Drug. Monit. *10*, 1–7.
105. Kahan, B. D., Shaw, L. M., Holt, D., Grevel, J. & Johnson, A. (1990) Consensus Document: Hawk's Cay Meeting on Therapeutic Drug Monitoring of Cyclosporine. Clin. Chem. *36*, 1510–1516.
106. Shaw, L. M., Yatscoff, R. W., Bowers, L. D., Freeman, D. J., Jeffery, J. R., Keown, P. A. & McGilveray, I. J. (1990) Canadian Consensus Meeting on Cyclo-

sporine Monitoring: Report of the Consensus Panel. Clin. Chem. *36*, 1841–1846.

107. Yatscoff, R.W. (1991) Cyclosporine Monitoring: Consensus Recommendations and Guidelines. Can. Med. Assoc. J. *144*, 1119–1120.
108. International Consensus Conference on Immunosuppressive Drugs, Lake Louise, Alberta, May 05-07, 1995.
109. McBride, J.H., Kim, S.S., Rodgerson, D.O., Reyes, A.F. & Ota, M.K. (1992) Measurement of Cyclosporine by Liquid Chromatography and Three Immunoassays in Blood from Liver, Cardiac, and Renal Transplant Recipients. Clin. Chem. *38*, 2300–2306.
110. Morris, R.G., Saccoia, N.C., Ryall, R.G., Peacock, M.K. & Sallustio, B.C. (1992) Specific Enzyme-multiplied Immunoassay and Fluorescence Polarization Immunoassay for Cyclosporine Compared with cyclotrac (^{125}I) Radioimmunoassay. Ther. Drug Monit. *14*, 226–233.

1.3 Special Aspects of Forensic Toxicology

H. Brandenberger

1.3.1 History

As has been pointed out in the introduction to chapter 1.1 of this book, analytical toxicology can function in the service of medicine and health or in the service of law. In the first case, we talk about clinical or emergency toxicology, the second service is called legal or forensic toxicology. While the beginning of forensic toxicology dates back to the 18th century, clinical toxicology is of much more recent origin. This may have different reasons:

1. Up to the middle of the 19th century, chemical analyses for the presence of poisons in body fluids or body tissues were very time-consuming. In most cases they also required large biological samples which could not always be provided by the clinicians. Let us consider – as an example – the quantitative analysis for mercury. The method of Stock (1, 2), the only available mercury determination prior to 1950, consists of isolating all mercury in the biological sample as droplets of free metal with quantification by determining the drop sizes under a microscope. Such an analysis can only be accurate and reliable in the µg-range. And since mercury concentrations down to 10 µg/kg must be measured, biological samples weighing at least 100 g are needed. The decomposition and analysis of such a sample takes about 2 working days. This also is not always acceptable in clinical investigations.

2. It is only relatively recently that clinicians became aware of the support they (respectively their patients) can obtain from the information about a possible presence or absence of certain poisons or poison groups. Less than 20 years ago, the head of a clinical emergency ward told this author that he never asks for help from chemical toxicology, since his ward is anyway always using symptomatic therapies, regardless of the possibilities of exogenous intoxications. He refused to consider a change of policy, because of the low percentage of the ward's mortality rate, so he said.

Since forensic toxicological investigations – and this holds especially for the past – are much less restricted than clinical analyses by a lack of large biological samples as well as by the need to provide results in the nick of time, laboratories for toxicological analyses have first been founded by institutions involved in forensic work, either in departments of forensic medicine (in most of Europe, US and Japan), laboratories dealing with criminalistics (England) or, less often, in pharmaceutical institutes (Benelux countries). However, at the beginning, the toxicological analyses carried out in the institutes for forensic medicine were often delegated to people with insufficient chemical training, i. e. to medical assistants and medical technicians.

In one case we have knowledge of, arsenic determinations according to the method of Marsh (3) were carried out by the janitor of the building, who had been sent to the near-by chemical institute to learn a technique that he could not fully comprehend.

But due to the call for more precise analytical information, originating especially from cross-examinations in court, this stage of "laymen chemistry" did not last very long. The institutes of forensic medicine had to provide a solid base also for their chemical laboratories and entrusted them to people with chemical training, to chemists and chemical assistants familiar with analytical investigations and aware of the problems connected with this work. This development was furthered by the advances in the field of chemical analysis, especially by the introduction of new instrumental methods of separation (chromatographic techniques), as well as the successive establishment of the various spectroscopic identification methods no chemical toxicologist would want to do without today.

While the toxicological chemists and chemical technicians, as late as 1960, were often regarded and treated by the medical staff as "Messknechte" (servants charged with simple measurements), the situation has – in most places – changed by now. In most institutes of legal medicine, chemists and their assistants are regarded as partners and their help is appreciated. In some places, the forensic institutions have branched out into a medical and a chemical department, working under separate technical management, but still in close collaboration. This author believes that both partners have benefited from such a separation. It has often led to improvements in the validity of chemical results and furthered scientific development in the field of chemical toxicology.

Slowly, some clinicians and MDs in other medical departments, even private medical doctors, became aware of the possibilities inherent in a toxicological analysis for poisons. They started to send blood and urine for chemical investigations. Sometimes, they requested a check for a specific poison, in other cases they asked for a group analysis (i.e. an analysis for heavy metals, barbiturates or hypnotics), and occasionally they even wanted a general search for "all" drugs or poisons. Often, such a service worked successfuly. At times, it was hampered due to an unsatisfactory exchange of information, or – since clinicians and chemists do speak different languages – also due to misunderstandings. I remember such a case from my first year as a chemical toxicologist. Urine was sent by mail from the nearby hospital (mailed in the post office located just beside our laboratory) with the handwritten request "Analyze for Mo". I selected an analytical procedure for molybdenum determination and had a technician adapt it for use with urine. After 2 days, just as the analysis of the sample was under way, I received a telephone call from the young intern who had sent the request. She had expected an analysis for morphine, which was at that time – 1961 in Switzerland – not one of the common problems as it is today. As a newcomer in the field, I was not yet familiar with medical slang.

To analyze for a specific poison is usually simple. To search for a group of poisons can sometimes be quite a bit more complicated, but it is usually feasible in a moderate amount of time and with a reasonable effort. But to search for an unspecified poison or "all" poisons in a small biological specimen is a sizable undertaking which requires chemical know-how, excellent instrumentation, skill and an abundance of time. The chemical toxicologist is therefore happy to obtain all the information which can give him a hint as to the nature of the poison(s) which may be involved: all available anamnestic data about the patient, his job, his hobbies, a descriptions of the most important symptoms, a list of all medicaments he may have taken. Powders, pills, bottles, glasses and other suspect materials which may have been involved in the intoxication, should also be submitted to the chemical investigator.

Some clinicians refuse to furnish anamnestic or clinical information to a person outside the hospital. They believe that medical data should not be released to non-medical people. They also do not realize how complicated a chemical investigation may be. This is a very serious draw-back for an efficient and successful collaboration. By submitting the pertinent information, the extremely difficult task of a general search for poisons may be reduced to the much easier job of a group analysis or even, if the anamnestic information contains a lucky hint, to the determination of a single compound.

A reluctance to furnish information and sometimes even biological samples from the hospital, misunderstandings and fear that a forensic institution may not be capable of working fast, may have helped in the establishment of sections devoted to toxicological analysis in the departments for clinical chemistry of the larger hospitals. Such a development was greatly furthered by the appearance of immunochemical assays. They are very simple to carry out. They can be automated and performed in large series and resemble in this respect other analytical determinations a clinical laboratory must carry out daily. They even may be assigned to technicians without professional education. Since conclusive results were first expected from such simple tests, the step into toxicological analysis was made without hesitation. It took some time and a number of complaints and contested results, until the clinical chemists became aware that, in order to exclude mistakes, today's immunotests must be confirmed by a second non-immunochemical technique (4). Now the hospital laboratories had to introduce also analytical methods needed to confirm the results of the immunotests: chromatographic separation and spectroscopic identifications. Today, there is no longer much difference in the equipment of forensic and clinical toxicological laboratories. Furthermore, both types of laboratories are staffed by personnel with similar training, know-how and experience. Differences, however, do exist. They are described in the next part of this chapter.

1.3.2 Differences between Forensic and Clinical Toxicology

1.3.2.1 Goals of the analytical investigations

Clinical toxicology is a diagnostic assistance for the medical profession. It should help to heal a patient by detecting one or several poisons which may be the cause of the illness. It has to assist in the efforts to save the patient's life.

Forensic toxicology, on the other hand, is assistance in a legal investigation of an accident or a crime. It must verify if a poisons is responsible for an event or may have influenced it. Further aims are to clarify if the poison found may have been administered deliberately or accidentally, and to shed light on the time sequence of the events.

In both fields, the results of the chemical investigation may further be used as a basis for taking proper measures to prevent a repetition of the accident or crime.

1.3.2.2 Mandators of the analytical investigation

In clinical toxicology, the mandators are MDs, especially clinicians and among them mainly the doctors in charge of medical emergency wards.

In forensic toxicology, the mandators are people involved in the preservation of law and order, i.e. judges charged with the investigation of an accident or crime, in some countries also police officiers, state or district attorneys respectively prosecuters. In accordance with a request from a legal authority, the pathologist charged with the autopsy of a deceased person may also act as mandator of a forensic-toxicological investigation.

Intentional intoxications are classed in most countries as offences requiring public prosecution. An investigator is therefore entitled to extend a toxicological search, if he is convinced that additional information of service to the law may be found. This is at least this author's opinion, who has always met with approval from the legal authorities for additional investigations he had decided to carry out.

1.3.2.3 Specimens submitted for analysis

In clinical toxicology, blood serum and urine samples are the specimens usually submitted for toxicological analysis, less often whole blood, and in recent times but still seldom, also hair or saliva. Between 5 and 10 ml of the serum or urine specimens are usually available. That is a small amount, if several determinations have to be carried out, as is the case in a group analysis or in a general search for poisons.

In forensic toxicology, the entire stomach content or stomach wash, whole blood and the total content of the bladder are the most important biological specimens. Quite often, brain tissue (i.e. for alcohol analysis, if death has occurred after vomiting and aspiration, which can falsify an alcohol analysis carried out with heart blood), liver tissue (specially important if not much blood is available), kidney (essential in the absence of urine), gall bladder content (i.e. for morphine analysis) or the intestinal contents may also be submitted together with the body fluids. For special purposes, hair can be of great value, i.e. for determining intoxications with arsenic, other toxic metals and semi-metals, as well as drugs of abuse. Hair analysis, if properly carried out (simultaneously grown segments should be analyzed separately), can not only indicate that an intoxication has occurred; it may also pin down the time period of exposition.

The pathologist should know that in forensic cases, he must secure not only small samples, but the entire content of stomach and bladder. Unfortunately, this is not always observed, and the situation is aggravated, if the chemist does not get informed. It may also be that the pathologist decides to use a part of the stomach or urine for other purposes than chemical analysis. Then he must measure the total volumes of the materials very carefully and forward the information to the chemist. For stomach, the total poison contents and not the concentration at the time of death is needed for the interpretation of the case. It must also be remembered that the stomach content has to be homogenized before aliquots can be removed. But before homogenization, it must be carefully screened for residues of tablets or food materials, evidences which are destroyed by the homogenization.

In forensic investigations, non-biological materials which may be connected with the accident or crime should always be forwarded to the chemist, together with the biological specimens or directly by the police who has confiscated them. They can then be investigated already before the biological samples arrive from the pathology. It is not easy to make a choice of materials to be secured. Experienced police officiers may have a flair for such an investigation. An example for a first unsuccessful and then successful search by the police follows:

A young soldier died while on weekend leave in the home of his parents. The police found a number of pharmaceuticals and even a forbidden drug among the belongings the soldier had left in the barracks of his service town. They were submitted to us. None of these drugs or their metabolites could be found in the body fluids of the soldier. On the other hand, we could establish the presence of Rohypnol and serveral of its metabolites. Since it was the first time that we attributed a death to the sole action of this hypnotic (then fairly new on the market), and since no other lethal intoxications with Rohypnol (at the time considered as safe as Valium by the medical toxicologists) had so far been reported, our findings were contested. A police woman was assigned to further investigate the case. She solved the problem in one day: The father of the soldier had received Rohypnol from his personal MD, but had not used it. The police woman found the almost empty package in the night table in the father's bedroom.

1.3.2.4 Containers, labeling, sample delivery

In a search for organic poisons, compounds liberated from most plastic materials (plasticizers such as tributylphosphate and phthalates) and other omni-present materials (solvent traces, diethylene glycol) can greatly impede toxicological analyses. This author used to request that specimens collected for the analysis of solvents, organic pesticides or drugs, particularly body fluids, be submitted in glass containers. He also insisted that body fluids for metal analysis should always be submitted in plastic bottles, certainly not in lead glass flasks. In a forensic institution, the chemist may be successful with such requests, although some pathologists hate to be told that they must deal with as simple things as containers. Mandators from the outside are usually eager to put their analytical problems in good hands and therefore willing to comply with such rules, especially if the containers are provided free of charge. It probably should be added that our specifications for container materials are not shared by all toxicologists (5).

A hospital laboratory can have more difficulty with the container problem. In clinical chemistry, specimens are often secured in plastic tubes, and an MD may not easily understand that for toxicological investigations carried out in the same laboratory tract, glass should be used. He may also not realize that for a complex toxicological investigation, a few ml of body fluid may be insufficient, and that some problems cannot be solved with serum as an analytical specimen. It happened several times that we received serum samples for determining carbon monoxide. In one instance, when I tried to tell the young medical intern that such an analysis requires whole blood and not serum, she cut me short by commanding: "You have to run the sample, the head of the clinic has given the order."

Utmost care must be taken that all specimens are properly labeled, in clinical as well as in forensic investigations. We consider it criminal to risk that analytical results can be assigned to the wrong person because of sheer negligence. In forensic cases, the wrong person may be inculpated, i.e. as a drunken driver or a drug dealer. In clinical cases, a medical treatment may be given to the wrong patient while the man in need might go without help. We can see no reason why the problem of labeling should not be given the same attention in clinical as in forensic toxicology.

Sample delivery is another point of importance, especially in clinical emergency toxicology. But this does not mean that in forensic work, slow ways of sample delivery should be permitted. If the clinical ward and the laboratory are housed in the same building complex, fast delivery should be (but is not always) secured. Smaller hospitals have to send their samples out, to a large central hospital in the nearby town or to a laboratory in a different town. With proper organization, this should also not pose a major problem. In most countries, mail delivery has become slow. Valuable time is lost, unrefrigerated samples can deteriorate. For a fast delivery, special messengers with cars or motorcycles should be used. We recommend that delivery be made against a timed receipt.

The author's laboratory at Zürich University used to work for 14 Swiss states (Kantone). Samples were submitted over distances of several hundred km. In default of a special messenger, we recommended that the hospitals hand the specimens over to a ticket man of the next express train to Zürich. A laboratory staff member would wait for the material on arrival time at Zürich train station. This proved to be an efficient delivery method, even free of charge.

1.3.2.5 Analytical investigations

In clinical toxicology, speed is of primary importance. Preliminary tests have to be followed at once by fast extractions. Time-consuming purifications of extracts may be neglected in favor of getting fast preliminary results. There is no time to confirm every result by 2 different methods, as is often requested in forensic investigations. A chemist may even have to take calculated risks, since the analysis is often a race against the progress of the illness, against death. In our laboratory, several technicians used to work on such emergency cases; the responsible person kept a steady watch over all experiments. Until useful information could be transmitted to the emergency ward, all other activities (research, work on forensic problems) had to be relegated.

Forensic investigations must be handled differently. The methods best suited for a problem have to be carefully selected. Care must be taken that the results are sufficiently conclusive to be used as evidence in court. It is always desirable that results be corroborated by a second independent method. And it is always helpful if the analytical data are instrumentally recorded, either by analog recorders (in GC and HPLC, UV, FS and IR) or, even better, digitalized by a computer (in GC-MS, LC-MS, GC-FTIR and NMR). Only exceptionally (i.e. if a possible prosecution is based on the outcome of laboratory trials) should not yet completely verified preliminary results be reported by telephone. In spite of a later rectification, they may stay on record in the case documentation and are very difficult to erase. It is always preferable that a forensic search be finished before a report is handed out.

1.3.2.6 Book keeping and reporting of results

As mentioned above, analytical results in the service of emergency toxicology have to be reported as soon as possible by telephone. A written report is of course also requested, but that can follow later in a relatively short form: materials received for analysis, analytical methods used and results. An interpretation of the results can be helpful for the clinician, who has little time to run to the library for obtaining the necessary references.

A forensic report must contain much more information. The specimens received have to be described. The analytical pathways should be listed, all analytical steps and results reported. Not all the chemical information may always be understood by the lawyers. This is why a report should contain a summary in plain words every layman can understand. Still, the chemical part of the report is important, since an involved party may try, helped by a chemical consultant, to object to the conclusions. The chemist must be able to back up his results by a complete analytical documentation. It does not have to be included in the report, but it should be kept on file.

In forensic toxicology, bookkeeping is of extreme importance. The chemist is responsible for all specimens received and must account for them. In general, he is not permitted to use them up; a part (i.e. half) must be kept for further investigations. A forensic toxicologist who has to admit that he has used all submitted materials is not in a good position, if his work is contested. He loses credibility.

Every chemist and chemical technician must keep a laboratory notebook and record all experiments in chronological order. This should be a rule in all chemical research, but is often neglected. In forensic work, the notebooks must receive special attention. Forensic investigations not reported in laboratory notebooks should not be accepted.

1.3.2.7 Fate of analytical specimens

In clinical toxicology, analytical specimens may be discarded at the end of an investigation. In forensic toxicology this is an offense. The remaining specimens have to be kept, the body fluids and tissues in the frozen state, for a relatively long time which should be specified by the legal authorities of each country or state. Failure to comply with this obligation can be considered as destruction of legal evidence, since the authorities lose the possibility to request additional evidence, be it from the same or from a second laboratory.

1.3.3 Instrumentation of a Laboratory for Forensic Toxicology

In 1979, the German Research Society (DFG) appointed a senate commission to establish guidelines for the promotion of clinical toxicology in Germany. It was stated that the existing laboratory facilities were not adequate to deal with the roughly 200000 intoxicated patients admitted per year to West-German hospitals, often as emergency cases. The commission recommended that 3 different types of toxicological laboratories are needed (5):

- very simply equipped so-called A-type labs which every hospital handling intoxications should install,
- better-equipped B-type labs for larger city hospitals, distributed over the whole country, and
- about a dozen C-type labs possessing all the equipment needed for sophisticated work.

In the A-type laboratories, a qualitative search for the most common drugs and poisons in blood and urine should be feasible by simple color reactions, thin-layer chromatography and immunochemical assays. The B-type laboratories should possess, additionally, also the know-how and equipment for the qualitative and quantitative determination of less common drugs and poisons, including toxic metals. Their equipment must therefore contain instrumentation for GC, HPLC, UV, FS and AA. In the largest C-type laboratories, difficult investigations such as a search for unsuspected poisons and for very seldom occurring poisons, as well as trace detections of organic and inorganic compounds, should also be possible. The review of the existing facilities showed that the requirements requested for a C-type laboratory could be met in part or completely only by the toxicological departments of the institutes for forensic medicine.

What specific instrumentation has to be available in ready-to-use condition in such a forensic or C-type clinical toxicology laboratory?

1. Equipment for carrying out preliminary informative tests, which may furnish guidelines for the subsequent analytical strategy, such as:
   - gas detectors and equipment for color or spot tests,
   - equipment for TLC,
   - instrumentation and reagent kits for running immunochemical assays.
2. Equipment for carrying out steam distillations and organic extractions of body fluids and homogenized tissue samples, as well as the means for enzymatic pre-treatment of the biological samples.
3. Instrumentation for all kinds of separation techniques such as sublimation, TLC, GC and HPLC, if possible also for supercritical fluid chromatography (SFC) and capillary electrophoresis (CE).
4. Recording spectrophotometers for obtaining UV, IR and fluorescence spectra (FS), permitting work with micro-samples (micro cells for UV, beam condensers or FTIR-technique for IR). In addition, access to a nearby NMR instrument should be possible.

5. One or two mass spectrometers must be part of the installation. At least one of them should permit also CI-MS and negative ion MS. Access to a high resolution MS and an isotope ratio MS should be possible.
6. The equipment listed under points 4 and 5 must be capable of working independently and on-line with chromatographic techniques, permitting the following combinations: TLC-UV, HPLC-UV, HPLC-FS, HPLC-MS, GC-MS, GC-FTIR, CE-UV, CE-MS.
7. For quantitative metal analysis, AA is indispensable. The qualitative search for metallic poisons may be carried out in different ways: by special HPLC (ion chromatography) or CE, using electrochemical or fluorescence trace detection methods, or – best – by an ICP-MS combination (Plasma-MS).

Such an installation is expensive, especially if it has to be purchased not only for a forensic laboratory, but also for a near-by laboratory for clinical toxicology. Furthermore, one must consider, that in addition to all these instruments and a lot of everyday laboratory equipment, considerable literature must be available, such as methodological books, periodicals, handbooks (6–9), and compilations of UV-, FS-, IR- and mass spectral data (10–16). And last but not least, the instruments need laboratory space to be conveniently placed. Overcrowded rooms constitute a danger for cross-contaminations.

1.3.4 Staff of a Laboratory for Forensic Toxicology

A chemical toxicologist has to be an analytical all-rounder. Nowadays, such people are hard to find. Today's university education in chemistry, especially chemical analysis, breeds specialists. They may be tops in their field, be it AA, IR, NMR or MS, but they are hardly masters of many different analytical techniques. The staff of an average toxicological laboratory is usually small. It may number up to 10 people, seldom more, and it is therefore not possible to hire a specialist for each technique. Flexibility is required. Specialists desiring to "play" with a technique they know best are often more interested in "analytical cosmetics" than in chemical analysis (17). They are less useful in a toxicological laboratory than all-rounders, who do not mind switching methods according to the requirements of the case under investigation.

An additional difficulty to find laboratory staff may be the fact that it must be prepared to do night work. It is therefore not only the expenses for instrumentation and laboratory space, but also the problems of finding suitable staff, which must be considered in answering the question: Should forensic and clinical toxicology be treated in separate institutes or in a single department dedicated to both services. For the second solution speaks also the fact that clinical toxicological cases very often lead to forensic investigations: "After medicine has done her work, the law can take over".

1.3.5 Concluding Remarks

The author of this chapter ran for over a quarter of a century a university toxicology laboratory, which was initially established for the needs of forensic science. But during all these years, the institution has also served a large number of Swiss hospitals as a laboratory for clinical-, including emergency-toxicology. He is convinced that both sides of the work greatly benefited from the experiences gained by this dual obligation.

Forensic investigations must be extremely precise and detailed. Since time is not always of primary concern, the investigators may be tempted to get lost in details and dig too deep. Clinical toxicology, on the other hand, has to provide results fast. Due to this urgency, the quality of the chemical investigation may suffer. Time does not always permit to select the top methodological approach and the most favorable conditions. Not seldom, short cuts must be taken in extraction work or chromatography. If a chemist does this every day, the accuracy of his general working habits may be affected.

We have made the experience that clinical toxicology, especially emergency toxicology, animates a laboratory crew. The occupation with forensic problems – on the other hand – develops the sense for accuracy and completeness. Both qualities should be furthered in chemists and chemical assistants. Both should be promoted. This is possible by assigning – to the same staff member – once a forensic problem and another time a clinical emergency case. The staff of a toxicological laboratory with such a dual function (for forensics and emergency toxicology) has a more interesting work basis and will receive a more complete education than the personnel of an institution devoted to only one kind of service job. – The chemist in charge must of course always be aware of the different requirements between an analysis for forensic and one for clinical purposes. The dual function will broaden also his horizon. The spectrum of poisons and medicaments he is confronted with will be larger than in pure forensic or pure clinical work, and the different requirements for reporting results should be a welcome change from routine.

References

1. Stock, A., Lux, H., Cucuel, F. & Köhle, H. (1933) Zur mikrometrischen Bestimmung kleinster Quecksilbermengen. Z. für Angew. Chem. *46*, 62–65.
2. Stock, A., Cucuel, F. & Köhle, H. (1933) Die Bestimmung kleinster Quecksilbermengen in organischem Material. Z. für Angew. Chemie *46*, 187–191.
3. Leifheit, H.C. (1961) Arsenic in biological material. In: Standard Methods of Clinical Chemistry *3*, 23–34. Academic Press, London and New York.
4. Brandenberger, H. & Maas, R.A.A. (1988) Empfehlungen der Deutschen Forschungsgesellschaft zur klinisch-toxikologischen Analytik. Folge 1: Einsatz von immunochemischen Testen in der Suchtmittelanalytik. VCH, Weinheim.
5. Geldmacher-von Mallinckrodt, M., Bäumler, J., Brandenberger, H., Büttner, J., v. Clarmann, M., van Heijst, A.N.P., Heyndrickx, A., Ibe, K., Machata, G., Maes, R.A.A., Moll, H., Stamm, D., Stoeppler, M., de Zeeuw, R.A. & Fabricius, W. (1983) Klinisch-toxikologische Analytik; Denkschrift der Deutschen Forschungsgesellschaft. VCH, Weinheim.
6. The Merck Index, 12th Ed. (1996) An Encyclopedia of Chemicals, Drugs and Biologicals. Merck & Co., Inc., Rahway, N.J.
7. Clarke's Isolation and Identification of Drugs in pharmaceuticals, body fluids and

post-mortem materials (1986), 2. Ed. (Moffat, A.C., Jackson, M.S., Moss, M.S. & Widdop, B., eds.). The Pharmaceutical Press, London.
8. Schütz, H. (1982) Benzodiazepines – A Handbook, Vol. I. Springer-Verlag, Berlin, Heidelberg & New York.
9. Schütz, H. (1989) Benzodiazepines – A Handbook, Vol. II. Springer-Verlag, Berlin, Heidelberg, New York, London, Paris & Tokyo.
10. Organic Electronic Spectral Data, Volumes *1* (1960) to *29* (1993). Covers UV-spectral data (maxima with log-values) reported in the literature since 1946 until 1988, further volumes in preparation. John Wiley and Sons, Chichester, West Sussex.
11. Sadtler Standard Spectra and Sadtler Reference Spectra: A large selection of different libraries with UV-, FTIR- and NMR-spectra from different technical fields such as pharmaceuticals, drugs of abuse, pesticides and agriculture chemicals, priority pollutants, solvents, compounds of forensic importance, and others. Collected by Sadtler Research Laboratories Inc., Philadelphia, Pa., USA, and distributed by the Bio-Rad Laboratories.
12. The Wiley/NBS Registry of Mass Spectral Data (1989). A registry of over 150000 mass spectra from over 112000 compounds, collected by F.W. McLafferty & D.B. Stauffer, 7 volumes. John Wiley and Sons, Chichester, West Sussex.
13. McLafferty, F.W. & Stauffer, D.B. (1991) Important Peak Index of the Registry of Mass Spectral Data. A derivative of reference 12 containing over 400000 entries. 3 volumes. John Wiley and Sons, Chichester, West Sussex.
14. The NIST/EPA/MSDC Mass Spectral Database. A mass spectral collection of 54000 different compounds for computer retrival, which is constantly enlarged. It consists almost entirely of complete mass spectra and includes structural drawings. The disk version is obtained from the suppliers of MS instrumentation and HD Science Limited, Nottingham, England. It is considered today's standard mass spectral reference data base.
15. Pfleger, K., Maurer, H.H. & Weber, A. (1992) Mass spectral and GC data of drugs, poisons, pesticides, pollutants and their metabolites. 2. Ed., VCH-Verlagsgesellschaft, Weinheim.
16. Drug Identification Atlas, Department of Agriculture. Canada. Vol. 1 to 5 (to be continued). Contians UV-, IR- and mass spectra, as well as data on TLC and GC on all drugs listed in the race track supervision regulations.
17. Brandenberger, H. (1984) Chemical analysis or analytical cosmetics. Proceedings of the 21. International Meeting of TIAFT in Brighton, England (Dunnett, N. & Kimber, K.J., eds.). Published by "The International Association of Forensic Toxicologists".

1.4 Analytical Aspects of Doping in Sports

D. de Boer, T.J.A. Seppenwoolde-Waasdorp and R.A.A. Maes

1.4.1 Introduction

The use of special food and agents to enhance physical performance and to influence psychological well-being dates as far back as thousands of years. Nowadays, several of these agents of the past, such as ethanol, ephedrine and cocaine, are still used. Modern sports, as a special kind of competitive physical activity, encouraged men to look for even more effective compounds. As soon as the scientific progress in the 19th and in the 20th century made it possible to develop these compounds, the phenomenon of using agents for sport enhancement was called doping.

The origin and the meaning of the word "doping" in connection with performance is not clear. The word "dope" in this context appeared in an English dictionary in 1889 as a stimulating narcotic mixture for race horses (1). From a historical point of view, drug testing in sport started in horse racing (2). In 1910 the Russian chemist Bukowski is said to have realized a method to confirm the presence of heroin and cocaine in saliva of horses. The first horse doping cases were proved in Austria in 1912 by Fränkel, also in saliva. In those days however, doping was not considered as a main issue in human sports. It was not until the sixties, when several deaths were associated with doping, that drug testing was introduced in that kind of sports.

Analytical techniques are of course essential in the doping control procedures and often limit the possibilities of drug testing in sports. Nowadays, the legal defensibility of these techniques is of utmost importance and determines the choice of analytical procedure and the assessment of quality control. Moreover, analytical techniques also play a role in the development of doping epidemics (3). This review describes the analytical techniques applied in drug testing in sports and discusses current practical problems.

1.4.2 Doping Control Procedure

The doping control procedure starts at the sport event itself, where an athlete or animal is subjected to strict procedural guidelines. In order to provide a medium to detect doping, a biological specimen is collected.

At the collection site the sample is divided into two portions, the test (A) and reserve (B) sample. The respective specimen containers are sealed and transported to a qualified laboratory. The test sample is used for the analysis and the reserve sample is stored under strict conditions (sealed, 4°C or less in a secured storage room). If the test sample contains a forbidden compound, the athlete is suspected to have applied doping. The reserve sample is available for a second analysis.

1.4.3 Selection of Biological Specimen

The selection of biological specimens depends on many aspects. At the moment, urine is most frequently analyzed in human sport doping control. The main advantages of using urine are its non-invasive way of collection and the relative high concentrations of the substances to be analyzed. The use of blood, however, may provide a better correlation with the pharmacokinetics and pharmacodynamics of substances of interest and for some compounds a blood sample is even the only way to detect their abuse. It is obvious that the detection of blood doping requires blood specimens (4).

The disadvantages of the collection of urine are the possibilities to manipulate and the so-called humiliating experience. As an invasive technique, the collection of blood also presents a number of restrictions. The amount of specimen is limited, handling requires special precautions and some religious groups may refuse to donate blood.

Recent developments in the field of drugs of abuse have shown that saliva and hair may also be specimens of interest (5, 6). When hair is analyzed it may not only reveal the identity of the drug, but also the moment of administration, especially when the drug has been cleared from other biological fluids and tissues. Saliva may be of interest as it may reflect the blood concentration of a drug. For doping control however, both drug testing in hair and saliva still have to be evaluated. Because much experience with urine and blood is already available and sampling of both specimens provides complementary possibilities to detect doping, it may be expected that urine and blood will be collected together in the very near future. In horse doping control both blood as well as urine specimens are already collected frequently, also because most of the 'ethical' disadvantages mentioned are ignored.

1.4.4 Doping List

Several attempts to define "doping" in one sentence did not lead to a satisfactory definiton and probably such an approach will never be satisfactory. A more pragmatic approach comes from the International Olympic Committee (IOC), who started in the late sixties with a list of forbidden substances. After several updates, the list resulted in the one as we know by now (table 1). By their definition doping can be referred to as the use of prohibited classes of drugs and of banned methods as mentioned on a defined list. Fortunately, most international and national sport federations comply to the IOC rules.

Table 1. List of forbidden substances and methods of the International Olympic Committee (7)

I. Doping classes	
	A. Stimulants
	B. Narcotic analgesics
	C. Anabolic agents
	D. Diuretics
	E. Peptide hormones and analogs
II. Doping methods	
	A. Blood doping
	B. Pharmacological, chemical und physical manipulation
III. Classes of drugs subject to certain restrictions	
	A. Alcohol
	B. Marijuana
	C. Local anesthetics
	D. Corticosteroids
	E. Beta-blockers

The IOC-list is based on substances classified on their pharmacological actions and/or structure. Every class is ended by "and related compounds" and the active agents mentioned on the list are merely meant to be examples. The advantage is that every drug, which belongs to one of the classes and sometimes may be especially designed for doping purposes, is banned automatically.

Some sport federations have their own specific doping list. The International Amateur Athletic Federation (IAAF) for example, does not consider the use of beta-blockers and diuretics as doping (8). In athletics the use of beta-blockers is indeed probably no matter of importance. The diuretics, however, could be of concern if those are used with the intention to dilute urine specimens. In such a case those substances can be considered as prohibited. In some sports even more substances than mentioned on the IOC-list are forbidden. In modern pentathlon the use of all sedatives, and drugs with sedative constituents are forbidden (9). In the USA the society is concerned about the effects of alcohol, tobacco and street drugs. The National Collegiate Athletic Association (NCAA) has therefore implemented the possibility to analyze for alcohol, nicotine, cannabinoids and heroin in their doping control program (10).

1.4.4.1 Stimulants

The group of stimulants on the IOC-list is a heterogenous collection of compounds, which according to their pharmacological action can be divided in subclasses, although each of these classifications is subject to discussion (11). The most best-known subclass in doping analysis concerns the psychomotor stimulant drugs such as the amphetamines (I), pemoline (II) and cocaine (III). The designer drugs like the 3,4-methylenedioxy-amphetamines show that although not marketed legally, several kinds of *N*-alkyl amphetamine-like compounds may be dealt with in drugs of abuse and thus also in doping analysis (for a selection see table 2).

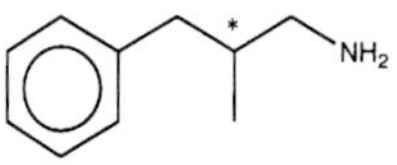

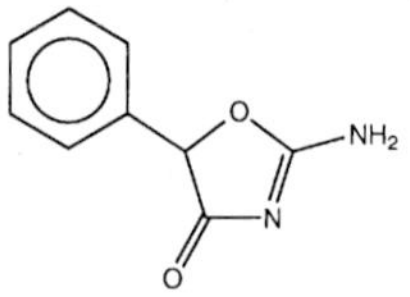

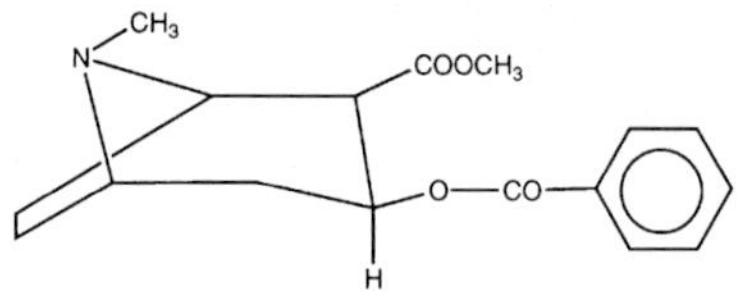

I II III

The sympathomimetic ephedrines show a less pronounced central action than the amphetamines. Their use has been led to interpretation problems, because of the presence of ephedrine (IVa; R = CH_3) or norephedrine (IVb; phenylpropanolamine; R = H) in ginseng extracts and formulations, which are commonly used as over-the-counter cough and cold preparations. To avoid such problems, the IOC considers only a concentration (total of the ephedrines and norephedrines) of more than 10 μg/ml urine as a violation of the IOC-list.

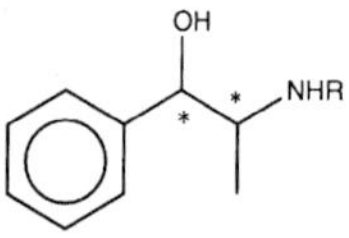

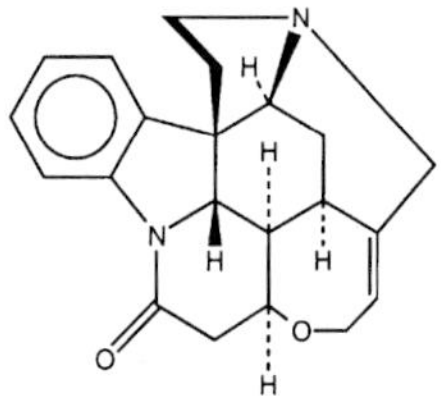

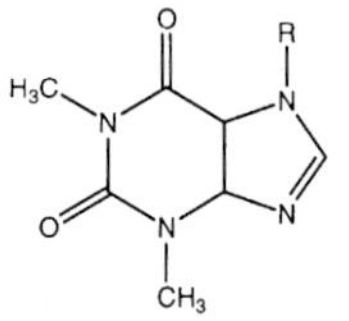

IV V VI

Analeptica such as strychnine (V) and the xanthine caffeine (VIa; R = CH_3) also belong to the stimulants. The socially tolerated stimulant caffeine has been used for centuries by millions of people consuming several kinds of food or beverages. Only large doses of caffeine are banned, assuming a dose-response curve for the ergogenic effects of caffeine. Currently, a urinary concentration > 12 μg/ml of caffeine is defined as doping by the IOC (7). It is estimated that recreational use of coffee will result in urine levels of 3 to 6 μg/ml (25). For theophylline (VIb; R = H), another xanthine and mainly used as a bronchodilator, no limit exists. It is assumed that up until now theophylline demonstrates not a main problem in sports. Mesocarb (VII) and amineptine (VIII) are structurally unique stimulants, being a meso-ionic compound (26) and a tricyclic antidepressant-like drug (27), respectively.

Many stimulants can be detected in urine as their parent compounds, although some of the stimulants are subject to extensive metabolism. It must be mentioned, that a number of different amphetamine-like compounds have shown to be metabolized to *N*-dealkylated compounds, which are not specific metabolites (table 2). In such cases the presence of any possible precursor must be checked. Urinary concentration ratios of different metabolites or the presence of characteristic metabolites

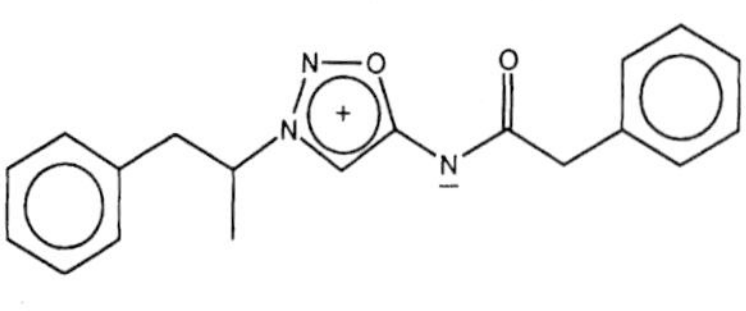

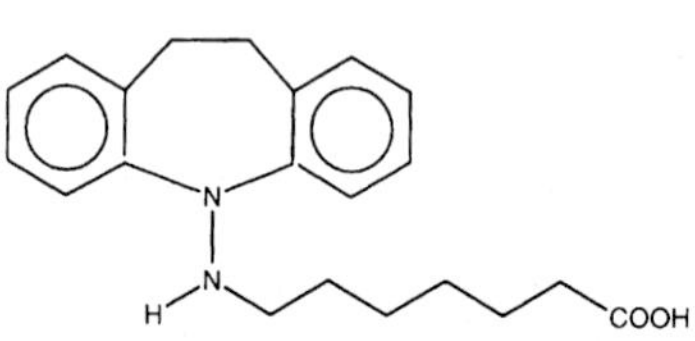

VII VIII

Table 2. Amphetamine-like compounds, examples of their trademarks and certain non-specific metabolites

amphetamine-like compounds	example of trademark	possible non-specific metabolite	reference(s)
deprenyl	Eldepryl®	(meth)amphetamine	(12)
dimethylamphetamine	–	(meth)amphetamine	(13, 14)
fencamine	Sicolor®	(meth)amphetamine †	(15)
furfenorex	Frugalan®	(meth)amphetamine	(13)
methamphetamine	Pervitin®	(meth)amphetamine	(13, 14)
N-acetyl-methamphetamine	–	(meth)amphetamine	‡
ethylmethamphetamine	–	ethyl- and (meth)amphetamine	(14)
ethylamphetamine	Apetinil®	(ethyl)amphetamine	(13, 14)
isopropylmethamphetamine	–	isopropyl- and (meth)amphetamine	(14)
isopropylamphetamine	–	(isopropyl)amphetamine	(14)
benzphetamine	Didrex®	benzyl- and (meth)amphetamine	(13, 14)
benzylamphetamine	–	(benzyl)amphetamine	‡
amphetamine	Benzedrine® $	amphetamine	(13, 14)
N-acetyl-amphetamine	–	amphetamine	‡
N-formyl-amphetamine	Formetorex®	amphetamine	‡
fenproporex	Fenorex®	amphetamine	(16)
mefenorex	Pondinil®	amphetamine	(17)
mesocarb	Sydnocarb®	amphetamine	(18)
prenylamine	Reocorin®	amphetamine	(19)
ephedrine	Ephetonin® $	(nor)ephedrine	(13, 20)
methylephedrine	Pholcomed®	(nor)ephedrine	(13, 20)
etafedrine	Nethaprin®	(ethyl)(nor)ephedrine	(13, 20)
amfepramone	Tepanil®	(ethyl)norephedrine	(21)
norephedrine	Propadrine® $	norephedrine	(13, 20)
mephentermine	Wyamine®	phentermine	(22)
phentermine	Ionamin®	phentermine	(13, 14, 22)
phendimetrazine	Prelu-2®	phenmetrazine	(13)
phenmetrazine	Preludin®	phenmetrazine	(13, 23)
fenfluramine	Ponderal®	norfenfluramine	(13)
norfenfluramine	–	norfenfluramine	(13)
3,4-methylenedioxy-ethylamphetamine	'Eve'	3,4-methylenedioxy-amphetamine	‡
3,4-methylenedioxy-methamphetamine	'XTC', 'Adam'	3,4-methylenedioxy-amphetamine	(24)
3,4-methylenedioxy-amphetamine	'MDA'	3,4-methylenedioxy-amphetamine	‡

$ many preparations under various trademarks are on the market
† metabolism described, but no definite identification of the metabolite reported
‡ metabolism assumed, but no identification of the metabolite reported

may also be indicative. The relative concentration of 4-hydroxy-methamphetamine to methamphetamine for example, might be characteristic for methamphetamine administration (28). Metabolism of methamphetamine and amphetamine either as parent compounds or as intermediate metabolites may result in the (nor)ephedrines as well, but the observed urinary concentrations are very low and insignificant (29). GC/MS confirmation of methamphetamine in general must also proceed with caution in the presence of ephedrine because of analytical complications (30). Of the parent compounds as mentioned in table 2, the use of deprenyl or prenylamine is not forbidden.

1.4.4.2 Narcotic analgesics

The use of narcotic analgesics in general is forbidden in sports. Morphine (IXa) is a well known example. Codeine (IXb) is a narcotic of which the medical use is allowed only recently by the IOC. It is applied as a cough suppressant and has probably been interpreted wrongly in former days as a sport-enhancing drug. Buprenorphine (X) and fentanyl (XI) are more potent analgesics than morphine, and are examples of which the detection of their abuse will be challenged due to the relatively low doses (< 1 mg) (31).

Metabolism of narcotics primarily occurs through *N*- and *O*-dealkylation and subsequent conjugation with glucuronic acid (28). If no [monoacetyl-] (IXc) or ethylmorphine (IXe) is found (fig. 1), the calculation of certain ratios and concentrations

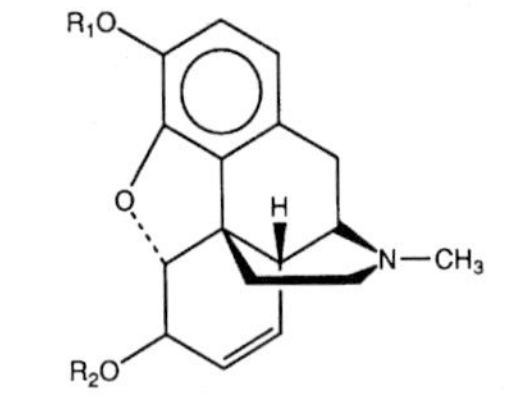

IXa	R_1 = H;	R_2 = H
IXb	R_1 = CH_3;	R_2 = H
IXc	R_1 = H;	R_2 = CO - CH_3
IXd	R_1 = CO - CH_3	R_2 = CO - CH_3
IXe	R_1 = C_2H_5;	R_2 = H

IX

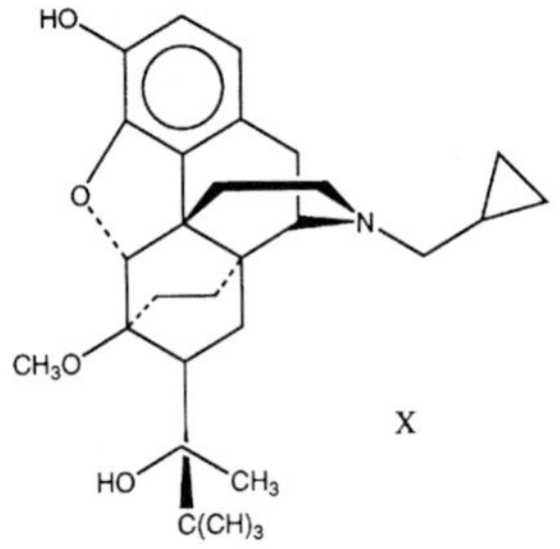

X

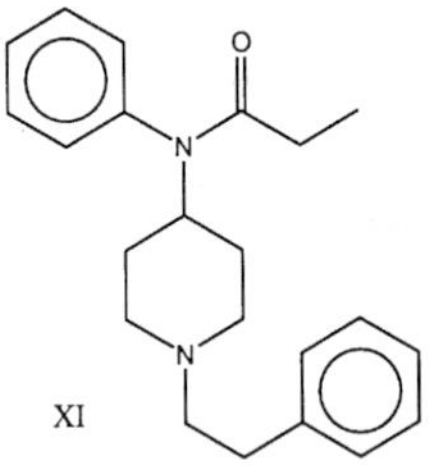

XI

can be useful for an interpretation of the presence of morphines. Since morphine may also originate from codeine, heroin (IXd), ethylmorphine (IXe) (28) or even poppy seeds (32), quantitative analysis is required, if morphine has been identified. Concentrations of free morphine < 1 [λg/ml] urine are not reported by the IOC. The medical use of codeine is allowed. Although no concentration limit exists, quantitative analysis may be indicative for codeine abuse (33). Of the *O*-alkylmorphines, benzylmorphine is obsolete.

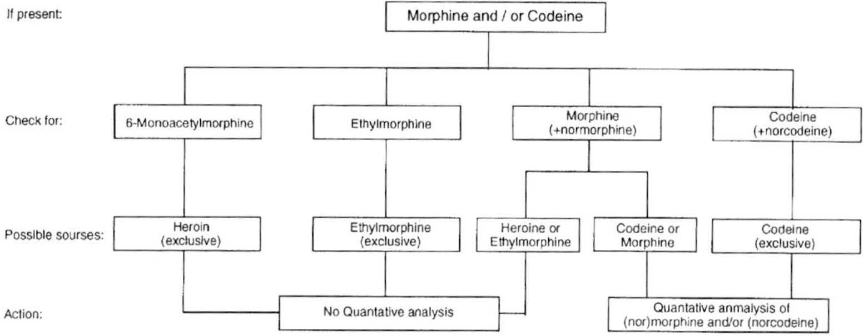

Figure 1. Decision diagram if during screening morphine and/or codeine is found in a urine sample.

1.4.4.3 Anabolic agents

Anabolic agents are divided in two subclasses, namely the anabolic androgens and the β2-agonist clenbuterol. The anabolic androgens may be classified into androstane (e.g. XII–XIV) and estrane (e.g. XV) derived steroids and heterocyclic steroids (e.g. XVI–XVII).

The detection of the use of testosterone (XII) is problematic since it is an endogenous steroid. Qualitative analysis is therefore useless. Quantitative analysis is also not satisfactory, because after doping control untimed samples are collected and the ranges of 'normal' and 'abnormal' concentrations are difficult to distinguish. The ratio of the glucuronides of testosterone and an other endogenous steroid, epitestosterone (XIII), provides however a specific indicator (34). Despite of all kinds of endocrinologic aspects and problems using such a ratio, it still has proven to be very useful (3). Additional information regarding the possible abuse of testosterone can be obtained by follow-up studies of a suspected case (35) or by using other ratios (36, 37). The more potent anabolic agent dihydrotestosterone (XIV) may become a future problem, especially as it is also an endogenous steroid. Research for dihydrotestosterone-sensitive specific indicators is in progress (38).

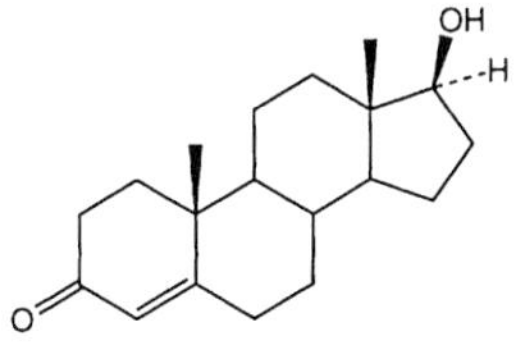

XII 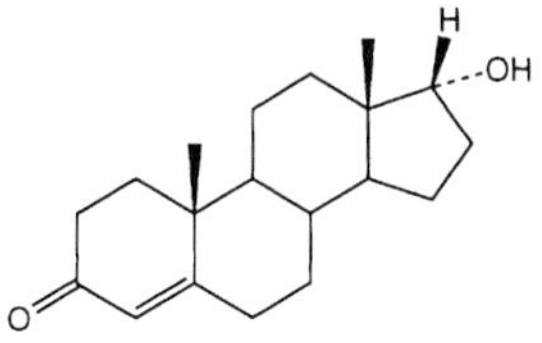 XIII 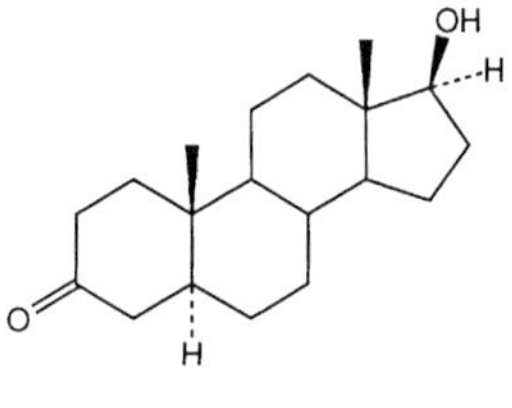 XIV

Although nandrolone (XV) is still considered as an exogenous steroid, it is actually an endogenous one. During pregnancy it is formed by the placenta (39), and in follicular fluids (40) a nandrolone metabolite can be found. However until now, due to the low endogenous concentrations, a detection of nandrolone metabolites in urine samples during doping control is considered to be of exogenous origin (41).

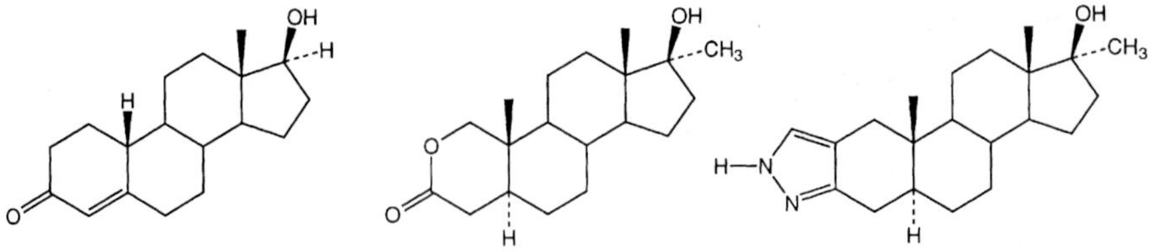

Heterocyclic anabolic androgens, such as oxandrolone (XVI) and stanozolol (XVII), are exclusively exogenous.

In general anabolic androgens are extensively metabolized and the detection is usually based on metabolites only (42). Hydroxylations, oxidations and reductions are the most common metabolic pathways. Metabolites are excreted as free substances or as conjugates with glucuronic acid or sulphate. In certain cases metabolites of anabolic androgens may originate from different precursors and the presence of a more specific metabolite must be checked (table 3). For example, besides nandrolone itself, also some oral contraceptives are possible sources of the main metabolite 19-norandrosterone (59). Other examples are some methylandrostanediol metabolites, which may originate from several parent compounds. Some of them result in the 17α-methyl-5α-androstan-isomer or the 5β-isomer mainly and others in a mixture of both isomers (table 3). 17-Methylepimer metabolites are also formed, but they are less important and probably are formed in urine *in vitro* (49, 56). The 1α- and 2α-methylandrostanediol metabolites may originate from two kinds of precursors.

Table 3. Non-specific and specific metabolites of some oral contra-ceptives and anabolic androgens

steroid	possible non-specific metabolites						specific metabolite	reference(s)
	A	B	C	D	E	F		
nandrolone	**	–	–	–	–	–	none	(43)
oral contraceptives								
ethynodiol	*	–	–	–	–	–	17α-ethynyl-5β-estran-3α,17β-diol	(44)
lynestrenol	*	–	–	–	–	–	17α-ethynyl-5β-estran-3α,17β-diol	(45)
norethisterone	*	–	–	–	–	–	17α-ethynyl-5β-estran-3α,17β-diol	(44)
norethynodrel	*	–	–	–	–	–	17α-ethynyl-5β-estran-3α,17β-diol	(44)
anabolic 1-methyl(en)-androgens								
mesterolone	–	–	–	–	–	*	1α-methyl-5α-androstan-3α-ol-17-one	(46)
methenolone	–	–	–	–	–	†	1-methyl-5α-androst-1-en-3α-ol-17-one	(47)
anabolic 2-methyl(en)-androgens								
drostanolone	–	–	–	–	–	*	2α-methyl-5α-androstan-3α-ol-17-one	(46)
stenbolone	–	–	–	–	–	†	2-methyl-5α-androst-1-en-3α-ol-17-one	(48)
anabolic 17α-methyl-androgens								
mestanolone	–	**	–	*	–	–	none	(49, 50)
oxymesterone	–	**	–	*	–	–	parent compound	(49)
oxymetholone	–	**	–	*	–	–	seco acidic metabolites	(51, 52)

Table 3. (continued)

steroid	possible non-specific metabolites						specific metabolite	reference(s)
	A	B	C	D	E	F		
4-chloromethan-dienone $	–	–	‡	–	–	–	6β-hydroxy-4-chloromethandienone	(53)
methandienone	–	*	**	–	*	–	6β-hydroxy-methandienone	(50, 54)
methandriol	–	‡	**	–	–	–	none	(49, 55)
methyltestosterone	–	*	**	*	*	–	parent compound	(49, 50)

A 19-norandrosterone
B 17α-methyl-5α-androstan-3α,17β-diol
C 17α-methyl-5β-androstan-3α,17β-diol
D 17β-methyl-5α-androstan-3α,17α-diol
E 17β-methyl-5β-androstan-3α,17β-diol
F 1α- or 2α-methyl-5α-androstan-3α,17β-diol
** relative important metabolite
* relative less important metabolite
– not present or not detected
$ dehydrochloromethyltestosterone
† metabolism described, but no definite identification of the metabolite reported
‡ metabolism assumed, but no identification of the metabolite reported

Clenbuterol (XVIII) has only been added recently to the IOC-list. Although its so-called anabolic effects are questionable, athletes may use this β2-agonist as an anabolic agent.

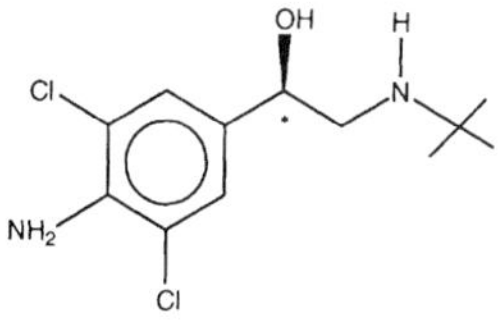

XVIII

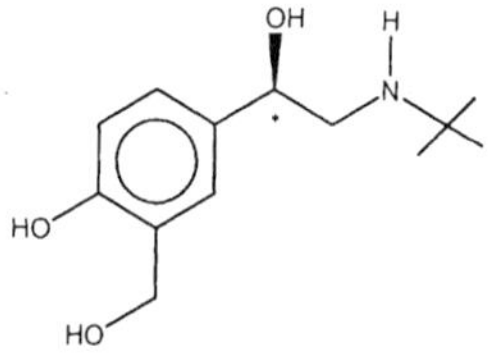

XIX

Structurally related β2-agonists may even become a new doping epidemic as observed in the breeding of cattle (57). Problems however, will arise as some β2-agonists are commonly applied as β2-receptors stimulants, e.g. salbutamol (XIX; albuterol), for the treatment of asthma.

1.4.4.4 Diuretics

Diuretics, therapeutically used for hypertension and edema, increase the urine excretion, thereby diluting the concentration of other possible forbidden substances. They are also used to reduce body-weight quickly in order to qualify for lower weight classes in certain sports, e.g. judo, weightlifting and boxing. The group of diuretics includes carbonic anhydrase inhibitors (e.g. acetazolamide; XX), benzothiazides (e.g. hydrochlorothiazide; XXI), high ceiling (loop) diuretics (e.g. furosemide; XXII), aldosterone antagonists (e.g. spironolactone; XXIII) and potassium-sparing diuretics (e.g. triamterene; XXIV). The IOC-list also contains mercurial diuretics, although those have disappeared from clinical practice (e.g. mersalyl; XXV).

Several of the diuretics are excreted as their parent compounds. Spironolactone is excreted as canrenone, an active diuretic itself (58).

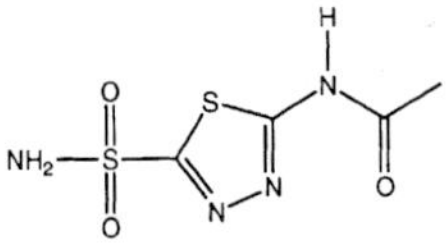

XX 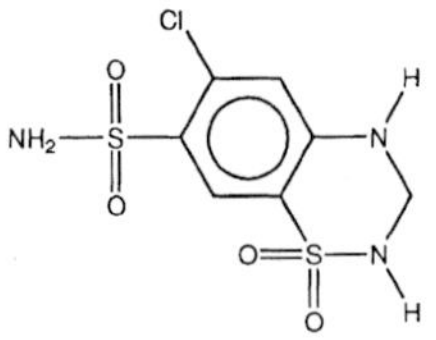 XXI 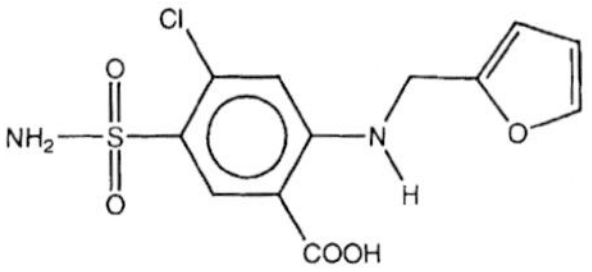 XXII

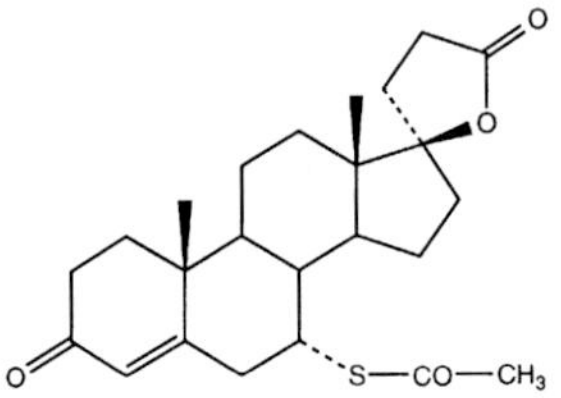

XXIII 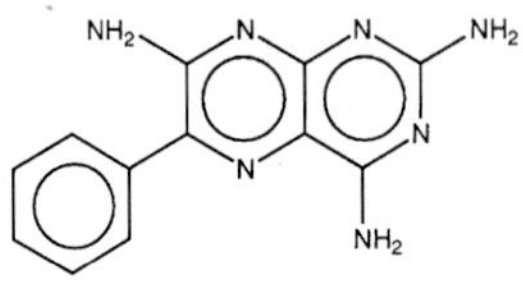 XXIV 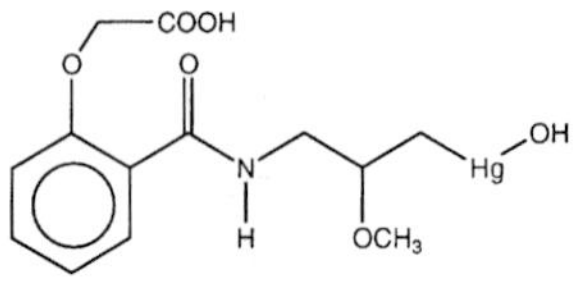 XXV

1.4.4.5 Peptide hormones and analogs

In the eighties, recombinant-DNA techniques made it possible to synthesize certain peptide hormones, for instance Human Growth Hormone (HGH) (60) and Erythropoietin (EPO) (61), in virtually unlimited quantities. This development was welcomed by the athletes, since these hormones are supposed to have a potential enhancement effect, and the routine testing systems for other doping agents had become more sophisticated. For athletes, another peptide hormone, Human Chorionic Gonadotrophin (HCG) extracted from urine of pregnant women, was already available and was used as well. Nowadays, peptide hormones like corticotrophin (ACTH) and tetracosactrin are also frequently applied. In the former German Democratic Republic, within the period of 1984 to 1988, experiments with the small peptides oxytocine, substance P and lypressin were performed aiming to improve central nervous system reactions (62). In table 4 some characteristics of the peptide hormones are summarized.

The group of peptide hormones and analogs is a classification based on their structures rather than their pharmacological effects. Regarding the effects and their use in sports, specific peptides may replace certain other non-peptide substances as mentioned on the doping list. The use of peptide hormones may therefore be considered as a new development, and in doping control this group will be a main problem in the future.

Table 4. Some characteristics of peptide hormones relevant for doping control

name	abbreviation	number of subunits	amino acid residues	carbo-hydrate	total MW (daltons)
Corticotrophin	ACTH	1	39	0%	4500
Erythropoietin	EPO	1	165	30%	34000
Chorionic Gonadotrophin	CG	2	α-92 β-145	31%	38000
Growth Hormone †	GH	1	191	0%	22150
Lypressin	–	1	9	0%	1000
Luteinizing Hormone	LH	2	α-89 β-115	16%	30000
Oxytocin	–	1	9	0%	1000
Substance P	–	1	11	0%	1350
Tetracosactrin ‡	–	1	24	0%	2900

† Growth hormone may consist of 2 variants, which can form di- and oligomers (63)
‡ Synthetic polypeptide identical with the first 24 of the 39 amino acids of ACTH

1.4.4.6 Blood doping

Blood doping refers to the practice of intravenously infusing blood into an individual in order to induce erythrocythemia (4). The procedure may be autologous (one's own blood) or homologous (donated blood). Nowadays however, it can be expected that blood doping will probably be replaced by the use of erythropoietin.

1.4.4.7 Pharmacological, chemical and physical manipulation

The use of substances and of methods which alter the integrity and validity of urine samples used in doping control are banned. Examples of such methods are catheterization, urine substitution and/or tampering and inhibition of renal excretion by probenecid (XXVI) (64) and related compounds such as carinamide (XXVII) and sulfinpyrazone (XXVIII).

In combination with catheterization, local anesthetics may be used. Because these anesthetics may contaminate an urine sample, if some of the drug remains in the urethra after catherization, it is useful to check for the parent compound and metabolites of local anesthetics. In the case of catheterization, only the parent compound of local anesthetics will be found and no metabolites are seen at all.

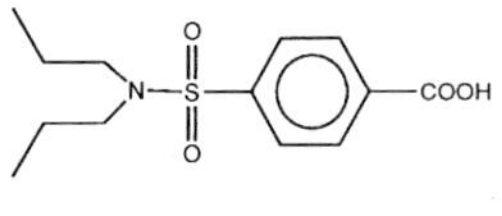

XXVI 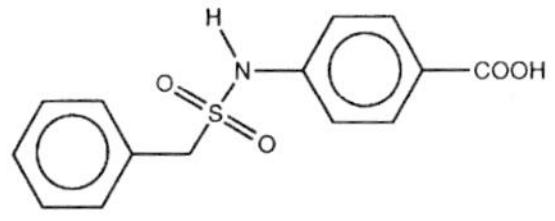 XXVII 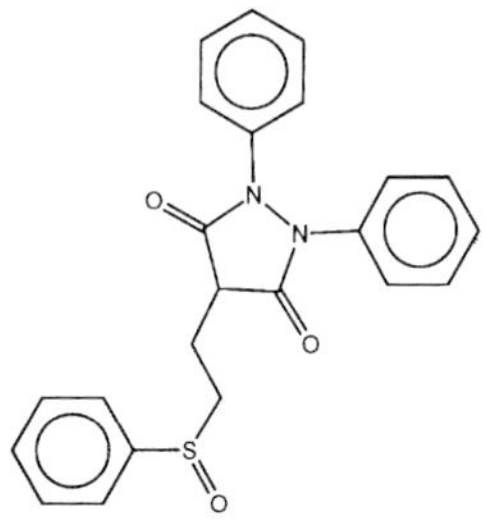 XXVIII

1.4.4.8 Alcohol

Alcohol (CH_3-CH_2-OH; ethanol) is a general central nervous system depressant and is not considered as a main problem in sport, at least not as a sport-enhancing drug. The IOC has therefore classified alcohol as a restricted compound (table 1). Although a legal and tolerated drug, it is recognized in for example the National Football League in the USA as the most abused drug (65).

1.4.4.9 Marijuana

Although marijuana is not considered as an ergogenic aid, it is one of the most frequently used illicit drugs. The principal active constituent is Δ^9-tetrahydrocannabinol (XXIX). 11-Nor-Δ^9-tetrahydrocannabinol-9-carboxylic acid is one of the main urinary metabolites (XXX) (66).

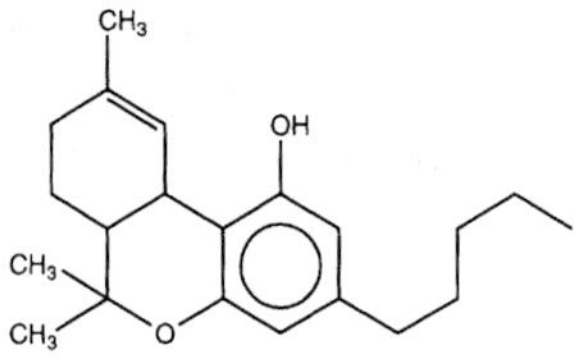

XXIX

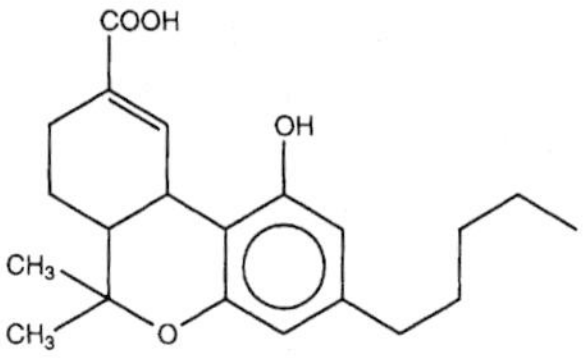

XXX

1.4.4.10 Local anesthetics

Certain local or topical injectable anesthetics are permitted under restricted conditions. Their use is medically justified only if an athlete is able to continue the competition without potential risk to his or her health. Examples are procaine (XXXI) and lidocaine (XXXII). Metabolism of the amide-linked local anesthetics involves *N*-dealkylation and subsequent hydrolysis (67).

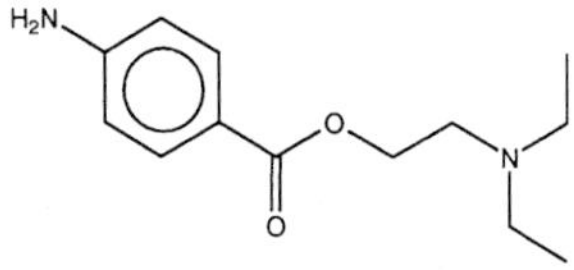

XXXI

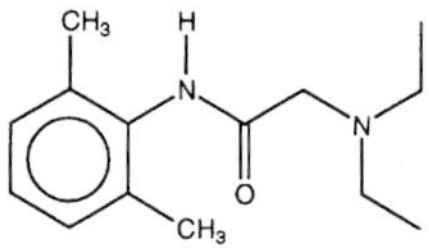

XXXII

1.4.4.11 Corticosteroids

Of the corticosteroids only the glucocorticosteroids are of interest for doping control. The naturally occurring and synthetic corticosteroids are mainly used as anti-inflammatory drugs which also relieve pain. Hydrocortisone (XXXIII) is an endogenous steroid and prednisolone (XXXIV) and triamcinolone acetonide (XXXV) are synthetic ones. As for anabolic androgens, research of endogenous corticosteroid ratios as specific indicators for the use of those steroids is in progress (68).

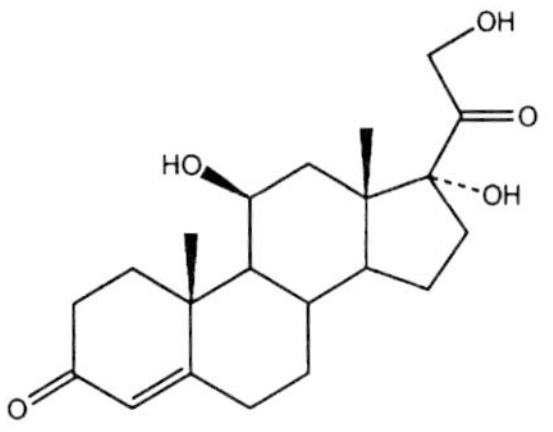

XXXIII 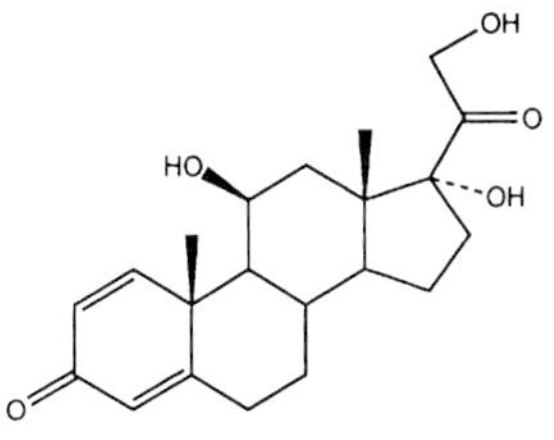 XXXIV 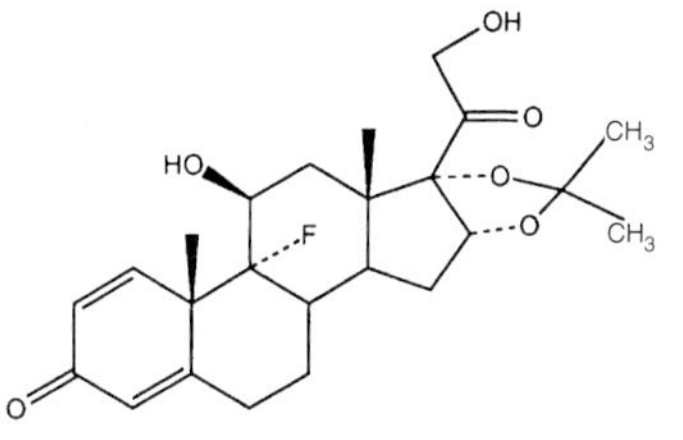 XXXV

1.4.4.12 Beta-blockers

Beta-adrenergic blocking drugs refer to a group that provoke a blockade of both β1- and β2-receptors (non-selective blockers) or only the β1-receptors (selective blockers). The major area of abuse of beta-blockers lies in those events in which motor skills can be affected by muscle tremor caused by anxiety or exercise. Propranolol (XXXVI) and nadolol (XXXVII) are examples of the non-selective blockers and metoprolol (XXXVIII) of a selective one. Labetolol (XXXIX) is an unique adrenergic blocking drug as it blocks both α- and β-receptors (69).

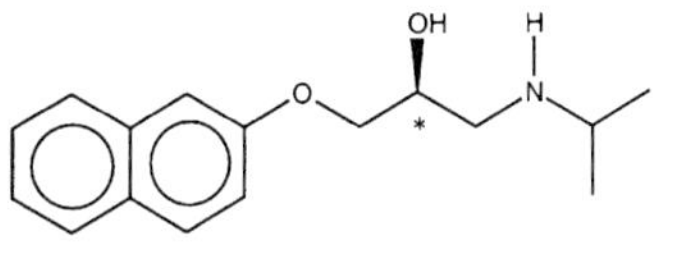

XXXVI 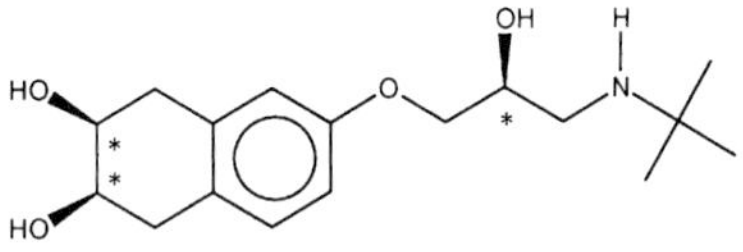 XXXVII

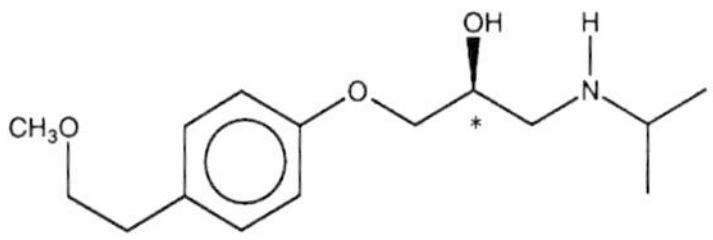

XXXVIII 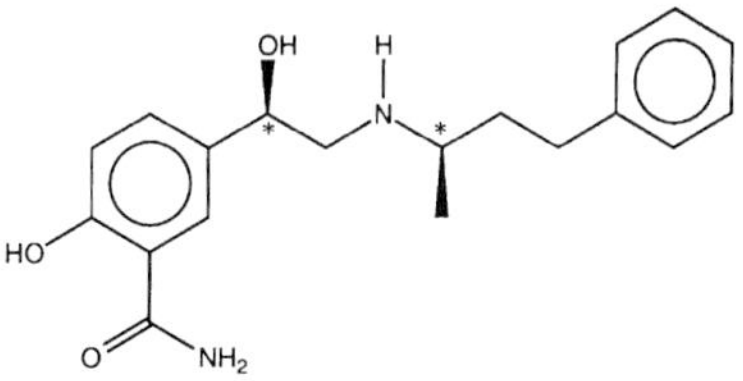 XXXIX

1.4.4.13 Sedatives

At the moment sedatives, such as barbiturates (XL)and benzodiazepines (XLI), are not on the list of the IOC. However, it could make sense, that if hand and arm steadiness is important, such as in riflery and archery, testing for sedatives is considered for these kinds of sports.

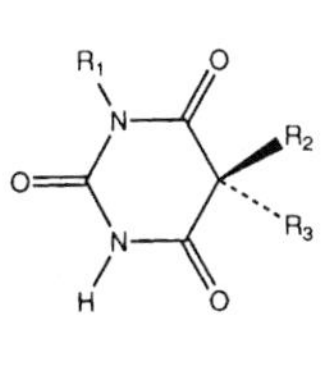

XL 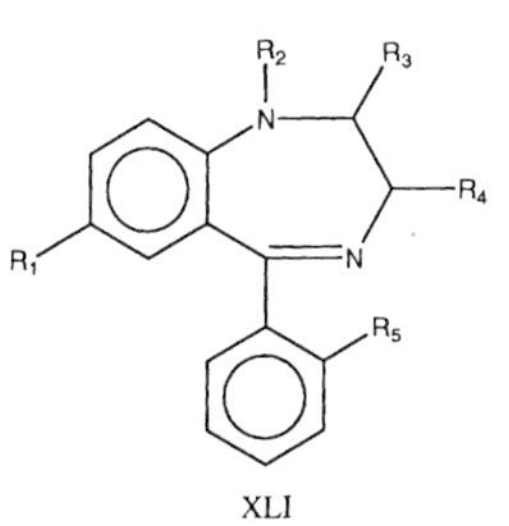 XLI 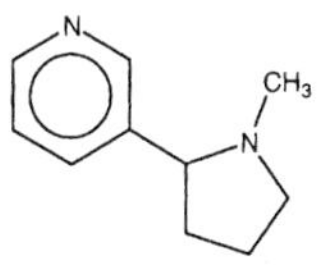 XLII

1.4.4.14 Nicotine

The use of nicotine (XLII) is not forbidden by the IOC, but the NCAA may conduct screening for nicotine or its main metabolite cotinine for nonpunitive research purposes (10).

1.4.4.15 Miscellaneous

Currently, the use of several drugs in sports, which are not mentioned on any doping list in human sports, but which may have potential sport-enhancing effects or should be applied under restricted conditions, is being reviewed critically every year. The effect of H_1-receptor antagonists or antihistamines on exercise performance for example is still uncertain and should be investigated (70).

The treatment of sporting injuries by non-steroidal anti-inflammatory drugs (NSAID's) is sometimes of concern. In the athlete with a mild soft tissue or bone injury, NSAID-use during competition may be regarded as ergogenic (71). In acute or chronic injuries, its use could even be harmful. During training however, it could hasten recovery (72) and should therefore even be recommended.

1.4.5 Analytical Procedures

1.4.5.1 Group analysis

Considering the doping list, an analytical laboratory is faced with a large number of compounds with totally different structures and chemical properties. Covering as many compounds and their metabolites as possible, the testing procedures should also meet criteria such as sensitivity and specificity. It is therefore impossible for all compounds to be isolated and detected by one procedure, although one should try to keep the number of methods as small as possible. An analysis starts with relatively simple and quick screening procedures, in which immunochemical procedures, chromatographic methods and hyphenated techniques play an indispensable role. In table 5 an overview is given of the analytical methods commonly applied to detect forbidden groups of compounds in doping analysis. The required sensitivities are also shown.

Any sample in which the presence of a drug and/or metabolite(s) is suspected has to be re-analyzed. The drug and/or metabolites have to be identified by mass spectrometry. Principally most analyses are performed qualitatively, although in some cases a quantitative determination is required (e.g. caffeine, ephedrines and morphines).

1.4.5.2 Sample preparation

The immunoassays applied require in general no sample preparation. In contrast, no chromatographic methods without sample preparation are used, although for the interpretative purposes a direct quantitative analysis of caffeine in urine through

Table 5. Overview of the analytical methods commonly applied to detect forbidden groups of compounds in doping analysis [73, 74]

Screening procedure	Chemical classification	Sample preparation			Analytical method	
		Hydrolysis	Extraction	Derivatization	Separation and detection technique	Sensitivity (ng/ml)
I	Volatile nitrogen containing compounds excreted free in urine; e.g. amphetamines and ephedrines	no	diethyl ether basic pH	no	GC/NPD	100
II	Non-volatile nitrogen containing compounds excreted as conjugates with sulfate or glucuronic acid; e.g. phenolalkylamines, betablockers, morphines	yes	diethyl ether/ t-butyl alcohol	no	GC/NPD	10
				O-TMS-N-TFA or TFA	GC/MS	10
IIIa	Xanthines, pemoline and corticosteroids	no	diethyl ether/	no	HPLC/DAD	100† 10‡
IIIb	Acidic compounds; e.g. diuretics and masking agents such as probenecid	no	diethyl ether/ acid pH	no	HPLC/DAD	100
			extractive methylation/SM-7 clean-up		GC/MS	100
IVa	Anabolic agents excreted free in urine; e.g. oxandrolone and clenbuterol	no	XAD-2/ diethyl ether	O-TMS-N-HFB or TMS-enolTMS	GC/MS	1
IVb	Anabolic androgens excreted as conjugates with sulfate or glucuronic acid; e.g. nandrolone	yes	XAD-2 diethyl ether	TMS-enolTMS	GC/MS	1
V	Miscellaneous non-peptides for which antibodies are available	no	no	no	immunochemical	100
VI	Peptide hormones	no	no	no	immunochemical	$

XAD-2 = polystyrene resin; SM-7 = copolymer resin; TFA = trifluoracetyl; TMS = trimethylsilylation; GC = gas chromatography; NPD = nitrogen phosphorous detection; MS = mass spectrometry; HPLC = high performance liquid chromatography; DAD = photodiode array ultra violet detection

† xanthines and pemoline
‡ corticosteroids
$ sensitivity required depends on the kind of peptide hormone

high performance liquid chromatography (HPLC) using ultra violet (UV) detection is recommended (75). All chromatographic methods applied include preparation of a sample by extraction and if appropriate combined with a hydrolysis and/or a derivatization step (table 5).

1.4.5.2.1 Extraction

Liquid-liquid extractions are still generally used (table 5). Solid phase extractions (SPE) are becoming popular owing to the relative simplicity of the procedure, their time-saving character, and the minimal use of solvents and manipulations. XAD-2 and SM-7 SPE are used in specific applications, but those columns still have to be prepared manually just before extraction. XAD-2 is a polystyrene resin, which adsorbs organic compounds in general and makes it thus possible to remove inorganics. SM-7 is a macroreticular acrylic copolymer resin, which is more specific and is applied to remove coextracted salts of tetrahexylammonium after methylation of diuretics (76). Commercially available SPE columns are gaining importance. Examples are the isolation of beta-blockers and steroids by use of ethyl-columns (77) and octadecyl columns (42), respectively.

1.4.5.2.2 Hydrolysis

Several compounds of interest in doping analysis are excreted in urine as conjugates with sulphate and/or glucuronic acid. Hydrolysis can be performed by a chemical or enzymatic procedure. In general enzymatic hydrolysis is preferred, because it is non-destructive. The choice of the enzymes and experimental conditions are however critical and should be optimalized in each case. Examples for the optimalization of the enzymatic hydrolysis of steroid conjugates (78–80) and of metabolites of probenecid (81) and sulfaphenazole (82) have been described. A rapid chemical hydrolysis using methanolysis is reported for the sulphate and glucuronyl conjugates of steroids (83).

In some cases destruction by chemical hydrolysis may be an advantage, especially if in contrast to its destruction product, the compound itself can not be detected by gas chromatographic methods. Pemoline, for example, may be hydrolyzed to 5-phenyl-2,4-oxazolidinedione using 1 N hydrochloric acid in the confirmation procedure of this stimulant (84). Another example is the detection of the use of mesocarb. Hydrolysis with 6 N hydrochloric acid and a basic extraction convert the *p*-hydroxymetabolite in *N*-(1-methyl-2-phenylethyl)-*N*-nitroso-α-aminoacetamide. After derivatization and gas chromatographic analysis, a pyrolysis product can be detected, although a more thermo-stable compound can be obtained after specific derivatization with fluoroacyl anhydride (figs. 2, 3).

1.4.5.3 Immunoassays

The applications of immunoassays can be divided in methods detecting non-peptides and peptides. Regarding the non-peptides, immunological methods sometimes can be considered as advantageous in doping control, because of rapid turnover time, low cost and high sensitivity (85). Commercially available procedures, e. g. EMIT®,

TDx®/ADx®, may cover the screening of a selection of amphetamines, barbiturates, benzodiazepines and opiates and more specifically for cannabinoids, the cocaine metabolite, methadon and phencyclidine.

Enzyme immunoassay (EIA) kits are available for compounds which are difficult to implement, such as trenbolone (86). Trenbolone is an anabolic androgen and derivatization using the common methodology applied for gas chromatography/mass spectrometry (GC/MS) of anabolic androgens in doping analysis leads to many unwanted derivatization products (87). Because of low therapeutical doses and the analytical sensitivity required, EIA techniques for the screening of clenbuterol are useful (88, 89). Sensitive confirmation methods are also available (90).

For peptides, immunoassays are the only available detection techniques, which can be performed routinely. Currently, human Chorionic Gonadotrophin (91) and Luteinizing Hormone (36) are the peptides screened in doping analysis. The detection can be performed by various immunoassays (91–93). For confirmation, complementary assays combined with ultra-filtration are recommended (94), although mass spectrometry is still required by the IOC.

1.4.5.4 Thin layer chromatography

Thin Layer Chromatography (TLC) is usually not performed in doping analysis in human sports. This technique has been considered "inadequate and obsolete" in the context of drugs of abuse in general (95). Reasons are that this kind of chromatography, despite its simplicity and low-cost, is time-consuming, relatively non-specific and insensitive.

1.4.5.5 High performance liquid chromatography

High Performance Liquid Chromatography (HPLC) using photodiode array ultraviolet detection (DAD) is applied in the quantitation of caffeine, thermolabile compounds (e. g. corticosteroids) and polar compounds, which are difficult to derivatize

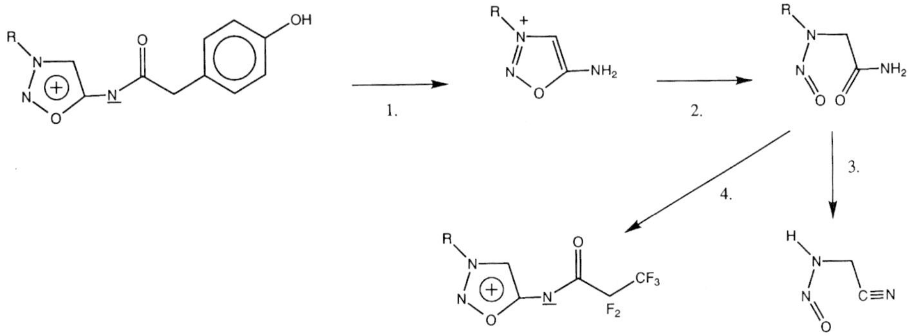

Figure 2. Conversions of the mesocarb metabolite during analysis (for complete structure of mesocarb metabolite see mesocarb VII); 1. acidic hydrolysis; 2. basic extraction; 3. pyrolysis of thermo-labile product during certain derivatizations and/or GC analysis; 4. derivatization with pentafluoropropionyl anhydride yielding a thermo-stable product.

(e.g. diuretics) (table 5). Detection methods using electron capture or fluorescence are not used routinely.

1.4.5.6 Gas chromatography

The combination of gas chromatography (GC) with Nitrogen Phosphorous Detection (NPD) is very important in the screening for *N*-containing stimulants and narcotic analgesics (table 5). Examples of the combination with fourier transform infrared (FTIR) have been reported (96), but until now have not been implemented in routine doping analysis. Although very specific, the draw back for FTIR-detection is the lack of sensitivity. Detectors based on flame ionization and electron capture are not applied. Derivatization in order to improve gas chromatographic properties remains essential. Coupling of an extra group to a polar site of a molecule not only results in an amelioration of volatilization during gas chromatographic injection, but also thermo-stability and peak symmetry, if using a column with an apolar phase, can be improved.

1.4.5.7 Mass spectrometry

The presence of any forbidden drug and/or its metabolite(s) has to be identified by mass spectrometry (MS). For confirmation, often sophisticated quadrupole mass spectrometers in combination with GC and the electron impact (EI) or chemical ionization (CI) mode (87), or in combination with HPLC and thermospray, are used (97). The detection of the use of mesocarb by analysis of 3-(1-methyl-2-phenylethyl)-*N*-pentafluoropropionyl-sydnone imine (figs. 2, 3) is a typical example of confirmation by GC/MS in the negative ion CI mode.

For screening, mass spectrometric analysis is mainly performed in combination with GC using low-cost quadrupole instruments in the EI mode. Derivatization

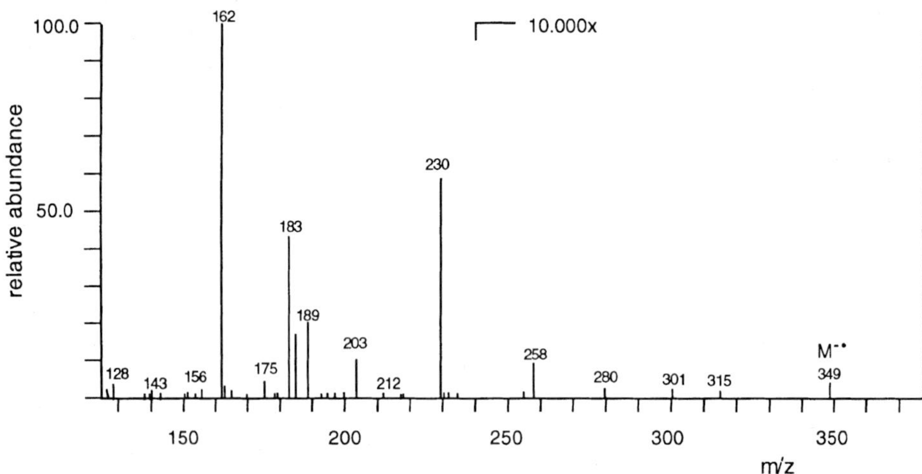

Figure 3. Negative ion chemical ionization spectrum of 3-(1-methyl-2-phenylethyl)-*N*-pentafluoropropionyl-sydnone imine.

may not only improve gas chromatographic properties, but also can make it possible to guide mass spectrometric fragmentation and to perform the detection of typical compounds of interest more selectively.

1.4.5.8 Hematological techniques

At the moment, only the international skiing federations perform blood doping analysis. The goal is to evaluate blood testing in order to detect autologous and homologous blood doping and the use of erythropoietin (EPO).

For the detection of autologous blood doping several factors must be considered (4). Blood samples must be taken after a resting period of at least 12 hours. Due to considerable variation within the 'normal' population, a single parameter alone is not sufficient and therefore two or more parameters (increase in haemoglobin (Hb) and serum iron and bilirubin, decrease in serum EPO) are necessary. At the present there are indications that this kind of blood doping can be detected in about 50 % of blood doped athletes throughout the first two weeks. The suggested method is based on analysis of differences in Hb and serum EPO levels between two blood samples taken with a minimal interval of one to two weeks. Both parameters may also be used to screen exogenous EPO (98). Research and data collection of reference values is however still in progress.

The detection of heterologous (foreign) red blood cells after homologous blood doping is less difficult and is based on the statistical probability of finding a difference in blood group factors between the recipient and the transfused heterologous red blood cells (98).

1.4.6 Horse Doping Control

In horse sport events of the Fédération Equestre Internationale (FEI) the use of almost every drug is not allowed (99). Among other compounds, the use of all xanthines (including theobromine), all NSIADs and local anesthetics, anabolic androgens, corticosteroids and also camphor play an important role in horse doping control.

1.4.7 Chirality

Chirality and the implicit stereoselectivity in pharmacology and toxicology will gain more importance in general (100) and in doping analysis in particular (101). The (*S*)- and (*R*)-forms[1] of the stimulant amphetamine for example, show different potent effects (11). A more striking difference can be observed for the four isomers

[1] In the literature various terms are used to describe the same stereoselectivity; the prefixes *d*-, D-, (*S*)-, (+)- and dex(-tro)- refer to the different enantiomers as does *l*, L, (*R*)-, (−)- and lev(o)-prefixes. The Cahn-Ingold-Prelog system (102), which employs the description R and S, is now more usually used for the identification of the configuration.

of (*1'RS,1''RS*)-labetalol (69). The (*1'R,1''R*)-form has a strong β-blocking effect, the (*1'S,1''S*)-form weakly and the (*1'S,1''R*)-form a pronounced α-blocking activity. The (*1'R,1''S*)-form of labetalol is virtually inactive. Perhaps in relevant cases, an ergogenic effect should be coupled quantitatively to a specific enantiomer.

Separation of enantiomers in doping analysis can be essential for a correct interpretation. The first example is the detection of the use of deprenyl (selegiline), which is a selective monoamine oxidase (MAO) inhibitor. Therapeutically, the (*R*)-form of deprenyl is used, because it is low-toxic, has less unwanted side-effects and is a 500 times more potent MAO inhibitor than the (*S*)-form. The metabolites desmethyldeprenyl, methaphetamine and amphetamine are therefore also in the (*R*)-form. Desmethyldeprenyl is relatively short lived and detection of the amphetamines is possible after the desmethyldeprenyl is eliminated. Detection of the (*S*)-form of the amphetamines in that case would indicate that deprenyl was not the parent compound. Nevertheless, recently a sensitive detection method for desmethyldeprenyl has become available using GC/MS in the negative ion CI mode (12) and providing a more specific confirmation procedure.

Another item concerns the (*S*)- and (*R*)-forms of the morphinan series (101). (*R*)-Methorphan is a potent narcotic analgesic, while (*S*)-methorphan is not a narcotic and is only used for its antitussive effects. (*S*)-Methorphan is found in cough syrups, tablets and capsules as the hydrobromide salt. It is metabolized to (*S*)-3-hydroxy-*N*-methylmorphinan and (*S*)-3-hydroxymorphinan (104). The (*R*)-form of 3-hydroxy-*N*-methylmorphinan is the only commercially available opioid agonist of the morphinan series and its use in sports is forbidden.

Diastereoisomers possess different physicochemical properties and can be separated by non-chiral chromatographic techniques. In this was the separation of the diastereomeric pairs of (*1RS, 2RS*)- and (*1RS, 2SR*)-(nor)ephedrines (the pseudo-forms) can be achieved by HPLC/DAD (105) or as their *N*-trifluoroacetyl-*O*-trimethylsilyl derivatives by GC/MS (96).

The separation of enantiomers can be achieved through a chiral derivatization step to form a pair of diastereomers. Examples of derivatization reagents are *N*-trifluoroacetyl-(*S*)-prolyl (103) or *N*-heptafluorobutyryl-(*S*)-prolyl chloride (106) and (*S*)-α-methoxy-α-(trifluoromethyl)-phenylacetyl chloride (MPTA) (107). The chiral purity of the reagent is one of the most critical points. Donike and Shin have successfully analyzed the (*RS*)-enantiomers of amphetamine and methamphetamine, and (*R*)-3-hydroxy-*N*-methylmorphinan through conversion to either the *N*-MPTA or *O*-MPTA derivatives by GC/MS in the EI mode (101).

Non-derivatizing methods require chromatographic techniques using a chiral selector, which may be part of the stationary phase or present as a mobile phase additive in the case of HPLC (107). Nowadays, several chiral HPLC columns are commercially available and to a lesser extent also some GC columns.

1.4.8 Quality Control

The accuracy of testing has been one of the most questioned and controversial aspects of doping control. Accreditation of laboratories and quality control of the analytical procedures (Good Laboratory Practice) however, have increased the analytical standards significantly.

1.4.9 Analytical Developments

1.4.9.1 Historical overview

Although the first analytical procedure for horse doping control was already described in 1910 (see the introduction), it was until the sixties, that analytical methods were developed for human sport doping control. Beckett et al. were the first to apply sensitive gas chromatographic testing procedures after an athletic event in 1972 (16). During the Summer Olympic Games of 1974 Donike et al. introduced the GC/NPD to screen for *N*-containing stimulants and narcotics (108). Radioimmunoassays (RIA's) developed by Brooks et al. made it possible to screen for anabolic androgens in 1976 during the Summer Olympic Games (109). In 1980, the RIA's for anabolic androgen screening disappeared, because of the unacceptable number of false negatives and because of the necessity to measure testosterone and epitestosterone specifically. The introduction of low cost mass spectrometric detectors (MSD's) in the early eighties made it possible to replace the RIA's.

Table 6. Analytical equipment and techniques used during the last 3 Summer Olympic Games by laboratories accreditated by the IOC (110–112).

Year	1984	1988	1992
Screening procedure	Analytical techniques and number of equipment		
I	5 GC/NPD	6 GC/NPD	4 GC/NPD
II	5 GC/NPD	2 GC/MSD	3 GC/MSD
IIIa	1 HPLC/UVD	3 HPLC/DAD	2 HPLC/DAD
IIIb		3 HPLC/DAD	3 HPLC/DAD
IVa	6 GC/MSD	4 GC/MSD	2 GC/MSD
IVb		4 GC/MSD	4 GC/MSD
V	RIA	FPIA	FPIA EIA
VI		EIA	EIA
Confirmation procedure	1 GC/MSD	1 GC/MS 1 HPLC/MS	1 GC/MS 1 HPLC/MS RIA

GC = gas chromatography; NPD = nitrogen phosphorous detection; MSD = low cost mass spectrometric detector; HPLC = high pressure liquid chromatography; UVD = ultra violet detection; DAD = diode array detection; RIA = radio immunoassay; FPIA = fluorescence polarization immunoassay; EIA = enzyme immonoassay

The development in analytical techniques applied to doping analysis since 1984 is summarized in table 6. Single wave length ultraviolet detection has been replaced by DAD. For confirmation purposes, sophisticated mass spectrometric techniques combined with either GC or HPLC have become available. In 1984, immunoassays were re-introduced. The number of applied immunological techniques is still increasing, although compared to other analytical fields their frequency of application is still low.

1.4.9.2 Group analysis

At the moment, IOC accreditated laboratories use up to six screening procedures to cover all substances belonging to the IOC-list. A more comprehensive multi-analyte approach has been achieved by Kraft (113), who combined the screening for stimulants, narcotic analgesics, beta-blockers and anabolic androgens. From the chemical point of view it was possible to combine some of the existing screening procedures using GC/MS. However, such an application will only be feasible under routine conditions, if the data-evaluation is performed by a powerful data system. Up to now, such a sophisticated system is not available.

1.4.9.3 Immunological techniques

The relative low frequency of the use of immunological techniques is caused by the large number of substances to be tested. Receptor-assays as an alternative for immunoassays can be a more successful approach, but also their application is limited.

Immunoaffinity chromatography may enhance the detection limit of an analytical method by reduction of the biological background (114). For the confirmation of the presence of steroids and β2-agonists, with concentrations in complicated biological matrices which are often low, this isolation techniques is of special interest (114, 115).

1.4.9.4 Mass spectrometry

Other hyphenated techniques such as LC/MS with electrospray ionization (116) and capillary zone electrophoresis (CZE) combined with MS (117) will be useful for the analysis of peptide hormones. For the detection of their use in doping analysis however, typical clinical chemical parameters (e.g. hematocrit, blood levels measured by immunological techniques, or urine values expressed per unit creatinine) will be necessary, as these hormones are eliminated fast, while their effects last longer. Therefore, it may be expected, that MS will not continue to be the only required identification technique.

The use of ion trap (IT) technology will increase (118) in general and in doping analysis in particular (119). The low cost MS detectors, which are commercially available, are based on either conventional quadrupole (CQ) or ion trap (IT) technology. The difference between CQ and the IT mass spectrometry is that ionization and ion separation are separated by space or time, respectively. A promising advantage of IT technology is that MS^n possibilities are relatively easy and cheap to achieve, because the main adaptations have to be realized in the software. In CQ instrumentation these possibilities require comprehensive hardware, as for MS^n mass spectrometry 2n-1 quadrupoles are needed, making the technique rather expensive. Moreover, IT shows more efficient abilities to collect ions, making it the more sensitive MS^n technique. An MS^3 analysis using IT technology has been described to detect the metabolites of 3,4-methylenedioxy-methamphetamine (120)

Tandem mass spectrometry (MS/MS) is a powerful analytical tool (121), in which the different scan modes offer several possibilities. Daughter ion scan consists of selecting a parent ion, fragmentation of that ion and scanning to obtain a daughter spectrum. The advantage of using this technique is an increase in the signal to noise ratio and selectivity (87, 122). For trace analysis, selected reaction monitoring (SRM), i.e. selecting a limited number of parent ion-daughter ion pairs, may result in a further increase in sensitivity, although at the expense of some selectivity (87). In parent scan, a daughter ion is selected and the parent ions are scanned. Neutral loss scan is scanning with a constant difference in mass, which has been useful for screening of barbiturates (123).

An increase in signal to noise ratio can also be achieved by high resolution mass spectrometry (HRMS). The analysis of drugs and/or metabolites in blood samples in general and full scan analysis of anabolic androgenic metabolites in particular require such enhancements of detection limits. The quantitative analysis of steroids by GC/MS has been improved by isotope dilution mass spectrometry (IDMS) and will be implemented into urine steroid profiling (124).

1.4.10 Development of Doping Epidemics and the Role of the Analytical Tools

During this century several doping epidemics have been observed (fig. 4) (3). Before World War II, mainly alkaloids were used by athletes. In the thirties, amphetamines were discovered as more effective stimulants than some of the alkaloids (11). This kind of stimulants were given in World War II to soldiers deprived from sleep to improve their endurance and attentiveness (125). As soon as addicted soldiers introduced these stimulants in their society, the athletes applied these compounds in the hope to improve their performances. The postwar stimulant epidemic in sport started. In the early seventies, the development of GC/NPD was an important progress for the stimulant testing system (108). It contributed significantly to the suppression of the stimulant epidemic.

In World War II, anabolic androgens were given to a small number of German soldiers in an attempt to increase aggressiveness (126). It was in 1976 that the anabolic androgens were added to the list of forbidden substances and during the Olympics of that year some positive cases were found (127). After the introduction of the low cost mass spectrometric detectors in the early eighties, it was possible to test for anabolic androgens successfully on a large scale. However, despite the sophisticated testing system the use of anabolic androgens is still widespread nowadays and it is not limited to athletes. At the high schools in the USA, these steroids are also used by non-athletes (128). Some deaths are also associated with these doping agents (129–139).

As mentioned, another problem started when recombinant-DNA techniques made it possible to synthesize certain peptide hormones in the eighties. Rumors about athletes using these peptide hormones have not been confirmed by drug testing yet. Therefore, it can be stated that the abuse of peptide hormones is the latest doping epidemic in sports.

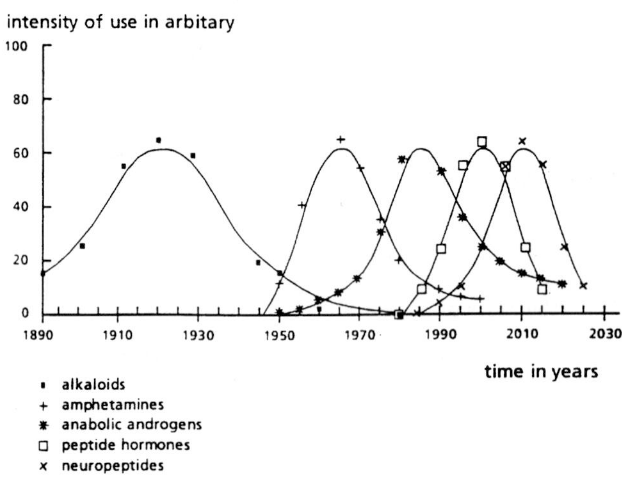

Figure 4. Developments of doping epidemics in sports.

The ingredients for the next doping epidemics seem to be present. In the former German Democratic Republic, in the period of 1984 to 1988, experiments with neuropeptides were already performed aiming to improve central nervous system reactions (62). Up until now, the results of these experiments were not very successful, but they could be a first step into a new direction. Another doping epidemic may be the β2-agonists, such as clenbuterol and related compounds. It may be clear that despite all analytical developments, the shift towards new groups of doping agents can be expected.

References

1. Prokop, L. (1970) The struggle against doping and its history. J. Sports Med. Phys. Fitness *10*, 45–48.
2. Hoberman, J.M. (1992) Mortal engines. The Free Press, Macmillan Inc. USA.
3. De Boer, D. (1992) Profiling anabolic androgens and corticosteroids in doping analysis. Utrecht, the Netherlands, University of Utrecht, Thesis.
4. Berglund, B. (1988) Development of techniques for the detection of blood doping in sport. Sports Med. *5*, 127–135.
5. Schramm, W., Smith, R.H. & Craig, P.A. (1992) Drugs of abuse in saliva: a review. J. Anal. Tox. *16*, 1–9.
6. Moeller, M.R. (1992) Drug detection in hair by chromatographic procedures. J. Chromatogr. *580*, 125–134.
7. International Olympic Committee (IOC) (1993) In: International Olympic Commit-

tee Medical Commission list of doping classes and methods. Lausanne.

8. International Amateur Atheletic Federation (IAAF) (1992) In: Procedural guidelines for doping control. IAFF London.
9. Rockstroh, W. (1990) Domestic anti-doping programmes and activities: Union Internationale de Pentathlon Moderne et Biathlon. In: Official Proceedings IInd. International Athletic Foundation (IAF) world symposium on doping in sport. Bellotti et al. (eds.), 179–185. IAF London.
10. National Collegiate Athletic Association. The NCAA drugtesting program 1988–1989. (1989) In: Drugs and the athlete. Ryan (ed.), 272–285. FA Davis company Philadelphia, PA.
11. Hoffmann, B.B. & Lefkowitz, R.J. (1990) Catecholamines and sympathomimetic drugs. in: The pharmacological basis of therapeutics. 8th ed. Goodman, Gilman et al. (eds.). 187–220. Pergamon Press Inc.
12. Reimer, M.J.L., Mamer, O.A., Zavitsanos, A.P., Siddiqui, A.W. & Dadgar, D. (1993) Determination of amphetamine, methamphetamine and desmethyldeprenyl in human plasma by gas chromatography/negative ion chemical ionization mass spectrometry. Biol. Mass. Spectrom, *22*, 235–242.
13. Beckett, A.H., Tucker, G.T. & Moffat, A.C. (1967) Routine determination and identification in urine of stimulants and other drugs, some of which may be used to modify performance in sports. J. Pharm. Pharmacol., *19*, 273–294.
14. Vree, T.B. & Van Rossum, J.M. (1970) Kinetics of metabolism and excretion of amphetamines in man. In: Amphetamines and related compounds. Proceedings of the Mario Institute for Pharmacological Research. Milan, Italy. Costa and Garattini (eds.), 165–190. Raven Press, New York, NJ.
15. Mallol, J., Pitarsch, L., Coronas, R. & Pons, A. (1974) Determination of dl-fencamine in rat and human urine. Arnzeim.-Forsch. *24*, 1301–1304.
16. Beckett, A.H., Shenoy, E.V.B. & Salmon, J.A. (1972) The influence of replacement of the *N*-ethyl group by the cyanoethyl group on the absorption, distribution and metabolism of (±)-ethylamphetamine in man. J. Pharm. Pharmacol. *24*, 194–202.
17. Blum, J.E. (1969) Experimentelle Untersuchungen mit dem neuen Appetithemmer *N*-(3-chloropropyl)-1-methyl-2-phenyl-äthylamin-hydrochlorid (Ro 4-5282). Arzneim.-Forsch. *19*, 748–755.
18. Shumin, Y., Changjiu, Z. & Kairong, C. (1993) Detection of mesocarb in urine by HPLC and GC-MSD. In: Proceedings of the 10th Cologne workshop on dope analysis 7th to 12th June 1992. Donike et al. (eds.), 159–161. Sport und Buch Strauß, Cologne.
19. Remberg, G., Eichelbaum, M., Spiteller and Dengler, H.J. (1977) Metabolism of DL-(^{14}C)Prenylamine in man. Biomed. Mass. Spectrom. *4*, 197–304.
20. Wilkinson, G.R. & Beckett, A.H. (1968) Absorption, metabolism and excretion of the ephedrines in man. I. The influence of urinary pH and urine volume output. J. Pharm. Exp. Ther. *162*, 139–150.
21. Testa, B. & Beckett, A.H. (1972) Studies on the metabolism of diethylpropion. I. Analytical procedure. J. Chromatogr. *71*, 39–54.
22. Beckett, A.H. & Brookes, L.G. (1970) The effect of chain and ring substitution on the metabolism, distribution and biological action of amphetamines. In: Amphetamines and related compounds. Proceeding of the Mario Institute for Pharmacological Research. Milan, Italy. Costa & Garattini (eds.), 109–120. Raven Press, New York.
23. Franklin, R.B. & Dring, L.G. (1977) The metabolism of phenmetrazine in man and laboratory animals. Drug. Metab. Disp. *5*, 223–233.
24. Lim, H.K. & Foltz, R.L. (1989) Identification of metabolites of 3,4-(methylenedioxy)methamphetamine in human urine. Chem. Res. Toxicol. *2*, 142–143.
25. Wadler, G.I. & Hainline, B. Caffeine (1989) In: Drugs and the athlete. Ryan (ed.), 107–113. FA Davis company, Philadelphia, PA.
26. Ramsden, C.A. (1979) Meso-ionic heterocycles. In: Heterocyclic compounds. Samnes (ed.), 1171–1228. Vol. 4: Comprehensive organic chemistry. Barton & Ollis (eds.), Pergamon Press, London.
27. Lachatre, G., Piva, C., Riche, C., Dumont, D., Defrance, R., Mocaer, E. & Nicot, G. (1989) Single-dose pharmacokinetics of amineptine and its main metabolite in healthy young volunteers. Fundam. Clin. Pharmacol. *3*, 19–26.
28. Tilstone, W.J. & Stead, A.H. (1986) Pharmacokinetics, metabolism, and the interpretation of results. In: Clarke's isolation and identification of drugs. Moffat et al. (eds.), 276–305. The Pharmaceutical Press, London.

29. Smith, R. L. & Dring, L. A. (1970) Patterns of metabolism of β-phenylisopropylamines in amn and other species. In: Amphetamines and related compounds. Proceedings of the Mario Institute for Pharmacological Research. Milan, Italy. Costa & Garattini (eds.), 121–139. Raven Press, New York, NJ.
30. Wu, A. H. B., Johnson, K. G., Ballatore, A. & Seifert, W. E. (1992) The conversion of ephedrine to methamphetamine and methamphetamine-like compounds during and prior gas chromatographic/mass spectrometric analysis of CB and HFB derivatives. Biol. Mass. Spectrom. *21*, 278–284.
31. Martindale. Opioid analgesics. (1989) In: Martinadale. The extra pharmacopoeia 29th Reynolds (ed.). The Pharmaceutical Press, London.
32. Hayes, L. W., Krasselt, W. G. & Mueggler, P. A. Concentrations of morphine and codeine in serum and urine after ingestion of poppy seeds. Clin. Chem. *33*, 806–808.
33. ElSohly, M. A. & Jones, A. B. (1989) Morphine and codeine in biological fluids: approaches to source differentiation. Forensic. Sci. Rev. *1*, 14–22.
34. Donike, M., Bärwald, K. R., Klostermann, K., Schänzer, W. & Zimmermann, J. (1983) The detection of exogenous testosterone. In: Sport: Leistung und Gesundheit. Heck et al. (eds.), 293–300. Deutscher Ärzte-Verlag, Cologne.
35. Oftebro, H. (1992) Evaluating an abnormal urinary steroid profile. The Lancet *339*, 941–942.
36. Kicman, A. T., Brooks, R. V. Collyer, S. C., Cowan, D. A., Nanjee, M. N., Southan, G. J. & Wheeler, M. J. (1991) Criteria to indicate testosterone administration. Br. J. Sp. Med. *24*, 253–264.
37. Carlström, K., Palonek, E., Garle, M., Oftebro, H., Stanghelle, J. & Björkheim, I. (1992) Detection of testosterone administration by increased ratio between serum concentrations of testosterone and 17α-hydroxyprogesterone. Clin. Chem. *38*, 1779–1784.
38. Southan, G. J., Brooks, R. V., Cowan, D. A., Kicman, A. T., Unnadkat, N. & Walker, C. J. (1992) Possible indices for the detection of the administration of dihydrotestosterone to athletes. J. Steroid Biochem. Molec. Biol. *42*, 87–94.
39. Reznik, Y., Herrou, M., Dehennin, L., Lemaire, M. & Leymarie, P. (1987) Rising plasma levels of 19-nortestosterone throughout pregnancy: determination by radioimmunoassay and validation by gas chromatography/mass spectrometry. J. Clin. Endocr. Metab. *64*, 1086–1088.
40. Dehennin, L., Silberzahn, P., Reiffsteck, A. & Zwain, I. (1984) Présence de 19-norandrostènedione et de 19-nortesosterone dans les fluides folliculaires humain et équin. Path. Biol. *32*, 828–829.
41. De Boer, D., Bensink, S. N., Borggreve, A. R., Van Ooijen, R. D. & Maes, R. A. A. (1993) Profiling 19-norsteroids. III. GC/MS/MS analysis of 19-norsteroids during pregnancy. In: Proceedings of the 10th Cologne workshop on dope analysis 7th to 12th June 1992. Donike et al. (eds.), 121–122. Sport und Buch Strauß, Cologne.
42. Massé, R., Ayotte, C. & Dugal, R. (1989) Studies on anabolic steroids. I Integrated methodological approach to the gas chromatographic/mass spectrometric analysis of anabolic steroid metabolites in urine. J. Chromatogr. *489*, 23–50.
43. Massé, R., Laliberté, C., Tremblay, L. & Dugal, R. (1985) Gas chromatographic/mass spetrometric analysis of 19-nortestosterone urinary metabolites in man. Biomed. Mass. Spectrom. *12*, 115–121, 1985.
44. Ranney, R. E. (1977) Comparative metabolism of 17α-ethynyl steroids used in oral contraceptives. J. Toxicol. Environ. Health *3*, 139–166.
45. Yasuda, J., Honjo, H. & Okada, H. (1984) Metabolism of lynestrenol: characterization of 3-hydroxylation using rabbit liver microsomes in vitro. J. steroid. Biochem. *21*, 777–780.
46. De Boer, D., De Jong, E. G., Maes, R. A. A. & Van Rossum, J. M. (1992) The methyl-5α-dihydrotestosterones mesterolone and drostanolone; gas chromatographic/mass spectrometric characterization of urinary metabolites. J. Steroid. Biochem. Molec. Biol. *42*, 411–419.
47. Goudreault, D. & Massé, R. (1990) Studies on anabolic steroids 4. Identification of new urinary metabolites of methenolone acetate (primobolan®) in human by gas chromatography/mass spectrometry. J. Steroid. Biochem. Molec. Biol. *37*, 137–154.
48. Goudreault, D. & Massé, R. (1991) Studies on anabolic steroids – 6. Identification of urinary metabolites of stenbolone acetate (17β-acetoxy-2-methyl-5α-androst-1-en-3-one) in human by gas chromatography/mass spectrometry. J. Steroid. Biochem. Molec. Biol. *38*, 639–655.

49. Schänzer, W., Opfermann, G. & Donike, M. (1992) 17-Epimerization of 17α-methyl anabolic steroids in humans: metabolism and synthesis of 17α-hydroxy-17β-methyl steroids. Steroids *57*, 537–550.
50. Massé, R., Bi., H., Ayotte, C., Du, P., Gélinas, H. & Dugal, R. (1991) Studies on anabolic steroids. V. Sequential reduction of methandienone and structurally related steroid A-ring substituents in humans: gas chromatographic-mass spectrometric study of the corresponding urinary metabolites. J. Chromatogr. *562*, 323–340.
51. Bi, H., Du, P. & Massé, R. (1992) Studies on anabolic steroids – 8. GC/MS characterization of unusual seco acidic metabolies of oxymetholone in human urine. J. Steroid. Biochem. Molec. Biol. *42*, 229–242.
52. Bi, H., Massé, R. & Just, G. (1992) Studies on anabolic steroids. 10. Synthesis and identification of acidic urinary metabolies of oxymetholone in a human. Steroids *57*, 453–459.
53. Dürbeck, H.W., Büker, L., Scheulen, B. & Telin, B. (1983) GC and capillary column GC/MS determination of synthetic anabolic ssteroids. II. 4-Chloro-methandienone (oral turinabol) and its metabolites. J. Chromatogr. Sci. *21*, 405–410.
54. Schänzer, W., Geyer, H. & Donike, M. (1991) Metabolism of metandienone in man: identification and synthesis of conjugated excreted urinary metabolites, determination of excretion rates and gas chromatographic-mass spectrometric identification of bis-hydroxylated metabolites. J. Steroid. Biochem. Molec. Biol. *38*, 441–464.
55. Schänzer, W. & Donike, M. (1993) Metabolism of anabolic steroids in man: synthesis and use of reference substances for identification of anabolic steroid metabolites. In: Proceedings of the 10th Cologne workshop on dope analysis 7th to 12th June 1992. Donike et al. (eds.), 97–100. Sport und Buch Strauß, Cologne.
56. Bi, H., Massé, R. & Just, G. (1992) Studies on anabolic steroids. 9. Tertiary sulfates of anabolic 17α-methyl steroids: synthesis and rearrangement. Steroids *57*, 306–312.
57. Leyssens, L., Driessen, C., Jacobs, A., Czech, J. & Raus, J. (1991) Determination of β2-receptor agonists in bovine urine and liver by gas chromatography-tandem mass spectrometry. J. Chromatogr. *564*, 515–527.
58. Karim, A., Hribar, J., Aksamit, W., Doherty, M. & Chinn, L.J. (1975) Spironolactone metabolism in amn studied by gas chromatography-mass spectrometry. Drug Metab. Disp. *3*, 467–478.
59. De Boer, D., De Jong, E.G., Maes, R.A.A. & Van Rossum, J.M. (1988) The problems of oral contraceptives in dope control of anabolic steroids. Biomed. Environ. Mass Spectrom. *17*, 127–128.
60. Von Kley, H.K. (1985) Influence on performance by peptide hormones: growth hormone, an anabolic agent? Deutsch Z. Sportmed. *4*, 113–118.
61. Schmidt, W. (1990) Erythropoietin and sports – facts and opinion. Deutsch. Z. Sportmed. *4*, 304–312.
62. Berendonk, B. (1991) Tabelle 2: Die gebräuchlichen Dopingmittel im DDR-Sport. In: Doping Dokumente. Von der Forschung zum Betrug. 78–79. Springer-Verlag, Berlin–Heidelberg.
63. Lewis, U.J., Singh, R.N.P., Tutweiler, G.F., Sigel, M.B., Vanderlaan, E.F. & Vanderlaan, W.P. (1980) Human growth hormone: a complex of proteins. Recent Prog. Horm. Res. *36*, 477–508.
64. Geyer, H., Schänzer, W. & Donike, M. (1993) Probenecid as masking agent in dope control – inhibition of the urinary excretion of steroid glucuronides. In: Proceedings of the 10th Cologne workshop on dope analysis 7th to 12th June 1992. Donike et al. (eds.). 141–151. Sport und Buch Strauß, Cologne, Germany.
65. Wadler, G.I. & Hainline, B. (1989) Alcohol. In: Drugs and the athlete. Ryan (ed.). 124–134. FA Davis Company, Philadelphia, PA.
66. Karlsson, L., Jonsson, J., Aberg, K. & Roos, C. (1983) Determination of Δ^9-tetrahydrocannabinol-11-oic acid as its pentafluoropropyl-pentafluorpropionyl derivative by GC/MS utilizing negative ion chemical ionization. J. Anal. Tox. *7*, 198–202.
67. Benowitz, N.L. & Meister, W. (1978) Clinical pharmakinetics of lignocaine. Clin. Pharmacokinet. *3*, 177–201.
68. Park, S.-J., Kim, Y.-J., Pyo, H.-S. & Park, J. (1990) Analysis of corticosteroids in urine by HPLC and thermospray LC/MS. J. Anal. Tox. *14*, 102–108.
69. Hoffmann, B.B. & Lfkowitz, R.J. (1990) Adrenergic receptor antagonists. II β-Adrenergic receptor antagonists. In: The pharmacological basis of therapeutics. 8th ed. Goodman Gilman et al. (eds.), Vol. 11. 235–236. Pergamon Press Inc.

70. Montgomery, L. C. & Deuster, P. A. (1993) Effects of antihistamine medications on exercise performance. Sports Med. *15*, 179–195.
71. Hasson, S. M., Daniels, J. C., Divine, J. G., Niebuhr, B. R., Richmond, S., Stein, P. G. & Williams, J. H. (1993) Effect of ibuprofen use on muscle soreness, damage, and performance: a preliminary investigation. Med. Sci. Sports Exerc. *25*, 9–17.
72. Noble, C. (1981) Fenbufen in the treatment of overuse injuries in log distance runners, S.A. Med. J. *5*, 387–392.
73. Donike, M. (1990) Overview on present analytical procedures in dope analysis. In: Official Proceedings IInd. International Athletic Foundation (IAF) world symposium on doping in sport. Bellotti et al. (eds.) 83–92. IAF London.
74. Segura, J., Pascual, J. A., Ventura, R., Ustraran, J. I., Cuevas, A. & González, R. (1993) The experience of antidoping control at the XI Panamerican Games in Havana, Cuba. In: Proceedings of the 10th Cologne workshop on dope analysis 7th to 12th June 1992. Donike et al. (eds.). 325–343. Sport und Buch Strauß, Cologne.
75. Gotzman, A. & Donike, M. (1993) Quantitative caffeine determination by direct injection of urine. In: Proceedings of the 10th Cologne workshop on dope analysis 7th to 12th June 1992. Donike et al. (eds.). 281–284. Sport und Buch Strauß, Cologne.
76. Lisi, A. M., Kazlauskas, R. & Trout, G. J. (1993) Diuretic screening in human urine by gas chromatography mass spectrometry. Part 2. In: Proceedings of the 10th Cologne workshop on dope analysis 7th to 12th June 1992. Donike et al. (eds.). 213–219. Sport und Buch Strauß, Cologne.
77. LeLoux, M. S., De Jong, E. G. & Maes, R. A. A. (1989) Improved screening method for beta-blockers in urine using solid-phase extraction and capillary gas chromatography-mass spectrometry. J. Chromatogr. *488*, 357–367.
78. Shackleton, C. H. K. (1986) Profiling steroid hormones and urinary hormones. J. Chromatogr. *379*, 91–156.
79. Clausnitzer, C. & Grosse, J. (1985) Analytische Probleme bei der Ermittlung und Interpretation des Testosterone-Epitestosterone-Quotienten im Rahmen von Dopingkontrollen. Med. u. Sport *25*, 212–215.
80. Houghton, E., Grainger, L., Dumasia, M.C. & Taele, P. (1992) Application of gas chromatography/mass spectrometry to steroid analysis in equine sports: problems with enzyme hydrolysis. Org. Mass. Spectrom. *27*, 1061–1070.
81. Vree, T. B. & Beneken Kolmer, E. W. J. (1990) Direct measurement of probenecid and its glucuronide conjugate by high pressure liquid chromatography in plasma and urine of humans. Pharm. Weekbl. (Sci) *14*, 83–87.
82. Vree, T. B. Beneken Kolmer, E. W. J. & Hekster, Y. A. (1990) High pressure liquid chromatographic analysis and preliminary pharmacokinetics of sulphenazole and its N2-glucuronide and N4-acetyl metabolites in plasma and urine of man. Pharm. Weekbl. (Sci) *12*, 51–59.
83. Tang, P. W. & Crone, D. L. (1989) A new method for hydrolyzing sulfate and glucuronyl conjugates of steroids. Anal. Biochem. *182*, 289–294.
84. Vermeulen, N. P. E., De Roode, D. & Breimer, D. D. (1977) Gas chromatographic determination of pemoline as 5-phenyl-2,4-oxazolidinedione in human urine. J. Chromatogr. *137*, 333–342.
85. De la Torre, R., Badia, R., Gracia, M., González, G., Farré, M., Lamas, X. & Segura, J. (1993) Screening of some dope agents by immunological methods. In: Proceedings of the 10th Cologne workshop on dope analysis 7th to 12th June 1992. Donike et al. (eds.). 285–302. Sport und Buch Strauß, Cologne.
86. Garle, M., Ocka, R. & Palonek, E. (1993) One year experience of trenbolone measurement by ELISA. In: Proceedings of the 10th Cologne workshop on dope analysis 7th to 12th June 1992. Donike et al. (eds.) 303–306. Sport und Buch Strauß, Cologne.
87. De Boer, D., Gainza Bernal, M. E., Van Ooijen, R. D. & Maes, R. A. A. (1991) The analysis of trenbolone and the human urinary metabolites of trenbolone acetate by gas chromatography/mass sepctrometry and gas chromatography/tandem mass spectrometry. Biol. Mass. Spectrom. *20*, 459–466.
88. Ploum, M. E., Haasnoot, W., Paulussen, R. J. A., Van Bruchem, G. D., Hamers, A. R. M., Schilt, R. & Huf, F. A. (1991) Test strip immunoassays and the fast screening of nortestosterone and clenbuterol resudues in urine samples at the parts per billion level. J. Chromatogr. *564*, 413–427.
89. Meyer, H. H., Rinke, L. & Dürsch, I. (1991) Residue screening for the β-agonists clen-

buterol, salbutamol and cimaterol in urine using enzyme immunoassay and high performance liquid chromatography. J. Chromatogr. *564*, 551–556.

90. Polettini, A., Ricossa, M.C., Groppi, A. & Montagna, M. (1991) Determination of clenbuterol in urine as its cyclic boronate derivative by gas chromatography-mass spectrometry. J. Chromatogr. *564*, 529–535.
91. Brooks, R.V., Collyer, S.C., Kicman, A.T., Southan, G.J. & Wheeler, M.J. (1990) HCG doping in sports and methods for its detection. In: Official Proceedings IInd. International Athletic Foundation (IAF) world symposium on doping in sport. Bellotti et al. (eds.), 37–45. IAF, London.
92. Kiburis, J. (1990) The need for standardization of the hCG assays in doping control. In: Official Proceedings IInd. International Athletic Foundation (IAF) world symposium on doping in sport. Bellotti et al. (eds.), 31–36. IAF, London.
93. De Boer, D., De Jong, E.G., Van Rossum, J.M., Maes, R.A.A., Thomas, C.M.G. & Seegers, M.F.G. (1991) Doping control of testosterone and human chorionic gonadotrophin: A case study. J. Int. Sports Med. *12*, 46–51.
94. Cowan, D.A., Kicman, A.T., Walker, C.J. & Wheeler, M.J. (1991) Effect of administration of human chorionic gonadotrophin on criteria used to assess testosterone administration in athletes. J. Endocrinol. *131*, 147–154.
95. Finkle, B.S. (1987) Drug-analysis technology: Overview and state of the art. Clin. Chem. *33 (11B)*, 13B–17B.
96. De Jong, E.G., Keizers, S. & Maes, R.A.A. (1990) Comparison of the GC/MS ion trapping technique with GC-FTIR for the identification of stimulants in drug testing. J. Anal. Tox. *14*, 127–131.
97. Ventura, R., Nadal, T., Pretel, M.J., Solans, A., Pascual, J.A. & Segura, J. (1993) Possibilities of liquid chromatography-mass spectrometry for the identification of labile compounds. In: Proceedings of the 10th Cologne workshop on dope analysis 7th to 12th June 1992. Donike et al. (eds.), 231–248. Sport und Buch Strauß, Cologne.
98. Videman, T. (1990) Experiences in blood doping testing at the 1989 world cross-country ski championship in Lahti, Finland. In: Official Proceedings IInd. International Athletic Foundation (IAF) world symposium on doping in sport. Bellotti et al. (eds.), 5–12. IAF, London.
99. Fédération Equestre Internationale (FEI). (1990) In: Veterinary regulations. Bern.
100. Ariëns, E.J. (1993) Nonchiral, homochiral and composite chiral drugs. TiPS *14*, 68–75.
101. Donike, M. & Shin, H.-S. (1993) Chirality in doping analysis. In: Proceedings of the 10th Cologne workshop on dope analysis 7th to 12th June 1992. Donike et al. (eds.), 159–161. Sport und Buch Strauß, Cologne.
102. Cahn, R.S., Ingold, C.K. & Prelog, V. (1966) Specifications of molecular chirality. Angew. Chem. Int. Ed. *5*, 385–415.
103. Cody, J.T. (1992) Determination of methamphetamine enantiomer ratios in urine by gas chromatography. J. Chromatogr. *580*, 77–95.
104. Park, Y.H., Kullberg, M.P. & Hinsvark, O.N. (1984) Quantitative determination of dextromethorphan and three metabolites in urine by reverse-phase high performance liquid chromatography. J. Pharm. Sci. *73*, 24–29.
105. Imaz, C., Carreras, D., Navajas, R., Rodriguez, C., Rodriguez, A.F., Maynar, J. & Cortes, R. (1993) Determination of ehedrines in urine by HPLC. In: Proceedings of the 10th Cologne workshop on dope analysis 7th to 12th June 1992. Donike et al. (eds.), 269–280. Sport und Buch Strauß, Cologne.
106. Lim, H.K. & Foltz, R.L. (1993) Stereoselective disposition: enantioselective quantitation of 3,4-(methylenedioxy)-methamphetamine and three of its metabolites by gas chromatographic/electron capture negative ion chemical ionization mass spectrometry. Biol. Mass. Spectrom. *22*, 403–411.
107. Taylor, D.R. & Maher, K. (1992) Chiral separations by high-performance liquid chromatography. J. Chromatogr. Sc. *30*, 67–84.
108. Donike, M. & Stratmann, D. (1974) Temperaturprogrammierte gaschromatographische Analyse stickstoffhaltiger Pharmaka: Die Reproduzierbarkeit der Retentionszeiten und der Mengen bei automatischen Injektionen (II) (Die Screeningprozedur für fluchtige Dopingmittel bei den Spielen der XX. Olympiade München 1972.) Chromatographia *7*, 182–189.
109. Brooks, R.V., Firth, R.G. & Sumner, N.A. (1975) Detection of anabolic steroids by radioimmunoassay. Br. J. Sport. Med. 9, 89–92.

110. Catlin, D.H., Kammerer, R.C., Hatton, C.K., Sekiera, M.H. & Merdink, J.L. (1987) Analytical chemistry at the Games of the XXIIIrd Olympiad in Los Angeles, 1984. Anal. Chem. *33*, 319–327.
111. Park, J., Park, S., Lho, D.-S., Choo, H.P., Chung, B., Yoon, C., Min, H. & Choi, M.J. (1990) Review of the Seoul laboratory activities – Seoul Olympic Games 1988. In: Official Proceedings IInd. International Athletic Foundation (IAF) world symposium on doping in sport. Bellotti et al. (eds.), 47–61. IAF, London.
112. Pasqual, J.A., Ewin, R.R. & Segura, J. (1993) Automated control of doping samples and their analyses preparing for Barcelone '92. Part I. Development of a new laboratory information management system (LIMS) for doping control. In: Proceedings of the 10th Cologne workshop on dope analysis 7th to 12th June 1992. Donike et al. (eds.), 345–367. Sport und Buch Strauß, Cologne.
113. Kraft, M. (1990) Approach to combine the present analytical methods for the detection of dope agents to a comprehensive screening procedure with GC/MS detection. In: Official Proceedings IInd. International Athletic Foundation (IAF) world symposium on doping in sport. Bellotti et al. (eds.), 93–106. IAF, London.
114. Van Ginkel, L.A. (1991) Immunoaffinity chromatography, its applicability and limitations in multi-residue analysis of anabolizing and doping agents. J. Chromatogr. *564*, 363–384.
115. Schänzer, W., Delahaut, P., Völker, E. & Donike, M. (1993) Immunoaffinity chromatograhy in isolation of anabolic steroids. In: Proceedigs of the 10th Cologne workshop on dope analysis 7th to 12th June 1992. Donike et al. (eds.), 307–320. Sport und Buch Strauß, Cologne.
116. Smith, R.D., Loo, J.A., Edmonds, C.G., Baringana, C.J. & Udseth, H.R. (1990) New developments in biochemical mass spectrometry: electrospray ionization. Anal. Chem. *62*, 882–899.
117. Nielsen, R.G., Sittampalam, G.S. & Rickard, E.C. (1989) Capillary zone electrophoresis of insulin and growth hormone. Anal. Biochem. *177*, 20–26.
118. Cooks, R.G., Glish, G.L., McLucky, S.A. & Kaiser, R.E. (1991) Ion trap mass spectrometry. Chem. Engin. News. *69*, 26–41.
119. De Boer, D., De Jong, E.G. & Maes, R.A.A. (1991) Mass spectrometric characterization of different norandrosterone derivatives by low-cost mass spectrometric detectors using electron impact and chemical ionization. Rappid, Commun. Mass. Spectrom. *4*, 181–185.
120. Lim, H.K. & Foltz, R.L. (1991) *In vivo* formation of aromatic hydroxylated metabolites of 3,4-(methylenedioxy)methamphetamine in the rat: identification by ion trap tandem mass spectrometric (MS/MS and MS/MS/MS) techniques. Biol. Mass. Spectrom. *20*, 677–686.
121. Johnson, J. & Yost, R.A. (1985) Tandem mass spectrometry in trace analysis. Anal. Chem. *57*, 758–768.
122. Edlund, P.O., Browns, L. & Henion, J. (1989) Determination of methrandrostenolone and its metabolite in equine plasma and urine by coupled-column liquid chromatography with ultraviolet detection and confirmation by tandem mass spectrometry. J. Chromatogr. *487*, 341–356.
123. Brzezinka, H., Bold, P. & Budzikiewicz, H. (1993) A screening method for the rapid detection of barbiturates in serum by means of tandem mass spectrometry. Biol. Mass. Spectrom. *22*, 346–350.
124. Donike, M. (1993) Steroid profiling in Cologne. In: Proceedings of the 10th Cologne workshop on dope analysis 7th to 12th June 1992. Donike et al. (eds.), 47–62. Sport und Buch Strauß, Cologne.
125. Strauss, R.H. & Curry, T.J. (1987) Magic, science and drugs. In: Drugs & performance in sports. 3–9. Strauss et al. (eds.). WB Saunders Co, Philadelphia, PA.
126. O'Shea, J.P. (1978) Anabolic steroids in sport: A biophysiological evaluation. Nutr. Rep. Int. *17*, 607–627.
127. Dirix, A. (1978) A general summary of dope testing. Farm. Tijdschr. Belg. *3*, 29–36.
128. Windsor, R. & Dumitru, D. (1989) Prevalence of anabolic steroid use by male and female adolescents. Med. Sci. Sports. Exerc. *21*, 494–497.
129. Strauss, R.H. (1987) Anabolic steroids. In: Drugs & performance in sports. Strauss et al. (eds.), 59–68. WB Saunders Co, Philadelphia, PA.
130. Ferenchick, G.S. (1990) Are androgenic steroids thrombogenic? N. Engl. J. Med. *322*, 476.

1.5 Pharmacokinetics as Applied to Drug and Doping Research

T. B. Vree

1.5.1 Introduction

Pharmacokinetics is the science which quantifies as a function of time the absorption, distribution, metabolism and excretion of drugs in the body. It studies and describes the fate of chemicals in a living organism. As drug concentrations cannot be read in the body one has to take at different and consecutive times samples from the respective body fluids in order to measure these concentrations and to reconstruct afterwards concentration-time curves. However, it is a matter of debate what concentration-time curve must be reconstructed; this depends severely on the scientific interests of the researcher, the aim of the investigation, whether a total mass balance of the drug is aimed, and of course on the availability of the samples. Drug metabolites must also be considered when they have pharmacologic activities that contribute to clinical responses. Doping research and doping control are two extremes of the scale and deserve a different approach for solving the problems and aims. Using pharmacokinetic functions, it is possible to evaluate the importance to man of processes which have been shown to occur in animals by tissue analysis and studies of metabolism.

Pharmacokinetic principles and the equations describing the one-, two- and multicompartment model are now well-known (1, 2), but only describe the observed rise and decline of a plasma concentration of a drug in time and do not tell the mechanisms of absorption and elimination. After absorption has taken place, it is the elimination which governs the practical actions to be taken in order to reconstruct the pharmacokinetic picture of the drug in question.

1.5.2 Elimination

Elimination of a drug takes place by two processes: a) metabolism (transformation of the molecule into a different structure) and b: excretion. One has to realize that metabolism is not elimination in the strict sense; it changes the parent drug to metabolites, but it does not bring these compounds outside the body. Excretion eliminates the drug and its metabolites from the body and is the final elimination step. The liver is the principal site of biotransformation and the kidney is primarily responsible for the excretion of most non-inhaled toxicants and their metabolites.

Metabolism
Phase I metabolism of a drug takes place by the cytochrome P450 isoenzyme complex or family of isoenzymes which is able to reduce and to oxidize exogenous compounds. The composition of these isoenzymes depends strongly upon the species which it serves. It maintains the metabolic homeostasis of the body to which it belongs, e. g. the enzyme complex of a crocodile fulfills the needs of the crocodilian body, while that in man fulfills the needs of the human body. That means that compound A can be metabolized differently by crocodiles and man, depending on the isoenzymes available to attack a certain exogenous compound. The metabolic picture of drug A in for instance a mouse or a spider cannot be transformed entirely to humans; the metabolites formed may be similar but the yields in which they are formed are different. The same holds true for phase II metabolism by conjugation performed by families of glucuronyl-, acetyl-, glycinyl-, and glutathionetransferases which are present in the different animal species. Within one species the quantity of each isoenzyme will show a Gaussian distribution, while also a specific enzyme may be not present (deficient metabolism).

Excretion
Excretion of the drug or its metabolites may take place by the kidneys, the bile, the lungs, the saliva, the sweat, etc. Excretion of drugs with a relatively low molecular weight takes place by the kidney, while those with a high molecular weight (> 500) take place by the bile and feces.

Renal excretion takes place by a combination of the following passive and active mechanisms: glomerular filtration followed by passive tubular reabsorption, glomerular filtration followed by active tubular secretion or active tubular reabsorption. These mechanisms determine the concentration of the drug in urine and the renal clearance of a compound, that is the amount of blood cleared from the drug per unit of time. The urine flow determines the final concentration of the drug in a urine sample. All that means that diseases of the liver and the kidney will alter the elimination of many drugs, mostly causing an increase in drug concentration and in elimination half-life. For example, diazepam metabolism is retarded in hepatic failure and the parent drug will persist.

Biliary excretion proceeds via similar mechanisms, though the bile flow is low in comparison to urine flow.

1.5.3 Sampling

Blood. Blood is relatively easily accessable for sampling though one needs help by the insertion of an indwelling catheter or by performing a vene puncture; sampling time may therefore be limited by office hours. Fingertip puncture can be performed by the subject him/herself and samples can be taken at any time required to construct a correct plasma concentration-time curve. Fingertip puncture limits the sample size to 2 ml blood and 1 ml plasma respectively.
Urine. Urine production and collection is a continuous process, urine voiding fortunately is a discontinue process. Untimed or spontaneously voided urine samples must be collected in order to reconstruct a renal excretion rate-time curve.

Feces. Feces production is a continuous process, defecation a discontinue one.
Saliva. Drugs appear in saliva and are swallowed again giving rise to recirculation. Saliva samples may be used to detect drugs.
Sweat. Drugs appear in sweat and sweat samples may also be used to detect drugs (3).

1.5.4 Number of Samples

Blood. The number of samples depends on the elimination process under investigation. First, in each case at least 3–5 blood samples are needed for accurately describing a pharmacokinetic phase in a one-, two- or multicompartment model. Second, the time course of a sampling period should be 5–7 times the expected elimination half-life of the drug and its metabolites under investigation. Half-lives of drugs depend on their molecular structure and vary between 1 and 300 h.
Urine. Each untimed urine sample should be collected over a period of 7 × half-life. The shorter the urine collection time the smoother the renal excretion rate-time profile will be. Timed urine samples (0–2 h, 0–12 h, 0–24 h) allow crude approximations of the renal excretion rate-time profile.

1.5.5 Analysis

The collected blood, urine samples or body fluid must be analyzed for their concentration of drug and metabolite(s). The analytical method must be able to discriminate between parent drug and each known and available metabolite in order to avoid cross reaction or interference of compounds in for example chromatographic peaks. Drug conjugates should be measured as such and not hydrolyzed to the parent drug or aglycon (4, 5). Treatment of a plasma or urine sample with a β-glucuronidase to liberate the drug under study depends on the effectivity of the selected β-glucuronidase and the molecular structure of the glucuronide. When not the right β-glucuronidase is selected, hydrolysis is minimal and the drug or aglycon is missed in the analysis or screening.

Reference compounds.
For the development of chromatographic analysis or screening, the reference compounds (drug plus their metabolites) must be present and available. Drug glucuronide conjugates are hard to synthesize and are usually not available via commercial routes. These compounds must be identified in the chromatogram and thereafter isolated by preparative chromatography. An effective β-glucuronidase can be selected and calibration curves for the glucuronides can be constructed as follows:

$$(\text{Drug-glucuronide}) = d(\text{Drug}) \times M_{\text{Drug-gluc}}/M_{\text{Drug}}$$

where d(Drug) is the difference in concentration of the typical drug before and after deconjugation and M is the relative molecular mass (6, 7, 8, 9, 10, 11).

Isolation of a drug-glucuronide from the matrix (urine) is necessary when two or more possible glucuronides or conjugates can be formed from one compound containing for instance an alcohol- and a carboxy moiety (12, 13). However, it is not an absolute requirement when only one conjugate is formed (6, 7, 8, 9, 10, 11).

Tentative identification of drug glucuronides can be carried out by selective β-glucuronidase treatment, by using the group contribution factor of the glucuronide or conjugate group to the chromatographic process (ratio k'glucuronide/k'aglycon), by constructing a pharmacokinetic curve of the conjugate in which the concentration rises and declines parallel to that of the parent drug. Absolute identification must be carried out by LS-MS, and/or LC-MS^2 depending upon the availability of that specific instrumentation (14).

1.5.6 Doping Research

The *pharmacokinetics* of each drug suspected to be used in doping must be investigated first in healthy volunteers. The appropriate number of blood samples must be collected in addition to all untimed urine samples over a period of seven half-lives. The *plasma concentration-time curve(s) and renal excretion rate-time profile(s)* of parent drug and each metabolite should be constructed. A renal excretion rate is required rather than urine concentrations because the renal excretion rate-time profile (μg/min) runs parallel to the plasma concentration-time curve with the renal clearance (Clr = ml/min) as the proportionality constant. The renal excretion rate (μg/min) is obtained by multiplying the measured urine concentration (μg/ml) with the average urine flow (ml/min) over that specific urine collection period. For this purpose one has to control the time (min) and the total volume of the void (ml). That urine flow figure of one sample represents the average urine flow (ml/min) over that period, and so the renal excretion rate is the average value over the same period. In the pharmacokinetic graph, the renal excretion rate should be expressed as a histogram-like curve, with horizontal lines representing the average renal excretion rate during a certain collection period (fig. 1, 2, 4, 5, 6). The renal clearance (Clr = ml/min) can be calculated for each urine collection period by dividing the renal excretion rate (μg/min) by the plasma concentration (μg/ml) corresponding to the midst of the urine collection period. When tubular reabsorption is the dominant process in the renal excretion process, the renal clearance of the drug or the metabolite(s) will strongly depend upon the urine pH and the urine flow. When the pKa or pKb of the drug differs 2–3 units from the physiological pH (7.4) the urine pH may affect the renal clearance of a drug as demontrated for amphetamine (pKb 9.9), salicylic acid (pKa 3.5), sulphamethoxazole (pKa 4.9), fencamfamine (pKb 8.7), in general strongly acidic or alkaline drugs (fig. 1). When the urine pH is kept acidic (pH 5.0–5.5) by the daily intake of 2–4 gram ammoniumchloride, or alkaline (pH 7–8) by the daily intake of 5–10 gram of sodiumbicarbonate, then the influence of the urine flow on the renal clearance becomes visible (15, 16). Active tubular reabsorption takes place with NSAID's; with acidic urine the renal clearance is negligible, with alkaline urine the excreted acyl glucuronide is labile, hydrolyzes to the parent compound and isomerizes to a stable isoglucuronide. That means that the virtual increase in renal clearance by alkaline urine pH of a parent drug, which forms an acyl glucuronide, may in part be due to hydrolysis of the acyl glucuronide conjugate.

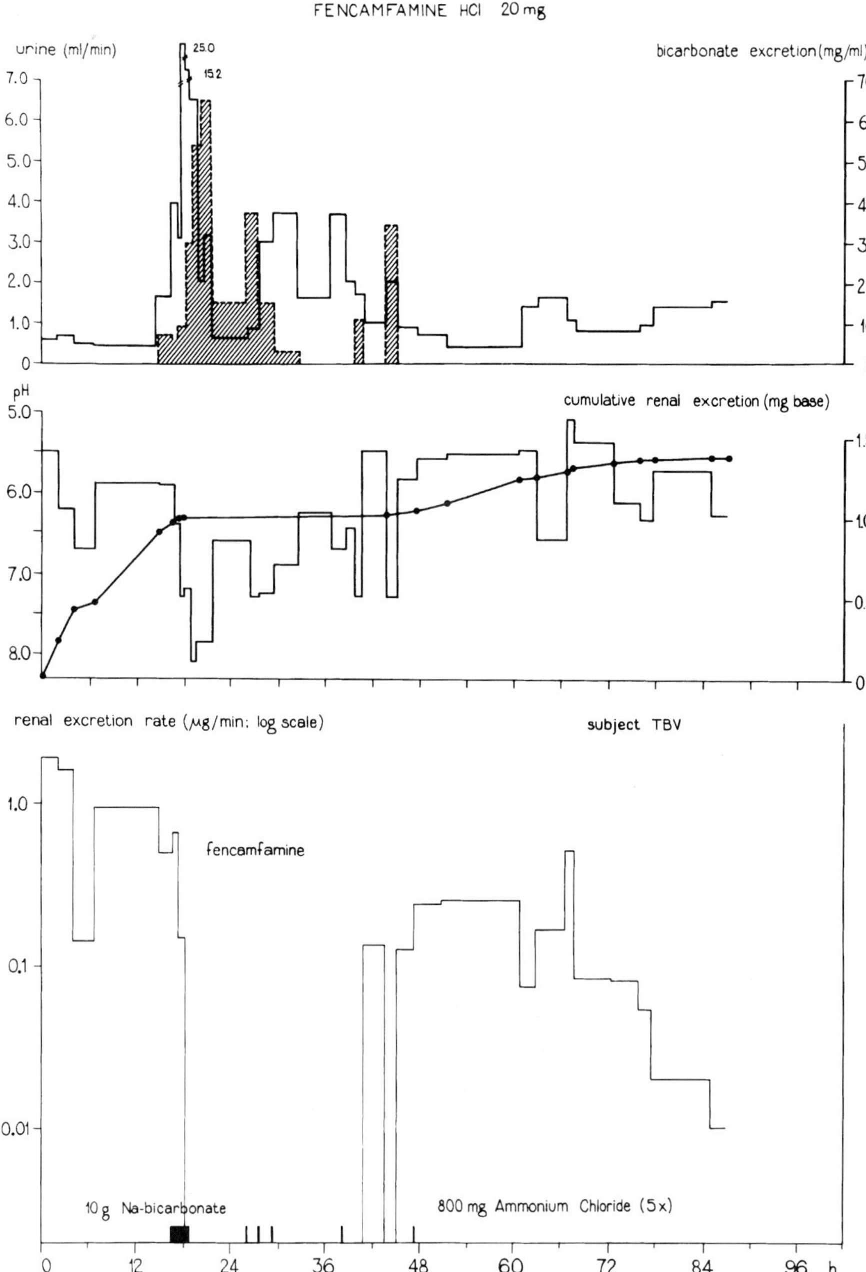

Figure 1. Renal excretion rate-time profile of fencamfamine under acid and alkaline urine conditions. Administration of $NaHCO_3$ makes the urine alkaline, stimulates the passive tubular reabsorption and reduces or inhibits the renal excretion (15, 16).

A fast indication about the renal clearance value is obtained when the total amount of drug and metabolite excreted (mg) is divided by the AUC (mg.L.h^{-1}). This value is the average over the total elimination period and masks the effects of urine pH and urine flow. Long urine collection periods or a 24 h urine sample must be avoided in pharmacokinetic studies. The 24 h urine collection period is inherited from the creatinine clearance estimation (steady state situation) using endogenous creatinine as marker for the kidney function.

Doping research leads to the respective reference values and pharmacokinetic knowledge which are absolutely essential in routine doping control. Furthermore the study of the pharmacokinetics of the xenobiotics involved may contribute considerably to the right interpretation of the results obtained from doping analysis.

1.5.7 The Site of Metabolism

'Normally' the liver metabolizes the drug and the kidney excretes drug and metabolites. For instance sulphonamides are acetylated and the N-acetyl conjugates are excreted by both glomerular filtration and active tubular secretion (6, 8). That system operates very effectively, it produces a metabolite that is excreted with the highest possible rate. Drug glucuronides are excreted by the same mechanism. That phenomenon can be easily understood when the drug and its metabolites are present in both plasma and urine. For instance ether glucuronidation of codeine results in the formation of codeine-6-glucuronide which is present in high concentrations in plasma as well as in urine (fig. 2) (17, 18).

When the kidney produces the same metabolites as the liver and excretes them in the urine, this process will be unnoticed and the calculated or apparent renal clearance will be too high. For instance capacity limited glycination of salicylic acid to salicyluric acid takes place in both the liver and the kidney (19).

When the metabolites are present in the urine in high concentration but are not detectable in plasma, the following situations can be distinguished:

First, the plasma concentration of the metabolite is below the quantitation limit of the analytical method due to an extremely high renal clearance. When the metabolite is ingested, its intrinsic elimination half-life is much smaller than that of the parent drug (naproxen t1/2 20 h, O-desmethylnaproxen t1/2 1 h; (12, 20)).

Second, the metabolite is synthesized by the kidney and immediately excreted in the urine. This situation is not easily recognized, because the first possibility is the most logical one. Studying the drug probenecid for example, we found an indication that the kidney is able to glucuronidate the parent compound.

Probenecid

Probenecid was developed for the inhibition of the carrier system in the active tubular secretion of anions (21). In that way it inhibits secretion and extends the half-life of drugs in the body. Occasionally it has been reported that probenecid interfered with and inhibited the glucuronidation of drugs (5, 20).

Probenecid is mainly conjugated by glucuronic acid in humans and probenecid acyl glucuronide is the main metabolite in the urine (fig. 3) (22, 23). When tracking

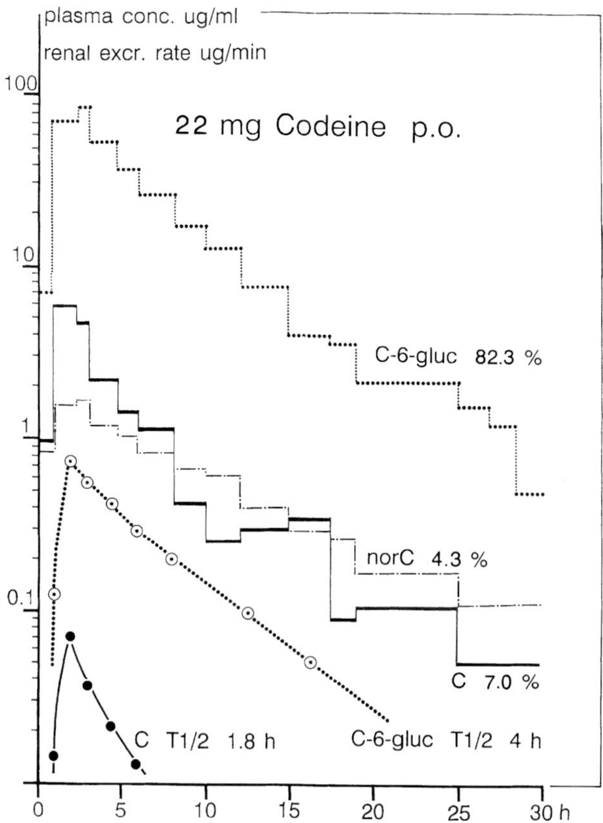

Figure 2. Plasma concentration-time curves and renal excretion rate-time profiles of Codeine (C), its metabolite norcodeine (norC), and the glucuronide codeine-6-glucuronide (C-6-gluc), in a volunteer after an oral dose of 30 mg codeine phosphate (22 mg codeine). The volunteer is unable to O-dealkylate codeine into morphine and lacks the cytochrome P450 IID6 isoenzyme. The apparent renal clearance of codeine and its glucuronide is approximately 100 ml/min (17, 18).

both renal excretion rate-time profiles and plasma concentration-time curve of probenecid and its metabolites after the oral intake of 1000 mg of the drug, it was noticed that the renal excretion rate was constant and maximal for 10 h (fig. 4). The duration of the maximum was dose dependent, and the height of the plateau value was subject (kidney) dependent (22, 23). This capacity-limited renal excretion was not observed in the plasma elimination curve. Probenecid glucuronide was not present in plasma, but it must have been even it had the highest renal clearance (700 ml/min). When it is accepted by now that probenecid is glucuronidated by the human kidney, more drugs may follow the same pattern and interference at renal conjugation may become a reality.

Probenecid forms an acyl- or ester glucuronide which is labile under alkaline conditions (24, 25). The pharmacokinetic studies and metabolic inhibitions under probenecid co-medication must therefore be carried out under acidic urine conditions (pH 5–5.5) in order to stabilize the formed acyl glucuronide conjugates. The anti-

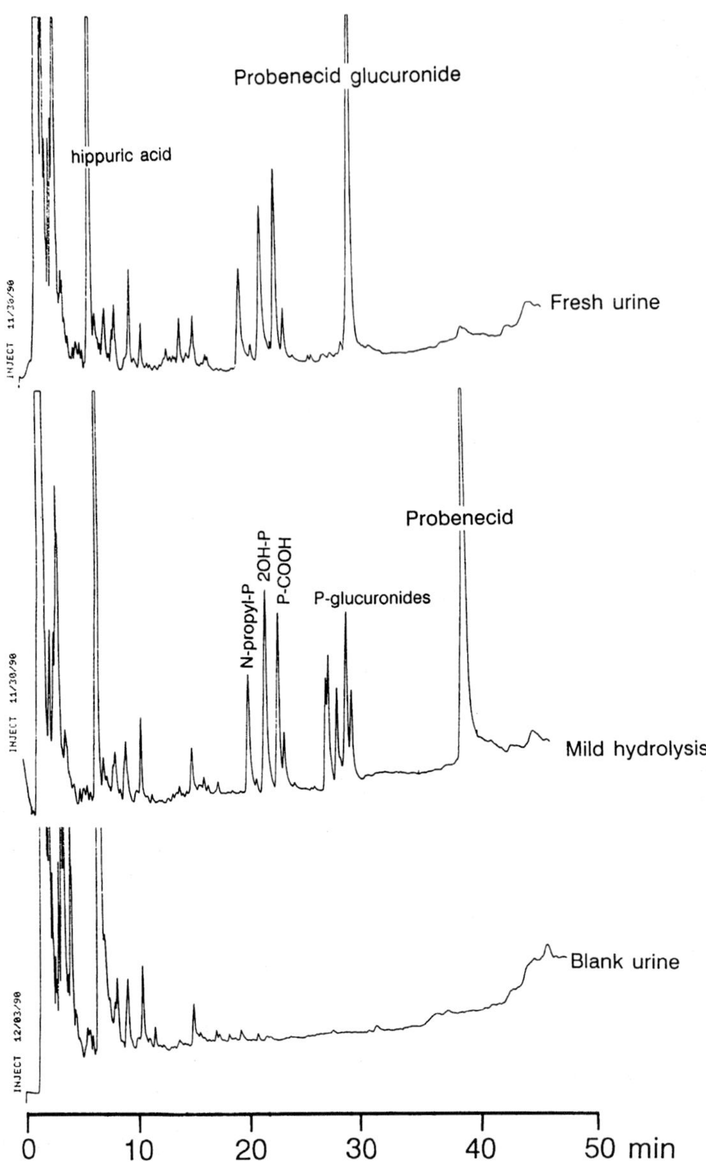

Figure 3. Chromatograms of probenecid, probenecid acyl glucuronide, and its phase I metabolites in fresh urine injected immediately after voiding, and after mild hydrolysis at pH 6.0. Mild hydrolysis leads to the formation of the positional isomers and hydrolysis of probenecid 1-O-glucuronide (acyl gluc) (22).

microbial drug nalidixic acid, (t1/2 3 h) and the NSAID indomethacin (t1/2's 3 + 10 h) may be thought to be glucuronidated by the human kidney, as their acyl glucuronides are not present in plasma. Probenecid comedication interferes and inhibits the formation of the acyl glucuronide of nalidixic acid (20, 27) and that of

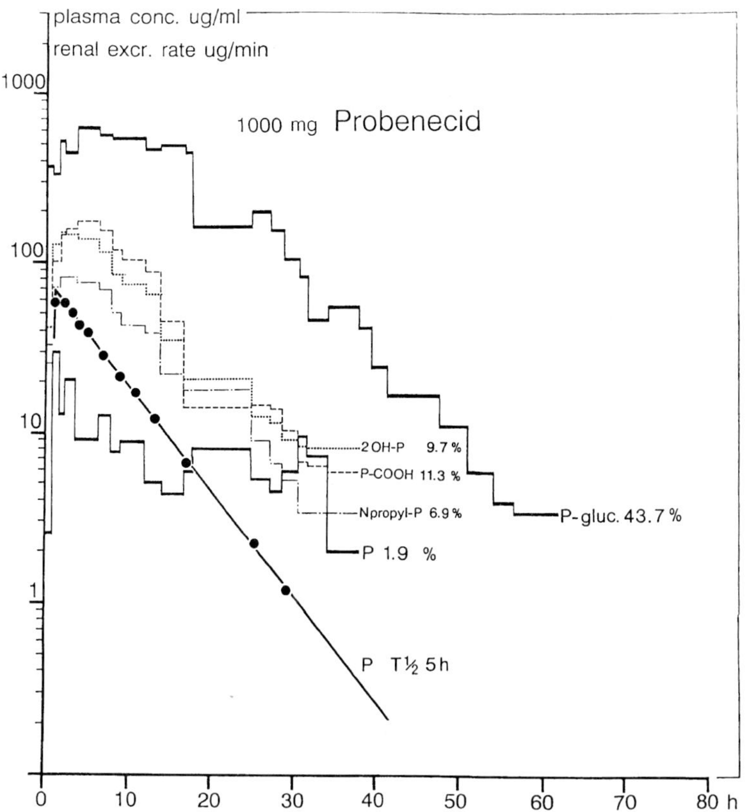

Figure 4. Plasma concentration-time curve and renal excretion rate-time profiles of probenecid (P), its metabolites 2-hydroxyprobenecid (2 OH-P), 2-carboxyprobenecid (P-COOH), N-monopropylprobenecid (Npropyl-P) and probenecid acyl glucuronide (P-gluc) in a human volunteer after administration of one oral dose of 1000 mg Probenecid. Note the plateau value in the renal excretion curve of P-gluc (23).

indomethacin (20). The degree of inhibition will depend on the intrinsic affinity of each structure for binding at the specific glucuronyl-transferase isoenzyme.

1.5.8 Reconstruction of the Pharmacokinetic Process

When do we need the reconstruction of the pharmacokinetic process of a certain drug? The following situations can be distinguished:

a the pilot and volunteer study
b a comparison of bioavailability of two drugs
c therapeutic drug monitoring
d after an overdose or toxic dose
e traffic drug control
f doping control
g drug of abuse policies

In situations a, b and c, the identity of the drug is known and according to good pharmacokinetic principles the required number of samples can be obtained. In situation d, the identity of the drug may be known or retrieved and plasma and urine samples can be obtained as soon as the patient is hospitalized. In situation e, police officers are entitled to take and analyse a breath or blood sample on the spot when somebody is suspected of alcohol abuse. In situations f and g, the identity of the drug used is unknown and the number of samples will be limited to only one due to practical reasons.

The rules of the committee organizing the sporting event force a certain number of competitors to deliver a urine sample produced just after the finish of the game. A blood sample could be more appropriate than a urine sample. Blood samples

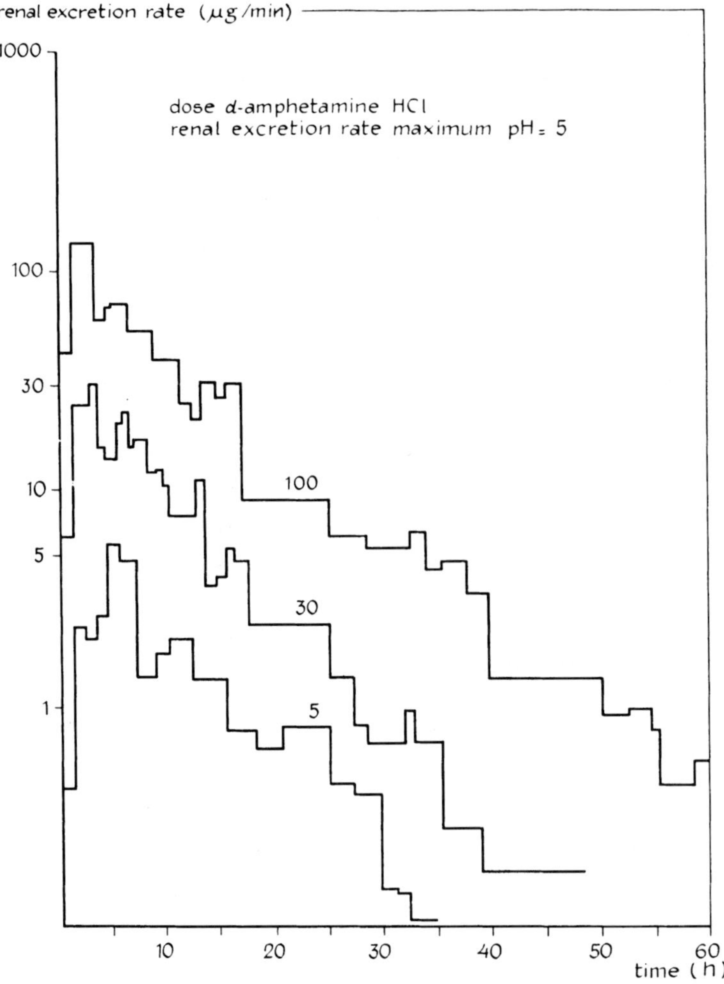

Figure 5. Renal excretion rate-time profiles under acidic urinary conditions of amphetamine after different dosages. Note that the maximum in the renal excretion rate in µg/min corresponds with the dose in mg (HCl salt). This coincidence may be used in doping control (16).

can be easily obtained, but the procedure violates the integrity of the competitor. Urine samples can be produced after some time when sufficient fluid has been taken. Urine samples are usually concentrated due to the low urine flow during and after the sporting event, because of alternative fluid loss by sweating. They are acidic (pH 5–5.0) after a sporting event, which may stimulate or minimize renal excretion of stimulants (fig. 5) (16).

1.5.9 Doping Control

The limitation to just one urine sample with the uncertainty about the collection time period, makes a reconstitution of the renal excretion rate impossible. At least

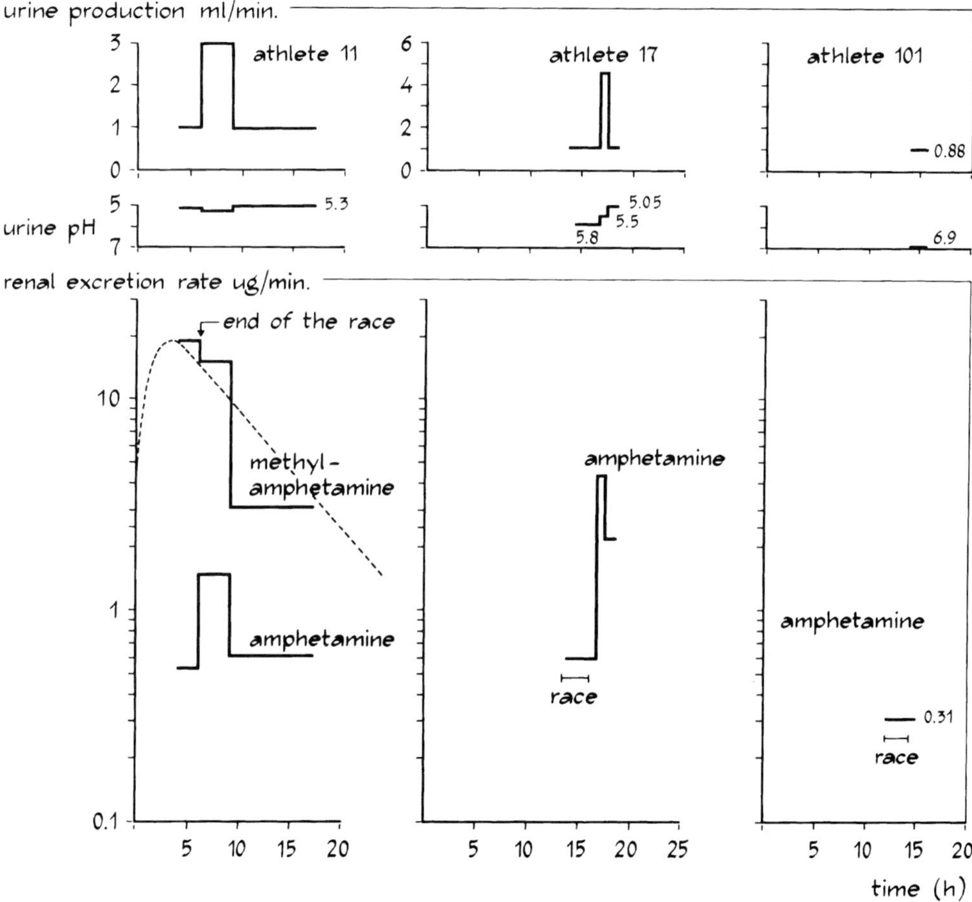

Figure 6. A) An example of a positive result in the doping control when it is performed as proposed in this chapter. The dose, time of administration and the compound can be identified.
B) Just 5 mg of amphetamine was taken before the event.
C) With only one urine sample, the compound can be identified, but important parameters as, renal excretion rate, half-life and dosage must be ignored (16).

three untimed urine samples (volume and time noted) are required to reconstruct a part of the excretion curve as shown in figure 6 (16).

From this time course and that of the metabolite it is possible to estimate the time of administration of the drug and to judge whether the drug has been taken just before or during the competition with the aim to falsify the result. When a certain renal excretion rate of a doping compound can be measured and calculated, it is possible to give a certain threshold value. With values found higher than the threshold dosis and excretion rate, the drug is taken at the day of the competition, and with values lower than the threshold, the drug is taken days before the day of the competition. It has been proposed in 1973 (16) and became effective in 1985 (14). This type of judgement is possible when the pharmacokinetic picture of each doping drug is known to each doping control laboratory and medical commission officers of a sporting organisation.

Cheating the doping control

There are compounds devoid of stimulating properties but which are taken to dilute, to mask or inhibit the excretion of known doping products. One typical example of this kind of drugs is *probenecid.*

Healthy human subjects who participate in probenecid inhibition studies know by experience that the effects of probenecid can be tolerated for one day. Probenecid not only inhibits the active tubular secretion of drugs but must also inhibit more active transport processes in the body, resulting in (extreme) tiredness (22, 23). Therefore it is difficult to understand why athletes should attempt to take probenecid, since it makes the condition for the match even worse. It is claimed that probenecid inhibits the renal excretion of anabolic steroids. If probenecid inhibits the renal clearance of these steroids, a dose of 1–2 gram/day of probenecid is needed to reduce renal excretion (4). If (where) probenecid inhibits renal excretion, it only inhibits the active tubular secretion and never the glomerular filtration. Reduction of the excretion of steroid glucuronides to approximately 10% of the pre-test or initial value overrates the contribution of tubular secretion in the overall excretion process.

Anabolic steroids in urine are measured after a deconjugation step (5, 14). Not the conjugate and its aglycon are measured but only the aglycon. Using that procedure, it is not certain whether under probenecid co-medication, the renal excretion of the steroid is inhibited, the renal excretion of the steroid conjugate is inhibited, or the renal conjugation/glucuronidation of the steroid is inhibited.

The experiments which nalidixic acid and indomethacin under probenecid co-medication show that probenecid is able to inhibit and to reduce the renal glucuronidation of these drugs, as shown by the direct HPLC analysis of the glucuronides of the drug (13, 20, 26, 28). This approach must also be followed for the analysis of steroid conjugates in urine, though the concentrations will be much lower and may be the limiting factor. A logical explanation of the effect of probenecid on the renal excretion of steroids may be that probenecid inhibits the renal conjugation of the steroids and not only the renal excretion.

1.5.10 Conclusion

Pharmacokinetics as applied to drug and doping research describes the research investigations and experiments. Depending on the aims of the investigations and personal interests of the researcher (27), the study is designed and the study protocol written. Doping control operates at the extreme other end of the scale, in absolute terms it detects compounds in body fluids. This detection is an art in itself for those who love to do this work and who love to operate at this particular end of the scale. When one likes to operate at the other or opposite end of the scale, doping control is a tedious job not satisfying scientific questions. As with all opposites, they belong to each other and cannot exist without each other. The discussion about the social need for a doping control is beyond the scope of this article.

So the art of doping control needs the art of drug/doping science in order to operate at the highest scientific level.

References

1. Jusko, W.J. (1986) Guidelines for collection and analysis of pharmacokinetic data. In: Applied Pharmacokinetics, principles of therapeutic drug monitoring. (Evans, W.E., Schentag, J.J., Jusko, W.J., & Harrison, H., eds.). 9–54. Applied Therapeutics Inc, Spokane, WA 99210.
2. Van Rossum, J.M., van Ginneken, C.A.M., Henderson, P.Th., Ketelaars, H.C.J. & Vree, T.B. (1977) Pharmacokinetics of biotransformation. In: Kinetics of Drug Action. Handbook of Experimental Pharmacology, Vol. 47 (Van Rossum, J.M., ed) 125–167, Springer Verlag, Heidelberg.
3. Vree, T.B., Muskens, A.Th.J.M. & van Rossum, J.M. (1972) Excretion of amphetamines in human sweat. Archiv. Int. Pharmacodyn. *199*, 311–317.
4. Geyer, H., Schänzer, W. & Donike, M. (1993) Probenecid as masking agent in dope control-inhibition of the urinary excretion of steroid glucuronides. In: Proceedings 10th Cologne workshop on dope analysis, june 1992. (Donike, M., Geyer, H., Gotzmann, A., Mareck, U., Engelke, A. & Rauth, S., eds.) 141–151, Sport & Buch Strausz Edition Sport, Köln.
5. Ryu, J.C., Kwon, O.S., Song, Y.S., Yang, J.S. & Park, J. (1992) The effects of probenecid on the excretion kinetics of stanozolol, an anabolic steroid, in rats. J. Appl. Toxicol. *12*, 385–391.
6. Vree, T.B., Beneken Kolmer, E.W.J., Martea, M., Bosch, R. & Shimoda, M. (1990) High-performance liquid chromatography of sulphadimethoxine and its N_1-glucuronide, N_4-acetyl and N_4-acetyl-N_1-glucuronide metabolites in human plasma and urine. J. Chromatogr. Biomed. Appl. *526*, 119–128.
7. Vree, T.B., Beneken Kolmer, E.W.J., Martea, M., Bosch, R., Hekster, Y.A. & Shimoda, M. (1990) Pharmacokinetcis, N_1-glucuronidation and N_4-acetylation of sulfadimethoxine in humans. Pharm. Weekbl. (Sci). *12*, 71–75.
8. Vree, T.B., Beneken Kolmer, E.W.J., Hekster, Y.A., Shimoda, M., Ono, M. & Miura, T. (1990) Pharmacokinetics, N_1-glucuronidation and N_4-acetylation of sulfa-6-monomethoxine in humans. Drug Metab. Dispos. *18*, 852–858.
9. Vree, T.B., Beneken Kolmer, E.W.J. & Hekster, Y.A. (1990) High pressure liquid chromatographic analysis and preliminary pharmacokinetics of sulfaphenazole and its N_2-glucuronide and N_4-acyl metabolites in plasma and urine of man. Pharm. Weekbl. (Sci) *12*, 243–247.
10. Vree, T.B., Steegers-Theunissen, R.P.M., Baars, A.M. & Hekster, Y.A. (1990) Direct high-performance liquid chromatographic analysis of p-hydroxyphenyl-phenylhydantoin glucuronide, the final metabolite of phenytoin, in human serum and urine. J. Chromatogr. Biomed. Appl. *526*, 81–89.
11. Vree, T.B., Baars, A.M. & Wuis, E.W. (1991) Direct high-performance liquid chromatographic analysis and preliminary pharmacokinetics of enantiomers of oxazepam and temazepam with their corresponding glucuronide conjugates. Pharm. Weekbl. (Sci) *13*, 83–91.

12. Vree, T.B., van den Biggelaar-Martea, M. & Verwey-van Wissen, C.P.W.G.M. (1992) Determination of naproxen and its metabolite O-desmethylnaproxen with their acyl glucuronides in human plasma and urine by means of direct gradient HPLC analysis. J. Chromatogr. Biomed. Appl. *578*, 239–249.
13. Vree, T.B., van den Biggelaar-Martea, M. & Verwey-van Wissen, C.P.W.G.M. (1993) Determination of indomethacin, its metabolites with their acyl glucuronides in human plasma and urine by means of direct gradient HPLC analysis. J. Chromatogr. Biomed. Appl. *616*: 271–202.
14. De Boer, D. (1992) Profiling anabolic androgens and corticosteroids in doping analysis. Academic PhD thesis University of Utrecht, SSN Nijmegen.
15. Vree, T.B. & van Rossum, J.M. (1969) Suppression of renal excretion of fencamfamine in man. Eur. J. Pharmacol 7, 227–230.
16. Vree, T.B. (1973) Dope and Doping control. In: Pharmacokinetics and metabolism of amphetamines. Academic PhD thesis, University of Nijmegen, 184–192. Drukkerij Brakkenstein, Nijmegen.
17. Vree, T.B. & Verwey-van Wissen, C.P.W.G.M. (1992) Pharmacokinetics and metabolism of codeine in humans. Biopharm. Drug Dispos. *13*, 445–460.
18. Verwey-van Wissen, C.P.W.G.M., Koopman-Kimenai, P.M. & Vree, T.B. (1991) Direct determination of codeine, norcodeine, morphine and normorphine with their corresponding O-glucuronide conjugates by HPLC with electrochemical detection. J. Chromatogr. Biomed. Appl. *570*, 309–320.
19. Wan, S.H. & Riegelman, S. (1972) Renal contribution to the overall metabolism of drugs. II. Biotransformation of salicylic acid to salicyluric acid. J. Pharm. Sci. *61*, 1285–1287.
20. Vree, T.B., Hekster, Y.A. & Anderson, P.G. (1992) Contribution of the human kidney to the metabolic clearance of drugs. Ann. Pharmacother. *26*, 1421–1428.
21. Beyer, K.H. (1950) Functional characteristics of renal transport mechanisms. *2*, 227–280.
22. Vree, T.B. & Beneken Kolmer, E.W.J. (1992) Direct measurement of probenecid and its glucuronide conjugate by means of high performance liquid chromatography in plasma and urine of humans. Pharm. Weekbl. (Sci) *14*, 83–87.
23. Vree, T.B., Beneken Kolmer, E.W.J., Wuis, E.W. & Hekster, Y.A. (1992) Capacity limited renal glucuronidation of probenecid by humans. A pilot Vmax finding study. Pharm. Weekbl. (Sci). *14*, 325–332.
24. Eggers, N.J. & Doust, K. (1981) Isolation and identification of probenecid acyl glucuronide. J. Pharm. Pharmacol. *33*, 123–124.
25. Faed, E.M. (1984) Properties of acyl glucuronides: implications for studies of the pharmacokinetics and metabolism of acidic drugs. Drug Metab. Rev. *15*, 1213–1249.
26. Vree, T.B., van den Biggelaar-Martea, M., van Ewijk-Beneken Kolmer, E.W.J. & Hekster, Y.A. (1993) Probenecid inhibits the renal clearance and renal glucuronidation of nalidixic acid. Pharm. World. Sci. *15*: 165–170.
27. Vree, T.B. (1991) Pharmacist-physician interaction: a battle of genes. Ann. Pharmacother. *25*, 1132.
28. Vree, T.B., van den Biggelaar-Martea, M., Verwey-van Wissen, C.P.W.G.M., van Ewijk-Beneken Kolmer, E.W.J. (1994) Probenecid inhibits the glucuronidation of indomethancin and O-desmethylidnomethacin in humans. Pharm. World. Sci. *16*, 22–26.

1.6 The Future of Immunochemical Drug Analysis

T. Takatori and M. Nagao

1.6.1 Introduction

Some drugs and other chemicals are well known causes of acute poisoning and hypersensitivity in the host (1, 2). In order to evaluate the conditions of patients with acute poisoning or drug allergy, monitoring and detecting drugs and chemicals in specimens is required.

The conventional methods for detection of drugs are generally instrumental procedures such as gas chromatography, high-performance liquid chromatography and gas chromatography-mass spectrometry. Although the sensitivity of these methods is sufficiently high, these procedures are complicated for untrained researchers. Analytical procedures for monitoring drugs or chemicals in biological materials should be simple enough for untrained researchers to perform rapidly and exactly. Immunoassays meet this criterion and have proven to be effective in detecting and monitoring clinically important compounds.

In this section, we describe the developments of immunoassay systems of psychotropic drugs and paraquat, which is widely used as a herbicide, and we also present the toxicological application of antibodies against these compounds.

1.6.2 Immunochemical Analysis of Psychotropic Drugs

Diazepam or haloperidol is among the most widely used benzodiazepine and butyrophenone derivatives, respectively. These psychotropic drugs have often been used as induction drugs for suicide. Therefore, in forensic and clinical fields, it is often necessary to detect trace amounts of them in biological specimens.

1.6.2.1 Radioimmunoassay for diazepam (3)

Several radioimmunoassays (RIAs) for benzodiazepine derivatives have been reported by Peskar and Spector (4) and Dixon and Crew (5). We describe here the development of a RIA for diazepam using temazepam (oxydiazepam)-3-hemisuccinate conjugated bovine serum albumin, an immunogen not reported previously (4, 5).

Preparation of the immunogen

Figure 1 shows the method of synthesis of the immunogen temazepam (oxydiazepam)-3-hemisuccinate (TZP-3-HS)-BSA. In order to prepare TZP-3-HS, 300 mg (1.0 mmol) of temazepam and 150 mg (1.5 mmol) of succinic anhydride containing 37 kBq of (1,4-^{14}C) succinic anhydride in 5 ml of dry pyridine were refluxed for two hours. After the solvent was removed by evaporation *in vacuo* in a water bath, the reactants were separated by thin layer chromatography on plates coated with silica gel 60 H using a solvent system of chloroform-methanol (9 : 1, by vol.). The TZP-3-HS eluted from silica gel 60 H with pyridine was then crystallized from methylene choride-hexane.

The diazepam hapten, TZP-3-HS, was coupled to BSA according to the method of Erlanger et al. (6) with a slight modification. For the estimation of the amounts of TZP-3-HS coupled to the carrier protein, the radioactivity of the dialyzate as the antigen was measured, the degree of conjugation being calculated to be 32.5 mol of the hapten per one mol of BSA (mol. wt. 68,000).

Immunization

A 0.5 ml volume of TZP-3-HS-BSA solution in saline containing 1.0 mg protein, was emulsified with an equal volume of complete Freund's adjuvant. Male albino domestic rabbits, weighing 2.5–3.0 kg, were injected subcutaneously in several different sites at the back. Rabbits received 1.0 ml of the emulsion twice the first month, then once every 3 weeks for 6 months as a booster injection. Blood was collected from a carotid artery ten days after the final injection and allowed to clot at 4 °C. The serum was separated by centrifugation and served as the source of antiserum.

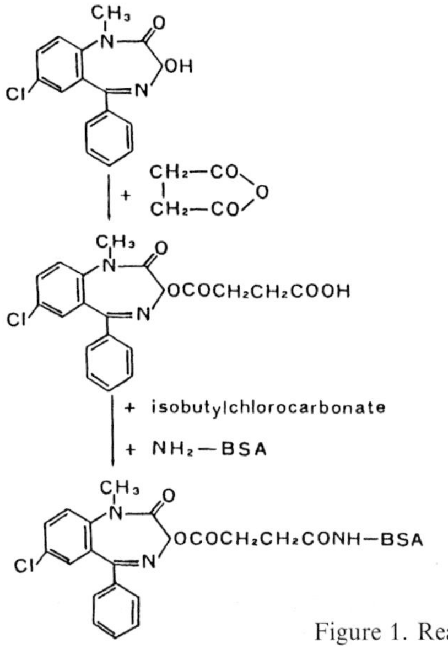

+ isobutylchlorocarbonate

+ NH_2—BSA

Figure 1. Reaction Scheme for the Preparation of TZP-3-HS-BSA.

Radioimmunoassay for diazepam
The RIA is based on competitive binding of unlabelled diazepam to the specific anti-diazepam antibodies present in rabbit antisera. After incubation, antiserum-bound (^{3}H)diazepam was separated from free by the salting-out method as described previously (7).

Each assay tube contained the following components: 0.1 ml of (^{3}H)diazepam (specific activity 2.88 TBq/mmol; about 5,000 dpm), 0.1 ml of diluted antiserum (1 : 10,000), 0.1 ml of 1 % bovine serum gamma globulin which was added to obtain sufficient ammonium sulfate precipitates, 0.1 ml of either unlabelled diazepam, other ligand solutions, or samples and 0.1 ml of 0.05 mM phosphate-buffered saline (pH 7.5, PBS). The unlabelled ligands were first dissolved with dimethyl sulfoxide, and each working solution was prepared to make 0.1 % dimethyl sulfoxide with PBS. The incubation mixture was allowed to stand for two hours at room temperature. After the addition of 0.5 ml of saturated ammonium sulfate solution to the mixture, the precipitate containing diazepam bound to the antibody was collected and washed twice with 0.5 ml of 50 % saturated ammonium sulfate solution, and its radioactivity was measured in a liquid scintillation spectrometer.

Tissue samples were prepared as follows. The organs were minced with scissors and homogenized with 10 volumes of 0.1 N HCl using a universal homogenizer. The whole homogenate was centrifuged at 14,000 xg for 20 min at 4 °C; an aliquot of the supernatant was used for this assay.

In this RIA procedure, the antiserum-bound (^{3}H)diazepam was competitively and linearly released with increasing amounts of unlabelled diazepam up to 128 pg as shown in Figure 2. In the absence of unlabelled diazepam, the antiserum of 1 : 10,000 dilution could bind 50 % of (^{3}H) labelled diazepam by this assay. As shown in Figure 2, a level as low as 1 pg of diazepam could be detected, and 32 pg resulted in a 50 % inhibition.

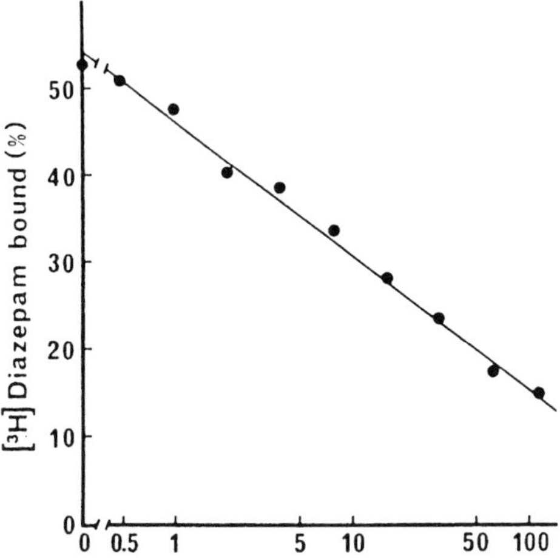

Figure 2. Displacement of (^{3}H)diazepam Bound to anti-TZP-3-HS-BSA Antiserum by Unlabelled Diazepam.

Table 1 shows the specificity of anti-diazepam antiserum. These results indicate that the antiserum is not specific for a hydroxy group at position-3 of a 1,4-benzodiazepine skelton, but is specific for a methyl group at position-1 of the skeleton and for a keto group at position-2 of the ring.

Table 1. Cross-Reactivity of Anti-Diazepam Antiserum with Benzodiazepine Derivatives.

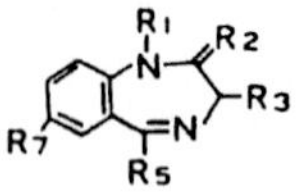

Ligands	R1	R2	R3	R5	R7	Cross-reactivity (%)
Diazepam	CH_3	O	H		Cl	100
Temazepam	CH_3	O	OH		Cl	100
Fludiazepam	CH_3	O	H	F	Cl	16
Prazepam	CH_2-◁	O	H		Cl	0.1
Medazepam	CH_3	H_2	H		Cl	0.08
N-Desmetyl-diazepam	H	O	H		Cl	0.025
Nitrazepam	H	O	H		NO_2	0.016
Oxazepam	H	O	OH		Cl	<0.0025
Lorazepam	H	O	OH	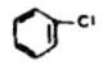 Cl	Cl	<0.0025
Bromazepam	H	O	H	N	Br	<0.0025
Clobazam	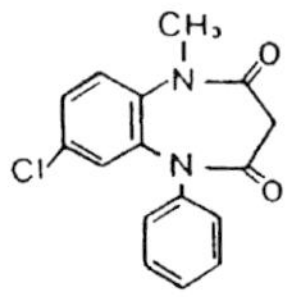					16
Chlordiazepoxide	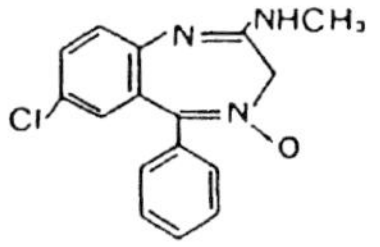					<0.0025

The antiserum produced by immunization with TZP-3-HS-BSA is highly specific to diazepam, and the RIA described herein is highly sensitive and reliable. For the determination of diazepam from trace amounts of biological materials, this RIA system will be advantageous in clinical and forensic fields (8).

1.6.2.2 Radioimmunoassay for haloperidol (9)

Production of anti-haloperidol antiserum

The haloperidol hapten, 4-fluoro-4-(4-hemisuccinate-4-*p*-chlorophenylpiperidino) butyrophenone (HPD-HS) was prepared by the method of Rubin et al. (10). HPD-HS was coupled to BSA according to the method of Erlanger et al. (6). The immunization procedure was the same as that for producing anti-diazepam antiserum.

RIA procedure and characterization of antiserum

The RIA procedure for haloperidol is almost the same as that for diazepam. The labelled haloperidol and unlabelled ligands were first dissolved with methanol and each working solution was prepared to make 30% methanol with PBS. (^{3}H)Haloperidol (specific activity 758.5 GBq/mmol; about 5,000 dpm in 0.1 ml of PBS solution) was added in an assay tube. In this RIA procedure, the antiserum-bound (^{3}H)haloperidol with increasing amounts of unlabelled haloperidol was competitively and linearly challenged up to 250 pg of haloperidol as shown in Figure 3. In the absence of unlabelled haloperidol, the antiserum of 1 : 1,000 dilution could bind 50% of (^{3}H) labelled haloperidol by this assay. As shown in Figure 3, as little as 30 pg of haloperidol could be detected, and the amount of haloperidol causing a 50% inhibition was 93 pg.

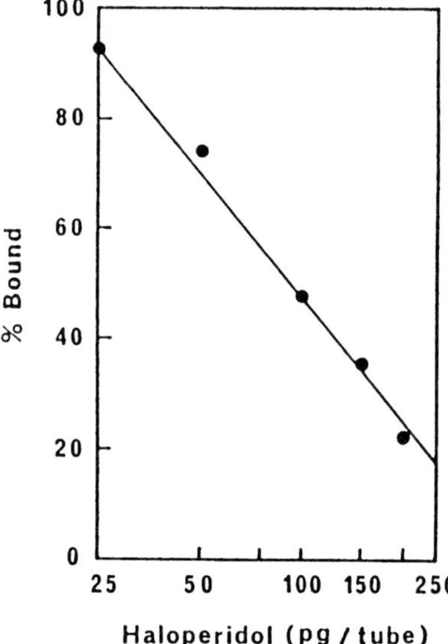

Figure 3. Displacement of (^{3}H)haloperidol Bound to Anti-HPD-HS-BSA Antiserum by Unlabelled Haloperidol.

The specificity of the anti-haloperidol antiserum was evaluated by cross-reactive studies with butyrophenone derivatives and their metabolites (Table 2). Substitutions at the carbon-4 of the piperidine ring of butyrophenone derivatives, such as timiperone, floropipamide and spiperone, showed a remarkably low cross-reactivity with

Table 2. Cross-Reactivity of Anti-Haloperidol Antiserum with Butyrophenone Derivatives and Their Metabolites.

Generic name	Structure	50% Inhibition(ng)
Haloperidol	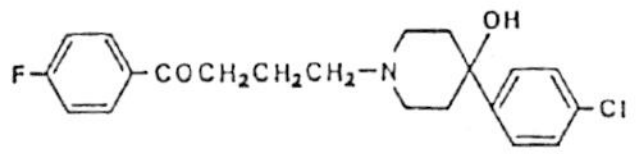 F– $COCH_2CH_2CH_2$–N, OH, Cl	0.093
Bromperidol	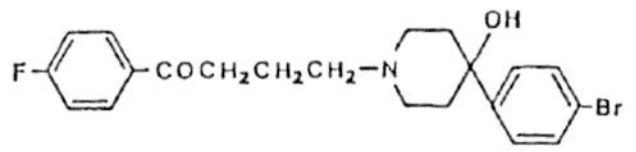 F– $COCH_2CH_2CH_2$–N, OH, Br	0.75
Moperone	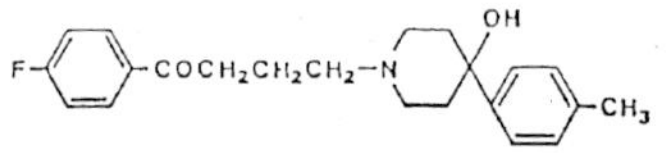 F– $COCH_2CH_2CH_2$–N, OH, CH_3	1.4
Timiperone	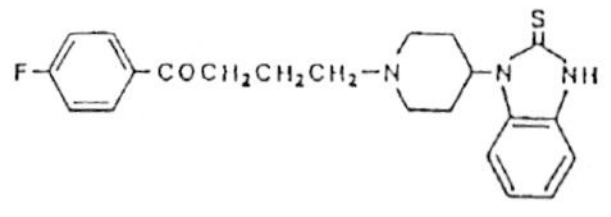 F– $COCH_2CH_2CH_2$–N, N, S, NH	2.5
Floropipamide	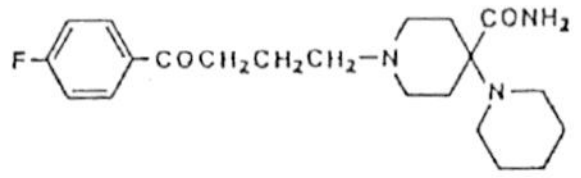 F– $COCH_2CH_2CH_2$–N, $CONH_2$, N	55
Spiperone	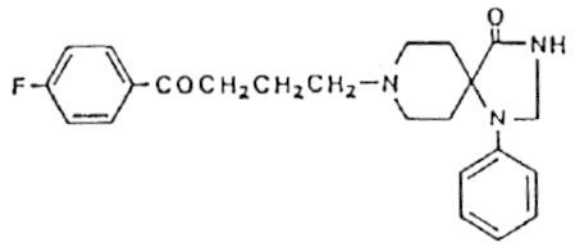 F– $COCH_2CH_2CH_2$–N, O, NH, N	45

Generic name	Structure	50% Inhibition(μg)
Reduced Haloperidol	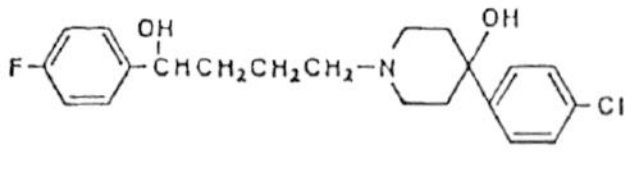 F– OH, $CHCH_2CH_2CH_2$–N, OH, Cl	0.24
p-Fluorobenzoyl-propionic acid	 F– $COCH_2CH_2COOH$	>500
p-Fluorophenyl-aceturic acid	 F– $CONHCH_2COOH$	>500
p-Fluorophenyl-acetic acid	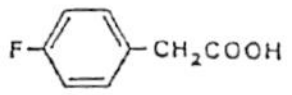 F– CH_2COOH	>500

the antibody. The binding affinities of bromperidol and moperone for the antibody were relatively decreased, suggesting that the minor change from the chloro group of haloperidol to the bromo or methyl group was also discriminated by the specific antibody. The antibody showed little binding affinity for other metabolites of haloperidol. These results indicate that the antibody recognizes not only the *p*-fluorobenzoylpropyl group but also the 4-*p*-chlorophenyl-4-hydroxypiperidine ring.

1.6.3 Immunochemical Analysis of Paraquat

Paraquat (1,1′-dimethyl-4,4′-bipyridinium) is widely used as a herbicide. In recent years there have been many cases of poisoning following ingestion of paraquat, and deaths from poisoning have been reported (11, 12). As there is no specific antidote for paraquat, a treatment for paraquat poisoning involves eliminating the herbicide from blood or other biological fluids as quickly as possible.

Polyclonal and monoclonal antibody production

A paraquat hapten, 1-methyl, 1′-hexanoic acid -4,4′-bipyridinium (MHBP) was synthesized according to the method of Niewola et al. (13) with a slight modification, and coupled to BSA, gelatin and keyhole limpet hemocyanine (KLH) with carbodiimide (CDI) (Fig. 4). Polyclonal antibodies were produced by immunizing rabbits (14) with almost the same procedure as described for diazepam, and monoclonal antibodies were obtained by immunizing BALB/c mice (15).

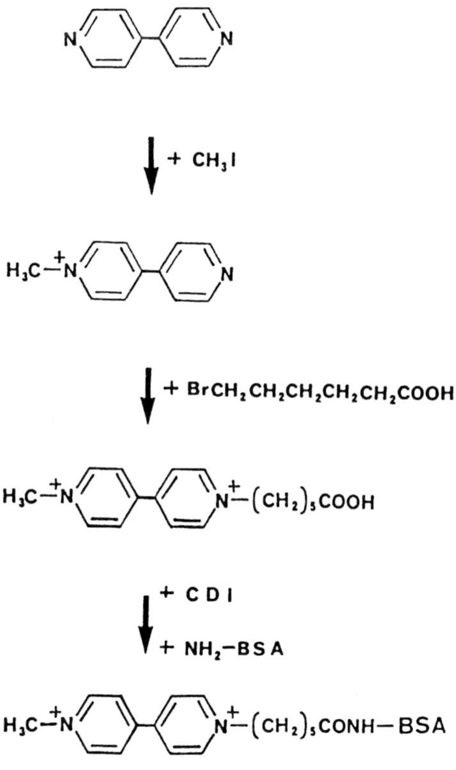

+ CH_3I

+ $BrCH_2CH_2CH_2CH_2CH_2COOH$

H_3C-N^+ … $N^+-(CH_2)_5COOH$

+ CDI

+ NH_2-BSA

H_3C-N^+ … $N^+-(CH_2)_5CONH-BSA$

Figure 4. Reaction Scheme for the Preparation of MHBP-BSA.

Hybridomas producing antibody were screened by an enzyme-linked immunosorbent assay (ELISA). In this screening procedure, the hydridoma cells which are not only positive against MHBP-gelatin used as a solid phase in the ELISA but also negative against BSA, were selected as the antibody producing hydridomas. Three hydridoma cells from positive wells were subcloned twice by a limiting dilution to ensure their monoclonal origins. Finally these three clones were establihed as anti-paraquat monoclonal antibodies (APM-1, 2 and 3).

Immunochemical analysis of paraquat

The RIA procedure for paraquat has been described previously (14). (^{3}H)Paraquat (specific activity 92.5 GBq/mmol; about 5,000 dpm in 0.1 ml of PBS solution) was added to an assay tube. In this RIA procedure, the antiserum bound (^{3}H)paraquat was competitively and linearly challenged with up to 32 ng of unlabelled paraquat as shown in Figure 5. In the absence of unlabelled paraquat, the antiserum of 1 : 100 dilution could bind 50% of (^{3}H) labelled paraquat by this assay. As can be seen in Figure 5, as little as 0.5 ng of paraquat could be detected, and the amount of paraquat causing a 50% inhibition was 3.8 ng.

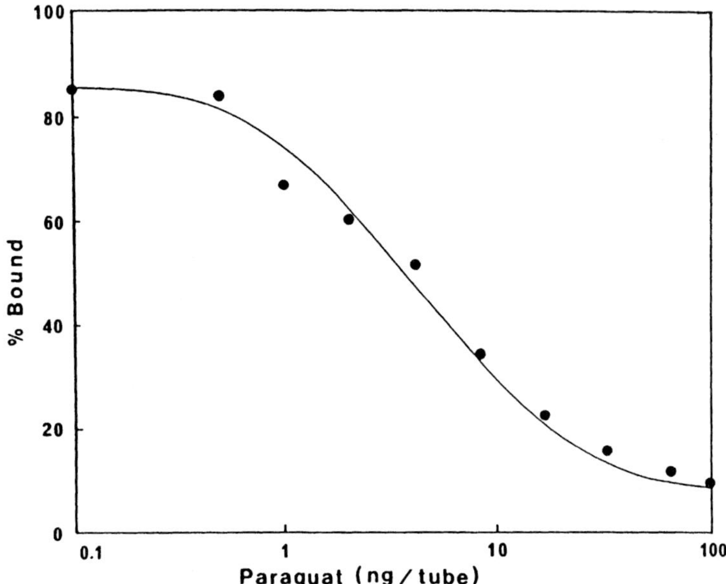

Figure 5. Displacement of (^{3}H)paraquat Bound to Anti-paraquat Antiserum by Unlabelled Paraquat.

Table 3 shows the specificity of anti-paraquat antiserum. Monoquat and diethyl paraquat showed relatively significant cross-reactivities with this antibody. Diquat, morfamquat, MPTP, and reduced paraquat could not bind this antibody, and MPP, as a similar congener to monoquat bound weakly. These findings suggest that both a bipyridyl ring and a methyl group at either the 1- or 1′-position of paraquat are strongly recognized by this anti-paraquat antibody.

Table 4 shows the concentrations of paraquat dichloride in tissues of the seven poisoning cases which were determined by this RIA system. All paraquat levels in fixed tissues were lower than those in the unfixed tissues. This indicates that during fixation the paraquat incorporated into the tissues is easily released into the fixative. The high sensitivity of the RIA system presented here even allowed the detection of the low paraquat levels in fixed tissues.

Table 3. Cross-Reactivity of Anti-Paraquat Antiserum with Paraquat, Bipyridyl Derivatives and Similar Analogs to Paraquat.

Generic name	Structure	Cross-reactivity(%)
Paraquat	$H_3C-\overset{+}{N}$ ⟨bipyridyl⟩ $\overset{+}{N}-CH_3$	100
Monoquat	$H_3C-\overset{+}{N}$ ⟨bipyridyl⟩ N	16.5
Diethyl-paraquat	$H_5C_2-\overset{+}{N}$ ⟨bipyridyl⟩ $\overset{+}{N}-C_2H_5$	5.59
Diquat	N^+ ^+N, H_2C-CH_2	0.0038>
Morfamquat	CH_3, CH_3, $NOCH_2C-\overset{+}{N}$ ⟨bipyridyl⟩ $\overset{+}{N}-CH_2CON$, CH_3, CH_3	0.0038>
Reduced Paraquat	H_3C-N ⟨⟩ $N-CH_3$	0.38>
MPTP	H_3C-N ⟨⟩	0.0038>
MPP+	$H_3C-\overset{+}{N}$ ⟨⟩	0.76

Table 4. Concentrations of Paraquat in Tissues smong Seven Paraquat-Poisoned Cadavers Measured by the Present RIA.

No.	Brain	Liver	Kidney	Heart muscle	Lung
1	21[a)]	8.4	5.6	12	10
2	0.31	0.48	0.23	0.33	0.40
3	0.19	0.24	0.68	0.62	0.20
4	0.15	0.40	0.16	0.30	0.24
5	1.3	0.54	0.25	0.29	0.29
6	0.47	0.30	2.0	0.17	0.34
7	0.32	0.25	0.31	0.09	0.13

a) µg/g wet weight

ELISA procedure and characterization of anti-paraquat monoclonal antibodies
The ELISA procedure has been described previously (16). Figure 6 shows an affinity status of antiserum for paraquat; the absorbance decreased with increasing amounts of paraquat dichloride in the reaction mixture. When the MHBP-KLH or the diazo-coupled paraquat-KLH (16) was used as the solid phase, the decrease in the color development in this ELISA system was almost linear in the range of 10 ng to 1,000 ng and 1 ng to 10 ng, respectively; the amounts of paraquat dichloride causing 50% inhibition with antiserum were 43 ng and 3 ng, respectively. The polyclonal antibody against paraquat described here is thought to contain antibodies from a number of clones which may recognize the spacer as well as the carrier protein, and some of these antibodies will bind to the spacer of MHBP-KLH to cause an increase in non-specific reaction. Therefore, the use of MHBP-KLH as a solid phase was not adequate in this ELISA system. On the other hand, the diazo-coupled paraquat-KLH has a different spacer and a different carrier protein from the immunogen (MHBP-BSA). Little non-specific reaction was observed in this ELISA system using the diazo-coupled paraquat-KLH as a solid phase, suggesting that the antibody selectively and strongly recognizes the portion of paraquat in the diazo-coupled paraquat-KLH. As a result, the sensitivity of this ELISA system was shown to be more than 14 times higher than the solid phase system (Figure 6). In an ELISA

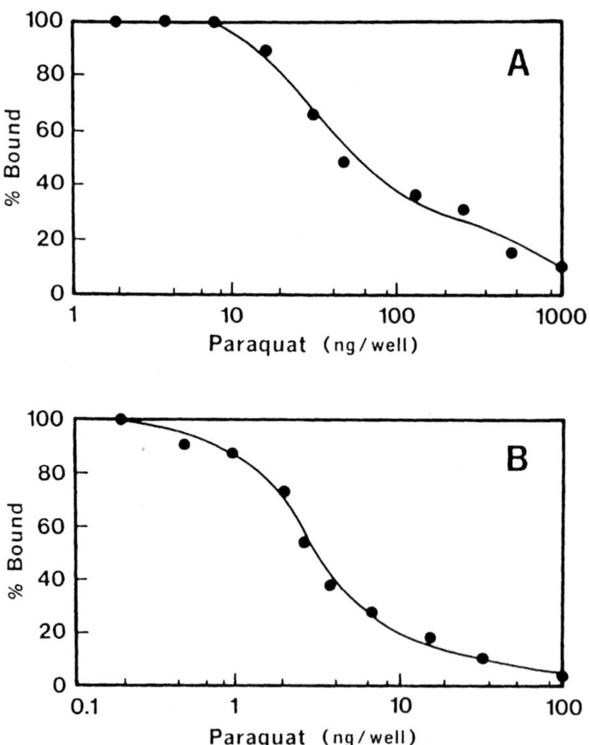

Figure 6. a) Inhibition of Binding of Antiserum to MHBP-KLH in the ELISA by Paraquat Dichloride. b) Inhibition of Binding of Antiserum to Diazo-Coupled Paraquat-KLH in the ELISA by Paraquat Dichloride.

system, a spacer and its binding mode between haptens and carrier protein are of great importance in general to raise sensitivity of the ELISA and to obtain a steep inhibition curve.

The competition ELISA by use of three monoclonal antibodies was also able to detect nanogram amounts (1 to 10 ng) of paraquat, or as sensitive as the RIA. As shown in Tables 3 and 5, the cross-reactivities of the anti-paraquat monoclonal antibodies are similar to those of anti-paraquat antiserum. It is thought that of many different IgGs in anti-paraquat polyclonal antiserum many have specificities similar to those of the present monoclonal antibodies which recognize only the paraquat portion of MHBP as the hapten.

Table 5. Cross-Reactivity of Three Different Anti-Paraquat Monoclonal Antibodies with Paraquat, Bipyridyl Derivative and Similar Analogs to Paraquat.

Reagent	Structure	Cross-reactivity(%)		
		APM-1	APM-2	APM-3
Paraquat	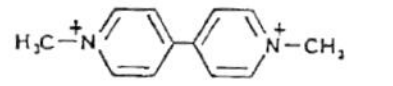 H_3C-N^+ … N^+-CH_3	100	100	100
Monoquat	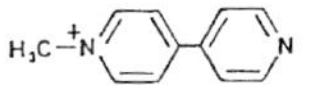 H_3C-N^+ … N	4.3	2.0	1.0
Diethyl paraquat	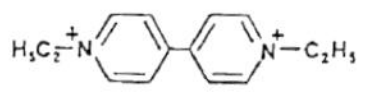 $H_5C_2-N^+$ … $N^+-C_2H_5$	22.0	8.1	11.8
Diquat	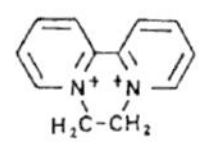 N^+ ^+N H_2C-CH_2	4.4	0.3	1.4
Morfamquat	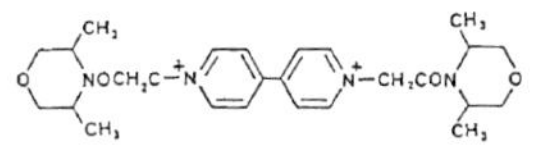 CH_3, O, $NOCH_2C-N^+$ … N^+-CH_2CON, O, CH_3	0.004>	0.002>	0.004>
MHBP[a]	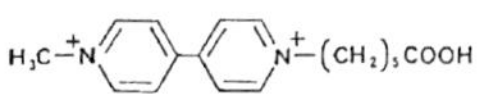 H_3C-N^+ … $N^+-(CH_2)_5COOH$	36.9	20.1	19.9

[a]1-Methyl,1'-hexanoic acid-4,4'-bipyridinium

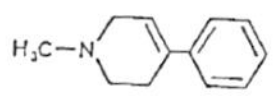

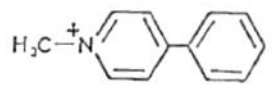

Reagent	Structure	Cross-reactivity(%)		
		APM-1	APM-2	APM-3
MPTP	H_3C-N …	0.08	0.0009>	0.03
MPP+	H_3C-N^+ …	0.5	0.9	0.4

1.6.4 Conclusions

In this section, we describe the analytical uses of anti-hapten antibodies against drugs and poisons. In order to develop a sensitive immunoassay system, it is important to produce hapten-specific antibodies which are not affected (or little affected) by spacers and carrier proteins. The development of anti-drug monoclonal antibodies which selectively recognize the hapten portion will satisfy this criterion.

A hapten-specific antiserum was also reported using biodegradable carboxymethyl-chitin as a carrier (17). CM-chitin will be a useful carrier in the future.

Recently anti-drug antibodies have been used for immunohistochemical studies (18, 19). In the future, anti-drug antibodies will also be useful tools for toxicological studies of localization and dynamics of drugs in the cell or interations between drugs and microorganelles in the cell.

References

1. Parker, C.W., Shapiro, J., Kern, M. & Eisen, H.N. (1962) Hypersensitivity to penicillenic acid derivatives in human being with penicillin allergy. J. Exp. Med. *115*, 821–838.
2. Halpern, B.N., Hotzer, A., Liacopoulos, P. & Meyer, J. (1958) Allergy to pyrazolone derivatives (aminopyrine) with evidence of a reaginic type antibody. J. Allergy. *29*, 1–12.
3. Tomii, S. & Takatori, T. (1984) Production and characterization of antibodies reactive with diazepam. Jpn. J. Clin. Chem. *13*, 281–285.
4. Pesker, B. & Spector, S. (1973) Quantitative determination of diazepam in blood, plasma and saliva by radioimmunoassay. J. Pharmacol. Exp. Ther. *186*, 167–172.
5. Dixon, W.R., Earley, J. & Postma, E. (1975) Radioimmunoassay of chlorodiazepoxide in plasma. J. Pharm. Sci. *64*, 937–939.
6. Erlanger, B.F., Borex, F., Beiser, S.M. & Lieberman, S. (1957) Steroid-protein conjugates. I. Preparation and characterization of conjugates of bovine serum albumin with testosterone and with cortisone. J. Biol. Chem. *228*, 713–727.
7. Yamaoka, A. & Takatori, T. (1979) Determination of phenobarbital by radioimmunoassay. J. Immunol. Methods *28*, 51–57.
8. Takatori, T., S. Tomii, K. Terazawa, M. Nagao, M. Kanamori, Y. Tomaru (1991) A comparative study of diazepam levels in bone marrow versus serum, saliva and brain tissue. Int. J. Legal. Med. *104*, 185–188.
9. Takatori, T., Kanamori, M., Fukuda, H. & Terazawa, K. (1986) Development and forensic application of radioimmunoassay for haloperidol. Jpn. J. Legal Med. *40*, 361–365.
10. Rubin, R.T., Tower, B.B., Hays, S.E. & Poland, R.E. (1982) In: Methods in Enzymology (Langone, J.J. Vunakis, H.V., eds.) *84*, pp. 532–542, Academic Press, New York · London.
11. Carson, D.J.L. & Carson, E.D. (1976) The increase of paraquat as a suicidal agent. Forens. Sci. *7*, 151–160.
12. Haley, T.J. (1979) Review of the toxicology of paraquat (1,1′-dimethyl-4,4′-bipyridinium chloride). Clin. Toxicol. *14*, 1–46.
13. Niewola, Z., Walsh, S.T. & Davies, G.E. (1983) Enzyme linked immunosorbent assay (ELISA) for paraquat. Int. J. Immunopharmacol. *5*, 211–218.
14. Nagao, M., Takatori, T., Terazawa, K., Wu, B., Wakasugi, C., Masui, M. & Ikeda, H. (1989) Development and application of immunoassay for paraquat: Radioimmunoassay. J. Forens. Sci. *34*, 547–552.
15. Nagao, M., Takatori, T., Wu, B., Tertazawa, K., Gotouda, H. & Akabane, H. (1989) Development and characterization of monoclonal antibodies reactive with paraquat. J. Immunoassay *10*, 1–17.
16. Nagao, M., Takatori, T., Wu, B., Terazawa, K., Gotouda, H., Akabane, H. & Wakasugi, C. (1989) Enzyme-linked immunosorbent assay (ELISA) for paraquat and its toxicological application. Clin. Chem. Enzym. Comms. *1*, 221–227.
17. Tokura, S., Hasegawa, O., Nishimura, S., Nishi, N. & Takatori, T. (1987) Induction of methamphetamine-specific antibody using biodegradable carboxymethylchitin. Anal. Biochem. *161*, 117–122.
18. Nagao, M., Takatori, T., Inoue, K., Shimizu, M., Terazawa, K. & Akabane, H. (1990) Immunohistochemical localization and dynamics of paraquat in small intestine, liver and kidney. Toxicol. *63*, 167–182.
19. Nagao, M., Takatori, T., Inoue, K., Shimizu, M. & Terazawa, K. (1991) In: ACS symposium series (Vanderlaan, M., Stanker, L., Watkins, B.E. & Roberts, D.W. eds.) *451*, 264–271. American Chemical Society, Washington D.C.

1.7 The Future of Instrumental Analysis in Toxicology

H. Brandenberger

1.7.1 History

Since many years, the strategy of toxicological analysis consists of three basic steps:

- preliminary orientation respectively screening tests,
- isolation and identification of poison(s),
- quantification(s).

Before the development of immunochemical assay methods, color reactions and related spot tests have been used to obtain a first orientation, i.e. the Reinsch test (1) as an indication of the presence of heavy metals, the Zwicker reaction (2) as a screening test for barbiturates, the FPN-reagent (2) for detecting phenothiazines or the Fujiwara test (2) for revealing trichloro-compounds, just to name a few. During the last two decades, such tests have been to a large part substituted by new immunochemical assay methods (3–5).

The actual chemical analysis for organic poisons usually involves a separation from the biological matrix, followed by isolation, purification and identification of any poison(s) present. After World War II, new instrumental methods began to take hold of chemical analysis. Highly efficient chromatographic separation techniques were developed, as well as various spectroscopic identification methods. But not all of these new instrumental possibilities have immediately found an open door in the toxicological laboratories. TLC on paper and on silica gel sheets was used fairly early, since about 1955 (6–8). The first applications of GC for poison analysis in the body have been made before 1960 (9, 10), of HPLC some 10 years later. Up to now, newer separation techniques like supercritical fluid chromatography (SFC) and capillary electrophoresis (CE) have received only little attention in the field of analytical toxicology.

Among the spectroscopic identification methods, UV, FS and IR have entered analytical toxicology around 1960. A great impact had, after 1965, the adoption of MS, already in combination with GC (11–13). In inorganic toxicology, the introduction of AA constituted a great advance. Clinical and forensic toxicologists were very quick to grab this technique and even contributed to its further development, not only in flame AA, but also in flameless AA (see chapter 2.10).

At present, the general procedure for organic toxicological investigations consists of

- preliminary testing with immunochemical assays,
- extraction of possibly present drugs and poisons,

- chromatographic separation of the extracts,
- spectrophotometric identification of the isolated compounds,
- eventually followed by quantifications.

It has also become customary to combine chromatographic separation techniques with spectrophotometric identification procedures to on-line methods, thus eliminating the delicate and time-consuming handling of isolated compound traces.

Especially in the US, but recently also in Europe, the strict rule has been established that an analysis for drugs of abuse has to be performed by immunochemical screening, followed by GC-MS analysis to confirm positive immunochemical results. Such a rule may be practical for the moment. However, we fear that it will hamper further developments in toxicological analysis.

1.7.2 Outlook on Future General Strategy

The general conviction that today's commercially available immunochemical assays can only have the status of screening methods, providing indications but not final verdicts about the presence or absence of a specific drug or poison, is certainly fully justified. It implies that all positive immunochemical results must be confirmed by a specific non-immunochemical method (5). Some of the available tests possess high degrees of reliability (but not 100%), others show poorer performances. Their specificity is often low, and cross reactions result in a considerable number of false positive results, or their sensitivity is relatively low and responsible for many wrong negative results. Two examples are given as illustrations:

- The so-called immunochemical test for barbiturates does not detect all members of this class of hypnotics. It has been developed for revealing the few barbiturates sold in the US, and is not able to see some of the barbiturates on the market in other countries (i.e. diethylbarbituric acid).
- The immunochemical test for benzodiazepines has very variable sensitivities for different members of this large class of tranquilizers and hypnotics. Most are able to detect low concentrations of diazepam (Valium), but only very high concentrations of the important hypnotics flunitrazepam (Rohypnol) and bromazepam (Lexotanil) (see chapter 3.2, tables 6 and 7).

To use such immunotests alone, without the knowledge of possible cross-reactions and the lack of sensitivity towards some of the drugs to be searched for, is extremely dangerous. Confirmation by chemical methods is a must. But all this concerns today's commercially available immunochemical assays. It would be wrong to believe that also the immunotests of the future will possess the same drawbacks. They will very likely be based on monoclonal antibodies and show improved specificity and sensitivity. Future immunochemical methods may be useful not only for a preliminary screening, but also for obtaining final data.

Another blind in the planning of the future strategy in toxicological analysis is the belief that screening methods, if not immunochemical, must always consist of very simple color reactions or related spot tests, which require practically no equipment. The new sophisticated instrumentation, which is indispensable to any modern toxicological laboratory, can be used also to carry out fast and simple screening reactions. Once such a procedure is developed, it can very well compete with immunochemical assays with respect to speed and experimental ease. In addition, such tests have the advantage that they do not require expensive reagents. The next subchapter gives an example for such a screening test on instrumental basis.

The modern instrumental development of analytical chemistry has started in the US around 1945 and reached the toxicological laboratories about 10 years later. Every two or three years, a new method has appeared. This gave the chemist some time to learn how to use it, before a new technique stood before his door. We have heard repeatedly that advances in instrumental analysis will soon have reached a culmination point and that not much further progress can be expected in the future, but can not agree with such a prediction. Technical innovations will keep appearing. For the next few years, we expect specifically advances in the field of extraction, as well as in the combination of identification techniques, designed and constructed to increase analytical speed and to facilitate the interpretation of data. It is not the analytical innovator who has slowed down, but it may be that the toxicological laboratories have become saturated. They are usually confronted with a heavy case load and there is a little time to study and introduce all the new techniques. In addition, the relatively small toxicological laboratories have only limited floor space. This makes it more and more difficult to cope with the increasing supply of new methods and corresponding instrumentation.

In a previous chapter (1.3), we stated that a trend exists to separate the services for forensic and clinical toxicology. This implies that two different laboratories must be equipped with identical instruments and staffed with people of similar background. We have further outlined that a laboratory staff can only profit by working simultaneously for clinical and forensic needs. To dispense with a separation of the 2 services would not only ensure a more interesting work basis and a broader training for the staff, but also permit to assign additional floor space, more funds for equipment and a larger staff to cope with the increasing technical demands.

1.7.3 Screening Test for Dithiocarbamates as an Example for a Preliminary Test on Instrumental Basis

In food analysis and agriculture chemistry, the search for dithiocarbamate pesticides consists of decomposing the biological materials with strong sulfuric acid and measuring the released carbon disulfide with a color reaction. We have worked out and successfully used for many years the following screening test for pesticides and pharmaceuticals with dithiocarbamate structure (14, 15): A sample (1 g) of the biological material is decomposed with strong sulfuric acid in a serum vial (5–20 ml) capped with a silicone (not rubber) membrane. Aliquots of the head space gas in the vial (0.5–2 ml) are then transferred to a GC-MS combination. The chromatogram is controlled by monitoring the appearance of the molecular ion of carbon disulfide

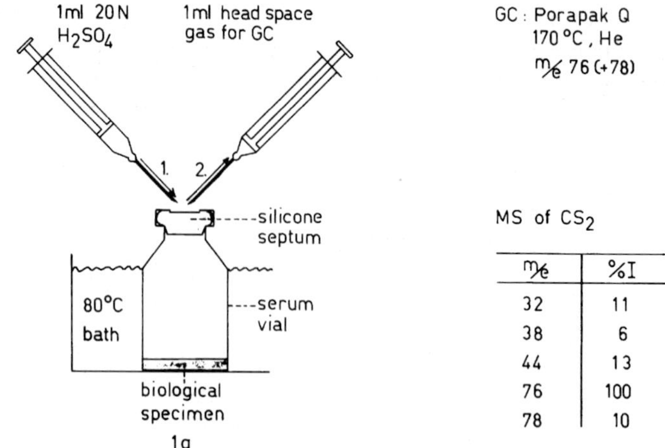

m/e	%I
32	11
38	6
44	13
76	100
78	10

Figure 1. Screening Test for Dithiocarbamates.
Decomposition and Sampling Procedure, GC Conditions and Mass Spectrum of CS_2.

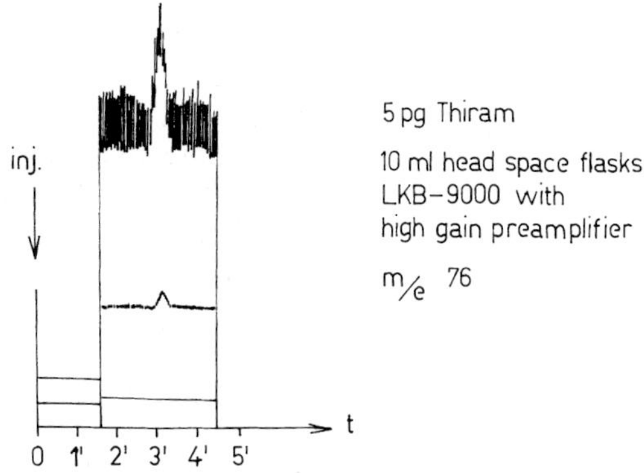

Figure 2. Screening Test for Dithiocarbamates.
CS_2 Peak from 5 pg of Thiram by Mass Specific Recording of m/z 76 (M^+).

with mass 76. For increased specificity, the isotopic molecular ion with mass 78 and/or the fragment ions with masses 44 and 46 can be recorded simultaneously.

Figure 1 illustrates the decomposition and sampling procedure. It also shows the tabulated mass spectrum of CS_2. Figure 2 is a reproduction of a chromatogram from the decomposition of only 5 pg (20 fmol) of Thiram. Figure 3 finally shows a calibration curve: the intensity of the m/e 76 peak is recorded for the analysis of Thiram samples between 0.5 and 100 ng. A similar curve is obtained by detection by photoionization, which is relatively specific for sulfur-compounds such as CS_2. The photoionization detector is a simple tool, but not as sensitive as a mass spectrometer. With selected ion detection, the screening test can reveal fmol-quantities of dithiocarbamates (fig. 2), while a photoionization detector can only measure quantities down to the upper pmol-range.

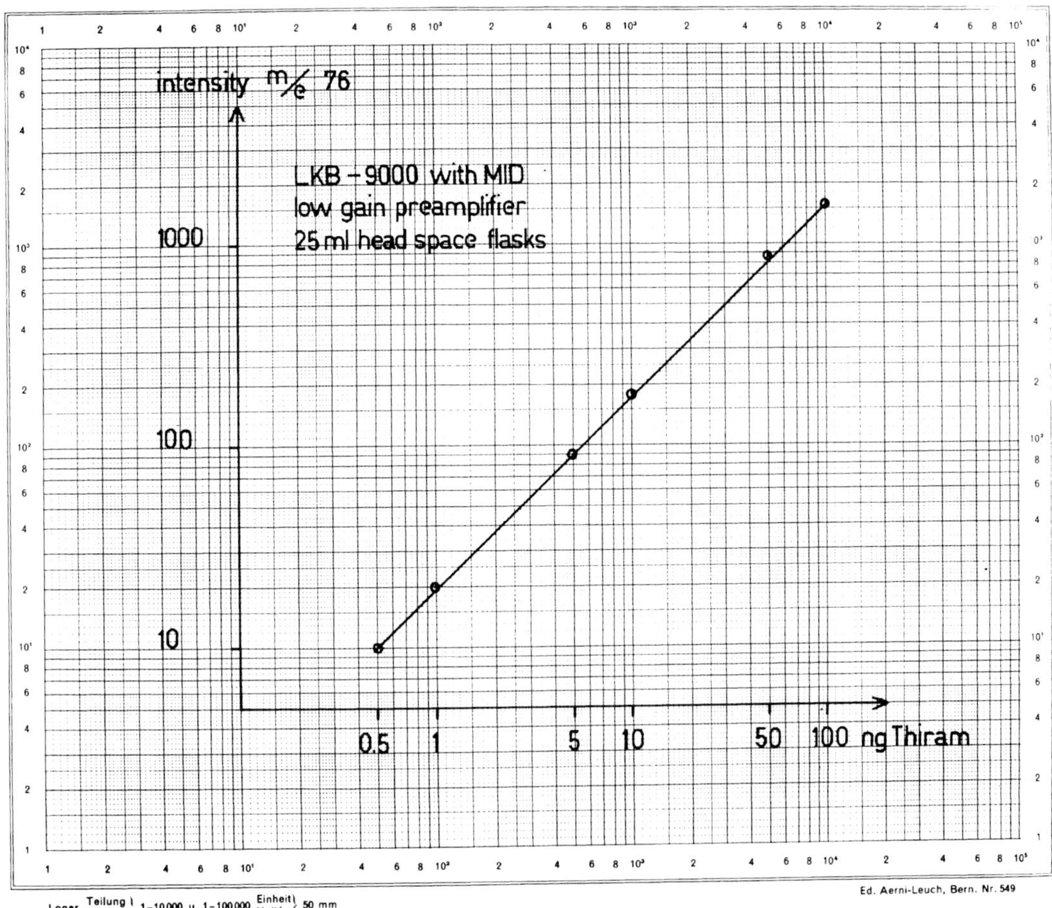

Figure 3. Screening Test for Dithiocarbamates.
Calibration Curve for Quantification of Thiram by Mass Specific Recording of M^+ of released CS_2.

Only a few minutes are needed for an analysis. Positive cases must be followed up by an identification of the dithiocarbamate. If only one is present, its concentration can be calculated from the GC-peak of the screening test.

1.7.4 New Extraction Techniques

1.7.4.1 Solid phase extractions

For urine samples, the solid phase extraction methods, consisting of adsorption (or absorption) of the analyte on a solid support (usually in column form, but also as disk or thread) with subsequent elution, are becoming more and more popular.

The instrumentation provided for carrying out such extractions is quite satisfactory; but the solid phases could be improved. Columns should be of glass or teflon and not of soft plastic which releases impurities. The present column materials may be satisfactory for the bulk of the analytical work, directed to the detection of one specific compound or a specific compound class. But we do not dare to recommend their use in a search for completely unknown poisons or drugs.

1.7.4.2 Supercritical fluid extractions

Supercritical fluid extraction is not new. It is used in food chemistry and technology, in pharmaceutical chemistry and related fields (16, 17). Its advantages are high efficiency (due to high diffusibility), good selectivity (can be tuned) and short extraction times. It does not involve thermal decompositions, nor does it introduce disturbing solvent impurities. But toxicologists have so far shied away from the method. Among all the papers presented in the meetings of the International Association of Forensic Toxicologists), only a single contribution has treated this subject (18). A reason for the lack of interest may be that the presently available equipment is intended for extracting solids, not liquids or suspensions. But this will change. We are convinced that extracting drugs and pesticides from biological fluids and tissues by compressed gases instead of organic solvents will become a welcome help in analytical toxicology.

1.7.4.3 Extractive dialysis

Many years ago, we have presented a new extraction technique called "Extractive Dialysis" (19). It was intended for extracting tissue suspensions, as is often required in forensic toxicology. Since it involves little manual labor, we have also used it successfully for extracting whole blood, stomach wash and occasionally even urine, if the time factor permitted.

The aqueous suspensions are placed into dialysis bags (cellulose or cellulose acetate tubing, knotted on both ends), and "dialyzed" against organic solvents. The semipermeable membrane permits the transfer of the soluble components into the organic phase, acts as a filter and also retains the higher molecular weight substances. The pH of the aqueous phase, solvent, solvent volumes and number of extraction steps can be selected exactly as in the commonly used liquid-liquid extraction methods. Clean extracts are obtained, with at least as good yields as with separatory funnel or solid phase extractions, as illustrated in tables 1 and 2 (agitation in extractive dialysis effected by gentle rolling of the bottle with blood in a dialysis bag placed in the solvent ether).

Of special interest is the possibility to combine enzymatic hydrolysis with extractive dialysis, that is to start dialysis and enzymatic treatment together. This makes up for the slowness of the dialysis process. We have had good results by adding the proteolytic enzyme Subtilisin or hydrolyzing enzymes to the aqueous phase at the start of extraction.

A possibility to reduce the volumes of organic solvents and therefore also the dilution consists of placing a relatively small dialysis bag containing the solvent into the recipient with the aqueous phase and extracting "into" the bag. Such a

Table 1. Extraction of Methaqualone from Blood

Initial pH 9.0 Solvent = Ether	
Method	*Recovered*
Separatory Funnel (3 steps)	1.21 mg per 100 ml
Extractive Dialysis (2 steps of 2 hours duration)	1.22 mg per 100 ml

Table 2. Recovery Experiment with Salicylic Acid Added to Whole Blood

Initial pH 1.7 Solvent = Ether Salicylic Acid Added = 0.25 mg		
Extraction Method		
Extractive Dialysis		92%
(2 steps of 2 hours)		96%
Separatory Funnel		86%
(2 steps)		
	New	77%
Extrolut Column		
	Used	70%

"reversed extractive dialysis", however, gives lower yields, unless the number of extraction steps (and therefore also the total extraction time) is increased.

For the extraction of tissue suspensions in forensic investigations, extractive dialysis can be a nearly ideal solution. For clinical toxicology it is too slow. But it seems possible that similar techniques based on semipermeable membranes may be able to eliminate this disadvantage.

1.7.5 Fractionation Methods

1.7.5.1 Established techniques

Toxicologists can hardly complain about a lack of available separation methods. In this field, the instrumental development during the past 40 years has been overwhelming. TLC, GC and HPLC are used in most toxicological laboratories.

The possibility to identify the separated spots directly on the sheet, using color tests, UV- or FS-analysis, has always been a great asset of TLC. It is interesting to see, that despite the triumphant advances of GC and HPLC, innovations in TLC are continually reported in the literature. Of special interest are the formation of fluorescent derivatives (20), as well as the efforts to combine TLC (off-line) with MS (21, 22).

GC column technology underwent an impressive development. I still remember how my first GC-column was made, way back in 1960: A red brick was procured from the next building under construction, ground to a fine powder in a simple laboratory mortar, moistened with Apiezon L grease and packed into a metal tube. We used this column to separate the fatty acids of tea, as free acids and methyl esters (23). Today, packed and capillary columns are sold by a large number of suppliers, and it is not easy to keep fully informed on all the materials on the market in order to make the best choice.

For a considerable time, some GC-users kept the packed columns, while others shifted to very narrow high performance capillaries (down to 0.1 mm inner diameter). Such long capillaries with very high plate numbers give excellent separations also of very complex mixtures. But they have to be used with inlet split and that is not exactly what a toxicologist likes. His extract seems too precious to be wasted into the air. Furthermore, in on-line with an identification device such as FTIR or MS, too small column effluents require make-up gas before entering the spectrophotometer. Larger "Megabore" capillaries of 0.53 mm inner diameter have now eliminated these inconveniences. They permit direct injection into the column, and their gas flow of several ml/min is compatible with most detectors. Their resolution is sufficient for most toxicological needs. For the direct coupling (without interface) to a fast scanning mass spectrometer, 0.32 mm columns with a gas flow of near 1 ml/min may even be a better choice.

In HPLC, the large 4 mm diameter columns did hold their place for a long time. There has been a lot of reluctance to switch to smaller semimicro- and microcolumns (2 mm, 1 mm and 0.5 mm inner diameter) or even capillaries. This author believes that in this case, the chemists have been unreasonably conservative. For analytical purposes, there seems no valid argument to retain the initial 4 mm tubes. – The choice of stationary phases is a more difficult problem. In liquid chromatography, these materials are still being improved. The initial silica-bound phases are more and more replaced by polymeric resins.

Some toxicologists have preferences, for TLC, for GC or HPLC. They have selected one favored technique and use it for most problems. We think that all the separation methods have their place, since all have advantages and drawbacks. They should be used side by side. A rule of thumb for selecting a method for a given toxicological problem may be: Whenever it is possible to carry out a separation by gas chromatographic means, GC should be given the preference. Thermolabile compounds or compounds with insufficient volatility (too polar, too high molecular weight) should be tackled by HPLC, unless the problem requires better resolving power. In such cases, supercritical fluid chromatography, capillary electrophoresis or capillary electrochromatography should be selected (24).

1.7.5.2 Supercritical fluid chromatography (SFC)

SFC combines the advantages of gas chromatography and liquid chromatography, because density and solvating power of supercritical fluids approach those of liquids, while the viscosity is similar to that of gases (25). SFC is also a logical consequence of SFE. A combination of these two methods can completely eliminate organic solvents, unless they are used for sample transfer or as organic modifiers to increase the fluid polarity. The solvation power of supercritical fluids can be varied by changes in density (pressure) or temperature, and by adding modifiers (i. e. water or alcohols). The preferred fluid is liquid CO_2, but liquid N_2O, Xe, propane, NH_3 and Freons have also been successfully used. GC-as well as HPLC-detectors can be adapted to this technique.

During the first years, both packed columns and capillaries were used in SFC. But the capillaries have taken over, except for preparative work. Diffusion coefficients are smaller than in GC, so that very narrow capillaries with inner diameters below 100 μm can be recommended. With 50 μm columns, efficiencies of 5000 plates per meter can be obtained. Due to the low viscosity, the pressure drop across the columns is small, and long capillaries can be used. This all ensures optimal resolution, comparable to that of GC and certainly better than that of HPLC.

Up to now, the method has not been used much in toxicological analysis. But this will change since on-line combinations of supercritical fluid chromatography with spectrometric identification techniques became available. In this connection, it should be pointed out that the combination of SFC with FTIR gives superior results to the (so far not satisfactorily solved) HPLC-FTIR combination.

1.7.5.3 Capillary electrophoresis (CE)

CE or, with other names, capillary zone electrophoresis (CZE) or open-tubular zone electrophoresis (26) is another separation technique which has not yet found wide acceptance in the field of toxicology, in spite of the fact that its resolution power is superior to that of chromatographic techniques. The separation is based on the charges of the analyte particles, as in other electrophoretic procedures. With respect to technology, however, the method can be compared to capillary GC or micro-HPLC in capillaries, except that neither pumping nor gas pressure is needed for effecting the analyte flow.

Separation occurs in a buffer medium of a fused-silica capillary (glass and teflon can also be used) with 25 to 100 μm inner diameter. It can be regulated by the choice of buffer and of voltage differences applied at the column ends (as a rule between 15 and 30 kV), and also by coatings on the inner wall of capillaries.

In a capillary without wall effect (fused silica with inside coating, teflon), positively charged particles migrate to the cathode, negatively charged particles to the anode, while the uncharged molecules do not move and do not separate. In unmodified fused silica capillaries, the silanol groups of the capillary surface bear a negative charge and attract the hydrated positive ions of the buffer. This creates an electro-osmotic flow of the entire liquid toward the cathode. Its velocity is a function of voltage, capillary diameter and buffer properties. The flow is nearly uniform across the capillary, unlike in a pumped HPLC-system with its parabolic flow effect. This reduces band broadening and improves resolution. The electro-osmotic flow is usually faster than the electrophoretic mobility of all analytes; all net movement is therefore directed toward the cathode. The separated cations reach the cathode first, the neutral components follow together with the speed of the electro-osmotic flow, the separated anions make up the rear. With the help of additions, i.e. detergents, it is possible to separate also the neutral components. They form so-called micelles with the analyte molecules, which are individually retarded in the electro-osmotic flow.

Detection possibilities are so far limited in CE. On-column detection with UV- or fluorescence light as well as electrochemical detection have mainly been used in the past, also for the separation of inorganic cations and anions (27–29). It is only relatively recently that CE-units with scanning UV-detectors are sold, and that the combination with MS has been realized, using electrochemical ionization at atmospheric pressure (30).

Capillary electrophoresis has initially been developed for separating high molecular weight biomolecules. But its field of application is much larger. It includes low molecular weight compounds: acids, bases, neutral molecules, inorganic cations and anions. A considerable number of separations of drugs and pesticides have been published, but seldom applied to real life problems, respectively case work.

1.7.5.4 Capillary electrochromatography (CEC)

By the introduction of capillary electrochromatography, a combination of HPLC and CE, an important improvement of the resolution possibilities of capillary HPLC can be expected in the near future. CEC can be carried out with CE-instrumentation, fitted with packed capillaries containing specially tailored particles on silica basis. The retention mechanism is the same as in HPLC, but the solvent flow a loan from CE. In place of hydraulic pressure from a pumping system, the electro-osmotic flow created by the potential difference and the negative surface charges on silica particles and column walls causes a transport of the solvent toward the cathode. Much higher separation efficiencies can be expected, since

- due to the absence of back pressure, the column length is not limited and column fillings with much smaller particle sizes can be used,
- the linear profile of the solvent front, obtained with the electro-osmotic flow (in contrast to the parabolic profile in a hydrostatic system), reduces band broadening and leads to sharper and more symmetrical peaks.

1.7.5.5 Chiral separations

In the forgoing, we did not mention a development which will soon become extremely important in analytical toxicology, the problem of optical activity or chirality. Chiral components form an intergral part of all biological systems. They constitute a chiral environment in which the enantiomers of biologically active compounds such as drugs or pesticides behave as totally different entities (31). It is common knowledge that the D-form of tyrosine has a sweet-, but the L-form a bitter taste, and that the R-form of limonene has an odor of oranges, the S-form the odor of lemon. Similar differences exist among enantiomers of drugs and pesticides. While (+)-propoxyphene has analgesic activity, (–)-propoxyphene is an antitussive and respiratory depressant. The pharmaceutical action is usually tied to one of the enantiomers only, the other one may be inactive or even responsible for unwanted dangerous effects. A therapeutically useless enantiomer of a racemic drug must be considered as an impurity, as 50% ballast, which may not always be tolerated easily by the body. The sedative thalidomide is an example for the disastrous consequences of using a racemic mixture.

There are direct and indirect methods of separating enantiomers by chromatography. The direct method consists of a separation of the optical isomers on a chiral stationary phase. Natural compounds such as oligosaccharides, proteins or amino acids linked to solid supports, cellulose derivatives as well as optically active synthetic compounds have been used. As an alternative, the chiral separation can be carried out on an achiral stationary phase in presence of a chiral additive to the mobile solvent. The indirect method (31) consists of converting the enantiomers into diastereomers by reaction with optically active reagents, followed by separation of the derivatives on a normal achiral phase.

Already today, a considerable percentage of the pharmaceuticals are pure enantiomers. This percentage will grow fast. The interest in the individual drug enantiomers will give a boost to the development of better chiral separation methods by GC, TLC, HPLC, SFC, CE and CEC. At high GC-temperatures, there is of course a danger of racemization. We therefore believe that HPLC and SFC, probably also CE and CEC, are more likely to become successful chiral separation techniques than GC.

1.7.5.6 Identification by retention indicies or with on-line spectrometric methods

In the last years, there has been a trend to rely, for the purpose of identification, largely on retention data. Impressive compilations of Rf-values and GC RI-values have appeared (32, 33). They can be a great help for primary orientations or as additional identification criteria. However, compound identification by retention data should not replace the unequivocal structural identification by the combined application of spectroscopic techniques. New compounds of toxicological interest, such as drugs and pesticides, are continually appearing on the market. A chemist looking for an unspecified poison or drug will meet such substances in his case work before collectors of data have had the chance to incorporate them into the files. – In addition, the many impurities of non-endogeneous and endogeneous nature, present in biological extracts, will always endanger a correct interpretation on the sole basis of retention data. Identification based on retention values is a fine method for analyzing series of homologues. For systems as complex as the extracts of biological materials, we strongly recommend relying on on-line combinations of separation techniques with spectrometric identification methods.

1.7.6 Trace Analysis by Mass Specific Detection in the Negative Ion Mode

For compounds with high electron affinity, GC with detection by electron capture has been the method of choice since the late 1960's. Already many years ago, we have shown that negative ion MS is not only a more specific but also a much more sensitive detection method. Negative ion MS is a CI-method. If used with a high reagent gas pressure (around 1 torr or at atmospheric pressure), the molecular or the $[M-H]^-$ quasi-molecular anions are usually base peaks and not much fragmentation occurs. However, at a low ion source pressure, near 10^{-2} torr, an interesting fragmentation can be observed (34–38). With many chlorine-containing compounds, Cl^- becomes base anion and can be traced by monitoring masses 35 and 37. Figure 4 gives an example: 5 pg (17 fmol) of hexachlorocyclohexane are revealed by mass specific recording of the 2 chloride anions (upper traces) and by a good electron capture detector (lower trace). The signals are amplified to give similar peak intensities. The signal to noise ratio of the specific mass recordings is much better. This is by far the more sensitive method than detection by EC (from ref. 38).

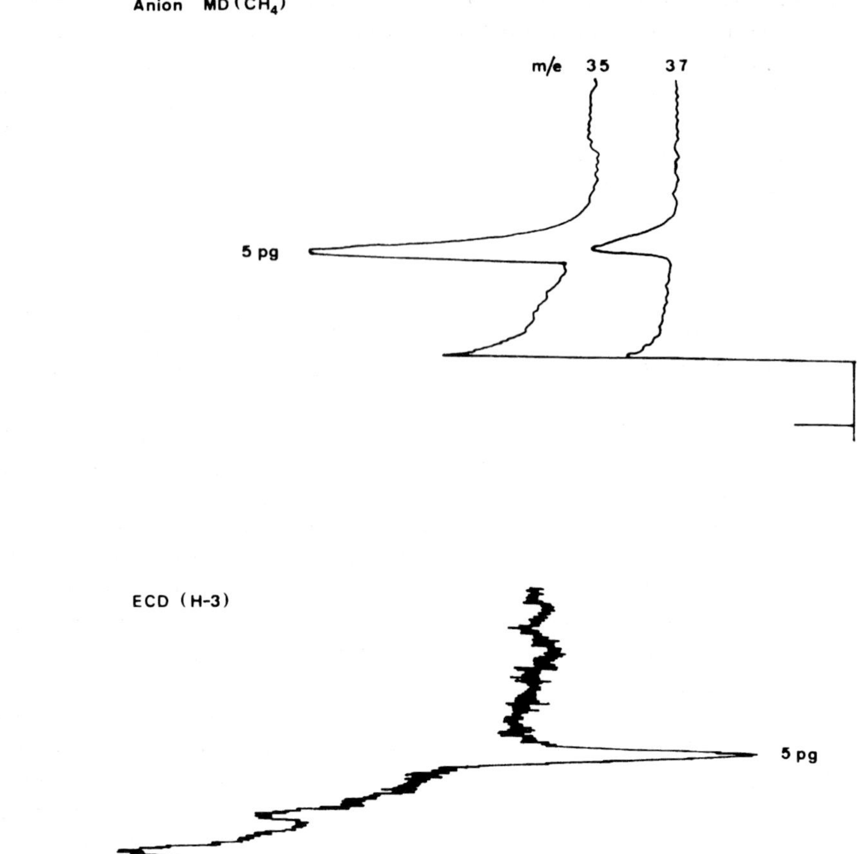

Figure 4. Trace Analysis of Lindane by GC with low pressure Negative Ion CI and with ECD. Upper trace: recording of chloride anions, lower trace: recording by ECD.

1.7.7 Chromatographic Separations with On-line Identification Methods

Today, spectrometric identification techniques in on-line combinations with chromatographic separations can be found in almost every larger toxicological laboratory. The leading combination is GC-MS, with the spectrometer used in the EI- or CI-mode, in order to obtain both structural and molecular mass information. Some laboratories are also equipped with a mass spectrometer which can be used as on-line detector for HPLC, SFC, CE and CEC (30). On the other hand, only few service laboratories of analytical toxicology are working with GC-FTIR combinations. This may change. Since there will be a strong emphasis on the separation of optical enantiomers, and since MS is usually unable to distinguish them, IR or

NMR will have to be used, especially FTIR, since NMR does not yet possess the sensitivity required for a chromatographic detection system.

Double on-line combinations such as GC-FTIR + MS are extremely powerful identification tools (39). With two different spectrometric techniques in parallel, chances for identifying unknown compounds are drastically improved, as long as the two methods yield complementary information and not a duplication of results. For IR and MS, this is the case.

1.7.8 "Dual-MS", a Combination of Positive EI-MS with Negative CI-MS at Low Source Pressure

Another possibility to obtain complementary information is a mass spectrometric technique we have developed some years ago and called "Dual-MS" (40–42). It consists of running the mass spectrometer at an ion source pressure near 10^{-2} torr, using selected reagent gases, and recording, quasi-simultaneously, the positive EI and the low-pressure negative CI-spectra. Our dual-MS method yields not only a conventional (positive) EI fragmentation picture, but also a negative ion spectrum, which often indicates the molecular mass, but also yields – due to the low source pressure used – a negative ion fragmentation which is basically different from that shown by EI-MS.

Figure 5 illustrates why positive EI- and low-pressure negative CI-spectra can be recorded simultaneously in the same chromatographic run or same direct solid inlet experiment. The reagent gases which have so far been most useful to us in low pressure negative ion work are CH_4 for electron attachment, N_2O for negative charge transfers from negative reagent ions (mostly $[NO]^-$ and O^-) to sample molecules, and CO_2 as a practical reagent gas which will favor both types of ionization mechanisms (electron attachment and charge transfer from O^-) and the subsequent fragmentations.

We have used the "Dual-MS" technique with excellent success for many years in our toxicological case work, for the identification of volatile anaesthetics, drugs and drug metabolites, pesticides, as well as unexpected impurities.

Figure 6 shows the dual-MS recording of an impurity often present in biological extracts. As long as we ran only EI-spectra, we just called it the "impurity with masses 45 and 75". Later, when working with low-pressure negative ion CI-MS with a different spectrometer and different GC-columns, we could observe a peak which we called the "impurity with negative mass 105". Only after combining the 2 ionization techniques, we realized that the peaks originated from the same compound, we could now easily identify on the basis of the 2 spectra as diethylene glycol. Figure 7 is a dual-MS recording of another impurity, the plasticizer tributylphosphate. Neither of the two spectra indicates the molecular mass 266. However, the different fragmentation schemes certainly help to identify the compound.

As a further illustration of the complementary nature of the information by positive EI-MS and low-pressure negative CI-MS, we refer to chapters 3.1 (Figure 10) and 3.10 (Figures 5 & 6).

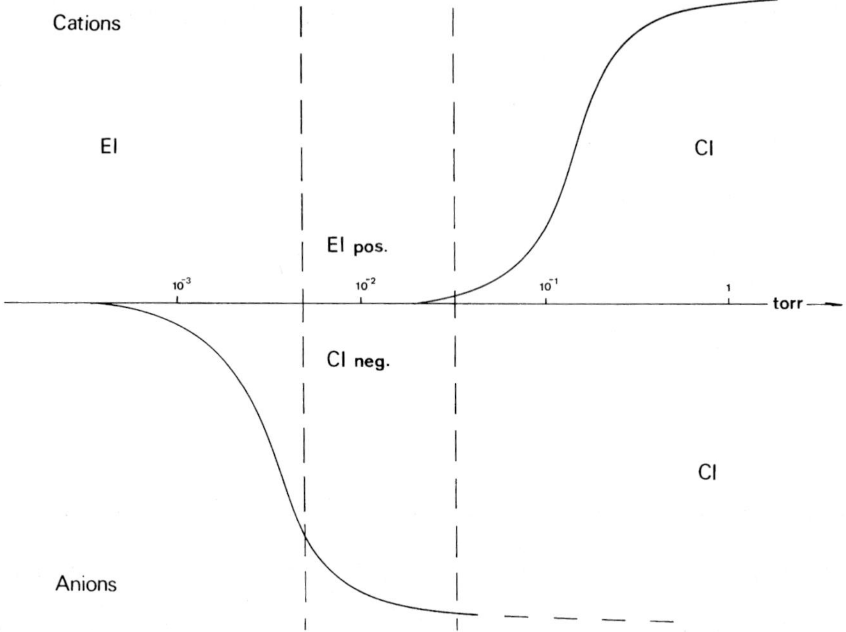

Figure 5. Positive and Negative Ion Mass Spectrometric Ionization Efficiency as a Function of Pressure in the Ion Source.
The practical absence of negative EI permits simultaneous recording of negative CI- and positive EI-spectra in the pressure range of 10^{-3} to over 10^{-2} torr.

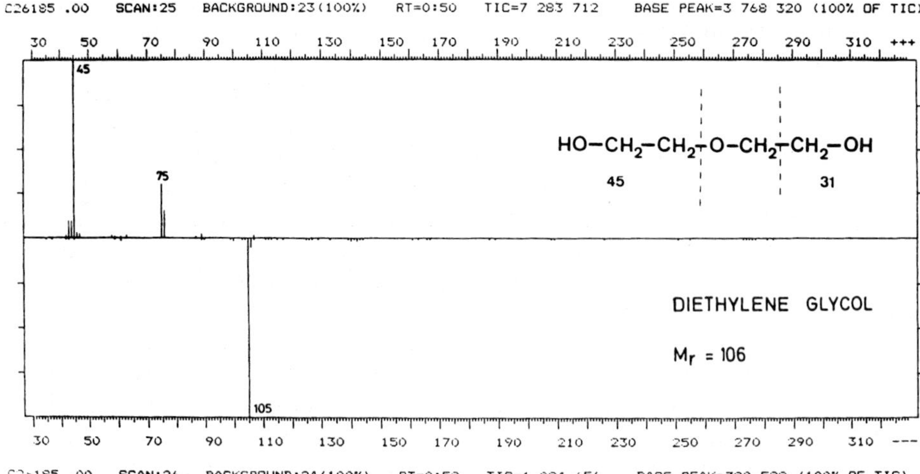

Figure 6. Dual-MS Recording of Diethylene Glycol.
Positive EI shows the fragments m/z 45 and 75,
Negative CI the quasi-molecular ion $[M-H]^-$.

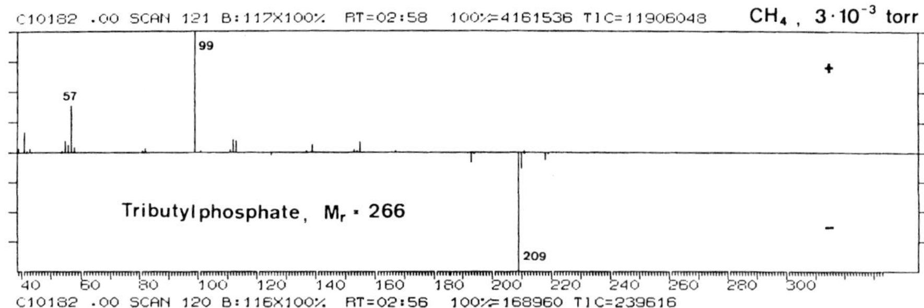

Figure 7. Dual-MS Recording of Tributylphosphate.
The joint interpretation permits identification.

1.7.9 GC-FTIR and The Need for Derivatization

It can often be heard that the sensitivity of GC-FTIR is too low for toxicological analysis. We can not agree. The light-pipe method, the simpler and cheaper of the two combinations, can identify compounds in the lower ng-range. The more complex quasi on-line combinations with elimination of carrier gas can even detect pg-quantities. Partly responsible for a low opinion about the sensitivity of the light pipe technique may be the fact that it is ignored that hydrogen stretching bands of hydroxy and especially amino groups show only very low intensity in the gas phase. Just as nobody would try to analyze aliphatic hydrocarbons by UV, primary and secondary amino groups should not be searched for by gas phase IR. Such groups must be derivatized.

Many biogenic primary amines and phenylethylamine type drugs can be converted spontaneously to isothiocyanates by adding a few ml of CS_2 to the organic extract (43, 44). The IR absorption of the isothiocyanate group is approximately 1000 times stronger than that of a primary amino group. Figure 8 shows the gas phase IR spectrum of phenylethylamine, recorded by GC-FTIR with the light-pipe technique. As evident from the aliphatic and aromatic C–H bands in the region of 3000 cm^{-1}, the concentration must be very high, but still no NH_2-band is visible. Figure 9 shows the IR-spectrum of the CS_2-derivative phenylethylisothiocyanate. The low intensity of the C–H bands illustrates that only about 1 % of the quantity used for the chromatogram of the amine has been injected. Nevertheless, the isothiocyanate band at 2070 cm^{-1} is extremely strong (45).

In the face of today's sophisticated physical equipment, the toxicologist often forgets that he can also use chemical reactions in his analytical work. Before HPLC was introduced, he had to derivatize many compounds to increase their volatility and make them suitable for GC. Now, HPLC has eliminated this need. But derivatization reactions should not be forgotten. They can be used to improve the visibility of compounds by our spectrometric instrumentation. The best example is fluorescence spectrometry. Derivatization with fluorogenic reagents could certainly be used more often in toxicological analysis, not only in TLC.

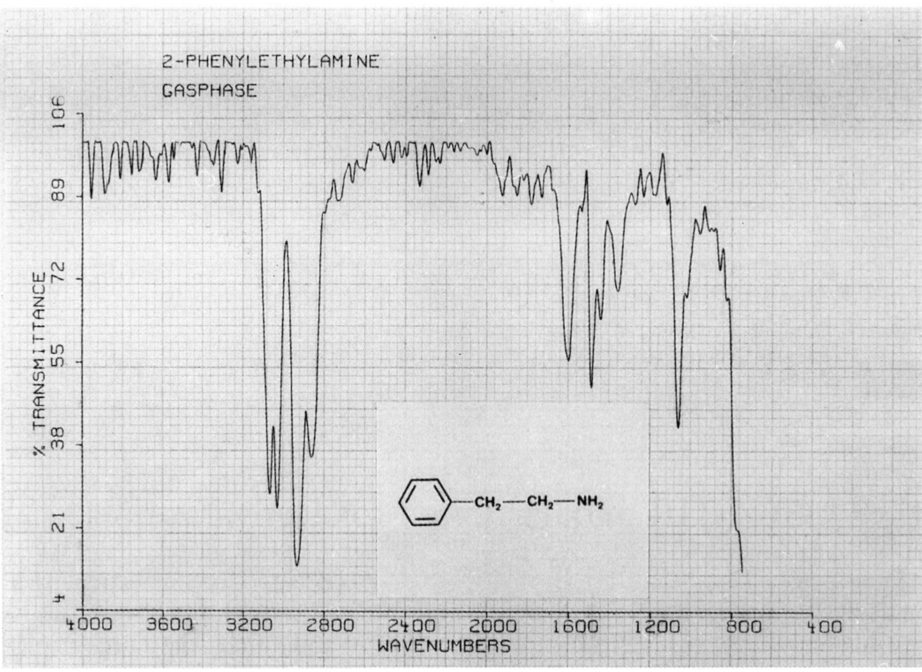

Figure 8. Gas Phase IR Spectrum by on-line GC-FTIR from 2 μg Phenylethylamine.

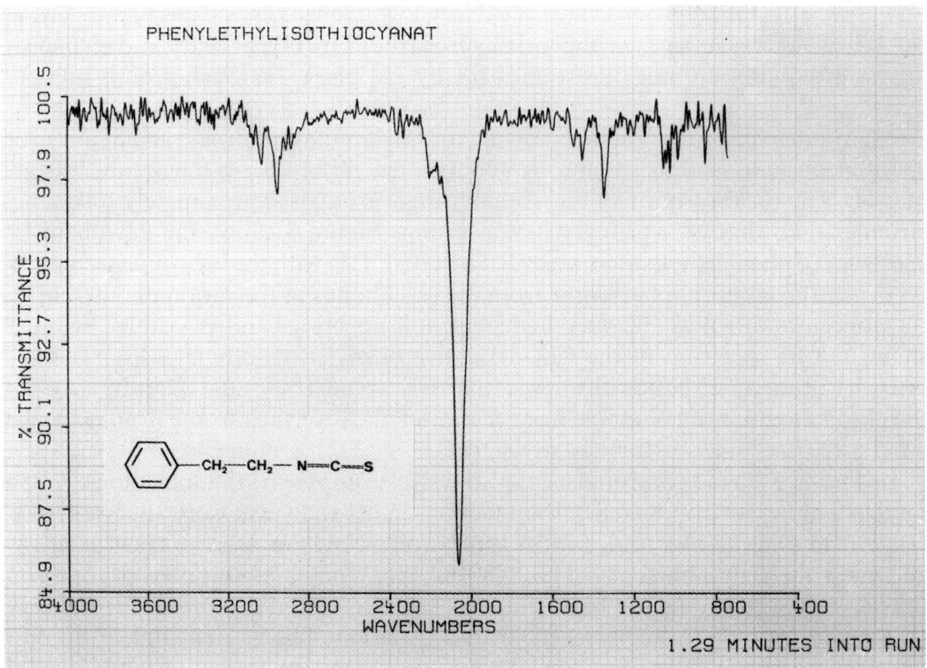

Figure 9. Gas Phase IR Spectrum by on-line GC-FTIR from 20 ng Phenylethylisothiocyanate.

1.7.10 Summary and Concluding Remarks

Instrumental innovations have not come to a stop. A number of new developments are on the market, but have not yet found entrance into the toxicological laboratories. This is mainly due to a saturation effect. The relatively small laboratories are too short of space, funds and staff to make use of all inovations. Which methods deserve to be adopted by the analytical toxicologists?

- Supercritical Fluid Extraction and Extractive Dialysis seem welcome additions to the classical liquid-liquid and the newer solid-liquid extractions.
- The chromatographic techniques must be supplemented by Capillary Electrophoresis and Capillary Electrochromatography, maybe also by Supercritical Fluid Chromatography. One reason for such a demand is that the separation of enantiomers is going to receive increased attention, and that GC is not an ideal method for this purpose.
- Among the established spectrophotometric techniques, Fluorescence Spectrometry has been sadly neglected by many toxicologists. It should be remembered that for fluorescent compounds, SF-sensitivity can be up to 100 times better than UV-sensitivity. Since fluorescence emission spectra are qualitatively almost identical to UV spectra, they can be helpful in screening very dilute extracts.
- Only few toxicological laboratories use Negative Ion MS. This is not easy to comprehend, since many of today's mass spectrometers are capable of recording negative ions. We have particularly pointed out the advantages of "Dual-MS", consisting of a quasi-simultaneous combination of positive EI and low-pressure negative CI. In addition, negative ion MS is also the best trace detector for compounds with high electron affinity, superior to any ECD.
- On-line MS-methods should be supplemented with On-Line FTIR. GC-FTIR is ready to be used, other combination will follow. IR can often complement information from MS. It can distinguish between structural isomers and enantiomers.
- A number of newer methods which involve MS deserve the attention of toxicologists, i.e. Isotope Ratio MS, which has considerable potential in forensic investigations, as well as ICP-MS for qualitative and quantitative inorganic analysis.

New instrumental methods are new tools and often better tools. But they do not replace chemical reasoning. Our greatest concern for the future of toxicological analysis is the danger that all the sophisticated laboratory equipment will side-track the toxicologist and seduce him to become a specialist. A toxicologist must try to stay a chemical all-rounder. He should aim at incorporating all new instrumental possibilities into his chemical strategy and choose for each job the best suited chemical and instrumental approach. He should not shy away from new instrumental techniques, but must also not forget established analytical methods such as derivatization reactions.

It is extremely interesting to try to get the very best performance out of an old or a new tool. But it is certainly a lot of fun to have all the chemical and instrumental possibilities working side by side to solve the many problems which crop up in a forensic or clinical toxicological laboratory. Let's try to stay master of the whole laboratory instead of becoming servant of one specific method or instrument.

References

1. Yeoman, W.B. (1986) Metals and anions. In: Clarke's Isolation and Identification of Drugs, 2nd Ed. (Moffat, A.C., Jackson, J.V., Moss, M.S. & Widdop, B., eds.), 57. The Pharmaceutical Press, London.
2. Stevens, H.M. (1986) Colour Tests. In: Clarke's Isolation and Identification of Drugs, 2nd Ed., pp. 128–147. The Pharmaceutical Press, London.
3. Hunter, W.M. & Corrie, J.J. (eds.) (1983) Immunoassays for Clinical Chemistry, 2nd Ed. Churchill Livingstone, Edinburgh.
4. Stewart, M.J. (1986) Immunoassays. In: Clarkes Isolation and Identification of Drugs, 2nd Ed., 148–159. The Pharmaceutical Press, London.
5. Brandenberger, H. & Maes, R.A.A. (1988) Empfehlungen zur klinisch-toxikologischen Analytik, Folge 1: Der Einsatz von immunochemischen Testen in der Suchtmittelanalytik. Mitteilung X der Senatskommission für Klinisch-toxikologische Analytik der DFG. VCH, Weinheim.
6. Vidic, E. (1955) Die Anwendung papierchromatographischer Methoden beim forensischen Suchtmittelnachweis. Arzneimittel-Forschung *5*, 291–297.
7. Machata, G. (1960) Dünnschichtchromatographie in der Toxikologie. Mikrochim. Acta *1960*, 79–86.
8. Bäumler, J. & Rippstein, S. (1961) Die Dünnschichtchromatographie als Schnellmethode zur Analyse von Arzneimitteln. Pharm. Acta Helv. *36*, 382–388.
9. Weinig, E. & Lauterbach, L. (1958) Die Gaschromatographie als neue Methode in der forensischen Toxikologie und Kriminalistik. Arch. f. Krimin. *122*, 11–17.
10. Cadman, W.J. & Johns, T. (1960) Application of the gas chromatograph in the laboratory of criminalistics. J. of Forensic Science *5*, 369–385.
11. Hammar, C.G., Holmstedt, B. & Ryhage, R. (1968) Mass fragmentography – Identification of chlorpromazine and its metabolites in human blood by a new method. Analytical Biochem. *25*, 532–548.
12. Hammar, C.G., Hanin, I., Holmstedt, B., Kitz, R.J., Jenen, D.J. & Karlén, B. (1968) Identification of acetylcholine in fresh rat brain by combined gas chromatography – mass spectrometry. Nature *220*, 915–917.
13. Brandenberger, H. (1982) Fünfzehn Jahre Massenspektrometrie in der forensischen Toxikologie. In: Entwicklung und Fortschritte der Forensischen Chemie (Arnold, W. & Püschel, K., eds.). Dieter Helm, Heppenheim.
14. Brandenberger, H. (1979) Trace detection and dosage of dithiocarbamate pesticides and pharmaceuticals by acid decomposition and analysis of the evolved carbon disulfide by GC, using a photoionization detector or mass specific detection systems. Meeting of the International Association of Forensic Toxicologists, August 1979, Glasgow.
15. Brandenberger, H. (1992) The tools of the toxicological chemist. Proceedings of the 30th International Meeting of the International Association of Forensic Toxicologists, October *1992* in Fukuoka, Japan, 7–13 (Nagata, T., ed.). Yoyodo Printing Kaisha.
16. McHugh, M.A. & Krukonis, V.J. (1986) Supercritical Fluid Extraction Principles and Practice. Butterworths, Boston.
17. Hawthorne, S.B. (1990) Analytical scale supercritical fluid extraction. Anal. Chem. *62*, 633A–642A.
18. Brandenberger, H. & Schneider, P. (1986) Extractions with liquid carbon dioxide. In: Proceedings of the 23rd Meeting of the International Association of Forensic Toxicologists, August *1986* in Ghent (Heyndrickx, B., ed.), 212–215. State University of Ghent.
19. Brandenberger, H. & Bucher, M. (1980) Extractive Dialysis. In: Forensic Toxicology (Oliver, J.S., ed.), 104–108. Croom Helm, London.
20. Seiler, N. (1993) Fluorescent derivatives. In: Handbook of Derivatives for Chromatography, 2nd ed. (Blau, K. & Halket, J., eds.), 175–213. John Wiley, Chichester, New York, Brisbane, Toronto & Singapore.
21. Durden, D.A., Juorio, A.V. & Davis, B.A. (1980) Thin-layer chromatographic and high resolution mass spectrometric determination of -hydroxyphenylethylamines in tissues as dansyl-acetyl derivatives. Anal. Chem. *52*, 1815–1820.

22. Kubis, A.J., Somayajula, K.V., Sharkey, A.G. & Hercules, D.M. (1989) Laser mass spectrometric analysis of compounds separated by thin-layer chromatography. Anal. Chem. *61*, 2516–2523.
23. Brandenberger, H. & Müller, S. (1962) A gas chromatographic separation of the volatile fatty acids of black tea. J. of Chromatog. *7*, 137–141.
24. Brandenberger, H., Engelhardt, H. & Haerdi, W. (1989) Empfehlungen zur klinisch-toxikologischen Analytik No. 3: Einsatz der Hochleistungsflüssigchromatographie in der klinisch-toxikologischen Analytik. Mitteilung XII der Senatskommission für Klinisch-toxikologische Analytik der DFG. VCH, Weinheim.
25. Markides, K.E. & Lee, M.L. (1988) Supercritical Fluid Chromatography Applications. Brigham Young University Press, Provo, Utah.
26. Engelhardt, H., Beck, W., Kohr, J. & Schmitt, T. (1993) Capillary Electrophoresis: Methods and Scope. Angew. Chem., Int. Ed. Engl. *32*, 629–649.
27. Jones, W.R. & Jandik, P. (1992) Various approaches to analysis of difficult sample matrices of anions using capillary ion electrophoresis. J. Chromatog. *608*, 385–393.
28. Beck, W. & Engelhardt, H. (1992) Separation of non UV-absorbing cations by capillary electrophoresis. J. Anal. Chem. *346*, 618–621.
29. Beck, W. & Engelhardt, H. (1992) Capillary electrophoresis of organic and inorganic cations with indirect UV detection. Chromatographia *33*, 313–316.
30. Huang, E.C., Wachs, T., Conboy, J.J. & Henion, J.D. (1990) Atmospheric pressure ionization mass spectrometry – Detection for the separation sciences. Anal. Chem. *62*, 713A–725A.
31. Skidmore, M.W. (1993) Derivatization for chromatographic resolution of optically active compounds. In: Handbook of Derivatives for Chromatography, 2nd ed. (Blau, K. & Halket, J., eds.), 215–252. John Wiley & Sons, Chichester, New York, Brisbane, Toronto, Singapore.
32. DFG & TIAFT (1992) Thin-Layer Chromatographic Rf Values of Toxicologically Relevant Substances on Standardized Systems. Report XVII of the DFG Commission for Clinical-Toxicological Analysis (de Zeeuw, R.A., Franke, J.P., Degel, F., Machbert, G., Schütz, H. & Wijsbeek, J., eds.). VCH, Weinheim.
33. DFG & TIAFT (1992) Gas Chromatographic Retention Indices of Toxicologically Relevant Substances on Packed or Capillary Columns with Dimethylsilicone Stationary Phases. Report XIII of the DFG Commission for Clinical-Toxicological Analysis (de Zeeuw, R.A., Franke, J.P., Maurer, H.H. & Pfleger, K., eds.). VCH, Weinheim.
34. Brandenberger, H. & Ryhage, R. (1978) Possibilities of negative ion mass spectrometry in forensic and toxicological analyses. In: Blood drugs and other analytical challenges (Reid, E., ed.). Ellis Horwood, Chichester.
35. Brandenberger, H. (1978) Negative ion mass spectrometry, a new tool for the forensic toxicologist. In: Instrumental Applications in Forensic Drug Chemistry. International Symposium organized by the US Department of Justice Drug Enforcement Administration (Klein, M., Kruegel, A.V. & Sobol, S.P.), 48–59. US Government Printing Office, Washington, D.C..
36. Brandenberger, H. & Ryhage, R. (1978) New developments in the field of mass spectrometry and GC-MS and their applications to toxicological analysis. In: Recent Developments in Mass Spectrometry in Biochemistry and Medicine, Vol. I (Frigerio, A., ed.) 327–341. Plenum Publ. Corp., New York.
37. Brandenberger, H. (1979) Present state and future trends of negative ion mass spectrometry in toxicology, forensic and environmental chemistry. In: Recent Developments in Mass Spectrometry in Biochemistry and Medicine, Vol. II (Frigerio, A. ed.) 227–255. Plenum Publ. Corp., New York.
38. Brandenberger, H. (1980) Negative ion mass spectrometry by low-pressure chemical ionization. In: Recent Developments in Mass Spectrometry in Biochemistry and Medicine (Frigerio, A. & McCamish, M., eds.) 391–404. Elsevier, Amsterdam.
39. Wilkins, Ch.L. (1987) Linked gas chromatography-infrared-mass spectrometry. Anal. Chem. *59*, 571A–581A.
40. Brandenberger, H. (1983) Negative ion mass spectrometry by low pressure chemical ionization: inherent possibilities for structural analysis. Intern. Journ. of Mass Spectrometry and Ion Physics *47*, 213–216.
41. Brandenberger, H. (1982) Negative ion mass spectrometry by low pressure chemical ionization and its combination with conventional positive ion mass spectrometry by electron impact. Abstracts of the 30th Annual Conference on Mass Spectrometry and al-

lied topics, June 1982, Honolulu, 468–469. American Society for Mass Spectrometry.

42. Brandenberger, H. (1987) Low pressure negative chemical ionization mass spectrometry and dual mass spectrometry: a review of the work since 1977. In: TIAFT 22, Reports on Forensic Toxicology (Proceedings of the International Meeting of The International Association of Forensic Toxicologists, August 1985, Rigi Kaltbad (Brandenberger, H. & Brandenberger, R., eds.). Branson Research, CH-6356 Rigi Kaltbad.
43. Brandenberger, H. & Hellbach, E. (1967) Über die gaschromatographische Bestimmung von Amphetamin als (1-Phenyl-isopropyl)-isothiocyanat. Helv. Chim. Acta *50*, 958–962.
44. Brandenberger, H. & Schnyder, D. (1972) Toxikologischer Spurennachweis von biologisch wirksamen Aminen (Amphetamine, Catecholamine, Halluzinogene) durch Gas-Chromatographie mit massenspezifischer Detektion (GC-MD). Z. Anal. Chem. *261*, 297–305.
45. Brandenberger, H. (1987) The role of FTIR in chemical toxicology. Increased sensitivity of the on-line combination with GC by Derivatization. Proceedings of the 24th International Meeting of "The International Association of Forensic Toxicologists", Banff, Canada (Jones, G. R. & Singer, P. P., eds.). 330–339. Univ. of Alberta, Alberta.

Part 2
Chapters Concerned with Special Poison Classes

2.1 Natural Gases and Vapors

H. Brandenberger

2.1.1 General Remarks and Classification

Occasionally, a toxicologist is confronted with intoxications or lethal accidents caused by gases or vapors which occur in nature. In table 1, we have tried to classify them as reduction and oxidation products of the non-metallic elements C, N, S, P

Table 1. Natural Gases of Toxicological Importance

Elemental buildings blocks	(C)	N_2 – 196° – PG, PhA	(S)	(P)	Cl_2 – 35° Co, Od TL 1
H_2 – 253° – PG	CH_4 – 161° neutral – PG, PhA	NH_3 – 33° basic Od PG TL 100	SH_2 – 60° acid Od PG TL 20	PH_3 – 87° basic (Od) (PG) TL 0.2	HCl – 85° acid Od TL 10
O_2 – 183°	CO – 191° – TL 100	NO – 151° Od TL 5 ↓	SO_2 – 10°C Od TL 5 ↓	(P_2O_3)	Cl_2O + 2°C Co, Od TL 1
	CO_2 – 78° – PG, PhA	NO_2 + 22° Co, Od TL 5	SO_3 + 45° Od TL 5	(P_2O_5)	ClO_2 + 10° Co, Od TL 0.1
O_3 – 111° Od TL 0.1	$COCl_2$ + 8° (Od) TL 0.2	NOCl – 6° Co, Od TL 1		AsH_3 – 55° (Od) TL 0.05	HCN + 26° Od TL 10

Formulas in brackets indicate that the compound is not a gas, but a liquid or a solid.

Co = Color
Od = Odor } If set in brackets, the intensity in not sufficiently strong to serve as a warning signal.

PG = important Natural Putrefaction Gas
PhA = dangerous Physical Asphyxiant

TL (with number) = Toxicity Level (appearance of harmful effects may be expected), in ml/m^3 (ppm). These figures are as a rule identical with the two-fold MAK-values or lower.

and halogens. Under each of the elements (Cl represents all halogens), the hydrogen derivatives are listed first (CH_4 stands in place of a large number of hydrocarbons), followed by the oxygen derivatives (the partially oxidized ones on the upper and the final oxidation products on the lower line). The volatility of the gases is illustrated by their boiling points, and the danger potential of the toxic compounds is shown by quoting the level above which toxic effects can be expected. Additional remarks point out the gases which can originate by putrefaction, the potential physical asphyxiants, as well as the compounds with a (warning) color or odor. For the reduction products, the acidity is also mentioned. At the bottom of the table, some structurally-related gases have been added, which do not exactly fit into our scheme. They must nevertheless be mentioned; their high toxicity has been responsible for many lethal intoxications.

2.1.2 The Elemental Gases

2.1.2.1 Hydrogen and nitrogen

Among the elemental gases, H_2 and N_2 can be formed by putrefaction of organic matter. Both gases can act as asphyxiants. Many years ago, the presence of concentrated N_2 (respectively a lack of O_2 in the air) has had lethal consequences for a large group of miners in an underground construction camp in the Swiss mountains. The absence of warning properties (odor or color) increases the danger potential of the 2 gases. It is smaller for the light gas H_2 than for N_2, since H_2 rises to the ceiling of the rooms and, if possible, escapes. Sub-lethal concentrations of N_2 can lead to mental confusion and incoordination, probably due to its slight anesthetic action.

2.1.2.2 Oxygen

The atmosphere should contain between 16 and 21 % of the life-sustaining O_2. Concentrations below 16 % require increased respiration and may result in a slight diminution of coordination. Below 12 %, the ability to think clearly will be lost. Concentrations below 10 % lead to unconsciousness and death.

Physical asphyxias can be due to a mechanical obstruction in the respiratory tracts, to a reduction of air pressure or to a dilution of air with another gas. Chemical asphyxias, in contrast, inhibit access of O_2 to the tissues by blocking hemoglobin or inactivating a respiratory enzyme.

Increased O_2 concenrations up to 40 % are well-tolerated. However, much higher concentrations (over 70 %), inhaled for longer periods, can irritate the lungs.

2.1.2.3 Ozone

Ozone (O_3) originates in nature in volcanic eruptions, under the influence of the UV-rays of the sun and in the electrical discharges of lightning. Today, O_3 has also become an industrial pollutant. It is formed by electric contacts of motors, during welding, by high intensity mercury lamps and by the action of UV-light (sun) on the nitrous oxides and hydrocarbons which are emitted by modern traffic. It is also produced in various industrial processes and by modern appliances such as laser printers or photocopiers. O_3 is produced synthetically and used to sterilize air and water (i.e. swimming pools).

O_3 is a much more powerful oxidizing agent than O_2. It has a high degree of acute and chronic toxicity. If set free in workshops, an adequate ventilation should be provided. However, the technically-produced O_3 does not significantly raise the O_3 level in the free air, which is determined by the much larger quantities originating from photochemical reactions. Natural O_3 and the O_3 created by the sun's action on the products of car exhaust constitute the bulk of O_3 in today's atmosphere.

While there is considerable concern about the temporary high O_3 levels in our atmosphere, the reduction of the O_3 content in the stratosphere probably represents a more serious danger. The O_3 layer protects us from the UV-rays of the sun. Its disappearance may contribute to a global warming, to an increase of human skin cancer and to changes in our vegetation. It may, finally, endanger our life on earth.

O_3 has a strong odor (warning signal), which can mask other odors. This may sometimes be falsely attributed to an air purification effect of O_3.

2.1.2.4 Halogens (fluorine and chlorine)

F_2 and Cl_2 are highly corrosive gases. They irritate the skin, and damage especially the respiratory tracts and lungs. Their toxic action can produce lung edemas. Chronic exposure leads to bronchitis, in the case of F_2 also to osteosclerosis. Both gases are yellow and possess a strong smell. But the color of F_2 is not sufficiently intensive to serve as warning signal, since it can only be detected at concentrations well over the toxic level. While F_2 is a typical gas, Cl_2 boils just at $-35°$. The other halogens are liquid (Br_2, with boiling point 50 °C) or solid (I_2, with melting point 113° and boiling point 184 °C). Their vapors also affect the respiratory tracts and lungs.

F_2 is emitted by aluminum plants and has been held responsible for damage caused to the vegetation of the surroundings. The heavy Cl_2 is an even more important industrial gas. It is produced in electrolytic processes and used as a chlorinating and bleaching agent and, in low concentrations, as a water disinfectant, for example for swimming pools.

2.1.3 The Products of Reduction

2.1.3.1 General remarks

The first three or even four members among the reduction products are typical putrefaction gases. Their acid respectively basic properties can be explained by their tendency to acquire the stable electron structure of the noble gases. NH_3 and PH_3, as proton acceptors, have basic character. HF, HCl and (to a lesser extent) SH_2, as proton donators, are acids. The bases and acids are corrosive and can damage body tissues.

2.1.3.2 Methane (and other hydrocarbons)

Neutral CH_4 is not toxic; but it must be classed among the asphyxiants. It can be found in mines, caves and pits. This is reflected in its German names "Erdgas" and "Grubengas". Here again, the lack of color and odor increases its danger potential. The symptoms which the lowest member of the hydrocarbon class can provoke are at most a slight depression of the central nervous system. In the higher saturated aliphatic hydrocarbons, these symptoms can be stronger. But these substances have no liver- or kidney-toxicity and are not cancerogenic like the unsaturated, cyclic and especially the aromatic hydrocarbons.

2.1.3.3 Ammonia

The strong base ammonia (NH_3) is ejected by volcanos ("volcanic gas") and formed in appreciable concentrations in sewers ("sewer gas") as a result of putrefaction of animal proteins. Its corrosive action on lung tissue and its local corrosive action on skin are especially dangerous. Chronic exposures to NH_3 may lead to inflamation of the eyes and of the bronchia. Exceptionally, the digestion can also be affected. It seems that one can become accustomed to NH_3, which has a pronounced warning odor. Its industrial use is widespread. Large amounts of the gas are synthesized and used in the production of plastics, fibers, colors, fertilizers and many other products.

2.1.3.4 Hydrogen sulfide

The slightly acid hydrogen sulfide (H_2S) has a strong odor, which resembles that of putrified eggs. It is a volcanic gas and is formed in sewers just like NH_3. It has therefore been given the same names. In contrast to most other putrefaction gases, H_2S is a strong poison, which inactivates many metallo-enzymes. Longer exposures to concentrations over 0.3‰ lead to nausea, respiratory difficulties and narcosis (formation of sulfhemoglobin), followed by cramps and delirium. Concentrations over 1‰ can be lethal within minutes or less.

Our laboratory has been confronted with a few H_2S poisonings: children in primitive toilets, farmhands in a stable, sailors in a freighter. In most cases, H_2S acted in combination with other putrefaction gases.

2.1.3.5 Phosphine and arsine

Phosphine (PH_3) may also be classed among the putrefaction gases. Its odor, not strong enough to serve as a warning, resembles putrefying fish. PH_3 is an extremely strong metabolic and nerve poison. Even low concentrations produce nausea and cramps. Intoxications with PH_3 have been reported in submarines carrying P-containing torpedos.

Arsine (AsH_3) is a nerve poison with a similar action to PH_3. In addition, it is also a potent blood poison. The excretion of red or even brown urine can serve as an indication for an exposure to AsH_3. Its distinctive but weak garlic odor is not sufficiently strong to be a warning signal. AsH_3 is discussed in chapter 2.10.17.

2.1.3.6 Hydrogen halogenides

The three hydrogen halogenides hydrofluoric acid (HF), hydrochloric acid (HCl) and hydrobromic acid (HBr) are very strong acids with corresponding corrosive properties. But except for HF, lower concentrations of these gases or vapors can be tolerated fairly well by the body, even if inhaled. The gases are very soluble in water. HCl and HBr dissociate immediately and completely to the ions. HF attacks glass and must be kept in teflon containers. Its dissociation is less complete, which explains its lipo-solubility and better ability to penetrate biological tissues. A contact of skin, mucous membranes or the respiratory tract with HF is much more dangerous than a contact with HCl or HBr.

2.1.4 The Products of Oxidation

2.1.4.1 The oxides of carbon

Carbon monoxide (CO) originates in all incomplete oxidations of carbon-containing materials. It is often formed simultaneously with soot. Thus, in the presence of soot, an occurrence of CO in the air can be suspected.

In many countries, CO is the poison responsible for the largest number of intoxications. We must remember that it has been for a long time (at least in Europe) the main constituent of cooking gas, and that it was (and often still is) a main component of car exhaust. Many accidental and intentional killings with CO (suicides and homicides) have occurred in the past and still occur today. During a long period, in Switzerland, the CO of the cooking gas was the preferred suicide weapon among women and the CO from car exhaust the suicide poison of choice among men. Many accidental CO intoxications result from insufficient air supply to gas stoves or from fires in closed rooms, i.e. airplanes.

The danger due to an elevated CO concentration in air during our daily occupations is often not realized. It does affect the well-being of a person and can lower the reaction time (i.e. while driving). In view of the importance of this poison, this book contains a separate chapter (2.2) on CO and its determination in the body.

While CO can be called a chemical asphyxiant, since it blocks the transport sites of oxygen in the body, carbon dioxide (CO_2) is the most dangerous of the physical asphyxiants. This results not only from its complete lack of warning properties that it shares with CO, but also from its high specific weight. Since it is 1.5 fold heavier than air, it accumulates in pits, mines, sewers, silos, boat-holds and wine cellars. It displaces the air and can cause immediate death of people and animals by asphyxia.

Our air usually contains close to 0.03 % CO_2. The concentration has been slowly increasing during this century, a further factor which is held responsible for a recent global warming. Expelled air (human breath) contains up to 4 % CO_2. Air contents from 8 % CO_2 upwards can provoke toxic symptoms (respiratory difficulties, rise of blood pressure, staggering). Death can occur at concentrations over 20%. Such high levels may result from evaporation of liquid CO_2 (used as extraction fluid) or solid CO_2 (dry ice). They can also be caused by fire.

Joint intoxications by CO and CO_2 are not uncommon. They correspond to a combined action of a chemical and a physical asphyxia. This explains why the carboxy-hemoglobin saturation level of fire victims is often below 50%, that is below the mark with lethal consequences.

2.1.4.2 The oxides of nitrogen (NO_x)

The two most important oxides of nitrogen are nitrogen monoxide (NO) and nitrogen dioxide (NO_2). They emanate from the combustion of nitrogen-containing compounds, the oxidation of the N_2 in air at very high temperatures (soldering, traffic vehicles with efficient gasoline motors), the action of nitric acid on metals and organic compounds (nitration), and also biologically in grain- or corn silos. The nitrogen oxides are definitely environmental poisons for which industry and modern traffic are responsible. They have a good lipid solubility and enter the alveoles easily. Symptoms of NO and NO_2 intoxications are cough, lung edemas, respiratory pains and cyanosis. Both oxides possess strong odors, NO_2 also a brown warning color.

NO acts fast. It leads also to the formation of methemoglobin (ferri-hemoglobin). NO_2 usually acts only after a latent period of 6 to 12 hours. But after every exposure, even if no symptoms are manifest, absolute rest is indicated, if possible also hospitalization. Chronic exposure to low concentrations of NO_x provokes a catarrh of the respiratory tracts.

2.1.4.3 The oxides of sulfur

At room temperature, sulfur dioxide (SO_2) is a gas, but sulfur trioxide (SO_3) a liquid which combines with water to sulfuric acid. Both oxides are formed by the combustion of sulfur and sulfur-containing compounds. They are also emitted in volcanic eruptions. But the biological formation of SO_2 and SO_3 is insignificant when compared to the large pollution by industry and traffic. Sulfur oxides in our environment stem mainly from thermic power plants, oil heating systems and gasoline-powered vehicles. Together with CO and O_3, they are the most dangerous environmental poisons.

Sulfur oxides are strongly irritating for eyes, the respiratory tract and lungs. They cause bronchitic symptoms and can damage the respiratory organs so badly that an asphyxia results. They also produce, especially in highly industrialized areas, immense damage to buildings and cultural objects. It has been estimated that the restoration costs of such damage amount to many 100 million francs yearly just for a small country like Switzerland, and 10 times more for Germany. Elimination of SO_2 from plant smoke as well as from the exhaust of household heating systems is possible and of primary importance.

2.1.4.4 The oxides of the halogens

Few accidents caused by halogen oxides have been reported. Three compounds may need mentioning: dichloro-monoxide (Cl_2O), monochloro-monoxide (ClO) and chlorine dioxide (ClO_2). They all decompose to Cl_2 and have therefore properties similar to this elemental gas. ClO_2 is used industrially as a bleaching agent. Its boiling point is relatively high (10 °C). The three Cl-oxides possess a strong yellow color as well as a (warning) odor.

2.1.5 Phosgene and Nitrosyl Chloride

The heavy phosgene ($COCl_2$) originates from the combustion of chlorohydrocarbons (solvents, grease- and paint removers, dry-cleaning agents). It is a very strong lung poison which has been used as a war gas. If inhaled, it decomposes to HCl. The weak odor of $COCl_2$ (reminiscent of moldy hay) is not a sufficient warning signal, already unperceptible concentrations are toxic. The effects have been extensively described (1, 2).

Nitrosyl chloride (NOCl), an industrial gas, is even more toxic than NO and NO_2. It combines the toxic properties of its constituents Cl_2 and NO_x. Its odor and brown color are better warning signals than in the case of phosgene.

2.1.6 Hydrocyanic Acid (Hydrogen Cyanide, Prussic Acid) and Related Compounds

In contrast to HCl and HBr, hydrocyanic acid (HCN) owes the toxicity not to its acidity, but to its inhibition of tissue oxidation. It prevents the use of O_2 carried by the blood. Unlike CO, HCN is a protoplasmic poison which also kills insects and other low forms of animal life, but not bacteria.

HCN has the (false) reputation of being the most powerful of the commonly used poisons. Its abuse goes back to ancient times. It has been used for murder and for suicides, including mass killings. Its role in the extermination of Jews in German concentration camps during World War II, as well as its use for a mass suicide in Guayana in 1987, are still in people's minds. HCN is by far not one of the strongest, but one of the fastest acting poisons. In sufficiently high concentrations, it kills almost instantly. On the other hand, sublethal doses of HCN are quite rapidly detoxified by conversion to thiocyanate. It seems therefore unlikely that chronic HCN intoxications play a significant role.

HCN originates in nature from the decomposition of the glycoside prunasin, which is present in the seeds and cores of many prunus species. The gas is released slowly by the enzyme emulsin, when the plant withers or gets injured. Linseeds, apricot stones and bitter almonds can contain high concentrations of such HCN-releasing glycosides. It is estimated that the oral ingestion of 60 (in children already 5 to 10) bitter almonds can be lethal (3), and it has been reported that in ancient times, death sentences were carried out by forcing the convicts to eat such almonds until death occurred.

HCN has been used as a fumigating agent for plant protection and disinfection of rooms, whole buildings (warehouses) and boats. We had to deal repeatedly with intoxications connected with such applications. They could be avoided by proper precautions. HCN is also produced in the combustion of N-containing materials (plastics, i.e. fires in modern airplanes). Such fumes can incorporate CO, CO_2 and HCN, three different types of asphyxiants. Often, CO and CN^- are found only in sublethal concentrations in the bodies of the victims, which indicates a simultaneous presence of high levels of CO_2.

Especially rescue people should remember that HCN, if present in high concentrations, can also be absorbed through the skin, and that gas masks do not yield 100% protection. The gas possesses a distinctive strong warning odor, similar to that of nitrobenzene. It can usually be detected immediately in the victims of HCN intoxications (expired air, vomit, stomach content).

The alkali salts of HCN are widely used in the manufacture of noble metals (Au, Ag), steel hardening processes, electroplating (galvanization), photography and as pesticides. Oral ingestions of such salts can be lethal, since HCN is set free by the acidity of the stomach and diffuses into the body. The action is less rapid than a death by direct confrontation with HCN gas. It can take many minutes until death occurs after oral poisoning with a cyanide salt such as potassium cyanide. Toxicity and analysis of cyanide anion will be treated in chapter 2.9.

Cyanogen respectively dicyan, $(CN)_2$, can also be a product of fires in presence of polyurethanes and other plastics with high nitrogen content. While HCN boils at +26 °C, $(CN)_2$ is more volatile; it boils at −20 °C. The physiological effects of the two compounds are identical.

A related toxic gas is cyanogen chloride (CNCl) with boiling point +13 °C. It is strongly irritating, especially to the respiratory tract. Its toxicity resembles that of $COCl_2$ and HCN.

2.1.7 Chemical Investigation of Gas Samples

In cases of gas poisoning, not only the biological specimens, but also samples of gas from the place of exposure (a tunnel, pit, mine, tank, living- or working room or a motor vehicle) are often submitted to the chemist. The toxicologist may also be asked to collect the gas samples, just after the event or during its reconstruction. Gas testing tubes from the Dräger Co. (4) and similar gas testing equipment can be very helpful for such inspections. They permit qualitative or semiquantitative tests for the presence of a large number of gases. But in addition to this rapid screening, the collection of a gas sample for an exact laboratory analysis must be recommended, since this permits not only to determine the concentration of a poison, but also the composition of the atmosphere at the site of the accident.

A simple way of collecting a gas sample is by emptying, in rooms which have to be investigated, water-filled vessels which can be tightly closed. We use containers with a seal, through which samples of the trapped gas can be removed with a gas syringe for analysis. Such flasks are on the market. But mineral water or beer bottles with pressure caps sealed by rubber rings or, for small gas samples, serum vials, are equally helpful.

GC is, of course, the preferred method for analysis. Aliquots are removed from the collection flask with a gas-tight syringe for direct injection into the GC column. In the past, we have used the following adsorbents: Molecular sieve 5A (not for CO_2 and H_2O), silica gel, and/or polyaromatic resins such as Porapak Q or Chromosorb 101 to 106. The corresponding retention data have been reported (5). Recently, porous carbon molecular sieves (Carboxenes) have been introduced as packed micro columns and as open tubular capillaries (6). They give excellent separations. Since neither the normal components of air nor the noncombustible gases can be detected by flame ionization, thermoconductivity detectors with He or H_2 as carrier gases are often used. If He is chosen, it should be remembered that the thermoconductivity coefficient of H_2 is lower than that of He; small amounts of H_2 will therefore be indicated by a negative signal. Significant improvements in sensitivity can be obtained by using an electron capture detector in the search for halogenated compounds or a thermoionic detector in the search for N-containing molecules. By far the best detector for a general trace analysis of gases is the mass spectrometer. It reveals all gases, including the incombustibles, with a sensitivity which is lower by several orders of magnitude than that of a thermoconductivity detector.

2.1.8 Procedures for Investigating Biological Materials

There are three possibilities for the toxicological diagnosis of gas poisoning:

- The gas is expelled from the biological material, collected and directly determined, i.e. by GC.
- The gas is detected by its chemical action in the body. Thereby, not the gas itself, but the altered biological constituent is detected and measured.
- The gas is determined in the exhaled air (breath) and, if the correlation with the blood level has been previously established, converted to the blood concentration (only for on-site controls immediately after exposure). This is, i.e., a possibility for determining blood CO saturation values (7).

Steam distillation is often used to remove a gas from biological materials. To remove volatile acids, the sample should be acidified (we use sulfuric-, phosphoric- or tartaric acid). To isolate volatile bases, the sample must be made alkaline (we recommend carbonate). As long as the nature of the gas to be collected is unknown, condensation or absorption of vapors may be incomplete. This is one reason why we prefer to work with the much simpler "head-space" technique:

A biological homogenate is placed in a serum vial and hermetically sealed. If an acid or base in needed to liberate the gas or vapor, it can be injected through the membrane. A raise in temperature increases the displacement of the volatiles into the gas phase of the vial. After equilibration, aliquots of the head space gas are transferred by a gas-syringe from the vial (through the membrane seal) directly to the chromatographic column. If needed, the constituents of air can be previously expelled from the vial by rinsing with GC carrier gas.

For the analysis of biological fluids, we proceed inversely: The acid (preferably a solid such as sodium bisulfate) or alkali (a solid carbonate) is weighed into the vial. Upon rinsing with carrier gas, the flask is sealed and the body fluid injected. After equilibration, usually at an elevated temperature, gas aliquots are transferred from the gas phase of the vial to the chromatographic column.

Spectrometric methods are often used to analyze the reaction products of toxic gases with body constituents, which usually possess altered spectras. But changes in other physical properties can also be measured. An example is carboxyhemoglobin. In contrast to oxyhemoglobin, it does not precipitate when blood in acetate buffer of pH 5 is heated for 5 minutes to 55 °C (8).

2.1.9 Examples for the Analysis of Biological Materials

2.1.9.1 Determination of hydrocyanic acid (respectively cyanide) by head space GC (9)

Serum vials (5 to 25 ml) are loaded with 1 g of solid $NaHSO_4$ and capped with the membranes. 1 or 2 ml of body fluid (stomach content, blood or urine) are injected with a syringe and the vials placed into a water bath (40 to 80 °C). After temperature equilibrium is reached, 1 or 2 ml of head space gas are injected into the GC column (i.e. Porapak Q at 100 °C) and the HCN peak measured and compared with those obtained from external standards. For low HCN concentrations, a thermionic nitrogen detector may be used, but better sensitivity is obtained by mass specific detection of the molecular ion with mass 27 in the normal EI-mode, or the cyanide ion with mass 26 in the negative low pressure CI-mode.

For the analysis of tissues, a 1 or 2 g sample is weighed into the serum vial. After capping, 1 or 2 ml of a 1 + 1 (v + v) dilution of phosphoric acid with water is injected, and the analysis is continued as above. Small serum vials give better sensitivity (for trace detection); larger vials permit repeated injections of head space gas (for quantification).

2.1.9.2 Determination of the blood hemoglobin saturation index with carbon monoxide by GC (10)

A 25 ml serum vial is rinsed with He and capped with the membrane. 5 ml of a hemolyzed blood solution (prepared by adding 15 ml of a 3 % aqueous solution of saponin to 5 g of blood) is injected, 4 ml of a 30 g/l solution of K_3 $Fe(CN)_6$ in water added, and the vial agitated (10 minutes). Aliquots of the head space gas are then transferred to a column (i.e. molecular sieve 5 A at 100 °C. The chromatogram is recorded with a thermoconductivity or mass spectrometric detector. The peaks for O_2, N_2 and CO are quantitatively measured (surface or, in case of very narrow peaks, peak height). Since it can be assumed that the N_2 stems from residual air,

25% of this value is deducted from the O_2 peak to obtain the O_2 liberated from the oxyhemoglobin. Comparison of the corrected value for O_2 with the CO peak yields the % CO saturation of hemoglobin.

We can recommend this method particularly for measuring CO in denatured blood. It is also helpful for the analysis of blood containing methemoglobin and cyano-methemoglobin. But for CO saturation values below 10%, the procedure lacks sensitivity.

2.1.9.3 Determination of the blood hemoglobin saturation index with carbon monoxide by atomic absorption analysis of Fe (11)

1 to 2 ml of blood are mixed with 0.5 ml 0.1 mol/l NH_3 and diluted to 5 ml with water. After 15 minutes hemolysis, the stroma is removed by centrifugation. 4 ml of the supernatant are diluted to 20 ml with sodium acetate buffer of pH 5 (ionic strength 2.25) and the solution distributed into an even number of glass centrifuge tubes. Half of them are placed for exactly 5 minutes into a 55 °C water bath, the rest kept at 0 °C. Then, all samples are rigorously centrifuged, and the solutions sprayed directly (without decantation from the sediments) into an air-acetylene flame. The Fe contents are compared. If the absorbance signal of the unheated solutions is adjusted to 100%, the absorbance signal of the heated samples indicates directly the percentage of non-precipitated hemoglobin and, after a negative correction of 3 to 4%, the % CO saturation of the blood.

This method is better suited for diagnosing highly toxic than sub-toxic CO concentrations, since the inaccuracy of the correction is less important. With denatured blood samples, i.e. in case of fire victims, the method should not be used. Also, some hemoglobin derivatives other than oxy- and carboxy-hemoglobin may interfere. If such compounds must be expected, it is advisable to check first whether they precipitate on heating, or whether they undergo reduction under the conditions used in many molecular spectrophotometric procedures, and to choose the analytical method accordingly.

For the physical properties of hemoglobin derivatives, we refer to the excellent old monograph of Schwerd (12). Since the formation of methemoglobin and other "abnormal" hemoglobin derivatives is not uncommon in gas intoxications, a simultaneous use of two methods based on different analytical principles, as discussed in the next chapter, is recommended. Information on the formation of "abnormal" hemoglobin derivatives by the action of anions, as well as on the determination of methemoglobin, can be found in chapter 2.9.

References

1. Hardy, E.E. (1982) In: Kirk-Othmar Enzyclopedia of Chemical Technology, Vol. 17, 416–425. Wiley-Interscience, New York, 3rd ed.
2. Clayton, G.D. & Clayton, F.E. (1982) In: Patty's Industrial Hygiene and Toxicology, Vol. 2C, 4126–4128. Wiley Interscience, New York, 3rd ed.
3. Moeschlin, S. (1980) Klinik und Therapie der Vergiftungen: Cyanwasserstoff. 6th ed., p. 252. Georg Thieme, Stuttgart & New York.

4. Drägerwerk AG (1994) Dräger-Röhrchen Handbuch: Boden-, Wasser- und Luftuntersuchungen sowie technische Gasanalyse. LN-Druck, Lübeck, 9th ed.
5. Brandenberger, H. (1974) Toxicology: Investigation of gas samples. In: Clinical Biochemistry, Principles and Methods (Curtius, H.Ch. & Roth, M., eds.) pp. 1427–1429. Walter de Gruyter, Berlin & New York.
6. Betz, W. (1994) Analyse von Permanentgasen und C2- und C3-Kohlenwasserstoffen mit Carboxen mikrogepackten Säulen und Porous Layer Open Tubular (PLOT) Kapillarsäulen. The Reporter: Recent Developments in Separation Technology *13/4*, 1–4. Published by Supelco, Gland.
7. Schütz, H. & Machbert, G. (1988) Photometrische Bestimmung von Carboxy-Hämoglobin im Blut (Report VIII of the Senate Commission for Clinic-toxicological Analysis of the German Research Society), Vorproben-mit Atemluft als Untersuchungsmaterial, pp. 93–94. VCH, Weinheim.
8. Wolff, E. (1947) Ann. méd. légale et criminol. police sci. et toxicol. *27*, 221–227.
9. Brandenberger, H. (1974) Toxicology: Determination of cyanide in body fluids. In: Clinical Biochemistry, Principles and Methods (Curtius, H.Ch. & Roth, M., eds.) p. 1431. Walter de Gruyter, Berlin and New York.
10. Brandenberger, H., unpublished method, used for over ten years for the investigation of carbon monoxide poisonings.
11. Brandenberger, H. (1974) Toxicology: Determination of Carbon Monoxide in Blood as Carboxyhemoglobin. In: Clinical Biochemistry, Principles and Methods (Curtius, H.Ch. & Roth, M., eds.) pp. 1432–1434. Walter de Gruyter, Berlin & New York.
12. Schwerd, W. (1962) Der rote Blutfarbstoff und seine wichtigsten Derivate. Max-Schmidt-Römhild. Lübeck.

2.2 Carbon Monoxide

T. Kojima

2.2.1 Introduction

Carbon monoxide (CO) gas is highly poisonous, odorless, colorless, tasteless, nonirritating, lighter than air, and a ubiquitous product of incomplete combustion. As a result of these properties it is referred to as the "silent killer" (1–3).

Toxicity of CO is due to its combination with hemoglobin to form carboxyhemoglobin (COHb), which cannot carry oxygen. If a person is exposed to air containing 0.1 % CO, the COHb saturation would rise to be around 50 %, as the affinity of hemoglobin for CO is approx. 220 times greater than that for oxygen. COHb in blood reduces its oxygen-carrying capacity, and has an inhibitory influence on the dissociation of oxygen from hemoglobin in tissues (3). CO also binds to muscle myoglobin and to intracellular cytochrome oxidases which can contribute to tissue and cellular hypoxia (3–6).

COHb saturation has been reported to range from 0.25 to 2.1 % in normal nonsmoking persons living in cities under conditions of minimal exposure to CO, and from 0.7 to 6.5 % in normal tobacco smokers. Occasionally, COHb saturations as high as 12 % have been found in heavy cigarette smokers immediately after smoking

Table 1. Correlation between COHb Saturations and Signs and Symptoms (2, 3, 6)

% COHb	Signs and symptoms
0– 5	No symptoms
5	In healthy individuals, impairs driving skills and decreases exercise tolerance
10–20	Headache: fatigue; dilatation of cutaneous blood vessels
20–25	Increased lactic acid with compensated metabolic acidosis
20–30	Severe headache; weakness; dizziness; dimness of vision; syncope; nausea; vomiting; diarrhea; interference with motor dexterity-mild symptoms
30–40	Syncope; increased respiration, increased heart rate; nausea; vomitting; confusion
40–50	Coma; convulsions; confusion; increased respiration; increased heart rate
50–60	Coma, convulsions; Cheyne-Stokes respiration; depressed respiration; depressed cardiovascular status-severe
60–70	Coma; convulsions; cardiorespiratory depression; bradycardia; hypotension-often fatal
70–80	Respiratory failure and death
80–90	Death in less than an hour
90–	Death within a few minutes

(5–7). Signs and symptoms at various COHb saturations are shown in table 1. When COHb levels rise to the 30 to 50% range, effects of CO become increasingly severe (2, 3, 6).

About 3,000 to 4,000 accidental or intentional deaths from acute intoxications are reported each year in Japan, and more than 50% of them are due to acute CO poisoning (8).

2.2.2 Determination of CO

Excellent reviews by Maehly (1) and Feldstein (9) have thoroughly covered the literature of methods for the analysis of CO in air, breath, and blood samples. According to these reviews the various procedures for the analysis of CO in gas samples may be roughly classified into infrared spectrometry, gas chromatography, colorimetry, oxidation methods, indicator tube methods, gas volumetric methods, and others. Selection of the method depends on the equipment available, the required sensitivity, and the quantity of sample. The methods are based on the chemical reactions of CO, the absorption properties of CO or on the separation of CO on a gas chromatographic column. In a complex mixture containing other compounds in relatively high concentrations, there may be interferences. The use of infrared spectrometry provides the specificity desired for the unequivocal identification of CO (9), and Brandenberger (10) recommends gas chromatography and gas chromatography-mass spectrometry.

A variety of methods have been developed for determining the COHb saturation in blood (1, 9). They include microdiffusion methods, colorimetry, infrared spectrometry, gas chromatography, gas volumetric methods, spectrophotometry, and others. Fourier transformation infrared spectrometry (11), highly sensitive gas chromatography in which CO is converted into methane and detected by a hydrogen flame ionization detector (12), derivative spectrophotometry (13), and automated differential spectrophotometry (CO-Oximeter) (14, 15) have been developed. The choice of method is often dictated by the equipment available, desired speed of analysis and identification ability (9). The microdiffusion method using the Conway cell can be chosen as a simple screening test (16). The CO-Oximeter IL 182 (14) evaluated by Maas et al. (17) is recommended as a quantitative test (7), and the CO-Oximeter IL 282 (15) for medicolegal use was studied by Okada et al. (18). It seems that the CO-Oximeter is useful in the clinical field. The gas chromatographic method using a microthermal-conductivity detector is reported as an elegant, precise procedure (19), and gas chromatography has been developed for determining CO in tissue (20, 21).

In the investigation of a death associated with a fire, the presence or absence of COHb is used as an indication of whether the victim was alive and died during the fire or whether the victim was dead prior to the conflagration. Therefore, a precise method for determining the COHb saturation in blood is extremely important for forensic purposes.

The spectrophotometric method, which was originally developed by Fetwurst and Meinecke (22) and later modified by Fukui (23) and then by Sakai et al. (24,

25), and the gas chromatographic method (CO-Total Hb method) developed by Kojima et al. (26, 27) have been used for the determination of COHb saturation in blood samples collected from cadavers in our department.

The spectrophotometric method: A small quantity of blood sample is diluted with 0.1 % sodium carbonate solution containing 0.3 % sodium hydrosulfite to convert hemoglobin derivatives into COHb and reduced Hb. Then the hemoglobins are denatured by sodium hydroxide. After maximum absorbancies at around 530 nm (A) and 560 nm (B) are determined, the COHb saturation is obtained by plotting the ratio of B/A against the calibration curve, obtained by analyzing standard COHb blood. The maximum absorbance at around 530 nm (A) should be between 0.3 and 0.4.

The CO-Total Hb method: A 5-ml glass syringe with a blood collection needle is used as a CO releasing apparatus. One gram of blood sample, three drops of 4% Triton-X 100, and three drops of n-octanol are poured into the tube of the syringe containing a small ball as a stirrer. After washing the sample with 0.5 ml of He, the gas phase is expelled. A half ml of He and 2 µl (20 µl, if COHb saturation is more than 10%) of methane as an internal standard are added to the syringe, and 0.3 ml of a 20% potassium ferricyanide solution as a CO releasing agent. After shaking for 5 min, the He phase is analyzed by gas chromatography, and the CO level is calculated by the peak area ratio of CO to the internal standard using the calibration curve. The gas chromatograph is a Shimadzu GC-3BT with a thermal conductivity detector and a Molecular Sieve 5 A column with a length of 2.1 m. The column temperature is set at 100 °C and He with a flow rate of 24/min is used

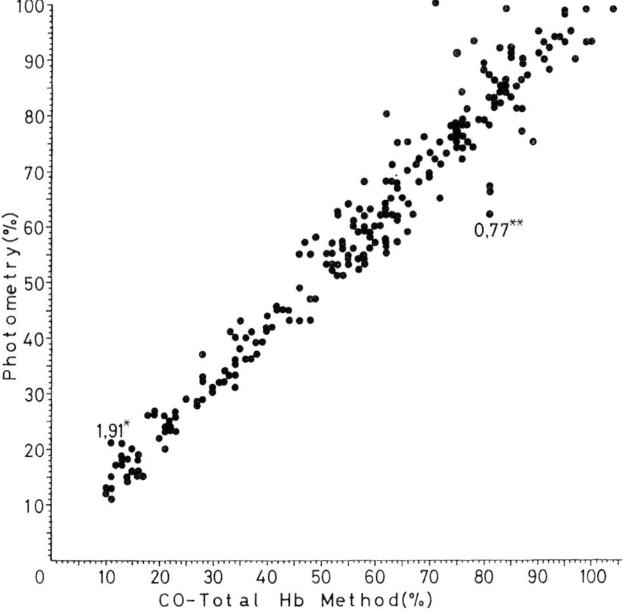

Figure 1. Comparison of COHb Saturation Values in Blood Samples Determined with the Spectrophotometric Method and by the CO-Total Hb Method (COHb Saturations Using the CO-Total Hb Method ≧ 10%). * The Highest and **the Lowest Ratio of the COHb Saturation Using the Spectrophotometric Method to that Using the CO-Total Hb Method.

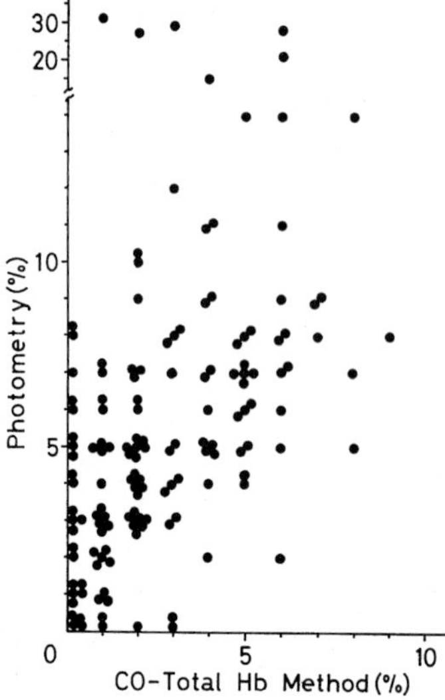

Figure 2. Comparison of COHb Saturation Values in Blood Samples Determined with the Spectrophotometric Method and by the CO-Total Hb Method (COHb Saturations Using the CO-Total Hb Method < 10%).

as a carrier gas. Total Hb level is determined by the modified international CNmHb method. COHb saturation is calculated by the ratio of content and CO-binding capacity.

In 232 out of 377 blood samples analyzed, COHb saturations were found to be 10% or more using the CO-Total Hb method. The correlation coefficient between COHb saturations using the spectrophotometric and the CO-total Hb method was 0.978. Ratios of the COHb saturation using the spectrophotometric method to that using the CO-Total Hb method ranged from 1.91 to 0.77. The average was 1.05 and the coefficient of deviation was 12.4% (Figure 1). In 145 out of 377 blood samples, COHb saturations were less than 10% when using the CO-Total Hb method. No correlation was observed between the COHb saturation using the CO-Total Hb method and that using the spectrophotometric method (Figure 2). According to these results, spectrophotometry, which is simpler and quicker than gas chromatography, is recommended for clinical use and for screening tests to determine whether death has been caused by CO poisoning.

2.2.3 Formation of CO in Humans

It has been reported that approx. 10 ml/day of endogenous CO, arising mainly from the catabolism of heme proteins, is formed in a normal adult, and that COHb saturations range from 0.4% to 1.6% (5–7, 28–31).

CO is also produced after death (27, 32–34). When the saturation level of COHb in blood samples and in reddish discolored body cavity fluids of 32 cadavers not exposed to fire or CO was analyzed, the maximum saturation of COHb in blood samples and in body cavity fluids was 6.2% and 83.7% respectively. These results indicate that the interpretation of COHb saturations in the blood could not be affected significantly by the postmortem formation of CO, and that body cavity fluids should not be used for CO determination (34). The bacterial activity supported by hemin is suspected to be responsible for the postmortem production of CO (35–37).

2.2.4 Interpretation of COHb Saturations in Blood from the Left and Right Heart Cavities of Cadavers

COHb saturations in samples collected at medicolegal autopsies from cadavers both exposed and not exposed to fire or CO gas were analyzed by the gas chromatographic method. The ratio of COHb in blood samples from the left heart cavity to that in blood samples from the right heart cavity (L/R ratio) was calculated (38, 39). Petroleum ingredients in the blood samples were analyzed by gas chromatography-mass spectrometry after extraction, using organic solvents both with and without an Extrelut® column (40).

In 20 victims not exposed to fire or CO gas, COHb saturations in blood samples from the left and right heart cavities ranged from 0.15% to 4.31%, and L/R ratios ranged from 0.96 to 1.33 (average 1.12).

In three victims exposed to fire after death, COHb saturations in blood samples from the left and right heart cavities ranged from 0.21% to 2.30%, and L/R ratios from 0.88 to 1.29.

In 89 victims exposed to fire or CO gas prior to death, COHb saturations in blood samples from the left and right heart cavities ranged from 0.15% to 103.9%. In 72 out of 89 cases, COHb saturations in blood samples from the right heart cavity were 10% or more, and L/R ratios ranged from 0.88 to 1.45 (average 1.09). In only one case was the L/R ratio higher than 1.33. In 17 out of 89 cases, COHb saturations in blood samples from the right heart cavity were less than 10%. L/R ratios ranged from 0.85 to 2.58. In four out of the 17 cases, no soot could be detected in the airway by the naked eye. Petroleum ingredients in blood samples from the left heart cavity were detected in nine of ten cases in which inflammable substances had been used.

In five victims in which it was not known whether death had occurred before or after exposure to fire, no soot in the airway was detectable by the naked eye. In two cases in which use of inflammable substances was suspected, none were detected. COHb saturations in blood samples from the left and right heart cavities ranged from 0.27% to 5.68%, and L/R ratios from 0.94 to 1.79. In two of the five cases, L/R ratios were higher than 1.33. It seemed likely that death had occurred after burning in these two cases.

According to these results, determination of the COHb saturation ratio between blood samples from both the left and right heart cavities appears to be useful for establishing whether death has occurred before or after exposure to fire. This is especially true in cases where no soot can be detected in the airway by the naked eye, no inflammable substances found in blood samples from the left heart cavity, and the COHb saturation of blood sample lies within the level considered normal for tobacco smokers.

2.2.5 Treatment of Acute CO Intoxication

The obvious and specific antagonist to CO is oxygen. The half-recovery time in terms of blood COHb levels for resting adults breathing air at 1 atm is 320 min. Oxygen (100%) reduces this time to 80 min, and hyperbaric oxygen (3 atm) further to 23 min (41).

Hyperbaric oxygen enhances the elimination of COHb and increases the amount of dissolved oxygen in plasma, but this therapy has complications such as decompression sickness and oxygen toxicity (2, 6). Therefore, a decision to use hyperbaric oxygen requires a clinical evaluation of the potential benefits and complications (6).

References

1. Maehly, A.C. (1962) In: Methods of Forensic Science, Vol. I (Lundquist, F., ed.) 539–592, Interscience Publishers, a division of John Wiley & Sons, New York · London.
2. Gossel, T.A. & Bricker, J.D. (1984) Principles of Clinical Toxicology, 84–90, Raven Press, New York.
3. Klaassen, C.D. (1990) In: Goodman and Gilman's The Pharmacological Basis of Therapeutics, 8th ed. (Gilman, A.D., Rall, T.W., Nies, A.S. & Taylor, P., eds.) 1615–1639, Pergamon Press, New York/Oxford/Beijing/Frankfurt/Sao Paulo/Sydney/Tokyo/Toronto.
4. Turino, G.M. (1981) Carbon monoxide toxicity: Physiology and biochemistry. Circulation *63*, 253A–259A.
5. Smith, R.P. (1986) In: Casarett and Doull's Toxicology, 3rd ed. (Doull, J., Klaassen, C.D. & Amdur, M.O., eds.) 223–244, Macmillan Publishing, New York.
6. Ellenhorn, M.J. & Barceloux, D.G. (1988) Medical Toxicology, 820–829, Elsevier, New York · Amsterdam · London.
7. Freireich, A., Landau, D., Dubowski, K.M. & Jain, N.C. (1975) In: Methodology for Analytical Toxicology (Sunshine, I. ed.) 67–69, CRC Press, Cleveland.
8. National Research Institute of Police Science (ed.) (1986–1990) Annual Case Reports of Drug and Toxic Poisoning in Japan (in Japanese). Vol. 28–33, Tokyo.
9. Feldstein, M. (1967) In: Progress in Chemical Toxicology, Vol. 3 (Stolman, A. ed.) 99–119, Academic Press, New York/London.
10. Brandenberger, H. (1974) In: Clinical Biochemistry. Principles and Method, Vol. 2 (Curtius, H.C. and Roth, M.) 1425–1467, Walter de Gruyter, Berlin/New York.
11. Kijewski, H., Seefeld, K.P. & Pöhlmann (1985) Eine neue Methode zur Carboxyhämoglobinbestimmung in flüssigem und getrocknetem Blut. Z. Rechtsmed. *95*, 67–74.
12. Collison, H.A., Rodkey, F.L. & O'Neal, J.D. (1968) Determination of carbon monoxide in blood by gas chromatography. Clin. Chem. *14*, 162–171.
13. Sanderson, J.H., Sotheran, M.F. & Stattersfield, J.P. (1978) A new method of carboxyhaemoglobin determination. Br. J. Industr. Med. *35*, 67–72.
14. Freireich, A.W. & Landau, D. (1971) Carbon monoxide determination in postmortem clotted blood. J. Forensic Sci. *16*, 112–119.
15. Brown, L.J. (1980) A new instrument for the simultaneous measurement of total hemoglobin, % oxyhemoglobin, % carboxyhemoglobin, % methemoglobin and oxygen content in whole blood. IEE Trans Biomed. Eng. *27*, 132–138.
16. Williams, L.A. (1975) In: Methodology for Analytical Toxicology (Sunshine, I. ed.) 64–66, CRC Press, Cleveland.

17. Maas, A.H.J., Hamelink, M.L. & De Zeeuw, R.J.M. (1970) An evaluation of the spectrophotometric determination of HbO_2, HbCO and Hb in blood with the CO-Oximeter IL 182. Clin. Chim. Acta *29*, 303–309.
18. Okada, M., Okada, T. & Ide, K. (1985) Studies on the utilization of a CO-oximeter in the medico-legal field (in Japanese with abstract in English). Jpn. J. Legal Med. *39*, 318–325.
19. Finkle, B.S. (1975) In: Methodology for Analytical Toxicology (Sunshine, I. ed.) 64–66, CRC Press, Cleveland.
20. Blackmore, D.J. (1970) The determination of carbon monoxide in blood and tissue. Analyst *95*, 439–458, 1970.
21. Iffland, R., Balling, P. & Eiling, G. (1988) Optimierung des Kohlenmonoxidnachweises im biologischen Material. Beitr. Gerichtl. Med. *47*, 87–94.
22. Fretwurst, F. & Meinecke, K.H. (1959) Eine neue Methode zur quantitativen Bestimmung des Kohlenoxydhämoglobins im Blut. Arch. Toxicol. *17*, 273–283.
23. Fukui, M. (1972) In: Saiban Kagaku, 2nd Ed. (in Japanese, Fukui, M. et al. eds.) 239–246, Hirokawa, Tokyo.
24. Sakai, K. & Kojima, T. (1978) Comparison of three spectrophotometric methods for determination of carbon monoxide hemoglobin in blood (in Japanese with English summary). Jap. J. Legal Med. *32*, 1–6.
25. Kojima, T., Yashiki, M. & Une, I. (1983) Selection of specimen in the determination of carboxyhemoglobin saturation by spectrophotometry. Hiroshima J. Med. Sci. *32*, 87–91.
26. Kojima, T., Nishiyama, Y., Yashiki, M. & Une, I. (1981) Determination of carboxyhemoglobin saturation in blood and body cavity fluids by carbon monoxide and total hemoglobin concentrations (in Japanese with English summary). Jap. J. Legal Med. *35*, 305–311.
27. Kojima, T., Nishiyama, Y., Yashiki, M. & Une, I. (1982) Postmortem formation of carbon monoxide. Forensic Sci. Int. *19*, 243–248.
28. Sjöstrand, T. (1949) Endogenous formation of carbon monoxide in man under normal and pathological conditions. Scand. J. Clin. Lab. Invest. *1*, 201–214.
29. Coburn, R.F., Williams, W.J., White, P. & Kahn, S.B. (1967) The production of carbon monoxide from hemoglobin in vivo. J. Clin. Invest. *46*, 346–356.
30. Landaw, S.A. (1969) Endogenous production of carbon monoxide: The human body as a cause of air pollution. Ann. intern. Med. *70*, 1275–1276.
31. Geldmacher-von Mallinckrodt, M. (1975) In: Gerichtliche Medizin, 2nd Ed. (Mueller, B., ed.) 832–855, Springer-Verlag, Berlin · Heidelberg · New York.
32. Kojima, T., Yashiki, M., Une, I. & Nishiyama, Y. (1980) Post-mortem formation of carbon monoxide in a drowned body. Jap. J. Legal Med. *34*, 163–168.
33. Kojima, T., Yashiki, M. & Une, I. (1983) Experimental study on postmortem formation of carbon monoxide. Forensic Sci. Int. *22*, 131–135.
34. Kojima, T., Okamoto, I., Yashiki, M., Miyazaki, T., Chikasue, F., Degawa, K., Oshida, S. & Sagisaka, K. (1986) Production of carbon monoxide in cadavers. Forensic Sci. Int. *32*, 67–77.
35. Engel, R.R., Matsen, J.M., Chapman, S.S. & Schwartz, S. (1972) Carbon monoxide production from heme compounds by bacteria. J. Bacteriol. *112*, 1310–1315.
36. Hayashi, A., Tauchi, H. & Hino, S. (1985) Production of carbon monoxide by bacteria of the genera Proteus and Morganella. J. Gen. Appl. Microbiol. *31*, 285–292.
37. Hino, S. & Tauchi, H. (1987) Production of carbon monoxide from aromatic amino acids by Morganella morganii. Arch. Microbiol. *148*, 167–171.
38. Kojima, T., Miyazaki, T., Yashiki, M. & Chikasue, F. (1991) Interpretation of COHb saturations in the left and right heart blood of cadavers. In: Proceedings of the 29th International Meeting of The International Association of Forensic Toxicologists (Kaempe, B., ed.), 257–261.
39. Miyazaki, T., Kojima, T., Yashiki, M. & Chikasue, F. (1992) Interpretation of COHb saturations in the left and right heart blood of cadavers. Int. J. Legal Med. *105*, 65–68.
40. Kojima, T., Yashiki, M., Chikasue, F. & Miyazaki, T. (1990) Analysis of inflammable substances to determine whether death has occurred before or after burning. Z. Rechtsmed. *103*, 613–619.
41. Peterson, J.E. & Stewart, R.O. (1970) Absorption and elimination of carbon monoxide by inactive young men. Arch. Environ. Health *21*, 165–171.

2.3 Determination of Alcohol Levels in the Body

H. Brandenberger, R. Brandenberger and K. Halder

2.3.1 Introduction

Among all drugs, ethyl alcohol or ethanol has had the largest impact on human life. It is used and abused in every part of the world and accountable for a tremendous number of acute and chronic intoxications with dangerous side effects: accidents in the home, at work and especially in connection with traffic. If we add the fatalities due to drunken driving to the list of ethanol intoxications with lethal consequences, this single component is responsible for the largest number of deaths, at least in the industrialized world.

In most countries, driving under the influence of alcohol is considered an offense. Recently, this rule has been extended also to traffic participants other than motor vehicle drivers. In addition, certain job holders are forbidden to ingest alcohol before and during working hours. For this reason, a great number of control analyses involving much medical and administrative work is requested. Alcohol determinations take a high position among the obligations of a laboratory for forensic toxicology. In many countries, official regulations specify reasons and justifications for requesting an alcohol analysis, for determining what specimens must be analyzed and how they should be collected, for prescribing how many analytical determinations have to be carried out and which methods must be used, and finally also how the data should be interpreted. Since the results of blood alcohol determinations are often contested, much more frequently than other work of a chemical toxicologist, this part of his many obligations may take an unreasonable amount of his time and efforts. It seems therefore justified to devote an entire chapter to the problems of alcohol analysis.

Ethyl alcohol is usually ingested as a constituent of alcoholic drinks and therefore also called "Trinkalkohol" (drinking alcohol). Normal beer contains around 4% (weight/volume) alcohol, wine between 7 and 12%, but usually 10%, liquors and spirits from 25 to 50% and occasionally even more, but often near 40%. These alcoholic beverages of different strength are served in glasses with volumes which are inversely proportional to the alcohol content of the beverages. Or, in other words, since the volume of every distinct drink is adjusted to its alcoholic strength, each drink, regardless of its alcohol concentration, introduces on the average the same amount of pure ethanol into a human body, as a rule between 8 to 10 g.

So-called alcohol-free beverages are permitted to contain up to 0.5% ethanol. Such a content can hardly have a clinical significance, since an adult would have to drink several liters of such a beverage to bring his body alcohol level into a range which might have dangerous consequences.

Some pharmaceutical preparations also contain ethanol. But in spite of their sometimes high alcohol concentrations, medicines can not be held responsible for elevated alcohol levels in a body; too little of the pharmaceutical is ingested. Only in the case of infants can alcoholic medicines have significant effects.

Ethanol dissolves well in body liquids and most tissues, but not in bones and only very little in body fat. Thus, the ingested ethanol distributes itself on the average in only 70% of the human body. According to the build of a person, this value may of course be higher or lower, in extreme cases as high as 80% or as low as 55%. As a very rough rule of thumb, it can therefore be concluded that each alcoholic drink (8–10 g of pure ethanol) is capable of raising the blood alcohol level of a person weighing 60 to 70 kg (of which only 40 to 50 kg are accessible to ethanol) by about 0.2‰, respectively 0.2 g/kg.

2.3.2 Toxic Effects of Ethanol

2.3.2.1 Acute intoxications

In adults with a certain alcohol tolerance, ethanol blood concentrations up to 0.5‰ (g/kg) seldom have significant effects. Higher levels decrease the capacity for mental observation and concentration and can lead to an euphoric state with a strong urge to speak and act. They may affect the equilibrium, slow down the reactivity of the pupils and lead to nystagmus. Above 1.5‰, these effects become very pronounced. Vision and walking capability, as well as the power of judgement and reasoning, are disturbed. Over 2.5‰, finally, a person's memory is badly affected, as well as her ability to speak correctly and to walk upright. At blood alcohol levels of 3.5‰ and more, a person may lose all reflexes and fall into a state of narcosis. Her life is in danger. She may aspirate her vomit and die from suffocation, she may fall in a cold place and, in the absence of help, freeze to death. Or she may die from respiratory arrest.

Lethal ethanol blood levels for an adult vary between 3.5 and 5‰. They depend strongly on the general state of health and on the tolerance to ethanol that a person has developed. Some cases have been reported in which people with blood alcohol levels near 6‰ have survived. Children, of course, tolerate much less alcohol than adults. For them, concentrations of 2‰ can already be fatal.

2.3.2.2 Chronic intoxications

A habitual abuse of alcoholic drinks over a prolonged period of time will affect many body constituents; some of the alterations are irreversible. The following illnesses have been attributed to chronical ethanol abuse:

- Gastritis and pancreatitis with formation of carcinomas in the esophagus, stomach and pancreas.
- Liver cirrhosis, leading to hepatitis and liver carcinomas.
- Affection of blood vessels with disturbances of the clotting mechanism, leading to bleeding and to infarct.

- Inhibition of respiration, leading to apnea and hypoxia.
- Atrophy of muscle tissues.
- Disturbances of the central nervous system with tremor, epileptic fits, delirium tremens and encephalopathy.

2.3.2.3 Mixed intoxications of ethanol and drugs

It is well known that ethanol can increase and also qualitatively influence the toxicity of many pharmaceuticals and that the effects of the common action may not merely be additive, but even exponential. It is extremely dangerous to consume alcohol before, simultaneously or shortly after taking drugs such as analgesics, antipyretics, sedatives, hypnotics and narcotics. Combinations of ethyl alcohol with tranquillizers, muscle relaxants, local anesthetics, antihistamines and antidiabetics should be avoided. Every investigation for drugs and other poisons should also involve a determination of the body alcohol concentration. This value is needed for a correct interpretation of the analytical data from the drug analyses. Unfortunately, the effects of the interaction of ethanol with many pharmaceuticals are not very well known.

2.3.3 Biological Specimens for Ethanol Analyses

In a living person, the biological specimens collected for ethanol analysis will always be **body fluids**: whole blood, plasma, serum, urine and, very seldom, sputum. Every country which has established a control procedure for ethyl alcohol, be it for traffic or work place control, has defined the specimens which must be collected. Since the results must be expressed as the alcohol content of blood (in ‰ respectively g ethanol per kg blood), we consider that **blood** is the analytical specimen of choice. The only inconvenience is that coagulated blood must be homogenized before removing a sample for analysis. This problem, however, can be easily solved by using a high speed homogenizer with a narrow head, which can be introduced directly into the sample tube.

The best sampling procedure is to collect 8 to 10 ml of venous blood with a "vacutainer", an evacuated glass vial fitted with a steel needle. So-called "venules" or "Waterman" tubes with sterile needles and absolutely dry syringes can of course also be used. Skin disinfection requires solvent-free disinfectants. The personnel in charge of collecting the blood (MD's, nurses) must constantly be reminded to check their disinfectants. It is advisable that blood vials (vacutainers) and disinfectant be provided by the analytical laboratory. Immediately before or after the blood collection, the vials must be labeled with the name and birth date of the donor as well as the date and time of collection. If the laboratory provides the containers, an anticoagulant (i.e. fluoride) can be added to prevent coagulation and eliminate the need for homogenization.

In Germany and some other countries, only the liquid part of the blood, **plasma** or **serum**, is used for the analysis. The separation from the solid particles respectively the coagulate requires centrifugation. Since the alcohol concentration in plasma and serum is higher than in the erythrocytes or the coagulum, the result from an analysis of blood liquids must be corrected to obtain the concentration in full blood. This is done by dividing the analytical results by 1.20, which is of course only an approximation. With the analytical methods of the past, it may have been justified to analyze blood liquid and not full blood, and to correct the results. But with the present possibilities, such a sidetrack seems no longer defensible, but rather a source of error.

In some countries, i.e. Sweden, both blood and **urine** must be collected for ethanol analyses. The analytical determination for urine can be identical to blood alcohol analysis, except that a homogenization step is not needed. Attempts have been made to correlate the alcohol concentrations of the two fluids. But we would like to warn against trying to calculate blood concentrations from urine results, using such "correlations". To us, the profit from a urine analysis seems to be restricted to the possibility of finding ethanol in urine after it can no longer be detected in blood, since ethanol disappears later from urine than from blood. In addition, comparing the ethanol concentrations of blood and urine may in some cases help to shed some light on the time of alcohol consumption.

As an alternative to blood alcohol analysis, the ethanol concentration can also be measured in the **sputum**. It has been reported that a quite good correlation exists between sputum and blood ethanol levels. But it may not always be easy to collect sputum in a proper way from an accused driver. It should not be sampled right after the end of a drinking period and/or the mouth must be rinsed with water beforehand.

A better alternative seems to be the control of the alcohol content of **exhaled air**. Breath analysis yields a fairly good picture of the state of intoxication. The correlation with blood alcohol levels is well established (1–4). The distribution of alcohol between alveolar air and blood is approximately 1 to 2100 (or slightly higher). Even if the accuracy of breath alcohol measurements may not always be high, they are used in many countries as preliminary tests to decide whether a collection of blood for a legally relevant alcohol analysis is indicated. New Zealand has discarded blood alcohol control in favor of a specific analysis of breath alcohol levels (5).

Many methods for breath alcohol analysis are based on the reducing power of ethanol, which is controlled colorimetrically (i.e. by reduction of chromic acid or permanganate) or electrochemically (fuel cells). Semiconductor detectors, infrared absorption instrumentation measuring the radiation at 2950 cm^{-1} (3.39 μ), as well as simple gas chromatographs are also used for this purpose (5, 6). Since this chapter deals specifically with the analysis of blood and other body fluids, we must refer the reader interested in breath analysis to the extensive literature quoted in our references (5) and (6).

When death is preceded or caused by aspiration of vomit, the alcohol concentration of blood, especially heart blood, may be falsified. Alcohol-containing vomit in the lung may cause diffusion of ethanol into heart blood. Therefore, if aspiration is observed or suspected, a body tissue, preferably **brain** or, if no autopsy is carried out, **muscle tissue**, should also be secured, besides blood. These tissue samples must be placed immediately in not too large, air-tight glass containers to minimize evaporation losses. For the same reason, the samples should be as large as possible. It has been claimed that the alcohol concentrations in tissues are lower than in blood. On the basis of our own data (7), we suspect that evaporation losses in the autopsy room, during storage (too small samples, too large containers) and/or during preparation of the samples for analysis may be responsible for some of the differences quoted. Our recommendations for measuring the alcohol content of tissue samples are given in 2.3.7.

2.3.4 Review of Methods for Blood Alcohol Analysis

2.3.4.1 Interferometric analysis

This is one of the 2 methods which has been used in Switzerland and other places up to 1961, in some laboratories even up to 1968 (8). It illustrates the lack of specificity of toxicological ethanol determinations in the past:

> 5 g of blood is diluted with 50 ml of 0.02 m phosphoric acid and submitted to distillation from a flask with a short reflux condenser. 25 ml of distillate is collected and its refractive index measured interferometrically. The reading is converted directly into alcohol concentration with the help of a calibration curve.

The method is very sensitive, but lacks specificity. Neutral and acid volatiles passing into the distillate (alkaline substances are retained by the acid) can produce false positive readings. The method is definitely outdated.

2.3.4.2 Chemical oxidation

Chromic acid solution is used to oxidize ethanol to actaldehyde and (in part) acetic acid. The strong yellow oxidizing agent (Cr^{+6}) is reduced to green Cr^{+3} salt. The oxidation can be controlled colorimetrically (loss of yellow or appearance of green color) or titrimetrically. The alcohol content is given by the reduced chromic acid. This principle has been used in many variations. The best known is the procedure of Widmark (9):

> 1 g or 0.1 g blood is placed in one compartment and 1 ml of chromic acid solution into the second compartment of a diffusion cell at 60 °C. All the ethanol diffuses into the chromic acid and reduces part of the Cr^{+6} to Cr^{+3}. The remaining chromic acid is measured by titration or photometry.

In 1962, when the requests for blood alcohol determinations in our laboratory began to increase, we modified this oxidation method to permit its application to a large number of samples (10):

> Duplicate 5 ml aliquots of blood distillates (described in 2.3.4.1) are pipetted into colorimetric tubes (17 mm light path). Equal volumes of chromic acid solution (17 g $K_2Cr_2O_7$ in 400 ml H_2O added to 3600 ml concentrated sulfuric acid) are added by means of an automatic burette. The initial reaction temperature is 96 °C, the reaction time a few minutes. After cooling to room temperature, the green Cr-III-salt formed in the redox process is determined colorimetrically at 600 nm and recorded on a paper strip which indicates directly the alcohol concentration in ‰ (g/kg). As an alternative, the disappearance of the yellow color

can also be measured at 350 nm, but the signal to noise ratio is rather lower despite the higher specific absorption.

The determination of ethanol by chemical oxidation has been used for many years, in some countries until quite recently, regardless of the fact that its specificity is not very good. All volatile reducing substances can be oxidized by the acid chromate solution, yielding false positive results. As a matter of fact, the same principle is also used in food analysis to measure the strength of aroma distillates.

2.3.4.3 Enzymatic oxidation

In the presence of nicotinamide-adenine-dinucleotide (NAD), the enzyme alcohol dehydrogenase (ADH) oxidizes ethanol to acetaldehyde, while NAD is reduced to its dihydro-derivative NADH. NADH, but not NAD, possesses a large UV absorption band with maximum at 339 nm, which serves to measure the redox reaction and calculate the ethanol (11, 12). In place of the 339 nm absorption reading, which requires a spectrophotometer with monochromator, the absorption can also be controlled, with a little loss in sensitivity, by using a simple filter photometer at 365 nm. The method of Bonnichsen and Theorell (11), as modified by Bucher and Redetzki, is shortly described (12):

> 0.5 g blood are added to 4 ml perchloric acid and the mixture is cleared by centrifugation. 0.1 ml of the supernatant is added to 5 ml of a reagent mixture containing ADH and DPN in a buffer solution. The formation of DPNH is measured photometrically after an incubation period of 70 minutes at 25 °C.

A large number of different procedures and modifications for blood alcohol analyses by ADH have been published or described by individual authors and commercial firms. Since it would take too much space to discuss them, we refer the reader to a recent publication which also lists the pertinent literature (13).

Enzymatic blood alcohol determinations are usually carried out with plasma or serum, but can also be applied to whole blood. In our hands, the best results were obtained using blood distillates.

Commercially available ADH is not 100% specific for ethanol. This fact has often been overlooked. ADH can also oxidize (in decreasing order) propanol, butanol and isopropanol. Several higher alcohols can retard the reaction (14). Nevertheless, the method is still widely used today. In some countries, it is even one of the 2 official methods to be used in blood alcohol analysis.

2.3.4.4 Gas chromatography

Before the availability of flame ionization detectors, the toxicological determination of ethanol by GC with heat conductivity detection was rather complicated. Already in 1958, Cadman and Johns (15, 16) proposed a procedure, which we have adapted and used since 1962 as a (third) specific method for all cases which gave divergent results with the two unspecific routine techniques in use at the time (10):

To 1 g of blood, 1 g of K_2CO_3 and 1 ml of n-butyl acetate are added. The vials are shaken for 1 min, then centrifuged at 4000 rpm for 10 minutes. 40 μl of the organic extract is injected (duplicate determinations) into a column held at 100 °C (Beckman GC-2 with heat conductivity detector and He as carrier gas). Ethanol is separated from other volatiles such as acetone, methanol and isopropanol, which often cause elevated results by the chemical oxidation method.

With the appearance of the much more sensitive flame ionization detector which permits work with aqueous solutions, gas chromatographic ethanol analysis became much simpler. Methods were proposed which use direct injection of body fluids into the columns (17–19). Shortly afterwards, "head-space gas" procedures were described, in which the gas over a blood sample is collected and injected (20–23). We joined the new trend and replaced all our ethanol determinations with 2 different gas chromatographic procedures, one with direct injection of full blood, the other with the head-space sampling approach. In order to comply with the official regulations asking for 2 different methods, gas-solid chromatography was used for one and gas-liquid chromatography for the second approach (24–26).

These 2 methods have been in use side by side since 1965. Every few years, they have been adapted to the technical progress and modernized, especially with respect to data collection and processing, which was first done manually, then with computing integrators and off-line connection to the central university computer, and finally with a small dedicated laboratory computer (27–29). But the principle of the method could be retained for more than a quarter of a century. That speaks for the quality of the methodological combination.

A detailed description of the procedures, as well as the computer program in basic (which has not been published elsewhere), will be presented in the following paragraphs of this chapter.

2.3.4.5 Other methods

HPLC with electrochemical detection has been used for ethanol determination in serum. It is diluted with phosphate buffer pH 7.4 and the ethanol separated by HPLC on a silicic acid cation exchange column. The effluent passes into an enzyme reactor containing immobilized ADH. The H_2O_2 formed in the reaction of ethanol with ADH is determined in the through-flow mode by an electrochemical detector. Methanol and propanol can also be detected, but with low sensitivity. Since they are separated by HPLC, they cannot falsify the ethanol results (for literature see ref. 13).

An alcohol oxidase **enzyme sensor** has also been developed. It consists of an electrode covered with a ADH-containing gel. Serum is diluted with buffer solution and the sensor is introduced into the mixture. The electrode measures the O_2-consumption of the enzymatic oxidation, which takes 6 minutes. The advantage of the easy sample preparation is outbalanced by the lack of specificity inherent in the enzyme test and the aging of the electrode.

An **immunochemical test** for ethanol has also been developed (EMIT). It seems to yield reliable results only in the low alcohol concentration range (up to 1‰). In view of the instrumental possibilities available, we do not consider that the use of an immunochemical assay is justified.

The "**Lion Alcometer**" uses the head space sampling approach to analyze every kind of body fluid. About 0.5 ml are placed into a reaction cell. A part of the head space gas diffuses into an electrolytic cell. Ethanol is oxidized to acetic acid. The electric current is amplified and measured. The measurements show a good linearity, but ethanol levels below 0.1‰ are not detected and other reducing substances can falsify the results.

2.3.5 Criteria for Selection of Analytical Methods

The overwhelming part of the requests for toxicological alcohol analyses results from traffic control. High expectations are placed on the reliability of results. They must satisfy the obligation to help enforce traffic laws, but also protect the public from unjust accusations. The number of blood specimens submitted for analysis is usually high. Our laboratory at Zürich University, for example, had to analyze about 6000 blood samples a year, collected in half of Switzerland, most in connection with traffic delicts (26). Since they do not arrive evenly spread over the working days, the laboratory had to be able to handle at least 40 blood samples a day and secure every result by two different and independent methods.

When dealing with a large number of samples, the possibility of mix-ups always exists. The most important prerequisite for high-quality work is therefore the establishment of a system which excludes such mistakes with the highest possible security. Provisions should be made to trigger warning signals for possible errors during registration and analysis of specimens, processing of data and reporting of results. This is facilitated by carrying out the two requested methods completely independently. The dosage of the analytical samples, for example, must be done at different times, so that a mistake can not affect both determinations.

The second point of importance is the choice of methods. Priority must be given to the claim for specificity. Since specific methods are available, we can see no reason for retaining unspecific procedures, even if their precision should be superior. Bearing this in mind, we must select chromatographic separation techniques, and, since GC is best suited for the analysis of volatiles, this is the method of choice. Gas chromatographic techniques can be based on different separation principles such as gas-solid adsorption or gas-liquid partition. This permits the use of two different gas chromatographic methods side by side.

The analytical accuracy has to be constantly controlled in order to detect systematic deviations at an early stage. In this respect, the use of two independent methods is of much help. The two chosen methods should not only yield highly accurate results with the control samples, but the corresponding data from all analyses must not show a statistical deviation between the two methods. If one of them should constantly yield higher results than the other, a bug is in the system and must be removed.

Much attention is usually given to the problem of precision. In Germany and other countries, recently also in Switzerland, not only two different methods are requested, but also at least two analytical runs with each of these methods. This seems an unnecessary complication. With the present methods, a lack of precision is seldom a problem. The results from precision controls coincide usually much better than the results from accuracy tests. It would be much better to devote the time needed for running double analyses with each method to check and improve analytical accuracy.

A last point, finally, which speaks strongly for a simultaneous use of two independent and different gas chromatographic procedures, is the possibility to detect also additional volatile components. A knowledge of their presence can be important for the evaluation of the event. By including such an additional goal into the methodology, routine testing for blood alcohol is upgraded to a general solvent analysis. All specimens submitted for clinical or forensic investigations can be subjected to such an analytical procedure. Unexpected interesting results are often obtained by such solvent screening with a minimum of effort and time.

In the light of these observations, it seems absolutely justified to select GC as technique for both requested methods. It could be objected that the directive to use two different procedures would be better served by using combinations of GC with HPLC or capillary electrophoresis. This is correct. But considerable additional expenses for instrumentation and chemicals would be involved, as well as more complications in the technical work and data acquisition. No better accuracy or precision would result. And is not one important principle for choosing a method also the aim to produce the best results with the least effort and costs as possible?

2.3.6 Detailed Procedures for Blood Alcohol Analysis by two Different Gas Chromatographic Methods

In this place, we will describe the procedure used at Zürich University for over twenty years. It consists of two different and independent gas chromatographic methods working side by side. One uses a column with a solid adsorbent, onto which an aliquot of the biological liquid is directly injected, the second a partition column with a head space gas injection technique. For quantification, both methods rely on **internal standarization**. For adsorption chromatography, acetonitrile has been chosen as internal standard, for partition chromatography 1,4-dioxane. Both solvents are nonbiological and toxic. Nobody would dare to ingest them prior to supplying blood, in order to falsify the analyses (an increase of the standard peaks lowers the results for the alcohol content).

Simple single column gas chromatographs with flame ionization detectors are used in both methods. They are identical except for the **column fillings**. Columns of 160 cm length and 3 mm ∅ are used. They contain:

- a cross-linked polystyrene resin (Porapak Q or Chromosorb 102) for adsorption chromatography,
- liquid Carbowax 20 M on Chromosorb (300 g/kg) for partition chromatography.

Column temperatures are about 110 °C, detector temperature 150 °C. Nitrogen (40 ml/min) is used as carrier gas.

The analytical specimen is homogenized (if needed) with a high frequency rod (Polytron or Ultraturrax, 0.8 mm ∅ head) which reaches into the collection tube. Transferals must be avoided. Homogeneous mixtures are obtained in 2 or 3 sec. The sampling for the first analysis is done immediately after homogenization, the sampling for the second analysis only later, completely independent from the first operation.

For **adsorption chromatography**, 1.00 g of the analytical specimens are weighed into 6 ml glass vials with PVC-stoppers and diluted with constant volumes (ca. 1 ml) of an aqueous solution of acetonitrile (1 g/l). Aliquots of the mixture (0.5 to 1 µl) are injected into the column which contains, in the injection region, a metal wire or a glass thread insert for collecting the solid particles and non-volatiles. It is replaced after every 10 blood injections. Eluting times are ca 200 sec for ethanol and 270 sec for acetonitrile. To detect a possible presence of solvents with longer retention times, the chromatographic run is extended to or over 600 sec.

For **partition chromatography**, 1 g of the analytical specimens are weighed into 20 ml serum flasks and diluted with constant volumes of an aqueous solution of 1,4-dioxane (4 g/l). The serum flasks are capped with silicone or rubber stoppers, then equilibrated in a 40 °C water bath. 0.5 to 1.0 ml aliquots of the gas phases in the serum flasks are injected into the GC. Reasonable elution times are 120 sec for ethanol and 270 sec for 1,4-dioxane. To detect a possible presence of additional volatiles, the chromatographic run is extended to 400 or more sec.

In order to obtain optimum accuracy in the legally most important concentration range of 1 ‰, the dosages of the internal standard solutions, made with automatic dispensers, are adjusted to yield peaks of comparable size to those of a 1 ‰ ethanol solution. The peaks for ethanol and the internal standard are measured and divided by computing integrators. These ratios, multiplied by correction factors, give the ethanol concentration in g/kg. The factors are determined by injecting at the start and the end of the daily runs, as well as repeatedly inbetween, aqueous solutions containing 1 ‰ ethanol (from sealed ampules, freshly opened). The computer program presented at the end of this chapter takes care of all calculations and displays the results in a tabular form. The data obtained by both individual methods are compared, and a possible systematic deviation is signalized.

Independently from the main analytical investigation, aqueous ethanol solutions containing 0.5, 1.0, 1.5, 2.0 and 2.5 g/l are also injected to control the linearity of both methods.

2.3.7 Analysis for Ethanol in Tissue Samples

Tissue samples can also be analyzed by the two gas chromatographic methods, if ethanol (and other solvents) are first isolated by distillation. 25 g of tissue is added to 100 ml of 0.02 M phosphate buffer and homogenized. The first 25 ml of distillate are collected and submitted to the two gas chromatographic analyses. For less accurate results, homogenized tissue samples can be analyzed directly (without distillation) by the head space injection technique, but not by the direct injection method. The problem with this simple approach is the evaporation losses involved in handling such small tissue samples.

2.3.8 Identification of Additional Volatiles

Figure 1 (25) gives the relative retention times of about 60 volatile solvents on the two previously described columns (Porapak Q for adsorption, Carbowax 20 M for partition chromatography). The retention values (RRT = 1.00) of the two internal standards used in ethanol analysis (acetonitrile for adsorption and 1,4-dioxane for partition chromatography) determine the time axis. It is evident that by using two different columns, many more components can be identified than with just a single column. Due to the differences of the separation principles, some volatiles elute late on the adsorption column (i.e. no. 33) but fast on the partition column and vice versa.

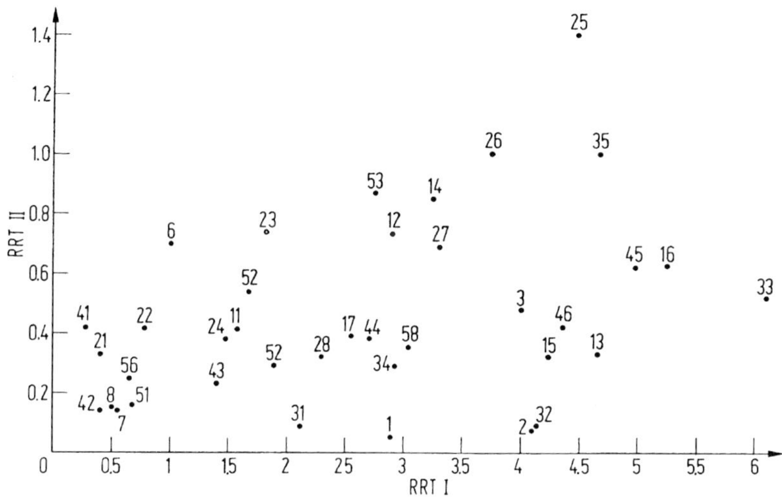

Figure 1. Relative retention times of some important solvents on 2 different GC-columns.

RRT I = relative retention to acetonitrile on Poropak Q
RRT II = relative retention to dioxane on Carbowax 20 M

1. $CH_3-(CH_2)_3-CH_3$
2. $CH_3-(CH_2)_4-CH_3$
3. C_6H_6

6. CH_3-CN
7. HCN
8. H_2S

11. CH_2Cl_2
12. $CHCl_3$
13. CCl_4
14. CH_2Cl-CH_2Cl
15. CCl_3-CH_3
16. $CCl_2=CHCl$
17. $CF_3-CHClBr$

21. CH_3-OH
22. CH_3-CH_2-OH
23. $CH_3-(CH_2)_2-OH$

24. $(CH_3)_2CH-OH$
25. $CH_3-(CH_2)_3-OH$
26. $(CH_3)_2CH-CH_2-OH$
27. $CH_3-CH_2-CH(CH_3)-OH$
28. $(CH_3)_3C-OH$

31. $CH_3-CH_2-O-CH_2-CH_3$
32. $(CH_3)_2CH-O-CH(CH_3)_2$
33. $CH_3-(CH_2)_3-O-(CH_2)_3-CH_3$
34. CH_2-CH_2 / O / CH_2-CH_2 (ring)

35. CH_2-CH_2 / O O / CH_2-CH_2 (ring)

41. HCHO
42. CH_3-CHO
43. $CH_3-CO-CH_3$
44. $CH_3-CH_2-CO-CH_3$
45. $CH_3-(CH_2)_2-CO-CH_3$
46. $(CH_3)_2CH-CO-CH_3$

51. HCOOH
52. CH_3-COOH
53. CH_3-CH_2-COOH

56. $HCOO-CH_3$
57. $CH_3-COO-CH_3$
58. $CH_3-COO-CH_2-CH_3$

Our computer program is capable of displaying, along with the alcohol content of the samples, the relative retention times of all additional peaks. The most common solvents are identified on the basis of their two relative retention times and communicated. The presence of volatiles not recognized by our computer program is signalized by stating their retention times. They are subsequently identified by combined GC-MS. For this purpose, we use different techniques (29):

1. Conventional EI-MS, the hard ionization mode which produces many fragments. It has the advantage that the spectra obtained can be identified with the help of mass spectral data bases.
2. Positive CI-MS. It will often furnish the molecular mass of an eluting compound, if the EI spectrum is not conclusive.
3. CI-MS in the negative ion mode, which we carry out at a low ion source pressure (around 10^{-2} torr) (30–32). It is an excellent method for detecting volatiles with high electron affinity, such as halogenated hydrocarbons or volatile anesthetics (see chapter 3.10 for examples).
4. Our so-called "Dual-MS" system, which is capable of recording quasi-simultaneously the positive EI- and the negative CI-spectra of an eluting compound (33–35). This combination can most likely identify every eluting compound (see chapter 3.10 for examples).

Communicating the presence of additional volatiles together with the results of an ethanol analysis can be of substantial value. It may yield information about a possible illness of the blood donor or be used as evidence for judicially important deductions. A few examples will serve as illustrations:

- The presence of considerable concentrations of acetone is an indication for a metabolic disorder. We could find quite high acetone concentrations in the blood of skiers involved in an accident, who had to wait in the cold until they could be transported (36).
- The presence or absence of methanol can help to verify the claims of the blood donor concerning type and quality of the ingested alcoholic beverages. Often, an accused driver pretends that his blood alcohol level is entirely or in part due to a specific strong liquor he claims to have drunk between accident and blood collection. Since many liquors (i.e. gin, or the Swiss "Kirsch" and "Williams") contain substantial concentrations of methanol which are detectable in the blood, such claims can be checked and often refuted.
- The presence of isopropanol (together with its oxidation product acetone) points to an abuse of a technical alcohol, i.e. a solvent used in cosmetics, such as a hair tonic.
- The presence of aliphatic ketones other than acetone is an indication that denatured alcohol has been ingested.
- The presence of several higher alcohols may indicate that the analytical specimen has undergone putrefaction and that the ethanol found is not necessarily all of exogenous origin.
- The presence of volatile anesthetics used in hospitals implies that the blood sample was taken after the start of a narcosis. In such cases it must be checked whether the patient had an infusion before the blood was collected, since this would have lowered his blood alcohol concentration.

- Additional volatiles found on the occasion of a simple request for alcohol analysis have helped us on several occasions to solve the problem of an unexpected death, whether of accidental nature or homocide or even murder. In one such case, a 16 year old girl was found dead in her bed. Neither the police investigation nor the autopsy revealed facts which could shed light on the case. No ethanol was detected in a blood sample submitted as a routine test, but another volatile was found and identified as ethyl chloride by EI-MS (see figure 1, chapter 3.10). Subsequent inquiries led to the arrest of the girl's stepfather. He later confessed to having used on several occasions this narcotic solvent, procured from his workshop, to depress his stepdaughter's feelings while abusing her.

2.3.9 Computer Program for the Collection and Processing of Data and Presentation of Results of Forensic Alcohol Analysis by two Different Gas Chromatographic Methods

The program was first written in PL-1 and later translated into Basic. This version is presented in the following.

2.3.9.1 Program description

This program evaluates the results of two different GC methods of blood alcohol analysis (channel A and B). The ethanol fraction of a sample is determined by the internal standard method. Solutions with a known amount of alcohol and standard (tests) are prepared to calibrate the method. The ethanol fraction of a sample is calculated by comparing the peak area of the alcohol peak with the peak area of the standard which has been added to each sample. The tests and samples are stored in arbitrary order on tape. The data of a single run consist of a header and the peak data as listed in table 1. Tests have run number smaller than 100, actual samples larger than 100. There is a maximum of 10 tests and of 35 samples per method. Each sample may contain up to 25 peaks. For each peak, retention time and peak area are printed. One of the peaks corresponds to the standard which is used to calibrate the area of the alcohol peak.

Calibration:

1. Determination of the alcohol and standard peaks in the test data. Of the two peaks with the largest area, the earlier is the alcohol peak, the later is the standard peak.
2. Calculation of the calibration factor as the ratio of the area of the alcohol peak divided by the area of the standard peak.
3. Evaluation of the mean values of the retention times of alcohol and standard peaks, and of the calibration factors. The mean values are determined using

Table 1. Data format on tape (depends on chromatographic data system used)

Header of a chromatogram

name	**variable**	**number of INPUT commands used to read the data from tape**	**type**	**description**
channel number	H1(1)	1	a number between 1 and 4	channel A-D
system	H2(1)-H2(3)	3	characters	not used
user number	H3(1)-H3(25)	25	characters	the first 4 characters are used to store the run number H.
option	H4(1)-H4(2)	2	characters	not used
flag	H5(1)	1	character	not used
sample number	H6(1)-H6(5)	5	characters	not used
sample name	H7(1)-H7(10)	10	characters	not used
number of peaks	H8	1	number	number of peaks in the chromatogram

Peak data (repeated for each peak in a chromatogram)

start of the peak	P(peak number,1)	1	number	not used
end of the peak	(peak number,2)	1	number	not used
retention time	P(peak number,3)	1	number	retention time of a peak
peak area	P(peak number,4)	1	number	peak area
peak height	P(peak number,5)	1	number	not used

only the tests, whose calibration factors do not deviate too much from the mean calibration factor. The mean relative retention time is the mean alcohol retention time divided by the mean standard retention time.

If there are less than three valid tests, the samples are not analyzed. There are various error messages:
"less than 2 peaks": the test contains less than two peaks.
"multiple peaks": it was not possible to identify the alcohol or standard peak.
"rejected": the deviation of the calibration factor from the mean is too large.

Evaluation of the samples:

1. Determination of the alcohol and standard peaks of the samples. The alcohol and standard peaks are identified as the largest peaks in the respective allowed retention time intervals. The intervals are given by the mean values determined

during the calibration and by the allowed deviation (a parameter which can be modified).

2. Calculation of the ethanol fraction of the sample. The fraction (in ‰) is given by the area of the alcohol peak devided by the area of the standard peak multiplied by the calibration factor. If there is no peak in the retention time window for alcohol, the result is given as 0‰.
3. The samples are sorted by channel in sequence of their run number according to the linked list method (37).
4. Determination of the average alcohol fraction of both channels. The allowed deviation of the fraction from the mean is ±X3 (or ±X4) for a fraction smaller than 1‰, and ± fraction × X3 (or X4) for a fraction larger than 1‰. X3 is chosen according to the external limits required by Swiss regulations, X4 according to the more stringent internal limits, selfimposed by our laboratory.
5. Determination of the sum and mean value of the average ethanol fraction of all samples of a channel.

Again, there are various error messages:

"no data": no data on this channel for the given sample number.

"multiple peaks aw": the alcohol peak cannot be identified (aw means alcohol window).

"multiple peaks sw": the standard peak cannot be identified (sw means standard window).

"no standard": no peak in the retention time window corresponding to the standard.

"rejected": the deviation of the alcohol fraction from the mean (of all the data for the given sample) is too large (see pt. 4 above).

Additional peaks:

A maximum of four additional peaks outside of the retention time intervals of alcohol and standard can be identified according to the relative retention times (and the allowed windows for these times).

2.3.9.2 Structure of the program

This chart structured according to (38) is given in Figure 2 (see p. 156).

2.3.9.3 Listing of Variables

A: average ‰ value of channels A and B.

A(0,0), A(0,5), A(0,6): number of analyses, number of analyses without errors, number of analyses not rejected.

A(0,1), A(0,3): sum of the ‰ values of all samples (channel A and B).

A(0,2), A(0,4): sum of the ‰ values of all samples (channel A and B) which were not rejected.

A(1–35,0): sample number

A(1–35,1), (A(1–35,2): ‰ values (channel A and B).

A(1–35,3), A(1–35,4): error code (channel A and B).

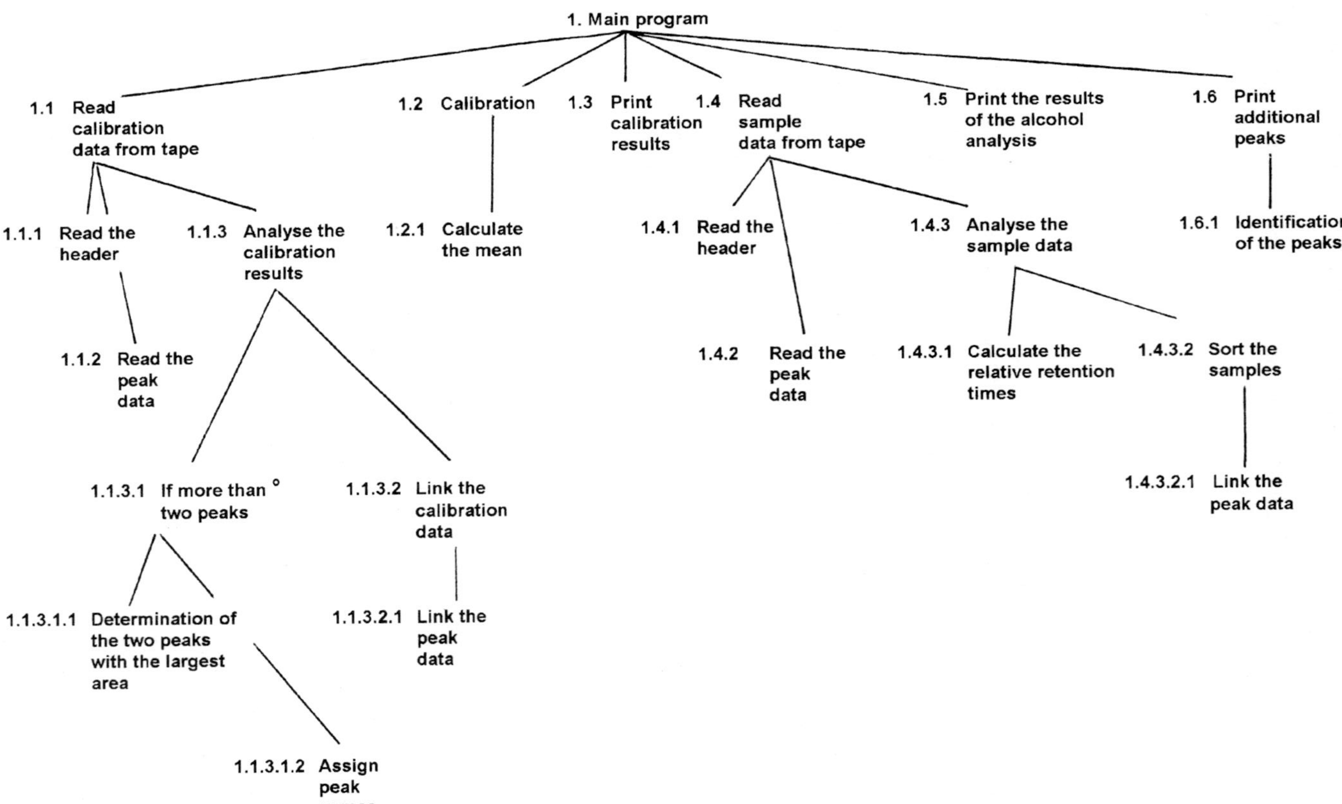

Figure 2. Structure of the Computer Program for Blood Alcohol Analysis with Identification of Additional Volatiles. The Program Is Executed in the Order of the Numbers Stated. The Names of the Subroutines Correspond to the Titles in the Progam Listing.

A(1–35,5)–A(1–35,8), A(1–35,9)–A(1–35,12): relative retention times of additional peaks (channel A and B).
B(1), B1–9(1), D(1), D1–9(1): relative retention times of known additional compounds (channel A and B).
B(2), B1–9(2), D(2), D1–9(2): allowed deviation of the above retention times (channel A and B).
C1, C2: error code for tests and samples respectively.
C$, F$, G$: text of the error message.
E1(0,0), E1(0,1), E2(0,0), E2(0,1): number of tests (tests without errors respectively) for channel A and B.
E1(0,2), E2(0,2): average calibration factor (channel A and B).
E1(0,3), E1(0,4), E2(0,3), E2(0,4): average retention times of alcohol and standard (channel A and B).
E1(0,5), E2(0,5): average relative retention of the alcohol (channel A and B).
E1(1–10,0), E2(1–10,0): test number (channel A and B).
E1(1–10,1), E2(1–10,1): calibration factor (channel A and B).
E1(1–10,2), E1(1–10,3), E2(1–10,2), E2(1–10,3): retention time of alcohol and standard (channel A and B).
E1(1–10,4), E2(1–10,4): error codes (channel A and B)
F: flag.
H: run number.
H1(1): channel (header).
H2(1)–H2(3): system (header).
H3(1)–H3(25): user number (header).
H4(1)–H4(2): option (header).
H5(1): flag (header).
H6(1)–H6(5): sample number (header).
H7(1)–H7(10): sample name (header).
H8: number of peaks (header).
I, J, K: counter.
M1, M2, M3: message codes for tests and samples respectively.
N, N1: lowest sample number in the linked list (37).
N1$, N2$: names of additional compounds (channel A and B).
O: upper bound on the search interval (in the linked list (37)).
P(0,1)–P(0,4): relative retention times of the additional peaks.
P(1–25,1): start of the peak.
P(1–25,2): end of the peak.
P(1–25,3): retention time of the peak.
P(1–25,4): peak area.
P(1–25,5): peak height.
P1, P2: peak number for alkohol and standard.
P3, P4: number of peaks in the alkohol and standard retention time intervals.
P(1–35): reverse pointer in the linked list (37).
T, T1, T2: auxiliary variables.
T(1): start of the peak.
T(2): end of the peak.
T(3): retention time of the peak.

T(4): peak area.
T(5): peak height.
U: lower bound on the search interval (in the linked list (37)).
V(1–35): forward pointer in the linked list (37).
X, X5, X6: auxiliary variables for the tolerance range
X1, X2: allowed deviation of the calibriation factor from the average (channel A and B).
X3, X4: allowed deviation of the ‰ value from the average (external and internal range).
X(1,1), X(1,2), X(2,1), X(2,2): allowed deviations of the retention times of alcohol and standard from the avarage (channel A and B).
Z: number of valid peaks.
Z4: number of additional peaks.

2.3.9.4 Program listing

```
10  REM Set initial parameter values
12  DIM T(5)
14  DIM H1(1),H2(3),H3(25),H4(2),H5(1),H6(5),H7(10)
16  DIM P(25,5)
18  DIM E1(10,5),E2(10,5)
20  DIM V(35),R(35)
22  DIM X(2,2)
24  DIM A(35,14)
25  DIM B(2),B1(2),B2(2),B3(2),B4(2),B5(2),B6(2),B7(2),B8(2),B9(2)
26  DIM D(2),D1(2),D2(2),D3(2),D4(2),D5(2),D6(2),D7(2),D8(2),D9(2)
30  M1=0,M2=0,M3=0
32  U=0,O=0,N=0
34  H8=0
36  REM Allowed deviation of retention time, calibration factor and ethanol con-
    tent from mean
40  X1=4,X2=4,X(1,1)=15,X(1,2)=20,X(2,1)=10;X(2,2)=15,X3=0.05,X4=
    0.03
100 REM Allowed deviation of rel. retention time of additional compounds (chan-
    nel A)
109 B(1)=1.295,B(2)=0.025,B1(1)=0.28,B1(2)=0.02,B2(1)=0.38,B2(2)=0.015
110 B3(1)=0,B3(2)=0,B4(1)=0,B4(2)=0,B5(1)=0,B5(2)=0,B6(1)=0,B6(2)=0
111 B7(1)=0,B7(2)=0,B8(1)=0,B8(2)=0,B9(1)=0,B9(2)=0
150 REM Allowed deviation of rel. retention time of additional compounds (chan-
    nel B)
159 D(1)=0.32,D(2)=0.01,D1(1)=0.35,D1(2)=0.01,D2(1)=0.39,D2(2)=0.01
160 D3(1)=0.73,D3(2)=0.01,D4(1)=1.15,D4(2)=0.02,D5(1)=0,D5(2)=0,
    D6(1)=0,D6(2)=0
161 D7(1)=0,D7(2)=0,D8(1)=0,D8(2)=0,D9(1)=0,D9(2)=0
```

```
200 REM Main program
210 OPEN IN#,UTIL,TAPE,RAN,"GLC RESULTS"
220 FOR I=1 TO 120          read tape until 120 runs or non valid channel number
230   INPUT# UTIL, H1(1)
240   IF H1(1)<1 GOTO 280
250   IF H1(1)>4 GOTO 280
260   GOSUB 500                                       read calibration data
270 NEXT I
280 CLOSE# UTIL
290 GOSUB 3100                                                    calibration
300 GOSUB 4200                                       print calibration results
310 OPEN IN#, UTIL,TAPE,RAN"GLC RESULTS"
320 FOR I=1 TO 120                                             read tape again
330   INPUT# UTIL,H1(1)
340   IF H1(1)<1 GOTO 380
350   IF H1(1)>4 GOTO 380
360   GOSUB 5500                                 read and analyse sample data
370 NEXT I
380 CLOSE# UTIL
390 GOSUB 9000                                           print sample results
395 GOSUB 9500                                    print additional compounds
400 STOP

500 REM Read calibration data from tape
510 GOSUB 600                                                     read header
520 GOSUB 800                                                  read peak data
530 IF H>=100 THEN RETURN                          number < 100 = calibration
535 IF H=0 THEN RETURN
540 GOSUB 1000                                      analyse calibration data
550 RETURN

600 REM Read the header
610 H=0
620 INPUT# UTIL, H2(1),H2(2),H2(3)
630 FOR J=1 TO 25
630   INPUT# UTIL, H3(J)                                   read number string
650 NEXT J
652 FOR J=1 TO 4
654   IF H3(J)<48 GOTO 660                             characters numerical?
655   IF H3(J)>57 GOTO 660
656 NEXT J
658 H=(H3(1)-48)*1000+(H3(2)-48)*100+(H3(3)-48)*10+(H3(4)-48)
                                                   convert string to number
660 INPUT# UTIL, H4(1),H4(2)
670 INPUT# UTIL, H5(1)
680 INPUT# UTIL, H6(1),H6(2),H6(3),H6(4),H6(5)
690 FOR J=1 TO 10
```

```
700      INPUT# UTIL, H7(J)
710   NEXT J
730   INPUT# UTIL, H8                                 number of peaks
740   RETURN

800   REM Read the peak data
810   Z=0
820   FOR J=1 TO H8
830      FOR K=1 TO 5
832         INPUT# UTIL, T(K)                         read peak data
834      NEXT K
837      IF T(3)=0 GOTO 860     reasonable retention time and peak area?
840      IF T(4)=0 GOTO 860
846      IF Z>24 GOTO 860
847      Z=Z+1                                        count peaks
849      FOR K=1 TO 5                    store max. 25 peaks/run
856         P(Z,K)=T(K)
857      NEXT K
860   NEXT J
870   RETURN

1000  REM Analyze the calibration results
1005  C1=0
1040  IF Z> =2 GOTO 1080
1050  C1=11                              error: less than two peaks
1060  GOTO 1150
1080  IF Z>2 GOTO 1130                                2 peaks
1090  P1=1                                            1.peak ethanol
1100  P2=2                                            2. peak standard
1110  GOTO 1150
1130  GOSUB 1400          more than 2 peaks: determine 2 largest peaks
1140  GOSUB 1800                                      assign peak name
1150  IF H1(1)=1 THEN GOSUB 1900    store calibration data (channel A)
1160  IF H1(1)=2 THEN GOSUB 2300    store calibration data (channel B)
1170  RETURN

1400  REM Determination of the two peaks with the largest area
1410  T1=P(1,4)                                       search largest peak
1420  P1=1
1430  FOR J=2 TO Z
1440     IF P(J,4)< =T1 THEN 1470
1450     T1=P(J,4)
1460     P1=J
1470  NEXT J
1480  T1=0                                 search peak with same area
1490  FOR J=1 TO Z
1500     IF J=P1 THEN 1540
```

```
1510     IF P(J,4)< >P(P1,4) THEN 1540
1520     T1=T1+1                              count peaks with same area
1530     P2=J
1540   NEXT J
1550   IF T1=1 THEN RETURN
                         one peak with same area: = second largest peak
1560   IF T1=0 THEN 1590
1570   C1=12                   error: more than one peak with same area
1580   RETURN
1590   T1=0                                   search second largest peak
1600   FOR J=1 TO Z
1610     IF J=P1 THEN 1650
1630     IF P(J,4)< =T1 THEN 1650
1640     T1=P(J,4)
1640     P2=J
1650   NEXT J
1660   T1=0
1670   FOR J=1 TO Z
1680     IF J=P1 THEN 1710
1690     IF J=P2 THEN 1710
1700     IF P(J,4)=P(P2,4) THEN T1=T1+1        count peak with same area
1710   NEXT J
1720   IF T1>0 THEN 1740
1730   RETURN
1740   C1=13                   error: one or more peaks withs same area
1750   RETURN

1800   REM Assign peak names
1810   IF P1<P2 THEN RETURN
1820   T1=P1                                            1. peak=ethanol
1830   P1=P2                                          2. peak = standard
1840   P2=T1
1850   RETURN

1900   REM Link the calibration data (channel A)
2000   IF E1(0,0)> =10 GOTO 2180                        store max. 10 tests
2010   FOR J=1 TO E1(0,0)
2020     IF E1(J,0)=H GOTO 2120
2030   NEXT J
2040   E1(0,0)=E1(0,0)+1              new test: increase number of tests
2050   J=E1(0,0)
2060   IF C1< >0 GOTO 2090
2070   GOSUB 2700                                no error: store peak data
2075   E1(0,1)=E1(0,1)+1                     increase number of valid tests
2080   RETURN
2090   E1(J,0)=H                         error: store number and error code
2100   E1(J,4)=C1
```

```
2110 RETURN
2120 IF C1<>0 GOTO 2145                        test repeated, no error:
2125 IF E1(J,4)<>0 THEN E1(0,1)=E1(0,1)+1
                           if old test non valid increase number of tests
2130 GOSUB 2700                                           store peak data
2140 RETURN
2145 IF E1(J,4)=0 THEN E1(0,1)=E(0,1)-1
                           error: if old test valid decrease number of tests
2150 E1(J,0)=H                               store number and error code
2160 E1(J,4)=C1
2170 RETURN
2180 M1=2                                               more than 10 tests
2190 RETURN

2300 REM Link the calibration data (channel B)
     ... like channel A lines 1900-2190
     E2 instead of E1
     GOSUB 2900 instead of GOSUB 2700
     M2 instead of M1
     adjust line numbers of GOTO commands
2590 RETURN

2700 REM Link the peak data (channel A)
2720 E1(J,0)=H                                                store number
2730 E1(J,1)=P(P1,4)/P(P2,4)       calculate and store calibration factor
2740 E1(J,2)=P(P1,3)                         store retention time ethanol
2750 E1(J,3)=P(P2,3)                        store retention time standard
2760 E1(J,4)=0                                                    no error
2770 RETURN

2900 REM Link the peak data (channel B)
     ... like channel A lines 2700-2770
     E2 instead of E1
2970 RETURN

3100 REM Calibration
3120 F=0,T=0,T1=0,T2=0,X5=0                                      channel A
3125 IF E1(0,1)<3 GOTO 3250
3130 GOSUB 3600                        more than 3 tests: calculate mean
3135 X5=(E1(0,2)/100)*X1     calculate possible deviation from mean factor
3140 E1(0,1)=0
3150 FOR I=1 TO E1(0,0)
3160   IF E1(I,4)<>0 GOTO 3220
3165   T1=ABS(E1(0,2)-E1(I,1))
3170   IF X5>T1 GOTO 3210
3172   F=1                              deviation from mean factor too large
3174   IF T1<=T GOTO 3210
```

```
3176      IF T2=0 GOTO 3185
3178      E1(T2,4)=0              delete error code of tests with smaller deviation
3180      E1(0,1)=E1(0,1)+1
3185      T2=I
3190      E1(I,4)=14                          mark test with largest deviation
3195      T=T1
3200      GOTO 3220
3210      E1(0,1)=E1(0,1)+1
3220   NEXT I
3230   IF E1(0,1)> =3 GOTO 3270
3250   M1=1                                            error: less than 3 test
3260   GOTO 3280
3270   IF F=1 GOTO 3120   recalculate mean without test with largest deviation
3280   F=0,T=0,T1=0,T2=0,X6=0                                        channel B
       ... like channel A lines 3120-3270
       X6 instead of X5
       E2 instead of E1
       X2 instead of X1
       GOSUB 3900 instead of GOSUB 3600
       M2 instead of M1
       adjust line numbers of GOTO commands
       RETURN instead of GOTO 3280 line 3260
3440   RETURN

3600   REM Calculate the mean (channel A)
3620   E1(0,2)=0
3630   E1(0,3)=0
3640   E1(0,4)=0
3660   FOR I=1 TO E1(0,0)
3670      IF E1(I,4)< >0 GOTO 3720
3690      E1(0,2)=E1(0,2)+E1(I,1)
3700      E1(0,3)=E1(0,3)+E1(I,2)
3710      E1(0,4)=E1(0,4)+E1(I,3)
3720   NEXT I
3730   E1(0,2)=E1(0,2)/E1(0,1)                                 mean factor
3740   E1(0,3)=E1(0,3)/E1(0,1)                 mean retention time ethanol
3750   E1(0,4)=E1(0,4)/E1(0,1)                mean retention time standard
3760   E1(0,5)=E1(0,3)/E1(0,4)       mean relative retention time ethanol
3770   RETURN

3900   REM Calculate the mean (channel B)
       ... like channel A lines 3600-3770
       E2 instead of E1
       adjust line numbers of GOTO commands
4070   RETURN
```

```
4200 REM Print calibration results
4220 PRINT " "
4221 PRINT " "
4222 PRINT " "
4223 PRINT " "
4224 PRINT " "
4225 PRINT "RESULTS TESTS:"
4226 PRINT "--------------"
4227 PRINT " "
4270 PRINT "RUNS CHANNEL A: ";E1(0,0),"REJECTED: ";E1(0,0)-E1(0,1)
4280 PRINT "RUNS CHANNEL B: ";E2(0,0),"REJECTED: ";E2(0,0)-E2(0,1)
4290 PRINT " "
4300 IF E1(0,0)<>0 THEN PRINT "CHANNEL A:";TAB(29);"TEST";
     TAB(49);"RT ALC";TAB(67);"RT STAND";TAB(89);"REL RT";
     TAB(109);"FACTOR"
4305 IF E1(0,0)<>0 THEN PRINT "-------------"
4310 FOR I=1 TO E1(0,0)
4320   IF E1(I,4)=0 GOTO 4390
4330   IF E1(I,4)=11 THEN PRINT TAB(28);E1(I,0)[4.0];TAB(40);"LESS
       THAN 2 PEAKS"
4340   IF E1(I,4)=12 THEN PRINT TAB(28);E1(I,0)[4.0];TAB(40);"MUL-
       TIPLE PEAKS"
4350   IF E1(I,4)=13 THEN PRINT TAB(28);E1(I,0)[4.0];TAB(40);"MUL-
       TIPLE PEAKS"
4360   IF E1(I,4)=14 THEN PRINT TAB(31);E1(I,0);TAB(37);"REJECTED";
       TAB(47);E1(I,2)[4]; TAB(67); E1(I,3)[4];TAB(106);E1(I,1)[4.3];
4380   GOTO 4400
4390   PRINT TAB(28);E1(I,0)[4.0];TAB(47);E1(I,2)[4.0];TAB(67);E1(I,3)[4.0];
       TAB(106);E1(I,1)[4.3]
4400 NEXT I
4415 IF M1=1 GOTO 4440
4420 PRINT TAB(29);"------------------------------------------------"
4425 PRINT " "
4430 PRINT TAB(23);"MEAN";TAB(47);E1(0,3)[4.0];TAB(67);E1(0,4)[4.0];
     TAB(87);E1(0,5)[4.2];TAB(106);E1(0,2)[4.3]
4440 PRINT " "
4442 PRINT " "
4444 PRINT " "
4450 IF E2(0,0)<>0 THEN PRINT "CHANNEL B:";TAB(29);"TEST";
     TAB(49);"RT ALC";TAB(67);"RT STAND";TAB(89);"REL RT";
     TAB(109);"FACTOR"
```

… *like channel A lines 4300–4430*
E2 *instead of* E1
M2 *instead of* M1
GOTO 400 *instead of* GOTO 4440 *in line number 4415*
adjust line numbers of GOTO-*commands*

```
4585 IF M1=1 GOTO 400
```

```
4590 RETURN

5500 REM Read sample data from tape
5510 GOSUB 600                                          read header
5520 GOSUB 800                                          read peak data
5530 IF H<=100 THEN RETURN                              number >100=sample
5550 GOSUB 5700                                         analyse samples
5570 RETURN

5700 REM Analyse the sample data
5710 C2=0,P1=0,P2=0,P3=0,P4=0
5720 IF H1(1)=1 GOTO 5750
5730 IF H1(1)=2 GOTO 6110
5740 RETURN
5750 FOR J=1 TO Z                                       channel A
5760   IF P(J,3)<(E1(0,3)-X(1,1)) GOTO 5930
                                     check: peak within ethanol window?
5770   IF P(J,3)>(E1(0,3)+X(1,1)) GOTO 5850
5780   P3=P3+1
5790   IF P3>1 GOTO 5820
5800   P1=J                                             1 peak: peak=ethanol
5810   GOTO 5930
5820   IF P(J,4)<=P(P1,4) GOTO 5930
                                     more than 1 peak: largest peak=ethanol
5830   P1=J
5840   GOTO 5930
5850   IF P(J,3)<(E1(0,4)-X(1,2)) GOTO 5930
                                     check: peak within standard window?
5860   IF P(J,3)>(E1(0,4)+X(1,2)) GOTO 5940
5870   P4=P4+1
5880   IF P4>1 GOTO 5910
5890   P2=J                                             1 peak: peak=standard
5900   GOTO 5930
5910   IF P(J,4)<=P(P2,4) GOTO 5930
                                     more than 1 peak: largest peak=standard
5920   P2=J
5930 NEXT J
5932 IF P3>1 GOTO 5940
5934 IF P4>1 GOTO 5940
5936 GOTO 6070
5940 FOR J=1 TO Z        more than 1 peak within ethanol or standard window
5950   IF P(J,3)<(E1(0,3)-X(1,1)) GOTO 6060
5960   IF P(J,3)>(E1(0,3)+X(1,1)) GOTO 6010
5970   IF P1=J GOTO 6060             additional peak within ethanol window
5980   IF P(J,4)<>P(P1,4) GOTO 6060
5990   C2=21                 error: additional peak same size as ethanol peak
```

```
6000    GOTO 6060
6010    IF P(J,3)<(E1(0,4)-X(1,2)) GOTO 6060
6020    IF P(J,3)>(E1(0,4)+X(1,2)) GOTO 6070
6030    IF P2=J GOTO 6060                    additional peak in standard window
6040    IF P(J,4)<>P(P2,4) GOTO 6060
6050    C2=22                  error: additional peak same size as standard peak
6060 NEXT J
6070 IF P4>0 GOTO 6085                            check: standard peak exists?
6080 C2=23                                             error: no standard peak
6085 GOSUB 8850                                   calculate relative retention
6090 GOSUB 6700                                insert sample into linked list
6100 RETURN
6110 FOR J=1 TO Z
     ... like channel A lines 5750-6100
     E2 instead of E1
     X(2,1) instead of X(1,1)
     X(2,2) instead of X(1,2)
     GOSUB 8950 instead of GOSUB 8850
     adjust line numbers of GOTO-commands
6450 RETURN

6700 REM Sort the samples
6710 IF A(0,0)>0 GOTO 6760
6715 A(0,0)=A(0,0)+1                               1. sample: count samples
6720 J=A(0,0)                                          store sample number
6722 IF H1(1)=1 THEN GOSUB 8500                store peak data (channel A)
6725 IF H1(1)=2 Then GOSUB 8700                store peak data (channel B)
6730 N=A(0,0)                                 number=smallest number
6740 RETURN
6760 IF H<=A(A(0,0),0) GOTO 7080                             other samples:
6770 U=A(0,0)                           new number higher than last number:
6780 GOTO 6860         reset lower limit until right place in sorted list found
6840 IF H<=A(V(U),0) GOTO 6920                     or highest number reached
6850 U=V(U)
6860 IF V(U)<>0 GOTO 6840
6862 IF A(0,0)>34 GOTO 7490                            end of the list reached
6865 A(0,0)=A(0,0)+1                                         count samples
6870 V(A(0,0))=0                                insert number in linked list
6875 V(U)=A(0,0)
6880 R(A(0,0))=U
6885 J=A(0,0)                                          store sample number
6900 IF H1(1)=1 THEN GOSUB 8500                store peak data (channel A)
6905 IF H1(1)=2 THEN GOSUB 8700                store peak data (channel B)
6910 RETURN
6920 IF H<A(V(U),0) GOTO 7000
6930 J=V(U)                          overwrite previous data: store number
6940 IF H1(1)=1 THEN GOSUB 8500                store peak data (channel A)
```

```
6945 IF H1(1)=2 THEN GOSUB 8700          store peak data (channel B)
6950 RETURN
7000 IF A(0,0)>34 GOTO 7490              right place in sorted list found
7010 A(0,0)=A(0,0)+1                     count samples
7015 O=V(U)                              insert number in linked list
7020 V(A(0,0))=O
7025 V(U)=A(0,0)
7030 R(A(0,0))=U
7040 R(O)=A(0,0)
7050 J=A(0,0)                            store number
7060 IF H1(1)=1 THEN GOSUB 8500          store peak data (channel A)
7065 IF H1(1)=2 THEN GOSUB 8700          store peak data (channel B)
7070 RETURN
7080 IF H<A(A(0,0),0) GOTO 7170
7090 J=A(0,0)                            overwrite previous data: store number
7100 IF H1(1)=1 THEN GOSUB 8500          store peak data (channel A)
7105 IF H1(1)=2 THEN GOSUB 8700          store peak data (channel B)
7110 RETURN
7170 O=A(0,0)                            new number lower than last number
7175 GOTO 7270        reset upper limit until right place in sorted list found
7250 IF H>=A(R(O),0) GOTO 7340           or lowest number reached
7260 O=R(O)
7270 IF R(O)<>0 GOTO 7250
7275 IF A(0,0)>34 GOTO 7490              end of list reached
7280 A(0,0)=A(0,0)+1                     count samples
7285 V(A(0,0))=O                         insert number in linked list
7290 R(A(0,0))=0
7295 R(O)=A(0,0)
7300 N=A(0,0)                            reset lowest number
7310 J=A(0,0)                            store number
7320 IF H1(1)=1 THEN GOSUB 8500          store peak data (channel A)
7325 IF H1(1)=2 THEN GOSUB 8700          store peak data (channel B)
7330 RETURN
7340 IF H>A(R(O),0) GOTO 7420
7350 J=R(O)                              overwrite previous data: store number
7360 IF H1(1)=1 THEN GOSUB 8500          store peak data (channel A)
7365 IF H1(1)=2 THEN GOSUB 8700          store peak data (channel B)
7370 RETURN
7420 IF A(0,0)>34 GOTO 7490              right place in sorted list found
7430 A(0,0)=A(0,0)+1                     count samples
7435 U=R(O)                              insert number in linked list
7440 V(A(0,0))=O
7445 V(U)=A(0,0)
7450 R(A(0,0))=U
7455 R(O)=A(0,0)
7460 J=A(0,0)                            store number
7470 IF H1(1)=1 THEN GOSUB 8500          store peak data (channel A)
```

```
7475  IF H1(1)=2 THEN GOSUB 8700              store peak data (channel B)
7480  RETURN
7490  M3=1                                    error: more than 35 samples
7500  RETURN

8500  REM Link the peak data (channel A)
8520  A(J,0)=H                                store number
8530  IF C2>0 GOTO 8570
8540  A(J,1)=P(P1,4)/(P(P2,4)*E1(0,2))        sample o.k.; store ethanol content
8542  FOR K=1 TO Z4
8545    A(J,K+4)=P(0,K)                       store relative retention
8547  NEXT K
8550  A(J,3)=1                                error code=1
8560  RETURN
8570  A(J,3)=C2                               error: store error code
8580  RETURN

8700  REM Link the peak data (channel B)
      ... like channel A lines 8500–8580
      A(J,2) instead of A(J,1)
      E2 instead of E1
      A(J,K+8) instead of A(J,K+4)
      A(J,4) instead of A(J,3)
      adjust line numbers of GOTO commands
8780  RETURN

8850  REM Calculate the relative retention times (channel A)
8852  IF C2< >0 THEN RETURN
8856  Z4=0
8857  FOR J=1 TO Z
8860    IF P(J,3)<(E1(0,3)-X(1,1)) GOTO 8875       <retention ethanol
8862    IF P(J,3)>(E1(0,3)+X(1,1)) GOTO 8867
8865    GOTO 8880
8867    IF P(J,3)<(E1(0,4)-X(1,2)) GOTO 8875
                                  >retention ethanol + <retention standard
8870    IF P(J,3)>(E1(0,4)+X(1,2)) GOTO 8875       >retention standard
8872    GOTO 8880
8875    Z4=Z4+1                                    count additional peaks
8877    P(0,Z4)=P(J,3)/P(P2,3)                     calculate rel. retention
8878    IF Z4> =5 GOTO 8882                        store max. 4 peaks
8880  NEXT J
8882  RETURN

8950  REM Calculate the relative retention times (channel B)
      ... like channel A lines 8850–8882
      E2 instead of E1
```

```
      X(2,1) instead of X(1,1)
      X(2,2) instead of X(1,2)
      adjust line numbers of GOTO commands
8982  RETURN

9000  REM Print the results of the alcohol analysis
9001  N1=N                                            1.number in linked list
9003  PRINT " "
9005  PRINT " "
9010  PRINT "RESULTS SAMPLES:"
9015  PRINT "----------------"
9020  PRINT " "
9034  PRINT "SAMPLES: ";A(0,0)
9036  PRINT " "
9040  IF A(0,0)<>0 THEN PRINT TAB(1);"CASE";TAB(10);"CHANNEL
      A"; TAB(29);"CHANNEL B";TAB(48);"AVERAGE";TAB(61);"RANGE
      INT";
9041  IF A(0,0)<>0 THEN PRINT TAB (76);"REJECTED";TAB(90);
      "RANGE EXT";TAB(105); " "REJECTED";                    print title
9042  PRINT
9050  PRINT " "
9070  FOR I=1 TO A(0,0)
9080    C$="           ",G$="           ";A=0
9081    F$=" "
9090    IF A(N,3)<>1 GOTO 9250
9100    IF A(N,4)<>1 GOTO 9320
9105    A(0,1)=A(0,1)+A(N,1)                  sample on channel A and B o.k.
9110    A(0,3)=A(0,3)+A(N,2)
9115    A(0,5)=A(0,5)+1
9118    A=(A(N,1)+A(N,2))/2        calculate mean ‰ value of channel A and B
9119    A=(INT(0.40001+A*100))/100                 round (0,5 becomes 0.0)
9122    IF A>0.10 GOTO 9137                                 ‰ value <=0.1‰
9123    IF X4<ABS(A-A(N,1)) THEN F$="YES"
                                    deviation of ‰ value too large (intern)
9124    IF X3>=ABS(A-A(N,1)) GOTO 9130
9126    C$="YES          "          deviation of ‰ value too large (extern)
9128    GOTO 9135
9130    A(0,2)=A(0,2)+A(N,1)                      deviation of ‰ values o.k.
9132    A(0,4)=A(0,4)+A(N,2)
9134    A(0,6)=A(0,6)+1
9135    PRINT TAB(1);A(N,0)[4];TAB(14);A(N,1)[1.3];
        TAB(33);A(N,2)[1.3];TAB(51);A[1.2];TAB(66);"-";TAB(78);F$;
        TAB(95);"-";TAB(107)C$;                             print sample result
9136    GOTO 9420
9137    IF A>1.00 GOTO 9143                           0.1‰ <‰ value<=1‰
9138    X5=X4
9139    IF X5<ABS(A-A(N,1)) THEN F$="YES"
```

```
9140   X=X3
9141   IF X>=ABS(A-A(N,1)) THEN GOTO 9170
9142   GOTO 9155
9143   X5=A*X4                                              ‰ value>1‰
9144   X5=(NT(0.500001+X5*100))/100
                               calculate max. allowed deviation (intern)
9145   IF X5<ABS(A-A(N,1)) THEN F$="YES"
                                 deviation of ‰ value to large (intern)
9147   X=A*X3
9149   X=(INT(0.500001+X*100))/100
                               calculate max. allowed deviation (extern)
9151   IF X>=ABS(A-A(N,1)) GOTO 9170
9155   C$="YES        "          deviation of ‰ value too large (extern)
9160   GOTO 9239
9170   A(0,2)=A(0,2)+A(N,1)                 deviation of ‰ values o.k.
9180   A(0,4)=A(0,4)+A(N,2)
9190   A(0,6)=A(0,6)+1
9239   S$=STR$(A(N,0)[4])&"     "&STR$(A(N,1)[1.3])&"     "&STR$
       (A(N,2)[1.3])&"     "&STR$(A[1.2])&"  "
9240   PRINT TAB(1);S$;A-X5[1.2];"-";A+X5[1.2];TAB(78);F$;TAB(89);A-X
       [1.2];"- ";A+X[1.2];TAB(107);C$;
9245   GOTO 9420                                   print sample result
9250   IF A(N,3)=0 THEN C$="NO DATA      "         error on channel A
9260   IF A(N,3)=21 THEN C$="MULTIPLE PEAKS AW"
9270   IF A(N,3)=22 THEN C$="MULTIPLE PEAKS SW"
9280   IF A(N,3)=23 THEN C$="NO STANDARD     "
9290   IF A(N,4)<>1 GOTO 9392
9300   PRINT TAB(1);A(N,0)[4];TAB(10);C$;TAB(33);A(N,2)[1.3];
                                                   print sample result
9310   GOTO 9420
9320   IF A(N,4)=0 THEN G$="NO DATA      "         error on channel B
9340   IF A(N,4)=21 THEN G$="MULTIPLE PEAKS AW"
9350   IF A(N,4)=22 THEN G$="MULTIPLE PEAKS SW"
9360   IF A(N,4)=23 THEN G$="NO STANDARD     "
9380   PRINT TAB(1);A(N,0)[4];TAB(14);A(N,1)[1.3];TAB(29);G$
                                                   print sample result
9390   GOTO 9420
9392   IF A(N,4)=0 THEN G$="NO DATA   "     error on channel A and B
9394   IF A(N,4)=21 THEN G$="MULTIPLE PEAKS AW"
9396   IF A(N,4)=22 THEN G$="MULTIPLE PEAKS SW"
9398   IF A(N,4)=23 THEN G$="NO STANDARD   "
9400   PRINT TAB(1);A(N,0)[4];TAB(10);C$;TAB(29);G$;  print sample result
9420   N=V(N)                                 next number in linked list
9430 NEXT I
9435 PRINT " "
9436 PRINT " "
9437 IF A(0,0)=0 THEN RETURN
```

```
9438  IF A(0,5)=0 GOTO 9471
9440  PRINT "SUM AND AVERAGE
      CALCULATED RUNS CHANNEL A:            ";A(0,1)[6.2];" : ";A(0,5);
      "=";(A(0,1)/A(0,5))[6.2]:             mean of all runs channel A
9450  PRINT "SUM AND AVERAGE
      CALCULATED RUNS CHANNEL B:            ";A(0,3)[6.2];" : ";A(0,5);
      "=";(A(0,3)/A(0,5))[6.2]:             mean of all runs channel B
9455  PRINT " "
9456  IF A(0,6)=0 GOTO 9471
9460  PRINT "SUM AND AVERAGE -NOT REJECTED-
      RUNS CHANNEL A:                       ";A(0,2)[6.2];" : ";
      A(0,6);"=";(A(0,2)/A(0,6))[6.2]       mean of not rejected runs channel A
9470  PRINT "SUM AND AVERAGE -NOT REJECTED-
      RUNS CHANNEL B:                       ";A(0,4)[6.2];" : ";
      A(0,6);"=";A(0,4)/A(0,6))[6.2]:       mean of not rejected runs channel B
9471  PRINT " "
9472  PRINT " "
9473  PRINT " "
9480  RETURN

9500  REM Print additional peaks
9502  N=N1                                  1.number in linked list
9505  PRINT " "
9510  PRINT " "
9515  PRINT "ADDITIONAL COMPOUNDS:"
9520  PRINT "------------------ "
9525  PRINT " "
9530  IF A(0,0)<>0 THEN PRINT "CASE";TAB(36);"CHANNEL A";
      TAB(76); "CHANNEL B"
9535  IF A(0,0)<>0 THEN PRINT TAB(29);"RRT";TAB(43);"IDENTIFICA-
      TION";TAB(69);"RRT";TAB(83);"IDENTIFICATION";       print title
9540  PRINT " "
9545  FOR I=1 TO A(0,0)
9550    FOR J=1 TO 4
9555      N1$="      ",N2$="      "
9560      IF A(N,J+4)=0 GOTO 9605
9565      IF A(N,J+8)=0 GOTO 9640
9570      GOSUB 9700                        identify peaks channel A
9575      GOSUB 9850                        identify peaks channel B
9580      IF J>1 GOTO 9595
9585      PRINT TAB(1);A(N,0)[4.0];TAB(27);A(N,J+4)[4.2];TAB(43);N1$;
          TAB(67);A(N,J+8)[4.2];TAB(83);N2$;
9590      GOTO 9665
9595      PRINT TAB(27);A(N,J+4)[4.2];TAB(43);N1$;TAB(67);A(N,J+8)
          [4.2];TAB(83);N2$;

9600      GOTO 9665
```

```
9605      IF A(N,J+8)=0 GOTO 9667
9610      GOSUB 9850                                   identify peaks channel B
9615      IF J>1 GOTO 9630
9620      PRINT TAB(1);A(N,0)[4.0];TAB(67);A(N,J+8)[4.2];TAB(83);N2$;
                                                 print 1.peak with sample number
9625      GOTO 9665
9630      PRINT TAB(67);A(N,J+8)[4.2];TAB(83);N2$;
                                        print next peaks without sample numbers
9635      GOTO 9665
9640      GOSUB 9700                                   identify peaks channel A
9645      IF J>1 GOTO 9660
9650      PRINT TAB(1);A(N,0)[4.0];TAB(27);A(N,J+4)[4.2];TAB(43);N1$;
9655      GOTO 9665
9660      PRINT TAB(27);A(N,J+4)[4.2];TAB(43);N1$;
9665    NEXT J
9667    N=V(N)                                         next number in linked list
9670  NEXT I
9671  PRINT " "
9672  PRINT " "
9673  PRINT " "
9674  PRINT " "
9675  PRINT " "
9680  RETURN

9700  REM Identification of the peaks (channel A)
9702  IF B(2)<ABS(A(N,J+4)-B(1)) GOTO 9710              peak in aceton window
9705  N1$="ACETON    "
9707  RETURN
9710  IF B1(2)<ABS(A(N,J+4)-B1(1)) GOTO 9720        peak in methanol window
9712  N1$="METHANOL "
9715  RETURN
9720  IF B2(2)<ABS(A(N,J+4)-B2(1)) GOTO 9730
                                                  peak in acetaldehyd window
9722  N1$="ACETALDEH."
9725  RETURN
9730  IF B3(2)<ABS(A(N,J+4)-B3(1)) GOTO 9740
9732  N1$="XXXXXXXXXX"
9735  RETURN
9740  IF B4(2)<ABS(A(N,J+4)-B4(1)) GOTO 9750
9742  N1$="XXXXXXXXXX"
9745  RETURN
9750  IF B5(2)<ABS(A(N,J+4)-B5(1)) GOTO 9760
9752  N1$="XXXXXXXXXX"
9755  RETURN
9760  IF B6(2)<ABS(A(N,J+4)-B6(1)) GOTO 9770
9762  N1$="XXXXXXXXXX"
9765  RETURN
```

```
9770  IF B7(2)<ABS(A(N,J+4)-B7(1)) GOTO 9780
9772  N1$="XXXXXXXXXX"
9775  RETURN
9780  IF B8(2)<ABS(A(N,J+4)-B8(1)) GOTO 9790
9782  N1$="XXXXXXXXXX"
9785  RETURN
9790  IF B9(2)<ABS(A(N,J+4)-B9(1)) THEN RETURN
9792  N1$="XXXXXXXXXX"
9795  RETURN

9850  REM Identification of the peaks (channel B)
9852  IF D(2)<ABS(A(N,J+8)-D(1)) GOTO 9860          peak in aceton window
9854  N2$="ACETON    "
9857  RETURN
9860  IF D1(2)<ABS(A(N,J+8)-D1(1)) GOTO 9870        peak in ethrane window
9862  N2$="ETHRANE   "
9865  RETURN
9870  IF D2(2)<ABS(A(N,J+8)-D2(1)) GOTO 9880
9872  N2$="HALOTHANE "                               peak in halothane window
9875  RETURN
9880  IF D3(2)<ABS(A(N,J+8)-D3(1)) GOTO 9890
9882  N2$="CHLOROFORM"                               peak in chloroform window
9885  RETURN
9890  IF D4(2)<ABS(A(N,J+8)-D4(1)) GOTO 9900
9892  N2$="PENTHRANE"                                peak in penthrane window
9895  RETURN
9900  IF D5(2)<ABS(A(N,J+8)-D5(1)) GOTO 9910
9902  N2$="XXXXXXXXXX"
9905  RETURN
9910  IF D6(2)<ABS(A(N,J+8)-D6(1)) GOTO 9920
9912  N2$="XXXXXXXXXX"
9915  RETURN
9920  IF D7(2)<ABS(A(N,J+8)-D7(1)) GOTO 9930
9922  N2$="XXXXXXXXXX"
9925  RETURN
9930  IF D8(2)<ABS(A(N,J+8)-D8(1)) GOTO 9940
9932  N2$="XXXXXXXXXX"
9935  RETURN
9940  IF D9(2)<ABS(A(N,J+8)-D9(1)) THEN RETURN
9942  N1$="XXXXXXXXXX"
9945  RETURN
10000 END
```

2.3.9.5 Sample Output

An example of a daily data output is given in Figure 3.

COMPUTER OUTPUT OF BLOOD ALCOHOL DATA FROM JULY 18, 1978

Department of Forensic Chemistry, University of Zurich, Switzerland

CALIBRATION	RUNS	REJECTED	RT ALC	RT STAND	RRT ALC	FACTOR
METHOD 1	8	0	239	315	0.758	0.73
METHOD 2	8	0	98	207	0.473	0.48

CASE	METHOD 1	METHOD 2	AVERAGE	RANGE	REJECTED
3497	0.94	0.86	0.90	0.85 – 0.95	
3498	2.55	2.46	2.50	2.36 – 2.64	
3499	2.50	2.46	2.48	2.34 – 2.62	
3500	1.01	0.96	0.98	0.93 – 1.03	
3501	0.70	0.70	0.70	0.65 – 0.75	
3502	1.52	1.45	1.48	1.40 – 1.56	
3503	2.47	2.55	2.51	2.37 – 2.65	
3504	0.00	0.00	0.00	–	
3505	1.56	1.52	1.54	1.46 – 1.62	
3506	1.46	1.52	1.49	1.41 – 1.57	
3507	0.00	0.00	0.00	–	
3508	1.10	1.08	1.09	1.04 – 1.14	
3509	2.76	2.70	2.73	2.59 – 2.87	
3510	1.32	1.30	1.31	1.23 – 1.39	
3511	0.55	0.54	0.54	0.49 – 0.59	
3512	1.29	1.26	1.27	1.19 – 1.35	
3513	2.07	2.06	2.06	1.95 – 2.17	
3514	2.29	2.27	2.28	2.17 – 2.39	
3515	1.88	1.90	1.89	1.78 – 2.00	
3516	1.20	1.15	1.17	1.12 – 1.22	
3517	0.00	0.00	0.00	–	
3518	2.77	2.66	2.71	2.57 – 2.85	
3519	0.72	0.67	0.69	0.64 – 0.74	
3520	1.28	1.30	1.29	1.21 – 1.37	

ADDITIONAL COMPOUNDS	RRT1	IDENTIFICATION	RRT2	IDENTIFICATION
3499	0.000		0.280	ACETONE
3503	0.260	METHANOL	0.280	ACETONE
3505	0.393	ACETALD.	0.000	
3506	0.257	METHANOL	0.000	
3507	0.000		0.410	HALOTHANE
3507	0.000		0.280	ACETONE
3510	0.000		0.280	ACETONE

PROGRAM FINISHED

Figure 3. Example of a Daily Data Printout in Blood Alcohol Analysis.

References

1. Harger, R., Raney, B., Birdwell, E. & Kitchel, M. (1950) The ratio of alcohol between air and water, urine and blood: Estimation and identification of alcohol in these liquids from analysis of air equilibrated with them. J. Biol. Chem. *183*, 197–213.
2. Harger, R., Forney, R. & Barnes, H. (1950) Estimation of level of blood alcohol from analysis of breath. J. Lab. Clin. Med. *36*, 306–318.
3. Harger, R. & Forney, R. (1967) Aliphatic alcohols, III C, Blood/breath alcohol ratio. In: Progress in Chemical Toxicology, Vol. 3 (Stolman, A., ed.) pp. 10–13, Academic Press, New York & London.
4. Dubowski, K.M. (1979) The blood/breath ratio of ethanol. Clin. Chem. *25*, 1144.
5. Denney, R.C. (1980) Instrumental Advances in breath alcohol analysis. Chemistry in Britain *16*, 428–431.
6. Chamberlain, R.T. (1983) Medical legal aspects of breath alcohol analysis. American Association of Clin. Chemistry *5(4)*, 1–8.
7. Aufdermaur, M. & Brandenberger, H., unpublished data.
8. Bock, J.C. (1931) The interferometric determination of alcohol in blood. J. of Biol. Chem. *18*, 645–655.
9. Widmark, E.M.P. (1922) Eine Mikromethode zur Bestimmung von Aethylalkohol im Blut. Biochem. Z. *131*, 473–484.
10. Brandenberger, H. (1963) Forensische Chemie heute: Forensische Bestimmung von Aethylalkohol. Chimia *17*, 249–257.
11. Bonnichsen, R.F. & Theorell, H. (1951) An enzymatic method for the microdetermination of ethanol. Scand. J. Clin. Lab. Invest. *3*, 58–68.
12. Bucher, Th. & Redetzki, H. (1951) Eine spezifische photometrische Bestimmung von Aethylalkohol auf fermentativem Wege. Klin. Wochenschr. *29*, 615–616.
13. Gibitz, H.J. & Schütz, H. (1993) Bestimmung von Ethanol im Serum. Mitteilung XX der Senatskommission für klinisch-toxikologische Analytik der Deutschen Forschungsgemeinschaft. VCH, Weinheim.
14. Bernheim, J., Sachs, V. & Steigleder, E. (1962) Specificity of the alcohol dehydrogenase procedure (in relation to ethyl and other alcohols). Dtsch. Z. Ges. Ger. Med. *53*, 28–33.
15. Cadman, B.A. & Johns, Th. (1958) Gas chromatographic determination of ethyl alcohol and other volatiles in blood. Communication at the 9th Annual Pittsburgh Conference on Analytical Chemistry and Applied Spectroscopy. Abstract Handbook & special Beckman Report.
16. Cadman, B.A. & Johns, Th. (1960) Application of the gas chromatograph in the laboratory of criminalistics. J. of Forensic Sciences *5*, 369–385.
17. Maricq, L. & Molle L. (1959) Recherches sur la détermination de l'alcoolémie par chromatographie gazeuse. Bull. Acad. Méd. Belg. *24*, 199–230.
18. Maricq, L. & Molle, L. (1960) Détermination simultanée de l'éther-oxyde d'éthyle et de l'éthanol dans le sang et les viscères par chromatographie gazeuse. Ann. Pharmac. françaises *18*, 811–816.
19. Machata, G. (1962) Die Routineuntersuchung der Blutalkoholkonzentration mit dem Gaschromatographen. Mikrochim. Acta *1962*, 691–700.
20. Curry, A.S., Hurst, G., Kent, N.R. & Powell, H. (1962) Rapid screening of blood samples for volatile poisons by gas chromatography. Nature *195*, 603–604.
21. Goldbaum, L.R., Domanski, T.J. & Schloegel, M.T. (1964) Analysis of biological specimens for volatile compounds by gas chromatography. J. Forensic Sci. *9*, 63–71.
22. Goldbaum, L.R., Schloegel, E.L. & Dominguez, A.M. (1963) Application of gas chromatography to toxicology. In: Progress in Chemical Toxicology, Vol. 1 (Stolman, A., ed.), pp. 11–52. Academic Press, New York & London.
23. Machata, G. (1964) Ueber die gaschromatographische Blutalkoholbestimmung: Analyse der Dampfphase. Mikrochim. Acta *1964*, 262–271.
24. Brandenberger, H. (1968) Entwicklung und heutiger Stand der forensischen Alkoholanalytik am Gerichtlich-medizinischen Institut der Universität Zürich. In: Gerichtsmedizin – Bindeglied zwischen Medizin und Recht (Edition for the 70th birthday of F. Schwarz), pp. 76–85. Ed. Stämpfli, Bern.
25. Brandenberger, H. (1974) Toxicology. In: Clinical Biochemistry, Principles and Methods (Curtius, H.Ch. & Roth, M., eds.), pp. 1425–1467. Walter de Gruyter, Berlin & New York.
26. Brandenberger, H. (1983) Die Zürcher Blutalkohol-Analytik. Kriminalistik *37*, 568–571.

27. Brandenberger, R. & Brandenberger, H. (1975) Data processing in blood alcohol analysis. Forensic Sciences *5*, 110.
28. Brandenberger, H. (1978) Computer output of blood alcohol data. Bull. International Association of Forensic Toxicologists (TIAFT) *14(1)*, 3–4.
29. Brandenberger, H., Brunner-Miersch, M., Fluri, T., Frangi-Schnyder, D., Leuenberger-West, F. & Stoll, C. (1987) Identification of additional volatiles in blood samples submitted for alcohol analysis. In: "TIAFT 22", Reports on Forensic Toxicology (Proceedings of the 1985 International Meeting of the International Association of Forensic Toxicologists, Rigi Kaltbad, Switzerland (Brandenberger H. & R., eds.), pp. 437–453. Publ. by Branson Research, CH-6356 Rigi Kaltbad.
30. Brandenberger, H. (1979) Present state and future trends of negative ion mass spectrometry in toxicology, forensic and environmental chemistry. In: Recent Developments in Mass Spectrometry in Biochemistry and Medicine, Vol. 2 (Frigerio, A., ed.), pp. 227–255. Plenum Publ. Corp., New York.
31. Brandenberger, H. (1980) Negative ion mass spectrometry by low-pressure chemical ionization. In: Recent Developments in Mass Spectrometry in Biochemistry and Medicine, Vol. 6 (Frigerio, A. & McCamish, M., eds.), Elsevier Publ. Comp., Amsterdam.
32. Brandenberger, H. (1983) Negative ion mass spectrometry by low pressure chemical ionization; inherent possibilities for structural analysis. Journal of Mass Spectrometry and Ion Physics *47*, 213–216.
33. Brandenberger, H. (1982) Negative ion mass spectrometry by low pressure chemical ionization and its combination with conventional positive ion mass spectrometry by electron impact. In: Proceedings of the 30th Annual Conference on Mass Spectrometry and Allied Topics, Honolulu, pp. 468–469. Publ. by American Society for Mass Spectrometry.
34. West, F. B. Ch. & Brandenberger, H. (1984) Further applications of the dual mass spectrometric technique in forensic and toxicological analysis. In: Topics in Forensic and Analytical Chemistry. Analyt. Chemistry Symposia Series Vol. 20 (Proceedings of the European Meeting of the International Association of Forensic Toxicologists, Munich, 1983) (Maes, R. A. A., ed.) pp. 131–136. Elsevier, Amsterdam, Oxford, New York & Tokyo.
35. Brandenberger, H. (1987) Low pressure negative chemical ionization mass spectrometry and dual mass spectrometry: A review of the work since 1977. In: "TIAFT 22", Reports on Forensic Toxicology (Proceedings of the 1985 International Meeting of the International Association of Forensic Toxicologists, Rigi Kaltbad, Switzerland (Brandenberger, H. & R., eds.) pp. 437–453. Publ. by Branson Research, CH-6356 Rigi Kaltbad.
36. Geiger, R., Matter, P., Brandenberger, H. & Biener, K. (1977) Alkohol und Skilauf. Münch. med. Wschr. *119*, 109–110.
37. Wirth, N. (1976) Algorithms + data structures = programs, pp. 171–189. Prentice-Hall Inc., Englewood Cliffs, N. Y..
38. Jackson, M. A. (1975) Principles of Program Design. Academic Press, London.

2.4 Volatile Hydrocarbons

T. Nagata and K. Kimura

2.4.1 General Aspects

Organic volatile hydrocarbons include various kinds of solvents, which are commonly and widely used in factories, offices, hospitals and even individual homes. Following exposure to vapor containing high concentrations of volatile hydrocarbons, as can happen in the case of accidents or malpractice, acute poisoning occurs, sometimes resulting in death. Exposure to lower concentrations over a long period results in chronic poisoning. Poisoning can occasionally be caused by peroral intake, as seen in forensic cases such as homicide or suicide.

The problem of chronic poisoning arising from long-term exposure to a low concentration of volatiles is treated within the field of environmental hygiene. This chapter therefore is intended to deal solely with acute poisoning, describing toxicity, therapy and the analytical methods regarding body fluids or tissues. A few related volatiles are supplementarily described.

2.4.2 Toxicity

The toxicity of volatiles is to a large extent due to their volatility. They can readily contact the human body, and can be aspirated into the respiratory tract and/or partly absorbed through the skin, later being distributed into the body tissues via the blood stream. When orally ingested, volatiles are likely to be distributed into the body tissues via the walls of the digestive tract after death.

In cases of acute poisoning, toxic symptoms appear rapidly or during the early stage of exposure. This fact would suggest a quick distribution into the brain which is enriched with lipids. The initial symptoms arise from a "narcotic" effect on the central nervous system (CNS) causing dizziness, disorientation and paralysis of the tastebuds, and are followed by illusion, loss of consciousness and convulsions. Sudden death by cardiac arrest often occurs. From a clinical perspective, in cases of ingestion, any symptoms originating from the CNS suggestive of an anesthetic process should be taken into consideration prior to symptoms suggesting disturbance to the parenchymatous organs such as the liver or the kidney.

Acute toxicity is influenced by the concentration of the vapor and the length of time of exposure. The blood level of volatiles is important for estimating the grade of toxicity. Some of the fatal blood levels described in this chapter are cited from C.L. Winek, "Drug & Chemical Blood-Level Data", Fisher Scientific 1985, Pittsburgh, USA.

2.4.3 Therapy

The following emergency measures are taken in cases of acute poisoning, according to the situation: Fresh air or oxygen inhalation, artificial respiration and/or transfusion are used following exposure to vapor. Gastric irrigation following oral intake should be carried out under respiratory care using a trachea tube, but this treatment is often too dangerous in the case of corrosive agents because of the danger of perforation of the stomach wall. When the condition is serious, such as when poisons have been absorbed into the body, blood purification including hemodialysis or direct hemoperfusion is effective. The symptomatic treatments should be chosen and carried out according to the situation.

2.4.4 Chemical Analysis

Chemical analysis is the most decisive diagnostic method in cases of volatile poisoning. It should be carried out as early as possible. Since innumerable analytical methods have been described in other publications, only a few examples of the methods used in the authors' laboratory are introduced below.

2.4.4.1 Sampling

One to 5 ml of blood and/or urine should be collected as soon as possible, since most volatiles are excreted rapidly from the lung. The samples are placed into a 10 to 15 ml volume glass bottle which is then sealed with a tight silicone cap. Antiseptic agents are unnecessary, except for a small amount of anticoagulant such as heparin, since added chemicals could interfere with the analytical procedure and lead to false results. The container is warmed to 55–60 °C for 30 min, and 1 ml of the vapor phase is aspirated into a syringe and introduced into a gas chromatograph or a gas chromatograph/mass spectrometer. In cases where instant analysis is impossible, samples must be stored in a freezer at temperatures lower than − 20 °C until analysis.

2.4.4.2 Analytical methods

Gas chromatography (GC) is at present the most common and selective method of qualitative and quantitative determination for volatiles, since GC can excellently separate any mixture of chemicals when they are in a gaseous state. If strict identification is required, gas chromatography/mass spectrometry (GC/MS) should be employed.

High performance liquid chromatography (HPLC), ultraviolet spectrometry (UV) and infrared spectrometry (IR) are selectively used. Chemical tests are seldom used in practice. As far as volatiles are concerned, GC is the preferred method.

Gas chromatographs should be properly utilized according to the required purposes by the choice of appropriate columns and detectors, such as combining a packed or capillary column together with a flame ionization detector (FID) or an electron capture detector (ECD). In practice, it is possible to use a packed column to deal with volatiles, while a capillary column is commonly used at present for other chemical substances owing to its high resolving power. With respect to the detector, FID is generally used, ECD selectively for halogens. GC/MS is in most cases carried out in the electron impact mode (EI mode), and total ion chromatograms, mass chromatograms and mass spectra are recorded simultaneously. For sensitive quantitative determination, selected ion monitoring (SIM) is generally performed.

Model 1 (for propane etc.):
a) GC-FID. Column: 2 m × 3 mm i.d. glass tube packed with Porapak Q (80–100 mesh). Temperature: 120 °C at column, 150 °C at injection port and detector. Carrier gas: N_2 (80 ml/min).
b) GC/MS. Column: 2 m × 3 mm i.d. glass tube packed with Porapak Q (80–100 mesh). Temperature: 120 °C at column, 200 °C at injection port, 250 °C at separator and 270 °C at ion source. Carrier gas: He (30 ml/min). Ionization energy: 20 eV or 70 eV for mass scan.

Model 2 (for toluene etc.):
a) GC-FID. Column: 2 m × 3 m i.d. glass tube packed with 10 % PEG-1000 on Uniport HP (60–80 mesh). Temperature: 80 °C at column, 130 °C at injection port and detector. Carrier gas: N_2 (60 ml/min).
b) GC/MS. See Model 5c.

Model 3 (for halogenated hydrocarbons):
a) GC-ECD (electron source, Ni^{63}). Column: 1 m × 3 mm i.d. glass tube packed with 15 % MS 550 on Universal B (60–80 mesh). Temperature: 60 °C at column. Carrier gas: N_2 (75 ml/min).
b) GC/MS. Column: 1.5 m × 3 mm i.d. glass tube packed with 3 % OV-1 on Chromosorb W HP (80–100 mesh). Temperature: 27 °C at column, 150 °C at injection port, 230 °C at separator and ion source. Carrier gas: He (30 ml/min). Ionization energy: 20 eV or 70 eV for mass scan.

Model 4 (for chlorinated hydrocarbons):
a) GC-FID. Column: 1 m × 3 mm i.d. glass tube packed with Porapak P (80–100 mesh). Temperature: 155 °C at column, 200 °C at injection port and detector. Carrier gas: N_2 (60 ml/min).
b) GC/MS. Column: 1 m × 3 mm i.d. glass tube packed with Porapak P (80–100 mesh). Temperature: 155 °C at column, 200 °C at injection port, 250 °C at separator and 270 °C at ion source. Carrier gas: He (30 ml/min). Ionization energy: 20 eV or 70 eV for mass scan.

Model 5 (for gasoline):
a) GC-FID. Column: 2 m × 3 mm glass tube packed with 10 % OV-17 on Chromosorb W HP (80–100 mesh). Temperature: 100 °C at column, 170 °C at detector. Carrier gas: N_2 (40 ml/min).
b) GC/MS. Column: 1 m × 3 mm i.d. glass tube packed with Porapak P (80–100 mesh). Temperature: 180 °C at column, 220 °C at injection port, 250 °C at separator, 270 °C at ion source. Carrier gas: He (30 ml/min). Ionization energy: 20 eV.

c) GC/MS. Column: 2 m × 3 mm i.d. glass tube packed with 10% OV-17 on Chromosorb W HP (80–100 mesh). Temperature: 120 °C at column, 150 °C at injection port, 250 °C at separator, 270 °C at ion source. Carrier gas: He (30 ml/min). Ionization energy: 70 eV.
d) GC/MS for MS and SIM. Column (capillary): 10 m × 0.53 mm i.d., 2.65 μm film thickness coated with Hewlett Packard HP-1 cross-linked methylsilicone gum phase. Temperature: programmed from 50 °C (2 min) to 200 °C (10 °C/min) at column, 200 °C at injection port, 250 °C at separator and ion source. Carrier gas: He (15 ml/min). Ionization energy: 20 eV for MS and 70 eV for SIM.

2.4.5 Specific Hydrocarbons

2.4.5.1 Aliphatic hydrocarbons

Methane and ethane, gaseous in the normal phase, are considered to be nontoxic in practice, however hydrocarbons with carbon numbers higher than 3 are generally toxic if there is exposure to their dense vapor or gas. Propane, the main component of liquefied petroleum gas, is generally considered to be harmless, and yet it is toxic and sometimes fatal when present at concentrations from 40 to 50% in the air. In most cases, hypoxic or anoxic condition leads to fatality. Hydrocarbons with larger carbon numbers and unsaturated bonds have higher toxicity. For example, butane or propylene are more toxic than propane. Cases of single poisoning by pentane, hexane or hydrocarbons with larger carbon numbers are rarely seen in practice. Detection method: Model 1 (1, 2).

2.4.5.2 Aromatic hydrocarbons

Benzene C_6H_6. Benzene is known as one of the most common solvents and there have been numerous reports dealing with benzene poisoning. In the case of acute intoxication, the symptoms closely resemble those which occur with other aromatic hydrocarbons: narcotic effect on the CNS and occasionally myocardial disturbance. Gastric irrigation is limited only to cases where large amounts have been ingested. The chronic toxicity of benzene is characterized by the disturbance of enzymes and leukemia arising from chromosome interruption.

The MAC (Maximum Allowable Concentration) is 35 ppm., while exposure to 10,000 ppm and ingestion of 100 ml can be dangerous. The lethal blood level is reported to range from 1 to 30 μg/ml.

Detection method: Model 2 (3).

Toluene $C_6H_5CH_3$. Toluene, the main component of paint thinner with a concentration of from 60 to 70%, is one of the most important solvents in terms of the social problem of addiction among young people, often leading to juvenile delinquency. Accidental poisoning often occurs in paint factories or work places. The acute toxicity of toluene is greater than that of benzene. On the other hand, chronic toxicity of the former is much lower than that of the latter.

The main symptoms of toluene intoxication are irritation of the air passages, and mental disturbance such as headaches, euphoria, illusions, hallucinations, mental excitement, convulsions, and, finally, coma. In serious cases, patients die of respiratory disturbance.

Toluene is excreted partly by the air from the lung and partly by the urine as hippuric acid, which is a glycine conjugate.

The MAC is 200 ppm, while 5,000 ppm can be highly toxic, and the fatal blood level ranges from 10 to 20 μg/ml. In order to confirm toluene exposure, toluene needs to be detected by GC using Model 2. For quantitative analysis, an internal standard (I.S.) should be added to the sample. For this purpose, o-xylene is added for the analysis of toluene, and benzene or toluene are added for the analysis of xylenes. Hippuric acid should be additionally searched for, using GC and UV as described below.

GC(4): FID. Column: 1 m × 3 mm i.d. glass tube packed with 3% Dexsil 300 GC on Chromosorb W AW DMCS (80–100 mesh).
Temperature: 170 °C at column, 200 °C at injection port and detector. Carrier gas: N_2 (60 ml/min).

UV: Urine (0.2 ml) is diluted 10 times, and then 0.4% p-dimethylaminobenzaldehyde in pyridine (4 ml) and anhydrous acetic acid (0.5 ml) is added. The mixture is stirred and analyzed by a UV-spectrometer at 460 nm. The data are compared with the average level for healthy persons, 0.34 mg/ml.

Xylenes $C_6H_4(CH_3)_2$. The toxicity and metabolism of xylenes are similar to those of toluene. Xylenes are easily taken into the body and metabolized into methylhippuric acids, which always indicate xylene exposure in that they are not otherwise present in the body. The MAC is 200 ppm, and the lethal concentration is 20,000 ppm.

Detection method: Model 2.

2.4.5.3 Gasoline and kerosene

Poisoning after exposure to fuels such as gasoline or kerosene often occurs among those who work in engine rooms and is occasionally seen in abusers sniffing toluene or similar substances. Erythema and/or bulla are seen on the skin surface after direct exposure to dense vapor or direct contact. In the case of exposure, the symptoms are similar to those which occur with toluene. Irritation and inflammation of the mucous membrane of the air passages are common, and in serious cases, mental delusion, coma, convulsions etc. occur. Death is caused by arrhythmia or cardiac arrest. When ingested accidentally by children, gasoline causes erosion of the mucous membrane, and/or deglutition pneumonia, and often leads to death. Exposure to 10,000 ppm, and/or ingestion of 100 to 200 ml can be fatal. Although toxicity is related to largely unsaturated and aromatic hydrocarbons, the exact details remain unknown, since these fuel components are too numerous to define.

In case of oral intake, gastric irrigation should be carried out by means of tracheal catheterization, and active carbon often needs to be administered.

GC is used for screening, and reliable analysis of biological specimens should be carried out using GC/MS (5, 6). In GC, a number of peaks appear, but far less than with the original fuel vapor. Usually we can see aliphatic and aromatic hydrocarbons with carbon numbers of 5 to 8 in the case of exposure to gasoline, and aliphatic hydrocarbons with carbon numbers of 8 to 10 and the group of aromatic hydrocarbons with the carbon number of 9 after exposure to kerosene.

Detection method: Model 5.

2.4.6 Halogenated Hydrocarbons

The hydrocarbons containing halogen generally have higher toxicity. Some common solvents are described below.

Chloroform $CHCl_3$. Although chloroform is nowadays seldom used as an anesthetic in a clinical setting, unexpected poisoning or death often occurs accidentally or in cases of violence (7). When exposed to chloroform vapor, patients have narcotic symptoms at a concentration of 14,000 ppm, and fatality occurs at a concentration of over 16,000 ppm The MAC is 100 ppm. In the case of oral intake, a dose ranging from 10 to 200 ml is considered fatal. The blood levels in fatal cases have been reported to range from 30 to 95 μg/ml. Chloroform is likely to change to phosgene under conditions of sunlight or heat, and toxicity then increases. When inhaled, the vapor irritates the respiratory tract and acts on the CNS, causing a quickened pulse, arrhythmia, ventricular fibrillation, and often leads to sudden cardiac attack. When contact is via the skin, redness and abrasion are observed. Chloroform abuse is known. Exposure to low concentrations of chloroform leads to liver damage. In addition to the general treatment, calcium gluconate is given symptomatically.

No special findings are visually apparent at the time of pathological observation, however chloroform has a characteristic odor. Fujiwara's test is known to detect chlorinated hydrocarbons, since pyridine-sodium hydroxide becomes dark red in the presence of chloroform. However in practice, such chemical tests have now been replaced by GC.

Detection method: Model 3 (8), 4 or 1.

Methyl chloride CH_3Cl and methyl bromide CH_3Br. These methane derivatives are used as extinguishants, pesticides and coolants. In cases where the CNS is affected, the toxicity is likely due to the halogen ion and methyl alcohol set free within the body. Lung edema and insufficiency of circulation are characteristic in the case of methyl bromide.

The MAC is 100 ppm for methyl chloride and 20 ppm for methyl bromide.

Detection Method: Model 3 or 4.

Dichloroethylene $CHClCHCl$, **trichloroethylene** $CHClCCl_2$ and **tetrachloroethylene** CCl_2CCl_2. Among these ethylenes which are used as solvents, trichloroethylene is the one which is most widely employed in manufacturing, and moreover, it used to be employed as a narcotic under the name of Trilene or Trichloren. It acts mainly on the CNS and often results in addiction. When inhaled, narcotic symptoms appear following slight excitement, and in rare cases sudden cardiac attack can ensue. The greater part of trichloroethylene is metabolized into trichloroacetic acid and trichloroethanol, and excreted in the urine. Acute toxicity is rather low compared with other chlorinated hydrocarbons. The MAC is 50 ppm, and the fatal blood level reportedly varies from 3 to 110 µg/ml.

Dichloroethylene and tetrachloroethylene are less toxic than trichloroethylene. The MAC of dichloroethylene is 200 ppm, while that of tetrachloroethylene is 50 ppm.

Therapy is not difficult in cases of acute poisoning.

Detection method: Model 3, 4 or 5.

Carbon tetrachloride (tetrachlormethane) CCl_4 and **trichloroethane** (1,1,1-trichloroethane) CCl_3CH_3. Although carbon tetrachloride was once widely used domestically, its utilization is now limited to industry as one of the raw materials for manufactoring Freon. Carbon tetrachloride acts on the CNS, producing narcotic symptoms similar to those seen with chloroform, and in cases of serious poisoning, it damages the parenchymatous organs such as the liver or kidney. This agent changes in the body to phosgene, which is also toxic. The MAC is 5 ppm, the toxic level ranges from 10 to 20 ppm, and the fatal blood level ranges from 100 to 200 µg/ml. The lowest oral lethal dose is reportedly 5 ml.

Trichloroethane is less toxic than carbon tetrachloride, with a MAC of 350 ppm, and a toxic level of ca. 600 ppm. The concentration of 1,1,1-trichloroethane in the brain of a fatal case was reported to have been 3.9 mg/100 g (9). As for therapy in the case of carbon tetrachloride, liquid paraffin needs to be given to prevent absorption from the stomach, and this needs to be washed away by gastric irrigation. Fatty foods and drinks such as milk should be prohibited.

Detection method: Model 1, 3 or 4.

Fluorocarbons. A number of halogenated methanes and ethanes containing chlorine and fluorine are commercially called Freons, among them FC-11 (CCl_3F), FC-12 (CCl_2F_2), FC-22 ($CHClF_2$) and FC-113 (CCl_2FCClF_2) are common and FC-11 and FC-12 are widely used as aerosols, propellants or coolants. The MAC is 1,000 ppm. Despite their low toxicity, poisoning and occasional death occur in cases of abuse, where disturbances in the circulatory system can be observed. The blood level of a fatal case was reported to have been 18 mg/100 ml (10). The greater portion of fluorocarbons is excreted unchanged after absorption, without any apparent clinical symptoms.

Detection method: Model 3 or 4.

Halothane (Fluothane) $C_2HBrClF_3$. Halothane is widely used as a safe inhalational anesthetic at a concentration of 1.5 to 2.0%. Halothane can easily become distributed in adipose tissue and is mainly excreted unchanged from the lung. It often causes bradycardia, a reduction of blood pressure, and occusionally cardiac arrest.

The liver suffers from intoxication known as "hepatotoxicity". Death can occur after oral ingestion with the intention of suicide. The fatal blood levels are considered to range from 30 to 50 µg/ml.

Detection method: Model 3, 4 or 1.

2.4.7 Other Important Volatiles

The common volatiles, acetone, diethyl ether and formaldehyde are additionally described.

Acetone CH_3COCH_3. Exposure to a high concentration of acetone irritates the conjunctivae, and causes edema and inflammation of the respiratory passages, dyspnoe, and narcotic symptoms, with accompanying disturbance to the CNS. Small amounts of acetone are always found physiologically. The so-called ketone bodies, including acetone, acetoacetic acid and beta-hydroxybutyric acid, increase in cases of starvation, where combustion of hydrocarbon declines. The MAC is 500 ppm, while exposure to over 12,000 ppm or the oral intake of 50 to 100 ml are considered dangerous and can be fatal.

In serious cases, sodium bicarbonate should be administered against acidosis. Detection method: Model 1 or 2.

Diethyl ether $C_2H_5OC_2H_5$. Ether was once used as a surgical anesthetic. The symptoms following exposure to the vapor at a high concentration are irritation of the respiratory tract and a narcotic effect on the CNS. Most patients soon recover without any serious symptoms when placed in fresh air. Most of the ether is excreted by expiration, with a small part being excreted in the urine.

The MAC is 400 ppm, while exposure to 100,000 ppm or ingestion of 30 ml can be fatal.

Detection method: Model 1 or 2.

Formaldehyde HCHO. Formaldehyde is characterized by an intense irritating odor and protein-coagulating action. In practice, formaldehyde is used as a 35% solution, called formalin. Exposure to a high concentration of formaldehyde produces severe inflammation in the mucous membrane of the air passages such as the nose, larynx, trachea and bronchus, and sometime leads to pneumonia. Edema of the larynx is dangerous. The symptoms after peroral intake either by accident or with the intention of suicide are damage to the digestive canal including erosion, ulcer or perforation, acidosis, and kidney insufficiency with oliguria or anuria. If the patient survives, formaldehyde is oxidized into formic acid and carbon dioxide.

The MAC of formaldehyde is 1 ppm and an intake of 30 to 60 g of formalin is considered to be fatal.

Gastric irrigation is recommended using 0.2% ammonia, 1 to 2% ammonium bicarbonate, milk or lukewarm water. Patients in a serious condition should undergo hemodialysis in addition to oxygen inhalation.

Detection method: GC (11) and LC (12).

References

1. Kageura, M., Nagata, T., Hara, K. & Totoki, K. (1976) Gas chromatographic determination of LP (liquid petroleum) gas (mainly butane) in biological materials (in Japanese with English summary). Jpn. J. Legal Med. *30*, 1–5.
2. Kojima, T., Takeoka, T., Kobayashi, H., Hara, K., Kageura, M. & Nagata, T. (1979) Determination of liquefied petroleum gas (mainly propane) in biological materials by gas chromatography and gas chromatography-mass spectrometry, a case report. Jpn. J. Legal Med. *33*, 80–85.
3. Kato, K., Nagata, T., Kimura, K., Kudo, K., Imamura, K. & Noda, M. (1990) Components of paint thinner in body fluids clearly detected using the salting-out technique. Forensic Sci. Int. *44*, 55–60.
4. Mizuki, E., Kageura, M. & Nagata, T. (1981) Determination of urinary hippuric acid by gas chromatography with 4-benzamidobutanoic acid as an internal standard. Jpn. J. Legal Med. *35*, 42–47.
5. Kimura, K., Nagata, T., Hara, K. & Kageura, M. (1988) Gasoline and kerosene components in blood – a forensic analysis. Human Toxicol. *7*, 299–305.
6. Kimura, K., Nagata, T., Kudo, K., Imamura, T. & Hara, K. (1991) Determination of kerosene and light oil components in blood. Biol. Mass Spectrom. *20*, 493–497.
7. Allan, A. R., Blackmore, R. C. & Toseland, P. A. (1988) A chloroform inhalation fatality-an unusual asphyxiation. Med. Sci. Law. *28*, 120–122.
8. Davies, D. D. (1978) A method of gas chromatography using electron-capture detection for the determination of blood concentrations of halothane, chloroform and trichloroethylene. Br. J. Anaesth. *50*, 147–155.
9. Kojima, T., Kobayashi, H., Hara, K. & Nagata, T. (1978) A case of 1,1,1-trichloroethane poisoning by inhalation. Jpn. J. Legal Med. *32*, 205–208.
10. Standefer, J. C. (1975) Death associated with fluorocarbon inhalation: report of a case. J. Forensic Sci. *20*, 548–551.
11. Yasuhara, A. & Shibamoto, T. (1989) Formaldehyde quantitation in air samples by thiazolidine derivatization: factors affecting analysis. J. Assoc. Off. Anal. Chem. *72*, 899–902.
12. Tomkins, B. A., McMahon, J. M., Caldwell, W. M. & Wilson, D. L. (1989) Liquid chromatographic determination of total formaldehyde in drinking water. J. Assoc. Off. Anal. Chem. *72*, 835–839.

2.5 Volatile Halogenated Compounds

R. J. Flanagan

2.5.1 General Introduction

Chlorinated solvents and other volatile halogenated compounds (substances which are either gases or liquids exerting a significant vapour pressure at room temperature) have been widely used for many years. The Kellner-Solvay process for the production of caustic soda (sodium hydroxide) by the electrolysis of brine was introduced on a large scale in the 1920s. Elemental chlorine was a by-product and the search for a market led to investigation of the potential uses of trichloroethylene and other organochlorines (1). Such compounds and their fluorinated and/or brominated analogues have since been used extensively in industry, in hospitals, in the laboratory, and in the home (tables 1 and 2). However, since the mid 1970s concern as to the consequences of the release of massive quantities of organochlorine and organobromine compounds such as chlorofluorocarbons (CFCs) into the atmosphere has led to the planned phased withdrawal of many of the compounds in current use (Montreal Protocol, table 3). Deodorised 'butane' (liquified petroleum gas, LPG) and dimethyl ether (DME) have already largely replaced CFCs as propellants in aerosols, for example, in many countries. Only time will tell which volatile halogenated compounds will be in widespread use by the year 2000 although in the case of refrigerants the trend is to polyfluorinated compounds such as 1,1,1,2-tetrafluoroethane.

Table 1. Some Volatile Halogenated Compounds [Alternative names and Chemical Abstracts Service (CAS) numbers in brackets]

Bromochlorodifluoromethane (BCF, FC 12B1, halon 1211) (353-59-3)
Bromoform (tribromomethane, halon 1003) (75-25-2)
Bromomethane (methyl bromide, halon 1001) (74-83-9)
Bromotrifluoromethane (FC 13B1, halon 1301) (75-63-8)
Carbon tetrachloride (tetrachloromethane, halon 104) (56-23-5)
Chlorbutol (chlorbutanol, chlorobutanol, chloretone, 1,1,1-trichloro-2-methyl-propan-2-ol) (57-15-8)
2-Chloro-1,1-difluoroethane (FC 142) (75-68-3)
Chlorodifluoromethane (FC 22, halon 121) (75-45-6)
Chloroform (trichloromethane, halon 103) (67-66-3)
Chloromethane (methyl chloride, monochloromethane, halon 101) (74-87-3)
1-Chloro-1-nitropropane (600-25-9)
Chloropicrin (trichloronitromethane, nitrochloroform, vomiting gas) (76-06-2)
1-Chloroprop-2-ene (allyl chloride, 3-chloropropylene) (107-05-1)
1,2-Dibromoethane (ethylene dibromide) (106-93-4)
2,2′-Dichlorodiethyl sulphide (Yperite, sulphur mustard, mustard gas) (505-60-2)
Dichlorodifluoromethane (FC 12, halon 122) (75-71-8)
1,1-Dichloroethane (ethylidine dichloride) (75-34-3)

Table 1. (continued)

1,2-Dichloroethane (ethylene dichloride) (107-06-2)
1,2-Dichloroethylene (acetylene dichloride) (156-59-2)
Dichloromethane (methylene chloride, methylene dichloride, halon 102) (75-09-2)
1,2-Dichloropropane (propylene dichloride) (78-87-5)
1,2-Dichlorotetrafluoroethane (FC 114, cryofluorane) (76-14-2)
Difluoromethane (methylene fluoride, FC 32, halon 12) (75-10-5)
1,2-Difluorotetrachloroethane (FC 112) (76-12-0)
Enflurane (2-chloro-1,1,2-trifluoroethyl difluoromethyl ether) (13838-16-9)
Fluorotrichloromethane (FC 11, halon 113) (75-69-4)
Halothane (2-bromo-2-chloro-1,1,1-trifluoroethane) (151-67-7)
Isoflurane (1-chloro-2,2,2-trifluoroethyl difluoromethyl ether) (26675-46-7)
Methoxyflurane (2,2-dichloro-1,1-difluoroethyl methyl ether) (76-38-0)
Monochloroethane (chloroethane, ethyl chloride, halon 201) (75-00-3)
Perfluoropropane (FC 218, halon 38) (76-19-7)
Phosgene (carbonyl chloride, carbonic dichloride) (75-44-5)
Tetrachloroethylene (perchloroethylene) (127-18-4)
1,1,1,2-Tetrafluoroethane (FC 134a) (811-97-2)
1,1,1-Trichloroethane (methylchloroform, 'Genklene') (71-55-6)
Trichloroethylene (trichloroethene, 'trike', 'Trilene') (79-01-6)
1,2,3-Tichloropropane (glycerol trichlorohydrin) (96-18-4)
1,1,1-Trichlorotrifluoroethane (FC 113a) (354-58-5)
1,1,2-Trichlorotrifluoroethane (FC 113) (76-13-1)
Vinyl chloride (chloroethylene) (75-01-4)

Table 2. Some Past and Current Uses of Volatile Halogenated Compounds (N. B. Many of the compounds listed in Table 1 are also used as chemical intermediates, as laboratory solvents, etc.)

Use and/or product		Major volatile component(s)
Aerosol propellants		FCs 11, 12 & 22 (sometimes with 'butane' and/or dimethyl ether)
Anaesthetics/ analgesics:	Inhalational	Enflurane, halothane, isoflurane, methoxyflurane (rare)
	Topical	FCs 11 & 12, monochloroethane
Anthelmintic (veterinary)		Carbon tetrachloride
Antiseptic/preservative		Chlorbutol
Commercial dry cleaning and degreasing agents		Dichloromethane, 1,1,1-trichloroethane, tetrachloroethylene, trichloroethylene, FC 113 (rarely carbon tetrachloride, 1,2-dichloropropane)
Domestic spot removers and dry cleaners		1,1,1-Trichloroethane, tetrachloroethylene, trichloroethylene
Fire extinguishers		BCF, bromomethane, FCs 11 & 12
Fuel additive		1,2-Dibromoethane
Fumigants		Bromomethane, carbon tetrachloride, chloropicrin, 1,2-dibromoethane, 1,2-dichloroethane
Paint remover		Dichloromethane (usually with toluene)
Refrigerants		FCs 11, 12 & 114, chloromethane
Sedative		Chlorbutol
Surgical plaster/chewing gum remover		1,1,1-Trichloroethane, trichloroethylene
Typewriter correction fluids/thinners		1,1,1-Trichloroethane

Table 3. Revised Montreal Protocol (November 1992)

The Montreal Protocol is an international agreement on the manufacture and import/export of the volatile chlorinated and brominated compounds listed below. Originally it was planned to gradually reduce the use of these controlled substances to zero by the year 2000 (2005 in the case of 1,1,1-trichloroethane) with a 10-year extension granted for less developed countries. Manufacture of some controlled substances was still to be permitted for safety-critical and other 'essential' applications such as use in medical inhalers (bromofluorocarbon fire extinguishers, for example, were to be exempted) or for captive use as chemical intermediates. Not all countries have signed the agreement. A revised timetable (see below) was adopted in Copenhagen in November 1992. Cutbacks are from 1986 production. The European Community has advocated tighter controls, including the phase-out of chlorofluorocarbons containing at least 1 hydrogen atom (HCFCs) by 2014.

Controlled Substances (N.B. Details of fluorocarbon nomenclature are given in Section 1).

a. Chlorofluorocarbons (denoted CFCs in the protocol) – FCs 11, 12, 113, 114, 115 and all other fully halogenated chlorofluorocarbons with 1, 2 or 3 carbon atoms, including all isomers: 75% cut in production by 1.1.1994; complete phaseout by 1.1.1996.

b. Bromofluorocarbons (denoted halons in the protocol) – FCs 12B1, 13B1 and 114B2, including all isomers: complete phaseout by 1.1.1994.

c. Carbon tetrachloride: 85% cut in production by 1.1.1995; complete phase-out by 1.1.1996.

d. 1,1,1-Trichloroethane: 50% cut in production by 1.1.1994; complete phase-out on 1.1.1996.

e. Chlorfluorocarbons with 1, 2 or e carbon atoms containing at least one hydrogen atom (HCFCs, e.g. FC 21, FC 22, etc.): no increase in production from 1996; complete phase out by 2030.

f. Methyl bromide: cut in production (to 1991 levels) by 1995; further cuts likely in future.

Toxicity from exposure to the vapour of halogenated compounds is not a new problem in clinical medicine. Chloropicrin, mustard gas and phosgene (table 1) were all used as chemical warfare agents in World War I while, on the Home Front, many cases of jaundice occurred when 1,1,2,2-tetrachloroethane was used in 'doping' (varnishing) fabric-covered aircraft (1). Today, if anaesthesia is excluded, acute poisoning with halogenated solvents and indeed with other volatile substances usually follows deliberate inhalation of vapour in order to become intoxicated [volatile substance abuse (VSA)]. Patients who ingest solvents or solvent-containing products, either by accident or deliberately, and the victims of domestic and industrial mishaps, provide further groups which may suffer acute poisoning by these compounds. In addition, chloroform and other volatiles are still used in the course of crimes such as rape and murder (4) and a related compound, chlorbutol, still sometimes used as a sedative and as a preservative (5), has been employed in doping racing greyhounds (6). Toxicity due to fumigants such as bromomethane or to compounds used primarily as chemical intermediates, on the other hand, is more commonly associated with occupational exposure to these agents.

One source of confusion when discussing chlorinated solvents or halons (halocarbons, aliphatic hydrocarbons in which one or more hydrogens are replaced by halogen atoms), is that there are often several names in common use for even simple compounds (table 1). As an example of the confusion which can arise, a recent 'fact-sheet' published in the UK by Re-Solv (The Society for the Prevention of Solvent and Volatile Substance Abuse) listed dichloromethane, methylene chloride and methylene dichloride as separate compounds (7). Obviously attempts have been made by bodies such as IUPAC (International Union of Pure and Applied Chemistry) to produce standardised nomenclature for full chemical names but even then one system is not applied universally, many North American publications, for example, following the Chemical Abstracts system rather then the IUPAC system.

With the halons, much confusion is generated by the existence of two separate 'shorthand' numbering systems. The Montreal Protocol (table 3), for example, uses elements of both systems for different compounds within a single document when it would have been easy to stick to either one or the other. The simplest system was promulgated by the US Army Corps of Engineers in the late 1950s und uses the word 'halon' together with a number denoting (reading from *left to right*) the numbers of carbon, fluorine, chlorine and bromine atoms in the molecule; terminal zeros are omitted. The number of hydrogen atoms is calculated by difference. Clearly for anything but very simple molecules this notation gives a 'group' classification. The second numbering system was devised by the American Society of Refrigeration Engineers (ASRE) for methane, ethane and cycloalkane refrigerants (8) and has since been extended to include other fluoroaliphatics (9). In this system all fluorocarbons (FCs) have an identifying number, the first digit (reading from *right to left*) being the number of fluorine atoms in the molecule, the second *from the right* being the number of hydrogens plus 1, and the third *from the right* being the number of carbons minus 1 (omitted if zero). The number of chlorine atoms in the molecule is ascertained by difference. In unsaturated compounds the number of double bonds is shown by the fourth number from the right; bromine is indicated by a capital 'B' followed by a number; different isomers are indicated by lower case suffixes ('a', 'b', etc.) allocated in order of decreasing symmetry; and so on.

Clearly the ASRE fluorocarbon numbering system becomes unwieldy with complex molecules. A further complication is that the numerical values derived in this manner are sometimes used together with the words 'propellant' or 'refrigerant', or with a multiplicity of trade (brand) names including Arcton (ICI), Freon (Du Pont), Frigen (Hoechst), Genetron (Allied-Signal), Isceon (Rhône-Poulenc), Isotron (Pennsalt), KLEA (ICI) and Ucon (Union Carbide). (N.B. 'Halon' is strictly an Allied-Signal trade-name for polytetrafluoroethylene). These trade names are sometimes used with additional numbers not derived using the ASRE system to denote azeotropic mixtures of FCs. To add to the confusion numbers based on the FC numbering system are sometimes used with the suffixes 'C' and/or 'H' to denote the presence of chlorine or hydrogen, respectively, in the molecule. It would appear possible to denote nonfluorinated halons using the ASRE fluorocarbon system but this becomes a bit nonsensical – all (H) (C) FCs are halons, but all halons are not (H) (C) FCs...

2.5.2 Volatile Substance Abuse (VSA)

There are now reports of VSA (also known as 'glue sniffing', inhalent abuse or solvent abuse) from most parts of the world. Schoolchildren or adolescents, often from poorer communities, are those principally involved. Thus, a survey of 1,836 school-children aged 9–17 years in Sao Paulo, Brazil, found that 24% had abused volatile substances; 5% reported 'sniffing' within 30 days of the survey (10). Solvents from adhesives, notably toluene, typewriter correcting fluids and thinners (at present often 1,1,1-trichloroethane), other halogenated solvents, hydrocarbons such as those found in petrol (gasoline), aerosol propellants, volatile anaesthetics such as halothane and fuel gases such as cigarette lighter refills ('butane') are amongst the substances which may be abused (11). Abuse of monochloroethane has been reported recently from the US (12). Since it appears largely immaterial whether halogenated or nonhalogenated volatile substances are abused, it is appropriate to discuss this topic as a whole.

VSA is characterised by a rapid onset of intoxication and usually by an equally rapid recovery. However, a 'high' can be maintained for several hours by repeated 'sniffing'. As with the ingestion of alcohol (ethanol), euphoria, disinhibition and a feeling of invulnerability may occur, but higher doses often lead to less pleasant and more dangerous effects. Nausea and vomiting with the risk of aspiration can occur at any time. Changes in perception may precede bizarre and frightening hallucinations while tinnitus, ataxia, agitation and confusion are often reported; dangerous delusions such as those of being able to fly or swim may also occur (13). Flushing, coughing, sneezing and increased salivation have also been described.

Local chronic sequelae of VSA include recurrent epistaxis, halitosis, oral and nasal ulceration, conjunctivitis, chronic rhinitis and increased bronchial expectoration. Systemic toxicity results in anorexia, thirst, weight loss and fatigue. Loss of concentration, depression, irritability, hostility and paranoia are further reported complications. In addition, there have been many reports of chronic, sometimes permanent, damage to the central nervous system (CNS), heart, liver and kidney (11). Neuropsychological impairment is often present in volatile substance abusers with well-defined neurological abnormalities. Studies have also found that abusers without reported neurological abnormalities obtain lower psychometric test scores than non-abusers, but this may not be caused by VSA (14). The psychosocial aspects of VSA, for example the disruption caused to the families and friends of abusers, must not be neglected. At present only a minority of abusers progress to heavy alcohol or illicit drug use in the UK (11).

The major risk associated with VSA is that of sudden death. This is probably associated with the rapid onset of action of volatile compounds leading to difficulties in controlling the dose, especially in the case of an inexperienced user. A further factor is that, unlike virtually all other abused compounds, there is no obvious unit dose. Bass (15) reported 110 VSA-related deaths in the US from abuse of aerosol propellants and chlorinated solvents during the 1960s. Further series of fatalities have been noted again from the US (16, 17) and from Scandinavia (18, 19). In the UK, sudden deaths from VSA have been monitored systematically since 1982 and increased from about 3 per year in the early 1970s to 151 in 1990; there were 73 such deaths in 1993, the latest year for which figures are available (20, 21). Nevertheless, the death rate from VSA in the UK is relatively low given the number of abusers indicated by prevalence studies. Taking all UK VSA-related deaths, the age at death ranged from 9 to 76 years, but most deaths (73%) occurred in adolescents aged less than 20 years. In contrast to the results of prevalence studies which indicate equal numbers of male and female 'sniffers', most deaths (88%) occurred in males. In some 20–30% of cases there was no evidence that the deceased had a previous history of VSA. There are no published data on VSA-related deaths from other countries comparable to those available in the UK, although individual cases and small series of deaths are reported regularly (22, 23).

The compounds encountered in UK VSA-related deaths are: fuel gases, mainly 'butane' (ca. 30% of cases); solvents from adhesives (ca. 30%); other solvents, notably 1,1,1-trichloroethane (again ca. 30%); and aerosol propellants (FCs and/or 'butane'). There has, however, been a trend towards the use of 'butane' cigarette lighter refills, aerosol propellants and even halon-containing fire extinguishers in the last few years. The precise mode of death is seldom clear, but indirect effects such as trauma, aspiration of vomit and asphyxia associated with the use of a plastic bag are more frequent in deaths involving solvents from adhesives. On the other hand, four modes of 'direct' acute VSA-related death can be recognised: anoxia, vagal stimulation leading to bradycardia and cardiac arrest, respiratory depression and cardiac arrhythmia (24). Of these, cardiac arrhythmia leading to cardiac or cardiorespiratory arrest is probably the most common cause of death. Sudden alarm, exercise or sexual activity may precipitate an arrhythmia since VSA is thought to sensitise the heart to circulating catecholamines; in many VSA-related deaths the immediate ante-mortem event was fright and running (15, 16).

2.5.3 Clinical Toxicology of Volatile Halogenated Compounds

As with other substances the severity of poisoning with volatile halogenated compounds depends on the toxicity of the agent in question and on the magnitude, duration and (sometimes) the route of exposure. Factors such as age, the presence of disease and co-ingestion of alcohol may also be important. The toxicity of compounds used as fumigants and other noxious gases is related to their chemical structure as discussed below. The acute CNS depressant and cardiotoxic effects of relatively inert compounds such as many solvents and inhalational anaesthetics are similar, however, being related more to physical properties than to chemical structure (25). The clinical toxicology of these compounds is thus discussed in a subsequent section. Additional information on the toxicity of the compounds under consideration can be obtained from various monographs (26–28).

2.5.3.1 Fumigants and other toxic gases

2.5.3.1.1 Bromomethane

Bromomethane, a colourless, non-flammable gas at room temperature, is nowadays employed as a fumigant in warehouses, grain stores and ships' holds and also in soil. Earlier uses as a refrigerant and fire extinguishing agent have been limited because of the risk of toxicity. Skin contact with bromomethane can give rise to large blisters several hours post exposure. Atmospheric bromomethane concentrations above 5 ppm (20 $mg m^{-3}$) (UK Occupational Exposure Standard, OES) can be detected by smell and by irritation of the mucous membranes of eyes and respiratory tract. Headache, soreness of the eyes, loss of appetite and abdominal discomfort may occur. Paraesthesiae, affecting particularly the lower limbs, may also occur and may persist for several months. In severe cases anuria may result from renal tubular damage and loss of consciousness may be followed by the development of pulmonary oedema, which may prove fatal, up to two days post-exposure. Disturbances of the CNS may follow lower level exposure. Early features of toxicity include malaise, headache, smarting of the eyes and nausea. After a latent interval of up to 48 h more serious effects including difficulty in focusing, ataxia and incoherence, apathy, drowsiness, weakness (especially of the lower limbs) and convulsions may occur. Depression, irritability, insomnia, visual disturbances and impairment of concentration may persist for months or even years after the acute symptoms have subsided. It is thought that the toxicity of bromomethane arises from methylation of cell macromolecules. Inorganic bromide is a metabolite and measurement of serum bromide can be used to monitor exposure (Section 2.5.2.4).

2.5.3.1.2 Chloromethane

A colourless gas at room temperature, this compound has been used as a refrigerant and as a blowing agent in chemical foams. Exposure to chloromethane can cause drowsiness, dizziness, headache, slurred speech, confusion, staggering gait, nausea, vomiting, blurred vision, weakness, personality changes, cyanosis, convulsions, coma and death. Several months may be required to attain full recovery. Formic acid, carbon dioxide and inorganic chloride are thought to be metabolites of chloromethane in man. Early recognition of the toxicity of this compound lead to the introduction of dichlorodifluoromethane (FC 12) as a 'safe' refrigerant (29).

2.5.3.1.3 Chloropicrin

Sometimes used as a soil fumigant and disinfectant, chloropicrin is a powerful irritant of all body surfaces. Marked irritation of the eyes occurs at an atmospheric concentration of 1 ppm which should act as a warning of exposure. Smarting of the eyes with lachrymation and blepharo-spasm occur if exposure is continued. Inhalation can give rise to bronchospasm, pulmonary oedema, and vomiting. The UK OES for chloropicrin is 0.1 ppm (0.7 mgm^{-3}).

2.5.3.1.4 Vinyl chloride

Vinyl chloride is an important intermediate in plastics manufacture. Exposure to atmospheric vinyl chloride concentrations above 1000 ppm causes CNS depression, with dizziness, lightheadedness, nausea, dulling of the senses and headache. Deaths have followed exposure to high concentrations of vinyl chloride vapour for only a few minutes in an enclosed space. Chronic exposure may cause degenerative bone changes, circulatory disturbances, thrombocytopenia, splenomegaly, hepatomegaly, and hepatic fibrosis. Angiosarcoma of the liver has also been described. N-Acetyl-S-(2-hydroxyethyl)cysteine and thiodiglycolic acid are urinary excretion products of vinyl chloride in the rat (30) but urinary thiodiglycolate excretion is not a reliable measure of exposure to vinyl chloride in man (31).

2.5.3.1.5 War gases

Discussion of the riot control agents CS (1-chloroacetophenone) and CR (*o*-chlorobenzylidene malononitrile) is beyond the scope of this review since these agents have to be dispersed as liquids, powders or smokes (32). Lewisite (2-chlorovinyl-dichloroarsine) would also have to be dispersed as a liquid if its use were ever contemplated (32). Phosgene (table 1), a colourless gas at normal temperatures, was first used in war by the German Army near Ypres in December 1915 and is estimated to have caused some 85% of the deaths attributable to chemical weapons during World War I. It has a suffocating odour said to be reminiscent of mouldy hay. Irritation of the eyes, throat and respiratory tract, with lachrymation, coughing, choking, tightness of the chest, nausea, retching and vomiting are early features of phosgene exposure. After a latent period of between 0.5 and 24 h more serious features, which may be precipitated by exercise, can ensue, including rapid shallow breathing, painful cough and cyanosis. Increasing expectoration of frothy fluid may be followed by the development of pulmonary oedema and death from circulatory collapse (32). *In vivo* hydrolysis to hydrochloric acid may contribute to the toxicity of this agent. Nowadays phosgene is mainly used as a chemical intermediate but it is also produced by the partial combustion of many chlorinated solvents including dichloromethane and carbon tetrachloride, providing one reason why compounds such as carbon tetrachloride are no longer used in fire extinguishers. Phosgene is also an important metabolite of chloroform (table 4).

2,2′-Dichlorodiethyl sulphide (sulphur mustard, table 1) was first used by the German Army on 12 July 1917 at Ypres and gave rise to some 168,000 casualties by the end of 1918. The death rate was only 2–3%. Nevertheless, sulphur mustard is generally recognised as the most effective chemical warfare agent introduced in World War I and was employed extensively by the Italians during their invasion of Ethiopia in 1936 and, more recently, in the Iran-Iraq war. Sulphur mustard, so-called because of its odour, is an oily liquid at room temperature and is a potent alkylating agent. The vapour is extremely penetrating and irritant, and can produce, after a latent period of 2 h or so, nausea, retching, vomiting, fatigue, headache, eye inflammation, reddening of exposed body surfaces, lachrymation and photophobia amongst other features. Later features include inflammation of inner thighs, axillae, genitalia, and buttocks followed by blister formation. The blisters are large, filled with yellow fluid, and may be pendulous. In more severe cases bronchitis with expectoration of necrotic slough may become established and may progress to bronchopneumonia with death in severe cases. Severe bone marrow depression may also occur.The value of sulphur mustard as a chemical warfare agent thus relies upon its potential ability to contaminate large areas of land and to produce large numbers of casualties rather than killing rapidly as in the case of the organophosphate nerve agents (32).

2.5.3.2 Halogenated solvents and inhalational anaesthetics

2.5.3.2.1 General considerations

Clinical features recorded after acute poisoning with halogenated solvents and inhalational anaesthetics include dizziness, euphoria, confusion and drowsiness leading to visual disturbances, coma, convulsions, with respiratory depression and circulatory failure in severe cases (1, 33, 34). Convulsions associated with cerebral infarction (35), congestive cardiomyopathy (36) and pulmonary oedema (37) have been reported after trichloroethylene exposure. A grand-mal convulsion after abuse of monochloroethane has also been described (38). Stuber (39) recorded 284 cases of industrial trichloroethylene poisoning – loss of consciousness occurred in 117 cases and 26 patients died. CNS depression was also the most common finding in 384 industrial accidents involving inhalation of tetrachloroethylene, 1,1,1-trichloroethane or trichloroethylene; only 17 patients died but 168 of the survivors were unconscious when first examined (40).

Cardiac arrhythmias have often been reported after inhalation of halogenated compounds. Indeed, the ability of trichloroethylene to produce cardiac arrhythmias is well known from its former use as an anaesthetic. The arrhythmogenic properties of trichloroethylene may be attributed to metabolism to the well-known cardiotoxin 2,2,2-trichloroethanol, at least in part. Wodka & Jeong (41) described a 15-year-old boy who suffered a cardiorespiratory arrest after abusing typewriter correction fluid containing trichloroethylene and 1,1,1-trichloroethane. Defibrillation was performed and sinus rhythm was established. Subsequent examination revealed acute anteroseptal myocardial injury; coronary artery spasm may have been the mechanism of myocardial injury. Halogenated anaesthetics such as halothane may have synergistic toxic effects in patients previously heavily exposed to chlorinated solvents either occupationally or as a result of VSA (42).

Gastrointestinal symptoms such as nausea and vomiting may predominate after oral ingestion of halogenated solvents, but in such cases the later features of poisoning are similar to those following inhalation of vapour. However, there is a risk of chemical pneumonitis if aspiration occurred at the time of ingestion. Blood concentrations associated with particular clinical features of toxicity tend to be higher after ingestion as compared to inhalation, probably reflecting differences in the rate of distribution to the CNS (see Section 2.5.4).

2.5.3.2.2 Hepatorenal toxicity of chlorinated solvents

Hepatorenal toxicity after acute, acute-on-chronic or chronic exposure to halogenated solvents often results from metabolic transformation, at least in part (table 4). The early features of acute poisoning with carbon tetrachloride are similar to those associated with other solvents although cerebellar dysfunction has also been reported (44). However, absorption of as little as 5 to 10 ml can cause serious hepatorenal damage. As in acute poisoning with paracetamol (acetaminophen), early use of N-acetylcysteine (NAC) may be beneficial even though hepatic reduced glutathione (GSH) stores are not depleted in animals poisoned acutely with carbon tetrachloride (45). Exposure to chloroform (46) and to 1,2-dichloropropane (DCP) (47) may also cause hepatorenal damage and, with both compounds, experimental data suggest NAC may be beneficial. Phosgene is a major metabolite of chloroform and phosgene production is associated with hepatic GSH depletion and hepatorenal damage in laboratory animals (45). Similarly, DCP administration is associated with GSH depletion in rats, possibly via 1-chloro-2-hydroxypropane production (48). Hepatorenal damage has also been attributed to trichloroethylene (49–51), tetrachloroethylene (52) and to 1,1,1-trichloroethane (53). However, no toxicological analyses were performed in these cases and the toxicity observed could have been due to more toxic compounds (chloroform, carbon tetrachloride) present as impurities.

2.5.3.2.3 Halothane and related halogenated anaesthetics

Halothane remains widely used as an inhalational anaesthetic. However, some 25 % of patients show mild disturbance of hepatic function (increased serum aminotransferase or glutathione-S-transferase activities) following exposure to this compound. There are also reports of hepatotoxicity after occupational exposure (54). Of most concern, however, is the massive liver necrosis ('halothane hepatitis') with a mortality of more than 50 % which follows some 1 in 10,000 anaesthetic procedures involving halothane. The mechanism of 'halothane hepatitis' remains uncertain, but there is evidence of an immune-mediated reaction in the form of a high incidence of recent exposure to halothane or other volatile halogenated anaesthetics, eosinophilia, and the presence of serum autoantibodies (54).

Some 20 % of a dose of halothane is metabolized. In addition to the metabolites of halothane listed in table 4, a number of reactive intermediates capable of bonding to macromolecules are formed. The pattern of metabolism appears to be different under hypoxic conditions (55). In order to reproducibly develop hepatotoxicity after halothane exposure, rats require pretreatment with ethanol, phenobarbitone, isoniazid, or tri-iodothyronine; systemic hypoxia is also required in the first two instances. Pretreatment with NAC exerts a protective effect in one of these model systems (56). However, it would be impracticable to use NAC to prevent halothane hepatitis in man because the condition is both unpredictable and rare. Careful history-taking to identify risk factors before using halothane or related compounds is at present a better way of reducing the incidence of this phenomenon (54). Fulminant liver failure may also follow anaesthesia with methoxyflurane and enflurane, but less commonly than with halothane (26).

2.5.3.2.4 Dichloromethane

Dichloromethane is commonly used in paint removers and in other applications (table 2). It is thought likely that this compound will be used increasingly as a replacement for 1,1,1-trichloroethane and possibly for other chlorinated solvents (table 3). However, some 40 % of an absorbed dose of dichloromethane is metabolized to carbon monoxide and this is thought to be responsible for morbidity and possibly even mortality after chronic exposure (57, 58). Co-exposure to methanol, also a common constituent of paint-remover, is said to extend the half-life of carboxyhaemoglobin derived from dichloromethane (59). At present most deaths from acute dichloromethane poisoning occur in confined spaces and the combination of direct dichloromethane toxicity and hypoxia is usually clearly more important than any additional, secondary effect of carbon monoxide. However, it remains to be seen whether morbidity from dichloromethane exposure will become more apparent with the predicted increase in usage of this compound (3).

2.5.3.3 Occupational/Environmental toxicology

Nowadays volatile halogenated compounds are virtually ubiquitous in the environment and, although the Montreal protocol (table 3) aims to restrict manufacture of many hitherto common compounds, it seems impossible to banish all such substances for ever. Chloroform derived from chlorination of organic matter, for example, occurs commonly in potable water supplies and in swimming pools and continues to be the subject of studies such as that of Aggazzotti et al. (60) who measured plasma chloroform concentrations in various groups of swimming pool users. Dry-cleaning establishments are another source of chlorinated hydrocarbons such as tetrachloroethylene, not only after occupational exposure but also in those dwelling nearby (61, 62). Anaesthetists and other operating theatre personnel are a further group where the possible consequences of long-term exposure to the vapour of volatile halogenated compounds give cause for concern (table 2).

2.5.4 Pharmacokinetics of Volatile Substances

The physical properties and pharmacokinetic parameters of some common volatile halogenated compounds are summarised in table 5. Some knowledge of the pharmacokinetics of volatile compounds is important in understanding the rate of onset, the intensity and the duration of intoxication with these compounds, as well as the rate of recovery after inhalational exposure. Such an understanding is also helpful when attempting to interpret the results of toxicological analyses performed on biological samples from poisoned patients (Section 2.5.5.2). These parameters have been studied extensively as regards occupational exposure (65–67) and anaesthesia (63) but not in relation to VSA. These three situations are similar in that dosage (although sometimes variable, especially during occupational exposure and VSA) is prolonged.

2.5.4.1 Absorption, distribution and elimination

The major factors influencing pulmonary uptake of a given solvent during chronic (occupational) exposure include: (a) the concentration of the compound in inspired air, (b) the air : blood and blood : tissue partition coefficients, (c) pulmonary ventilation and blood flow (respiratory rate, fitness, exercise), (d) the proportion of body fat (this may be as high as 50% of body weight in obese individuals), (e) work practices (adequate workplace ventilation), (f) interaction with other inhaled compounds, drugs or alcohol, (g) addiction or aversion to the compound, and (h) individual variation in metabolic clearance (68). The potential complexity of the situation is illustrated by a report on the effect of ethanol ingestion on toluene uptake in II subjects: although the total uptake of toluene was decreased, the maximum blood toluene concentration was increased and the apparent clearance decreased (69).

Inhaled compounds may rapidly attain high concentrations in well-perfused organs (brain, heart) while concentrations in tissues with a relatively poor blood supply such as muscle and adipose tissue may be very low. Should death occur, this situation is 'frozen', but if exposure continues the compound will slowly accumulate in less accessible (poorly perfused) tissues, only to be slowly released once exposure ceases. Thus, the blood concentrations of some compounds may fall monoexponentially, while others may exhibit two (or more) phases of decline (half-lives). Published data on the elimination half-lives of volatile substances are not easily comparable, either because too few samples were taken or the analytical methods used did not have sufficient sensitivity to measure the final half-life accurately (table 5). The partition coefficients of a number of compounds between air, whole blood and various tissues have also been measured *in vitro* using animal tissues, and some *in vivo* distribution data have been obtained from postmortem tissue measurements in human fatalities (table 5). However, these latter data must be used with caution since there are many difficulties inherent in such measurements (sampling variations, analyte stability, etc., see Section 2.5.5.2).

2.5.4.2 Metabolism

Some volatile halogenated substances are partly eliminated unchanged in exhaled air and partly metabolised in the liver and elsewhere, the metabolites being eliminated in exhaled air, in urine or possibly sometimes in bile (table 4). The rate and extent of metabolism may be affected by many factors such as age, disease, dose and exposure to other drugs or solvents. After oral ingestion the extent of hepatic 'first-pass' metabolism may influence systemic availability. Xenobiotics including volatile halogenated compounds may be metabolised in a number of ways, a frequent result being the production of metabolites of greater polarity (water solubility) and thus lower volatility than the parent compound. This is achieved by so-called Phase I (usually oxidation, reduction or hydrolysis) and Phase II reactions (conjugation with glucuronic acid, sulphate, acetate or an amino acid).

The pharmacological activity and pharmacokinetics of any metabolite(s) are often different to those of the parent compound(s), primarily because the more hydrophilic metabolites pass through biological membranes less readily. Although metabolism may result in detoxification, enhanced toxicity may also result. This is especially true of chlorinated solvents and related compounds: some aspects of the toxicity of carbon tetrachloride, chloroform, dichloromethane, 1,2-dichloropropane, trichloroethylene and possibly of halothane can be attributed to the formation of toxic metabolites as discussed in Section 2.5.3.2. Many other compounds such as FCs 11 and 12, tetrachloroethylene and 1,1,1-trichloroethane are largely excreted unchanged (table 5). There is only one common example in which a volatile halogenated metabolite is produced from a relatively non-volatile parent compound, viz. the production of the cardiotoxin 2,2,2-trichloroethanol, which is also a major metabolite of trichloroethylene (table 4), from the sedative drugs chloral hydrate, dichloralphenazone and triclofos.

Table 4. Summary of the Metabolism of Some Common Volatile Halogenated Hydrocarbons (data from Refs. 28 and 43).

Compound	Principal Metabolites (% absorbed dose metabolised)	Notes
Carbon tetra-chloride	Chloroform, carbon dioxide, hexachloroethane & others	Trichloromethyl free radical (reactive intermediate) probably responsible for the marked hepatorenal toxicity of this compound.
Chloroform	Carbon dioxide (up to 50%), diglutathionyl dithiocarbonate (GS.CO.SG)	Phosgene (reactive intermediate) depletes reduced glutathione and causes hepatorenal toxicity (45).
Dichloro-ethane	Carbon monoxide (up to 40%)	Carboxyhaemoglobin half-life 13 h. Blood [carboxyhaemoglobin] useful index of chronic exposure.
Enflurane	Difluoromethoxydifluoro-acetate, inorganic fluoride	Hepatotoxicity is apparent in a very few patients given this drug.
Halothane	2-Chloro-1,1,1-trifluoro-ethane, 2-chloro-1,1-di-fluoroethylene, trifluoro-acetate, inorganic bromide	Toxic metabolism may be important in the aetiology of the hepatotoxicity which occurs in patients exposed to halothane (54).
Methoxy-flurane	Methoxydifluoroacetate, dichloroacetate, oxalate, inorganic fluoride.	Some patients develop nephrotoxicity when given this drug; hepatotoxicity is rare
Tetrachloro-ethylene	Trichloroacetate (< 3%)	Urinary trichloroacetate excretion serves only as a qualitative index of exposure.
1,1,1-Tri-chloroethane	2,2,2-Trichloroethanol (2%), trichloroacetate (0.5%)	Urinary metabolites give qualitative index of exposure only (see tetrachloroethylene).
Trichloro-ethylene	2,2,2-Trichloroethanol, (45%), trichloroacetate (32%)	Trichloroethanol (glucuronide) and trichloroacetate excreted in urine (blood half-lives ca. 12 and 100 h). Trichloroacetate excretion quantitative index of exposure.

Table 5. Physical Properties and Pharmacokinetic Data of Some Volatile Compounds (data from Refs. 28, 63 and 64).

Compound	F.Wt.	B.Pt. (760 mm Hg) (°C)	OES[1] (ppm)	Inhaled Dose Absorbed (%)	Proportion of Absorbed Dose: Eliminated Unchanged (%)	Proportion of Absorbed Dose: Metabolised (%)	$t_{1/2}$ (h)	Brain:Blood Distribution Ratio (Fatal Cases)	Partition Coefficient (Blood:Gas) (37 °C)
Bromochloro-difluoromethane	165.3	− 4	–	–	–	–	–	–	–
Bromomethane	94.5	4	5	–	–	–	–	–	–
Carbon tetrachloride	153.8	76–78	2	–	50?	50?	48	–	1.6
Chloroform	119.4	61	2	–	20–70 (8 h)	> 30	1.5	4	8
Chloromethane	50.5	− 24	50	–	–	–	1–1.5	–	–
Dichloromethane	84.9	39	100	–	?50	< 40	0.7	0.5–1	5–10
Enflurane	184.5	56–58	50	90+	> 80 (5 d)	2.5	36	–	1.9
FC 11	137.4	23.7	1000	92	89	< 0.2	1.5	2.5	0.87
FC 12	120.9	− 29.8	1000	35	99	< 0.2	–	1.4	0.15
FC 22	86.4	− 41	1000	–	–	–	–	1.9	–
Halothane	197.4	50	10	90+	60–80 (24 h)	< 20	0.5	2–3	2.4
Methoxyflurane	165.0	103–108	–	–	19 (10 d)	> 44	–	2–3	11
Tetrachloroethylene	165.9	118–122	50	60+	> 90	1–2	72	9–15	9–19
1,1,1-Tri-chloroethane	133.4	71–81	350	–	60–80 (1 w)	2	10–12	2	1–3
Trichloroethylene	131.4	85–88	100	50–65	16	> 80	30–38	2	9.0
Vinyl chloride	62.5	− 13	7	–	–	–	–	–	–

Key: 1 – UK Occupational Exposure Standard (long-term exposure limit – 8 h time-weighted overage reference period)
Abbreviation: $t_{1/2}$ = terminal phase elimination half-life

2.5.5 Diagnosis of Poisoning due to Volatile Substances

A diagnosis of acute poisoning due to volatile substances should be based on a combination of circumstantial, clinical and analytical evidence rather than on any one factor. Toxicological investigations should be considered mandatory in all VSA-related deaths (24), and such analyses can also be helpful, for example, in cases of unexplained abnormal behaviour, especially in adolescents. Toxicological analyses should be performed routinely if there are forensic implications or if publication of case details is contemplated.

2.5.5.1 Clinical and circumstantial evidence

The recognition of volatile substance abusers can be difficult because many of the manifestations of VSA may appear similar to normal problems of adolescence. However, the diagnosis should be suspected in children with 'drunken' behaviour, unexplained listlessness, anorexia and moodiness. The hair, breath and clothing may smell of solvent, and empty adhesive tubes or other containers, potato crisp bags and cigarette lighter refill or aerosol spray cans are often found. The smell of solvent on the breath is related to the dose and duration of exposure and may last for many hours. The so-called "glue-sniffers' rash" (perioral eczema) is caused by repeated contact with glue poured into a plastic bag. However, only 2 of 300 children who regularly abused glue were found to exhibit this feature (70).

2.5.5.2 Toxicological analyses

For the purposes of this review volatile halogenated compounds have been defined as either gases or liquids exerting a significant vapour pressure at room temperature. Coincidentally this means that headspace gas chromatography (GC) of blood or tissue digests using either packed (71) or capillary (72) columns often provides a convenient mode of analysis. Detection of exposure to reactive molecules, however, is limited to measurement of metabolites or decomposition products, measurement of blood bromide to assess bromomethane exposure being a prime example (Section 2.5.5.2.4).

2.5.5.2.1 Sample collection and storage

The likelihood of detecting exposure to volatile substances by headspace GC of blood is influenced by the nature of the compound(s) involved, the extent and duration of exposure, the time of sampling in relation to the time elapsed since exposure, and the precautions taken when collecting and storing the sample (73). In one series of suspected abusers, volatile compounds or metabolites were detected in 79 of 125 cases (74). In 69 (87%) of positive cases the samples were obtained within 10 h of the suspected exposure. Nevertheless, exposure can be detected using later samples. Thus, in separate cases, toluene was detected at 40 h and 2,2,2-trichloroethanol (from trichloroethylene) at 48 h (74). Analysis of urinary metabolites may extend the time in which exposure may be detected but, of the halogenated compounds commonly abused, only trichloroethylene has a suitable metabolite (table 4). On the other hand, direct mass spectrometry (MS) of expired air can detect many compounds several days post-exposure. However, the use of this technique is limited by the need to take breath directly from the patient (75).

Most volatile compounds are relatively stable in blood if simple precautions are taken. The container should be glass, preferably with a cap lined with metal foil; greater losses may occur if plastic containers are used. The tube should be as full as possible and, ideally, should only be opened when required for analysis and then only when cold (4 °C) (73). If the sample volume is limited, it is advisable to select the container to match the volume of blood so that there is minimal headspace. An anticoagulant (lithium heparin or EDTA) should be used. Specimen storage between 0 and 4 °C is recommended and, in the case of esters such as ethyl and methyl acetates, addition of 1% (w/v) sodium fluoride is advisable to minimise esterase activity. However, many samples submitted in far from ideal circumstances still give useful qualitative results. It is vital that any products thought to have been abused are packed and stored separately from biological specimens to avoid cross-contamination. In a suspected VSA fatality, analysis of tissues (especially fatty tissue such as brain) may prove useful since high concentrations of volatile compounds may be found even if very little is detectable in blood. Tissues should be stored before analysis in the same way as blood.

2.5.5.2.2 Screening for volatile substances by Headspace GC

Headspace sample preparation together with temperature programmed gas chromatography (GC) and split flame ionization/electron capture detection (FID/ECD) provides a simple method of screening for a wide range of volatiles in biological specimens. Ramsey and Flanagan (71) used a packed column [2 m × 2 mm i. d. 0.3% (w/w) Carbowax 20 M on Carbopack C] programmed from 35 to 175 °C. On-column septum injections of up to 400 μl of headspace could be performed and thus good sensitivity (of the order of 0.1 mg/l or better using 200 μl of sample) could be obtained. Moreover, most compounds of interest were retained without resort to sub-ambient operation and the system could be used isothermally at an appropriate temperature for quantitative analyses. Disadvantages included the poor resolution of some very volatile substances, the long total analysis time (40 min), and variation in the peak shape given by alcohols between different batches of column packing.

Bonded-phase wide-bore capillary columns permit relatively large volume septum injections and can offer advantages of improved efficiency, reproducibility and reliability. A 60 m × 0.53 mm i. d. fused silica capillary coated with the dimethylpolysiloxane phase SPB-1 (5 μm film thickness) (Supelco) with split flame ionization/electron capture detection offers many advantages over the packed column system described above (72). Good peak shapes are obtained for polar analytes such as ethanol and on-column injections of up to 300 μl of headspace can be performed with no discernable loss of efficiency. Sensitivity is thus as least as good as that attainable with packed columns. Most commonly-abused compounds, including many with very low boiling points such as BCF, *n*-butane, dimethyl ether, FC 11, FC 12, isobutane and propane, can be retained and differentiated at an initial column temperature of 40 °C followed by programming to 200 °C. The total analysis time is 26 min. The reductions in costs and in the time taken in recycling which arise directly from the use of this relatively high starting temperature are considerable, especially if liquid carbon dioxide cooling would otherwise have been necessary. Quantitative measurements can be performed isothermally or on a temperature program.

Table 6. Headspace Capillary GC: Composition of Qualitative Standard Mixture (see fig. 1)

(a) Qualitative standard mixture [prepared in 125 ml gas sampling bulb (Supelco 2-2146)]

Compound	Amount added (ml)*
BCF	0.005
n-Butane	#
Dimethyl ether	1.0
Isobutane	#
FC 11	0.02
FC 12	0.3
FC 113a	0.5
Propane	#

Key: * Volume of vapour phase in headspace vial
2.0 mL commercial 'butane' (purified liquified petroleum gas, a mixture of *n*-butane, isobutane and propane) added

(b) Liquid components mixture (add 10 μl to mixture of gaseous components in gas sampling bulb [see Table 6 (a)]

Compound	Amount added (ml)
Acetone	7.5
Butanone	5.0
Carbon tetrachloride	0.05
Chloroform	0.5
Ethanol	5.0
Ethylbenzene	2.5
Halothane	0.1
n-Hexane	5.0
Methyl isobutyl ketone	2.5
Propan-2-ol	5.0
Tetrachloroethylene	0.025
Toluene	2.5
1,1,1-Trichloroethane	0.25
1,1,2-Trichloroethane	1.0
Trichloroethylene	0.25
2,2,2-Trichloroethanol	0.015

The analysis of a qualitative standard mixture (table 6) on the temperature program on the SPB-1 capillary column is illustrated in figure 1. Note especially the peak shapes given by ethanol and propan-2-ol and the absence of a peak derived from the septum (71). No deterioration in peak shape has been observed over a

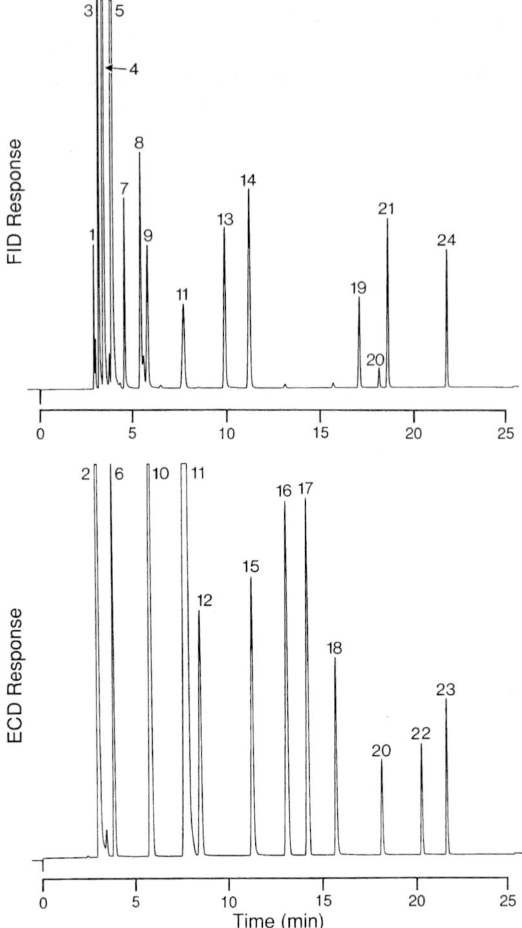

Figure 1 (a, b). Screening for Volatile Substances Using Headspace Capillary GC (reprinted with permission from Streete et al. (72)).
Analysis of the qualitative standard mixture (cf. table 6). Column: 60 m × 0.53 mm i.d. SPB-1 (5 µm film). Oven temperature: 40 °C (6 min), then to 80 °C at 5°/min, then to 200 °C at 10°/min. Injection: ca. 10 µl. Detector sensitivities (FSD): FID 3.2 nA, ECD 64 kHz (Hewlett-Packard 5890). Peaks: 1 = propane, 2 = FC 12, 3 = dimethyl ether, 4 = isobutane, 5 = *n*-butane, 6 = BCF, 7 = ethanol, 8 = acetone, 9 = isopropanol, 10 = FC 11, 11 = FC 113a, 12 = halothane, 13 = butanone, 14 = *n*-hexane, 15 = chloroform, 16 = 1,1,1-trichloroethane, 17 = carbon tetrachloride, 18 = trichloroethylene, 19 = methyl isobutyl ketone, 20 = 1,1,2-trichloroethane (internal standard), 21 = toluene, 22 = tetrachloroethylene, 23 = 2,2,2-trichloroethanol, 24 = ethylbenzene (internal standard).

five year period. Retention and detector response data for compounds in the standard mixture are given in table 7 – in all, data for 244 compounds are available (72). Of the commonly encountered compounds only isobutane/methanol and paraldehyde/toluene are at all difficult to differentiate. Compounds which did not elute during the program generally had boiling points (at atmospheric pressure) of 170 °C or above and retention indices (*n*-alkane) of 1000 or more. The retention data have proved reproducible in routine use over a five years period and should be applicable to other dimethylpolysiloxane-coated capillaries of similar dimensions and film thickness. However, the carrier gas flow might have to be adjusted to give retention data identical to those given in table 7.

Table 7. Retention Data for Compounds in the Qualitative Standard Mixture (cf. table 6; see legend to fig. 1 for GC conditions)

Compound	Retention Time (min)	Relative Retention Time
Propane	3.05	0.163
FC 12	3.18	0.170
Dimethyl ether	3.34	0.179
Isobutane	3.61	0.193
BCF	4.07	0.217
n-Butane	4.09	0.219
Ethanol	4.80	0.257
Acetone	5.66	0.303
Propan-2-ol	6.04	0.323
FC 11	6.13	0.327
FC 113a	8.01	0.428
Halothane	8.76	0.468
Butanone	10.17	0.545
n-Hexane	11.51	0.616
Chloroform	11.65	0.622
1,1,1-Trichloroethane	13.56	0.724
Carbon tetrachloride	14.70	0.785
Trichloroethylene	16.25	0.868
Methyl isobutyl ketone	17.60	0.942
1,1,2-Trichloroethane	18.72	1.000
Toluene	19.14	1.025
Tetrachloroethylene	20.89	1.116
2,2,2-Trichloroethanol	22.29	1.191
Ethylbenzene	22.38	1.198

The sample preparation procedure was that of Ramsey and Flanagan (71) except that 200 rather than 100 μl of internal standard solution were used and it was not necessary to use nitrogen-filled vials. Samples of solid tissues were analysed in the same way as blood after treatment with a proteolytic enzyme (subtilisin A). The analysis of a blood specimen from an adolescent who died after abusing 'butane' gas and vapour from a typewriter correcting fluid is shown in figure 2. The analysis of a blood specimen from a patient who died after inhaling vapour from an electrical component cleaner is illustrated in figure 3. It is clear that the concentrations of the components of interest were well above the limit of detection of the system. Indeed, although no formal studies have been performed, the sensitivity attainable appears to be similar to that obtained using an FID : ECD split ratio of 10 : 1 with the modified Carbopack packed column system, i.e. of the order of 0.01 mg/l for ECD-responding compounds and 0.1 mg/l for the remainder (71). Thus sensitivity enhancement either by "salting-out" or the use of purge-and-trap devices is unnecessary when working with clinical or forensic specimens.

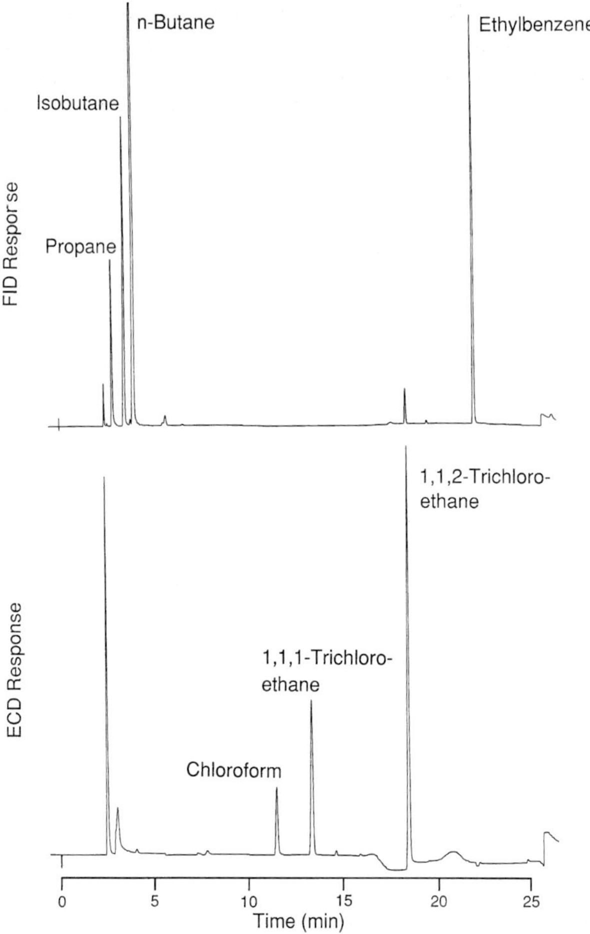

Figure 2. Screening for Volatile Substances Using Headspace Capillary GC (reprinted with permission from Streete et al. (72)).
Analysis of whole blood (200 µl) from a patient who died after abusing cigarette lighter refills and a typewriter correcting fluid containing 1,1,1-trichloroethane. GC conditions: as fig. 1. Injection: 300 µl headspace. Detector sensitivities (FSD): FID 80 pA, ECD 2 kHz. Whole blood 1,1,1-trichloroethane concentration 1.2 mg/l.

Workers using packed columns have emphasized the need to use retention data from two different columns before reporting results (76, 77). However, as in any toxicological investigation the results must never be considered in isolation from any clinical or circumstantial evidence. In addition, the use of an efficient capillary column together with two different detectors confers a high degree of selectivity, particularly for low formula weight compounds where there are very few alternative structures. If more rigorous identification is required, GC combined with MS or Fourier transform infra-red spectrometry (FTIR) may be used. However, GC-MS can be difficult when the fragments produced are less than m/z 40, particularly if the instrument is used for other purposes as well as solvent analyses. In particular, the available sensitivity and spectra of the low molecular weight alkanes renders them very difficult to confirm by GC-MS. GC-FTIR is more appropriate to the analysis of volatiles, but sensitivity is relatively poor particularly when compared with the ECD. In addition, interference, particularly from water and carbon dioxide in the case of biological specimens, can be troublesome.

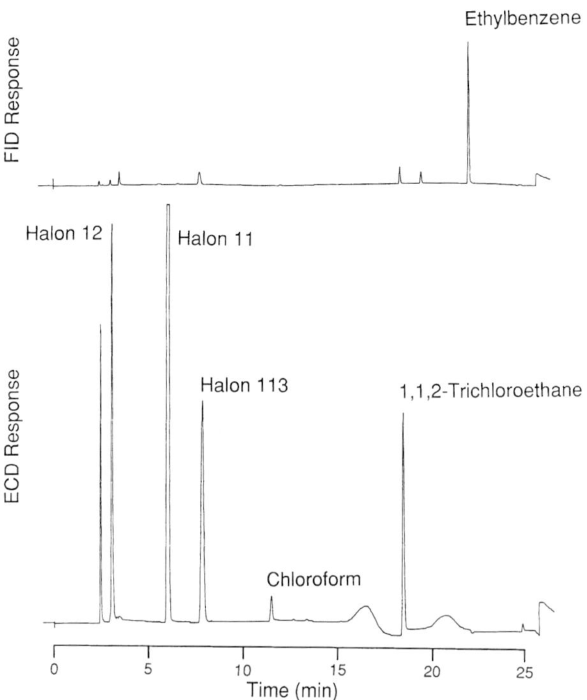

Figure 3. Screening for Volatile Substances Using Headspace Capillary GC (reprinted with permission from Streete et al. (72)).
Analysis of whole blood (200 µl) from a patient who died after abusing an aerosol used to clean electrical components and which contained FCs 11, 12 and 113a. GC conditions: as fig. 1. Injection: 150 µl headspace. Detector sensitivities (FSD): FID 160 pA, ECD 4 kHz. Whole blood FC 11 and 113a concentrations 3.2 and 1.0 mg/l, respectively.

2.5.5.2.3 Quantitative analyses by Headspace GC

Using the SPB-1 capillary column, quantitative measurements can be performed isothermally in duplicate at an appropriate temperature using the appropriate detector. Calibration solutions for liquid analytes were prepared by adding a known volume of the analyte to a volumetric flask containing 'blank' blood using a positive displacement pipette and ascertaining the exact amount added by weighing. Appropriate volume to volume dilutions were then performed, taking care to minimize losses of analyte by handling reagents and glassware at 4 °C and storing samples and standards at 4 °C with minimal headspace (73). Portions of the standards were transferred to headspace vials for analysis as described above and calibration graphs of peak height ratio of analyte to internal standard against analyte concentration prepared. In most cases either 1,1,2-trichloroethane or ethylbenzene could be used as the internal standard. The same calibration solutions were used in the analysis of blood and of tissue digests. Standard concentrations in the range 0.1 to 10 or 0.5 to 50 mg/l were usually adequate in cases of acute poisoning (figures 2 and 3).

After exposure to 350 ppm 1,1,1-trichloroethane (UK OES, table 5) for 1 h, the mean blood concentration was 2.6 mg/l (20 μmol/l) (78). Blood 1,1,1-trichloroethane concentrations ranged from 0.1 to 46 mg/l in samples from 66 patients, 29 of whom died as a result of VSA (74). There was a broad relationship between blood concentration and the severity of poisoning but there were large variations within each patient group. The absence of a strong correlation between blood concentration and clinical features of toxicity is probably due to rapid initial distribution into tissues. In addition, other compounds were present in many non-fatal cases, further complicating recognition of any dose-response relationship.

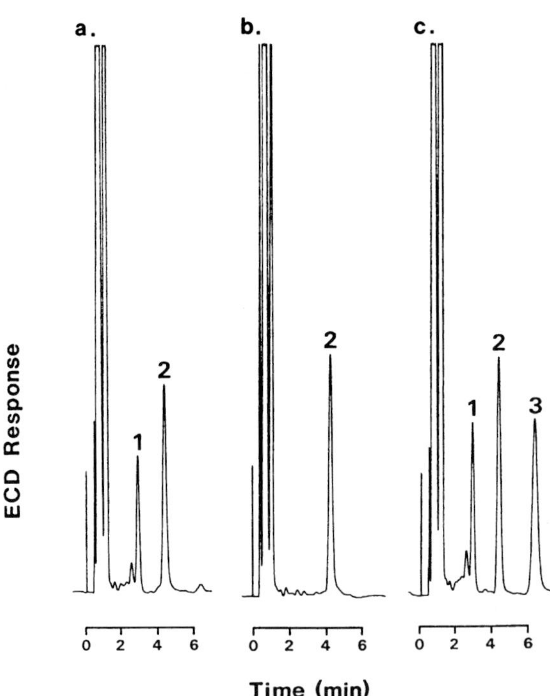

Figure 4. Chromatograms Obtained Using the GLC Trichloroacetic Acid (TCA) Assay (Ruprah (84)).
Column: 2 m × 2 mm i.d. glass packed with 8% (w/w) OV-225 on Chromosorb W HP, 100–120 mesh. Carrier-gas and purge (ECD) flow-rates: 30 & 60 ml/min, respectively. Oven temperature: 110 °C. Injections: 2 μl. Samples: (a) Derivatized extract of bovine plasma containing TCA (5 mg/l), (b) 'blank' human plasma, (c) plasma from a patient suspected of abusing trichloroethylene (TCA concentration 6 mg/l). Peaks: 1 = methylated TCA, 2 = 1,4-dichlorobenzene (internal standard), 3 = 2,2,2-trichloroethanol.

2.5.5.2.4 Analysis of metabolites

Inorganic bromide is a metabolite of bromomethane and may be measured in serum by colorimetry (79) or by chromatographic methods such as that of Goewie & Hogendoorn (80). The serum bromide concentrations associated with toxicity are much lower after bromomethane exposure than when inorganic bromide has been used in therapy. Thus, after therapeutic administration of inorganic bromide, toxicity is usually associated with serum bromide concentrations above 500 mg/l. On the other hand, after bromomethane exposure toxicity may occur with serum bromide concentrations above 30 mg/l and concentrations in the range 90–400 mg/l have been reported in fatal cases (28).

The urinary excretion of trichloroacetic acid (TCA), and to a lesser extent that of 2,2,2-trichloroethanol (TCE), can be used to monitor exposure to trichloroethylene (table 4). Methods based on the Fujiwara reaction lack specificity when applied to biological samples (81, 82). GC-ECD methods offer increased selectivity and the method of van der Hoeven et al. (83), in which TCA is methylated by heating with methanolic boron trifluoride at 80 °C, can be used to measure both TCA and total (free + conjugated) TCE (figure 4). Free TCE can be measured using GLC/ECD after extraction into toluene (figure 5). An automated headspace GC-ECD method for measuring trichloroethylene, TCE and TCA in blood and in urine which relies on thermal decarboxylation of TCA to chloroform has also been

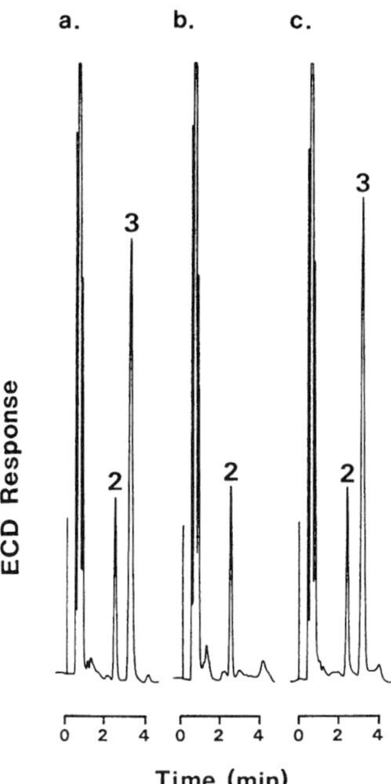

Figure 5. The Analysis of Free 2,2,2-Trichloroethanol (TCE) (Ruprah (84)).
GC conditions: as fig. 4 except column temperature 140 °C. Injection: 1 µl. Samples: (a) extract of bovine plasma containing TCE (3 mg/l), (b) 'blank' human plasma, (c) plasma from a patient suspected of abusing trichloroethylene (plasma free TCE concentration 3 mg/l, cf. fig. 4). Peaks: 2 = 1,4-dichlorobenzene (internal standard), 3 = TCE.

described (85). A similar GC-FID method for urine TCA has been reported (86). TCA has a long half-life in blood (table 4) and thus occupational trichloroethylene exposure is best monitored by collecting urine samples towards the end of the working week. TCA excretion is impaired by ethanol and it is therefore important that alcoholic drinks are avoided when urinary TCA is to be measured. Urinary TCA concentrations above 100 mg/g creatinine (70 μmol/mmol creatinine) are indicative of excessive trichloroethylene exposure.

Acknowledgements
I thank ICI Chemicals & Polymers (Runcorn), Fluorochem (Glossop), Mr. J.D. Ramsey (St George's Hospital Medical School) and Mr P.J. Streete and Dr. M. Ruprah (Poisons Unit) for help and advice.

References

1. Hunter, D. (1978) The Diseases of Occupations, pp. 570–621. Edition 6. Hodder & Stoughton, London.
2. Stevenson, R. (1990) CFCs RIP. Chem. Br. *26*, 731–733.
3. Stevenson, R. (1992) Goodbye carbon tet. Chem. Br. *28*, 208–209.
4. McGee, M.B., Jejurikar, S.G. & Van Berkom, L.C. (1987) A double homicide as a result of chloroform poisoning, J. Forensic Sci. *32*, 1453–1459.
5. Tung, C., Graham, G.G., Wade, D.N. & Williams, K.M. (1982) The pharmacokinetics of chlorbutol in man. Biopharmaceut. Drug Dispos. *3*, 371–378.
6. Smith, H. & Thorpe, J.W. (1978) Drugging racing greyhounds with chlorbutanol. J. Forensic Sci. *18*, 185–188.
7. Re-Solv (1992) Volatile Substance Abuse: Recognition of Products. Fact Sheet No. 4. Re-Solv, Stone.
8. ASRE (1957) Proposed ASRE Standard 34. Refrigerating Engineering *65* (February), 49–51.
9. PCR Inc. (1990) General Catalog. Gainsville, Florida: PCR Inc., 176–181.
10. Carlini-Cotrim, B. & Carlini E.A. (1988) The use of solvents and other drugs among children and adolescents from low socioeconomic background: a study in Sao Paulo, Brazil. Int. J. Addictions *23*, 1145–1156.
11. Flanagan, R.J., Ruprah, M., Meredith, T.J. & Ramsey, J.D. (1990) An introduction to the clinical toxicology of volatile substances. Drug Safety *5*, 359–383.
12. Hersh, R. (1991) Abuse of ethyl chloride. Am J. Psychiatr. *148*, 270–271.
13. Evans, A.C. & Raistrick, D. (1987) Phenomenology of intoxication with toluene-based adhesives and butane gas. Br. J. Psychiatr. *150*, 769–773.
14. Chadwick, O., Anderson H.R., Bland, M. & Ramsey, J. (1989) Neuropsychological consequences of volatile substance abuse: a population based study of secondary school pupils. Br. Med. J. *298*, 1679–1684.
15. Bass, M. (1970) Sudden sniffing death. J. Amer. Med. Assn. *212*, 2075–2079.
16. Carlton, R.F. (1976) Fluorocarbon toxicity: deaths and anaesthetic reactions. Ann. Clin. Lab. Sci. *6*, 411–414.
17. Garriott, J. & Petty, C.S. (1980) Death from inhalant abuse: toxicological and pathological evaluation of 34 cases. Clin. Toxicol. *16*, 305–315.
18. Edh, M., Selerud, A. & Sjoberg, C. (1973) Death and sniffing: a report on 63 cases. Svenska Lakartidningen *70*, 3949–3959.
19. Kringsholm, B. (1980) Sniffing associated deaths in Denmark. Forensic Sci. Internat. *15*, 215–225.
20. Anderson, H.R., Macnair, R.S. & Ramsey, J.D. (1985) Deaths from abuse of volatile substances: a national epidemiological study. Br. Med. J. *290*, 304–307.
21. Taylor, J.C., Norman, C.L., Bland, J.M., Anderson, H.R. & Ramsey, J.D. (1995) Trends in Deaths Associated with Abuse of Volatile Substances 1971–1993. Report 8. St George's Hospital Medical School, London.
22. Jacob, B., Heller, C., Daldrup, T., Bürrig, K.F., Barz, J. & Bonte, W. (1989) Fatal accidental enflurane intoxication. J. Forensic Sci. *34*, 1408–1412.

23. Walker, F.B. & Morano, R.A. (1990) Fatal recreational inhalation of enflurane. J. Forensic Sci. *35*, 197–198.
24. Shepherd, R.T. (1989) Mechanism of sudden death associated with volatile substance abuse. Human Toxicol. *8*, 287–291.
25. Clark, D.G. & Tinston, D.J. (1982) Acute inhalation toxicity of some halogenated and non-halogenated hydrocarbons. Human Toxicol. *1*, 239–247.
26. Ellenhorn, M.J. & Barceloux, D.G. (1988) Medical Toxicology. Diagnosis and Treatment of Human Poisoning. Elsevier, Amsterdam.
27. Proctor, N.H., Hughes, J.P. & Fischman, M.L. (1988) Chemical Hazards of the Workplace. J.B. Lippincott Co., Philadelphia.
28. Baselt, R.C. & Cravey, R.H. (1989) Disposition of Toxic Drugs and Chemicals in Man. Edition 3. Biomedical Publications, Davis, California.
29. Sayers, R.R., Yant, W.P., Chornyak, J. & Shoaf, H.W. (1930) Toxicity of dichlorodifluoromethane; a new refrigerant. Report No. 3013. United States Bureau of Mines, Washington.
30. Watanabe, P.G., McGowan, G.R., Madrid, E.O. & Gehring, P.J. (1976) Fate of (^{14}C) vinyl chloride following inhalational exposure to rats. Toxicol. Appl. Pharmacol. *37*, 49–59.
31. Draminski, W. & Trojanowska, B. (1981) Chromatographic determination of thiodiglycolic acid – a metabolite of vinyl chloride. Arch. Toxicol. *48*, 289–292.
32. Beswick, F.W. & Maynard, R.L. (1987) Poisoning in conflict. In: Oxford Text-book of Medicine, Edition 2, Ed. Weatherall, D.J., Ledingham, J.G.G. & Warell, D.A., pp. 6.59–6.65. Oxford University Press, Oxford.
33. McGee, M.B., Meyer, R.F. & Jejurikar, S.G. (1990) A death resulting from trichlorotrifluoroethane poisoning. J. Forensic Sci. *35*, 1453–1460.
34. Lerman, Y., Winkler, E., Tirosh, M.S., Danon, Y. & Almog, S. (1991) Fatal accidental inhalation of bromochloridifluoromethane (halon 1211). Human Exptl. Toxicol. *10*, 125–128.
35. Parker, M.J., Tarlow, M.J. & Anderson, J.M. (1984) Glue-sniffing and cerebral infarction. Arch. Dis. Childh. *59*, 675–677.
36. Mee, A.S. & Wright, P.L. (1980) Congested (dilated) cardiomyopathy in association with solvent abuse. J. Roy. Soc. Med. *73*, 671–672.
37. Seage, A.J. & Burns, M.W. (1971) Pulmonary oedema following exposure to trichloroethylene. Med. J. Aust. *2*, 484–486.
38. Nordin, C., Rosenqvist, M. & Hollstedt, C. (1988) Sniffing of ethyl chloride – an uncommon form of abuse with serious mental and neurological symptoms. Int. J. Addictions *23*, 623–627.
39. Stüber, K. (1931) Gesundheitsschädigungen bei der gewerblichen Verwendung des Trichloräthylens und die Möglichkeiten in der Verhütung. Arch. Gewerbepath. Gewerbehyg. *2*, 398–456.
40. McCarthy, T.B. & Jones, R.D. (1983) Industrial gassing poisonings due to trichloroethylene, perchloroethylene, and 1-1-1 trichloroethane, 1961–1980. Br. J. Industr. Med. *40*, 450–455.
41. Wodka, R.M. & Jeong, E.W.S. (1989) Cardiac effects of inhaled type-writer correction fluid. Ann. Intern. Med. *110*, 91–92.
42. McLeod, A.A., Marjot, R., Monaghan, M.J., Hugh-Jones, P. & Jackson, G. (1987) Chronic cardiac toxicity after inhalation of 1,1,1-trichloroethane. Br. Med. J. *294*, 727–729.
43. Toftgård, R. & Gustafsson, J-Å. (1980) Biotransformation of organic solvents: a review. Scand. J. Work Environmental Hlth. *6*, 1–18.
44. Johnson, B.P., Meredith, T.J. & Vale, J.A. (1983) Cerebellar dysfunction after acute carbon tetrachloride poisoning. Lancet ii, 968.
45. Flanagan, R.J. & Meredith, T.J. (1991) The use of N-acetylcysteine in clinical toxicology. Amer. J. Med. *91* (Suppl 3C), 131S–139S.
46. Storms, W.W. (1973) Chloroform parties. J. Amer. Med. Assn. *225*, 160.
47. Pozzi, C., Marai, P., Ponti, R., Dell'Oro, C., Sala, C., Zedda, S. & Locatelli, F. (1985) Toxicity in man due to stain removers containing 1,2-dichloropropane. Br. J. Industr. Med. *42*, 770–772.
48. Imberti, R., Mapelli, A., Colombo, P., Richelmi, P., Bertè, F. & Bellomo, G. (1990) 1,2-Dichloropropane (DCP) toxicity is correlated with DCP-induced glutathione (GSH) depletion and is modulated by factors affecting intra-cellular GSH. Arch. Toxicol. *64*, 459–465.
49. Baerg, R.D. & Kimberg, D.V. (1970) Centrilobular hepatic necrosis and acute renal failure in 'solvent sniffers'. Ann. Intern. Med. *73*, 713–720.

50. Clearfield, H. R. (1970) Hepatorenal toxicity from sniffing spot-remover (trichloroethylene). Digestive Dis. *15*, 851–856.
51. Litt, I. F. & Cohen, M. I. (1969) "Danger... vapour harmful": spot remover sniffing. N. Engl. J. Med. *281*, 543–544.
52. Meckler, L. C. & Phelps, D. K. (1966) Liver disease secondary to tetrachloroethylene exposure: a case report. J. Amer. Med. Assn. *197*, 662–663.
53. Halevy, J., Pitlik, S., Rosenfeld, J. & Eitan, B. D. (1980) 1,1,1-Trichloroethane intoxication: a case report with transient liver and renal damage. Review of the literature. Clin. Toxicol. *16*, 467–472.
54. Neuberger, J. M. (1990) Halothane and hepatitis: incidence, predisposing factors and exposure guidelines. Drug Safety *5*, 28–38.
55. Farrell, G. C. (1988) Mechanism of halothane induced liver injury. J. Gastroenterol. Hepatol. *3*, 465–482.
56. Keaney, N. P. & Cocking, G. (1981) N-Acetylcysteine protection against halothane hepatotoxicity: experiments in rats. Lancet ii, 95.
57. Stewart, R. D. & Hake, C. L. (1976) Paint-remover hazard. J. Amer. Med. Assn. *235*, 398–401.
58. Di Vincenzo, G. D. & Kaplan, C. J. (1981) Uptake, metabolism, and elimination of methylene chloride by humans. Toxicol. Appl. Pharmacol. *59*, 130–140.
59. Vale, J. A., Meredith, T. J. & Proudfoot, A. T. (1987) Poisoning from hydrocarbons. In: Oxford Text-book of Medicine, Edition 2, Ed. Weatherall, D. J., Ledingham, J. G. G. & Warell, D. A., pp. 6.41–6.53. Oxford University Press, Oxford.
60. Aggazzotti, G., Fantuzzi, G., Tartoni, P. L. & Predieri, G. (1990) Plasma chloroform concentrations in swimmers using indoor swimming pools. Arch. Environ. Hlth. *45*, 175–179.
61. Imbriani, M., Ghittori, S., Pezzagno, G. & Capodaglio, E. (1988) Urinary excretion of tetrachloroethylene (perchloroethylene) in experimental and occupational exposure. Arch. Environ. Hlth. *43*, 292–298.
62. Popp, W., Müller, G., Baltes-Schmitz, B., Wehner, B., Vahrenholz, C., Schmieding, W., Benninghoff, M. & Norpoth, K. (1992) Concentrations of tetrachloroethene in blood and trichloroacetic acid in urine in workers and neighbours of dry-cleaning shops. Int. Arch. Occup. Environ. Hlth., *63*, 393–395.
63. Fiserova-Bergerova, V. (1983) Modelling of Inhalation Exposure to Vapours: Uptake, Distribution and Elimination. Volumes I and II. CRC Press, Boca Raton.
64. Moffatt, A. C. (Ed). (1986) Clarke's Isolation and Identification of Drugs. Edition 2. Pharmaceutical Press, London.
65. Andersen, M. E. (1981) Pharmacokinetics of inhaled gases and vapours. Neurobehavioural Toxicol. Teratol. *3*, 383–389.
66. Astrand, I. (1985) Uptake of solvents from the lungs. Br. J. Industr. Med. *42*, 217–218.
67. Brown, W. D., Setzer, J. V., Dick, R. B., Phipps, F. C. & Lowry, L. K. (1987) Body burden profiles of single and mixed solvent exposures. J. Occup. Med. *29*, 877–883.
68. Gompertz, D. (1980) Solvents – the relationship between biological monitoring strategies and metabolic handling: a review. Ann. Occup. Hygiene *23*, 405–410.
69. Wallén, M., Naslund, P. H. & Byfalt Nordqvist, M. (1984) The effects of ethanol on the kinetics of toluene in man. Toxicol. Appl. Pharmacol. *76*, 414–419.
70. Sourindrhin, I. & Baird, J. A. (1984) Management of solvent abuse: a Glasgow community approach. Br. J. Addiction *79*, 227–232.
71. Ramsey, J. D. & Flanagan, R. J. (1982) Detection and identification of volatile organic compounds in blood by headspace gas chromatography as an aid to the diagnosis of solvent abuse. J. Chromatogr. *240*, 423–444.
72. Streete, P. J., Ruprah, M., Ramsey, J. D. & Flanagan, R. J. (1992) Detection and identification of volatile substances by headspace capillary gas chromatography to aid the diagnosis of acute poisoning. Analyst *117*, 1111–1127.
73. Gill, R., Hatchett, S. E., Osselton, M. D., Wilson, H. K. & Ramsey, J. D. (1988) Sample handling and storage for the quantitative analysis of volatile compounds in blood: the determination of toluene by headspace gas chromatography. J. Analyt. Toxicol. *12*, 141–146.
74. Meredith, T. J., Ruprah, M., Liddle, A. & Flanagan, R. J. (1989) Diagnosis and treatment of acute poisoning with volatile substances. Human Toxicol. *8*, 277–286.
75. Ramsey, J. D. (1984) Detection of solvent abuse by direct mass spectrometry on expired air. In: Drug determination in therapeutic and forensic contexts. Eds. Reid, E. & Wilson, I. D. New York: Plenum, pp. 357–362.

76. Franke, J.P., Wijsbeek, J., de Zeeuw, R.A., Möller, M.R. & Niermeyer, H. (1988) Systematic analysis of solvents and other volatile substances by gas chromatography. J. Analyt. Toxicol. *12*, 20–24.
77. Goebel, K.-J. (1982) Gas chromatographic identification of low boiling point compounds by means of retention indices using a programmable pocket calculator. J. Chromatogr. *235*, 119–127.
78. Mackay, C.J., Campbell, L., Samuel, A.M., Alderman, K.J., Idzikowski, C., et al. (1987) Behavioural changes during exposure to 1,1,1-trichloroethane: time-course and relationship to blood solvent levels. Amer. J. Industr. Med. *11*, 223–239.
79. Sunshine, I. (1975) Bromide. Type A procedure. In: Methodology for Analytical Toxicology, pp. 54–55. CRC Press, Cleveland.
80. Goewie, C.E. & Hogendoorn, E.A. (1985) Liquid chromatographic determination of bromide in human milk and plasma. J. Chromatogr. *344*, 157–165.
81. Tanaka, S. & Ikeda, M. (1968) A method for determination of trichloroethanol and trichloroacetic acid in urine. Br. J. Industr. Med. *25*, 214–219.
82. Gibitz, H.J. & Plochl, E. (1973) Oral trichloroethylene intoxication in a four and a half year old boy. Arch. Toxicol. *31*, 13–18.
83. van der Hoeven, R., Drost, R.H., Maes, R.A.A., Dost, F., Plomp, T.A. & Plomp, G.J.J. (1979) Improved method for the electron-capture gas chromatographic determination of trichloroacetic acid in human serum. J. Chromatogr. *164*, 106–108.
84. Ruprah, M. (1988) Development and application of simple analytical procedures to aid in the diagnosis and management of poisoning with volatile substances. Ph.D. thesis, University of London.
85. Christensen, J.M., Rasmussen, K. & Koppen, B. (1988) Automatic headspace gas chromatographic method for the simultaneous determination of trichloroethylene and metabolites in blood and urine. J. Chromatogr. *442*, 317–323.
86. Senft, V. (1985) Simple determination of trichloroacetic acid in urine using head space gas chromatography: a suitable method for monitoring exposure to trichloroethylene. J. Chromatogr. *337*, 126–130.

2.6 Pesticides

M. Geldmacher - v. Mallinckrodt and G. Machbert

2.6.1 General Remarks

2.6.1.1 Definition

A pesticide is defined as a chemical substance used for destruction of organisms detrimental to man or to some of his interests, excluding drugs or compounds meant for processed food, drugs, paints, varnish, tar and the like, nor does it include medicines designed for internal use, whether applied topically or systemically. Since $^3/_4$ of the 800,000 types of living creatures belong to the insect kingdom, their destructive power can be easily appreciated. Hence the common use of pesticides (1).

2.6.1.2 Classification of pesticides

There are three major classifications of pesticides, the use of which depends on the intended application (2).

2.6.1.2.1 Classification by usage

Examples for this classification are:

Acaricide
Aphicide
Bacteriostat (soil)
Fumigant
Fungicide, other than for seed treatment
Fungicide, for seed treatment
Herbicide
Insecticide
Insect growth regulator
Larvicide
Molluscicide
Miticide
Nematocide
Other use for plant pathogen
Plant growth regulator
Rodenticide
Repellant (species)
Synergist

2.6.1.2.2 Classification by chemical type

This chemical classification does not represent a recommendation on the part of the IPCS (International Programme on Chemical Safety) (2) as to the way in which pesticides should be classified. (IPCS is a joint venture of the United Nations Environment Programme (UNEP), the International Labour Organisation (ILO) and the World Health Organisation (WHO)). However, this classification helps to determine similarities in the mode of action and metabolism in the target organism, as well as in man. It is also a very suitable classification for analytical purposes. Furthermore it has some significance in the sense that pesticides may have a common antidote, or may be confused in the nomenclature with other chemical types (e.g. thiocarbamates are not cholinesterase inhibitors and do not have the same effects as carbamates).

It should be considered that some pesticides may fall into more than one group. Examples are:

Carbamate
Chloronitrophenol derivative
Organomercury compound
Organophosphorus compound
Organotin compound
Pyridyl derivative
Phenoxyacetic acid derivative
Pyrethroid
Triazine derivative
Thiocarbamate

2.6.1.2.3 Classification by hazard

This is the classification recommended by IPCS (2). A terminological distinction should be made between toxicity, which is an innate capacity of a substance to cause damage, and hazard, which is the risk of poisoning arising in practice. Hazard is based on the toxicity of the compound and on its formulations. In particular, allowance is made for the lesser hazards from solids as compared with liquids. Thus, toxicity and hazard are not synonyms, but hazard is also a function of two variables other than toxicity: contamination and time. This can be expressed in the form of an equation (3):

$$\text{Hazard} = \text{Toxicity} \times \text{Contamination} \times \text{Time}$$

where:
Hazard = risk of poisoning
Toxicity = ability to cause damage
Contamination = prerequisite for entering the body
Time = duration of contact with pesticide.

The classification is based primarily on the acute oral and dermal toxicity in the rat (LD_{50}). Four groups are proposed, see table 1 (2).

Substances with a lower LD_{50} than characterized in Class III are thought unlikely to present acute hazard in normal use.

Table 1. Classification of Pesticides by Hazard Recommended by WHO (2)

Class	LD_{50} for the rat (mg/kg body weight)			
	Oral		Dermal	
	Solids	Liquids	Solids	Liquids
IA Extremely hazardous	5 or less	20 or less	10 or less	40 or less
IB Highly hazardous	5–50	20–200	10–100	40–400
II Moderately hazardous	50–500	200–2000	100–1000	400–4000
III Slightly hazardous	Over 500	Over 2000	Over 1000	Over 4000

The terms "solids" and "liquids" refer to the physical state of the product or formulation being classified.

In compiling the last group, an oral LD_{50} of > 2000 mg/kg of solids and > 3000 mg/kg for liquids has been assumed.

However it should not be overlooked that in formulations of these technical products, solvents or vehicles may present a greater hazard than the actual pesticide and therefore classification of a formulation in one of the higher hazard classes may be necessary.

2.6.1.3 Toxicity for man

It is difficult to give reliable data on the toxic, acute toxic or lethal dose for man. The method of uptake (oral, percutaneous, by inhalation) plays an important role, as well as the nature of the commercial product (solvents, emulsifiers, absorption to granules, etc.), the age and health conditions of the exposed person etc.

A good approach to determine the highest nontoxic dose for oral intake is based on the so-called ADI values (4). ADI's (Acceptable Daily Intakes) are derived from measurements or estimates of the highest dietary level that does not cause significant changes in any measured variable. The most sensitive of these parameters, for example for organophosphorus or carbamate pesticides is the erythrocyte acetylcholinesterase or pseudocholinesterase activity in the blood.

Another approach to evaluate the toxicity of hazardous substances for man (uptake by inhalation or percutaneous) are the German MAK and BAT values (5, 6).

The MAK value (Maximum Allowable Concentration Value in the Workplace) is defined as the maximum permissible concentration of a chemical compound present in the air within a working area (as gas, vapor, particulate matter) which, according to present knowledge, generally does not impair the health of the employee or cause undue annoyance. As a rule, the MAK value is integrated as an average concentration over periods of up to one workday or one shift. For substances with systemic effects like organophosphorus pesticides, limitation of exposure peaks are given. In establishing the MAK values, scientifically based criteria for health protection, rather than technical or economical feasibility, are employed.

The BAT value (Biological Tolerance Value for a Working Material) is defined as the maximum permissible quantity of a chemical compound, its metabolites, or any deviation from the norm of biological parameters induced by these substances. For parathion, for example, this is the 4-nitrophenol excretion in urine and the inhibition of erythrocyte acetylcholinesterase activity in blood.

In the United States, the American Conference of Governmental Industrial Hygienists (ACGIH) has adopted similar limits, namely, the TLV's (Threshold Limit Values) and the BEI's (Biological Exposure Indices) (7).

In the following chapters ADI, MAK and BAT values for representatives of each group of pesticides discussed here are given.

According to their importance in human intoxications the following chemical pesticide groups will be discussed in detail:

Organophosphorus pesticides
Carbamate pesticides
Chlorinated hydrocarbon insecticides
Bipyridilium herbicides

2.6.2 Organophosphorus Pesticides

2.6.2.1 Chemical structure and physico-chemical properties

More than 100 anticholinesterase organophosphorus pesticides are known. They are normally esters, amides, or thiol derivatives of phosphoric, phosphonic, phosphorothioic, or phosphonothionic acids (3).

The essential structure of organophosphorus pesticides is

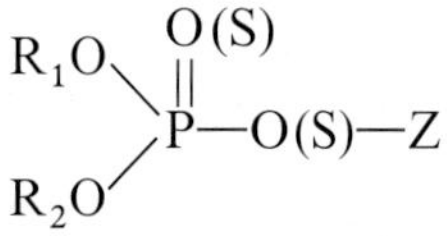

$$(R_1O)(R_2O)P(=O(S))\text{—}O(S)\text{—}Z$$

R_1 and R_2 are usually simple alkyl or aryl groups, both of which may be bonded directly to phosphorus (in phosphinates), or linked via $-O-$ or $-S-$ (in phosphates); or R_1 may be bonded directly and R_2 bonded via one of the above groups (phosphonates). In phosphoramidates, carbon is linked to phosphorus through an

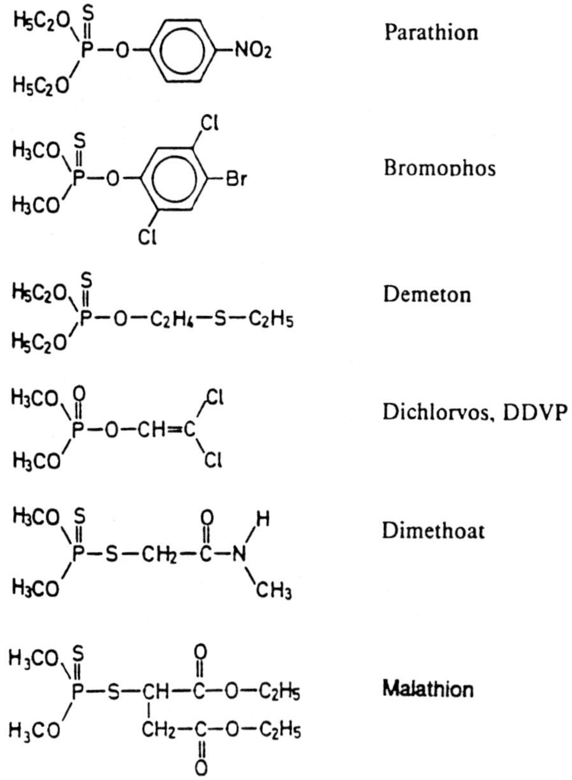

Parathion

Bromophos

Demeton

Dichlorvos, DDVP

Dimethoat

Malathion

Figure 1. Examples for Some Widely Used Organophosphorus Pesticides and Their Chemical Structure.

–NH group. The group Z, which is the leading group, can be any one of a wide variety of substituted and branched aliphatic, aromatic, or heterocyclic groups linked to phosphorus via a bond of some lability (usually –O– or –S–) (8). Figure 1 shows the chemical structures of some widely used organophosphorus pesticides.

Organophosphorus compounds representing a variety of chemical, physical, and biological properties are presently in commercial use. Most are only slightly soluble in water and have a high oil-water partition coefficient and a low vapor pressure (8, 9). Most, with the exception of dichlorvos, are of comparatively low volatility, and all are degraded by water, yielding water soluble products. Parathion, O,O-diethyl,O-4-nitrophenyl phosphorothioate, as an example, is freely soluble in alcohols, esters, ketones and aromatic hydrocarbons, but is practically insoluble in water (20 ppm) or in petroleum, kerosene, or spray oils.

Since all organophosphates are subject to degradation by hydrolysis to nontoxic products, the toxic hazard is essentially short-term in contrast to the persistent organochlorine pesticides.

2.6.2.2 Uses and route of exposure

Organophosphorus pesticides are used to control insect vectors which are found in food and commercial crops, and infestation in man or domestic animals.

(Organophosphorus compounds used as nerve gases are not discussed here.)

The compounds used for agricultural purposes are available mainly as emulsifiable concentrates or wettable powder formulations for reconstitution as liquid sprays, but also as granules. A limited number are also available as fogging formulations, smokes, impregnated resin strips for use indoors, or as animal or human pharmaceutical preparations.

Potential sources of exposure are product development, manufacture, storage, transport, and use. Food contamination, accidental poisoning in children, misuse, suicidal and criminal poisoning are known (10).

2.6.2.3 Absorption, distribution, metabolism, and elimination

Organophosphorus pesticides are absorbed by the skin as well as by the respiratory and gastrointestinal tracts. Metabolism occurs principally by oxidation, and hydrolysis by esterases and transfer of parts of the molecule to glutathione and glucuronic acid. Oxidation may lead to more or less toxic products. In general, phosphorothioates are not directly toxic but require oxidative metabolism to the proximal toxin (e.g. parathion, which is oxidized to the highly toxic paraoxon). Elimination of the metabolites derived from hydrolysis may proceed by urine or feces (8, 11, 12, 13, 57).

There is no evidence for prolonged storage of organophosphates in the human body. Yet, after the uptake of acute toxic high doses, they may be found in the fat for weeks.

2.6.2.4 Toxicology

2.6.2.4.1 Mechanism of action

The toxicity of organophosphorus pesticides in man is well understood: these compounds cause inhibition of red cell acetylcholinesterase, ACHE (EC 3.1.1.7) and plasma cholinesterase (pseudocholinesterase), CHE (EC 3.1.1.8), by phosphorylation of the enzyme, and accumulation of acetylcholine at susceptible receptors (14) (figure 2). The degree and timing of the effect varies with different compounds (12, 57).

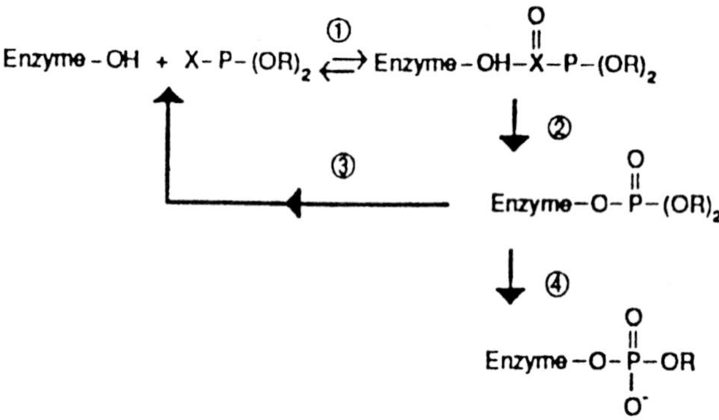

Figure 2. Inhibition of an Esterase System by Organophosphate Compounds.
1 = Reversible formation of Michaelis enzyme substrate complex; 2 = Phosphorylation of the enzyme – a specific serine residue in the protein is phosphorylated with loss of the group X; 3 = Reactivation reaction – may occur spontaneously at a rate which is dependent on the nature of the attached group and on the protein, but also on the pH and on the addition of oximes which may catalyse the reaction; 4 = 'Ageing' – involves cleavage of an R-OP bond.

In man, the neurotransmitter acetylcholine is present at the terminal endings of all postganglionic parasympathetic nerves (muscarinic receptors), at neuromuscular junctions (nicotinic receptors), and in the autonomic nervous system, sympathetic and parasympathetic ganglia (nicotinic receptors). The loss of activity of acetylcholinesterase results in excessive nervous system stimulation, which may lead to respiratory failure and death in severe cases. As a consequence, the following acute toxicity features from organophosphate poisoning can be seen (10):

Muscarinic features
- Increased bronchial secretion, broncheoconstriction cyanosis, pulmonary edema
- Excessive sweating, salivation and lachrimation
- Nausea, vomiting, diarrhea, abdominal cramp
- Urinary and fecal incontinence
- Bradycardia, hypotension, heart block
- Miosis (unreactive to light), blurred vision

Nicotinic features
- Muscle fasciculation, including diaphragmatic and respiratory muscles
- Generalized weakness
- Tachycardia, hypertension
- Hyperglycemia
- Pallor
- Mydriasis (rarely)

CNS (nicotinic) features
- Confusion, headache, restlessness, anxiety, poor concentration, tremor, ataxia, dysarthria
- Hypotension
- Respiratory depression
- Convulsions and coma

2.6.2.4.2 Toxicity

The toxicity of the various organophosphates varies, as can be seen in table 2.

The method of uptake (oral, by inhalation, percutaneous) plays an important role, as well as the nature of the commercial product (solvents, emulsifier, absorption to granules, etc.).

Table 2. LD_{50} (oral, rat) of Some Organophosphorus Pesticides Classified by IPCS 1996 (2)

Compound	Hazard Class	Physical state	LD_{50} mg/kg	
Azinophos-ethyl	IB	S	12	
Azinophos-methyl	IB	S	16	
Bromophos-ethyl	IB	L	71	
Carbophenothion	IB	L	32	
Chlorphenvinphos	IA	L	10	
Chlorpyrifos	II	S	132	
Demeton-O and -S	IA	L	2,5	
Demeton-S-methyl	IB	L	40	
Demeton-S-methyl-sulfon	IB	S	37	
Diazinon	II	L	300	
Dichlorvos	IB	L	56	volatile
Dimethoate	II	S	150	
Disulfoton	IA	L	2,6	
Ethion	II	L	208	
Fenitrothion	II	L	503	
Malathion	III	L	2100	
Mevinphos	IA	L	4	
Omethoate	IB	L	50	
Parathion-ethyl	IA	L	13	
Parathion-methyl	IA	L	14	
Thiometon	IB	oil	120	

It is difficult to give reliable data on the lethal dose for man.

For parathion, the ingestion or inhalation of 10–300 mg is assumed to be lethal for the adult. The estimated fatal dose of diazinon in man after oral ingestion is 25 g, the mean fatal dose of malathion about 60 g (13). Children have died after only 0.1 mg/kg of parathion (12, 15). The uptake of much higher doses can be survived following adequate therapy.

Taking all this into consideration, lethal doses for other organophosphate pesticides may be estimated very cautiously from the ADI, MAK, and TLV values given in table 3.

Table 3. Acceptable Daily Intake (ADI) (4), Maximum Allowable Concentration Value in the Workplace (MAK) (5) and Threshold Limit Values (TLV) (7) for Some Organophosphorus Pesticides

Compound	ADI (mg/kg bw)	MAK (mg/m^3)	TLV (mg/m^3)
Acephate	0.03	–	–
Azinophos-ethyl	no ADI	–	–
Azinophos methyl	0.005	0.2	0.2
Bromophos	0.04	–	–
Bromophos-ethyl	0.003	–	–
Carbophenothion	0.0005	–	–
Chlorphenvinphos	0.0005	–	–
Chlorpyrifos-methyl	0.01	–	0.2
Chlorthion	no ADI	–	–
Coumaphos	no ADI	–	–
Demeton	no ADI	0.1	0.11
Demeton-S-methyl and related	0.0003	–	0.5
Diazinon	0.002	1.0	0.1
Dichlorvos	0.004	1.0	0.9
Dimethoate	0.01	–	–
Disulfoton	0.0003	–	0.1
Ethion	0.002	–	0.4
Fenitrothion	0.005	–	–
Fenthion	0.007	0.2	0.2
Malathion	0.02	15	10
Mevinphos	0.0015	0.1	0.092
Omethoate	0.0003	–	–
Parathion-ethyl	0.004	0.1	0.1
Parathion-methyl	0.003	–	0.2
Pirimiphos-methyl	0.03	–	–
Thiometon	0.003	–	–

– = no entry

2.6.2.4.2.1 *Acute toxic effects*

The clinical picture of intoxications with organophosphorus insecticides results from the accumulation of acetylcholine at nerve endings. The symptoms can be summarized as muscarinic and nicotinic effects.

According to the severity of poisoning, the following signs and symptoms can occur (16):

Mild: anorexia, headache, dizziness, weakness, anxiety, substernal discomfort, tremors of the tongue and eyelids, miosis, and impairment of visual acuity.

Moderate: nausea, salivation, tearing, abdominal cramps, vomiting, sweating, slow pulse, and muscular fasciculations.

Severe: diarrhea, pinpoint and nonreactive pupils, respiratory difficulty, pulmonary edema, cyanosis, loss of sphincter control, convulsions, coma, and heartblock. Hypoglycemia and acute pancreatitis have also occurred.

2.6.2.4.2.2 Delayed clinical effects

In severe cases some of the diethyl phosphorothioates, being lipophilic, may remain in the body for days or even weeks, and provoke a recurrence of clinical effects after an initial period of apparent recovery. In contrast, dichlorvos, a dimethyl phosphate, and omethoate, a dimethyl phosphorothioate, are rapidly hydrolyzed to inactive products and are unlikely to cause late clinical effects.

2.6.2.4.2.3 Neuropathy

It has been stated (12) that the risk of developing a peripheral neuropathy following organophosphate poisoning is related to the inherent toxicity of the pesticide and the dose recieved, and may occur following single or chronic exposure (14). The organophosphates which have been noted to cause neuropathy are DEF (S,S,S-tributyl phosphorotrithioate), merphos and fenthions (17).

The phosphorylation, by organophosphates, of a nervous system target protein called neuropathy target esterase (NTE, neurotoxic esterase) has been considered to be a marker in animals of the subsequent development of a delayed neuropathy (18). A more accessive form of a similar human enzyme is found in peripheral blood leucocytes and platelets. At present, the monitoring of the peripheral form of the enzyme is being considered as a potentially predictive test in humans for the development of a delayed neuropathy (14, 18, 19, 20, 21).

2.6.2.4.2.4 Treatment of acute organophosphate poisoning

Therapy of acute poisoning can be divided into routine supportive care and specific antidote treatment for reversal of organophosphate toxicity (10, 16) (table 4).

When artificial ventilation is required, suxamethonium should be avoided, since it is normally rapidly metabolized by plasma cholinesterase (22).

Atropine is the antidote of choice and is useful in reversing the muscarinic features.

An initial "test dose" of atropine provides a measure of severity of organophosphate poisoning, since this is unlikely to be severe in the presence of the rapid development of signs of atropinization. Atropine has no effect on the rate of regeneration of inhibited acetylcholinesterase, which takes weeks. Oximes are able to reactivate cholinesterases inhibited by organophosphates. Some organophosphates cause rapid "aging" due to dealkylation of the enzyme-organophosphate complex, so that oximes are no longer effective. Some oximes have severe side effects, therefore the organophosphate should be identified before the application of oximes.

Table 4. Summary of Specific Antidote Treatment for Acute Organophosphate Poisoning (10)

Level of poisoning	Red cell ACHE (% normal)	Treatment
Subclinical	> 50%	Observation only
Mild	20–50%	Atropine 1 mg IV or IM (test dose) (child 0.01 mg/kg) Pralidoxime: 1 g IV or 7.5–10 mg/kg IM (child: 25 mg/kg over 15–30 min.)
Moderate	< 10–20%	Atropine 2 mg IV (IM) every 10–15 min (child: 0.02–0.05 mg/kg) Pralidoxime 1–2 g IV (max. rate 0.5 g/min) or over 15–30 min), or 7.5 mg/kg IM (child: 25–30 mg/kg over 15–30 min) repeatable after 1 h, than 8–12 hourly
Severe	10%	Atropine 4–5 mg IV (IM) every 10–15 min, then 0.08 mg/kg/h if needed (child: 0.02–0.05 mg/kg repeated) Pralidoxime 1–2 g IV (max 0.5 g/min, or over 15–30 min), repeated after 1 h, then 8–12 hourly (or 7.5–10 mg/kg IM, repeated, or up to 0.5 g/h by IV infusion) (child: 25–50 mg/kg IV over 15–30 min)

2.6.2.5 Biochemical and toxicological analyses

2.6.2.5.1 Biochemical analyses

2.6.2.5.1.1 Determination of serum cholinesterase (EC 3.1.1.8) activity

Toxicological analysis can most readily be performed by an indirect method, which measures the effect of the organophosphate on the level of serum (plasma) cholinesterase (CHE).

Spectrophotometric procedures based on Ellman's method (23) are recommended as a rapid screening test:

(1) Principle of test

CHE from plasma or serum hydrolyzes butyrylthiocholine to butyrate and thiocholine. With Ellman's reagent (DTNB = 5,5'-dithio-bis-2-nitrobenzoic acid) 5-thio-2-nitrobenzoic acid is formed (see Fig. 3). The enzyme activity is proportional to the rate of formation of 5-thio-2-nitrobenzoic acid, which can be measured photometrically at wavelengths between 400 and 440 nm.

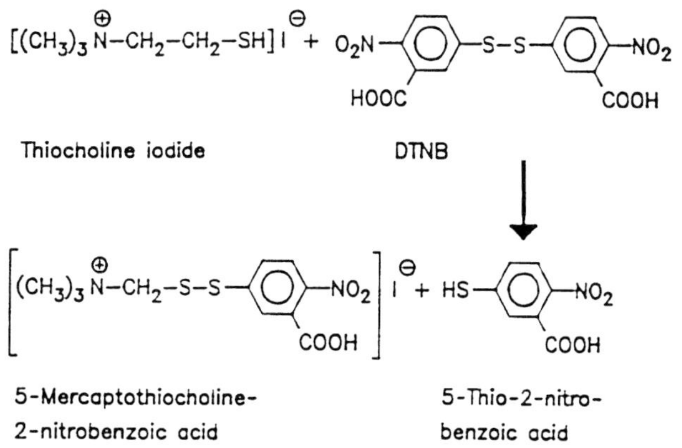

$$[(CH_3)_3 \overset{\oplus}{N}-CH_2-CH_2-SH]\,\overset{\ominus}{I} + \text{DTNB} \rightarrow \left[(CH_3)_3 \overset{\oplus}{N}-CH_2-S-S-C_6H_3(COOH)-NO_2\right]\overset{\ominus}{I} + HS-C_6H_3(COOH)-NO_2$$

Figure 3. Ellmans Reaction (23) for the Determination of Cholinesterase Activity. DTNB = 5,5′-dithio-bis-2-nitrobenzoic acid.

(2) Sampling
Serum or plasma (e.g. lithium heparinate; fluorides may not be added since they inhibit serum cholinesterase).

(3) Storage
At room temperature CHE is stable in serum or plasma of healthy persons for more than one week. However, CHE inhibited by organophosphates will be partly reactivated during storage, even at 4 °C (24).

(4) Chemicals and reagents
Test kits for the determination of CHE activity are commercially available (e.g. from Boehringer, D-68298 Mannheim).
Prepare reagents according to the advice of the manufacturer.
Control samples for quality control in Clinical Chemistry mostly include CHE.

(5) Equipment
- photometer (400–440 nm; Hg 405 nm)
- thermostat (cuvette holder) for 25, 30 or 37 °C
- cuvettes, 1 cm
- stop-watch
- blotter

or mechanized system (recommended).

(6) Procedure
Follow the instructions given by the manufacturer.

(7) Calibration procedure
Calculation of test results can be done using the molar absorption coefficient of 5-thio-2-nitrobenzoic acid or by comparison with a standard (e.g. serum multicalibrator from Boehringer, D-68298 Mannheim).

(8) Quality control
Quality control must be achieved by running control samples for monitoring imprecision and inaccuracy. At least one control sample with a strongly reduced CHE activity should also be used.

(9) Specificity
Butyrylthiocholine is hydrolyzed specifically by CHE. Acetylcholinesterase (ACHE) from erythrocytes does not interfere.

(10) Analytical assessment of the precision of the results:
Using a mechanized system (Hitachi 717, Boehringer, D-68 298 Mannheim) between-run coefficient of variation was 2.51 % at a mean activity of 2736 U/l.
Accuracy:
Mean absolute difference from the reference method value was 2.71 %.
Detection limit:
The minimal detectable extinction difference per 30 sec corresponds to an enzyme activity of about 200 U/l.

(11) Medical interpretation of the result
Reference values depend on substrate, age, sex, hormonal influences, and temperature (25), see table 5.
The correlation between serum cholinesterase activity and the clinical picture is weak. Signs of organophosphate poisoning can be expected if the CHE activity is below 10 % of the reference values.

Interpretation of low enzyme activities may be difficult because changes can be associated with many types of liver disease and also other disease states. Furthermore there are genetic influences (table 6) (10).

In practice, the response to atropine injection at the beginning of therapy gives the best information.

Inhibition of CHE can also be caused by uptake of anticholinesterase carbamates and alkylsulfates as well as certain drugs like cyclophosphamide, physostigmin and tubocurarin chloride.

Table 5. Reference Values for Serum/Plasma Cholinesterase (EC 3.1.1.8) Activity in Man (Substrate: butyrylthiocholine) (25)

Subjects	Reference values (U/l) 25 °C	37 °C*
Children; men, women older than 40 years	3500–8500	5400–13 200
Women (16–39 years old, not pregnant, no hormonal contraceptives)	2800–7400	4300–11 500
Women (18–41 years old, pregnant or intake of hormonal contraceptives)	2400–6000	3700– 9 300

* calculated (f = 1.553)

Table 6. Genetic Characteristics Affecting Plasma Cholinesterase Measurement (modified from (12))

Genetic variation		Laboratory investigations	
phenotype	% of population	dibucaine no. (%)	plasma CHE (% of normal)
Normal	96.865	79	100
Intermediate	3.106	62	35–74
Atypical	0.029	16	8–21

2.6.2.5.1.2 Determination of erythrocyte acetylcholinesterase (EC 3.1.1.7) activity

Acetylcholinesterase (ACHE) is structurally bound in membranes. It occurs in every organ and tissue in the human organism (26), especially in the CNS, the cholinergic synapses and in the motor end-plates in muscles. In addition, ACHE occurs in erythrocyte membranes.

Inhibition of ACHE in the neurons as a result of exposure to organophosphates, carbamates and sulfonate esters represents the real toxicological stress parameter. However, determination of ACHE in the neurons is not possible. The ACHE bound to the erythrocytes is, however, the correlate of the ACHE in the neurons. Thus inhibition of erythrocytic ACHE is determined as a measure of the stress parameter.

After isolation, washing and hemolysis, the ACHE of erythrocytes can also be measured by the colorimetric method of Ellmann, using acetylthiocholine as substrate (detailed procedure see (27)). Commercially available test kits can be used. In addition the hemotocrit value is measured in the original blood sample as well as in the washed erythrocyte suspension.

Inhibition of ACHE activity according to the severity of poisoning, see table 4.

Erythrocyte ACHE activities between 2900 and 4100 U/l were measured in healthy male adults using acetylthiocholine as a substrate. Physiological fluctuations and circadian rhythms can lead to variations of up to 30% for an individual reference value (27).

2.6.2.5.2 Toxicological analyses

2.6.2.5.2.1 Group reactions for organophosphorus pesticides

a) Bioassay with Drosophila melanogaster

This is a simple bioassay on substances with insecticidal properties, using adult Drosophila melanogaster (fruit fly, vinegar fly) as test objects. These insects are exposed either directly to a pulp of the sample to be tested (e.g. scene residues, stomach contents) or to a residue film left after the evaporation of the solvent from an extract solution. Mortality counts begin 24 h after exposure (28, 29). The detection limits shown in table 7 have been achieved in our laboratory.

It is always important to include a positive control, since resistance to a number of pesticides has developed in insects.

Specificity: other insecticides, like carbamates and organochlorine compounds, also yield a positive result.

Table 7. Detection Limits for Some Organophosphorus Pesticides Using the Bioassay with Drosophila Melanogaster

Compound	Detection limit (μg)
Dimethoate	10
Disyston	25
Metasystox	50
Metasystox R	100
Parathion	1
Thiometon	10

b) Test using gas detector tubes
For scene residues or stomach contents, gas detector tubes (e.g. from Draegerwerk, D-23558 Lübeck) can be used as simple screening tests. In the presence of organophosphorus pesticides a color change will be observed. This is due to suppression of cholinesterase by anticholinesterase pesticides in the reactive zone of the test tube, whereby butyrylthiocholine cannot be hydrolyzed.

The detection limit e.g. for dichlorvos in the head space area of a sample is about 0.05 ppm.

Other volatile substances which cause inhibition of cholinesterase, e.g. carbamate pesticides, will also give a positive result.

c) Thin layer chromatography in standardized systems (30, 31)
Organophosphorus pesticides are extracted with pentane. Aliquots of the extracts are chromatographed on silica plates together with a mixture of reference compounds. After the plates are developed the identification is based on

1. UV-absorption.
2. Color reaction with palladium chloride.
3. Release of 2-naphthol from 2-naphthylacetate by the action of cholinesterase. 2-Naphthol reacts with Fast Blue Salt BB, building a colored complex. Substances which inhibit cholinesterase can be recognized on the chromatogram as white spots on a rose-violet background (32, 33).
   This mode of detection is very sensitive. It can be made even more sensitive by treatment of the developed TLC plate with bromine water. The detection limit is about 0.1 to 1 ng/spot for most of the organophosphorus pesticides with a few exceptions.
   Anticholinesterase carbamates do interfere.

The use of standardized systems and reference compounds allows the identification of a substance via the corrected Rf-value (31) with a rather small error window.

Equipment and chemicals:
TLC plates: Silicagel with fluorescent marker
Solvent mixtures: see table 8
Reference compounds: see table 8

Table 8. Solvent Mixtures and Reference Compounds for Detection of Organophosphorus Compounds by TLC (31)

Solvent mixture[1]	Reference compounds[2]	
	Substance	hR_f-value
P1: n-hexane-acetone (80 + 20)	Triazophos	21
	Parathion-methyl	30
	Pirimiphos-methyl	50
	Bromophos-ethyl	69
P2: Toluene-acetone (95 + 5)	Carbofuran	20
	Azinophos-methyl	42
	Methidathion	56
	Parathion-ethyl	84

[1] Eluant composition: volume + volume. Saturated tanks are used.
[2] Reference compounds are available from:
Ehrenstorfer, D-86199 Augsburg
Riedel de Haen, D-30926 Seelze, (Pestanal[R]);
Supelco INC, Supelco Park, Belforte, PA 16823–0048, USA;
Sigma Chemical Company, 3050 Spruce Street, St. Louis, Missouri 63103, USA.
Solution of the reference compounds at a concentration of approximately 2 mg/ml of each substance

Enzymatic detection:
Chemicals:
Bovine liver (cholinesterase from horse serum is now available from Merck, D-64271 Darmstadt.)
β-Naphthylacetate
phosphate buffer pH 7.0; 0.02 M/l
Ethanol
Fast Blue Salt BB

Enzyme solution using bovine liver:
10 g bovine liver is homogenized with 90 m/l phosphate buffer (0.02 M; pH = 7,0), and then centrifuged 5 min at 3000 rpm. The supernatant is diluted with the phosphate buffer 1 : 4.
Liver and centrifuged homogenate can be stored frozen for many months without loss of activity.
Before use, the enzyme solution is diluted 1 : 10 with phosphate buffer.
β-Naphthylacetate solution:
Dissolve 125 mg β-napthylacetate in 100 ml ethanol.

Substrate solution:
Dissolve 20 mg Fast Blue Salt BB in 16 ml distilled water and add 4 ml of the β-napthylacetate solution. The substrate solution must be freshly prepared before spraying.

Enzymatic detection procedure:

1. The developed and dry TLC plate should be sprayed evenly with the enzyme solution.

Table 9. hRf Data in TLC systems P1 and P2, UV Absorption and Color Developing with Palladium Chloride for Some Organophosphorus Pesticides (31)

Substance	P1	P2	UV	$PdCl_2$
Acephate	0	0	–	yellow
Omethoate	0	0	–	yellow
Methamidophos	1	0	–	yellow
Demeton-S-Methylsulfone	1	3	–	yellow
Etrimfos	3	0	+	yellow
Trichlorphon/Trichlorphos	4	2	–	–
Dimethoate	4	4	–	yellow
Phosphamidon	7	2	+	yellow
Mevinphos/Phosdrin	12	10	+	–
Demeton	17	17	–	yellow
Oxydemeton-methyl	18	0	–	yellow
Heptenophos	20	18	–	–
Dichlorvos/DDVP	20	20	–	–
Azinphos-methyl	20	42	+	yellow
Triazophos	21	38	+	yellow
Temephos	23	75	+	yellow
Azinophos-ethyl/ Ethyl Guthion	24	48	+	yellow
Tetrachlorvinphos	25	29	+	–
Chlorphenvinphos	26	26	+	–
Coumaphos	27	61	+	yellow
Methidathion	29	56	+	yellow
Parathion-methyl	30	73	+	yellow
Malathion	31	53	–	yellow
Phosalone	31	67	+	yellow
Chlorthion	31	71	+	yellow
Pyrazophos	32	47	+	yellow
Fenitrothion	32	76	+	yellow
Ethoprofos/Ethiprop	33	28	–	yellow
Isophenfos	41	67	+	yellow
Fenthion	41	81	+	yellow
Parathion-ethyl	41	84	+	yellow
Phoxim	42	86	+	yellow
Dimpylate/Diazinon	47	50	+	yellow
Pirimiphos-methyl	50	75	+	yellow
Sulfotep	50	84	–	yellow
Etrimfos	52	67	+	yellow
Tolclofos-methyl	52	89	–	yellow
Chlorthiophos	55	91	+	yellow
Chlorpyriphos-methyl	56	89	–	black
Disulfoton	58	89	–	yellow
Bromophos	58	92	+	yellow
Fonofos	59	89	+	yellow
Terbufos	63	90	–	red
Chlormephos	64	91	–	yellow
Chlorpyriphos	64	95	+–	yellow
Bromophos-ethyl	69	93	+–	yellow
Demeton	83	81	–	yellow

2. After this the plate should be incubated for 60 min in a moisture saturated incubator at 37 °C.
3. Dry the plate for a short time with a hair dryer.
4. Spray the plate with the substrate solution. After a few minutes the background becomes red (reaction of enzymatically released β-naphthol with fast Blue Salt BB).
   Anticholinesterase compounds are recognized as white spots on the red background (inhibition of the esterase).
5. Calculate the corrected Rf-values of the spot.
6. Compare the corrected Rf-values of the spots wih the corrected Rf-values of organophosphate pesticides.

For corrected Rf-values for organophosphorus compounds, see table 9. This table also indicates whether the organophosphate absorbs UV, as well as the color reaction with palladium chloride.

d) Determination of alkylphosphate metabolites in urine
The basic method, published by Shafik (34), allows detection of the alkylphosphates dimethylphosphate (DMP), diethylphosphate (DEP), dimethylthiophosphate (DMTP), diethylthiophosphate (DETP), dimethyldithiophosphate (DMDTP) and diethyldithiophosphate (DEDTP) in urine. Residues of DMP and DEP are directly attributable to pesticide exposure; DMDTP and DEDTP are less directly associated with pesticide exposure and rapidly degrade to other alkyl phosphate metabolites, whereas nonpesticides may confound the interpretation of DMTP and DETP residues. Although the metabolic residues are known for many organophosphates, it is difficult to monitor exposure to a single compound, because many separate compounds share the same metabolic end points. In the case of an intoxication with organophosphorus compounds, these metabolites may be the basis for a useful global screening test. They can be detected in the urine for days.

In the original Shafik procedure (34) the analysis involves solvent extraction of the acidified urine sample, conversion of the alkylphosphates to volatile derivatives using diazopentane, fractionation, and cleanup of the extract with a silica gel column, then gas chromatographic analysis. Refinements have been developed (35, 36, 37, 38).

These methods have enjoyed wide application in studies of occupational pesticide exposure (39). Although the measurement of these urinary metabolites can be a valuable tool, its complexity must be noted. The result can give a qualitative answer, but quantitative evaluations are difficult. This is due to the variation in kinetics among different compounds, and partly very low recoveries of some of these metabolites. According to our experience, good recovery can be achieved following the screening procedure of Reid and Watts (37), which includes extraction of the urine sample with acetonitrile, derivatization with pentafluorobenzyl bromide, and determination by GC/NPD. Only the application of an isotope dilution method using deuterated alkyl phosphates may lead to reliable quantitative results (40).

e) Test for 4-nitrophenol metabolites in urine
Detection of nitrophenols in the urine can be used to assess exposure to organophosphorus pesticides like

parathion-ethyl (O,O-diethyl-O-(4-nitrophenyl)thionophosphate)
parathion-methyl (O,O-dimethyl-O-(4-nitrophenyl)thionophosphate)
EPN (O-ethyl-O-p-nitrophenyl-phenyl-thionophosphonate)
dicapthon (O,O-dimethyl-O-(2-chloro-4-nitrophenyl)-thionophosphate)
chlorthion (O,O-dimethyl-O-(3-chloro-4-nitrophenyl)thionophosphate),
folithion (O,O-dimethyl-O-(3-methyl-4-nitrophenyl)thionophosphate).

Most nitrophenol metabolites are excreted relatively rapidly. A simple screening test can be applied for the pesticides listed above:

The hydrolyzed urine sample is extracted with an organic solvent which in turn is reextracted with alkali. If nitrophenols are present, the yellow color of a phenolate is seen. The absorption curve of the phenolate is measured spectrophotometrically. For confirmation the alkaline solution is reacted with o-cresol and titanium trichloride. If 4-nitrophenols are present, a blue indophenol dye is formed whose absorption curve is also measured spectrophotometrically (41, 42, 43), see table 10. The detection limit for nitrophenol metabolites is about 2–10 mg/l.

Table 10. Nitrophenol Metabolites of Organophosphorus Pesticides, λ_{max} of Phenolates and Corresponding Indophenol Dyes (43)

Compound	Metabolite	λ_{max} phenolate (nm)	λ_{max} indophenol dye (nm)
Parathion-ethyl Parathion-methyl Paraoxon EPN	4-nitrophenol	400	610
Dicapthon	2-chloro-4-nitrophenol	395	600
Chlorthion	3-chloro-4-nitrophenol	390	600
Folithion	3-methyl-4-nitrophenol	394	636

One should consider that nitrophenols are also excreted in the urine as metabolites of e.g. nitrobenzene or 1-chlornitrobenzene. Furthermore, intact nitrophenolic herbicides like 2,4-dinitrophenol, dinitro-o-cresol, 2-sec-butyl-4,6-dinitrophenol or their metabolites are also excreted in the urine. However, the yellow di- and trinitrophenolates have different absorption curves and do not yield blue indophenol dyes.

Applying this reaction, a simple colorimetric method for the quantitative determination of 4-nitrophenol metabolites in urine is also possible (41, 42).

For cases of mild to severe intoxication the colorimetric method is sensitive enough. The BAT value for occupational parathion exposure (5) is based on the inhibition of erythrocyte ACHE (max. – 30% of the reference value) and the p-nitrophenol excretion in the urine (max. 500 μg/l). To achieve sufficient sensitivity 4-nitrophenol has to be determined by HPLC (44) (detection limit 50–100 μg/l).

2.6.2.5.2.2 *Special methods*

For confirmation and quantitative determination of organophosphorus pesticides, particularly in biological materials like blood, serum or stomach contents, advanced methods, including adequate extraction, have to be applied. Examples are given in the following paragraphs.

a) Gas chromatography

1. Gas chromatographic detection of organophosphorus pesticides using approved retention indices (45).

Extraction: with a suitable organic solvent, e.g. pentane (46)
Column: OV-1 (SE-30)
Detector: Halogen phosphorus detector
RI-values: see (45) (The RI-values have been stated using pure substances.)

Medical interpretation of organophosphate blood levels: It is difficult to give reliable toxic blood levels for organophosphate pesticides since the number of cases in which blood levels have been determined is low. Only a few can be found in (47) (table 11).

Table 11. Toxic Blood Levels in Organophosphate poisoning (47)

Compound	Blood level (mg/l)
Diazinon	0.05–1.0
Paraoxon	0.005
Parathion	0.01

We have observed two cases where a severe parthion intoxication was survived following adequate therapy. We found the following parathion blood levels (GC):

Pat. S.A., 32 years:

8.4.	Uptake of parathion
Blood levels:	
8.4./21.30 h	0,7 mg/l
23.30 h	0,4 mg/l
10.4./7.00 h	0,1 mg/l
11.4.	0.4 mg/l
12.4.	0,2 mg/l
18.4.	< 0,05 mg/l
20.4.	< 0,01 mg/l
21.–27.4.	not detectable

Pat. M.D., 25 years:

27.6./24.00 h	Uptake of parathion
Blood levels:	
28.6.	1,3 mg/l
29.6.	0,4 mg/l
30.6.	0,2 mg/l
2.7.	0,05 mg/l
4.7.	not detectable
5.7.	not detectable

Data derived from post mortem blood samples should be regarded with caution (48).

2. Gas chromatographic detection and determination of organophosphorus pesticides in plasma and urine (49)

The authors describe a simple and rapid method for the isolation of eleven organophosphate pesticides from human urine and plasma with Sep-Pak C18 Cartridges and their identification, using wide-bore capillary gas chromatography with flame ionization detection. The detection limits for most pesticides are excellent. A quantitative determination is possible.

3. Gas chromatographic detection of mevinphos and its metabolites in stomach contents, blood and urine (50)

Mevinphos in stomach contents, blood or urine can be extracted directly into chloroform. The solutions, concentrated under vacuum and diluted with acetone or hexane to a suitable concentration, are analyzed with a gas chromatograph equipped with a flame photometric detector operating in the phosphorus mode.

4. Gas chromatography/Mass spectrometry (GC/MS)

Procedure and GC/MS data for organophosphorus pesticides see (51).

b) Liquid chromatography (HPLC)

1. Determination of pirimiphos and five metabolites in urine and plasma (52)

Urine samples are shell-frozen, freeze-dried, and extracted with methanol. The extracts are filtered and injected directly to the column.
0.2 ml portions of plasma diluted with phosphate buffer (0.1 M; pH 7.0), are filtered and added to ice-cold methanol, shaken and centrifuged to remove precipitated protein. The supernatant is injected directly on the column.

2. Detection and quantitative determination of phospholan, mephisfolan, and related compounds in plasma (53)

The authors describe an HPLC method for qualitative and quantitative analysis of phospholan, mephisfolan and related compounds in plasma. The plasma samples were extracted with ethyl acetate. HPLC was done using C18 columns and UV detection at 254 nm.

2.6.3 Carbamate Pesticides

2.6.3.1 Chemical structure and physico-chemical properties

A large number of anticholinesterase carbamates are known. The carbamates used as insecticides and herbicides are synthetic derivatives of neutral esters of carbamic acid, $H_2N-COOH$. These carbamates have the common structure shown in figure 4. The insecticidal carbamates are usually mono-methyl-substituted and are often aryl esters as well. Some examples are shown in figure 5.

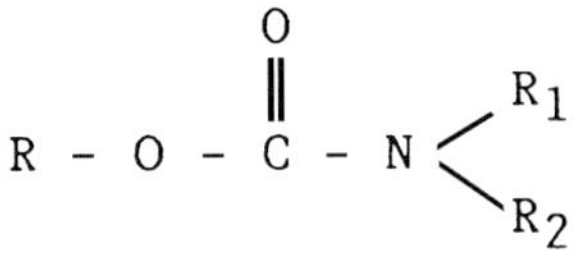

$$R - O - \overset{\overset{\displaystyle O}{\|}}{C} - N \begin{matrix} \diagup R_1 \\ \diagdown R_2 \end{matrix}$$

R generally a substituted benzene, naphthalene or other ring structure
R_1 usually only hydrogen
R_2 usually a methyl group

Figure 4. Common Chemical Structure for Insecticidal Carbamates.

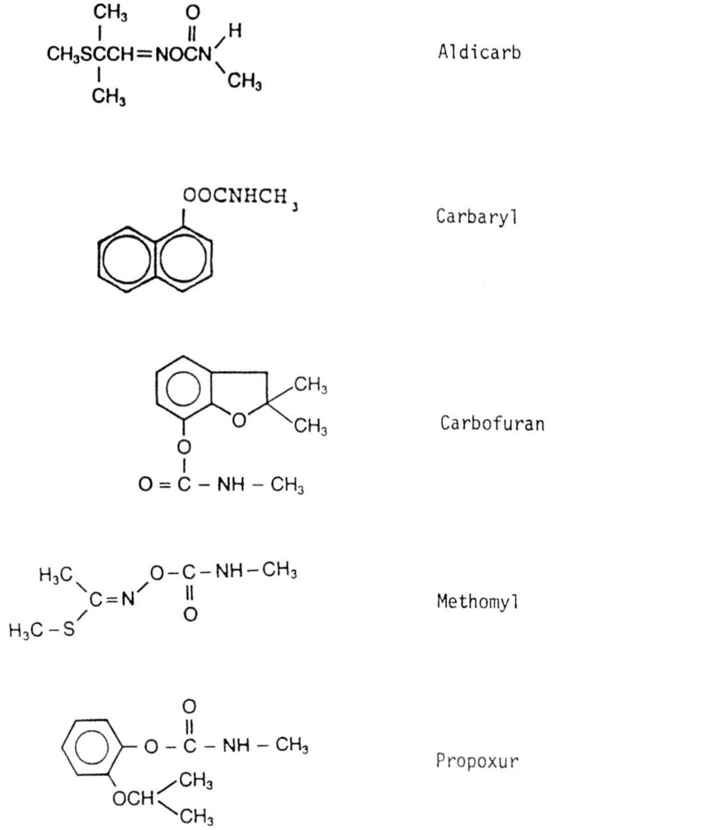

Figure 5. Structures of Some Widely Used Carbamate Pesticides.

Thiocarbamates and dithiocarbamates have not been included here because these compounds have a different mode of action.

Many carbamate pesticides representing a variety of chemical, physical, and biological properties are presently in commercial use. Most of them are slightly soluble in water and moderately soluble in solvents such as benzene, toluene, xylene, chloroform and 1,2-dichloromethane. In general, they are poorly soluble in nonpolar organic solvents, such as petroleum hydrocarbons, but highly soluble in polar organic solvents such as methanol, ethanol, acetone, dimethylformamide, etc. (9, 54).

Being esters, they are susceptible to hydrolysis. Most carbamate compounds are stable at acid pH values. However, under alkaline conditions, hydrolysis is rapid, the breakdown rate increasing approximately tenfold for each pH unit above seven. An increase of ten degrees will increase the hydrolysis rate approximately fourfold (9, 55). The instability with alkali is of use for decontamination and cleanup.

The vapor pressure of certain of the carbamates makes them particularly subject to the effects of temperature when they are sprayed on hot surfaces (56).

2.6.3.2 Uses and route of exposure

The anticholinesterase carbamates are mainly used in agriculture as insecticides, fungicides, herbicides, nematocides, or sprout inhibitors. In addition they are used as biocides for industrial or other applications and in household products.

Formulations containing less than 1 % to more than 95 % pure material are available. Compounds used for agricultural purposes are available commonly as wettable powders, dusts, granules, and emulsifiable concentrates. It is important to note that some formulations may contain carbamates in combination with other pesticides. This may be particularly important in the diagnosis and management of poisoning.

Agricultural and pesticide sprayers, crop harvesters and formulation plant employees may be subject to occupational exposure. Health hazards occur mainly from occupational overexposure to carbamate insecticides. The main routes of exposure are inhalation and dermal, but accidental and suicidal uptake (oral) is also known.

2.6.3.3 Absorption, distribution, metabolism and elimination

Anticholinesterase carbamates are well absorbed orally and by inhalation, more slowly via the skin. The oral route occurs mostly in people who purposely ingest the compound. In most cases of occupational poisoning, absorption occurs through the skin and the respiratory tract.

Carbamates are actively metabolized in the liver, and the degradation products are excreted by the liver and the kidneys.

All compounds have great similarities in their metabolism. However, no two compounds are identical (12, 57). Although hydrolysis occurs to some extent in all compounds, various oxidation steps that are catalyzed by mixed function oxidases also occur. The different behavior of different carbamates is illustrated by the oxidation by microsomal enzymes of 33 methyl and dimethyl carbamates and related compounds. Examples of N-demethylation, aromatic ring hydroxylation, O-dealkylation, alkyl hydroxylation, sulfoxidation, conversion of N-methyl to N-formamide or N-hydroxy methyl groups, and formation of a dihydrodihydroxy derivative of an aromatic ring were found (58). Those changes of a carbamate insecticide that fail to separate the ester bond usually produce cholinesterase inhibitors, and some of the products are more potent than the parent compound (12, 57).

Carbaryl, as an example, is known to be metabolized by ring hydroxylation, hydrolysis, and conjugation (glucuronic and sulfuric acids). The hydrolysis pathway results in the urinary excretion of free and conjugated 1-naphthol, which accounts for over 20% of an ingested dose (15), together with small amounts of conjugated 4-hydroxycarbaryl.

2.6.3.4 Toxicology

2.6.3.4.1 Mechanism of action

Carbamate insecticides act by combining with and inactivating acetylcholinesterase, but the combination is reversible with time. Thus the hazard is not increased by daily exposure to amounts less than those required to cause immediate symptoms. If symptoms develop, they usually do not persist for more than 8 h. Acute intoxication is due to inhibition of acetylcholinesterase, just as with organophosphates. Acetylcholinesterase is able to hydrolyze carbamate insecticides, although the rate of hydrolysis is not as fast as for the natural substrate acetylcholine. Thus carbamate insecticides are reversible inhibitors. Symptoms of intoxication develop when the amount of pesticide in the body is so large that the rate of carbamylation of acetylcholinesterase exceeds the rate of hydrolysis by the enzyme. Acetylcholine then accumulates in neuroeffector and synaptic regions, resulting in clinical signs similar to those of organophosphate poisoning. Carbamates do not cause delayed neuropathy (26, 54).

2.6.3.4.2 Toxicity

The acute toxicity of carbamate insecticides varies through a wide range (table 12). Most of the aromatic carbamate ester insecticides have low dermal toxicity, however one cannot generalize that carbamates are without dermal toxicity, as illustrated by the extreme toxicity of aldicarb by both oral and dermal routes.

Most insecticidal carbamates appear to lie in the toxicity range 50–500 mg as a probable lethal (oral) dose to humans (15).

Acceptable daily intakes (ADI's), MAK values and TLV's are given in table 13.

2.6.3.4.3 Acute toxic effects

Clinical effects will vary according to the amount ingested.

The following signs and symptoms listed in appropriate order of appearance begin within thirty to sixty minutes and are at a maximum at two to eight hours (59):

Mild: anorexia, headache, dizziness, weakness, anxiety, substernal discomfort, tremor of the tongue and eyelids, miosis, and impairment of visual acuity.

Table 12. LD_{50} (oral, rat) of Some Carbamate Pesticides Classified by IPCS (2)

Compound	Hazard class	Physical state	LD_{50} mg/kg
Aldicarb	IA	S	0.93
Aminocarb	IB	S	50
Bendiocarb	II	S	55
Carbaryl	II	S	300
Carbofuran	IB	S	8
Ethiofencarb	II	L	411
Methiocarb	II	S	100
Methomyl	IB	S	17
Oxamyl	IB	S	6
Propoxur	II	S	95

S = solid; L = liquid

Table 13. Acceptable Daily Intake (ADI) (4), Maximum Allowable Concentration Value in the Workplace (MAK) (5) and Threshold Limit Values (TLV) (7) for Some Anticholinesterase Carbamate Pesticides

Compound	ADI (mg/kg bw)	MAK (mg/m^3)	TLV (mg/m^3)
Aldicarb	0.003	–	–
Aminocarb	no ADI	–	–
Bendiocarb	0.004	–	–
Benomyl	0.1	–	–
Carbaryl	0.01	5	5
Carbendazim	0.03	–	–
Carbofuran	0.01	–	–
Chlorpropham	no ADI	–	–
Ethiofencarb	0.1	–	–
Methiocarb	0.001	–	–
Methomyl	0.03	–	2.5
Oxamyl	0.03	–	–
Pirimicarb	0.02	–	–
Propoxur	0.02	2	0.5

– = no entry

Moderate: nausea, salivation, tearing, abdominal cramps, vomiting, sweating, slow pulse, and muscular fasciculations.

Severe: diarrhea, pinpoint and nonreactive pupils, respiratory difficulty, pulmonary edema, cyanosis, loss of sphincter control, convulsions, coma and heart block. Hyperglycemia and acute pancreatitis have occurred.

2.6.3.4.4 Chronic poisoning

The hazard in environmental or occupational exposure is not increased by daily exposure to amounts less than those required to produce immediate symptoms. If symptoms develop, they do not persist for more than eight hours after the end of exposure.

2.6.3.4.5 Treatment of acute poisoning with anticholinesterase carbamates

Specific treatment is the use of atropine until signs of atropinization appear or complete reversal of symptoms occurs. It is important to keep the airway open and to prevent aspiration. In the case of ingestion, gastric lavage can be performed. The use of oximes is contradicted, see (16).

2.6.3.5 Biochemical and toxicological analyses

2.6.3.5.1 Biochemical analyses

2.6.3.5.1.1 Determination of serum cholinesterase (EC 3.1.1.8) activity

Serum cholinesterase (pseudocholinesterase) determination see 2.6.2.5.1.1.

Serum cholinesterase may be lowered in carbamate poisoning, whereas albumin concentration is unaffected. Serum cholinesterase activity returns to normal values within a few hours.

2.6.3.5.1.2 Determination of erythrocyte acetylcholinesterase (EC 3.1.1.7) activity

Procedure and reference values see 2.6.2.5.1.2.

In a patient who ingested propoxur (1.5 mg/kg), the lowest erythrocyte acetylcholinesterase level (27% of normal) was observed 15 min after ingestion. After 2h it reached 95% of the normal value again. Plasma cholinesterase remained entirely normal (60).

2.6.3.5.2 Toxicological analyses

2.6.3.5.2.1 Simple group reactions

a) Thin layer chromatography in standardized systems (see 2.6.2.5.2.1 c)

Principle of test:
Thin layer chromatographic separation of anticholinesterase carbamate pesticides (30, 31) and enzymatic detection using cholinesterase inhibition (33).

Sampling
Pure substance, suspect material, scene residues, stomach contents

Chemicals and reagents
see 2.6.2.5.2.1 c)

Sample preparation
Dissolve about 500 mg of the sample in a few ml of acetone or ethanol. Centrifuge or filter if necessary and use the clear filtrate.

Procedure
see 2.6.2.5.2.1 c)

Detection limits
Ezymatic detection is very sensitive. Detection limits (33) and corrected hRf values (30, 31) see tables 14 and 15.

Table 14. Detection Limits After Enzymatic Detection for Some Anticholinesterase Carbamate Pesticides (33)

Compound	Detection limit per spot (μg)
Carbaryl	3
Dimetan	80
Dimetilan	60
Isolan	10
Methiocarb	400
Minacide	6
Pyramat	200
Zectran	80

Table 15. Corrected hRf-Values in Solvent Mixtures P_1 and P_2 After TLC in Standardized Systems for Some Carbamate Pesticides (31)

Compound	Solvent P1 hR_f	Mixture P2 hR_f
Butocarboxim	15	11
Butocarboxim-sulphoxid	0	13
Butoxycarboxim	3	2
Carbaryl	18	25
Carbofuran	17	20
Carbosulfan	49	66
Ethiofencarb	21	25
Formetante	1	1
Methomyl	6	6
Phenmedipham	11	17
Pirimicarb	26	17
Propham	39	57
Propoxur	20	21
Thiofanox	20	17

b) Thin layer chromatography in the Toxi-Lab® system

On special request, instructions and reagents for the determination of 16 insecticidal carbamates and their metabolites can be obtained from the manufacturer (Toxi-Lab Inc., Goodyear 2, Irvine, CA 92718, USA). The detection limits (color reactions after dipping the TLC plates into the reagent mixtures) are about 0.5–1 mg/l.

2.6.3.5.2.2 Special methods

For confirmation and quantitative determination of anticholinesterase carbamates, particularly in biological materials like blood, serum, urine or stomach contents, special methods including adequate extraction have to be applied.

Examples are given in the following paragraphs.

a) Gas chromatographic screening (45)
Packed column OV-1 (OV-101)
Flame ionization detector (FID)
Retention indices see (45)

b) Screening by HPLC (62)
Column: 250 × 8 × 4 mm, RP18, 100-10, Nucleosil (Macherey-Nagel, D-52355 Düren)
Stationary phase: Reversed phase C18-particle size 10 μm
Mobile phase: acetonitrile/water
Detection: UV detector = 215 nm or 270 nm
Detection limit: 1–2 mg/l (215 nm)
5–60 mg/l (270 nm)

c) Quantitative determination of carbaryl in blood by gas chromatography (61)
Ball-mill extraction using acetone and methylene chloride, removing of lipid material by a freeze-out procedure, and florisil micro-column cleanup were employed. Derivatization of carbaryl with heptafluorobutyric anhydride in the presence of triethylamine allowed rapid processing for gas liquid chromatographic detection.
Column: 180 × 2 mm glass column containing 3% OV-17
on 80–100 mesh Chromosorb W-HP
Detector: ECD
Detection limit: 20 μg/l carbaryl for blood

d) Quantitative determination of 2-isopropoxyphenol, a metabolite of propoxur, in urine by gas chromatography (63)
2-isopropoxyphenol, available after hydrolysis of the conjugate in the free form, can be extracted with hexane and, after reaction with trichloroacetyl chloride, determined by gas chromatography.

Column: 3% DC-200 and 1.5% OV-17 on Chromosorb W (AW-DMCS), 80–100 mesh
Detector: ECD
Precision: 6.2%
Recovery; 92.7%
Detection limit: 0.5 μg/ml urine
Medical interpretation: the quantitative determination in urine may give important limits concerning the extent of exposure to propoxur.

e) Quantitative determination of total 1-naphthol, a metabolite of carbaryl, in urine (64)
Simple colorimetric determination of 1-naphthol, a metabolite of carbaryl, after hydrolysis in urine.

Medical interpretation (64):
Normal excretion level: 1.5–4 mg/l
Occupational exposure: 40% of the urine samples of 689 workers exposed to carbaryl contained in excess of 10 mg/l of total 1-naphthol.
Intoxication: in the urine sample of a 19 month old child collected 18 h after the uptake of carbaryl, 31 mg/l of 1-naphthol were found.

2.6.4 Chlorinated Hydrocarbon Insecticides

It is conventional to divide the chlorinated hydrocarbon insecticides into four groups:

a) DDT and its analogues
b) benzene hexachloride (BHC) or hexachlorocyclohexane (HCH)
c) cyclodienes and related compounds
d) toxaphene and related compounds

Cyclodienes and toxaphene and their related compounds, though at one time marketed or widely reported, are believed to be currently of little commercial interest in Europe (65). However, some are still widely used in other areas of the world. The compounds discussed in this chapter are very different in their usage, toxicology and treatment of poisoning, so, we think it is necessary to give more detailed information.

2.6.4.1 Chemical structure and physico-chemical properties

2.6.4.1.1 DDT and its analogues

The term DDT (p,p′-DDT) is generally understood throughout the world and refers to 1,1,1-trichloro-2,2-bis(4-chlorophenyl) (IUPAC) or 1,1′-trichloro-di-(4-chlorophenyl)ethane (CA). The structure of DDT permits several different isomeric forms. The chemical structure of DDT and some of its analogues is shown in Table 16. All isomers of the compound DDT are white, crystalline, tasteless, almost odorless with the empirical formula $C_{14}H_9Cl_5$ and a relative molecular weight of 354.5. The melting range of p,p′-DDT is 108.5–109.0 °C. DDT is soluble in organic solvents such as benzene, cyclohexanone, chloroform, and in petroleum solvents, and is practically insoluble in water (66).

Table 16. Structure of p,p′-DDT and Analogues

Name	Chemical Name	R	R′	R″
DDT	1,1,1-trichloro-2,2-bis-(*p*-chlorophenyl) ethane	$-Cl$	$-H$	$-Cl_3$
Dicofol (Kelthane®)	2,2,2-trichloro-1,1-bis-(*p*-chlorophenyl) ethanol	$-Cl$	$-OH$	$-Cl_3$
Ethylan (Perthane®)	1,1-dichloro-2,2-bis-(*p*-ethylphenyl) ethane	$-C_2H_5$	$-H$	$-HCl_2$
Methoxychlor	1,1,1-trichloro-2,2-bis-(*p*-methoxyphenyl) ethane	$-OCH_3$	$-H$	$-Cl_3$
Prolan®	2-nitro-1,1-bis-(*p*-chlorophenyl) propane	$-Cl$	$-H$	$-HNO_2-CH_3$
TDE[a)]	1,1-dichloro-2,2-bis-(*p*-chlorophenyl) ethane	$-Cl$	$-H$	$-HCl_2$

a) As an insecticide, this compound has the approved name of TDE; as a metabolite of DDT, it usually is called DDD.

2.6.4.1.2 Benzene hexachloride and lindane

Benzene hexachloride, a term firmly established by usage, is a misnomer for 1,2,3,4,5,6-hexachlorcyclohexane. It has eight separable steric isomers, the alpha-isomer exists in two enantiomorphic forms (figure 6). The acronym for benzene hexachloride is BHC. The acronym for the proper chemical name is HCH. Both BCH and HCH are recognized as common names in ISO recommendation R 1750. The International Organization for Standardization recognizes as distinct common names both Gamma-BHC (or Gamma-HCH) and lindane, which is defined as not less than 99% pure Gamma-BHC. All isomers of BHC have the empirical formula $C_6H_6Cl_6$ and a molecular weight of 290.80. The technical mixture is an off-white to brown powder with a persistent, musty odor. The mixture is moderately soluble in acetone, benzene and chlorinated hydrocarbons; it is poorly soluble in kerosene, only slightly soluble in fats and oils, and almost insoluble in water (66).

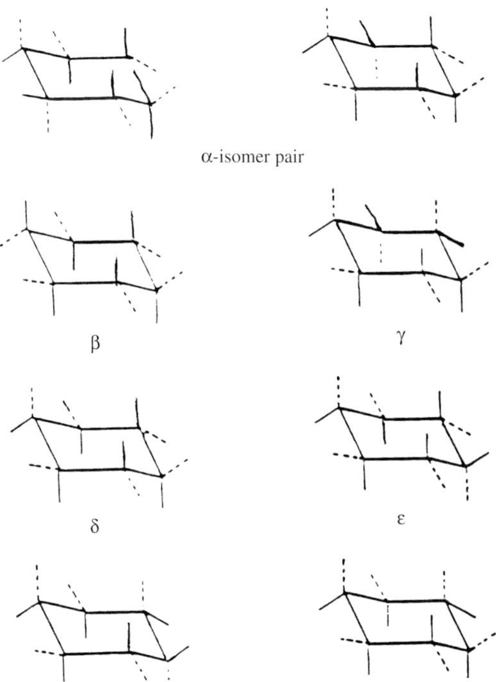

Figure 6. Structures of Isomers of BHC of which the α-, β-, γ-, and δ- are Relatively Strainless and Predominate in the Technical Product.
Solid lines indicate chlorine, and *dashed lines* indicate hydrogen. In the β-isomer all of the chlorine atoms are represented as being in the axial (*a*) position and all the hydrogens in the equatorial (*e*) position.

2.6.4.1.3 Cyclodienes and related compounds

The most important cyclodiene insecticides are shown in figure 7. Among them are two pairs of stereoisomers: aldrin and isodrin, dieldrin and endrin.

Chlordane has two isomers and the empirical formula $C_{10}H_6Cl_8$. The molecular weight is 409.80. The cis-isomer melts at 106 to 107 °C, the trans-isomer at 104 to 105 °C. Technical chlordane is a viscous, amber-colored liquid containing about 45 constituents. Chlordane is soluble in most organic solvents including petroleum oils, but is virtually insoluble in water.

Heptachlor has the empirical formula $C_{10}H_5Cl_7$ and a molecular weight of 373.35. The pure material forms white crystals, melting at 95–96 °C. Technical heptachlor is a soft wax melting in the range of 46 to 74 °C. It contains about 72 % heptachlor and 28 % related compounds. It is soluble in acetone, benzene, carbontetrachloride, cyclohexanone, xylene, kerosene, and alcohol.

Aldrin has the empirical formula $C_{12}H_8Cl_6$ and a molecular weight of 364.93. The pure compound is an odorless, white, crystalline solid melting at 104 °C. Technical aldrin contains not less than 90 % aldrin defined as a mixture, that is, not less than 85.5 % of the main ingredient, not less than 4.5 % of insecticidally active, related compounds and not more than 10 % of other compounds. Technical aldrin is a tan to dark brown solid with a mild chemical odor and a melting point in the range of 49 to 60 °C. It is moderately soluble in paraffins, aromatics, halogenated solvents, esters and ketones, but sparingly soluble in alcohols.

Dieldrin has the empirical formula $C_{12}H_8Cl_6O$ and a molecular weight of 380.93. In most countries, the common name stands for the technical product. Pure dieldrin is a white, crystalline, odorless solid that melts at 176 to 177 °C. Technical dieldrin contains not less than 95 % of a mixture with 80.75 pure dieldrin, not less than 14.25 % of insecticidally active, related compounds and not over 5 % other compounds. It is slightly soluble in minerals oils, other aliphatic hydrocarbons, and

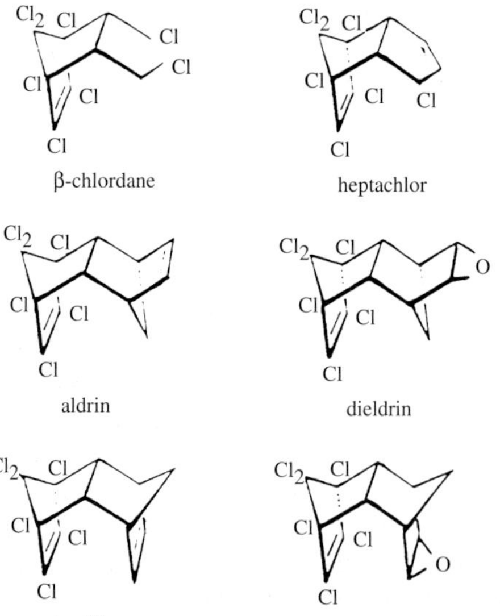

Figure 7. Structural Formulas of β-Chlordane, Heptachlor, Aldrin, Dieldrin, Isodrin and Endrin.

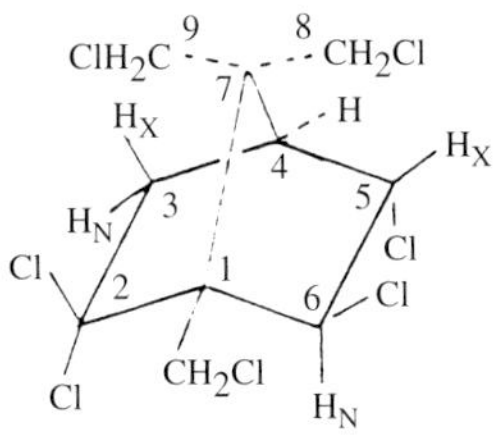

B

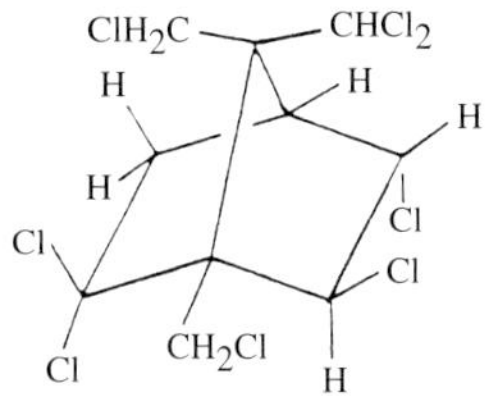

A-1

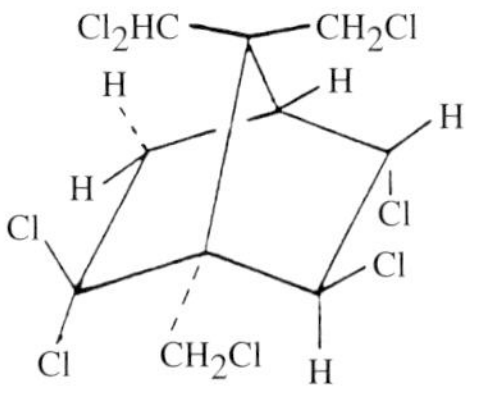

A-2

Figure 8. Toxaphene Toxicants A-1, A-2, and B.

alcohols; moderately soluble in acetone; soluble in aromatic and halogenated solvents (66).

2.6.4.1.4 Toxaphene and related compounds

Toxaphene is a reaction mixture of chlorinated camphenes containing 67 to 79% chlorine. Although this mixture contains more recognized compounds than most other technical pesticides, there is no evidence that toxaphene produced in the USA is very different in composition. The empirical formulae (figure 8) and molecular weights of toxicants A and B are $C_{10}H_{10}Cl_8$ (413.79) and $C_{10}H_{11}Cl_7$ (379.35), respectively. Technical toxaphene is a yellow or amber, sticky wax with a slight pine odor. Toxaphene is easily soluble in organic solvents including petroleum oils, but is soluble in water only to the extent of about 3 ppm (66).

2.6.4.2 Uses and route of exposure

2.6.4.2.1 DDT and its analogues

DDT is a potent non-systemic ingested and contact insecticide which is persistent on solid surfaces. DDT came to widespread attention because it dramatically controlled typhus and malaria at the time of World War II. Knowledge that DDT and its metabolites are stored in essentially everyone in the world kept DDT in the spotlight. Later it was implicated in the injury of a wide variety of wildlife. Under these circumstances DDT has been studied more throroughly than any other pesticide and in more diverse relationships than any other drug. As the situation now stands, DDT is still used extensively, both for agriculture and vector control, in some tropical countries. Information apparently is not available on how much of the agricultural use involves food protection or how much loss of food protection would result if the use of DDT were discontinued. Commerical products of DDT consisted predominantly of p,p'-DDT, 77.1 %; o,p'-DDT, 14.9 %; p,p'-TDE, 0.3 %; o,p'-TDE, 0.1 %, p,p'-DDE, 4.0 %, o,p'-DDE, 0.1 %, and unidentified compounds, 3.5 %. DDT was available as emulsifiable concentrate, wettable powder and dustable powder. Apparently all commercial products have been discontinued (66, 67).

2.6.4.2.2 Benzene hexachloride and lindane

Gamma-HCH acts as an ingested insecticide by contact and has some fumigant action. It is effective against a wide range of soil-dwelling and phytophagous insects hazardous to public health, other pests and some ectoparasites. Formulations are: suspension concentrate, solution for seed treatment, wettable powder and smoke generator.

BHC has anthelminthic properties and has been explored for possible practical use. Even for lindane, dosage levels that are dependably anthelminthic may produce serious side effects; the compound offers no advantage over available drugs and was never adopted for use. The situation is entirely different in connection with the use of lindane to control scabies caused by the mite Sarcoptes scabiei. Use against scabies is now standard and generally trouble-free. As a scabicide, lindane is applied locally, usually as a cream or ointment, but it may be formulated as an emulsion, solution, aerosol or shampoo. The formulation is usually 1 %, but 0.3 % is recommended for babies to avoid excessive absorption. A 0.3 % solution may be used to impregnate clothing to avoid infestation. Formulations similar to those for mites may be used to combat human lice, but their efficiency is increased by adding an enzyme to help dislodge the nits (66).

2.6.4.2.3 Cyclodienes and related compounds

Chlordane has been used for control of various insects, and extensively for control of household pests. **Heptachlor** has been used for control of cotton insects, grasshoppers, soil insects, and certain crop pests. **Aldrin** has been used mainly against insects attacking field, forage, vegetable and fruit crops. **Dieldrin** is used in tropical countries as a residual spray on the inside walls and ceilings of homes for the control of vectors of diseases, mainly malaria.

2.6.4.2.4 Toxaphene and related compounds

Toxaphene is a non-systemic insecticide with some action against mites. It is not toxic against plants except squash and melons. It is valued not only for its effectiveness but also because of its limited persistence in the environment and its rapid excretion by mammals. It has been used extensively for cotton insects and to lesser degrees for a wide range of pests of beef cattle, goats, sheep and swine, and for insects attacking vegetables, small grains, and soybeans.

2.6.4.3 Absorption, distribution, metabolism and elimination

All chlorinated hydrocarbon insecticides can be absorbed through the skin as well as by the respiratory and oral routes. The importance of dermal absorption varies greatly for the different compounds. This is partly because some of them, such as methoxychlor, have such a low toxicity that a small amount absorbed by any route is of no importance. More importantly, the efficiency of dermal absorption varies considerably for the different compounds. For example, **DDT** is poorly absorbed by the skin from solutions, and the solid is so poorly absorbed that it is difficult or impossible to measure either the absorbed compound or its effect. On the contrary, solid dieldrin, if very fine ground, is absorbed so effectively by the skin that it is about half as toxic when applied to the skin as when administered by mouth. Chlorinated hydrocarbon insecticides may be absorbed from the lung if they reach the respiratory epithelium in the form of solid or liquid aerosols of appropriate particle size. They seldom reach unpermissible levels in the air in the form of vapors (66).

The overall rate of absorption of technical **BHC** administered in food for 14 days was essentially identical (94, 95), but average absorption of the isomers differed somewhat (68). It is claimed that more rapid oral absorption of BHC occurs if it is administered with an alkyl surfactant.

The absorption of lindane through human skin was demonstrated by Feldmann and Maibach (69). Within five days this absorption constituted 9.3 % of the applied dose. Lindane is metabolized through cytochrome P-450 dependent enzymes. A large number of metabolites which are still imperfectly identified are produced by the liver. There are four possible reactions: (a) dehydrogenation (= gamma-hexachlorocyclohexene), (b) dehydrochlorination (= gamma-pentachlorocyclohexene (c) dechlorination (= gamma-tetrachlorocyclohexane) and (d) hydroxylation (= hexachlorocyclohexanol). In man, after acute exposure, lindane shows a short serum half-life before redistribution to fat and other tissue storage sites (70). In one clinical case described by the authors, a peak serum lindane level of 1.3 µg/ml was measured 12 hours after ingestion. It decreased to 0.8 µg/ml 36 hours after ingestion.

The cyclodiene insecticides aldrin, dieldrin, endrin and heptachlor are readily absorbed by the skin as well as by the lungs and the gastrointestinal tract. Absorption through the skin amounted to 7–8 % of the applied dose in volunteers. Up to 50 % of the inhaled aldrin vapor is absorbed and retained in the body. The percentage of an ingested dose that is actually absorbed has not been determined. After absorption, it is rapidly distributed over the organs and tissues, and a continuous exchange between blood and other organs takes place. Aldrin is readily converted to dieldrin, mainly in the liver, where two hours after ingestion of aldrin it is possible to determine the presence of dieldrin. After some hours, the presence of dieldrin is confirmed in the lipid tissues. Some of the ingested aldrin and dieldrin passes unabsorbed through the intestinal tract and is eliminated. Part is excreted unchanged from the liver into the bile, stored unchanged in the organs and tissues, particulary in the adipose tissue, and part is metabolized in the liver to more polar and hydrophilic metabolites, which are excreted via the bile in the feces. Aldrin and dieldrin are biodegraded to the same metabolites.

Experimentally, the major metabolites are the 9-hydroxy derivative, the trans-6,7-dihydroxy derivative, the dicarboxylic acid derived from the dihydroxy compound and the bridged pentachloroketone. Only the 9-hydroxy compound has been demonstrated in the feces of humans. The conversion of aldrin to dieldrin by mixed-function monooxygenases (aldrin-epoxidase) in the liver and the distribution and the subsequent deposition of dieldrin, mainly in lipid tissues, proceed much faster than the biodegradation and ultimate elimination of unchanged dieldrin and its metabolites.

2.6.4.4 Toxicology

2.6.4.4.1 Mechanism of action

2.6.4.4.1.1 Effects on the central nervous system

There is considerable evidence that DDT acts by changing the electrophysiologically associated enzymatic properties of nerve cell membranes, especially axonal membranes. The other chlorinated hydrocarbon insecticides have been studied less in this regard, but may have some basic similarity in action. Whatever the exact mechanism is, nonconvulsant doses of chlorinated hydrocarbon insecticides increase the susceptibility of animals to convulsion caused by many other poisons. One study of these relationships concluded that the convulsant effects of dieldrin may be mediated by effects on the hippocampus and other limbic systems (71).

Fever may be a specific result of poisoning of the temperature control center in the brain. It has been observed in BHC, dieldrin and endrin poisoning. The effect may be more common than has been assumed. What has been recognized in a few human cases is high fever of sometimes late but sudden onset, frequently followed promptly by death.

Fortunately, high fever of central origin is rare, but because it is such a grave sign, it is essential to distinguish it from other kinds of fever that may be the result from poisoning. Fever may accompany convulsions in man or larger animals simply because it may be impossible to dissipate heat as rapidly as it is generated by the violent activity, which is certainly muscular and perhaps also metabolic. Fever of this origin has no special prognostic significance beyond that of the convulsions that give rise to it. Regardless of the exact cause, a moderate increase in body temperature during the early course of illness carries no serious implications (72). However, unless fever subsides promptly after convulsions are controlled, some other basis for it must be sought. Fever may also be a response to chemical pneumonitis following aspiration of solvents or other chemical irritiants. Of course, fever of this origin may occur after a formulation of any chlorinated hydrocarbon insecticide has been aspirated. It may depend in part on secondary infection. Usually it is delayed about 12 hours or more, and in relatively mild cases may not appear until the patient has recovered from neurological manifestations.

2.6.4.4.1.2 Effects on the liver

There is no doubt that DDT and a number of other chlorinated hydrocarbon insecticides cause marked changes in the livers of various rodents and that these changes progress to tumor formation in some species, notably the mouse. There is no clear evidence that they do so in any other species. DDT promotes its own metabolism in some species of laboratory animals.

That the same is true in man is indicated by the fact that storage of DDT is relatively less at higher dosages (figure 9). However, the metabolism and subsequent excretion of DDT can be promoted even more by other agents. Patients who received phenobarbital or, more especially, phenytoin, stored much less DDT than other persons with similar exposure to DDT (73, 74, 75). This result concerning phenytoin was confirmed by McQueen et al (76) who also showed that other drugs produced a smaller but still highly significant reduction in DDT storage.

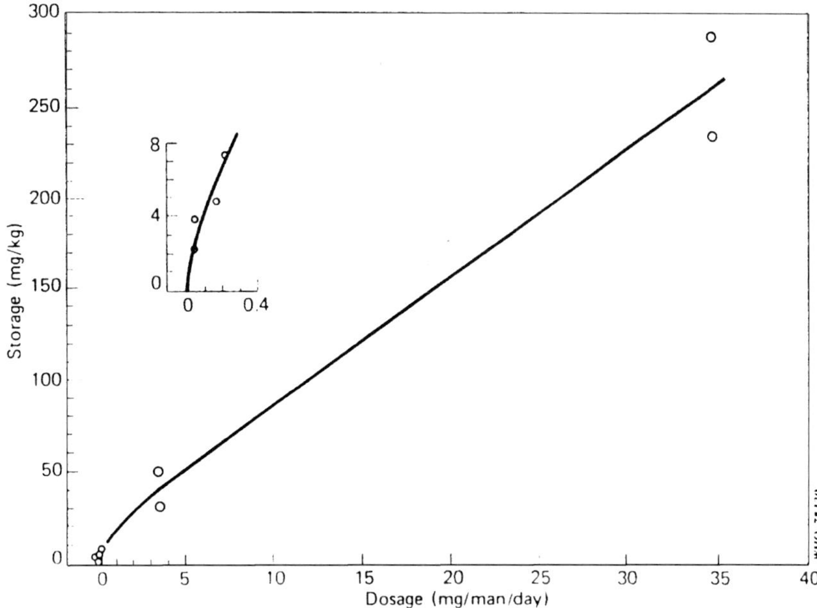

Figure 9. Concentrations of DDT in Body Fat Plotted Against Daily Dosages (102).

Establishment of a reduced equilibrium appeares to require about 2 months. Within this period, the regression of the level of DDT plus DDE on duration of treatment with phenytoin was highly significant (P 0.001).

At the end of nine months' treatment, the body fat of nonepileptic volunteers given phenytoin at a rate of 300 mg/man per day contained an average of 25% of the DDT and 39% of the DDE concentrations originally present before administration of the drug (77).

The same was true for workers whose exposure was greater than that of the general population. Maintenance doses of phenobarbital, phenytoin, or a combination of the two kept the storage levels of several organochlorine insecticides in epileptic workers as low as, or lower than, levels in the general population (78, 79).

2.6.4.4.2 Toxicity

The toxicity (LD_{50} and hazard class) of the chlorinated hydrocarbon insecticides can be seen in table 17. The acceptable daily intake (ADI), the maximun allowable concentration value in the workplace (MAK) and threshold limit values (TLV) are given in table 18.

Table 17. LD_{50} (oral, rat) of Some Chlorinated Hydrocarbon Insecticides Classified by IPCS 1996 (2)

Compound	Hazard Class	Phys. state	LD_{50} mg/kg
Aldrin	Ib	S	27
Benzene Hexachloride	see Lindane or BHC/HCH		
BHC/HCH	II	S	100
Camphechlor	II	S	80
Chlorobenzilate	III	S	700
Chlordane	II	L	460
DDT	II	S	113
Dieldrin	Ib	S	37
Dicofol (Kelthane)	III	S	690
Endosulfan	II	S	80
Endrin	Ib	S	7
Heptachlor	II	S	100
Hexachlorcyclohexane	see Lindane or BHC/HCH		
Lindane	see Lindane or BHC/HCH		

Table 18. Acceptable Daily Intake (ADI) (4), Maximum Allowable Concentration Value in the Workplace (MAK) (5) and Threshold Limit Values (TLV) (7) for Some Chlorinated Hydrocarbon Insecticides

Compound	ADI (mg/kg bw)	MAK (mg/m^3)	TLV (mg/m^3)
Aldrin	0.0001	0.25	0.25
Benzene Hexachloride	see Lindane or BHC/HCH		
BHC/HCH	no ADI	–	–
Camphechlor	no ADI	–	–
Chlordane	0.0005	0.5	0.5
DDT	0.02	1	1
Dieldrin	0.0001	0.25	0.25
Dicofol (Kelthane)	0.002	–	–
Endosulfan (endosulfan A)	0.006	–	0.1
Endrin	0.0002	0.1	0.1
Heptachlor	0.0001	0.5	0.5
Hexachlorcyclohexane	see Lindane or BHC/HCH		
Lindane	0.008	0.5	0.5
Methoxychlor	0.1	15	10

– = no entry

2.6.4.4.3 Acute toxic effects

The earliest symptom of poisoning by **DDT** is hyperesthesia of the mouth and lower part of the face. This is followed by paresthesia of the same area and of the tongue and then by dizziness, an objective disturbance of equilibrium, paresthesia and tremor of the extremities, confusion, malaise, headache, fatigue, and delayed vomiting. The vomiting is probably of central origin and not due to local irritation. Convulsions occur only in severe poisoning. Onset may be as soon as 30 minutes after ingestion of a large dose or as late as 6 hours after small but still toxic doses.

Nearly all the clinical findings reported in nonfatal poisoning by **BHC** were encountered among a group of 11 persons who drank Nescafé accidentally prepared with lindane in the place of sugar in such a way that each person received approximately 0.60 g of lindane or about 86 mg/kg. The interval of ingestion to onset of symptoms averaged about an hour, but varied from 20 minutes (two cases) to four hours (one case). Initial symptoms included malaise, faintness, and dizziness followed by collapse and convulsions, sometimes preceded by screaming and accompanied by foaming at the mouth and biting of the tongue. Nausea and vomiting occurred in many cases. The patients were presumably unconscious during convulsions and certainly unconscious afterwards. Loss of consciousness lasted from 0.25 to 3 hours (66).

Chlordane has not been a common cause of poisoning. All established cases have been associated with gross exposure. In most instances, including those with full recovery, convulsions appeared within 0.5 to 3 hours after ingestion (80, 81, 82) or after dermal exposure involving spillage (83). Following ingestion, some patients experienced nausea and vomiting before signs of central nervous system overactivity appeared. However, as often as not, a convulsion was the first clear indication of illness. Convulsions often last about a minute and may recur at intervals of about five minutes. Convulsions are usually accompanied by confusion, incoordination, excitability, or, in some instances, coma. In one instance, convulsions were so violent that the patient suffered compression fractures of varying severity of dorsal vertebrae four through nine, as revealed by an x-ray examination made after recovery from acute poisoning. During the acute episode, the same man experienced a brief episode of oliguria with proteinuria, hematuria, and mild hypertension, all of which returned to normal (84).

Severe poisoning by **endrin** involves repeated, violent, epileptiform convulsions, each lasting several minutes and followed by semiconsciousness or coma for 15 to 30 minutes unless the next fit occurs sooner. Fits may become almost continuous. Usually, there are few or no warning symptoms in severe poisoning, and even in moderate poisoning there may be no warning before the first fit (85, 86). A less common but very ominous feature of serious poisoning is hyperthermia (41 °C or more). The high fever is often followed by decerebrate rigidity. Deaths have occurred as little as half an hour after ingestion, and most suicidal cases are dead within an hour or two (86). Death usually occurs in accidental cases within the first 12 hours (87).

Serious poisoning by **aldrin** usually involves convulsions and is entirely similar to poisoning by dieldrin (88, 89), but a 3-year-old girl died following collapse and coma after only brief excitation and ataxia at onset (90). Hematuria and/or albuminuria have been noted in several cases, including some that were occupational in origin with no solvent to complicate the situation (91, 92). There have been at least three outbreaks of poisoning involving a total of 53 people and caused by the consumption of seed grain treated with aldrin, sometimes in combination with other pesticides. None of the cases in Kenya involved convulsions, but five of the most severe ones did involve unequal pupils with the smaller pupil unreactive to light and the larger reacting only sluggishly (93).

Ingestion of unknown volumes of a 5% **dieldrin** formulation prepared from an emulsifiable concentrate killed a 2-year-old girl and poisoned her 4-year-old brother.

The girl died before a physician arrived. In the boy, convulsions began within 15 minutes and became continuous for a time. He survived even though his temperature rose to 39.5 °C (94). The fact that his temperature was reduced by means of an ice mattress and chlorpromazine hydrochloride may have contributed to his survival, and administration of large doses of barbiturates almost certainly contributed. Analytical results in this case are discussed under laboratory findings. In another case, the dosage was also unknown but might have been smaller, since the onset was delayed about six hours. Then the baby suddenly lost consciousness, became dyspneic, and suffered a convulsion. Convulsions were finally controlled by treatment with phenobarbital, chloral hydrate, and chlorpromazine, but the baby remained unconscious. The temperature rose to 40 °C; cyanosis and tachycardia increased, and the child died 20 hours after exposure (95). Death from dieldrin is often associated with a massive dose as the result of an accidental or especially suicidal ingestion. In one example, 2 g dieldrin was recovered from the stomach contents alone (96).

2.6.4.4.4 Treatment of acute poisoning

The treatment of poisoning by a chlorinated hydrocarbon insecticide must be based mainly on general principles. It depends on the condition of the patient, whether first attention is given to removal of the poison or to sedation. Activated charcoal can speed fecal excretion (presumably by partially interrupting the enterohepatic circulation). The anticonvulsants that have been used in treating poisoning by chlorinated hydrocarbon insecticides are pentobarbital and phenobarbital. Furthermore, it might be necessary to continue dosage with barbiturates at a high rate for two weeks or more to increase metabolism. However, there is a growing acceptance for the use of diazepam for the control of convulsions. In at least two cases, paralysis (succinylcholine) combined with artificial respiration proved effective when anticonvulsants had failed.

Differential Diagnosis: Diagnosis might be difficult if exposure is unrecognized and the illness so mild that no convulsions occur. However, any such illness would be brief and without sequel, so that a failure of diagnosis would not be too serious. If the fact of exposure is unrecognized in a case involving one and more convulsions, the differential diagnosis must involve a) poisoning by chlorinated hydrocarbon insecticides, b) poisoning by some other kind of compound, including numerous drugs, c) epilepsy, d) convulsions secondary to infection, e) convulsions due to toxemia of pregnancy. Convulsion caused by strychnine involve far more tonic spasm and opisthotonus than ordinarily seen in poisoning by chlorinated hydrocarbon insecticides. Most convulsions associated with organic phosphorus insecticides occur late in the course of illness and are anoxic in origin. In the case of severe intoxication with chlorinated hydrocarbon insecticides, the convulsions and vital signs, especially respiration, are of poor quality. The presence of significant febrile illness before the onset of convulsions tends to point to a diagnosis of infection. Fever can occur in connection with poisoning by chlorinated hydrocarbon insecticides, but it tends to start after convulsions, not before (66).

2.6.4.5 Biochemical and toxicological analyses

2.6.4.5.1 Biochemical analyses

Apart from the nervous system, the liver is the only other organ significantly affected by DDT. A number of enzymes of intermediate metabolism are either stimulated or moderately inhibited by toxic doses of DDT. The possibility that these changes are the result, rather than the cause, of poisoning has not been excluded.

2.6.4.5.2 Toxicological analyses

2.6.4.5.2.1 Group reactions

a) Bioassay with drosophila melanogaster
The simple bioassay (see 2.6.2.5.2.1 a) can be used for chlorinated hydrocarbon insecticides as well as for carbamate and organophosphorus pesticides. The detection limits are shown in table 19.

Table 19. Detection Limits for Some Chlorinated Hydrocarbon Insecticides Using the Bioassay with Drosophila Melanogaster

Compound	Detection limit µg
Aldrin	0.5
Dieldrin	1
Chlordan	1
Endrin	2
Lindan	3
Toxaphen	30
DDT	150

b) Tests using gas detector tubes
Gas detector tubes for DDT, BHC/HCH, cyclodienes and taxophene are not available from Draeger or Kitagawa.

c) Thin layer chromatography in standardized systems (30, 31)
Chlorinated hydrocarbon insecticides are extracted with pentane. Aliquots of the extracts are chromatographed on silica plates together with a mixture of reference compounds. After the plates are developed, the identification is based on

1. UV-absorption
2. Color reaction with silver nitrate
3. Color reaction with N,N-dimethyl-p-phenylenediamine

Equipment and chemicals: TLC plates: Silicagel with fluorescent marker, UV-lamp (366 nm), silver nitrate, N,N-dimethyl-p-phenylenediamine hydrochloride and sodium ethylate.
Solvent mixtures: see table 20
Reference compounds: see table 20

Table 20. Solvent Mixtures and Reference Compounds for Detection of Chlorinated Hydrocarbon Insecticides by TLC (30, 31)

Solvent mixture	Reference Compounds	
	Substance	hR_f-value
P1 n-Hexane-Acetone (80 + 20)	Endosulfan	40
	Dieldrin	65
	Aldrin	89
P2 Toluol-Acetone (95 + 5)	Endosulfan	77
	Aldrin	98

Reagents: (a) Dissolve 1 g $AgNO_3$ in 5 ml distilled water and 2.5 ml ammonia (25%). Dilute with acetone to 100 ml.
(b) Dissolve 0.1 g N,N-dimethyl-p-phenylendiamine hydrochloride in 10 ml ethanol. Mix with 10 ml sodium ethylate just before spraying.

Procedure: After development in solvent mixtures P1 and P2
a) dip the plate in the reagent silver nitrate for about 3 min and dry for 15 min at 150 °C. A variety of colors may be seen against a brown background.
b) irradiate the plate with UV light of 366 nm for 10 min and spray homogeneously with the reagent N,N-dimethyl-p-phenylenediamine. Irradiate again with UV light of 366 nm for 10 min and observe the spots under the UV light against a gray to black background.

Corrected Rf-values of chlorinated hydrocarbon insecticides are given in table 21 as well as the color reactions with silver nitrate and N,N-dimethyl-p-phenylenediamine hydrochloride.

2.6.4.5.2.2 *Special methods: Gas Chromatography*

a) Gas chromatographic detection of organophosphorus pesticides using approved retention indices (45)

Extraction: with a suitable organic solvent, e.g. hexane (97)
Column: OV1 (SE-30)
Detector: Halogen phosphorus detector
RI-values: see Ref. 45 (The RI-values have been stated using pure substances.)

b) Determination of selected chlorinated hydrocarbon insecticides and some metabolites (98)

The most widely used method for analyzing blood involves hexane extraction (99). This modified method has been widely used to generate chlorinated pesticide blood residue data for the National Health and Nutrition Examination Survey II (100). The method of BURSE et al has less deleterious effect on the gas chromatographic system: The method can used to determine the residue level of certain chlorinated pesticides in the presence of polychlorinated biphenyls in serum. The method involves the following: 1. Extraction of denaturated serum (Methanol) with hexane/ethyl ether; 2. elution of the organic extract through micro-Fluorisil columns to obtain two fractions; 3. acid treatment of the less polar Florisil fraction and subsequent elution through activated silica gel to obtain two fractions; and 4. analysis of all fractions using gas-liquid chromatography with electron capture detection. The authors use two different separating columns: 3% SE-30 on 80/100 mesh Gas Chrom Q and 1.5% SP-2250/1.95% SP-2401 on 100/120 mesh Supelcoport. The method yielded in vitro recoveries for 8 pesticides (Gamma-HCH, Beta-HCH, Oxychlordane, Heptachlorepoxide, p,p′-DDT, p,p′-DDE, Dieldrin, Endrin), spiked in the range of 1–10.7 ppb, of 50.4% to 121.6% and in the range of 4.98-21 ppb, recoveries ranged from 47.7 to 112.6%.

c) Gas chromatographic detection of Gamma-Hexachlorcyclohexane in plasma (101).

Gamma-Hexachlorcyclohexane is extracted from plasma with hexane and analyzed using gas-liquid chromatography on a capillary column (HP-1.25 m × 0.32 mm) with Ni 63 electron capture detector. Oven temperature is programmed from 65 °C to 235 °C.

2.6.5 Bipyridylium Herbicides

2.6.5.1 Chemical structure and physico-chemical properties

Paraquat and diquat (figure 10) are bipyridylium herbicides. The term paraquat has been applied to two technical products: 1,1′-dimethyl-4,4-bipyridylium dichloride ($C_{12}H_{14}N_2Cl_2$) and 1,1′-dimethyl-4,4-bipyridylium disulfate ($C_{12}H_{14}N_2(CH_3SO_4)_2$). Both compounds are white crystalline solids and insoluble in hydrocarbons, slightly soluble in the lower alcohols, and very soluble in water. Diquat is the dibromide of 1,1′-ethylene-2,2′-bipyridylium ($C_{12}H_{12}N_2\ Br_2$). It is slightly soluble in alcohol, and practically insoluble in non-polar organic solvents (102).

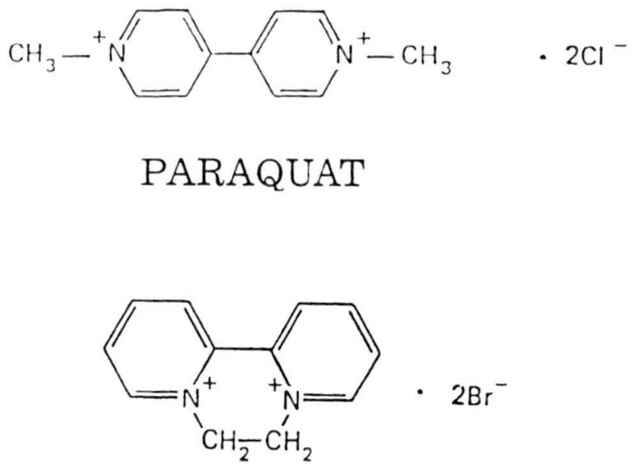

Figure 10. Structures of Paraquat and Diquat.

2.6.5.2 Uses and route of exposure

Paraquat and diquat are quick-acting herbicides which destroy green plant tissue by contact action and some translocation. They are used as plant desiccants for preharvest of cotton and potatoes. Paraquat and diquat are inactivated rapidly by the soil. They are formulated as aqueous solutions with surface-active agents and used in the form of an aqueous spray, which means that potential human exposure may occur as a result of their presence in air, on plants, in soil, or in water (103).

2.6.5.3 Absorption, distribution, metabolism and elimination

Although toxic amounts of paraquat may be absorbed after oral ingestion, the greater part of the ingested paraquat is eliminated unchanged in the feces. Paraquat can also be absorbed through the skin, particularly if it is damaged. Absorbed paraquat is distributed via the bloodstream to practically all organs and tissues of the body, but no prolonged storage takes place in any tissue. The lung selectively accumulates paraquat from the plasma by an energy-dependent process. Consequently, this organ contains higher concentrations than other tissues. Since the removal of absorbed paraquat occurs mainly via the kidneys, an early onset of renal failure following uptake of toxic doses will have a marked effect on paraquat elimination and distribution and on its accumulation in the lung (102).

Although paraquat and diquat have similar chemical, physical, and herbicidal properties, only paraquat has been shown to damage the lung. According to Sharp et al. (104), diquat concentrations in lung and muscle are much lower than the levels attained with equal doses of paraquat.

2.6.5.4 Toxicology

2.6.5.4.1 Mechanism of Action

The mechanism of the toxic effects of paraquat is largely the result of a metabolically catalyzed single-electron reduction-oxidation reaction, resulting in depletion of cellular NADPH and the generation of potentially toxic forms of oxygen such as the superoxide radical.

2.6.5.4.2 Toxicity

Apparently the smallest dose leading to death was 1 g ingested by a 23-year-old woman (105), indicating a dosage of about 16.7 mg/kg. This is consistent with the view (106) that anything over 5 ml of a 20% solution (about 14 mg/kg) is likely to be fatal. In one instance, a man ingested "only a mouthful" of (20%) liquid, most of which was said to have been rejected immediately" (107). A mouthful is about 50 ml. Therefore, if the patient's account was accurate, he may have swallowed about 20 ml or a dosage of about 57 mg/kg. The dose was fatal in 15 days. Many other fatal dosages were apparently larger. Solfrank et al. (108) expressed the view that 3 g of paraquat (about 43 mg/kg) was the largest dose allowing a chance for survival. However, there are exceptions to all such rules. A 40-year-old man survived a dose estimated at 7.2 g (109).

The LD_{50} values for paraquat and diquat are 150 and 231 mg/kg, respectively. Both compounds are classified as class II moderately hazardous (2).

The current acceptable daily intakes (ADI's) for paraquat ion and diquat ion are 0.04 and 0.08 mg/kg body weight, respectively (4).

2.6.5.4.3 Toxic effects

Two types of fatal poisoning can be distinguished: acute fulminate poisoning leading to death within a few days, and a more protected form that may last for several weeks, resulting in fatal pulmonary fibrosis. Depending on the severity of the poisoning, there may be involvement of kidneys, liver and other organs. Extensive damage to the oropharynx and the oesophagus are usually seen in cases of ingestion of liquid concentrate.

2.6.5.4.4 Treatment of poisoning

After ingestion, speed is imperative in commencing emergency treatment, and it should be noted that this can start before arrival of the patient at a hospital. Because there is no specific antidote for paraquat and diquat, treatment involves eliminating the herbicides from the body as quickly as possible. The use of adsorbents, e.g. Fuller's earth or bentonite, prevents the absorption of paraquat (110). The gut lavage should be continued until paraquat in blood or urine is not detectable (111).

The response to treatment of paraquat poisoning is very disappointing and the mortality rate remains high. In less severe cases, without lung damage, recovery has always been complete.

The prognosis in acute paraquat poisoning is largely determined by the time between ingestion and treatment, and by plasma paraquat concentrations before treatment.

Ikebuchi (112) reported the toxicological significance of paraquat concentration by means of the multivariate analysis method. The discrimination function (D) to separate the survival and fatal cases could be best described by the following equation:

$D = 1.3114 - 0.1617 \ln T - 0.5408 \ln [\ln (C \times 1000)]$, where T is time from ingestion (h); and C the plasma paraquat concentration (μg/ml). Ikebuchi et al. (113) introduced a new assessment of severity of paraquat poisoning using a toxicological index of paraquat (TIP). Figure 11 shows the distribution of D scores of 128 patients. The D score ranged from -0.086 to 0.437 in survivors. Fatal cases ranged from -0.489 to 0.066. The right and left parts of the figure refer to the classification by discriminant function as survival and death, respectively. To assess the severity of paraquat poisoning with higher reliability, the authors designate the obtained D score as TIP. The TIP can be divided into three types: TIP 1 type is characterized by $D > 0.1$; TIP 2, by $-0.1 < D < 0.1$; and TIP 3, by $D < -0.1$. TIP 1 and 3 permit the prediction of outcome with 100% probability in an early stage of acute paraquat poisoning. In TIP 2, the patient should be treated vigorously in the hope that the balance may be tipped in favor of survival. The TIP not only allows a more accurate assessment in patients, but also provides a more reliable method for judging the success of new treatments for paraquat poisoning.

2.6.5.5 Toxicological analyses

2.6.5.5.1 Group color reaction for paraquat and diquat (114)

(1) Principle of test
Paraquat and diquat develop a characteristic color after addition of sodium dithionite to alkaline urine, due to the formation of radicals.
(2) Sampling
Urine 10 ml
(3) Equipment
Pipets, test tubes
(4) Reagents
Dissolve 0.1 g sodium dithionite in 10 ml sodium hydroxide (the solution is stable for several hours when stored in ice water).
Paraquat stock solution: dissolve 0.1 g paraquat in 1 l water (freshly prepared)
(5) Negative control:
Urine of a nonexposed person.
(6) Positive control:
Mix 50 µl of paraquat stock solution with 950 µl of the negative control.
(7) Procedure
To 1000 µl of each sample (patient urine, negative control and positive control) add 1500 µl sodium dithionite solution. In a further test tube, mix 1000 µl patient urine with 1500 µl sodium hydroxide. Shake the mixtures for a short time. Blue color indicates the presence of paraquat. Diquat gives a slight yellow to green color.
(8) Analytical performance
Detection limit: The minimal detectable amount is 1.4 mg/l paraquat.
(9) Medical interpretation of the result
False positive results are not to be expected. Negative results are seen after intake of small amounts of paraquat and, in urine samples, if the intake is some time back. The test is not recommended to exclude an intake of diquat since this color reaction is not sensitive enough.

2.6.5.5.2 Special methods: Gas Chromatography

Van Dijk et al. (115) have reported a method for determination of paraquat in plasma which involves protein precipitation using trichloroacetic acid, followed by a reduction of the herbicides to ether – extractable compounds and subsequent gas chromatographic quantitation.

Extraction and reduction procedure
To 3.0 ml of plasma, add 1.0 ml internal standard containing 10.0 µg of ethyl viologen ion (1,1′-diethyl-4,4′-dipyridilium) per ml. Mix throroughly and add 1.0 ml of a 25% solution of trichloroacetic acid in distilled water. After reduction with sodium borohydride at pH 10 at 60 °C the sample is extracted with ether. The organic phase is removed, the residue reconstituted in methanol and injected in the gas chromatograph.

Column: 3% Poly A 135 on 80/100 mesh 01-1895 Supelcoport (Supelco).

Detector: Alkali Flame Ionization Detector.
The detection limit is 0.030 mg/l.

References

1. Mgeni, A.Y. (1991) Toxic effects of pesticides – Africa's health dilemma. In: Proceedings of the African Workshop on Health Sector Managements in Technological Disasters, Addis Ababa 26.–30.11.1990 (Holopainen, M., Kurttio, P. & Tuomiso, P., eds.) pp. 88–108, National Public Health Institute, Kuopio.
2. IPCS (International Programme on Chemical Safety) (1996) The WHO recommended classification of pesticides by hazard and guidelines to classification 1996–1997. WHO/PCS/96.4, Geneva.
3. Plestina, R. (1984) Prevention, diagnosis and treatment of insecticide poisoning. World Health Organisation, WHO/VBC/84.889, Geneva.
4. IPCS (International Programme on Chemical Safety) (1995) Summary of toxicological evaluations performed by the Joint FAO/WHO Meeting on Pesticide Residues (IMPR) 1995, WHO/PCS/95.50, Geneva.
5. DFG (Deutsche Forschungsgemeinschaft) (1995) Maximum concentrations at the workplace and biological tolerance values for working materials. VCH Verlagsgesellschaft Weinheim.
6. DFG (Deutsche Forschungsgemeinschaft) (1986) Biologische Arbeitsstoff-Toleranz-Werte (BAT-Werte), Arbeitsmedizinisch-toxikologische Begründungen. (Henschler, D. & Lehnert, G., eds.), VCH Verlagsgesellschaft Weinheim.
7. ACGIH (American Conference of Governmental Industrial Hygienists) 1991–1992 Threshold Limit Values for chemical substances and physical agents and Biological Exposure Indices.
8. Schrader, G. (1963) Die Entwicklung neuer insektizider Phosphorsäure-Ester. Verlag Chemie, Weinheim.
9. Industrieverband Agrar e.V. (1990) Wirkstoffe in Pflanzenschutz- und Schädlingsbekämpfungsmitteln. Physikalisch-chemische und toxikologische Daten. BLV Verlagsgesellschaft, München.
10. Minton, N.A. & Murray, S.G. (1988) A review of organophosphate poisoning. Medical Toxicol. **3**, 350–375.
11. Fest, Ch. & Schmidt, K.-J. (1970) Insektizide Phosphorsäureester. In: Chemie der Pflanzenschutz- und Schädlingsbekämpfungsmittel Vol. 1 (Wegler, R., ed.) pp. 246–453, Springer Berlin, Heidelberg, New York.
12. Hayes, W.J. (1982) Pesticides studied in man. Williams and Wilkins, Baltimore.
13. Baselt, R.C. & Cravey, B.S. (1989) Disposition of toxic drugs and chemicals in man, 3. ed. Year Book Medical Publishers Chicago, London, Boca Raton, Littleton, Mass.
14. WHO (World Health Organisation) (1986) Organophosphorus inecticides: a general introduction. Environmental Health Criteria 63, WHO, Geneva.
15. Gosselin, R.E., Hodge, H.C., Smith, R.P. & Cleason, M.N. (1984) Clinical toxicology of commercial products, 5th ed. Williams and Wilkins, Baltimore.
16. Ellenhorn, M.J. & Barceloux, D.G. (1988) Medical toxicology; Diagnosis and treatment of human poisoning. Elsevier, New York, USA.
17. Centers for Disease Control (1985) Neurological findings among workers exposed to fenthion in a veterinary hospital – Georgia. Morbidity and Mortality Weekly Report *34*, 402–403.
18. Lotti, M. & Johnson, M.K. (1978) Neurotoxicity of organophosphorus pesticides-predictions can be based on in-vitro studies with hen and human enzymes. Arch. Toxicol. *41*, 215–221.
19. Bertocin, D., Russolo, A., Caroldi, S. & Lotti, M. (1985) Neurotoxic esterase in human lymphocytes. Arch. Environ. Health *40*, 139–144.
20. Lotti, M. & Loretto, A. (1986) Inhibition of lymphocytic neuropathy target esterase predicts the development of organophosphate polyneuropathy in man. Human Toxicol. *5*, 114.
21. Maroni, M. & Bleeker, M.L. (1986) Neuropathy target esterase in human lymphocytes and platelets. J. Appl. Toxicol. *6*, 1–7.
22. Selden, B.S. & Curry, S.C. (1987) Prolonged succinylcholine-induced paralysis in

organophosphate insecticide poisoning. Ann. Emerg. Med. *16*, 215–217.

23. Ellman, G.L., Courtney, K.D., Andres, V. & Featherstone, R.M. (1961) A new and rapid colorimetric determination of acetylcholinesterase activity. Biochem. Pharmacol. *7*, 88–95.
24. Geldmacher-v. Mallinckrodt, M., Lindorf, H.H., Petenyi, M., Rabast, U. & Weiß, H. (1974) Zur Bewertung der Serum-Cholinesteraseaktivität in Leichenblut bei Verdacht auf eine E 605-Vergiftung. Z. Rechtsmed. *75*, 191–199.
25. den Blaauwen, D.H., Poppe, W.A. & Tritschler, W. (1983) Cholinesterase (EC 3.1.1.8) mit Butyrylthiocholinjodid als Substrat: Referenzwerte in Abhängigkeit von Alter und Geschlecht unter Berücksichtigung hormonaler Einflüsse und Schwangerschaft. J. Clin. Chem. Clin. Biochem. *21*, 381–386.
26. Goodman and Gilman's The pharmacological basis of therapeutics, 8. ed. (1992) (Goodman Gilman, A., Rall, T.W., Nies, A.S. & Taylor, P., eds.), pp. 131–149, McGraw-Hill, New York, St. Louis, San Francisco et al.
27. Lewalter, J., Domik, C. & Schaller, K.H. (1991) Acetylcholinesterase (ACHE; Acetylcholine-Acetylhydrolase EC 3.1.1.7) and Cholinesterase (CHE; Acylcholine-acylhydrolase EC 3.1.1.8). In: Analyses of hazardous substances in biological materials, Vol. 3, (Angerer, J. & Schaller, K.H., eds.), VCH Verlagsgesellschaft Weinheim.
28. Needham, P.H. (1960) An investigation into the use of bioassay for pesticide residues in foodstuffs. Analyst *85*, 792–809.
29. Schoof, H.F. (1964) Laboratory cultures of Musca, Fannia, and Stomoxys. Bull. World Hlth. Org. *31*, 539–944.
30. Erdmann, F., Brose, C. & Schütz, H. (1990) A TLC screening program for 170 commonly used pesticides using the corrected R_f value (R_f^c value). Int. J. Leg. Med. *104*, 25–31.
31. DFG/TIAFT (Deutsche Forschungsgemeinschaft / The International Association of Forensic Toxicologists) (1992) Thin-layer chromatographic R_f values of toxicologically relevant substances on standardized systems. VCH Verlagsgesellschaft Weinheim.
32. Geike, F. (1969) Dünnschichtchromatographisch-enzymatischer Nachweis und Wirkungsmechanismus von Chlorkohlenwasserstoff-Insektiziden. J. Chromatogr. *44*, 95–102.
33. Geike, F. (1970) Dünnschichtchromatographischer Nachweis von Carbamaten. I. Nachweis insektizider Carbamate mit Rinderleber-Esterase. J. Chromatogr. *53*, 269–277.
34. Shafik, T., Bradway, D.E., Enos, H.F. & Yobs, A.R. (1973) Human exposure to organophosphorus pesticides. A modified procedure for the gas-liquid chromatographic analysis of alkyl phosphate metabolites in urine. J. Agr. Food Chem. *21*, 625–629.
35. Lores, E.M. & Bradway, D.E. (1977) Extraction and recovery of organophosphorus metabolites from urine using an anion exchange resin. J. Agr. Food Chem. *25*, 75–79.
36. Blair, D. & Roderick, H.L.R. (1976) An improved method for the determination of urinary dimethylphosphate. J. Agr. Food Chem. *24*, 1221–1223.
37. Reid, S.J. & Watts, R.R. (1981) A method for the determination of dialkylphosphate residues in urine. J. Anal. Toxicol. *5*, 125–132.
38. Bradway, D.E., Moseman, R. & May, R. (1981) Analysis of alkyl phosphates by extractive alkylation. Bull. Environ. Contam. Toxicol. *26*, 520–523.
39. Coye, M.J., Lowe, B.S. & Maddy, K.J. (1986) Biological Monitoring of agricultural workers exposed to pesticides: II. Monitoring of intact pesticides and their metabolites. J. Occup. Med. *28*, 628–636.
40. Machbert, G., Distler, F. & Zhuo Xianyi (1994) Quantitative determination of alkylphosphates and alkylthiophosphates in urine using an isotope dilution method, to be published.
41. v. Eicken, S. (1954) Zur Ausscheidung von p-Nitrophenol im Urin nach Einwirkung von Pflanzenschutzmittel E 605. Angew. Chem. *66*, 551–564.
42. Elliot, J.W., Walker, K.C., Penick, A.E. & Durham, W.F. (1960) A sensitive procedure for urinary p-nitrophenol determination as a measure of exposure to parathion. J. Agr. Food Chem. *8*, 111–117.
43. Geldmacher-v. Mallinckrodt, M. & Herrmann, A. (1969) Gruppenreaktion zur Erfassung von p-Nitrophenolen, p-Aminophenolen und p-Phenylendiaminen im Harn. Z. Klin. Chem. Klin. Biochem. *7*, 34–37.
44. Michalke, P. (1982) Empfindlicher Nachweis von p-Nitrophenol in Blut und Urin mit HPLC nach E 605-Intoxikation. Z. Rechtsmed. *88*, 195–202.

45. DFG/TIAFT (Deutsche Forschungsgemeinschaft / The International Association of Forensic Toxicologists) (1992) Gas chromatographic retention indices of toxicologically relevant substances on packed or capillary columns with dimethylsilicone stationary phases. 3rd. ed., VCH Verlag Weinheim.
46. Sunshine, I. (1975) Methodology for analytical toxicology. CRC Press Cleveland.
47. DFG (Deutsche Forschungsgemeinschaft) (1990) Orientierende Angaben zu therapeutischen und toxischen Konzentrationen von Arzneimitteln und Giften in Blut, Serum oder Urin. pp. 11–33, VCH Verlag Weinheim.
48. Jones, G.R. & Singer, P.P. (eds.) (1988) Proceedings of the 24th International Meeting of the International Association of Forensic Toxicologists. Alberta Society of Clinical and Forensic Toxicologists, Edmonton, Alberta.
49. Liu, J., Susuki, O., Kumazawa, T. & Seno, H. (1989) Rapid isolation with the Sep-Pak C18 cartridges and widebore capillary gas chromatography of organophosphate pesticides. Forens. Sci. Int. *41*, 67–72.
50. Lewin, J.F. & Love, J.L. (1974) A death caused by the ingestion of mevinphos. Forens. Sci. *4*, 253–255.
51. Pfleger, K., Maurer, H.H. & Weber, A. (1992) Mass spectral and GC data of drugs, poisons, pesticides, pollutants and their metabolites. VCH Verlag Weinheim.
52. Brealey, C.J. & Lawrence, D.K. (1979) High-performance liquid chromatography of pirimiphos methyl and five metabolites. J. Chromatogr. *168*, 461–469.
53. Abou Donia, S.H. & Abou Donia, M.B. (1982) High-performance liquid chromatography of phosfolan, mephosfolan and related compounds. J. Chromatogr. *240*, 532–538.
54. Böcker, E. & Draber, W. (1970) Carbamate. In. Chemie der Pflanzenschutz- und Schädlingsbekämpfungsmittel, Bd. 1 (R. Wegler, ed.) pp. 220–245, Springer Berlin, Heidelberg, New York.
55. Mühlmann, R. & Schrader, G. (1957) Hydrolyse der insektiziden Phosphorsäureester. Z. Naturforsch. *12B*, 196–208.
56. Hayes, W.J. (1971) Studies on exposure during the use of anticholinesterase pesticides. Bull. WHO *44*, 277–278.
57. Hayes, W.J. jr. & Laws, E.R. (1991) Handbook on pesticide toxicology. Academic Press, San Diego, New York, Boston.
58. Oonnithan, E.S. & Casida, J.E. (1968) Oxidation of methyl and dimethylcarbamate insecticide chemicals by microsomal enzymes and anticholinesterase activity of the metabolites. J. Agr. Food Chem. *16*, 28–44.
59. Dreisbach, R.H. & Robertson, W.O. (1987) Handbook of poisoning, 12th ed. Lange Medical Pub., Los Altos, Calif.
60. Vandekar, M., Plestina, R. & Wilhelm, K. (1971) Toxicity of carbamates for mammals. Bull. WHO *44*, 241–249.
61. Mount, M.E. & Oehme, F.W. (1980) Microprocedure for determination of carbaryl in blood and tissues. J. Anal. Tox. *4*, 286–292.
62. Fleischmann, Ch. (1990) Extraktion und Nachweis von insektiziden Carbamaten aus dem Blutserum mittels HPLC-Entwicklung eines Screeningverfahrens. Thesis Erlangen.
63. Eben, A. & Barchet, R. (1978) 2-Isopropoxiphenol im Harn. In: Analysen in biologischem Material (Henschler, D. ed.), Verlag Chemie, Weinheim.
64. Best, E.M. & Murray, B.L. (1962) Observation on workers exposed to sevine insecticides: a preliminary report. J. Occup. Med. *10*, 507–517.
65. Worthing, Ch. R. & Hance R.J. (1991) The pesticide manual, a world compendium. 9th ed. The British Crop Protection Council, Old Walking, Surrey.
66. Hayes, W.J. Jr. (1982) Chlorinated hydrocarbon insecticides In: Pesticides studied in Man (Hayes W.J. ed.) pp 172–283, Williams & Wilkins, Baltimore/London.
67. WHO (1979) Environmental Health Criteria 9, DDT and its derivatives. United Nations Environment Programme and the World Health Organization, Geneva.
68. Albro, P.W. & Thomas, R. (1974) Intestinal absorption of hexachlorbenzene and hexachlorcyclohexane isomers in rats. Bull. Environ. Contam. Toxicol. *12*, 289–294.
69. Feldmann, R.J. & Maibach, H.J. (1974) Percutaneous penetration of some pesticides and herbicides in man. Toxicol. Appl. Pharmacol. *28*, 126–132.
70. Swanson, K. & Woolley, D. (1978) Neurotoxic effects of dieldrin. Toxicol. Appl. Pharmacol. *45*, 339.
71. Sunder-Ram-Rao, C.V., Shreenivas, R. & Singh, V. (1988) Disseminated intravascular coagulation in a case of fatal lindane poisoning. Vet.Hum.Toxicol. *30*, 132–134.

72. Osuntokun, B.O. (1964) "Gammelin 29" poisoning: A report of two cases. West Afr.Med. J. *13*, 207–210.
73. Davies, J.E., Edmundson, W.F., Carter, C.H. & Barquet, A. (1969) Effect of anticonvulsant drugs on dicophone (DDT) residues in man. Lancet *2*, 413–423.
74. Edmundson, W.F., Davies, J.E., Maceo, A. & Morgade, C. (1970) Drug and environmental effects on DDT residues in human blood. South. med. J. *63*, 1440–1441.
75. Watson, M., Gabica, J. & Benson, W.W. (1972) Serum organochlorine pesticides in mentally retarded patients on differing drug regimens. Clin. Pharmacol. Ther. *13*, 186–192.
76. McQueen, E.G., Owen, D. & Ferry, D.G. (1972) Effect of phenytoin and other drugs in reducing serum DDT levels. N.Z.Med.J. *75*, 208–211.
77. Davies, J.E., Edmundson, W.F., Maceo, A., Irvin, G.L., III, Cassady J. & Barquet, A. (1971) Reduction of pesticide residues in human adipose tissue with diphenyl hydantoin. Food Cosmet. Toxicol. *9*, 413–423.
78. Schoor, W.P. (1970) Effect of anticonvulsant drugs on inseticide residues, Lancet *2*, 520–521.
79. Kwalik, D.S. (1971) Anticonvulsants and DDT residues. J. Am. Med. Assoc. *215*, 120–121.
80. Micks, D.W. (1954) Potential health hazards of organic insecticides Tex. State J. Med. *50*, 148–153.
81. Curley, A. & Garrettson, L.K. (1969) Acute chlordane poisoning. Arch. Environ, Health *18*, 211–215.
82. Aldrich, F.D. & Holmes, J.H. (1969) Acute chlordane intoxication in a child. Arch. Environ. Health *19*, 129–132.
83. Derbes, V.J., Dent, J.H., Forrest, W.W. & Johnson, M.F. (1955) Fatal chlordane poisoning, J.A.M.A. *158*, 1367–1369.
84. Stranger, J. & Kerridge, G. (1968) Multiple fractures of the dorsal part of the spine following chlordane poisoning, Med. J. Aust. *1*, 267–268.
85. Davies, G.M. & Lewis, I. (1956) Outbreak of food poisoning from bread made from chemically contaminated flour. Br.Med.J. *2*, 393–398.
86. Reddy, D.B., Edward, V.B. & Rao, K.V. (1966) Fatal endrin poisoning. A detailed autopsy, histopathological and experimental study. J. Indian Med. Assoc. *46*, 121–124.
87. Weeks, D.E. (1967) Endrin food poisoning. A report on four outbreaks caused by two seperate shipments of endrin contaminated flour. Bull. WHO *37*, 499–512.
88. Pitsch, R.L. Finklea, J.F. & Ecotr, W.L. (1969) Stolen pesticides. J.S.C. Med. Assoc. *65*, 127–238.
89 Kazantzis, G., McLaughlin, A.I.G. & Prior, P.F. (1964) Poisoning in industrial workers by the insecticide aldrin. Br. J. Ind. Med. *21*, 46–51.
90. WHO (1959) Report of fatal case of aldrin poisoning in an African child. World Health Organization Information Circular on the Toxicity of Pesticides to Man, No. 3, p1, World Health Organization, Geneva.
91. Spiotta, E.J. (1951) Aldrin poisoning in man. Report of a case. Arch. Ind. Hyg. Occup. Med. *4*, 560–566.
92. Nelson E. (1953) Aldrin poisoning, Rocky Mt. Med. J. *50*, 483–486.
93. WHO (1958) Note by Secretariat on aldrin poisoning in Kenya. Information Circular on the Toxicity of Pesticides in Man, No 1, p 3. World Health Organization, Geneva.
94. Garrettson, L.K. & Curley, A. (1969) Studies in a poisoned child. Arch. Environ. Health *19*, 814–822.
95. AMA (1960) American Association Committee on Toxicology Report on: Occupational dieldrin poisoning, J.A.M.A. *172*, 2077–2080.
96. Weinig, E., Machbert, G. & Zink, P. (1966) Proof of dieldrin in dieldrin poisoning. Arch. Toxicol. *22*, 115–124.
97. Sunshine, I. (1975) Methodology for analytical toxicology. CRC Press, Cleveland, Ohio 81–82.
98. Burse, V.W., Head, S.L., Korver, M.P., McClure, P.C., Donahue, J.F. & Needham, L.L. (1990) Determination of selected organochlorine pesticides and polychlorinated biphenyls in human serum. J. Anal. Toxicol. *14*, 137–142.
99. Dale, W.E., Curley, A. & Cueto, C. Jr. (1966) Hexane extractable chlorinated insecticides in human blood. Life Sci. *5*, 47–54.
100. Cook, B.T., Settergren, S.K. & Carra, J. (1976–1980) Laboratory procedures for the analysis of human blood serum and urine from the second National Health and Nutrition Examination Survey (NHANES) II. Office of pesticides and toxic substance, U.S. Environmental Protection Agency, Washington, D.C.
101. Drummond, L., Gillanders, E.M. & Wilson, H.K. (1988) Plasma Gamma-Hexa-

chlorocyclohexane concentrations in forestry workers exposed to lindane. Brit. J. Ind. Med. *45*, 493–497.

102. WHO (1984) Environmental Health Criteria 39, Paraquat and diquat. United Nations Environment Programme and the World Health Organization, Geneva.
103. Hayes, W. J. Jr. (1982) Herbicides In: Pesticides studied in Man (Hayes W.J. ed.) pp 543–562, Williams & Wilkins, Baltimore/London.
104. Sharp, C.W., Ottolenghi, A. & Posner, H.S. (1972) Correlation of paraquat toxicity with tissue concentrations and weight loss of the rat. Toxicol. Appl. Pharmacol. *22*, 241–251.
105. FAO/WHO (1973) Paraquat. In: 1972 Evaluations of some pestizide residues in food, Rome, Food and Agriculture Organization of the United Nations.
106. Binnie, G.A.C. (1975) Paraquat. Lancet *1*, 169–170.
107. Bullivant, C.M. (1966) Accidental poisoning by paraquat: Report of two cases to man. Br. Med. J. 1, 1272–1273.
108. Solfrank, G., Mathes, G., Clarman, v.M. & Beyer, K.H. (1977) Haemoperfusion through activated charcoal in paraquat poisoning. Acta Pharmacol. *41*, 91–101.
109. Mayama, S., Haneda, T., Shimazu, W., Sato, H., Makino, E., Yamjima, T., Kumaktura, M., Obata, H. & Hattori, T. (1979) A case of acute intoxication due to paraquat dichloride. J. Jpn. Soc. Int. Med. *68*, 559–560.
110. De Kort, W.L.A.M., Sangster, B., Douze, J.M.C., Van Dijk, A. & Gimbrère, J.S.F. (1984) Intoxicaties door paraquat in Nederland, 1972–1982. Ned. Tijdschr. Geneeskd. *128*, 2441–2444.
111. Ellenhorn, M.J. & Barceloux, D.G. (1988) Medical toxicology. Diagnosis and treatment of human poisoning. Elsevier, New York – Amsterdam – London.
112. Ikebuchi, J. (1987) Evaluation of paraquat concentration in paraquat poisoning. Arch. Toxicol. *60*, 304–310.
113. Ikebuchi, J., Proudfoot, A.T., Matsubara, K., Hampson, E.C.G.M., Tomita, M., Suzuki, K., Fuke, C., Ijiri, I., Yuasa, I. & Okada, K. (1993) Toxicological index of paraquat: A new strategy for assessment of severity of paraquat poisoning in 128 patients. Forsenic Sci. Int. *59*, 85–87.
114. Daldrup, Th. (1994) Paraquat In: Einfache Methoden im klinisch-toxikologischen Laboratorium (H.J. Gibitz ed.) VCH Verlagsgesellschaft Weinheim.
115. Diijk van, A., Ebberink, R., de Groot, G., Maes, R.A.A., Douze, J.M.C. & van Heyst, A.N.P. (1977) A rapid and sensitive assay for the determination of paraquat in plasma by gas-liquid chromatography. J. Anal. Tox. *1*, 151–154.

2.7 Glycols

M. Roth

The importance of glycols in toxicology is related to the wide use of ethylene glycol and propylene glycol as automotive antifreeze agents. Occasionally severe intoxications due to accidental or intentional ingestion of ethylene glycol occur.

2.7.1 Ethylene Glycol (1,2 Ethanediol)

The toxic symptoms elicited by ethylene glycol are not due to the compound itself, but mainly to its metabolites glycolic acid and oxalic acid (1, 2) (fig. 1). The effects are a severe metabolic acidosis together with known effects of oxalate intoxication, namely hypocalcemia and urinary oxalate crystals.

Laboratory results of diagnostic importance are not only the concentration of ethylene glycol in blood, but also the anion gap and plasma bicarbonate. In cases where the identity of the poison is unknown, the assessment of metabolic acidosis is of great help as a first step in the search for the causal agent. The most frequent causes of toxic acidosis are the ingestion of ethylene glycol and of methanol. Salicylate intoxication also produces metabolic acidosis, but to a smaller degree, as the doses involved are generally lower. Large doses of ethanol produce an acidosis due to keto acids and lactic acid (5).

$HO-CH_2-CH_2-OH$ ethylene glycol

$HO-CH_2-C(=O)-OH$ glycolic acid

$HO-C(=O)-C(=O)-OH$ oxalic acid

Figure 1. Ethylene glycol and its principal acidic metabolites.

2.7.1.1 Plasma bicarbonate

A low plasma bicarbonate may be indicative of either metabolic acidosis or hyperventilation, two different conditions, both of which may be encountered in clinical toxicology. In combination with the analyses of sodium, potassium and chloride, bicarbonate serves to calculate the anion gap, an elevation of which indicates metabolic acidosis.

In certain hospitals, total CO_2 is routinely determined with automatic analyzers, for example by enzymatic assay (3).

Total CO_2 is the sum of bicarbonate, carbonic acid and dissolved CO_2. The latter two components account for only a small fraction of the total, so that total CO_2 may be taken as a good approximation of bicarbonate. In intensive care units, blood gas analyzers are used to determine pCO_2 and pH, both of which decrease in metabolic acidosis. Bicarbonate concentration may be calculated by use of the equation of Henderson-Hasselbalch:

$$\mathrm{pH} = \mathrm{pK} + \log \frac{[\text{bicarbonate}]}{[\mathrm{pCO_2}]}$$

For blood, pK (the dissociation constant of carbonic acid) equals 6.091.

2.7.1.2 The anion gap

This is defined as $([Na^+] + [K^+]) - ([Cl^-] + [HCO_3^-])$.

In a solution, the sum of cations equals the sum of anions. A number of circulating ions such as Ca^{2+} and proteins are not taken in consideration in the above formula, and since there are more non-measured anions than cations, the reference values for the anion gap of human plasma are positive (10–18 mmol/l). An anion gap above the reference interval indicates the presence of extraneous anions, such as keto acids in diabetes, phosphate, sulfate and different organic anions in the nephrotic syndrome and, in certain intoxications, the toxic agent or its metabolite(s) (4, 5).

In ethylene glycol intoxication, the elevated plasma glycolate is highly correlated to the anion gap (1).

In laboratories routinely performing the analyses of sodium, potassium, chloride and total CO_2, determination of these four analytes followed by calculation of the anion gap provides a fast information on a possible toxic acidosis. Some blood gas analyzers utilized in intensive care units, such as the NOVA stat profile 5 (NOVA Biomedical, PO Box 9141, Waltham, MA 02254-9825, USA) are capable of measuring sodium, potassium, chloride, pCO_2 and pH, and inferring the anion gap by calculation.

If one suspects an intoxication in presence of an elevated anion gap, one should first exclude the possibility of diabetic acidosis (a strip test for ketone will reveal this) and of uremia. The most likely causes of a toxic acidosis are the ingestion of either methanol or ethylene glycol by virtue of their respective metabolites formate and glycolate.

Gas chromatography usually serves to identify the alcohol involved.

Treatment of methanol and ethylene glycol poisoning involves administration of ethanol, which acts as competitive inhibitor of alcohol dehydrogenase, the enzyme responsible for the production of the metabolites formate and glyoxylate. Successful treatment results in a return of the elevated anion gap to normal values.

2.7.1.3 Gas chromatography of ethylene glycol

Because of the polar nature of ethylene glycol, its direct injection has posed problems to many investigators who observed peak tailing and poor sensitivity. For this reason, a number of methods use derivatization with phenylboronic acid followed by gas chromatography on packed columns filled with e.g. OV–101 (6,7). More recently, however, Aarstad et al (8) were able to design a direct injection method using a capillary column coated with Chromosorb 101 and a flame-ionization detector. The method distinguishes ethylene glycol from the less toxic compounds 1,2- and 1,3-propanediols and is well suited to the monitoring of ethylene glycol elimination upon treatment by dialysis.

2.7.1.4 Other methods

Ethylene glycol has been assayed by an enzymatic technique using glycerol dehydrogenase from **Enterobacter aerogenes** (9) and by HPLC after derivatization by p-methoxybenzoyl chloride (10).

Glycolate, the compound essentially responsible for ethylene glycol toxicity, has been determined by HPLC after derivatization with O-p-nitrobenzyl-N, N^1-diisopropyl urea (11), by isotachophoresis (1), by gas chromatography of the methyl ester (12) and by colorimetry (12). The gas chromatographic method seems convenient for following the elimination of glycolate upon hemodialysis (12).

2.7.2 Other Glycols

Propylene glycol (1,2-propanediol) is used as vehicle in some pharmaceutical formulations. Contact dermatitis has been reported as a side-effect of its application.

Analysis of propylene glycol in serum and wine may be done by gas chromatography (8).

References

1. Jacobsen, D., Ovrebo, S., Ostborg, J. and Sejersted, O. M. (1984) Glycolate causes the acidosis in ethylene glycol poisoning and is effectively removed by hemodialysis. Acta Med. Scand. *216*, 409–416.
2. Jacobsen, D., Akesson, I. Shefter, E. (1982) Urinary calcium oxalate monohydrate crystals in ethylene glycol poisoning.
3. Forrester, R.L., Wataji, L.J., Silverman, D.A. and Pierre, K.J. (1976) Enzymatic method for determination of CO_2 in serum.
4. Emmet, M. and Narins, R.G. (1977) Clinical use of the anion gap. Medicine *56*, 38–54.
5. Jacobsen, D., Bredesen, J.E., Eide, I. and Ostborg, J. (1982) Anion and osmolal gaps in the diagnosis of methanol and ethylene glycol poisoning. Acta Med. Scand. *212*, 17–20.
6. Flanagan, R.J., Dawling, S. and Buckley, B. M. (1987) Measurement of ethylene glycol (ethane-1,2-diol) in biological specimen using derivatisation and gas-liquid chromatography with flame ionisation detection. Ann. Clin. Biochem. *24*, 80–84.
7. Balikove, M. and Kahlicek, J. (1988) Rapid determination of ethylene glycol at toxic levels in serum and urine. J. Chromatography *434*, 469–474.
8. Aarstad, K., Dale, O., Aakervik, O., Ovrebo, S. and Zahlsen, K. (1993) A rapid gas

chromatographic method for determination of ethylene glycol in serum and urine. J. Anal. Toxicol. *17*, 218–221.

9. Standefer, J., and Blackwell, W. (1991) Enzymatic method for measuring ethylene glycol with a centrifugal analyzer. Clin. Chem. *37*, 1734–1736.
10. Matsumoto, N., Nagamatsu, M. and Ogata, H. (1992) Determination of alcohols in biological fluids, urine and pharmaceutical preparations by high-performance liquid chromatography after derivatization with p-methoxybenzoyl chloride. Iyakuhin Kenkyu *23*, 339–346.
11. Hewlett, T.P., Ray, A.C. and Reagor, J.C. (1983) Diagnosis of ethylene glycol (antifreeze) intoxication in dogs by determination of glycolic acid in serum and urine with high pressure liquid chromatography and gas chromatography-mass spectrometry. J. Assoc. Off. Anal. Chem. *66*, 276–283.
12. Fraser, A.D. and MacNeil, W. (1993) Colorimetric and gas chromatographic procedures for glycolic acid in serum. The major toxic metabolite of ethylene glycol. J. Toxicol. Clin. Toxicol. *31*, 397–405.

2.8 Inorganic Anions

M. Roth

2.8.1 Nitrate (NO_3^-) and Nitrite (NO_2^-)

We shall discuss both anions together as the toxicity of nitrate is due to its conversion *in vivo* to nitrite, so that the toxic manifestations are the same in both cases. Intoxications by nitrate have been observed in particular in rural areas where the concentration of nitrate in well water was abnormally high due to the increased use of nitrogenous fertilizers (1). Vegetables are another source of nitrate exposure. Nitrate itself is hardly toxic, but after oral ingestion, part of it is reduced to nitrite by intestinal bacteria. Absorbed nitrite then reacts with blood hemoglobin by oxidizing the constituent Fe^{2+} to Fe^{3+}, producing methemoglobin which cannot bind or transport oxygen. Normally, some methemoglobin is produced continuously in the red cells, but this is rapidly reduced back to hemoglobin by enzymes such as the cytochrome b_5 reductase. In nitrite intoxication, the quantities of methemoglobin produced overwhelm the capacity of the reduction systems.

In nitrate intoxication, a high concentration of nitrate in blood is observed (2). Determination of nitrate in biological fluids can be done by reduction of nitrate to nitrite followed by a chromogenic (3) or fluorigenic (4) reaction. Omission of the reduction step provides a measure of nitrite. In emergency toxicology, however, the most useful information is provided by the rapid determination of methemoglobin, as described under 2.8.3.1. A qualitative test for plasma nitrite using commercial reagent strips (Nephur-Test from Boehringer, Mannheim and N-Multistik from Ames) has also been reported (5).

Sodium nitrite has been the cause of accidental intoxications. A mistake in the preparation of a laxative solution, which contained sodium nitrite instead of sodium sulphate, caused a fatal methemoglobinemia in two patients (6). Soup prepared with water contamined by sodium nitrite caused an outbreak of methemoglobinemia in a school in USA (7).

Nitrite is a food additive, for example in sausages. The low concentrations are generally not toxic, but may induce methemoglobinemia in infants.

2.8.2 Chlorate (ClO_3^-) and Chlorite (ClO_2^-)

Chlorates are oxidizing agents which, if ingested, produce methemoglobinemia (see 2.8.3). Potassium chlorate is used in some matchheads and sodium chlorate in certain herbicides. In two cases of suicidal attempt with sodium chlorate, the treatment was exsanguino-transfusion (8).

Sodium chlorite, too, produces methemoglobinemia (9).

2.8.3 Methemoglobin as a Toxicity Index

Nitrites and chlorates are not the only compounds producing methemoglobinemia. Many organic compounds have the same effect, including the organic nitro derivates metabolized to nitrite. The intoxication mechanism is the same, producing a similar clinical picture. The emergency diagnosis rests on the laboratory assessment of methemoglobinemia. Table 1 lists a number of substances known to induce methemoglobinemia. The severity of symptoms correlates with the proportion of hemoglobin denaturated to methemoglobin (normal $> 1.5\%$; mild symptoms 15–20%; marked cyanosis; moderate symptoms 20–45%; severe symptoms 45–70%; $> 70\%$ usually lethal). The symptoms are those of hypoxia and cyanosis; the latter, due to the brown color of methemoglobin, has been referred to as "chocolate cyanosis". Methylene blue can be administrated as an antidote.

Table 1. Substances inducing methemoglobinemia

Substance	References
a) inorganic compounds	
nitrate	1
nitrite	6, 7
chlorate	8
chlorite	9
b) organic nitro compounds	
amyl nitrite	10, 11
butyl nitrite	10
isobutyl nitrite	10, 12, 13
methyl nitrite	14
nitrogylcerin	15
nitrobenzene	16
nitroethane	17
c) arylamine derivatives and metabolites	
dapsone	18, 19
phenylhydroxylamine	20
benzocaine	21
aniline	20
aminophenol	20

2.8.3.1 Determination of methemoglobin

2.8.3.1.1 Manual methods

2.8.3.1.1.1 Qualitative (1)

In severe methemoglobinemia, blood drawn from the patient has a brownish color. A simple test is to put a drop of blood on filter paper and to compare its chocolate brown color with that of a drop of normal blood.

2.8.3.1.1.2 Quantitative (22, 23)

The absorption spectrum of methemoglobin in the visible range is distinct from the other forms of hemoglobin, showing a maximum at 631 nm in slightly acidic solution. This maximum disappears upon conversion into cyanmethemoglobin. Sulfhemoglobin, which is rarely present, absorbs in the same spectral range (max. 618 nm), but its absorbance does not disappear upon addition of cyanide (24).

Blood is diluted 50-fold by volume in 20 mmol/l potassium phosphate buffer, pH 6.6. The absorbance of a 5 ml aliquot of the hemolyzate is measured at 631 nm before (A_1) and after the addition (A_2) of 5 mg potassium cyanide. A second aliquot is completely converted into methemoglobin (ca. 5 mg $K_3Fe(CN)_6$) and the absorbance again read before (A_3) and after (A_4) the addition of cyanide.

$$\frac{A_1 - A_2}{A_3 - A_4} = \text{mass fraction of methemoglobin in total hemoglobin}$$

The mass fraction is normally inferior to 0.015 (1.5%). Signs of intoxication appear above 0.15.

2.8.3.1.2 Automated method for the simultaneous determination of methemoglobin, hemoglobin and carboxyhemoglobin

Specially designed photometers for the differential determination of oxyhemoglobin and deoxyhemoglobin are used since many years in hospitals. The anticoagulated blood is injected into the instrument, which provides for the appropriate dilution and measures the absorbance of the hemolyzate at two or more wavelengths. The more recent instruments usually measure the absorbance at four to six different wavelengths and incorporate a microprocessor computing the concentrations of oxyhemoglobin, deoxyhemoglobin, carboxyhemoglobin and methemoglobin based on a series of equations (25, 26). The results are available in less than one minute after the injection of the sample. Such instruments are operated in a stat mode in intensive care units and central laboratories of clinical chemistry and represent an invaluable means of measuring methemoglobin and carboxyhemoglobin in cases of emergency.

2.8.3.1.2.1 Automated determination of methemoglobin

Commercial instruments capable of determining methemoglobin include the IL 482 CO-oxymeter, the Corning 2500 CO-oxymeter, the Radiometer OSM3 hemoxymeter and the AVL 912 CO-oxylite (27). The results are reported as a fraction of total hemoglobin. Sulfhemoglobin, if present in large quantities, interferes, but its presence is generally indicated by a warning signal.

2.8.3.1.2.2 Automated determination of carboxyhemoglobin

The analysis of carbon monoxide is discussed in chapter 2.2 of this book. Mahoney et al. (27) have compared five commercial CO-oxymeters for their performance in the determination of carboxyhemoglobin. They found them to compare favorably with spectrophotometric and gas chromatographic reference methods, except for concentrations inferior to 2.5% where the accuracy was unsatisfactory.

2.8.4 Bromate (BrO_3^-)

Bromates are not as potent as chlorates for inducing methemoglobinemia. The clinical manifestations of intoxication include acute renal failure and deafness. A cause of poisoning has been the ingestion of hair neutralizers used in permanent wave preparations (28, 29).

Determinations of plasma urea and creatinine are among the first laboratory tests to be done, for both diagnosis and treatment. Methods for the assay of total bromine are mentioned under 2.8.5.3.2.

2.8.5 Halogenides

2.8.5.1 Fluoride (F^-)

Fluoride, for example as sodium fluoride, is used for the prevention of dental caries. For this purpose, small concentrations may be incorporated in drinking water and toothpastes. At much higher concentrations, fluoride is a toxic agent altering calcium availability to tissues and causing cardiac dysfunction.

Chronic intoxication induces bone fractures and calcification of periarticular structures. Plasma fluoride (reference interval: 0.5–10.5 μmol/l) and urinary fluoride excretion (reference interval: 10–58 μmol/l) are elevated (30). Acute intoxication has been described in a group of patients who underwent hemodialysis. The cause was a high fluoride concentration in the dialysis water due to a faulty operation of the deionization system (31). The intoxicated patients had an increased plasma fluoride and a decreased plasma calcium. Other effects of acute fluoride intoxication are hyperkaliemia and hypomagnesemia (32). Hydrofluoric acid produces the same symptoms, with burns in addition (33).

Plasma and urine fluoride are conveniently determined by use of specific electrodes (34–36).

2.8.5.2 Chloride (Cl^-)

The toxicity of most metal chlorides is essentially determined by their cationic moiety, and is therefore dealt with in chapter 2.10. However, we choose to discuss the toxicity of sodium and potassium chlorides in the present section.

2.8.5.2.1 Sodium chloride

Salt is necessary for life, but if ingested at high doses it is toxic. Intoxications are rare in adults, since the taste prevents one from ingesting excessive amounts. This is different in young babies, who depend on their parents for the choice of food and drink. Salt poisoning, either accidental or intentional, has been reported in a number of children, generally from 1 to 9 months old.

Sodium, and to a lesser extent chloride, are part of the standard panel of analyses performed in the blood of hospitalized patients. Hypernatremia is observed in different sorts of electrolyte disturbances. In salt poisoning, it is generally the first indice leading to the correct diagnosis. Plasma sodium may then be as high as 200 mmol/l (reference interval 135–148 mmol/l) and chloride may reach 180 mmol/l (reference range 96–107 mmol/l). Accordingly, plasma osmolality then exceeds 400 mmol/kg (reference range 285–295 mmol/kg) (37, 38). If the renal function is not impaired, urinary sodium and chloride excretion are raised.

In hospitals, sodium has long been determined by flame emission spectrophotometry, and more recently with sodium-sensitive electrodes. Both methods are precise and accurate. The electrodes have the advantage of being better suited to automation. Sodium and potassium are usually assayed together (39). Chloride may be determined by mercurimetric titration, coulometric-amperometric titration, spectrophotometry and ion-sensitive electrodes (39). Automatic analyzers use either ion-sensitive electrodes or spectrophotometry.

2.8.5.2.2 Potassium chloride

Potassium chloride is used therapeutically as a potassium supplement and by the general public as a salt substitute. Overdose of potassium chloride produces a hyperkaliemia, which has to be distinguished from the other more frequent forms of hyperkaliemia observed in clinical practice (40). The reference range for heparinized plasma potassium is between 3.1 and 4.4 mmol/l. Determination of potassium in plasma, serum and urine is made with the same methods as for sodium (39).

2.8.5.3 Bromide (Br^-)

2.8.5.3.1 Intoxication by bromide

Bromides are known since the last century as anticonvulsants. In spite of their side-effects and the development of newer drugs, potassium bromide is still prescribed today as an antipileptic. Bromide is also a constituent of some drugs such as dextromethorphan hydrobromide (41) and pyridostigmine bromide (42). It is even contained in some over-the-counter drugs.

The therapeutic serum concentration of bromide in anticonvulsant therapy is 0.6–1.2 mmol/l. Confusion and lethargy occur above 6 mmol/l and a concentration of 40 mmol/l may be fatal (43).

Unsuspected bromide intoxication may be discovered in patients having an unexplained hyperchloridemia. Since most electrodes used for the assay of chloride in clinical laboratories respond to bromide, and the spectrophotometric method using mercuric thiocyanate also measures bromide, bromide intoxication produces a falsely elevated plasma chloride, together with a low or even negative anion gap. The non-specificity of these methods actually represents an advantage over more specific methods such as coulometery (41, 44, 45) for the detection of unsuspected bromide intoxication.

Millet et al. (46) reported two cases of intoxication by bromide-containing drugs. One patient, a child, had an apparent chloridemia of 194 mmol/l and up to 40 mmol/l of bromide as determined by a specific method.

2.8.5.3.2 Analysis

Numerous assay methods of bromide in blood have been published over the past seventy years. Many of them involve a time-consuming preparation of the sample. Some older colorimetric methods are not specific enough and overestimate the bromide concentration of serum.

The colorimetric method of Goodwin uses a specific reaction between bromine and rosaniline as the final step (47). The method actually measures the sum [bromide + bromine + bromate] but is indeed intended for the determination of bromide as the predominant form. Miller and Cappon used a protein-free ultrafiltrate of serum which they injected onto an ion-exchange chromatography column (48). The bromide concentration measured in human serum ultrafiltrates ranged from 44 to 125 μmol/l (3.5–10 mg/l). Another HPLC method has been described by Goewie and Hogendorn (49).

Head-space gas chromatography has been used to determine urine bromide (50). Bohn et al. published an ion-chromatographic method for the sumultaneous determination of bromide, nitrite, nitrate, phosphate and sulfate in urine (51). Total bromine has been determined in plasma and urine by ICP-MS (52).

2.8.5.4 Iodide (I^-)

2.8.5.4.1 Toxicity

Iodine (I_2), if ingested in large amounts, is very toxic, but few cases of acute intoxication have been reported. In the body, iodine is reduced to the iodide ion, which represents the form excreted in urine.

Iodide is required for the biosynthesis of thyroid hormones. If present in excess, it inhibits this synthesis, and hypothyroidism develops.

The repeated application of iodine solutions (used as a local disinfectant) to newborns has been a matter of concern in pediatry, as iodine rapidly crosses the skin of young babies. An increase in circulating total iodine and TSH has been reported under such conditions (53). Iodide poisoning has also been observed after long-term treatment with iodinated glycerol (54).

2.8.5.4.2 Analysis

An adequate method for determining iodide in blood is the gas chromatographic method devised by Doedens (55). The iodide is derivatized with methyl-isobutylketone, followed by extraction into hexane. A portion of the extract is analyzed by GC using a packed column and electron capture detection. Urinary iodide has been determined by HPLC with electrochemical detection (56).

Total iodine in urine may determined by colorimetry after ashing (57). Total iodine in serum, plasma and urine has been assayed by ICP-MS (52, 58).

2.8.6 Sulfide (S^{2-})

Whereas the sulfides of heavy metals are poorly soluble in water, sodium and potassium sulfides are soluble and generate hydrogen sulfide (SH_2) in acidic solution. Thus hydrogen sulfide is released in the stomach upon oral ingestion of sodium sulfide.

Hydrogen sulfide is a gas with an unpleasant odor of rotten eggs. If inhaled at high concentration, it is very toxic. It reacts with cellular respiratory enzymes such as cytochrome oxidase and rapidly produces cyanosis. In the blood, it reacts with methemoglobin by forming sulfhemoglobin, a green pigment which cannot transport oxygen but is otherwise devoid of toxicity. For this reason, nitrite, which produces methemoglobin, is used as an antidote for the treatment of acute SH_2 intoxication (59, 60).

Sulfides are metabolized to thiosulfate, which appears in the urine. Thiosulfate has been determined in urine by liquid chromatography (61) and in both urine and plasma by GC and GC-MS (62).

2.8.6.1 Determination of sulfhemoglobin

Sulfhemoglobin contains a sulfur atom within the porphyrin ring (63). Its presence in blood is rarely encountered in clinical practice. It is a typical product of SH_2 intoxication. A few drugs (phenacetin, dapsone (18)) may cause a delayed formation of sulfhemoglobin, probably by an intermediate mechanism occurring *in vivo*.

Some automatic CO-oximeters (see 2.8.3.1.2) give a warning signal if sulfhemoglobin is present. Sulfhemoglobin may be determined by differential spectrophotometry (25, 64–66).

2.8.7 Phosphate

Larson et al. (67) reported a case of laxative phosphate poisoning in a 4-month-old boy. Excess of inorganic phosphate in blood binds calcium, resulting in a dangerous hypocalcemia. Plasma calcium was 0.09 mmol/l (reference 2.20–2.52), phosphate 16.1 mmol/l (ref. int. 1.2–1.9). The anion gap was 44 mmol/l (ref. int. 10–18). The treatment consisted in intravenous calcium gluconate and parenteral rehydration. A similar case is reported by Craig et al. (68).

Phosphate is routinely assayed in clinical laboratories by spectrophotometry (69).

2.8.8 Cyanide (CN^-)

2.8.8.1 Cyanide intoxication

Cyanide is a deadly poison acting rapidly. Ingestion of salts such as KCN and inhalation of hydrocyanic gas HCN both result in poisoning. Cyanide binds with cytochrome oxidase, thus paralyzing cellular respiration. This results in cellular hypoxia and metabolic acidosis (70). Sodium and potassium cyanide have many industrial uses, especially in metallurgy. Strong acids convert them into HCN, a volatile liquid with bitter-almond odor.

Cyanide poisoning can be caused by the ingestion of cyanide salts, from accidental, suicidal or even criminal causes. Intoxication can also be due to the biochemical release of cyanide *in vivo* from a precursor such as laetrile (amygdalin obtained from the pits of apricots or peaches) (71), sodium nitroprusside, a drug used in the treatment of hypertension (72, 73), and acetonitrile, an organic solvent (74, 75). Pyrolysis of some nitrogen-containing synthetic polymers generates cyanide, and this is believed to contribute, in addition to carbon monoxide, to the toxicity of smokes inhaled during fires (76, 77).

Laboratory analyses helpful in the emergency diagnosis of cyanide intoxication are the determination of cyanide in plasma and gastric content, and the identification of metabolic acidosis by blood gas analysis, or plasma CO_2, pH and lactate.

If done promptly, treatment by an antidote is very effective. Sodium nitrite has been used as antidote (71). It produces methemoglobin which reacts by forming nontoxic cyanmethemoglobin. The disadvantage of this antidote is that it is detrimental to the capacity of hemoglobin to transport oxygen. It is not recommended in the treatment of fire victims (76). A much safer antidote is hydroxycobalamin, which binds cyanide by forming the nontoxic cyancobalamin (77–78). Thiosulfate as an antidote reacts slowly and is therefore not very helpful for the treatment of acute poisoning; on the other hand, it has been used in cases of chronic toxicity, for example during nitroprusside therapy.

Thiocyanate is a normal detoxification product of cyanide. The conversion is slow. Thiocyanate concentration in plasma is elevated in chronic cyanide toxicity (72).

2.8.8.2 Cyanide analysis

Various methods exist for assaying cyanide. Because of the high affinity of cyanide for methemoglobin, the concentration in whole blood is higher than that in plasma, and the analysis is usually done on whole blood. The microdiffusion principle of Conway has been widely used in the past. Anticoagulated blood is placed in the external compartment of the microdiffusion cell. After the addition of sulfuric acid, released HCN gas diffuses and is trapped by an alkaline solution placed in the internal compartment. The resulting solution is analyzed for cyanide by colorimetry (79, 80) or another procedure such as detection by an ion-specific electrode (81). A disadvantage of these methods is the time required for the diffusion. This can be accelerated by heating of the cell.

Head-space gas chromatography is a good means of assaying cyanide (82, 83). Using an electron capture detector, Odoul et al. obtained a detection limit of 5 μg/l.

Cyanide present in the erythrocytes as the non-toxic cyanmethemoglobin is released as HCN by strong acids such as H_2SO_4 or perchloric acid and therefore measured as cyanide in the above methods (84). Cyanocobalamin, on the other hand, does not contribute to the cyanide measured (80).

Cyanide measured in plasma or serum is in the free, toxic form. It correlates with whole blood cyanide during nitroprusside intoxication (72).

Cyanide specifically reacts with 2,3, naphthalene dialdehyde by forming a strongly fluorescing product. This is used in a promising HPLC method with precolumn derivatization (85).

Whole blood cyanide concentration in normal persons is inferior to 2 μmol/l. Smokers have higher concentrations than non-smokers. Severe toxicity is observed above 50 μmol/l, and 200 μmol/l may be lethal. If an antidote is given rapidly, a patient may survive after having had even higher concentrations.

References

1. Kross, B.C., Ayebo, A.D. and Fuortes (1992) Methemoglobinemia: nitrate toxicity in rural America. American Family Physician *46*, 183–188.
2. Hegesh, E. and Shiloah, J. (1982) Blood nitrates and infantile methemoglobinemia. Clin. Chim. Acta *125*, 107–115.
3. Schneider, N.R. and Yeary, R.A. (1973) Measurement of nitrite and nitrate in blood. Am. J. Vet. Res. *24*, 133–135.
4. Miski, T.P. Schilling, R.P. Salveminu, D., Moore, W.M. et al. (1993) A fluorometric assay for the measurement of nitrite in biological samples. Anal. Biochem. *214*, 11–16.
5. Soler Rodriguez, F., Miguez Santiyan, M.P. and Pedrera Zamorano, J.D. (1992) Evaluation of reagent strips for the rapid diagnosis of nitrite poisoning. J. Anal. Toxicol. *16*, 63–65.
6. Ellis, M., Hiss, Y. and Shenkman, L. (1992) Fatal hemoglobinemia caused by inadvertent contamination of a laxative solution with sodium nitrite. Israel J. Med. Sci. *28*, 289–291.
7. Askew, G.L., Finelli, L., Genese, C.A., Sorhage, F.E. et al. (1994) Boilerbaisse: an outbreak of methemoglobinemia in New Jersey in 1992. Pediatrics *94*, 381–384.
8. Granier, P., Danel, V., Barnoud, D., Lavagne, P. et al. (1985) Intoxication par le chlorate de soude. Presse médicale *14*, 1099.
9. Lin, J.L. and Lim, P.S. (1993) Acute sodium chlorite poisoning associated with renal failure. Re. Fail. *15*, 645–648.
10. Haley, T.J. (1980) Review of the physiological effects of amyl, butyl and isobutyl nitrite. Clin. Toxicol. *16*, 317–329.
11. Machabert, R., Testud, F. and Descotes, J. (1994) Methaemoglobinemia due to amyl nitrite inhalation: a case report. Human & Experimental Toxicology *13*, 313–314.
12. Wason, S., Detshy, A.S., Platt, O.S. and Lovejoy, F.H. (1980) Isobutyl nitrite by ingestion. Ann. intern. med. *92*, 637–638.
13. O'Toole, J.B., Robbins, G.B. and Dixon, D.S. (1987) Ingestion of isobutyl nitrite, a recreational chemical of abuse causing fatal methemoglobinemia. J. Forensic Sci. *32*, 1811–1812.
14. Wax, P.M. and Hoffman, R.S. (1994) Methemoglobinemia: an occupational hazard of phenylpropanolamine production. J. Toxicol. Clin. Toxicol. *32*, 299–303.
15. Bojar, R.M., Rastegar, H. Payne, D.D., Harkness, S.H. et al. (1987) Methemoglobinemia from intravenous nitroglycerin: a word of caution. Ann. Thorac. Surg. *43*, 332–334.
16. Schimelman, M.A., Soler, J.M. and Muller, H.A. (1978) Methemoglobinemia: nitrobenzene ingestion. JACEP *7*, 406–408.
17. Hornfeldt, C.S. and Rabe, W.H. (1994) Nitroethane poisoning from an artificial fingernail remover. J. Toxicol. Clin. Toxicol. *32*, 321–324.
18. Lambert, M., Sonnet, J., Mahieu, P. and Hassoun, A. (1982) Delayed sulfhemoglobinemia after acute dapsone intoxication. J. Toxicol. Clin. Toxicol. *19*, 45–50.
19. Hansen, D.G., Challoner, K.R. and Smith, D.E. (1994) Dapsone intoxication: two case reports. J. Emerg. Med. *12*, 347–351.
20. Harrison, J.H. and Jollow, D.J. (1987) Contribution of aniline metabolites to aniline-induced methemoglobinemia. Mol. Pharmacol. *32*, 423–431.
21. Rodriguez, L.F., Smolik, L.M. and Zbehlik, A.J. (1994) Benzocaine-induced methemoglobinemia: report of a severe reaction and review of the literature. Ann. Pharmacother. *28*, 643–649.
22. Winterhalter, K.H. Quantitative determination of various forms of hemoglobin in a mixture. In: H.C. Curtius and M. Roth (eds) Clinical Biochemistry, Principles and Methods. p.1312. W. de Gruyter, Berlin, New York, 1974.

23. Stahr, H.M. Analytical Methods in Toxicology. p. 5. J. Wiley & Sons, New York, 1991.
24. Brandenburg, R.O. and Smith, H.L. (1951) Sulfhemoglobinemia: a study of 62 clinical cases. Am. Heart J. *42*, 582–588.
25. Fogh-Andersen, N., Sigaard-Andersen, O., Lundsgaard, F.C. and Wimberley, P.D. (1987) Diode-array spectrophotometry for simultaneous measurement of hemoglobin pigments. Clin. Chim. Acta *166*, 283–289.
26. Ziljstra, W.G., Buursma, A. and Zwart, A. (1988) Performance of an automated six-wavelength photometer (Radiometer OSM3) for routine measurement of hemoglobin derivatives. Clin. Chem. *34*, 149–152.
27. Mahoney, J.J., Vreeman, H.J., Stevenson, D.K. and van Kessel, A.L. (1993) Measurement of carboxyhemoglobin and total hemoglobin by five specialized spectrophotometers (CO-oximeters) in comparison with reference methods. Clin. Chem. *39*, 1693–1700.
28. Warshaw, B.L., Carter, M.C., Hymes, L.C., Bruner, B.S. and Rauber, A.P. (1985) Bromate poisoning from hair permanent preparation. Pediatrics *76*, 975–978.
29. Gradus, D., Rhoads, M., Bergstrom, L.B. and Jordan S.C. (1984) Acute bromate poisoning associated with renal failure and deafness presenting as hemolytic uremic syndrome. Am. J. Nephrol. *4*, 188–191.
30. Raimondeau, J., Rodat, O., Nicolas, G., Boiteau, H.L. et al. (1993) Fluorose osseuse "médico-légale". A propos de deux cas de probable intoxication criminelle. Rev. Méd. Interne *14*, 263–267.
31. Arnow, P.M., Lee, A., Bland, M.A., Garcia-Houchins, S. et al. (1994) An outbreak of fatal fluoride intoxication in a long-term hemodialysis unit. Ann. Intern. Med. *121*, 339–344.
32. Stremski, E.S., Grande, G.A. and Ling, L.J. (1992) Survival following hydrofluoric acid ingestion. Ann. Emerg. Med. *21*, 1396–1399.
33. Bordelon, B.M., Saffle, J.R. and Morris, S.E. (1993) Systemic fluoride toxicity in a child with hydrofluoric acid burns. Case report. J. Trauma *34*, 437–439.
34. Fuchs, C., Dorn, D., Fuchs, C.A., Henning, H.V. et al. (1975) Fluoride determination in plasma by ionselective electrodes: a simplified method for the clinical laboratory. Clin. Chim. Acta *60*, 157–167.
35. Speaker, J.H. (1976) Determination of fluoride by specific ion electrode and report of a fatal case of fluoride poisoning. J. Forensic. Sci. *21*, 121–126.
36. Durst, R.A. and Sigaard-Andersen, O. *Electrochemistry*. In: C.A. Burtis & E.R. Ashwood (eds.) Tietz Textbook of Clinical Chemistry, 2nd edition, pp. 159–183. W.B. Saunders, Philadelphia 1994.
37. Saunders, N., Baffe, J.W. and Laski, M.D. (1976) Severe salt poisoning in an infant. J. Pediatr. *88*, 258–261.
38. Meadow, R. (1993) Non-accidental salt poisoning. Arch. Dis. Child. *68*, 448–452.
39. Tietz, N.W., Pruden, E.L. and Sigaard-Andersen, O. *Electrolytes*. In: C.A. Burtis & E.R. Ashwood (eds.) Tietz Textbook of Clinical Chemistry, 2[d] edition, pp. 1354–1374. W.B. Saunders, Philadelphia 1994.
40. Saxena, K. (1989) Clinical features and management of poisoning due to potassium chloride. Med. Toxicol. Adverse Drug Exp. *4*, 429–443.
41. Ng, Y.Y., Lin, B.C., Chen, T.W., Tsai, S.H. et al. (1992) Spurious hyperchloremia and decreased anion gap in a patient with dextromethorphan bromide. Am. J. Nephrol. *12*, 268–270.
42. Rothenberg, D.M., Berns. A.S., Barkin R. and Glantz, R.H. (1990) Bromide intoxication secondary to pyridostigmine therapy. JAMA *263*, 1121–1122.
43. Buchanan, J.J. in: K.R. Olson, ed. Poisoning and Drug Overdose pp. 98–99 Prentice-Hall International, London 1990.
44. Woody, R.C., Turley, C.P. and Brewster, M.A. (1990) The use of serum electrolyte concentrations determined by automatic analyzers to indirectly quantitate serum bromide concentration. Ther. Drug. Monit. *12*, 490–492.
45. Emancipator, K. and Kroll, M.H. (1900) Bromide interference. Is it really better? Clin. Chem. *36*, 1470–1473.
46. Millet, V., Samperiz, S., Mancini, J., Charrel, M. et al. (1993) Hyperchlorémie néonatale révélatrice d'une intoxication au brome. Pédiatrie *47*, 785–787.
47. Goodwin, J. (1971) Colorimetric measurement of serum bromide with a bromate-rosaniline method. Clin. Chem. *17*, 544–547.
48. Miller, M.E. and Cappon, C.J. (1984) Anion-exchange chromatographic determination of bromide in serum. Clin. Chem. *30*, 781–783.
49. Goewie, C.E. and Hogendoorn, E.A. (1985) Liquid chromatrographic determination of bromide in human milk and plasma. J. Chromatography *344*, 157–165.
50. Koga, M., Hara, K., Hori, H., Kodama, Y. et al. (1991) Determination of bromide ion concentration in urine using a head-space

gas chromatography and an ion chromatography-biological monitoring for methyl bromide exposure. Sangyo Ika Daigaku Zasshi *13*, 19–24.

51. Bohn, I., Niessen, K.H., Reineke, F. and Teufel, M. (1994) Simultane Bestimmung anorganischer Anionen in Körperflüssigkeiten. Eine neue Methode in der pädiatrischen Labordiagnostik; Ergebnisse einer Pilotstudie. Klin. Pädiatr. *206*, 95–99.
52. Allain, P., Mauras, Y., Dougé, C. and Jaunault, L. (1990) Determination of iodine and bromine in plasma and urine by inductively coupled plasma mass spectrometry. Analyst *115*, 813–815.
53. Charolle, J.P., and Rossier, A. (1978) Goitre and hypothyroidism in the newborn after cutaneous absorption of iodine. Arch. Dis. Child. *53*, 495–498.
54. Geurian, K. and Branam, C. (1994) Iodine poisoning secondary to long-term iodinated glycerol therapy. Arch. Int. Med. *154*, 1153–1156.
55. Doedens, D.J. (1985) Iodide determination in blood by gas chromatography. J. Anal. Toxicol. *9*, 109–111.
56. Rendl, J., Seybold, S. and Borner, W. (1994) Urinary iodide determined by paired-ion reversed-phase HPLC with electrochemical detection. Clin. Chem. *40*, 908–913.
57. Aumont, G. and Tressol, J.C. (1986) Improved routine method for the determination of total iodine in urine and milk. Analyst *111*, 841–843.
58. Vanhoe, H., van Allemeersch, F., Versiekc, J. and Dams, R. (1993) Effect of solvent type on the determination of total iodine in milk powder and human serum by inductively coupled plasma mass spectrometry. Analyst *118*, 1015–1019.
59. Stone, R.J., Slosberg, B. and Beacham, B.E. (1976) Hydrogen sulfide intoxication. Ann. Int. Med. *85*, 756–758.
60. Peters, W.J. (1981) Hydrogen sulfide poisoning in a hospital setting. JAMA *246*, 1588–1589.
61. Kangas, J. and Savolainen, H. (1987) Urinary thiosulfate as an indicator of exposure to hydrogen sulfide vapour. Clin. Chim. Acta *164*, 7–10.
62. Kage, S., Nagata, T. and Kudo, K. (1991) Determination of thiosulfate in body fluids by GC and GC/MS. J. Anal. Toxicology *15*, 148–150.
63. Park, C.M. and Nagel, R.L. (1984) Sulfhemoglobinemia. Clinical and molecular aspects. New Eng. J. Med. *310*, 1579–1584.
64. Fairbanks, V.F. and Klee, G.F. *Biochemical aspects of hematology* in: C.A. Burtis and E.R. Ashwood (eds) Tietz Textbook of Clinical Chemistry, 2nd ed., pp. 2026–2027. W.B. Saunders, Philadelphia, 1994.
65. Sigaard-Andersen, O., Nørgaard-Pedersen, B. and Rem, J. (1972) Hemoglobin pigments. Spectrophotometric determination of oxy-, carboxy-, met-, and sulfhemoglobin in capillary blood. Clin. Chim. Acta *42*, 85–100.
66. Zwart, A., Buursma, A., van Kampen, E.J., van der Ploeg, P.H.W. et al. (1981) A multiwavelength spectrophotometric method for the simultaneous determination of five hemoglobin derivatives. J. Clin. Chem. Clin. Biochem. *19*, 457–463.
67. Larson, J.E., Swigart, S.A. and Angle, C.R. (1986) Laxative phosphate poisoning: Pharmacokinetics of serum phosphorus. Human Toxicol. *5*, 45–49.
68. Craig, J.C., Hodson, E.M. and Martin, H.C. (1994) Phosphate enema poisoning in children. Med. J. Austr. *160*, 347–351.
69. Endres, B. and Rude, R.K. *Mineral and Bone Metabolism*. In: C.A. Burtis and E.R. Ashwood (eds), Tietz Textbook of Clinical Chemistry 2d ed. pp. 1908–1910. W.B. Saunders, Philadelphia 1994.
70. Singh, B.M., Coles, N., Lewis, P., Braithwaite, R.A. et al. (1989) The metabolic effects of fatal cyanide poisoning. Postgraduate Medical J. *65*, 923–925.
71. Hall, A.H., Linden, C.H., Kulig, K.W. and Rumack, B.H. (1986) Cyanide poisoning from laetrile ingestion: role of nitrite therapy. Pediatrics *78*, 269–272.
72. Vesey, C.J. and Cole, P.V. (1985) Blood cyanide and thiocyanate concentrations produced by long-term therapy with sodium nitroprusside. Br. J. Anaesth. *57*, 148–155.
73. Robin, E.D. and McCauley, R. (1992) Nitroprusside-related cyanide poisoning. Time (long past due) for urgent, effective interventions. Chest *102*, 1842–1845.
74. Caravati, E.M. and Litovitz, T.L. (1988) Pediatric cyanide intoxication and death from an acetonitrile-containing cosmetic. JAMA *260*, 3470–3473.
75. Turchen, S.G., Manoguerra, A.S. and Whitney, C. (1991) Severe cyanide poisoning from the ingestion of an acetonitrile-containing cosmetic. Am J. Emerg. Med. *9*, 264–267.
76. Kulig, K. (1991) Cyanide antidotes and fire toxicology. New Eng. J. Med. *325*, 1801–1802.

77. Tassan, H., Joyon, T., Lamaison, D. et al. (1990) Intoxication au cyanure de potassium traitée par l'hydroxycobalamine. Ann. Fr. Anesth. Réanim. *9*, 383–385.
78. Forsyth, J.C., Mueller, P.D., Becker, C.E., Osterloh, J. et al. (1993) Hydroxycobalamin as a cyanide antidote: safety, efficacy and pharmacokinetics in heavily smoking normal volunteers. J. Tox. Clin. Tox. *31*, 277–294.
79. Dunn, W.A. and Siek, T.J. (1990) A rapid, sensitive, and specific screening technique for the determination of cyanide. J. Anal. Toxicol. *14*, 256.
80. Laforge, M., Buneaux, F., Houeto, P., Bourgeois, F. et al. (1994) A rapid spectrophotometric blood cyanide determination applicable to emergency toxicology. J. Anal. Toxicol. *18*, 173–175. (see also *letter to the editor*, J. Anal. Toxicol. *18*, (1994), 424)
81. Yagi, K., Ikeda, S., Schweiss, J.F. and Homan, S.M. (1990) Measurement of blood cyanide with a microdiffusion method and an ion-specific electrode. Anesthesiology *73*, 1028–1031.
82. Odoul, M., Fouillet, B., Nouri, B., Chambon, R. et al. (1994) Specific determination of cyanide in blood by headspace gas chromatography. J. Anal. Toxicol. *18*, 205–207.
83. Brandenberger, H. *Determination of cyanide in body fluids*. In: H.C. Curtius and M. Roth, (eds) Clinical Chemistry, Principles and Methods, p. 1431. W. de Gruyter, Berlin, New York 1974.
84. Lundquist, P., Rosling, H. and Sörbo, B. (1985) Determination of cyanide in whole blood, erythrocytes, and plasma. Clin. Chem. *31*, 591–595.
85. Sano, A., Takimoto, N. and Takitani, T. (1992) High-performance liquid chromatographic determination of cyanide in human red blood cells by pre-column fluorescence derivatization. J. Chromatogr. *582*, 131–135.

2.9 Non-Metals of Group 16

M. Roth and M. Pelletier

2.9.1 Selenium

2.9.1.1 Introduction

Selenium is increasingly recognized as an important metalloid of the living world. It occurs in the earth's crust at concentrations between 50 and 90 µg/kg. The sources of pollution are essentially coal combustion and wastes from industries using selenium. Concentrations in soil and plants differ widely depending on regions, which explains why regional differences exist for the average concentrations in the blood of populations.

A small proportion of selenium occurs in the atmosphere as volatile organic derivatives and, mainly, in the form of dusts. Certain plants accumulate selenium and convert it into alkyl-selenium.

Selenium is an essential element for vertebrates. It is a constituent of glutathione peroxidase, an enzyme capable of eliminating hydrogen peroxide and lipid peroxides within living cells. A number of papers suggest that selenium depletion enhances the risk of cancer mortality (1, 2). In alcoholics and patients with hepatic insufficiency, low concentrations of selenium in plasma have been observed (3, 4). Ingestion of high doses of selenium causes atypical symptoms such as headache, skin inflammation, low appetite and gastrointestinal disorders (5, 6). Cases of intoxication occur almost exclusively in persons working in factories utilizing selenium. The only cases of environmental intoxication have been observed in regions where the soil, and in turn the vegetation, contain high quantities of selenium (5), as in a region of China where the amount ingested may be as high as 5 mg per day (7). People in this area have selenium concentrations in blood up to 3.2 mg/l (8).

Ingested quantities exceeding 1 mg daily per kg body weight may produce a chronic intoxication. Concentrations superior to 5 mg/kg in food or 0.5 mg/l in water or milk are considered dangerous (6, 9). Normally, the ingested quantities are far below these values, ranging from 28–30 µg/day in New Zealand to 326 µg/day in Venezuela. 60–300 µg/day is considered normal in adults (10, 11).

The average concentration of selenium in the blood of normal individuals is about 30–150 µg/l. There are great individual variations. At birth the concentration is higher, then it drops progressively up to 20 years of age (12, 13).

About half of the absorbed selenium is excreted in the faeces, mainly in mineral form and the other half in the urine, partly as alkyl-selenium (14). The concentration in normal urine is between 5 and 30 µl/l.

2.9.1.2 Analysis

Selenium analysis requires sensitive techniques in view of the low concentration found in biological samples. Precautions have to be taken in order to avoid contaminations from the biological matrix and losses due to the volatility of certain selenium compounds. Most methods use an oxidative digestion of the organic matrix yielding tetravalent or hexavalent selenium. The determination may be done by atomic absorption, fluorimetry, neutron activation, polarography or gas chromatography. Neutron activation is a highly sensitive technique providing good results, but few laboratories have access to the required instrumentation (6, 17). Tetravalent selenium reacts with various *o*-diamines, thus forming piazselenols which can be extracted into organic solvents and measured by gas chromatography or fluorimetry.

In the gas chromatography techniques, detection has been made by electron capture or microwave emission. A method for urine by gas chromatography-mass spectrometry has been described by Aggarwal et al. (18). The use of a ^{76}Se internal standard provides a convenient way of correction for losses which may occur during the preparation. The method by Ducros and Favier (13), which uses the same principle, is also applicable to serum and red blood cells.

Fluorimetric methods take advantage of the strong fluorescence exhibited by the compound formed upon reaction of tetravalent selenium with 2,3-diaminonaphthalene. The method devised by Koh and Benson (19) uses a digestion at 210 °C with a mixture of nitric and perchloric acids, reduction of Se^{+6} to Se^{+4} by HCl, extraction into cyclohexane and measurement of the fluorescence at 364 nm (excitation) and 523 nm (emission). The method has been successfully used for the analysis of large numbers of samples.

At present, one of the best techniques is atomic absorption. However, flame atomic absorption is not sensitive enough. The graphite furnace technique, which is sensitive, suffers from matrix interference at the low wavelength used (196 nm). Background correction by the Zeeman effect provides satisfactory results (21, 22). Hydride generation followed by atomic absorption spectrometry is the most reliable procedure (8, 15, 20). The sample (blood or urine) is heated for several hours with concentrated nitric acid in order to destroy most of the organic material. A complete mineralization is not necessary, as the goal is merely to avoid foaming during hydride generation. The sample is then mixed with a basic solution of sodium borohydride and diluted with hydrochloric acid. This generates selenium hydride which is then sent by a nitrogen flow through a heated quartz tube situated above the burner of an atomic absorption instrument, in the axis of the beam of the hollow cathode lamp.

We use the following procedure: 2–5 g of sample are mixed with 5 ml of suprapure nitric acid and placed for 5–6 h in a 25 ml Erlenmeyer at 100 °C. After cooling, the volume of the solution is completed to 10 ml with distilled water. 1 ml of the solution is mixed with 10 ml of 0.5 mol/l hydrochloric acid and placed into the hydride generation cell. After 30 seconds (allowing for the nitrogen stream to eliminate oxygen, which interferes at the wavelength of 196 nm), a 30 g/l solution of sodium borohydride in 0.25 mol/l potassium hydroxide is introduced into the cell. Potassium hydroxide is preferred to sodium hydroxide (often mentioned in the manufacturer's procedures) as it contains less selenium as an impurity. The maximum value of the absorbance is recorded. Since the nature of the sample and its degree of previous mineralization have a great influence on the speed of hydride generation (foaming), it is essential to use the technique of standard addition instead of a calibration curve. The above method is suitable for assaying selenium in blood, serum or urine down to concentrations of 10 μg/l.

Finally, we may mention the technique by Buckley et al. (22) which combines hydride generation with inductively coupled plasma mass spectrometry.

2.9.2 Tellurium

Tellurium is one of the rarest elements on earth. It is used in industry, for example in metallurgy as an additive to iron, steel and copper and as a semiconducting compound in microelectronic industry.

Few cases of poisoning are known. Müller et al. (24) described the case of a woman who had eaten meat that was contaminated by tellurium for an unknown reason. The tellurium concentration in blood serum was 27.6 μg/l (reference range <1 μg/l). The symptoms included nausea and hair loss, but no persistent health impairment developed.

One of the best indices for diagnosing tellurium intoxication is a characteristic garlic odor of breath, which may persist for months (24).

Analytical methods for the determination of tellurium bear some analogy with those for selenium, but since the reference range in human blood is much lower, a higher sensitivity is required.

Ashing procedures suffer from losses due to the volatility of tellurium. To overcome this, matrix modification (25) or isotope dilution GC-MS (26) have been used.

However, the most convenient method is probably hydride generation followed by atomic absorption spectrometry, in a way similar to the technique for selenium (27). For atomic absorption, the 214.3 nm line may be used (28).

References

1. Willet, W.C., Morris, J.S., Pressel, S., Taylor, J.O. et al (1983) Prediagnostic serum selenium and risk of cancer. Lancet II, 130–134.
2. Shamberger, R.J., Rukovena, E., Longfield, A.K., Tytko, S.A. et al (1973) Antioxidants and cancer. I. Selenium in the blood of normals and cancer patients. J. Natl. Cancer Inst. *50*, 863–870.
3. Dutta, S.K., Miller, P., Greenbery, L., Levanda, O.A. (1982) Evidence for selenium depletion in alcoholic subjects admitted for detoxification. Am. J. Clin. Nutr. *35*, 845.
4. Snook, J.T. (1991) Effect of ethanol use and other lifestyle variables on measures of selenium status. Alcohol (NY) *8*, 13–16.
5. NAS (1976) Medical and biological effects of environmental pollutants – Selenium. National Academy of Sciences, Washington DC.
6. Glover, J., Levander, O., Parizek, J. and Vouk, V. (1979) *Selenium* in: Friberg, L., Nordberg, G.F. and Vouk, V.B. (eds) Handbook on the toxicology of metals, pp 555–577. Elsevier-North Holland, Amsterdam.
7. Yang, G., Yin, S., Zhou, R., Gu, L., Yan, B. et al. (1989) Studies of safe maximal daily Se-intake in a seleniferous area in China. Part II: Relation between Se-intake and the manifestation of clinical signs and certain biochemical alterations in blood and urine. J. Trace Elem. Electrolytes Health Dis. *3*, 125–130.
8. Norheim, G. and Haugen, A. (1986) Precise determination of selenium in tissues using

automated wet digestion and an automated hydride generator-atomic absorption spectroscopy system. Acta Pharmacol. Tox. 59, suppl. 7, 610–612.
9. Newland, L.W. in Hutzinger, O. The Handbook of environmental chemistry, vol 3, Part B, pp 45–57. Springer-Verlag, Heidelberg, 1982.
10. Lo, M.T. and Sandi, E. (1980) Selenium: occurence in foods and its toxicological significance. A Review. J. Environ. Pathol. Toxicol. *4*, 193–218.
11. Kazantsis, G. (1981) Role of cobalt, iron, lead, manganese, mercury, platinum, selenium and titanium in carcinogenesis. Environ. Health Perspect. *40*, 143–161.
12. Fishbein, L. (1991) Selenium in: Ernest Merian, ed. Metals and their compounds in the environment, pp 1153–1190. Verlag Chemie, Weinheim, New York.
13. Ducros, Y. and Favier, A. (1992) Gas chromatographic-mass spectrometric method for the determination of selenium in biological samples. J. Chromatogr. *583*, 35–44.
14. Von Stockhausen, H.B. (1986) in Abstracts of international Workshop Trace element analytical chemistry in Medicine and Biology. Neuherberg.
15. Angeer, J. and Schaller, K.H. (1988) Selenium. In: Analyzes of hazardous substances in biological materials. Vol. 2, pp 231–247. VCH Verlagsgesellschaft, Weinheim, New York.
16. Einbrodt, H.J. and Michels, S. (1984) Selenium. In: Merian, E. (ed.) Metalle in der Umwelt, pp 541–554. Verlag Chemie, Weinheim.
17. Fishbein, L. (1985) In: Merian, E., Frei, R.W., Haerdi, W. and Schlatter, C. (eds) Carcinogenic and mutagenic metal compounds, pp. 96–112. Gordon & Breach, New York, London.
18. Aggarwal, S.K., Kinter, M., and Herold, David, A. (1992) Determination of selenium in urine by isotope dilution gas chromatography-mass spectrometry using 4-nitro-*o*-phenylenediamine, 3,5-dibromo-*o*-phenylenediamine, and 4-nitro-*o*-phenylenediamine. Anal. Biochem. *202*, 367–374.
19. Koh, T.S. and Benson, T.H. (1983) Critical re-appraisal of fluorometric method for determination of selenium in biological materials. J. Ass. Off. Anal. Chem. *66*, 918–926.
20. Welz, B. and Verlinden, M. (1986) IUPAC interlaboratory trial-Selenium determination in human body fluids using hydride generation atomic absorption spectrometry. Acta Pharmacol. Toxicol. *59*, suppl. 7, 577–580.
21. Nève, J., Molle, L. (1986) Direct determination of selenium in human serum by graphite furnace atomic absorption spectroscopy. Improvement due to oxygen ashing in graphite tube and Zeeman-effect background correction. Acta Pharmacol. Toxicol. *59*, 606–609.
22. Nève, J., Chamart, S., Molle, L. (1987) Optimization of a direct procedure for the determination of selenium in plasma and erythrocytes using Zeeman effect atomic absorption spectroscopy. In: Bratter, P., Schramel, P. (eds.) Trace element analytical chemistry in medicine and biology. W. de Gruyter, Berlin, pp. 349–358.
23. Buckley, W.T., Budac, J.J., and Godfrey, D.V. (1992) Determination of selenium by inductively coupled plasma mass spectrometry utilizing a new hydride generation sample introduction system.
24. Müller, R., Zschiesche, W., Steffen, H.M. and Schaller, K.H. (1989) Tellurium-Intoxication. Klin. Wochenschr. *67*, 1152–1155.
25. Newman, R.A., Osborn, S. and Siddik, Z.H. (1989) Determination of tellurium in biological fluids by means of electrothermal vaporization-inductively coupled plasma-mass spectrometry (ETV-ICP-MS) Clin. Chim. Acta *179*, 191–196.
26. Aggarwal, S.K., Kinter, M., Nicholson, J., and Herold, D.A. (1994) Determination of tellurium by isotope dilution gas chromatography/mass spectrometry using (4-fluorophenyl)magnesium bromide as a derivatizing agent and a comparison with electrothermal atomic absorption spectrometry. Anal. Chem. *66*, 1316–1322.
27. Schierling, P., Oeffele, C. and Schaller, K.H. (1982) Determination of arsenic and selenium in urine samples with the hydride AAS technique. Ärztl. Laboratorium *28*, 21–27.
28. Siddik, Z.H. and Newman, R.A. (1988) Use of platinum as a modifier in the sensitive detection of tellurium in biological samples. Anal. Biochem. *172*, 190–196.

2.10 Metals of Toxicological Significance

H. Brandenberger and M. Roth

2.10.1 Introduction

In most, but not all, cases of analysis for metals or semimetals, preliminary qualitative tests can be omitted and the material under investigation submitted directly to a quantitative determination of that element whose presence (in elevated concentration) is suggested by the case history and/or the clinical or pathological data. While, in the past, colorimetric, fluorimetric or polarographic procedures have usually been employed for such measurements, atomic absorption spectroscopy has now become the method of choice. For trace determination, neutron activation analysis is still used for a few elements, but plasma-mass spectrometry is becoming increasingly popular. Mass spectrometry combined with gas chromatographic separation can also be used successfully for measuring those semi-metals which can easily be converted to volatile hydrides. The use of ion selective electrodes for quantitative cation analysis is possible only in a few cases.

After a short discussion of the needs and possibilities for qualitative metal screening, the present chapter will review the clinical and toxicological significance of some metals and groups of metals, and recommend methods for their quantitative determination in biological materials.

A considerable number of metals are normal constituents, often trace constituents, of the human body. A lack of such an element may be as much of a health hazard as its presence in too high concentration. Other metals, not needed as body constituents, are nevertheless present at quite low levels without causing toxic effects. But at slightly elevated concentrations, such effects may become evident and endanger health as well as life. This can result from professional or environmental exposure, and also from accidental or intentional intake of a material containing or contaminated with metals. Since even small differences in the concentration may be responsible for toxic effects, highly accurate analytical methods, and for trace element analysis also highly sensitive methods, are acceptable as analytical tools. While, in a search for organic poisons such as drugs or pesticides, qualitative or semiquantitative analytical work may often have priority over exact quantitative determinations, the contrary is usually true in the toxicological control of the levels of metallic elements in body fluids and tissues.

At the time the present authors started their work in forensic and clinical toxicology, only a few metals were regarded as important toxic hazards. The interest was mainly focused on arsenic, lead, mercury and thallium. Today, we know that a much larger number of metals and semi-metals can cause toxic effects. For the present chapter, we had to set some limits. Just about two dozen elements are discussed, several of them rather briefly. To some, our selection may seem arbitrary. It is based on the intoxications we have been confronted with in many years of service work, and does not exclude that metals we have not discussed are also capable of producing toxic symptoms.

2.10.2 Qualitative Analysis for Metals in Biological Specimens

2.10.2.1 Preliminary tests

The Reinsch test (1, 2) is suitable as a qualitative screening method for some heavy metals in biological fluids. The sample is heated in solution or suspension with hydrochloric acid in presence of a clean copper plate or copper wire. A black deposit on the copper surface indicates a presence of arsenic, antimony or bismuth, a shiny metallic layer a presence of mercury or silver. The composition of such a deposit can easily be identified by qualitative tests, as well as by atomic absorption or other instrumental methods (i. e. flame or plasma emission analysis, electrochemical methods, ion chromatography or capillary electrophoresis).

2.10.2.2. Color reactions

Color reactions have been frequently used to detect metals in biological specimens. Some excellent tests are described in handbooks which are on most toxicologists' desks (2–4). In some cases, the metal ions must first be separated from the biological matrix by complex formation and extraction with organic solvents, or by conversion to a volatile derivative such as a hydride. The Gutzeit test for arsenic (1) is a good example of such an approach. But color reactions are usually tests for only a single metal or a group of metals, and are seldom helpful for a general screening.

2.10.2.3 Emission methods

Flame emission spectroscopy, an effective method for qualitative metal analysis, lacks the sensitivity needed for solving a large part of the toxicological problems. There are of course exceptions. For some elements of the Groups 1 to 3, the sensitivity of flame emission analysis is surprisingly good. But a general qualitative screening method should be able to detect a large spectrum of the metals of toxicological significance, if not all, and not just a few.

Inductive coupled plasma emission spectroscopy (ICP-AES) usually possesses superior sensitivity than flame emission analysis. But it still can not cope with all the requests of the toxicologist. The combination of inductive coupled plasma with mass spectrometry (ICP-MS) is certainly a better choice, both with respect to sensitivity and freedom from interferences. Tables 1 and 4 to 6 list the detection limits of the most common instrumental methods for the analysis of the metals discussed in the respective chapters. They show clearly the superiority of ICP-MS over ICP-AES for trace analysis. Recent innovations in the ICP-technique, such as ultrasonic nebulation, have permitted improvements in sensitivity of several factors of 10. Since background noise is fairly uniform across the mass range and since most elements are highly ionized in the plasma torch, detection limits on a molar basis are similar, except for the differences due to variations in the abundance of the major isotopes (5). This and the large linear range of the method (up to 5 orders of magnitude) are further facts which help to make ICP-MS the preferred screening technique. At present, the instrumentation is still expensive. But it can be expected that in a few years the costs will no longer exceed the budget of an average toxicological laboratory.

ICP-MS also provides a unique possibility for determining isotope ratios for elements in the ng-range. This can be used in quantitative analysis. The sample is spiked with an enriched isotope of the element of interest. The altered isotope ratio is measured and the original element concentration calculated from the ratio difference (5). Such an approach is not hampered by eventual losses of the analyte in sample preparation, nor by signal suppression or enhancement in the measuring process. ICP-MS is therefore not only a useful screening technique in toxicological metal analysis, it has also the potential of becoming a highly accurate quantitative method. In addition, the determination of isotope ratios may help to solve forensic problems.

2.10.2.4 Electrochemical methods

Polarographic or voltametric methods have once played a major role in metal analysis, also in the field of toxicology. But since the introduction of atomic absorption, their importance for the quantification of metals has faded. On the other hand, they are still used for screening purposes. Especially differential pulse voltametry (DPV) has recently been recommended (6). It can reveal up to 12 toxicologically important metals directly in urine, namely In, Tl, Sn, Pb, As, Sb, Bi, Cu, Zn, Cd, Co and Ni.

2.10.2.5 Thin-layer chromatography (TLC)

In the past, thin-layer chromatography has been used for separating and revealing metal ions. Particularly cellulose sheets were popular with, as moving phases, mixtures of dilute mineral acids and organic solvents such as methanol, ethanol or acetone. Visualization was accomplished by spraying with dithizone (which reveals As, Sb and Sn as yellow, red or purple spots), or treatment with ammonia vapor, followed by spraying with dithizone and then diphenylcarbazone (for the detection of Bi, Ca, Cr, Fe, Cu, Ni, Hg, Pb, Ti and Zn). For these and other metal analyses by TLC, we refer to the monograph by Stahl (7). In view of the more recent developments, their importance may be more historical.

2.10.2.6 Cation chromatography on columns

Ion chromatography has become a special branch of high performance liquid chromatography (8, 9). For trace cation work, an instrumentation without any metallic contacts should be used. This calls i.e. for special pumps, and if possible even for an all-plastic chromatographic unit. For most applications, polystyrene-based ion exchangers are chosen as stationary phases; they possess a pH-stability which extends from 1 to 13. The modern "Ion Pac" cation exchangers, for example, permit separation and detection of most inorganic cations in aqueous solutions in the concentration range of or below 1 ng/ml (10). The new self-regenerating suppressors continously remove the acid from the eluant. This improves the detection limits of conductivity detectors and greatly facilitates the work. However, direct and indirect UV absorption, as well as fluorescence analysis after derivatization, are also used.

2.10.2.7 Capillary electrophoresis

Capillary electrophoresis was initially conceived for the separation and detection of high molecular mass molecules such as proteins and nucleotides. Today, it is used increasingly for separating low molecular mass substances, recently even for inorganic anions and cations (11). Separation of the cations by electromigration is easily achieved by weak complexation to selectively reduce the mobility of the different cations. Detection of the non UV-absorbing cations is more problematic. An on-line conductivity detector (12) and a potentiometric detector (13) for capillary zone electrophoresis have been described and used for cation analysis. For routine work, these systems may not be too practical. Other possibilities are the analysis of the cations after complex formation with UV-absorbing molecules such as acetylacetone (14) or fluorescent compounds such as 8-hydroxyquinoline-5-sulfonic acid (15). Practical solutions are the use of indirect UV detection in the presence of strongly UV-absorbing background electrolytes (16, 17, 18) and indirect fluorescence detection (19). This field has not yet been sufficiently investigated. It can be expected that further progress will soon be made in capillary ion analysis. The method permits separation of the different cations in a few minutes. With respect to sensitivity it is, however, not yet competitive with some of the spectrometric techniques.

2.10.3 Lithium

2.10.3.1 Toxicity

Lithium carbonate is widely used in psychiatry, in particular for the treatment of maniac-depressive disorders. It has a narrow range of therapeutic plasma concentrations (0.6–1.5 mmol/l), and thus lithium determinations in blood serve to monitor therapy and prevent overdosage.

Severe intoxications, either deliberate or accidental, sometimes occur and a number of fatal cases have been reported. Toxic manifestations include renal and central nervous system effects (20–22). In particular, nephrogenic diabetes insipidus and dehydration may develop as a consequence of inhibition of the antidiuretic hormonal mechanism. The elimination is slow, and in severe intoxications treatment by hemodialysis is indicated (23, 24).

2.10.3.2 Lithium determination

For many years, the determination of lithium was mostly done by atomic absorption or flame emission spectrometry and was therefore restricted to large hospitals or specialized toxicology centers. More recently, the availability of instruments equipped with lithium-sensitive electrodes has made the determination accessible to smaller laboratories and clinics.

2.10.3.2.1 Atomic absorption spectrometers and flame emission

A comparison of atomic absorption at 668 nm with an air-acetylene flame and flame emission showed that both methods provided closely similar results (25). It is convenient to use diluted serum as material. Blood plasma or even whole blood hemolysate may be used, but in this case the anticoagulant should be lithium-free heparinate, for example ammonium heparinate (26, 27). However, in emergency situations, a lithium heparin concentration inferior to 30 IU per ml is acceptable. The determination may also be done on diluted urine.

2.10.3.2.2 Ion-sensitive electrodes (ISE)

The development of suitable lithium-sensitive electrodes has been delayed by the necessity of eliminating the interference of sodium, as the ratio of therapeutic concentrations of lithium to those of sodium in plasma is lower than 1:1000. In recent years, the availability of organic ionophores specific for lithium has enabled the development of the type of electrodes needed. They measure the activity of the lithium ion, and therefore may be used not only to analyze plasma, serum and urine, but also anticoagulated blood, irrespective of the presence of erythrocytes.

Comparative studies have shown that results with ISE correlate well with those by atomic absorption and flame emission spectrometry (28, 29). Some types of electrodes, however, are sensitive to changes in the pH of the sample (28), and if they are used, it is advisable to perform the analysis shortly after blood collection. Several commercial electrode instruments manufactured for the determination of sodium and potassium in intensive care units now incorporate also a lithium channel.

2.10.3.2.3 Comparison of results

The three methods, atomic absorption, flame emission and lithium-sensitive electrodes, are all convenient for the diagnosis and management of lithium intoxication. However, atomic absorption is the less practical one, whereas ion sensitive electrode systems are progressively gaining wide acceptance owing to their greater simplicity.

2.10.4 The Metals of the Group 2 (Alkaline Earths)

2.10.4.1 Beryllium

Beryllium has no known biological significance. But in recent times, it has acquired considerable toxicological importance, since it is increasingly used in industry, i.e. in steel manufacture, in atomic power plants and in the preparation of fluorescent powders. The use of Be in fluorescent light bulbs has led to a number of intoxications and had to be abandoned.

Only half of the ingested Be is excreted by way of the kidneys, the rest is deposited mainly in liver and bones. It can substitute for magnesium in the active site of enzymes and thus interfere with the action of alkaline phosphatases.

Chronic exposure to Be vapors or Be dust will lead to a so-called beryllium fever, to beryllium pneumonia, and to a formation of lung granulomes. The metal may also be carcinogenic, and individuals exposed to Be vapors or Be dust must therefore be submitted to analytical control.

Colorimetric and fluorimetric procedures for determining Be (i.e. with acetyl acetone or with morine) can be hampered by interferences and have been largely replaced by spectroscopic methods. Table 1 shows that flame AAS can detect as little as 1.5 μg Be per liter water, if the hot acetylene-nitrous oxide flame is used, and that flameless AAS in a graphite furnace can reveal almost 100 times less. Such a sensitivity equals that of ICP-MS, which we recommend as a screening procedure. For quantifications, AAS is preferred not only because of its good sensitivity, but also due to low costs, simplicity and high accuracy. For graphite furnace AA, matrix modification is recommended for the thermal stabilization of Be. It permits sufficiently high charring temperatures, which reduce background interferences. It can be done:

- by working in a medium of HNO_3 (30),
- by addition of $Mg(NO_3)_2$ (31),
- by diluting urine with a matrix containing $Mg(NO_3)_2$, HNO_3 and Triton X-100 (32), or
- by addition of $Al(NO_3)_3$ as a modifier (33).

For urine analysis, good results have been reported without addition of modifiers or other preliminary manipulations (34). 20 μl samples of urine can be directly added to the furnace. Urine samples from workmen in a beryllium plant contained up to 10 μg/l Be, but 120 random urine samples from individuals without known exposure between 0.4 and 0.9 μg/l Be. Such concentrations can still be measured by graphite furnace AAS.

Table 1. Determination of the Alkaline Earth Metals of the Group 2 (Beryllium, Magnesium, Calcium, Strontium and Barium) and the Group 1 Element Lithium. Detection Limits of 4 Instrumental Techniques in μg/l resp. ppb, in Aqueous Solution

Element	Li	Be	Mg	Ca	Sr	Ba
Inductive coupled plasma – emission spectrophotometry (ICP–AES)	1.5	0.1	0.15	0.15	0.08	0.15
Inductive coupled plasma – mass spectrometry (ICP–MS)	0.03	0.03	0.007	2*	0.001	0.002
Flame atomic absorption (F–AAS)	0.8	1.5	0.15	1.5	3	15
Graphite furnace atomic absorption (GF–AAS)	0.15	0.02	0.01	0.03	0.06	0.9

* using isotopic mass 44, since mass 40 is obscured by Ar

Most data taken from the Perkin Elmer Bulletin L-655F (The Perkin Elmer Co., USA), printed in June 1993: "The Guide to Technique and Applications of Atomic Spectroscopy". Several figures are rounded off. The quoted detection limits for water are usually a factor of 2 to 3 below those for urine. Considerably higher detection limits must be expected for serum, plasma or whole blood.

2.10.4.2 Magnesium

This element is an important body constituent. Human serum contains between 0.8 and 0.9 mmol/l Mg (19–25 mg/l) (35, 36). In the past, it has been assumed that the danger of Mg deficiency does not exist. It was believed that the human body could obtain all the Mg it needs with a normal diet. In recent times, however, doubts have arisen, and the existence of Mg deficiencies is no longer excluded, especially in connection with physical activities, i.e. sports.

But a more important problem is certainly the danger of Mg poisoning, since the element has gained high technical importance, especially in the construction of aircraft and other vehicles. MgO serves for the fabrication of fire resistant materials and for water neutralization, $MgCl_2$ as antifreeze. Mg salts can be found as antacids in dental preparations and similar products. Important is the medical use of $MgSO_4$ as prenarcotic agent, anticonvulsant, spasmolytic, as a medicament for improving blood circulation, as laxative (introduodenally in very high doses) and as antihelmintic. Such applications can involve a considerable poisoning danger, particularly in individuals with slow Mg elimination. 50 g of $MgSO_4$, given as a laxative, may be lethal.

The following intoxication symptoms have been described and correlated with serum Mg concentrations by J.A. Aita (1966). We quote them from a book (35), which contains a detailed but brief summary of the toxic effects of magnesium:

- lowering of blood pressure at 36 to 60 mg/l Mg in serum,
- loss of reflexes and ataxia at 85 to 120 mg/l Mg in serum,
- breath insufficiency at 120 mg/l Mg in serum,
- coma between 145 and 180 mg/l Mg in serum, and
- heart arrest at higher concentrations.

In the interpretation of analytical figures, it must be kept in mind that the serum Mg levels increase after death, usually to near 70 mg/l, occasionally even higher.

The daily urinary Mg elimination may vary from 10 to 150 mg (36). It can be very much elevated in Mg intoxications.

In order to measure such high Mg concentrations, trace methods are not necessary. Simple photometric procedures can be used. The corresponding reagent kits are commercially offered by several firms such as Merck, Pierce and others. However, if the metal does not have to be determined daily in series, it is more convenient to use flame atomic absorption.

The ease and high sensitivity with which Mg can be measured in body fluids by AA has had a decisive influence on the rapid introduction of this technique into clinical chemistry. It is largely due to the pionieering work of Willis shortly before and during 1960 (37–39). The work of Willis and other authors has been summarized and discussed by W. Slavin (36). More recent publications have not added significant improvements. Among all elements, Mg can be measured with the best sensitivity by flame AAS. The serum only needs to be diluted (1:50) and the solution injected into an air-acetylene flame. A small protein effect can be controlled by an addition of EDTA, HCl, lanthanum or strontium chloride. However, we have found that additions are hardly needed. In urine, Mg can be determined directly after 50-fold dilution with water. If the most sensitive resonance line at 285.2 nm is used, the optimal working range lies between 0.2 and 2 mg Mg/l (aqueous solution). This involves considerable dilutions of body fluids. By using the less sensitive line at 202.5 nm (in combination with a quartz-window hollow cathode lamp), or by rotating the burner head by 90°, or by a combination of both modifications, the analytical working range can be greatly extended to much higher concentrations (40).

2.10.4.3 Calcium and strontium

These two elements have very low toxicity and will seldom be of concern to a toxicologist, except perhaps for the diagnosis of internal and external burns, caused by the caustic action of their oxides or hydroxides. They can also cause skin eczemas, which are often found on the hands of bricklayers.

Calcium is an important body component. The daily requirement of an adult is approximately 1 g. Resorption is incomplete and depends on the availability of vitamin D. A deficiency causes rachitis, principally in children. But also adults should take care to eat plenty of Ca-containing aliments (milk, cheese) to prevent osteoporosis. This holds particularly for women after the menopause.

Serum Ca levels have been reported to lie near 100 mg/l. If they fall below 70 mg/l, tetanic cramps may occur. Therapeutic application of Ca salts is used for remineralization and for preventing inflammations. If given as injections, however, necroses may result as an undesirable side effect.

The colorimetric methods for Ca analysis are suited for automation and therefore widely used in clinical chemistry. However, the exactitude is better with flame AAS, and this is the method of choice in toxicology. It was again Willis, who has described this possibility at a very early stage in the development of the technique (39, 41, 42), and Slavin (36), who summarized and discussed the values of the procedure. While Willis recommended precipitation of serum proteins prior to injection into the flame, other authors reported satisfactory results by direct injection of 1:20 serum dilutions with water (43) or a dilute aqueous solution of EDTA (44). Urine Ca can also be determined after 20-fold dilution with water. The addition of lanthanum chloride has been recommended (36).

Strontium analysis by AAS is half as sensitive as Ca analysis. But its determination in body fluids is much more critical because of the low concentration at which it occurs. The interferences when using the acetylene-air flame can be eliminated by shifting to an acetylene-nitrous oxide flame with its much higher temperature (near 3000 °C) (36). But flame AAS can only measure body fluid Sr concentrations over 10 µg/l. For lower levels, graphite furnace AAS (45) or ICP-MS must be used (Table 1).

2.10.4.4 Barium

Soluble barium salts are highly toxic. They cause contraction of all muscles including the heart. The action is very similar to that of digitalis alkaloids. Oral poisoning by Ba salts such as Ba chloride, carbonate or nitrate leads to nausea, vomiting and diarrhea, at higher doses to paralysis, followed by exitus. 0.5 to 0.8 g of a soluble Ba salt may be lethal for an adult.

Soluble Ba salts are not used in medicine, in contrast to the non-soluble and non-toxic $BaSO_4$, which can function as an X-ray contrast medium.

The refractory element Ba cannot be measured by atomic absorption with the normally used acetylene-air flame, but only with the hot acetylene-nitrous oxide flame (36), as long as the levels do not lie below 20 µg/l. For lower concentrations, flameless AAS with graphite furnace (46, 47), ICP emission analysis (48), or ICP-MS must be used (table 1).

2.10.5 Transition Metals from Groups 5 to 10 (Old Nomenclature Groups Vb to VIIIb) with Technical and Toxicological Importance

2.10.5.1 Group information

Among the transition metals (Groups 3–12 according to the new, Groups IIIb to VIIIb according to the old nomenclature), iron, copper, silver and gold, zinc, cadmium and mercury are discussed in separate chapters, due to their importance as body constituents (iron) or their special toxicological significance. In this chapter, we have tried to summarize in tableform the toxicological aspects of some other technically important transition metals from the Groups 5 to 7, 9 and 10, namely vanadium, chromium, molybdenum, manganese, cobalt and nickel. They will occupy a toxicologist rather seldom. But his laboratory must still be capable of detecting and determining them in body fluids, should the need arise.

Table 2. Use of Some Technically Important Transition Metals of the Groups 5–10 (Vb–VIIIb), their Oxides and Salts

Group of periodic system							
new nomenclature	5	6	6	7	8	9	10
old nomenclature	Vb	VIb		VIIb		VIIIb	
Element (Transition metal)	V	Cr	Mo	Mn	Fe	Co	Ni
Steel manufacture	X	X	X	X	X	X	X
Electroplating		X					X
Battery component		X		X			X
Wood preservation	(X)	X		X			
Corrosion inhibitor		X					
Paint composite	X	X		X		X	X
Leather tanning		X					
Catalyst	X	X	X	X	X	X	X
Chemical reagent	X	X		X	X		
Agriculture use				X	X		(X)
Food additive or Pharmaceutical				X	X	(X)	
Essential biological element	X*	X	X	X	X	X	X*

X* shown so far only for other organisms, not for humans

Table 2 shows the main uses of these metals or their derivatives. It illustrates potential dangers of contamination. Table 3 lists some of the possible toxic effects

Table 3. Toxicological Injuries From Some Technically Important Transition Metals of the Groups 5–10 and their Salts

Element (Transition metal)	V	Cr	Mo	Mn	Fe	Co	Ni
Respiratory tract (irritation, bronchitis, pneumonia, oedema, respiratory arrest))	X	X	X	X		X	X
Gastrointestinal tract (irritation, corrosion, ulcers, carcinomas)	X	X		X	X	X	X
Skin (inflammations, allergies, eczemas, cancer)	X	X	X			X	X
Heart (hypertension, tachycardia, heart arrest)						X	
Blood vessels (dilation or vasoconstriction)	X					X	X
Blood cells (haemolysis, cyanosis, haemo-globinemia)		X					
Liver (oedemas)		X	X	X		X	X
Kidney (obstruction, anuria)		X	X				
Nervous system (tremor, parkinson, psycho-logical disturbances)				X	X	X	
Muscles				X	X		
Carcinogen		X			(X)	(X)	X

X = toxicological damages in humans have been demonstrated
(X) = toxicological damages only demonstrated in animal trials

on the human body. Table 4 compares, for 6 instrumental analytical methods, the detection limits which can be expected in the analysis of aqueous solutions. For the matrix urine, the figures may be twice as high in most cases, and considerably higher for plasma, serum or whole blood. But the table permits nevertheless a comparison of the methodological possibilities, at least with respect to analytical sensitivity. Except for iron, ICP-MS is, if available, certainly the best choice as a screening method for these metals. For quantifications, the graphite furnace AAS-technique can solve most problems. If better sensitivity is required, NAA may be used for determining traces of manganese and maybe also cobalt, but not for the other metals. In such cases, ICP-MS with isotope calibration can be employed.

Table 4. Detection Limits for the Analysis of Some Transition Metals from Groups 5 to 10 in µg/l resp. ppb, from Aqueous Solution

Transition metal	V	Cr	Mo	Mn	Fe	Co	Ni
Electrochemical analysis (DPV)	100	0.02	100	40	0.04	0.005	0.001
Inductive coupled plasma-emission spectrophotometry (ICP–AES)	3	3	7.5	0.6	1.5	3	6
Inductive coupled plasma-mass spectrometry (ICP–MS)	0.002	0.02	0.05	0.002	0.4*	0.001	0.005**
Flame atomic absorption (F–AAS)	60	3	45	1.5	5	9	6
Graphite furnace atomic absorption (GF–AAS)	0.3	0.1	0.2	0.1	0.3	0.4	0.8
Neutron activation analysis (NAA)	0.15	20	10	0.003	6	0.03	15

* using isotopic mass ^{54}Fe, since ^{56}Fe is obscured by Ar^{16}O
** using isotopic mass ^{60}Ni, since ^{58}Ni is obscured by Ar^{18}O

The data for DPV and NAA are approximations, taken from different publications and averaged. The data for emission and absorption spectrophotometry are taken from the Perkin Elmer Brochure "The Guide to Technique and Applications of Atomic Spectroscopy" (1993) (see footnote Table 1). We are greatly indebted to Dr. R. Funck, Perkin-Elmer AG, Rotkreuz, Switzerland, for providing us with the pertinent literature and for many helpful and stimulating discussions.

2.10.5.2 Vanadium

Vanadium is an essential trace element in some animals, perhaps also in man. It is increasingly used in the steel industry. Its dust and vapor constitute quite a hazard to exposed steel workers. 20 to 50 μg of V could be measured by GF-AAS in their daily urine samples. Poisonings by the pentoxide and by salts of vanadic acid have been reported. V_2O_5 may be the main cause for bronchitis and pneumonia of people in contact with "Thomasmehl" or "Thomasschlacke". V is present in crude oil and is held responsible for intoxications of workmen who have to clean oil tanks and oil pipes. The importance of V_2O_5 as a chemical catalyst, and of V salts in the color industry is stressed in Table 2. The toxicity of V salts is inferior to that of Cr salts, since the former are eliminated fairly rapidly (in a few days) through the kidneys. V concentrations in the blood of people without occupational exposition are usually inferior to 10 μg/l; very seldom do they reach 20 μg/l. They can be measured by graphite furnace AAS or by ICP-MS.

2.10.5.3 Chromium

The main uses of chromium are in steel composites (stainless steel can contain up to 30% of Cr) and electroplating. Up to 50 μg of Cr can be present in the daily urine of exposed workers, instead of the normal values of near 2 μg. Such levels can be measured easily by GF-AAS (49). Users of Cr paints must be aware that the metal can enter the body also by way of the skin. Its elimination is slow. Deposits of Cr have been measured in liver (up to 190 μg/kg), kidney (160 μg/kg) and bone marrow (100 μg/kg). Lethal oral poisonings have occurred by ingesting chromic acid (1–2 g) or potassium dichromate. The latter has been used for suicides and for inducing abortions. It is not common in the crime scene, since it can be recognised immediately by its intense color. The yellow oxidizing agent turns green on reduction to the Cr-III-salt.

Cr is an essential element. There is some concern that diabetes may be associated with Cr deficiency. But Cr intoxications caused by dust and vapors are a more important problem for the chemical toxicologist. The metal accumulates in the lung and can induce lung cancer.

A detection of Cr in stomach content, vomit or urine is possible by oxidation with hydrogen peroxide. Blue pentoxide (CrO_5) is formed and can be extracted with ether.

2.10.5.4 Molybdenum

Molybdenum is much less toxic than chromium. It is not easily resorbed. It has a function in the body as constituent of the enzyme xanthine oxidase. Its normal concentrations in different human tissues are reported to lie in the range of 0.14 (brain, muscle) to 3.2 (liver) ppm on a dry basis. The Mo levels in blood seldom lie above 5 μg/kg, usually around 1.5 μg/kg, evenly divided between plasma and red cells. Toxic levels of Mo can be recognized by direct GF-AAS, but for determining normal blood levels, either a concentration step or a shift to ICP-MS is indispensable. The urinary excretion of Mo can be drastically increased by sulfate ingestion. It has been determined that 11 g of potassium sulfate, taken orally, raises the daily Mo excretion from around 1 mg/l to over 20 mg (50). Such urinary eliminations can easily be controlled by GF-AAS.

2.10.5.5 Manganese

Manganese is another normal trace constituent of the body. The estimated total Mn content in an adult is 12 to 20 mg. The highest concentrations could be found in liver (approx 1.7 ppm on wet tissue basis), pancreas and kidney, as well as in the bones, the lowest values in muscle tissue. In blood, values in the range of 10 μg/l have been reported. Almost all the Mn is located in the red cells. Following an acute coronary occlusion, blood Mn levels can double.

Mn deficiencies have been observed and held responsible for skeletal abnormalities and impaired growth, as well as for defects in lipid and carbohydrate metabolism. A number of enzymes (hydrolases, decarboxylases, transferases) are activated by Mn. This metal is among the least toxic of the trace elements. Chronic poisonings have been reported among miners working with Mn ores. It can be assumed that the metal enters the body mainly as oxide dust via the lungs and the gastrointestinal tract. Reported poisoning effects are essentially of psychiatric nature (manifestations similar to Parkinson's disease and/or schizophrenia) (51).

Mn is mainly excreted in the bile flow and pancreatic juice, very little in the urine, unless chelating agents are administered. In the past, neutron activation analysis has been used for measuring Mn in body fluids and tissues. Today, considerably lower detection limits are obtained with ICP-MS. GP-AAS is useful for measuring blood or serum Mn in intoxications, but not for detecting Mn deficiencies.

2.10.5.6 Cobalt

Cobalt is another essential trace element. Its main physiological function is its central position in vitamin B-12. The body of an adult contains around 1 mg of this metal, half of it stored in muscle tissue. Different Co levels have been reported for human blood, from 0.1 to 60 μg/l. A careful study by flameless AAS (52) quotes a range from 0.1 to 1.2 μg per liter blood (mean 0.5 μg/l), and 0.1 to 2.2 μg per liter urine (control of 25 occupationally unexposed adults). The detection limit of GF-AAS lies in the lower range of these intervals. This is therefore not an ideal method for determining Co in body fluids, in particular not for tracing Co deficiencies. For such investigations, the analyst must resort to NAA or ICP-MS. On the other hand, GF-AAS possesses sufficient sensitivity for controlling Co intoxications, since Co toxicity is relatively low and high doses are usually needed to cause toxic effects. An oral dose of 0.5 g $CoCl_3$ can lead to vomiting and diarrhea. Injections of 10 to 50 mg of the metal salt can cause dizziness, rise of blood pressure, heart palpitations and impairment of breathing.

In the Canadian province Quebec, in the US-city Omaha, as well as in the Belgian town Liège, addition of Co salt to beer to stabilize its foam has caused severe heart muscle damage in heavy beer drinkers with a considerable number of fatalities (53, 54). The addition had to be abandoned.

2.10.5.7 Nickel

It is not yet certain that Nickel is also an essential trace element for man. Small quantities of this metal (0.3 to 0.5 mg) are ingested daily with normal food and excreted in urine and feces. Blood levels up to 120 µg/l (2 µmol/l) can still be considered normal (55). Serum concentrations are much lower, near 5 µg/l (56). The normal daily Ni elimination in urine lies between 15 and 100 µg (55).

Ni deficiencies have been reported for some animals (chicken, rats, pigs) but not yet for man. Ni is also an element with relatively low oral toxicity. During cooking, acid food often dissolves Ni from cooking utensils, but the metal is not well-resorbed and does not cause much damage. The inhalation of vapors or dust containing Ni respectively Ni-derivatives is much more dangerous. It affects the respiratory tracts and can induce cancer.

The most dangerous Ni derivative is $Ni(CO)_4$, a low boiling liquid used in the production of the pure metal. It decomposes in the body to Ni and CO. The early toxic symptoms are very similar to those of CO exposition (headache, dizziness, nausea and vomiting). They are followed by lung affections, gastrointestinal troubles and general weakness. In addition, the CNS may also be affected. A number of lethal intoxications by $Ni(CO)_4$ have been reported (57, 58). It must be kept in mind that a contact with Ni or Ni-derivatives can cause dermatitis, often observed on the skin of workmen in the galvanisation trade, but even on the hands of people working with Ni-plated tools or instruments. Also in regard to Ni dermatitis, the low boiling $Ni(CO)_4$ possesses the highest danger potential of all Ni derivatives.

Ni intoxications, especially by the tetra carbonyl derivative, can be easily detected by urine analysis with GF-AAS, because the daily urinary Ni eliminations are high, in lethal cases up to 1 mg/day. For measuring normal Ni concentrations, AAS is not sensitive enough, unless a concentration step is used. ICP-MS or electrochemical analysis (DPV) should be chosen.

2.10.6 Iron

2.10.6.1 Introduction

Following acid digestion in the stomach, ferric ions are liberated from the food and absorbed in the intestine by an active process involving reduction to the ferrous state. Once absorbed, iron is oxidized to the ferric form. In plasma, it is combined with the protein transferrin. In combination with the protein apoferritin, it also forms ferritin, which represents the form stored in tissues such as liver and spleen. The adult human body contains 50 to 70 mmol of iron. About 70% is present in the ferrous form (Fe^{2+}) in haemoglobin, up to 25% as Fe^{3+} in ferritin, and only about 0,1% circulates bound to transferrin. The remainder is present in myoglobin and cytochromes (59).

Acute iron poisoning most often occurs in children following the accidental ingestion of iron-containing pills. Occassionally, intentional overdosage is seen in adults. Chronic intoxication may occur in patients repeatedly transfused for treatment of thalassemia major or sickle cell anemia (60). The cause is the increased iron release from dying erythrocytes.

2.10.6.2 Symptoms of acute iron intoxication

Early signs appearing within six hours are gastrointestinal irritation, diarrhea and vomiting. After a period of relapse, shock and metabolic acidosis may occur. An increase of serum iron above the reference range (9–32 nmol/l) is characteristic, especially if the concentration is beyond the binding capacity of transferrin. In such cases, a treatment by the iron chelator desferrioxamine may be undertaken. As iron intoxication also affects the liver and the kidney, the concentrations in plasma of alanine aminotransferase (ALAT) and creatinine, respectively, may increase.

2.10.6.3 Laboratory assessment of iron poisoning

Determination of plasma iron is a useful emergency test, as concentrations above the reference range suggest an overload. However, since toxicity only occurs when the iron concentration is in excess of the binding capacity of transferrin, a better information is provided by the simultaneous determination of iron and transferrin, from which the saturation index can be calculated.

Desferrioxamine is a potent chelator of Fe^{3+}. Its chelating power is stronger than that of transferrin and apoferritin. The appearance of a *vin rosé* colour of the urine upon treatment by desferrioxamine indicates that the patient was actually intoxicated and confirms the effect of the therapy (59).

2.10.6.3.1 Serum or plasma iron by colorimetry

Many hospital laboratories use colorimetric methods in which iron is dissociated from transferrin by acid (e.g. hydrochloric or citric acids), reduced to the divalent state, and made to react with a compound producing a colored complex. In one type of method, the determination is made on a protein-free supernatant obtained after protein precipitation by trichloroacetic acid (61, 62). This has the advantage of eliminating any interference from turbid, icteric or lipemic samples. In another type of method, the color reagent is added directly to the serum and the absorbance is measured without precipitation of the proteins. The direct methods are simpler and well suited to automation, but require a reagent-free blank for each sample to correct for the absorbance of the serum (63, 64).

Among the numerous color reagents available, bathophenantholine sulfonate, ferrozine and Ferene S are the most commonly used. Addition of a copper-masking agent such as thiourea improves the specificity (65).

In emergency situations, iron may either be determined in urgent mode in a hospital laboratory, if this is convenient, or with a simple manual colorimetric method such as the Ferene-S method without deproteinization. The latter is commercially available, among others, as a kit (Sera-Pak) from Ames.

2.10.6.3.2 Plasma and urine iron by atomic absorption spectrometry

The suitability of atomic absorption for iron assay depends on the type of sample to be analyzed. The method is not recommended for the direct assay in plasma, as in contrast to colorimetric methods, it also measures haemoglobin iron, and thus overestimates iron in haemolyzed samples. On the other hand, it is not affected by the presence of desferrioxamine, which is the cause of low results with some colorimetric procedures (66, 67). Thus atomic absorption is well suited to the monitoring of patients under desferrioxamine treatment. It is also a good method to directly assay total iron in the urine of such patients (68).

2.10.6.3.3 Transferrin

For the assessment of iron overload, it is useful to know the iron concentration which is in excess of the total iron-binding capacity (69). Although a simple colorimetric method had been published for that purpose (70), the simultaneous determination of iron and transferrin is now a convenient way of obtaining this information. Serum transferrin may be assayed by immunoturbidimetric (71) or by immunonephelometric (72) procedures. It is convenient to express the concentrations as sites of iron fixation, on the basis that one mole of transferrin (M_r 79570) binds two moles of Fe^{3+}. Dividing the concentration of serum iron by that of serum transferrin then provides the transferrin saturation index, which is normally between 0.07 and 0.51. An index above 0.65 is already indicative of iron overload. Severe toxicity is associated with an index > 1.

2.10.6.3.4 Other analyses

Plasma ferritin increases together with iron in iron poisoning, reaching concentrations above the reference interval.

Other useful analyses include blood gas or CO_2 determination to monitor the metabolic acidosis consecutive to iron poisoning. After the first 24 hours, aminotransferase determinations serve to assess whether or not hepatic damage has occurred (73).

2.10.7 Copper

2.10.7.1 Toxicological importance

Metallic copper or copper oxide can only be dangerous if inhaled in form of dust or vapor. On the other hand, orally ingested water-soluble copper salts such as copper sulfate (constituent of dyes and wood preservative), copper acetate and oxychloride (plant protection agents) can produce severe or even lethal intoxications (74). It must also be kept in mind that copper may often be contaminated by arsenic or lead, and that such impurities can be partly responsible for the toxic manifestations. Symptoms due to copper poisoning after inhalation are fever with headaches, vomiting and often bloody diarrhea after oral ingestion. Severe oral intoxications will affect mainly blood and kidneys.

On the other hand, Cu is also an essential trace element. It plays an important role in carbohydrate and lipid metabolism and in the maintenance of heart and blood vessel activity. A daily Cu intake of 1.5 to 2 mg is essential. A Cu deficiency may be lethal just like a Cu poisoning.

2.10.7.2 Clinical significance

Only a small part of blood plasma Cu is present in the free or unspecifically bound state (i.e. as albumin complex). The main part is tied to the protein ceruloplasmin. Depending on the problem to be solved, either the total Cu concentration or only the ceruloplasmin Cu must be determined. The reference intervals for serum and blood Cu are 0.7–1.5 mg/l (11–23 μmol/l). Gravidity may nearly double these concentrations. Estrogens and contraceptives will also cause elevations. Since ceruloplasmin cannot pass the placenta, newborns show very low plasma Cu levels. Various pathological conditions can also raise or lower the ceruloplasmin content as well as the total Cu level of the plasma. Thus, ceruloplasmin is missing in most patients with Wilson's disease. Since their blood does not contain ceruloplasmin-bound Cu, the total serum Cu is reduced, even though the unbound serum Cu is elevated. The kidneys can eliminate free or albumin-bound Cu, but not ceruloplasmin Cu, since the high molecular weight protein can not pass the biological filter.

The daily Cu elimination in urine can vary. We have measured 10–40 μg (75), other authors 30–100 μg (35). We could show that urine Cu is present in an inorganic or at most only weakly complexed form. In patients with Wilson's disease, we have found daily copper eliminations of up to 5000 μg, occasionally even higher. A penicillamine treatment increases Cu elimination and prevents accumulation of the metal in the organs.

Cu analysis of liver biopsies is often requested for the diagnosis of Wilson's disease. In such cases, we have obtained concentrations of 30–60 μg per g tissue, instead of the normal 5–15 μg/g (75–77).

2.10.7.3 Detection possibilities

For a control of Cu metabolism and of possible Cu intoxications, a laboratory should be able to carry out 4 different analyses:

- determination of urine Cu,
- determination of total serum Cu,
- assay of ceruloplasmin-Cu in serum and
- assay of Cu in liver biopsies and other small tissue sections.

For a review of the older methods, we refer to readily available articles (70–80), which treat the use of colorimetric procedures, emission spectrophotometry and neutron activation, and also discuss the detection of the polyphenoloxidase ceruloplasmin on the basis of its enzyme activity. We will restrict ourselves to a description of atomic absorption, a technique especially well-suited for the assay of Cu in biological materials.

2.10.7.4 Copper analysis by atomic absorption

For the detection of elevated Cu concentrations in urine (patients with Cu intoxications or Wilson's disease) we recommend the use of flame-AA with direct injection of the urine sample (if necessary clarified by centrifugation or filtration). Quantification can be carried out by the method of inner standardization. Our analytical procedure has been described in detail (75–77).

Normal or decreased levels of urine copper can not be measured by this simple method; a concentration step is required. Berman (81) recommended wet-ashing of 25 ml urine with sulfuric and nitric acid, complexing the copper with diethyldithiocarbamate and extracting the complex with methylisobutyl ketone. We found that the decomposition step can be omitted (76):

> 40 ml urine are acidified with HCl to pH 1. 1 ml of 50 g/l diethyldithiocarbamate and 1 ml of 50 g/l EDTA are added, and the resulting solution is extracted with 2 ml of methylisobutyl ketone. To avoid a transfer of the extracts, wide centrifuge tubes (50 ml) with long necks (2 ml) are used. After centrifugation of the liquid layers, the organic phase is displaced into the neck of the centrifuge tube by addition of water, then sprayed directly into the flame. With such a twentyfold enrichment, Cu concentrations down to 0.5 μg/l urine can be determined.

In plasma, Cu is easier to measure than in urine, because the concentration is normally about 50-fold higher. Still, extraction procedures (81) or precipitation of proteins (82) have been described. However, in our hands, direct injection of serum samples after dilution with an equal volume of water (83, 84) gave better results. It is rapid and reduces the dangers of losses and contaminations.

Small tissue sections such as liver biopsies (usually not over 100 mg wet weight) have to be decomposed, i.e. by alternating treatment with nitric acid and hydrogen peroxide. The resulting solution (clear but not necessarily colorless) is standardized to a definite volume and directly atomized.

Cu in urine, serum, whole blood and tissue homogenates can also be measured directly by flameless AA in a graphite furnace. Such a procedure is particularly indicated for very small samples, since, due to the excellent sensitivity of flameless AA for copper analysis, it does not require more than 2 μg of a body fluid or a tissue homogenate (85). Flameless AA has also been used successfully to measure copper in hair (86).

For highest sensitivity, the resonance line at 424.7 nm of a copper hollow cathode lamp should be selected. But the line at 327.4 nm gives also good results. For unexpectedly high copper contents, we recommend a shift to the approximately twenty times less sensitive line at 249.2 nm.

2.10.8 The Noble Metals Silver and Gold

2.10.8.1 General remarks

The two noble metals gold and silver have no known vital function in man. But since they occur in nature, low traces are also present in the human body. Poisonings with noble metals have been reported, especially in connection with therapeutic applications, which are however seldom used in recent times. Poisonings due to professional or environmental contamination have also occurred; but they are rather rare.

2.10.8.2 Toxic effects of silver

If large amounts of metallic silver are inhaled as dust or vapor, injected as a colloid, or orally ingested over a longer period (old therapy for gastritis), the metal deposits as fine granules of Ag_2S and (under the influence of light) elemental Ag. This leads to a brown to black pigmentation of the inner organs and the skin (Argyria or Argyrose). Such a blackening of the skin can also result from external contact with Ag salt.

Acute intoxications by $AgNO_3$ have been reported. This corrosive salt has been medically used (and may still be used) as a disinfectant. Its strong acidic action can lead to tissue damage. The loss of eyesight from a 10% solution of $AgNO_3$ has occurred.

Environmental contamination with Ag is possible through food stored in Ag plated vessels and from Ag-foils used in decorating cakes and confectionary. Small amounts of Ag may also enter the organism from dental fillings or from contact with Ag batteries. However, such sources do not seem to represent real toxic dangers, even if they may be responsible for a "normal" level of Ag near 10 µg per kg blood (87).

2.10.8.3 Toxic effects of gold

Organic gold preparations have been used and may occasionally still be used for treating polyarthritis and bronchitis (asthma). Such applications are very risky. They can affect blood and spinal marrow. Albuminuria, hematuria, uremia, anuria and skin eczemas have been reported as toxic effects in a considerable percentage of the patients. Gold therapies must therefore never be conducted without a strict analytical control of the blood picture and of urine values, which must continue in the months after the end of the treatment (88).

2.10.8.4 Determination of noble metals in body fluids

Table 5 shows that ICP-MS is the most sensitive approach for determining Ag and Au in body fluids. In the case of Ag, GF-AAS, and in the case of Au, NAA, also possess sufficiently low sensitivities for detecting elevated levels.

Table 5. Detection Limits for the Analysis of some Transition Metals From Groups 11, 12 and 13 in µg/l resp. ppb, from Aqueous Solution

Metal	Cu	Ag	Au	Zn	Cd	Hg	Tl
DPV	0.002	0.1	10	0.02	0.0002	0.02	0.02
ICP–AES	1	1	10	1	1	1	30
ICP–MS	0.003	0.003	0.001	0.003	0.003	0.004	0.0005
F–AAS	1.5	1.5	10	1.5	1	300	15
GF–AAS	0.25	0.05	0.4	0.3	0.02	1.5*	0.4
NAA	0.1	2	0.005	2.5	1.5	0.03	40

* With cold vapor atomization method better than 0.01

2.10.9 Zinc

2.10.9.1 Physiological and toxicological importance

Zinc is a normal constituent of the human body and plays important roles in many enzyme systems. It is a key element for many body functions, i.e. muscle activity and wound healing. Our daily intake of Zn should reach at least 12–15 mg. But our present normal diet may not always furnish such an amount.

Massive ingestions of Zn salts, however, can produce intoxications with headaches, vomiting and diarrhea, the latter often bloody, as symptoms. The metal salts may stem from food kept in metallic containers which have been galvanized with Zn. But these intoxications are usually less severe than those created by Cu salts. Inhalation of Zn dust or Zn vapor can lead to fever (zinc fever), just like the inhalation of other metallic dusts or vapors (74).

The following reference intervals may be quoted for Zn body fluids and tissues:

13–15 µmol/l (0.8–1.1 mg/l) in serum (75, 89),
5–7 µmol/day (0.3–0.5 mg/day) in urine (75) and close to
0.8 µmol per g wet weight (50 µg/g) for liver (75).

Zn-levels have been used as a diagnostic tool. The many clinical conditions which are associated with hypozincaemia include myocardial infarction, acute infection, viral hepatitis, hepatitic cirrhosis and pernicious anaemia. They have been reviewed by Sunderman (89).

2.10.9.2 Zinc analysis with atomic absorption

Since the concentrations in body fluids are fairly high, and flame AA possesses excellent sensitivity for Zn, it can be used without pre-concentration steps, that is with direct injection of the fluid after 4- or 5-fold dilution with water (83, 84).

The absorbance line at 213.8 nm is used. At such a low wave length, light scattering effects may cause disturbances. But they can easily be eliminated with any of the available background compensation systems. Direct flame AA can detect less than 10 µg/l Zn in body fluids. The more sensitive but also more cumbersome flameless AA method in a graphite oven is therefore not needed, not even for the analysis of tissue sections. This eliminates the danger of evaporation losses during the drying steps in flameless AA. The pretreatment of tissues is identical to the one we have described for Cu analysis. For the determination of Zn in hair, however, flameless AA with the graphite furnace must be used (86).

2.10.10 Cadmium

2.10.10.1 Toxicological importance

Cadmium is an enzyme poison. It reacts with sulfhydryl groups. It can i.e. replace zinc from enzyme complexes and block enzymatic activity. Cd can be found in earth, water and air, and therefore in very low concentrations also in plant and animal organisms. Among human nutriments, some marine seafood such as oysters, as well as mushrooms, often also liver and kidney, possess considerably higher Cd levels than other tissues.

The element does not seem essential for life. Already relatively low concentrations are highly toxic for animals and humans.

Cd is more and more often used industrially: in batteries, metallic alloys, as coats for the protection of iron, in dyes and pigments, as a stabilizer in plastics. It can be found in silver and pot polishes and in shoe whiteners. Severe intoxications have been caused by Cd-plated metal containers or cooking and baking utensils, as well as by ceramic dishes with Cd-pigment decors. Acidic foods or drinks such as fruit juices can dissolve such Cd surfaces and become highly toxic.

Orally ingested Cd salt leads to stomach pains, vomiting and diarrhea. According to most authors (74, 90, 91), as little as 30–50 mg can already be lethal for an adult. Chronic exposures can produce rheumatic pains, followed by bone deformations and tendency to bone ruptures, since Cd can also replace Ca in the bones. The two ions have a similar size of the radii. Poisoning of the kidneys is indicated in abnormally high sugar and protein levels of the urine. Yellow rings on the base of the teeth may hint that a Cd contact has occurred (yellow CdS).

Very dangerous is also an inhalation of Cd dust or Cd vapors, which can evolve easily during industrial manufacture of the metal, which has a relatively high volatility (at ambient pressure, Cd melts at 321 °C and boils at 767 °C). The vapors produce coughs, headaches and abdominal pains. They also interfere with breathing. After a latent period of 1 to 2 days, a lethal lung edema may occur.

It should be kept in mind that the Cd content of tobacco smoke cannot be neglected. The body of an adult non-smoker may contain around 15 mg Cd, but that of a smoker 30 mg. A further important point is that Cd occurs very often as an impurity in Zn. Suspected cases of Zn poisoning with unclear symptoms should therefore also be tested for Cd.

The toxic potential of Cd has been neglected for a long time. The so-called Itai-Itai illness in Japan has reversed the situation and called world-wide attention to the extremely high toxicity of this element. Between 1945 and 1970, a large percentage of the population in the district of Toyama in Japan suffered from strange rheumatic affections, followed by severe deformations and ruptures of the bones. In addition, they developed gastrointestinal troubles as well as kidney, lung and heart affections. This illness was first diagnosed as a combined effect of malnutrition and lack of vitamin D. It was not until 1960 that possibilities of epidemic poisonings were also evaluated and not until 1968, after the illness had already claimed over 100 fatalities, that Cd could be identified as its main cause. The toxic metal originated from a Cd mine which sent the waste into the Jiutsu river. Since this river water is used to irrigate the rice fields of the region, the rice produced in the irrigated fields was shown to possess high Cd concentrations, up to 1 mg/kg.

The normal daily Cd intake, as well as the Cd excretion by way of kidneys and intestines, is estimated to be near 50 μg. Normal Cd blood levels are grouped around an average value of 3.5 μg/l. A professional or environmental exposure to Cd compounds can produce a thousandfold increase. The body eliminates the metal only slowly. Depositions in the liver, kidneys and other tissues occur. Their normal Cd levels, usually below 100 μg/kg, can also rise thousandfold (to 100 mg/kg), resulting in tissue damage. Cd intake can produce intestinal ulcers. It must be considered a carcinogen.

The bioaccumulation of Cd seems to be on the increase. Its concentration is especially high in the kidneys and liver of older animals, and it has therefore been recommended to refrain from eating too often kidney or liver from cows, pigs and especially horses.

2.10.10.2 Analytical possibilities

Due to the low levels of Cd in normal body fluids and tissues, most analytical methods require concentration steps. The use of colorimetric methods (dithizone), optical emission, neutron activation and atomic absorption for the determination of Cd in biological materials has been reviewed by Friberg (92). For the colorimetric methods, the book of Sandell (93) is also a good reference. The dithizone method is handicapped by the fact that over 15 metals react similarly to Cd; among them is lead. This is probably responsible for the fact that some Cd poisonings have been wrongly diagnosed in the past as caused by lead. AA can help to differentiate between the extracted elements. The organic phases with the dithizone complexes of the different metals can be measured separately by AA rather than by molecular spectrophotometry.

In a recent review of Cd distribution in nature, Cd toxicity and Cd analysis in biological materials (94), the analytical possibilities (spectrophotometric procedures, atomic emission including ICP-AES and ICP-MS, atomic absorption, electrochemical techniques, liquid chromatography and others) are evaluated and compared. From the detection limits quoted, it is evident that only ICP-MS, anodic stripping voltametry and electrothermal atomic absorption possess the sensitivity required for Cd determinations in biological fluids without preconcentration steps.

2.10.10.3 Cadmium determination by atomic absorption

Although flame AA has an excellent sensitivity for the element Cd (it is only exceeded or matched by that for Mg and Zn), it is still not able to determine the normally extremely low concentrations in blood or urine directly. Already in 1967, Berman (95) has described a good preconcentration procedure consisting of complexing Cd in urine or blood (after TCA precipitation of proteins) with diethyl dithiocarbamate, extracting the complex into a small volume of methylisobutyl ketone and analyzing the organic extract by flame AA using the 228.8 nm resonance line and an oxidizing flame. Her method can also reveal normal Cd levels in body fluids, not only elevated concentrations from acute or massive chronic exposures.

Many attempts for direct determination of Cd in normal body fluids by AA (without preconcentration) have been described. Some use the sampling boat technique (96), others the Delves cup method (97, 98). In both cases, a good compensation system for the elimination of unspecific background absorption is indispensable. Even then, the reproducibility of the methods was not ideal in our hands. The main difficulty is the low evaporation temperature of the metal, which does not lie sufficiently above the charring temperature of the matrix. For the same reason, we cannot recommend a direct determination of non-complexed Cd in undecomposed body fluids by graphite furnace atomization. However, with the help of matrix modification for the prevention of pre-atomization losses, such a procedure can yield good results. The addition of ammonium phosphate permits the use of higher ashing temperatures without significant Cd losses and reduces the disturbing molecular absorption signals during the subsequent atomization (99). If the normal argon gas flow in the furnace is replaced by O_2 during the first phase of the ashing step (but not during drying and atomization), ashing can be carried out at 600° C, followed, after a short delay, by atomization at 2100 °C. Another possibility to reduce the Cd losses during the ashing step is the addition of Triton X-100 to blood or urine to lower the ashing temperature below 500 °C (100). A Zeeman background compensation system and a L'vov platform in the pyrolytic tube have been used in this approach. It has the advantage of not requiring a change of gas flow during analysis.

2.10.11 Mercury

2.10.11.1 Introductory remarks

The unique liquid metallic element mercury, its volatility, its ability to dissolve many other metals (amalgamate), as well as its toxic properties have been known since ancient times. This may have been largely due to the fact that it occurs in nature in its native metallic form. Three different types of Hg intoxications must be distinguished. They result from exposure to three different states of Hg and, accordingly, possess also three different faces. We must deal with

- intoxications by metallic Hg vapor or Hg dust,
- oral poisonings by Hg salts (ionic Hg), and
- intoxications by organo-mercurial compounds.

These three danger potentials with their corresponding poisoning symptoms will be discussed independently.

2.10.11.2 Intoxications by metallic mercury

Liquid Hg has the lowest boiling point (357 °C) and highest vapor pressure of all metals. Already at ambient temperture, up to 15 mg Hg can accumulate in 1 m^3 air. The vapor is easily inhaled and diffuses readily across the alveolar membranes. Diffusion of the mono-atomic Hg within the body is facilitated by its lipid solubility (absence of charge). It reaches the brain fairly rapidly. Elemental Hg is only slowly oxidized to Hg ions. The distribution pattern obtained from exposure to Hg vapor is therefore different from one resulting from a direct administration of Hg ions (101, 102). The toxicity of Hg is generally attributed to the action of mercuric ions. Their distribution in body tissues, especially brain, determines the nature of the toxic effects and explains the differences in the toxic symptoms between atomic vapor exposure and oral intoxications by Hg salts.

Relatively few acute Hg vapor poisonings have been reported. The major symptom is an acute pneumonitis called mercury fume fever. It is accompanied by central nervous and gastrointestinal disturbances.

Chronic intoxications by Hg vapor, on the other hand, are very common, especially as a result of professional exposures. Serious poisonings have been observed in connection with gold plating (watch industry), production of mercury mirrors by amalgamation of tin (Murano island in Italy, city of Fürth in Germany), felt treatment in the hat industry, fabrication of thermometers, Hg batteries, Hg manometers and electronic gadgets with Hg switches, work with Hg-sealed stirrers or Hg diffusion pumps and other Hg-containing laboratory devices, electrolysis with Hg cells, all kinds of amalgamation processes (dentistry), and uses of Hg catalysts. Toxic symptoms connected with such occupations have been known for a long time, as we are told from old sayings, such as "crazy as a hatter", "the mad hatter", or "the hatter's shake". As early as 1700, Ramazzini (103) wrote that the workers involved in the mirror fabrication in Murano were able "to see in their own products the picture of their misery".

The main effects of chronic exposures to Hg vapor are psychological disturbances, fatigue, tremor and weight loss; side effects can be proteinuria, stomatitis and dermatitis.

Elemental liquid Hg is said to pass the intestinal tract without being resorbed and without causing damage. We have repeatedly been able, however, to diagnose severe intoxications by metallic Hg in cases of intestinal injury. In one, glass fragments from a Hg thermometer may have damaged the intestines. In a second case, a Hg-containing medical probe broke after introduction into the stomach tract. Due to a complete lack of intestinal activity, the Hg remained in the guts for many days. It must have been oxidized and subsequently resorbed.

2.10.11.3 Intoxications by mercury salts

Inorganic Hg salts have been used frequently as criminal poisons and also for suicides. Mercuric salts such as $HgCl_2$ are better soluble and therefore more toxic than mercurous salts, such as HgCl (calomel), with their much lower water solubility. Already a single oral dose of 0.5 g $HgCl_2$ may be lethal (104). The immediate effects of such an ingestion are gastro-intestinal irritation and corrosion of the exposed tissues. They are followed by intestinal hemorrhage and very serious kidney harm (anuria). Repeated or chronic exposures to smaller doses result in kidney damage and ulceration (105). While the brain seems to be the critical tissue in exposures to metallic Hg vapors, the kidneys are certainly the organs which suffer most from oral Hg salt intoxications.

Inorganic Hg salts are used in medicine as antiseptics, for the sterilization of medical instruments and, in dilute solutions, also for the disinfection of hospital rooms and beds. We had to deal repeatedly with poisonings by Hg disinfectants:

- A cleaning solution containing Hg salt had been used by the hospital staff without the recommended dilution.
- A Hg salt disinfectant (mercuric oxycyanide), which every Swiss military doctor carries in his medical kit, was used for a suicide attempt by the son of an MD.

- A hospital patient drank repeatedly the solution of the Hg disinfectant left on the table in his hospital room.

Hg dervatives (HgCl and HgO) have been used and may still be used as fungicides, i.e. in the treatment of potato-seed and lawns (golf courses), the red and yellow oxides as paint additives for coating ship hulls.

The world wide Hg production is estimated to exceed 10000 t a year. Perhaps one tenth contaminates, as waste, air, oceans and soil. It adds to the natural Hg levels and to the Hg pollution created by the combustion of fossil fuels. Today, the toxic danger potentials from metallic Hg and Hg salts seem to be fairly well under control. But we have good reasons to be more concerned about the intoxication risks from the organo-mercurial compounds with which we are also confronted.

2.10.11.4 Intoxications by organic mercury derivatives

The use of organic Hg derivatives as pharmaceuticals, i.e. diuretics (Mersalyl), has been abandoned, but not the medical application of some Hg-containing disinfectants such as phenylmercury acetate, phenylmercury chloride, phenylmercury borate (Merfen), phenylmercury nitrate, and o-carboxyphenyl-thioethyl-mercury (Thiomersal, Merthiolate, Merfamin). It is assumed that their toxicity is low, since they resorb only slowly. However, we have been confronted with several dangerous intoxications resulting from disinfection of large sections of burnt body tissue with such organo-Hg compounds. We wonder if their use can still be justified.

A much greater peril than the use of organo-mercurials in medicine is their large scale application in agriculture and industry. Hundreds of organic mercury derivatives have been designed as plant-protecting agents, mainly fungicides for the treatments of seeds. Their general structure is R–Hg–X. The nature of X is of secondary importance. It can be an inorganic or organic acid (acetate, benzoate, salicylate), an amide such as urea, thiourea or dicyandiamide, a phenolate or a mercaptide. This anionic part of the molecule can influence its vapor pressure, solubility and fungicide action. But with respect to toxicity, the substituent R of the molecule is the decisive factor. Usually, the organo-mercurials are therefore subdivided into alkyl Hg compounds (R = CH_3- or CH_3CH_2–), alkoxyalkyl derivatives (R = CH_3–O–CH_2–CH_2– or CH_3–CH_2–O–CH_2–CH_2–) and aryl Hg compounds (R = C_6H_5– or CH_3–C_6H_5–). If R is a methyl group, the R–Hg bond is very stable. If R is an aryl or alkoxyalkyl group, the bond can be broken by many microorganisms in soil or water. This means that aryl- and alkoxyalkyl-Hg and (slower) even ethyl-Hg compounds will finally decompose, permitting the freed Hg to volatilize. In addition, the molecules of methyl-Hg and ethyl-Hg are also very volatile.

Microorganisms can not only decompose organo Hg compounds, they have also the ability to methylate inorganic Hg to monomethyl-Hg salts and dimethyl-mercury. Such reactions occur especially in organic sediments of the sea and of fresh water. They are blocked by hydrogen sulfide, which binds Hg ions to insoluble HgS. The ability to methylate Hg is widespread. It is an important factor for comprehending the many aspects of Hg toxicity.

Chronic intoxications by methyl-mercury compounds start with sensory disturbances of the extremities and affection of hearing and eyesight, followed by tremor and ataxia. All parts of the central nervous system can be impaired. The prenatal intoxication danger is high. Children from exposed parents are mentally retarded and show motor disturbances and paralysis (106). The symptoms are rather non-specific and have for a long time not been connected to Hg contamination. The correlation was established with the help of some "epidemic" intoxications caused by these compounds.

2.10.11.5 Minamata disease

After 1950, in the fishing village Minamata on the west coast of Japan, an unusually high number of dead birds and cats was found. In 1956, a five-year old impaired girl, who had never learned to walk and speak correctly, died. First a few, then more and more residents of the village developed severe paralytic symptoms. In 1957, it was for the first time suggested that there might be a connection with the pollution caused by a nearby polyvinylchloride plant, which released its waste water into the Minamata river and bay. It took ten years until this possibility was investigated, and during this time, hundreds of people died and more were severely incapacitated. A similar "epidemic" poisoning in the city of Niigata with over hundred fatalities helped to shed light on the events. In both cases, factories had released Hg-containing waste, in Minamata into river and bay, near Niigata into the Agano river. The Hg settled in the sediments and was methylated to methyl mercury by microorganisms. This stable compound becomes concentrated in the natural carnivorous food chain. And since the main food of the residents of Minamata as well as the victims of Niigata consisted of fish, the population got poisoned by this highly toxic Hg derivative. Due to its good lipid solubility, it easily penetrates the placental barrier, which can explain the many prenatal intoxications.

2.10.11.6 Other "epidemic" intoxications with organo-mercurials

A large scale pollution with organo-Hg occurred also in Sweden. In the 1940's, seed-dressings containing methylmercury dicyandiamide were introduced and took over the market. This caused severe poisonings of seed-eating birds and resulted in a drastic decrease of the wild bird population. In 1966, Westöö (107, 108) called attention to high concentrations of Hg in Swedish eggs, meat and fish. It was present mainly as methylmercury.

Alkylmercury seed dressings (wheat, potatoes) have also caused several "epidemic" intoxications of humans in developing countries. Treated seeds, donated for agricultural use only, were erroneously used directly for human consumption.

2.10.11.7 Mercury levels in man

When evaluating the data on the concentrations of Hg in biological materials, one should keep in mind that the distribution of Hg in a body is strongly influenced by its chemical contact form (atomic vapor, mercury salt or organomercurial) (109). The analytical data should whenever possible be interpreted in consideration of the suspected contact form.

Many reference data for blood and urine levels of people without known Hg contacts have been published. Berlin (109) recommends that blood levels over 20 μg/kg Hg should be taken as an indication that some unspecific symptoms may be due to Hg exposure. He considers levels over 30 μg/kg as confirmation of a dangerous Hg contact and states that levels from 100 μg/kg up are always associated with some classical signs of mercurialism.

Between 1966 and 1977, our laboratory at Zürich University analyzed a large number of blood and urine samples of people with known or suspected exposures to Hg vapor, Hg salts or organic Hg derivatives, as well as a considerable number of reference cases with no known Hg contacts. On the basis of our results, we can conclude that without any extraordinary exposures, Hg-concentrations lie below 10 μg/kg blood and 20 μg/daily urine. In most cases with higher values, the existence of a special exposure to Hg in one or another form could be established.

In spite of the inconsistent data available for the Hg levels in tissues, it can safely be assumed that brain will show especially high Hg concentrations in cases of Hg vapor contact, the kidneys in oral poisoning cases with mercuric salts. For live people, Westermark and Ljunggren (110) recommend blood (separated into plasma and corpuscules) and hair as preferred sample materials. As normal values, he quoted 10 ng/g Hg for blood corpuscules and 1 μg/g for hair. In hair from Minamata patients, Hg concentrations between 16 and 760 μg/g could be found (111).

2.10.11.8 Methods for mercury analysis in biological samples

The old method of Stock (112, 113) was cumbersome and time consuming, but gave astonishingly reproducible results. Hg was collected from acid solution on a copper wire, sealed in a capillary, liberated from the copper by distillation, collected on the other end of the capillary as microdroplet, which was measured under a microscope. This method was used for more than 30 years. With the development of instrumental analytical methods, it could be replaced by more sensitive and more quantitative procedures such as:

- neutron activation analysis,
- radioactive tracer dilution techniques,
- colorimetric determination, especially as dithizone complex,
- atomic absorption,
- atomic fluorescence,
- plasma emission analysis with MS detection, and
- gas chromatography (for organo-mercurials).

But a good choice of the detection technique for the analysis of a given sample is by far no guarantee for analytical accuracy. Due to the high volatility of Hg and some of its derivatives, it is of primary importance that sample combustion and transfer to the detection system exclude any Hg losses. We can recommend the following sample decomposition techniques:

- For dry samples such as hair or nails, the oxygen flask decomposition of Schöniger (114) or also the decomposition in a stream of oxygen (115).

- For solids which contain over 30% water, decomposition by a mixture of concentrated nitric and sulfuric acids in the apparatus described by Bethge (116). The distillate collected during combustion contains the mercury and must be analyzed.
- For liquid samples such as blood and urine, destruction of the organic matrix by the combined action of sulfuric acid and permanganate, preferably in the cold (116). The excess permanganate can be removed by reduction with hydroxylammonium chloride.

Decomposition procedures usually dilute the analyte and must be followed by a concentration step. In colorimetric determinations of Hg as organic complex, such as in the dithizone procedures (117, 118), this can be accomplished by solvent extraction. A more efficient concentration is possible by amalgamation on copper or the noble metals silver and gold. This possibility has already been used by Stock. In 1967, we have established a combination of amalgamation on copper wire with flameless atomic absorption analysis after electrothermal desorption (119–122). In order to shorten the analysis, amalgamation is accelerated by electrolysis. If time is not an essential factor, the slower spontaneous amalgamation can also be used. Today, all noble metals have come into use for collecting Hg from aqueous solutions or gases by amalgamation.

2.10.11.9 Choice of analytical method

For a considerable number of years, neutron activation analysis was the method of choice in trace Hg analysis. It can be applied directly to a large selection of dry samples such as hair, nails or dry proteins (only mg-quantities are needed), but also to liquids or gases (123). But for such investigations, an average toxicological laboratory requires outside help. Today, the time involved and the costs of operation may not any longer be justified, since simpler and even more sensitive techniques have become available.

Lamm and Ruzicka have reviewed and compared the methods for determining traces of mercury in biological materials (124). They recommend the dithizone procedure only for higher concentrations (over 50 ng/g). They warn of interferences due to the formation of chelates other than $Hg(HDz)_2$, since they absorb light at similar wave lengths (near 485 nm) and falsify the results in trace analysis. They consider atomic vapor absorption the best choice for the analysis of samples with mercury concentrations between 5 and 1000 ng/g. For even lower concentrations, they recommend and describe radioisotope techniques, especially isotope dilution analysis and isotope exchange methods. Such applications are facilitated by the availability of the radioactive isotope ^{203}Hg with favorable properties (half life 47 days, easily measurable beta and gamma radiation). To differentiate between various types of organo-mercurials, GC seems to be the preferred method. It is able to detect and measure the most dangerous Hg poisons: methyl mercury, dimethyl mercury, phenyl mercury, as well as other organo-mercurials.

2.10.11.10 Mercury analysis by cold vapor atomic absorption

Flame AAS and graphite furnace AAS are not suitable for toxicological Hg determinations (Table 5). The measurement of Hg in its vapor state by flameless AAS, on the other hand, must be considered the most practical Hg analysis available today.

An important step in this direction was made already in 1959 by Lindstrom (125). He used a simple Hg vapor measuring apparatus to determine Hg in solutions such as urine. The liquid was atomized in a oxyhydrogen flame, the resulting gases were led through a system of filters into the gas chamber of the vapor meter, and the passage of the gas cloud measured by recording the absorption of the light of a Hg lamp. The procedure showed good sensitivity, but the instrumentation was somewhat complicated.

In 1967 and 1968, we described two simple methods for Hg analysis (119, 122), both with a detection limit of 100 pg Hg. In a first step, the cationic Hg in solution is collected by a spontaneous or electrolytically-aided amalgamation on a Cu spiral. The loaded wire is placed into a tube with quartz windows fitted into the light path of an AA spectrometer, the amalgamated Hg liberated by electrothermal heating, and the mono-atomic vapor measured by recording the absorbed Hg resonance light emitted by a Hg vapor or hollow cathode lamp. In a first stage of the development, we worked like Lindstrom with an open gas cuvette, through which the liberated Hg vapor was drawn. This "dynamic" method furnishes a time-dependent, peak-shaped signal, which must be graphically recorded (119, 120). But we soon changed over to a "static" vapor method (121, 122) by evaporating the Hg in a sealed tubular cuvette. The resulting vapor yields a steady static absorption signal which can be measured more accurately. Such a "static" procedure is less subject to error and easier to operate than the "dynamic" method.

The "dynamic" approach with the open absorption cell has been subsequently modified by many authors. The amalgamation of Hg on Cu, for example, was replaced by deposition on Ag or Au. In our opinion, this just raises the costs of the procedure. In addition, Ag often contains Hg impurities. A more relevant change was the replacement of the reduction by amalgamation by a reduction with $SnCl_2$ in solution (126–128). This principle, generally known as Hatch and Ott method, has been commercially exploited by the firms in the AAS market. Most of them sell Hg vapor phase accessories for their spectrometers, constructed for reduction in solution. In addition, independent Hg monitors, built exclusively for Hg vapor analysis, have also become available. With these instruments, down to 1 ng Hg can be measured. If 50 ml samples are analyzed, this corresponds to a concentration of 20 ng/l. Sources of error with these methods are that they furnish only fast transient signals and that the Hg cloud liberated from aqueous solutions is contaminated with humidity (vapor or micro-droplets) which must be removed with a desiccant or a filter in the gas flow line. This involves the risk of losses by adsorption, as well as a danger of contamination between samples. Furthermore, a presence of other easily reducible components can impair the reduction of Hg ions to elemental Hg. With $NaBH_4$ as reducing agent in place of $SnCl_2$, all hydride forming elements are also reduced, and the molecular absorption of the volatile hydrides can falsify the readings.

A good part of these sources of error can be eliminated by a background compensation system. Another approach to improve the method consists of collecting the Hg vapor from reduction in solution on a gold/platinum gauze and desorbing it subsequently by heat (129). This rather costly accessory has also been commercialized.

Laboratories which do not have to run large series of Hg determinations as daily routine tasks may not want to invest much money in a dedicated accessory to their AA spectrometer. For them, we can recommend our static Hg vapor technique after amalgamation (121, 122). The tubular optical cell can be contructed by any glass blower. The rest of the equipment is usually available in a laboratory. This makes for a low cost method, possessing a sensitivity which at least equals that of expensive commercial accessories. In an evaluation of methods, the amalgamation procedure was considered superior to the reduction in solution (130). It seems therefore justified to add a detailed description of this analytical procedure by reproducing a text already published in 1974 in Clinical Biochemistry (75), as well as a drawing of the simple equipment which is needed (fig. 1).

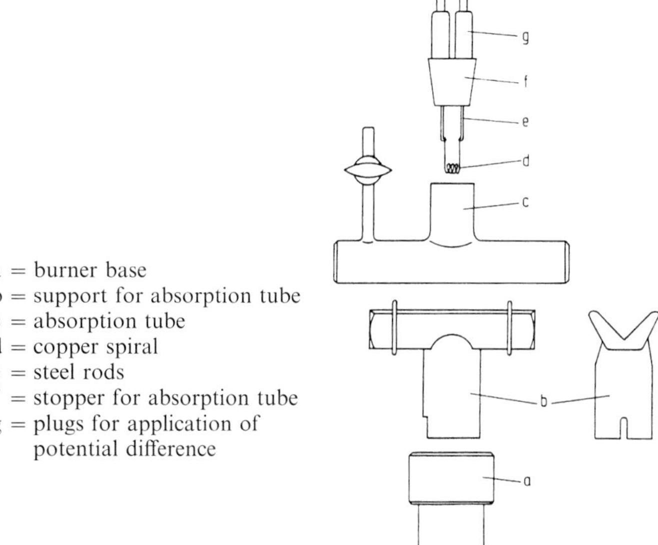

Figure 1. Simple Atomic Absorption Accessory Permitting "Static" Measurement of Mercury as Cold Vapor.

Figure 1 shows the burner base a, the holder b fitting into a, the absorption tube c to be placed on b, a silicon or rubber stopper f fitting into the neck of c and pierced by 2 small steel rods e which connect the ends of the copper spiral d to the plugs g. The spiral d is made from electrolytic copper wire of 0.1 mm diameter by twisting it around a 4 mm glass rod (approx. 20 cm wire, up to 10 windings, straight end pieces of 3 cm length for connection to e). Preferably, the absorption tube is made entirely of silica; however, a glass tube with planar silica windows can also be used. The length and the diameter of the absorption cell should be adapted to the light path of the instrument. The optimal dimensions can also be dependent upon the light source. They may lie between 15 and 20 mm diameter and around 10 cm length. The tube holder b permits focusing the absorption cell with the instrument-knobs provided for adjustment of the burner.

The spiral is attached to the stopper, cleaned with half concentrated nitric acid and rinsed with water. The electrolysis is carried out in 0.1 to 1 mol/l nitric acid in disposable plastic beakers. A small platinum wire serves as anode, and a plastic covered magnetic stirrer is used to mix the solution. We work with an electrolysis potential of 2 to 3 V, which is applied during 20 min for 5 ml samples and up to 60 min for 10 ml solutions.

The amalgamated copper cathode is rinsed with water and alcohol and introduced dry into the absorption cell. It should not reach into the light beam. A slight vacuum is applied using a water aspirator. The absorbance is set to zero, the valve to the vacuum closed, and the filament heated by applying a potential difference of 3 to 4 V (high enough to make the wire glow without burning it through, alternating or direct current). All the mercury evaporates instantaneously into the cell, and the absorption of the 253.7 nm resonance line can be measured immediately. The value should be constant for at least 30 seconds. Then, the mercury vapor is removed by the vacuum pump in order to avoid any condensation.

Evaporation of mercury into a non-evacuated tube is also possible. A slightly higher specific absorption value is even obtained. However, the signal is not as constant as with the evacuated system, since the mercury vapor distributes through the cell more slowly. After the measurement, the vapor must be displaced by an inert gas or air.

Depending on the size of the absorption cell, not over 100 to 200 ng of mercury should be deposited on the copper wire for one determination. Normally, the absorption tube can be re-used without cleaning. However, it should always be checked if its transparency for the resonance light has not diminished due to mercury deposition on the windows. A good absorption cell should not weaken the light intensity by more than 25%.

We just would like to add that we have also modified the reduction method with $SnCl_2$, so that a static absorption signal can be obtained (131). The Hg vapor is generated in a side arm of an evacuated and closed absorption cell. Since the volume of this side arm is larger than the neck of the cuvette used for desorption of the Hg from an amalgamated spiral, the sensitivity of this modification is slightly lower.

2.10.11.11 Differentiation between inorganic and various types of organic mercury

It is possible to differentiate between inorganic and organic bound Hg by analyzing a urine sample before and after combustion of the biological matrix. By carefully adjusting the pH of the solution, it is also possible to decompose phenyl-Hg, but not alkyl-Hg compounds (132). This means that three determinations are needed: one without acidification of the analyte solution (for inorganic Hg), the second after addition of sulfuric acid (for inorganic and phenyl-bound Hg), and a third after digestion with permanganate in presence of sulfuric acid (for total Hg). By using a similar combination of methods, we have been able to show that after exposure to elemental or ionic Hg, the urinary excretion of inorganic Hg slowly decreases, while that of organic Hg increases with time (133). This indicates that the human body is not only able to decompose, but also to synthesize, organo-mercurials.

GC has been used at an early stage (107, 108) for analyzing methyl-Hg in foodstuffs. Head space GC with microwave-induced plasma detection permits finding methyl-Hg compounds in water (134). With a combination of GC with AAS, ng-amounts of Me_2Hg, Et_2Hg, MeHgCl and EtHgCl can be determined in air (135). It must be expected that GC will soon be able to clear up many open problems in Hg toxicology. A combustion-free analysis of total Hg in biological specimens is possible (136), but the differentiation of the Hg-compounds needs some further studies.

2.10.12 Boron

2.10.12.1 Toxicological importance

Boric acid has been used to disinfect wounds for well over a hundred years. Its aqueous solution was for example held ready in special bottles in most laboratories for rinsing eyes in case of an occasional spill with aggressive chemicals. But the bactericidal and bacterostatic effects of such solutions are, at best, very limited. Boric acid solutions have also frequently been used for rinsing the bladder after prostate operations. They were considered harmless to the human body. However, around 1965 and later, it became evident that this belief was erroneous. A number of severe poisonings could be detected, and we can assume that many more have not been recognized. Along with oral ingestion, resorption by way of injured tissues and mucous membranes has led to intoxications. Between Fall 1966 and Spring 1968, we had to investigate three deaths after prostate operations. Their cause was the boric acid solution used for bladder rinsing. Concentrations beween 80 to 370 mg/kg boron could be found in blood and organs (137).

Moeschlin (138) considers boron blood concentrations up to 250 µg/l, Curry (3) even levels up to 800 µg/l as normal values. Concentrations up to 20 mg per liter blood must be regarded as subtoxic, higher levels as toxic. According to Moeschlin (138), concentrations over 80 mg/l may be fatal. For oral intakes, he estimates the following lethal doses: 2–3 g for infants, 5–6 g for children, 15–20 g for adults.

Resorbed boron is eliminated by the kidneys. In a poisoning case, we measured 25–50 mg boron per liter urine during the five days which elapsed between intoxication and death (137).

2.10.12.2 Detection possibilities

A qualitative recognition of boron by conversion to its boric acid ester, which burns with a green flame, can be used for its detection in biological materials: A small portion of the sample is placed in an evaporating dish (a watch glass), two drops of sulfuric acid and ten drops of ethanol or methanol are added and the mixture is ignited. A green border of the flame indicates the presence of boron (139). The element can also be detected by color reactions, i. e. with turmeric respectively curcuma paper, or with carminic acid (2, 3, 139).

2.10.12.3 Quantitative determinations

A quantitative colorimetric determination of boron in serum with carminic acid reagent is described (2), as well as a spectrometric determination with curcumin after its extraction with 2-methyl-pentane-2,4-diol in chloroform (140).

We have measured the element by flame atomic absorption at the resonance line of 249.8 nm of a boron cathode lamp (137) in the reducing (acetylene-rich) laminar flame of a 3-slit burner. The analytical sensitivities (better than 50 μg/ml) and detection limits (around 10 μg/ml) permit diagnosis of dangerous intoxications. Urine, serum, spinal fluid, dialysates as well as other homogenous solutions can be sprayed directly into the burner aggregate (137). Blood and tissues must first be decomposed. Koch (141) recommends dry ashing at 650° C after addition of Li_2CO_3. We have found that this method involves losses and prefer wet oxidation by nitric acid and hydrogen peroxide. The resulting clear but not necessarily colorless solutions are buffered with NH_3 to pH 2–3, brought to volume with water and sprayed directly into the flame of the AA spectrometer. For quantification, external as well as internal calibration can be used. For the determination of "normal" boron levels in biological specimens by flame-AAS, preconcentration procedures are indispensable.

Other methods than flame AAS are also available, some more sensitive, but in part also more cumbersome:

- graphite furnace AA (142, 143),
- flame emission spectrometry after conversion of boron to its methyl ester (144),
- ICP-AES (145, 146),
- direct current plasma atomic emission spectrometry (147) and
- ICP-MS combination.

2.10.13 Aluminium

2.10.13.1 Introduction

Aluminium is the most abundant metal in the earth's crust. It occurs in plant and animal tissues. Depending on its origin and preparation, domestic tap water may contain relatively high concentrations of aluminium.

In view of its widespread distribution and occurrence in aliments, aluminium has long been considered a non-toxic substance. There had been few reports on aluminium intoxication. However, in 1970, Berlyne et al. (148) reported elevated serum aluminium concentrations in patients with end-stage chronic renal failure, and subsequently showed that administration of aluminium to uremic rats produced toxic effects such as lethargy, periorbital bleeding and anorexia (149).

2.10.13.2 Causes of aluminium intoxication

Orally ingested aluminium is poorly absorbed by the intestinal tract. The proportion absorbed has been estimated to be of the order of 0.1 %. Absorption is increased by organic acids such as citrate. The poor intestinal absorption of orally ingested aluminium explains why normally no intoxication from the environment is observed. This is different if aluminium-containing drugs such as phosphate binders or antacids are ingested. Thus, daily administration of about 30 mg of aluminium per kg of body weight in the form of aluminium hydroxide results in significant increases of plasma aluminium in uremic patients (150–152).

From the *milieu intérieur*, aluminium is excreted by the kidneys. Uremic patients are particularly susceptible to aluminium intoxication because of the enhanced accumulation. Marked intoxication has been reported in a patient who received 3840 mg/day of aluminium hydroxide (153).

Whereas the intestinal epithelium functions as an efficient barrier against oral aluminium ingested at moderate doses, the parenteral administration is much more dangerous and represents the cause of most intoxications. Access may occur e.g. by haemodialysis, peritoneal dialysis, vesical application or dust inhalation. Thus, the relatively high aluminium concentration of some water supplies utilized to prepare dialysis baths has been responsible for a number of intoxications until it was realized that purified water should be used instead (150).

2.10.13.3 Clinical manifestations of aluminium intoxication

2.10.13.3.1 Encephalopathy

Patients dialyzed against a recipient fluid prepared from tap water rich in aluminium may develop speech disorders followed by dementia (154). The same symptoms may also develop in uremic patients treated by phosphate-binding aluminium salts. Acute reversible encephalopathy has been reported in a boy receiving vesical irrigations of aluminium (155).

2.10.3.3.2 Osteodystrophy

This has been observed in uremic patients given aluminium salts or dialyzed against aluminium-containing water and is most likely due to the hypophosphatemia induced by such treatment (156).

2.10.13.3.3 Microcytic anemia

Uremic patients dialyzed against aluminium-containing tap water develop a microcytic anemia unresponsive to iron treatment. This is probably due to disturbances in heme synthesis and porphyrin metabolism (157).

2.10.13.3.4 Alzheimer's desease

This disease is a common cause of dementia in the elderly. Histopathologic examination of the brain shows the presence of neurofibrillary tangles. Microprobe studies have shown neurofibrillary tangle-bearing neurons to have a significantly higher aluminium content than tangle-free ones (158, 159). Since aluminium normally does not pass the hemato-encephalic barrier, a cause different from dietary aluminium overload may exist to explain an increased aluminium access to the central nervous system in Alzheimer patients. The question whether aluminium is involved in the development of Alzheimer's disease is still open. More sensitive assay methods and determinations in such samples as cerebrospinal fluid will be helpful for investigating this problem.

2.10.13.4 Analysis of aluminium

Recognition of clinical manifestations related to aluminium toxicity had long been delayed by the absence of reliable methods of assaying traces of this metal in biological materials. The development of appropriate methods has played an important role in the progress of knowledge in this field. Not only should the procedures be sensitive, but care has to be taken to avoid any contamination, because of the abundance of aluminium in the environment.

A thorough discussion of possible sources of contamination during sample collection and analysis has been published by Savory and Wills (160). Needles, collections tubes, syringes, anticoagulants must be checked for aluminium contribution to the specimen by running analysis in which the sample is substituted by aluminium free water. The analytical materials such as glassware, pipets, tubes and the water utilized must be checked as well. The water should have a resistivity of at least 18 MΩ. A procedure for cleaning the glassware is described by Brown et al. (161).

The mostly used analytical technique is graphite furnace atomic absorption spectrometry, preferably with the L'vov platform and Zeeman-effect background correction (162–164). The 309.3 line, which is most sensitive, may be used. Determinations can be made in serum, heparinized plasma, tissues, water and dialysis fluid. The method is sufficiently sensitive to allow for assays in the plasma of healthy persons.

A fluorimetric procedure for determinations in serum has been published (165).

The required sensitivity is also achieved by inductively coupled plasma emission spectrometry (166). The method has been used for assays in plasma and skin (152).

2.10.13.5 Role of aluminium analysis in clinical toxicology

It is now well documented that the high aluminium concentration of some domestic water supplies was the cause of aluminium poisoning in dialyzed patients.

Nephrology centers located in areas where the water supply is subject to aluminium contamination now use purification procedures such as reverse osmosis to prepare the water of the dialysis baths. For home dialysis, deioniziation cartridges may be used. The water has to be checked periodically for aluminium content, the frequency being determined by the local situation. According to de Wolff et al. (167) a concentration of 5 μg/l (0.18 μmol/l) of aluminium in the dialysate is considered safe.

For the control of patients, aluminium determination in plasma or serum is the most convenient way, even if this does not necessarily reflect the content stored in body tissues. The approximate reference interval among healthy persons is below 0.37 μmol/l (> 10 μg/l) (167), whereas symptoms of aluminium toxicity have been observed with concentrations above 3.7 μmol/l (100 μg/l). In acute aluminium intoxication, plasma concentrations of 5 μmol/l or more have been reported (170).

Reviews on the biological monitoring of aluminium have been published (168, 169).

2.10.14 Thallium

2.10.14.1 Toxicological importance

Thallium is a highly toxic metal. It has no known function in the body and is therefore of no interest to the clinical chemist. On the other hand, Tl has been and still is of concern to the forensic as well as the environmental toxicologist. The widespread use of Tl (usually in form of its sulfate) as a rodent poison constitutes a real danger for adults, children and domestic animals. Tl sulfate has been a frequent weapon for homicides and suicides. Accidental oral poisonings have also occurred, especially in children. Since the toxic effects of Tl become evident only after a latent period of hours or days, an analytical control is the only means to confirm in time a suspected ingestion. For this purpose, a fast chemical method must be ready in toxicological laboratories. We have repeatedly been confronted with such situations, when small children were found playing with "Zelio" grains. They are impregnated with $TlSO_4$ and a warning color, and sold as mice or rat baits. But unfortunately, the color constitutes rather an attraction for children than a warning.

The estimated oral lethal dose for an adult is 1 g of $TlSO_4$. The salt is resorbed rapidly before the first toxic symptoms occur. After hours or even only after two or three days, gastrointestinal pains begin and lead to severe colics with obstipation. The symptoms can feign a twisting of the guts and have induced doctors to proceed with abdominal operations. Various neurological symptoms follow and may lead to paralytic affections. Since intoxications with lead or arsenic give similar poisoning effects (but usually less pronounced), a chemical analysis is indispensable as diagnostic tool. All too often, the existence of a Tl intoxication is only suspected ten or more days after an ingestion, when the hair begins to fall out in strands or bushels, sometimes over the whole body.

In the past years, it has become evident that Tl is also an environmental hazard. It can be found in the exhaust of coal power plants, as well as in the dust and smoke of cement factories and other plants. Thanks to the availability of more sensitive analytical methods, it has been shown that minute concentrations of Tl are present in the earth and in living organisms. Certain plants are able to concentrate Tl from the soil. Even humans who are not professionally exposed to Tl absorb traces of the metal by way of food and air (171, 172).

Contradictory information exists about "normal" Tl levels in urine. According to some authors, they are situated below 1 μg/l (173), while others claim that they extend up to 5 μg/l (172). Chronic exposures, as experienced by workers in cement plants, can elevate the urine Tl concentration to the 100 μg/l range. In lethal and sub-lethal oral ingestions, we have measured up to 100 mg Tl per 1 urine. While such high levels are easy to detect, the chemist may still have difficulties to pin down moderate chronic exposures, and distinguish them from the so-called "normal" values.

Tl accumulates in hair and nails. Especially hair is, besides blood and urine, the material of choice for a control of abnormal Tl exposures. It has the advantage that, by careful analysis of hair segments grown during different time periods, estimates of date and duration of an exposure are possible.

2.10.14.2 Determination by flame emission

Before the availability of atomic absorption, we have worked out a method for the determination of the metallic elements of the group 3 B (gallium, indium and thallium) by flame emission (174), and have applied it to the analysis of Tl in biological materials (174). The metals are extracted as halides (Tl as $TlBr_3$) with diethyl ether, the organic phase is concentrated, injected directly into the hydrogen-oxygen flame of an emission spectrometer, and Tl is measured by recording the emission lines at 377.8 and 535.0 nm. The shift from the solvent water to ether brings an over tenfold increase in sensitivity.

1.10.14.3 Determination by atomic absorption

Acute Tl intoxications can be detected in a few minutes by direct injection of saliva, blood or urine into the flame of an atomic absorption spectrometer. We recommend using the resonance line at 276.8 nm and repeating the measurement at 377.6 nm, where light-scattering interferences are less pronounced. This test is suitable for revealing Tl concentrations above 50 μg/l and can serve as a rapid check for a suspected oral ingestion (75).

Lower detection limits can be obtained by extracting Tl as pyrrolidine dithiocarbamate complex into a small volume of methyl isobutyl ketone. For urine, such an extraction is possible without previous decomposition, since urine Tl is present in the ionic state or as only weakly bound complex (75). By using a tenfold preconcentration, we have been able to measure urine Tl levels down to 5 μg/l. Berman (95) has described methods for measuring Tl in blood, urine, tissues and hair by chemical decomposition, chelation with dithiocarbamate, extraction with methyl isobutyl ketone, followed by analysis of extracts with flame AAS. This permits recognition of acute as well as massive chronic intoxications, but not the measurement of "normal" Tl levels in biological tissues. For such a task, access to GF-AAS or more sophisticated methods such as ICP-MS or NAA, is needed. A good combination is an extraction of Tl as bromide into ether (174) with subsequent analysis of the concentrated extract by GF-AAS (175). Similar combinations of organic extraction with flameless AA can also be used to measure Tl in hair.

2.10.14.4 Other methods for thallium analysis

Before the introduction of atomic absorption, a number of colorimetric determinations of Tl have been described and used, for example the formation of color complexes with the dyes Brilliant Green (absorption maximum at 630 nm) or Rhodamine B (absorption maximum at 560 nm), or the determinations with dithizone, sodium diethyldithiocarbaminate or thionalide. In the first two examples, Tl is first isolated as $TlBr_4$ or $TlCl_4$ with ether. All these methods suffer from unsatisfactory detection limits, some also from interferences by other metals.

To measure Tl in urine, differential pulse voltametry has often been used in the past (6). But by far the best sensitivity is obtained by ICP-MS. This method is able to reveal as little as 0.5 ng/l Tl in water (see table 5).

2.10.15 Germanium and Tin

2.10.15.1 Toxicological importance of germanium

This metal has gained only very limited interest among clinicians and toxicologists. It does not seem to be an essential element to man and has a low order of toxicity.

Oral doses of inorganic Ge are rapidly absorbed in the gastrointestinal tract and well excreted by way of the kidneys. It has been estimated that between 0.5 and 1.5 mg of Ge are ingested with the daily human diet. Almost all of it can be found in the daily urine excretion. Intoxications from "normal" exposition to inorganic Ge are not reported and not to be feared. It may be, however, that the increased use of Ge in the electronic industry will create new dangers of exposure in the future.

Since Ge respectively some of its organic derivatives have been connected with antitumor activity (176), the oral application of such compounds as antitumor drugs has gained some importance, first in Japan, later also in other countries. Especially the organo-Ge compound "Spirogermanium" (177), but also GeO_2 and the preparation "Sanumgerman", containing Ge as lactate and citrate, are being used. Overdosages are not uncommon. The normally prescribed single daily capsule containing 45 mg Ge may be tolerated by the body. But overdosages up to several g of Ge per day over months have led to fatal intoxications, particularly due to renal failure (178, 179). While "normal" Ge levels in human tissues are usually low, in the range of 100 µg/kg tissue and 1 mg/daily urine, they can attain mg-levels after ingestion of such high doses of Ge antitumor preparations.

2.10.15.2 Toxicological importance of tin

Metallic tin is used as a coat for sheet iron (i.e. in food cans), and in copper or iron cooking pans. It is more resistant to weak acids than zinc, and its oral toxicity is very low. Food out of Sn-coated cans contains usually between 20 and 50 mg of Sn per kg. It has no toxic effects, since only a small percentage of the ingested metal is resorbed, most is excreted with the faeces. Still, can food containing over 100 mg Sn per kg should be avoided. If the Sn concentration exceeds 250 mg/kg, the can food can impair the health.

A chronic inhalation of SnO_2 dust can damage the lungs, without being cancerogenic. Concentrations exceeding 10 mg Sn per kg of dry tissue could be found in such damaged lungs (180). The volatile tetrahydride, SnH_4, is the most dangerous of all inorganic Sn compunds. It affects the central nervous system very much like AsH_3 (arsine).

The organic compounds of Sn are usually very toxic. This holds especially for $Sn(C_2H_5)_4$, a nerve poison which penetrates the body through lung and skin. But also the di- and tri-alkyl Sn derivatives show toxic effects. They inhibit oxidative phosporylation and interfere with breathing. This should be kept in mind when handling Sn-containing fungicides and disinfectants.

An epidemical intoxication due to "Stalinone", a Sn-containing preparation prescribed against furuncles around 1954 in France, resulted in over 100 fatalities. It has not been established whether diethyl-tin-diiodide ($Sn(C_2H_5)_2I_2$), the main ingredient of the preparation, or a Sn-derivative present as impurity, such as $SnH(C_2H_5)_3$, was mainly responsible for the toxic action.

2.10.15.3 Analysis for germanium and tin in biological samples

Table 6 gives an idea of the detection limits which can be expected by analyzing aqueous solutions with different instrumental techniques. ICP-AES and F-AAS have only limited sensitivities, in contrast to ICP-MS or to the determination of the metals after conversion to their hydrides, that is with hydride AAS (181) for Sn, as well as with GC-MS, respectively GC with mass specific detection (Hy-MS) for both metals (182, see below). In addition, DPV is also a sensitive screening technique for Sn (6, 183). Data dealing with extraction of Ge followed by AAS have recently been summarized (184).

Table 6. Detection Limits for the Analysis of some Elements from the Groups 14 to 16 in μg/l resp. ppb, from Aqueous Solution

Element	Ge	Sn	Pb	As	Sb	Bi	Se	Te
DPV	–	0.03	0.001	0.2	0.1	0.1	0.02	0.06
ICP-AES	15	60	30	30	90	30	90	75
ICP-MS	0.01	0.002	0.001	0.006	0.001	0.001	0.06*	0.01
F-AAS	300	150	15	150	45	30	100	30
GF-AAS	0.2	0.5	0.15	0.5	0.4	0.6	0.7	1
Hy-AAS	?	0.8	–	0.03	0.15	0.03	0.03	0.03
Hy-MS	0.002	0.01	–	0.002	0.01	?	?	?
NAA	0.5	2	3000	0.05	0.2	20	1	0.5

* using isotopic mass ^{82}Se, since ^{80}Se is obscured by Ar_2

Hy-AAS = Atomic absorption analysis after hydride generation
Hy-MS = GC-MS analysis of hydrides using capillary column GC and specific mass detection (selected ion recording)

2.10.15.4 Analysis of volatile hydrides by GC-MS

In 1974, we have developed a simple and sensitive approach to the analyisis of Ge, Sn, As and Sb (182), which we have used successfully in our service work. Urine or digests from body fluids or tissue samples are sealed in small serum vials. A sodium borohydride solution is injected through the rubber or silicone septum. It reduces inorganic Ge to GeH_4, Sn to SnH_4, As to AsH_3, and Sb to SbH_3. The volatile hydrides accumulate in the head space gas room of the vials. They can be removed through the septum with a gas syringe and injected into the column of a GC-MS unit. The spectrometer is used as mass specific detector (selective ion recording mode).

Since Ge and Sn are isotopic mixtures, many masses corresponding to M^+ and the three possible ionic fragments of the isotopic hydrides (by successive losses of H) appear in the spectra. Their relative intensities depend slightly on the electron voltage used (20 or 70 eV). The tracer ions can be selected from the high intensity masses which we list below, with the corresponding relative intensities in brackets:

GeH_4, 20 eV: 77 (75), 76 (100), 75 (80), 74 (65), 72 (50),
70 eV: 77 (75), 76 (100), 75 (85), 74 (100), 72 (65).
SnH_4, 20 eV: 123 (100), 122 (75), 121 (80), 120 (78), 119 (75),
70 eV: 123 (89), 122 (70), 121 (80), 120 (100), 119 (65).

Internal standardization is easily feasible by adding an additional hydride-forming element to the mixture to be analyzed. For quantification of Ge, we add As, since the main fragments of AsH_3 (arsine) are m/z 76 and 78. Sb is the ideal internal standard for determining Sn, since m/z 122 is the mass of second intensity for SbH_3 (stibine).

This method is not only extremely simple and fast, it also possesses excellent selectivity (all high intensity ions of the hydrides can be recorded) and sensitivity (see Table 6). For a further discussion of the method, our chapter 2.10.17 on arsenic should also be consulted. Figure 2 describes the simple instrumentation we use for

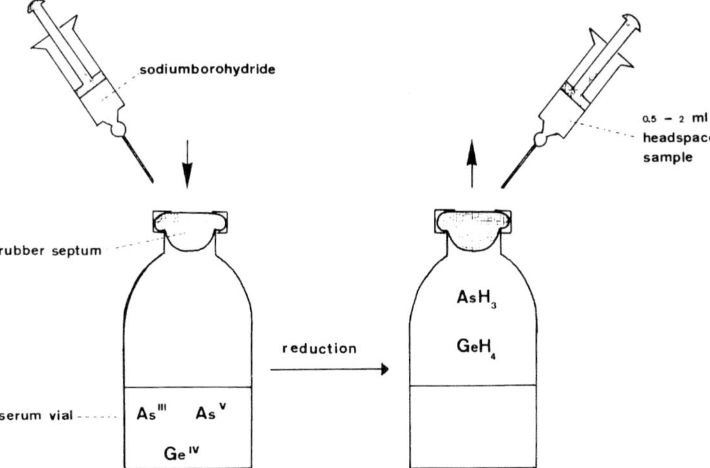

Figure 2. Procedure for the Reduction of Semi-Metals and Sampling of Resulting Volatile Hydrides.

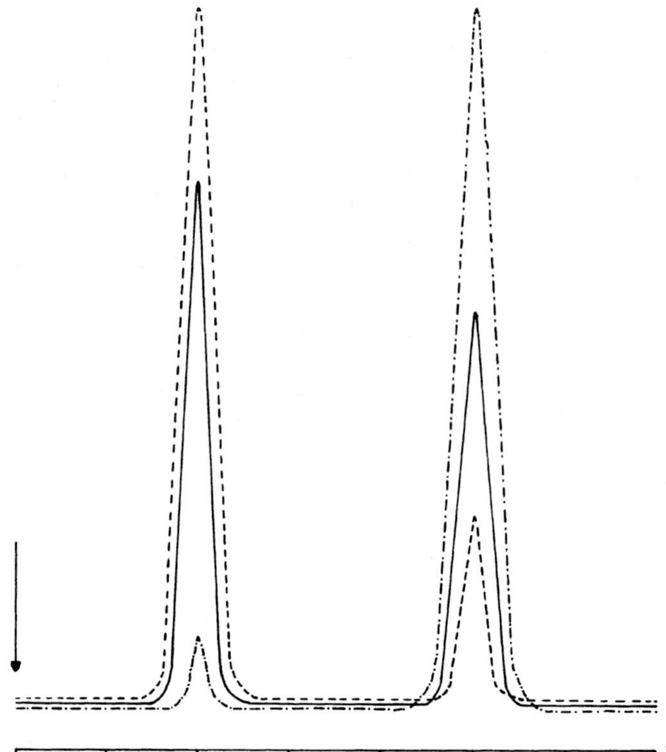

Figure 3. Trace Detection of Tin Hydride (first peak) and Antimony Hydride (stibine, second peak) by Gas Chromatography (on Tenax GC) with Mass Specific Recording of m/z 122 (middle trace, solid line), m/z 123 (highest trace in the first & lowest in the second peak), and m/z 124 (lowest trace in the first & highest in the second peak) at 20 eV.

reducing the semi-metals and for sampling the resulting hydrides. Figure 3 gives an example of a gas chromatographic hydride analysis by GC with mass specific recording of the ions with masses 122, 123 and 124. The relative peak heights must correspond to the relative peak intensities in the mass spectra of the hydrides.

2.10.16 Lead

2.10.16.1 Occurrence and possible function

Lead (Pb) is an ubiquitous element. It can be found in soil, air, water and living organisms. But it has not yet been clearly established whether it also has a vital function in the body and can be included in the list of essential elements. The total burden of Pb in adults ranges from 100 to 400 mg, with 90% in the skeleton. Wet tissues usually contain less Pb than 0.5 µg/g, hair 20 to 30 µg/g. Even people without a professional or environmental exposure ingest Pb with their daily nutriments. According to the WHO regulations, potable water should not contain over 100 µg/l

Pb. But normally the water levels are much lower, between 1 and 50 μg/l. Meat can contain up to 1 mg/kg (liver), plant food between 0.1 and 10 mg/kg. Luckily, only 5 to 10 % of orally ingested Pb is resorbed, the rest is excreted with the feces.

There is some disagreement about the "normal" blood Pb levels, since a large part of the population seems to suffer from "abnormal" environmental Pb exposure. On the basis of many analyses, carried out between 1960 and 1975, we consider concentrations between 100 and 200 μg/l blood as "normal" for an adult, and values up to 400 μg/l as acceptable in an urban and motorized environment. Higher concentrations are danger signals. Intoxication symptoms are to be expected from 800 μg/l on up. Only half of the stated values should be tolerated in children. Thus, the present "normal" and the toxic concentrations are not far apart; the safety margin is much smaller than with most other poisons.

The Pb concentration in urine is lower than in blood. The daily urinary elimination lies in the range of 15 to 100 μg. Higher values indicate dangerous exposures. In cases of chronic lead poisoning, daily excretions may not exceed 100 μg, but increase on administration of the complexing agent EDTA.

2.10.16.2 Toxicity of lead

In earlier years, lead was probably the most dangerous industrial poison. It caused severe intoxications in the painting trade, in factories producing batteries and other lead-containing products, as well as in the printing business. Metallic lead dust and vapor from heating the liquid metal over 1500 °C have been responsible for many intoxications. Today, most of the Pb poisoning sources are identified and could, thanks to a very strict analytical control of workmen and work places, be eliminated. But the metal still constitutes an appreciable environmental hazard, largely on account of the wide use of $Pb(C_2H_5)_4$ and $Pb(CH_3)_4$ as gasoline additives. Car exhaust contains the metal mostly as fine dust of $PbCl_2$. Although Pb-containing paints have been banned in many countries, the excellent white Pb cover paint with PbO has by far not been eliminated. It was never completely abandoned and is still used for painting houses, garden furniture or children's toys. This is dangerous especially for small children, who often chew on wooden toys or on the boards of the window frames.

Lead intoxications have also been caused by oral ingestion of lead with food or drink. In the Middle Ages, lead-salts ("lead sugar") have been used to sweeten wine. In the ceramic trade, colorful lead-containing glazings have become very popular. Not too long ago, such potteries could be found in Swiss households, often souvenirs from trips to southern countries. The glazings dissolve in vinegar and in acid drinks (i.e. wine, cider), if kept in such containers even for a short time.

Colics are the main symptoms of lead intoxication. This is reflected in the saying "the painter's colic". Other symptoms are paleness and general weakness. Chronic exposures over long periods can lead to encephalopathia with paralysis, especially in children. Statistical surveys indicate further that the probability of contracting tuberculosis is higher than the average in workmen with professional lead exposure.

Pb is an enzyme blocker. It interferes with the biosynthesis of heme by inhibiting the three enzymes aminolevulinate-dehydratase, coproporphyrin-decarboxylase and ferrochelatase (185).

The inhibition of aminolevulinate-dehydratase results in a decrease of its activity in erythrocytes and an increase of urinary delta-aminolevulinate. Inhibition of coproporphyrin-decarboxylase produces an elevation of urinary coproporphyrin, and inhibition of erythrocyte ferrochelatase has the effect of increasing erythrocyte protoporphyrin.

With the development of modern assay techniques, lead can be adequately determined both in blood and urine. Determination of blood Pb provides information on acute intoxications shortly after exposure. However, after some time, blood Pb concentration may appear normal, while toxic concentrations are still present in storage tissues, and the effects on heme biosynthesis are still evident. For identifying subjects who have been severely exposed in the past, and for monitoring chronic intoxication, it is necessary to assay the indicators of heme biosynthesis enzyme inhibition. The most sensitive of these are erythrocyte aminolevulinate dehydratase and erythrocyte protoporphyrin, the latter being the more practical to perform (186).

2.10.16.3 Indirect screening for lead intoxication

2.10.16.3.1 Urinary delta-aminolevulinate

Liquid chromatography with fluorescence detection represents a sensitive means of assaying this compound (187). The increase observed shortly after exposure correlates with that of blood Pb. Past exposure is not detected with this assay (186).

2.10.16.3.2 Delta-aminolevulinate dehydratase in erythrocytes

This is a very sensitive test to detect the inhibition produced **in vivo** by Pb. Even the intoxication by tetraalkyl lead, which has little effect on blood Pb and urinary delta-aminolevulinate, produces a significant decrease of the activity of this enzyme (187). Among others, an European standardized test may be cited (188). These assays, however, are not simple enough for screening purposes.

2.10.16.3.3 Erythrocyte protoporphyrin

Protoporphyrin X and its zinc chelate are fluorescent compounds occurring in increased concentration in the erythrocytes after lead intoxication. Their determination provides a reliable index of intoxication, even after past exposure.

A simple method consists of measuring the fluorescence of a diluted haemolysate in a spectrofluorimeter (189). It is also possible to measure the surface fluorescence of non-hemolyzed blood in a specially designed hemato-fluorimeter (190). These methods are quite practical and allow for the screening of large groups of subjects. They are not entirely specific, as other fluorescent compounds, for example in hyperbilirubinemia, may interfere. The most reliable method is the reverse-phase chromatography of porphyrins with fluorimetric detection of zinc protoporphyrin (191, 192).

Since the fluorescence of porphyrins is red, it is important to use a detector sufficiently sensitive in the red region of the spectrum. If the fluorimeter response is poor, it may be advantageous to substitute a red-sensitive photomultiplier for the standard one.

2.10.16.4 Direct chemical determination of lead

For a long time, Pb analyses were carried out colorimetrically after complexing the metal with dithizone and extraction with an organic solvent (193, 194) or by polarography (195). Today, these methods have generally been replaced by atomic absorption.

The detection limit of flame AA for Pb in urine lies between 50 and 100 µg/l. This permits revealing toxic concentrations, but not to determine subtoxic levels. A considerable number of methods using organic extraction of Pb complexes followed by AA have been described (196–198). But such concentration steps complicate an analysis and introduce additional error possibilities. Already before the development of the graphite oven technique, special "flame assisted microsampling" techniques have been introduced (199, 200). They permit Pb determination with 10 or 20 µl blood and have been used successfully, but have been superseded later by the more general flameless electrothermal evaporation method in graphite ovens (201–203). With a simple tenfold dilution with a Triton X–100 solution and by using Zeeman background compensation and the L'vov platforms, it has been possible to obtain a determination limit of about 1.5 µg/l with an excellent reproducibility.

Table 6 shows that better sensitivity than with GF-AAS can only be obtained by DPV (6) (for urine analysis only) and ICP-MS (204).

2.10.17 Arsenic

2.10.17.1 General remarks

Arsenic is one of the best known criminal poisons. It has been used for homicides already in antiquity, then in the Middle Ages and in modern times (205). Crimes committed with this poison have influenced the path of history. But the semi-metal constitutes also an environmental danger. Today, this toxicological aspect of As and its derivatives is probably a more important problem than its criminal abuse. Environmental and occupational contacts are possible with a number of arsenic compounds with basically different physical, chemical and toxicological properties. Some possess very high, others (some organo-As-derivatives) only low human toxicity.

The difficulties in the selection of the analytical method and the interpretation of results are the reasons why our knowledge of the human toxicity and metabolic transformation of some As compounds is far from complete, and that it has not always been sufficiently well established which forms of As are measured by a chosen analytical technique.

2.10.17.2 Sources and toxicity of arsenic

As is widely distributed in our biosphere. It occurs in normal soils at levels between 1 and 40 mg/kg, in seawater at levels of several µg/l, and also in public water supplies and in air. Most plants do not absorb much As from the soil. However, exceptions do exist. The average human foodstuffs contain less As than 0.2 mg/kg, but here also, we can find many exceptions. Fish from the sea may have As contents between 2 and 8 mg/kg fresh weight, oysters, shrimp and seaweed usually much more.

The daily As ingestion in humans depends largely upon the diet of a person (amount and proportion of sea food). It can also greatly be influenced by the As content of the water supplies. It is general knowledge that in the Austrian region of Steiermark, the As content in the bodies of the population is above the mean. We have also found a case in the Grisons region of the Swiss alps with well water containing an elevated As level. This was reflected in the As content in the bodies of the people using the water from this well.

The limit for As in the US and Canada drinking water has been set at 50 μg/l. It was assumed that the water contains mainly inorganic arsenite and arsenate. If such strict limits would be applied to seafood, most of it would be unfit for consumption, since it can contain 100 or 1000 times higher As concentrations. However, the main part of this As is present in relatively low-toxic organic forms such as arseno-choline and arseno-betaine (in crabs, lobsters, shrimp and fish), or as arseno-sugars (in seaweed). Arseno-betaine for example is excreted rapidly and unchanged with urine, and no toxic effects have been recorded (206, 207).

Industrial and agricultural uses of As compounds can also result in human exposure, especially in occupations such as mining, smelting, glass making and pesticide manufacture. Exposure to arsenate arises from its use as an insecticide and a wood preservative. Salts of the semi-metal are often found in paints, wallpapers, ceramics, weed killers and ant poisons. Exposures to monomethyl-arsonic and dimethyl-arsinic acids result from their roles as herbicides. In animal trials, these derivatives have shown much lower toxicity than inorganic As. An approximate listing of the As compounds in a descending order of toxicity may look as follows: arsine, arsenite, arsenate, monomethyl-arsonic acid, dimethyl-arsinic acid, arseno-choline and arseno-betaine (208, 209). For inorganic compounds, 100 mg of As_2O_3 may already constitute the minimal lethal dose for a human adult (104). However, for metabolically stable organic derivatives like arseno-betaine or arseno-choline, this dose may be thousand times higher.

Symptoms of acute As poisoning may start like a gastroenteritis with vomiting and bloody diarrhea, followed by a fall of blood pressure with general weakness. With lethal doses, convulsions, coma and death from respiratory failure follow. With highly toxic but sub-lethal doses, a paralysis may develop and persist over weeks or months.

Chronic As exposures may affect the central nervous system (polyneuritis), lead to gastrointestinal troubles (nausea, vomiting, abdominal cramps, liver cirrhosis), to dermatitis (edema, skin-bronzing), anemia and weight loss.

2.10.17.3 Arsenic in the human body

There is considerable confusion concerning "normal" arsenic contents of the human body. Values ranging from 0.01 to 0.6 μg/g have been reported for whole blood. In a large study by NAA, carried out with blood of English adults, a mean of 0.2 μg/g wet weight was established (210). Blood As seems to be concentrated mainly in the red cells.

Urinary excretion is one possibility to check for abnormal exposures to the poison. In our experience, daily As eliminations up to 100 μg can be considered as normal. Higher As levels indicate unusual exposures. Quite often, they can be traced to food intake or water supply. In exposed workmen, urinary excretions up to 5 mg/l could be found without any evidence of intoxication symptoms (211). It has been repeatedly established that a biotransformation of inorganic As to monomethyl-arsonic acid and dimethyl-arsinic acid occurs in the body. It has been called a detoxification (212). Typically, 60–80% of the arsenic in human urine is dimethyl-arsinic acid, the rest approximately half inorganic and half monomethyl-arsonic acid (213, 214).

There does not seem to be a marked accumulation of As in internal organs or tissues of humans (215). On the other hand, it accumulates in the skin and especially in hair and nails. In a study of over 1000 hair samples, taken at random from living subjects, As concentrations ranging from 0.03 to over 70 μg/g could be found with a mean of 0.81 μg/g (216). It has been suggested that As contents in hair of over 2 μg/g should be considered as suspect and further examined. It has furthermore been established that the content of As in hair roots increases immediately after intake of the element, and that the levels return to normal a few days after cessation of exposure(216). This makes hair an ideal material for detecting abnormal As contacts, especially since hair analysis allows determination of approximate time and duration of an exposure. For such an investigation, a bundle of hair must be carefully tied together, cut off near the skin or pulled out, divided into segments containing hair from the same growth periods, and each segment separately analyzed. Comparision of the concentrations in the different segments permits the estimation of time and length of exposure period(s).

2.10.17.4 Analytical possibilities

As mentioned in chapter 2.10.2, the Reinsch test can be used as a simple qualitative screening method for the presence of As in body fluids (1, 2). It does not require a previous combustion, as is necessary for the analytical determinations according to Gutzeit or Marsh (1). Decomposition of the organic materials can be carried out by wet-oxidation with nitric and sulfuric acids, without or with addition of perchloric acid or perhydrol. Dry decomposition after addition of $MgNO_3$ and MgO to prevent evaporation losses is also possible. Recently, microwave oven combustion has become the preferred method for decomposing As-containing organic materials (217). For hair and nails, oxygen flask combustion (182) can also be used.

Both Gutzeit and Marsh liberated As from the decomposition mixtures as arsine. Gutzeit led the gas over filter paper treated with $HgBr_2$, on which arsenic precipitates as a yellowish-brown mercury compound. Marsh isolated arsenic as a metallic mirror by thermal decomposition, while the arsine flows through a glass capillary. Both methods are at best only semi-quantitative, since the evaluation of results depends on visual inspection of color intensity or size of the metallic mirror. More quantitative results are obtained by absorbing AsH_3 in a solution of diethyldithiocarbamate and measuring the resulting color complex spectrophotometrically (218).

A direct assessment of As in urine samples, but not in other body fluids, is possible with DPV (6). In spite of its reasonable sensitivity (Table 6), this method does not seem to be widely used. Most of the reliable quantitative results were obtained by NAA (219), which has been for a long time the most sensitive method available for determining As. It may still be the method of choice, if only a single hair has to be analyzed, and if time and duration of a possible exposure to As must be determined.

The measurement of As by flame-AA is difficult. Arsenic resonance radiators are relatively poor and the absorption lines lie below 200 nm (189.0, 193.7, 197.2 nm), in a region with a strong self-absorption of air and flame gases. An improvement of flame transparency is possible by replacing the air-acetylene with an argon-hydrogen flame, but it lowers the sensitivity not over a factor of 5 to a value of approximately 0.2 μg/ml (table 6). This does not permit toxicological investigations. With graphite furnace AA, As concentrations down to 0.5 μg/l can be detected in aqueous solutions (220), but a much better choice is a combination of the hydride generation technique with flameless AA in a heated flow-through tube, which permits detection of even 10 times lower As concentrations (221, 222). This is the method most often used today in food and environmental analysis as well as toxicology.

2.10.17.5 Determination of arsenic and other hydride-forming elements by GC-MS

Already in 1974 (223), we have combined the hydride generation technique with an analysis of the AsH_3, released by direct reduction of urine or a solution obtained by decompostion of a body fluid or tissue, with GC-MS. The reaction vessel is a small serum flask. After introducing the body fluid and HCl, it is capped with a rubber or silicone septum. A sodium borohydride solution is injected as reducing agent. After several minutes, head space gas aliquots can be removed with a gas syringe and injected into the GC-column of the GC-MS unit (for example a column with Chromosorb 103). Figure 2 illustrates the extremely simple instrumentation which is used for this procedure. Figure 4 shows two mass spectra of AsH_3, recorded at 70 and 20 eV. Only four ions are formed. All can be incorporated in a mass specific detection system. If the relative heights of the four recorded peaks correspond to the relative peak intensities in the mass spectrum of AsH_3, its presence is confirmed with the same certitude as with a registration of the mass spectrum, but with a sensitivity gain of several factors of 10 (for sector instruments, less for quadrupole spectrometers and even less for ion traps). For internal standardization, a Ge salt can be used. It is reduced to GeH_4, which elutes (i.e. from a column with Chromosorb 103) shortly before AsH_3. Its mass spectrum (fig. 5) contains some of the same ions as that of AsH_3, but with different relative intensities, which is ideal for revealing the two hydrides side by side. We have extended the method also to the determination of the other hydride-forming semi-metals Ge, Sn and Sb (182), as described in chapter 2.10.15.4. If no mass spectrometer should be available, a search for AsH_3 and SbH_3 is also possible by GC with detection by photoionization (224). This combination can reveal As in the low ng-range, while GC-MS is able to detect As in the lowest pg-range.

ARSINE AsH_3

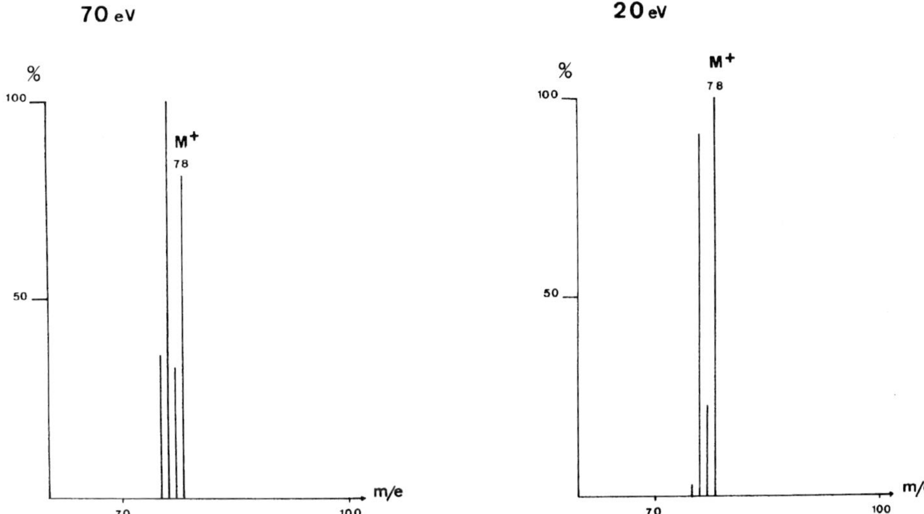

Figure 4. Mass Spectra of Arsine at 70 eV and 20 eV.

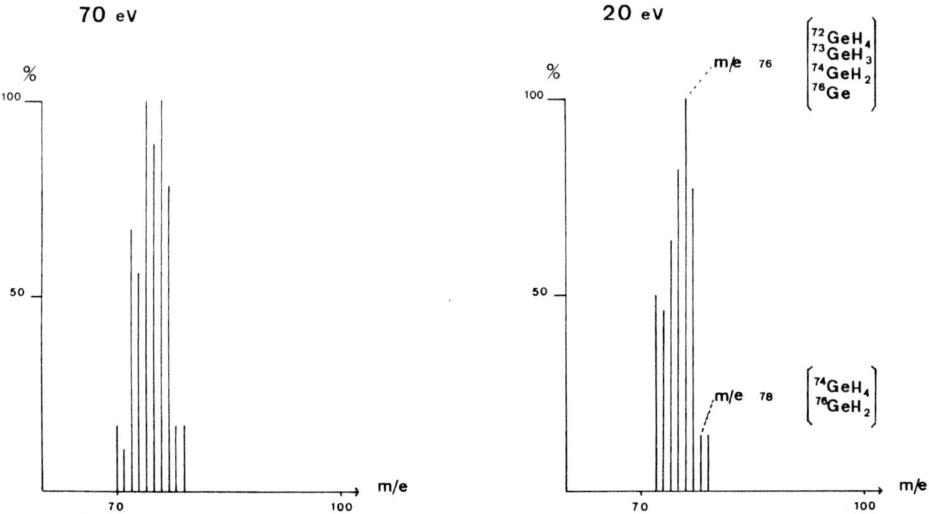

Figure 5. Mass Spectra of Germanium Hydride at 70 and 20 eV.

2.10.17.6 Differentiation of individual compounds of arsenic

A disadvantage of all methods discussed so far is that they do not permit to identify, side by side, different forms of As: inorganic As in different oxidation states and different types of organo-As-compounds. With the hydride generation techniques, for example, we can determine the sum of inorganic As plus methyl-arsonic acid, dimethyl-arsinic acid and some other urinary As metabolites which are reduced by sodium borohydride, but not specify them. The more stable arseno-betaine and arseno-choline are not reduced by the sodium borohydride solution. For a determination of the total As content, a combustion step is required. The problem could only recently be solved by combining HPLC with ICP-MS to an on-line detection method (225). This analytical combination permits revelation of the different urinary metabolites after ingestion of sea food such as crabs and shrimps, which contain arseno-betaine as main arsenic species, as well as seaweed, which contains mostly arseno-sugars. A recent investigation (226) aspires to the same goal by combining HPLC with hydride generation and ICP-MS. From our experiences with mass spectrometry of As derivatives, it must be expected that a combination of HPLC with MS could serve the same purpose.

2.10.18 Antimony

2.10.18.1 Occurrence, toxicological and medical importance

Like arsenic, antimony (Sb) can be considered as metal or metalloid. Elemental Sb is a silvery-grey material. It expands on solidifying. This property has been exploited in the lead-antimony alloys used in printing type. Sb can also be found in accumulators, ammunition and semiconductors. The trioxide, Sb_2O_3, is a white pigment which has been used in paints (227). Sb preparations play a considerable role in the pyrotechnique.

Toxic effects of antimony have been observed in miners and smelters. Inhalation of dust or exposure to vapors of heated alloys are the main causes of Sb intoxications. The effects are observed on the upper respiratory tract. Gastrointestinal irritation has also been reported. While orally ingested elemental Sb does not possess high toxicity, some of its salts do (228), sulfides more so than oxides, and Sb^{+3}-salts more than Sb^{+5}-salts. SbH_3 has a similar, but much weaker, action than AsH_3. Intoxications with Sb-salts resemble those of As-salts. It has been claimed (228) that antimony potassium tartrate (Brechweinstein, tartar emetic, Stibunal) gives the same toxic symptoms as As_2O_3, and that already 10 mg can be toxic and 100 mg lethal.

In medicine, Sb compounds have been used for centuries, tartar emetic and stibocaptate formerly for treating schistosomiasis. At present, pentavalent Sb, as Na stibogluconate or meglumine antimoniate, is the standard treatment for leishmaniases (229). The clinical symptoms observed on patients receiving organic Sb compounds are quite different from the effects of inorganic Sb salts. They consist in the induction of a pancreatitis, characterized by the elevation of plasma amylase and lipase, and they disappear upon discontinuing the treatment (229).

2.10.18.2 Analysis

Sb in the human lung can be determined non-invasively by X-ray spectrometry (227). In urine, Sb ions can be directly measured by DPV(6) respectively anionic stripping voltametry (230). Inorganic pentavalent Sb is extractable into organic solvents. It forms colored complexes with Rhodamine B (231, 232) or Brilliant Green (233). But they are not very specific; other elements form similar complexes which absorb in the same spectral range. The colorimetric procedures have therefore largely been replaced by GF-AAS, Hy-AAS, Hy-ICP-AES, Hy-ICP-MS and by Hy-MS (Table 6).

Prior to the determination by GF-AAS, Sb in biological materials must be concentrated. This is usually accomplished by extraction with organic solvents (234, 235). AAS of SbH_3 (221, 222) is somewhat more sensitive than GF-AAS. Therefore, in some cases, the preconcentration step may not be needed. An interesting approach is the combination of hydride generation with ICP-AES (236–238).

By using our Hy-MS approach, as in the determination of As, Ge and Sn (see corresponding sub-chapters and references 182, 223, 224), we have obtained excellent results with detection limits lower than those of other instrumental techniques except ICP-MS. The reduction of Sb ions to stibine (SbH_3) is carried out in presence of Sn ions for internal standardization. The gas phase of the reaction flask is transferred to a GC-MS unit and the gas chromatogram monitored by recording the ions with m/z 122, 123 and 124, originating from SbH_3 and SnH_4 (182). Alternatively, GC with detection by photoionization can also be used (224), as long as top sensitivity is not essential.

2.10.18.3 Body levels

The daily uptake of Sb by the human body (from food and dust) is 20 to 35 μg (239, 240), but higher values have also been reported for the US. Without a special exposure (Sb-containing dust, fumes or medicaments), the Sb concentration in urine is usually below 1 μg/l and not over 3 μg/l (240). With professional exposures, up to 1 mg/l could be found (241), up to 2 mg/l under the influence of Sb containing medicaments (242).

2.10.19 Bismuth

Bismuth has been used since the nineteenth century as an oral therapeutic agent for the treatment of gastrointestinal and other disorders. It is a constituent of certain alloys used in industry. The prolonged intake of some bismuth salts such as the subcitrate and the subgallate may produce severe neurological toxicity symptoms, and for this reason the use of the older bismuth-containing drugs has declined. More recently, new compounds such as bismuth subcitrate have proved effective in the treatment of peptic ulcer caused by the bacterium *helicobacter pylori*. These new drugs have the advantage of being poorly absorbed by the gastrointestinal tract, with the consequence of a lower toxicity (243).

Bismuth assay in blood and urine is therefore a valuable means to investigate the toxicity of bismuth-containing drugs and to assess the severity of accidental poisoning. After cessation of bismuth intake, the toxic effects are generally reversible.

Bismuth is usually assayed by atomic absorption. Methods differ as to the way of eliminating the matrix effect in the analysis of blood. The method using hydride generation (244) is a sensitive and reliable one. A method which does not require the equipment necessary for hydride generation has been proposed by Dean et al. (245). Protein precipitation and a palladium matrix modifier reduce the problem of foaming.

In laboratories equipped with inductively coupled plasma mass spectrometry, this represents a convenient way of determining bismuth in biological samples (246). The sensitivity is such that it allows for establishing the reference range in persons who do not ingest bismuth preparations (247).

2.10.20 Review and Preview

During the last half century, we could witness a spectacular development of the possibilities for analyzing metal ions in general and for the toxicological detection and quantitative determination of trace metals in biological materials in special. 50 years ago, wet chemical methods combined with colorimetric and electrochemical measurements dominated the field, while radioisotope and flame emission techniques were used only occasionally for special applications. The 25 years after 1960 have been characterized by a triumphant advance of atomic absorption analysis, first with flames as atomizing units, later with flameless techniques such as thermic atomization in furnaces, cold-vapor atomization for Hg, and hydride methods for some semi-metals. During the same period, flame emission, as a leading sequential method of metal analysis, was substituted by the more sensitive plasma emission techniques.

In the past 10 years, another development has been initiated. Metal analyses based on mass spectrometric techniques are rapidly gaining ground, especially (but not exclusively) in the form of ICP-MS. They hold a great deal of promise, since:

- they are not limited to the determination of metal ions, but also can measure most non-metals,
- they usually permit attainment of the lowest detection limits, even considerably below those of graphite furnace AAS,
- they can be used as sequential techniques for metal identification (metal screening), as well as for selective quantification of single elements,
- they open the field for the internal standardization by stable (non-radioactive) isotopes, the ideal internal standards that ensure highest accuracy,
- they are well suited (better than most other techniques) for the combination with separation methods like HPLC, capillary electrophoresis or GC, which is a prerequisite for differentiating between various species of a single metal,
- in some cases they are also capable, on the basis of the isotope distribution pattern, of shedding some light on the origin of a specific element (which can be important in the field of forensics).

It must be expected that also in the near future the clinical analysis of some metals which are present in relatively large concentrations (Ca, Mg and Fe) will still be carried out by the highly-automated colorimetric procedures in general use in clinical laboratories, maybe also by one or two measurements (Li) with ion sensitve electrodes, since these methods are inexpensive, fast and not labor-intensive. In typical toxicological laboratories, the practical AAS with its high accuracy will certainly remain for a long time the analytical tool of choice for single metal determinations. To obtain the lowest detection limits, however, as well as for sequential metal screening, ICP-MS and other mass spectrometric methods will enter the field. They will be used increasingly in combination with separation techniques, and this will extend our knowledge of the toxicity of the different species of single metals, and provide better information on the metabolic fate of elements which have entered the body by different pathways.

References

1. Leifheit, H.C. (1961) Arsenic in biological materials. In: Standard Methods of Clinical Chemistry *3*, 23–34. Academic Press, New York & London.
2. Yeoman, W.B. (1986) Metals and Anions, in: Clarke's Isolation and Identification of Drugs, 2nd ed. (Moffat, A.C., Jackson, J.V., Moss, M.S. & Widdop, B., eds.) pp. 55–69. The Pharmaceutical Press, London.
3. Curry, A.S. (1976) Poison Detection in Human Organs, 3rd ed. Charles C. Thomas, Springfield, Illinois.
4. Feigl, F. (1954) Spot Tests, Vol. I, Inorganic Applications, pp. 97, 99 & 348. Elsevier Publishing Company, Amsterdam, Houston, London & New York.
5. Douglas, D.J. & Houk, R.S. (1985) Inductively-coupled plasma mass spectrometry (ICP-MS). Prog. analyt. atom. Spectrosc. *8*, 1–18.
6. Daldrup, T. & Franke, J.P. (1993) Metallscreening aus Urin bei akuten Vergiftungen. Mitteilung XXII der Senatskommission für klinisch-toxikologische Analytik der DFG (Deutsche Forschungsgemeinschaft). VCH Verlagsgesellschaft, Weinheim.
7. Stahl, E. (1967) Dünnschichtchromatographie. Edition Springer, Berlin.
8. Fritz, J., Gjerde, D. & Pohlandt, Ch. (1982) Ion Chromatography. A. Huethig, New York.
9. Smith, F.C. & Chang, R.C. (1983) The Practice of Ion Chromatography, pp. 183–192. John Wiley & Sons, New York.
10. We are greatly indebted to Mr. R. Trost from Dionex, Olten, Switzerland, for furnishing advance information.
11. Engelhardt, H., Beck, W., Kohr, J. & Schmitt, T. (1993) Capillary electrophoresis: methods and scope. Angew. Chem. *32*, 629–649.
12. Huang, X., Pang, T.J., Gordon, M.J. & Zare, R.N. (1987) On-column conductivity detector for capillary zone electrophoresis. Anal. Chem. *59*, 2747–2749.
13. Haber, C., Silvestri, I., Röösli, S. & Simon, W. (1991) Potentiometric detector for capillary zone electrophoresis. Chimia *45*, 117–121.
14. Saitoh, K., Kiyohara, C. & Suzuki, N. (1991) Mobilities of metal-diketonato complexes in micellar electrokinetic chromatography. HRC *14*, 245–248.
15. Swaile, D.F. & Sepaniak, M.J. (1991) Determination of metal ions by capillary zone electrophoresis with on-column chelation using 8-hydroxyquinoline-5-sulfonic acid. Anal. Chem. *63*, 179–184.
16. Beck, W. & Engelhardt, H. (1992) Capillary electrophoresis of organic and inorganic cations with indirect UV detection. Chromatographia *33*, 313–316.
17. Beck, E. & Engelhardt, H. (1993) Separation of non UV-absorbing cations by capillary electrophoresis. Fresenius' Jour. of Anal. Chem. *346*, 618–621.
18. Beck, W. (1993) Verbesserung der Detektionsmöglichkeiten in der Kapillarelektrophorese. Diss. Univ. Saarbrücken.
19. Yeung, E.S. & Kuhr, W.G. (1991) Indirect detection methods for capillary separations. Anal. Chem. *63*, 275A–282A.
20. Goddard, J. and Bloom, S.R. (1991) Lithium intoxication. Brit. Med. J. *302*, 1267–1269.

21. Perrier, A., Martin, P.-Y., Favre, H. et al. (1991) Very severe self-poisoning lithium carbonate intoxication causing a myocardial infarction. Chest *100*, 863–865.
22. Friedman, B.C., Bekes, C.E., Scott, W.E. et al. (1992) ARDS following acute lithium carbonate intoxication. Intensive Care Med. *18*, 123–124.
23. Friedberg, R.C., Spycher, D.A. and Herold, D.A. (1991) Massive overdoses with sustained-release lithium carbonate preparations: pharmacokinetic model based on two case studies. Clin. Chem. *37*, 1205–1209.
24. Szerlip, H.M., Heeger, P. and Feldman G. (1992) Comparison between acetate and bicarbonate dialysis for the treatment of lithium intoxication. Am. J. Nephrol. *12*, 116–120.
25. Levy, A.L. and Katz, E.M. (1970) Comparison of serum lithium determination by flame photometry and atomic absorption spectrophotometry. Clin. Chem. *16*, 840–842.
26. Robertson, R., Fritze, K. and Grof, P. (1973) On the determination of lithium in blood and urine. Clin. Chim. Acta *45*, 25–31.
27. Frankfurt, E., van den Bergh, P.J., Drenth, J.P., et al. (1993) Erroneous diagnosis of lithium poisoning due to the use of lithium-heparin blood letting tubes. Ned. Tijdschr. Geneesk *137*, 1017–1018.
28. Okorodudu, A.O., Burnett, R.W., McComb, R.B. et al. (1990) Evaluation of three first-generation ion-sensitive electrode analyzers for lithium: systematic errors, frequency of random interferences and recommendations based on comparison with flame atomic emission spectrometry. Clin. Chem. *36*, 104–110.
29. Bertolf, R.L., Savory, G., Winborne, K.H. et al. (1988) Lithium determined in serum with an ion-selective electrode. Clin. Chem. *34*, 1500–1502.
30. Hurlbult, J.A. (1978) Determination of beryllium in biological tissues and fluids by flameless atomic absorption spectroscopy. Atomic Absorption Newsletter *17*, 121–124.
31. Slavin, W. & Carnrick, G.R. (1985) A survey of applications of the stabilized temperature platform furnace and Zeeman correction. Atomic Spectroscopy *6*, 157–160.
32. Paschal, D.C. & Bailey, G.G. (1986) Determination of beryllium in urine with electrothermal atomic absorption using the L'vov platform and matrix modification. Atomic Spectroscopy *7*, 1–3.
33. Luo, F., Li, X. & Wang, M. (1990) The study of modifiers for the determination of beryllium by graphite furnace atomic absorption spectrometry. Atomic Spectroscopy *11*, 78–82.
34. Grewal, D.S. & Kearns, F.X. (1977) A simple and rapid determination of small amounts of beryllium in urine by flameless atomic absorption. Atomic Absorption Newsletter *16*, 131–132.
35. Geldmacher-von Mallinckrodt, M. (1976) Einfache Untersuchungen auf Gifte im klinisch-chemischen Laboratorium, pp. 184–185. Georg Thieme, Stuttgart.
36. Slavin, W. (1968) Atomic Absorption Spectroscopy. Interscience (Division of John Wiley), New York, London, Sydney.
37. Willis, J.B. (1959) Determination of magnesium in blood serum by atomic absorption spectroscopy. Nature *184*, 186–187.
38. Willis, J.B. (1960) The determination of metals in blood serum by atomic absorption spectroscopy II: Magnesium. Spectrochim. Acta *16*, 273–278.
39. Willis, J.B. (1961) Determination of calcium and magnesium in urine by atomic absorption spectroscopy. Anal. Chem. *33*, 556–559.
40. Peterson, G. (1966) Low-sensitivity analysis for calcium and magnesium. Atomic Absorption Newsletter *5*, 117–118.
41. Willis, J.B. (1960) Determination of calcium in blood serum by atomic absorption spectroscopy. Nature *186*, 249–250.
42. Willis, J.B. (1960) The determination of metals in blood serum by atomic absorption spectroscopy I: Calcium. Spectrochim. Acta *16*, 259–272.
43. Trudeau, D.L. & Freier, E.F. (1967) Determination of calcium in urine and serum by atomic absorption spectrophotometry. Clin. Chem. *13*, 101–114.
44. Gimblet, E.G., Marney, A.F. & Bonsnes, R.W. (1967) Determination of calcium and magnesium in serum, urine, diet and stool by atomic absorption spectrophotometry. Clin. Chem. *13*, 204–214.
46. Bek, F., Janouskova, J. & Moldan, B. (1974) Determination of manganese and strontium in blood serum using the Perkin-Elmer HGA–70 graphite furnace. Atomic Absorption Newsletter *13*, 47–48.
46. Rollemberg, M.C. & Curtius, A.J. (1982) Flameless atomic absorption determination of barium in natural waters using the

technique of standard additions. Mikrochim. Acta *2*, 441–447.

47. Subramanian, K.S. & Méranger, J.C. (1984) A survey for sodium, potassium, barium, arsenic and selenium in Canadian drinking water supplies. Atomic Spectroscopy *5*, 34–35.
48. Zhuang, M. & Barnes, R.M. (1985) Determination of major, minor and trace elements in human serum by using inductively coupled plasma-atomic emission spectroscopy. Appl. Spectrosc. *39*, 793–796.
49. Schaller, K.H., Essing, H.G., Valentin, H & Schäcke, G. (1973) The quantitative determination of chromium in urine by flameless atomic absorption spectroscopy. Atomic Absorption Newsletter *12*, 147–150.
50. Dick, A.T. (1956) in: Inorganic Nitrogen Metabolism (W.D. McElroy & Glass, B., eds.) p.445. John Hopkins Press, Baltimore, Maryland.
51. Rosenstock, H.A., Simmons, D.G. & Meyer, J.S. (1971) Chronic manganism. J. Am. Med. Assoc. *217*, 1354–1358.
52. Lidums, V.V. (1979) Determination of cobalt in blood and urine by electrothermal atomic absorption spectrometry. Atomic Absorption Newsletter *18*, 71–72.
53. Kesteloot, H., Roelandt, J., Willems, J. & Claes, J.H. (1968) Inquiry into the role of cobalt in the heart disease of chronic beer drinkers. Circulation *37*, 854–864.
54. Wellmann, K.F. (1968) Bière au cobalt. Dtsch. Med. Wochenschr. *93*, 500–501.
55. Wirth, W. & Gloxhuber, Ch. (1981) Toxikologie, p.115. Georg Thieme, Stuttgart & New York.
56. Völlkopf, U., Grobenski, Z. & Welz, B. (1981) Determination of nickel in serum using graphite furnace atomic absorption. Atomic Spectroscopy *2*, 68–70.
57. Jones, C.C. (1973) Nickel carbonyl poisoning. Arch. Environm. Health *26*, 245–248.
58. Ludewigs, H.J. & Thiess, A.M. (1970) Arbeitsmedizinische Erkenntnisse bei der Nickelcarbonylvergiftung. Zbl. Arbeitsmed. Arbeitsschutz *20*, 329–339.
59. Proudfoot, A.T., Simpson, D. and Dyson, E.H. (1986) Management of acute iron poisoning. Med. Toxicol. *1*, 83–100.
60. Cohen, A. (1987) Management of iron overload in the pediatric patient. Hamatol. Oncol. Clin. North Am. *1*, 521–544.
61. Iron Panel of the International Committee for Standardization in Haematology (1990) Revised recommendations for the measurements of the serum iron in human blood. Brit. J. Haematol. *75*, 615–616.
62. International Committee for Standardization in Haematology (1978) Recommendations for measurement of serium iron in human blood. Brit. J. Haematol. *38*, 291–294.
63. Goodwin, J.F., Murphy, B. and Guillemette, M. (1966) Direct measurement of serum iron and binding capacity. Clin. Chem. *12*, 47–57.
64. Solem, E., Seiffert, U.B., Jalner, U., Kaul, M. et al. (1988) Automated determination of serum (plasma) and urine iron: a comparative investigation of chromogens. Improved tripyridyltriazine micromethod. J. Clin. Chem. Clin. Biochem. *26*, 697–703.
65. Artiss, J.D., Vinogradov, S. and Zak, B. (1981) Spectrophotometric study of several sensitive reagents for serum iron. Clin. Biochem. *14*, 311–315.
66. Gevirtz, N.R. and Wassermann, L.R. (1966) The measurement of iron and iron-binding capacity in plasma containing deferoxamine. J. Pediat. *68*, 802–804.
67. Helfer, R.E. and Rodgerson, D.O. (1966) The effect of deferoxamine on the determination of serum iron and iron-binding capacity. J. Pediat. *68*, 804–806.
68. Zettner, A. and Mansbach, L. (1965) Application of atomic absorption spectrophotometry in the determination of iron in urine. Am. J. Clin. Pathol. *44*, 517–519.
69. Beilby, J., Olynyk, J., Ching, S., Prins, A. et al. (1992) Transferrin index: an alternative method for calculating the iron saturation of transferrin. Clin. Chem. *38*, 2078–2081.
70. Ressler, N. and Zak, B. (1958) Serum unsaturated iron-binding capacity. Am. J. Clin. Pathol. *30*, 87–90.
71. Bergström, K. and Lefvert, A.K. (1980) An automated turbidimetric immunoassay for plasma proteins. Scand. J. Clin. Lab. Invest. *40*, 637–640.
72. Buffone, G.J., Lewis, S.A., Iosefsohn, M. and Hicks, J.M. (1978) Chemical and immunochemical measurement of total iron-binding capacity compared. Clin. Chem. *24*, 1788–1791.
73. Pootrakul, P., Josephson, B., Huebers, H.A. and Finch, C.A. (1988) Quantitation of ferritin iron in plasma, an explanation for non-transferrin iron. Blood. *71*, 1120–1123.
74. Moeschlin, S. (1980) Klinik und Therapie der Vergiftungen, 6th Ed., pp.139–142. Georg Thieme, Stuttgart & New York.

75. Brandenberger, H. (1974) Heavy- and semi-metals of clinical and toxicological importance. In: Clinical Biochemistry. Principles and Methods (Curtius, H.Ch. & Roth, M., eds.), pp. 1584–1603. Walter de Gruyter, Berlin & New York.
76. Brandenberger, H. (1967) Moderne Methoden im Dienste der praktischen Medizin. Praxis *56*, 1302–1307.
77. Brandenberger, H. (1967) Résolution de quelques problèmes de chimie toxicologique et légale par absorption atomique. Ann. Biol. clin. *25*, 1053–1062.
78. Sass-Kortsak, A. (1965) Copper metabolism. In: Advances in Clinical Chemistry *8*, 1–67. Academic Press, New York & London.
79. Rice, E.W. & Singer, W.H. (1963) Ceruloplasmin assay in serum: Standardization of ceruloplasmin activity in terms of international enzyme units. In: Standard Methods of Clinical Chemistry *4*, 39–46. Academic Press, New York & London.
80. Rice, E.W. (1963) Copper in serum. In: Standard Methods of Clinical Chemistry *4*, 57–63. Academic Press, New York & London.
81. Berman, E. (1965) Application of atomic absorption spectrometry to the determination of copper in serum, urine and tissue. Atomic Absorption Newsletter *4*, 296–299.
82. Olson, A.D. & Hamlin, W.B. (1968) Serum copper and zinc by atomic absorption. Atomic Absorption Newsletter *7*, 69–71.
83. Sprague, S. & Slavin, W. (1965) Determination of iron, copper and zinc in blood serum by an atomic absorption method requiring only dilution. Atomic Absorption Newsletter *4*, 228–233.
84. Salmela, S. & Vuori, E. (1984) Improved direct determination of copper and zinc in a single serum dilution by atomic absorption spectrophotometry. Atomic Spectroscopy *5*, 146–149.
85. Wawschinek, O. & Höfler, H. (1979) Diagnosis of Wilson's disease through determination of copper in serum and urine by flameless atomic absorption. Atomic Absorption Newsletter *18*, 97–98.
86. Harrington, D.E., Jones, J.S. & Bramstedt, W.R. (1983) Determination of aluminium, copper and zinc in human hair. Atomic Spectroscopy *4*, 176–178.
87. Underwood, E.J. (1977) Trace Elements in Human and Animal Nutrition. 4th Ed., p. 444. Academic Press, New York, San Francisco & London.
88. Möschlin, S. (1980) Klinik und Therapie der Vergiftungen. 6th Ed., pp. 134–139. Georg Thieme Verlag, Stuttgart & New York.
89. Sunderman, F.W. (1979) Trace elements. In: Chemical Diagnosis of disease (Brown, S.S., Mitchell, F.L. & Young, D., eds.), pp. 1009–1038. Elsevier, Amsterdam.
90. Auterhoff, H. (1978) Lehrbuch der Pharmazeutischen Chemie, 9th Ed., p. 94. Wissensch. Verlagsgesellschaft, Stuttgart.
91. Wirth, W. & Gloxhuber, Ch. (1981) Toxikologie für Aerzte, Naturwissenschaftler und Apotheker, 3rd Ed., pp. 119–120. Georg Thieme, Stuttgart & New York.
92. Friberg, L., Piscator, M. & Nordberg, G. (1971) Cadmium in the Environment. The Chemical Rubber Co., USA.
93. Sandell, E.B. (1959) Colorimetric Determination of Trace Metals, pp. 350–365. Interscience, New York.
94. Robards, K. & Worsfold, P. (1991) Cadmium: Toxicology and Analysis; a review. Analyst *116*, 549–568.
95. Berman, E. (1967) Determination of cadmium, thallium and mercury in biological materials by atomic absorption. Atomic Absorption Newsletter *6*, 57–60.
96. Hauser, T.R., Hinners, T.A. & Kent, J.L. (1972) Atomic Absorption determination of cadmium and lead in whole blood by a reagent-free method. Anal. Chem. *44*, 1819–1821.
97. Delves, H.T. (1970) A micro-sampling method for the rapid determination of lead in blood by atomic absorption spectrometry. Analyst *95*, 431–438.
98. Ediger, R.D. & Coleman, R.L. (1973) Determination of cadmium by the Delves cup technique. Atomic Absorption Newsletter *12*, 3–6.
99. Delves, H.T. & Woodward, J. (1981) Determination of low levels of cadmium in blood by electrothermal atomization and atomic absorption spectrophotometry. Atomic Spectroscopy *2*, 65–67.
100. Claeys-Thoreau, F. (1982) Determination of low levels of cadmium and lead in biological fluids with sample dilution by atomic absorption spectrophotometry using Zeeman effect background correction and the L'vov platform. Atomic Spectroscopy *3*, 188–191.
101. Magos, L. (1967) Mercury-blood interaction and mercury uptake by the brain after vapor exposure. Environ. Res. *1*, 323–337.

102. Berlin, M., Fazackerley, J. & Nordberg, G. (1969) The uptake of mercury in the brains of mammals exposed to mercury vapor and mercury salts. Arch. envir. Health *18*, 719–729.

103. Ramazzini, B. (1700) De morbis artificum diatriba. Editio novissima (1953). Ed. Pazzini, Rome.

104. Brandenberger, H. (1988) Wichtige exogene Vergiftungen und ihre differentialdiagnostische Erkennung. In: Differentialdiagnose Innerer Krankheiten, 16th Ed. (Siegenthaler, W., ed.), pp. 39.1–39.18. Georg Thieme, Stuttgart & New York.

105. Bidstrup, P. L. (1964) The toxicity of mercury and its compounds. Elsevier, Amsterdam, London & New York.

106. Lu, P.C., Berteau, P.E. & Clegg, D.J. (1972) The toxicity of mercury in man and animals. In: Mercury contamination in man and his environment. Technical Reports Series No. *137*, pp. 67–85. International Atomic Energy Agency, Vienna.

107. Westöö, G. (1966) Detection of mercury compounds in foodstuffs I. Methylmercury in fish, Identification and determination. Acta. Chem. Scand. *20*, 2131–2137.

108. Westöö, G. (1967) Detection of mercury compounds in foodstuffs II. Detection of methylmercury in fish, egg, meat and liver. Acta. Chem. Scand. *21*, 1790–1800.

109. Berlin, M. (1972) Present levels of mercury in man under conditions of occupational exposure. In: Mercury contamination in man and his environment. Technical Reports Series No *137*. International Atomic Energy Agency, Vienna.

110. Westermark, T. & Ljunggren, K. (1972) The determination of mercury and its compounds by destructive neutron activation analysis. In: Mercury contamination in man and his environment. Technical Report Series No. *137*, pp. 99–110. International Atomic Energy Agency, Vienna.

111. Saito, N. (1967) Levels of mercury in man. Paper presented at the IAEA Meeting on Mercury Contamination in Man and his Environment, Amsterdam.

112. Stock, A. & Zimmermann, W. (1928) Geht Quecksilber aus Saatgut-Beizmitteln in das geerntete Korn und in das Mehl über? Z. Angew. Chem. *41*, 1336–1337.

113. Stock, A. & Cucuel, F. (1934) Die Verbreitung des Quecksilbers. Naturwiss. *22*, 390–393.

114. Schöniger, W. (1955) Eine mikroanalytische Schnellbestimmung von Halogenen in organischen Substanzen. Mikrochim. Acta *1955*, 124–129.

115. Lidums, V. & Ulfvarson, U. (1968) Mercury analysis in biological material by direct combustion in oxygen and photometric determination of the mercury vapor. Acta chem. scand. *22*, 2150–2156.

116. Analytical Methods Committee (1965) The determination of small amounts of mercury in organic matter. Analyst (London) *90*, 515–530.

117. Iwantscheff, G. (1958) Das Dithizon und seine Anwendung in der Mikro- und Spurenanalyse. Verlag Chemie, Weinheim.

118. Stary, J. (1964) The solvent extraction of metal chelates. Pergamon Press, New York.

119. Brandenberger, H. & Bader, H. (1967) Die Bestimmung von Nanogramm-Mengen Quecksilber aus Lösungen durch ein flammenloses atomares Absorptionsverfahren. Helv. Chim. Acta *50*, 1409–1415.

120. Brandenberger, H. & Bader, H. (1967) The determination of nanogram levels of mercury in solution by a flameless atomic absorption technique. Atomic Absorption Newsletter *6*, 101–103.

121. Brandenberger, H. & Bader, H. (1968) The determination of mercury by flameless atomic absorption II. A static vapor method. Atomic Absorption Newsletter *7*, 53–54.

122. Brandenberger, H. (1968) Empfindlichkeitssteigerung der atomaren Absorptionsanalyse mittels flammenloser Atomisierung. Chimia *22*, 449–459.

123. Guinn, V.P. (1972) Determination of mercury by instrumental neutron activation analysis. In: Mercury contamination in man and his environment. Technical Report Series No. *137*, pp. 87–97. International Atomic Energy Agency, Vienna.

124. Lamm, C.G. & Ruzicka, J. (1972) The determination of traces of mercury by spectrophotometry, atomic absorption, radioisotope dilution and other methods. In: Mercury contamination in man and his environment. Technical Report Series No. *137*, pp. 111–123. International Atomic Energy Agency, Vienna.

125. Lindstrom, O. (1957) Rapid microdetermination of mercury by spectrophotometric flame combustion. Anal. Chem. *31*, 461–467.

126. Hatch, W.R. & Ott, W.L. (1968) Determination of sub-microgram quantities of mercury by atomic absorption spectrophotometry. Anal. Chem. *40*, 2085–2087.

127. Bailey, B.W. & Lo, F.C. (1971) Automated method for the determination of mercury. Anal. Chem. *43*, 1525–1546.
128. Kahn, H.L. (1971) A mercury analysis system. Atomic Absorption Newsletter *10*, 58–59.
129. Welz, B. & Melcher, M. (1984) Picotrace determination of mercury using the amalgamation technique. Atomic Spectroscopy *5*, 37–42.
130. Doherty, P.E. & Dorsett, R.S. (1971) Determination of trace concentrations of mercury in environmental water samples. Anal. Chem. *43*, 1887–1888.
131. Brandenberger, H. & Müller, S. (1975) Mercury in the body; its determination and fate. Forensic Sciences *5*, 109–110.
132. Campe, A., Velghe, N. & Claeys, A. (1978) Determination of inorganic, phenyl and total mercury in urine. Atomic Absorption Newsletter *17*, 100–103.
133. Brandenberger, H., unpublished results from the investigation of mercury intoxications.
134. Lansens, P., Meuleman, C., Leermakers, M. & Baeyens, W. (1990) Determination of methylmercury in natural waters by head space gas chromatography with microwave-induced plasma detection after preconcentration on a resin containing dithiocarbamate groups. Anal. Chim. Acta *234*, 417–424.
135. Jiang, G., Ni, Z., Wang, S. & Han, H. (1989) Determination of organomercurials in air by gas chromatography-atomic absorption spectrometry. J. Anal. At. Spectrom. *4*, 315–318.
136. Buneaux, F., Buisine, A., Bourdon, S. & Bourdon, R. (1992) Continous flow quantification of total mercury in whole blood, plasma, erythrocytes, and urine by inductively coupled plasma atomic emission spectroscopy. J. of Anal. Toxicol. *16*, 99–101.
137. Bader, H. & Brandenberger, H. (1968) Boron determination in biological materials by atomic absorption spectrophotometry. Atomic Absorption Newsletter *7*, 1–3.
138. Moeschlin, S. (1980) Klinik und Therapie der Vergiftungen, 6th Ed., pp. 206–208. Georg Thieme, Stuttgart & New York.
139. Feigl, F. (1954) Spot Tests, Vol. I, Inorganic Applications, pp. 310–315. Elsevier Publishing Company, Amsterdam, Houston, London and New York.
140. Azuarez, J. & Mir, J.M. (1984) Spectrophotometric determination of boron with curcumin after extraction with 2-methylpentane-2,4-diol in chloroform. Analyst *109*, 183–184.
141. Koch, O.G. & Koch-Dedic, G.A. (1964) Handbuch der Spurenanalyse, p. 353. Springer Verlag, Berlin.
142. Van der Geugten, R.P. (1981) Determination of boron in river water with flameless atomic absorption spectrometry (Graphite furnace technique). Z. Anal. Chem. *306*, 13–14.
143. Schramel, P., Lill, G. & Hasse, S. (1985) Mineral and trace elements in human urine. J. Clin. Chem. Clin. Biochem. *23*, 293–301.
144. Siemer, D.D. (1982) Determination of boron by methyl ester formation and flame emission spectrometry. Anal. Chem. *54*, 1321–1324.
145. Schramel, P. (1979) Determination of boron in milk and milk products by ICP-AES. Z. Lebensm. Unters. & Forsch. *169*, 255–258.
146. Manzoori, J.L. (1980) Inductive coupled plasma optical emission spectrometry: Application to the determination of Mo, Co and B in soil extracts. Talanta *27*, 682–684.
147. Urasa, I.T. (1984) Determination of As, B, C, P, Se & Si in natural waters by direct current plasma atomic emission spectrometry. Anal. Chem. *56*, 904–908.
148. Berlyne, G.M., Ben-Ari, Y., Pest, D., Weinberger, J. et al. (1970) Hyperaluminaemia from aluminium resins in renal failure. Lancet *2*, 494–496.
149. Berlyne, G.M., Yagil, R., Ben-Ari, Y., Weinberger, G. et al. (1972) Aluminium toxicity in rats. Lancet *1*, 564–567.
150. Kerr, D.N.S. and Ward, M.K. (1988) Aluminium intoxication: history of its clinical recognition and management. In: H. Sigel and A. Sigel (eds.) Metal ions in biological systems. Vol. *24*, Marcel Dekker Inc., New-York and Basel, pp. 217–258.
151. Salusky, I.B., Foley, J., Nelson, P. and Goodman, W.G. (1991) Aluminium accumulation during treatment with aluminium hydroxide and dialysis in children and young adults with chronic renal disease. New Engl. J. Med. *324*, 527–531.
152. Subra, J.F., Krari, N., Tirot, P., Mauras, Y. et al. (1991) Aluminium determination in the skin of patients with and without end-stage renal failure. Nephron *58*, 170–173.
153. Russo, L.S., Beale, G., Sandroni, S. and Ballinger, W.E. (1992) Aluminium intoxication in undialysed adults with chronic re-

nal failure. J. Neurol. Neuroswg. Psychiatry *55*, 697–700.

154. Alfrey, A.C., Mishell, J.M., Burks, J. Contiguglia, H. et al.. (1972) Syndrome of dyspraxia and multifocal seizures associated with chronic hemodialysis. Trans. Am. Soc. Artif. Inter. Organs *18*, 257–261.
155. Moreno, A., Dominguez, P., Dominguez, C. and Ballabriga, A. (1991) High serum aluminium levels and acute reversible encephalophathy in a 4-year-old boy with acute renal failure. Eur. J. Pediatr. *150*, 513–514.
156. Wills, M.R. and Savoy, J. (1988) Aluminium toxicity and chronic renal failure. In: H. Sigel and A. Sigel (eds) Metal ions in biological systems. Vol. *24*, Marcel Dekker inc., New-York and Basel, pp. 315–345.
157. McGonigle, R.J.S. and Parsons, V. (1985) Aluminium-induced anaemia in haemodialysis patients. Nephron *39*, 1–9.
158. Perl, D.P. (1988) Aluminium and Alzheimer's disease: methodologic approaches. In: H. Sigel and A. Sigel (eds) Metal ions in biological systems. Vol. *24*, Marcel Dekker Inc., New-York and Basel, pp. 259–283.
159. Perl, D.P. and Good, P.F. (1992) Aluminium and the neurofibrillary tangle: results of tissue microprobe studies. Ciba Found. Symp. *169*, 217–236.
160. Savoy, J. and Wills, M.R. (1988) Analysis of aluminium in biological materials. In: H. Sigel and A. Sigel (eds) Metal ions in biological systems. Vol. *24*, Marcel Dekker Inc., New-York and Basel, pp. 347–371.
161. Brown, S., Bertholf, R.L., Wills, M.R. and Savoy, J. (1984) Electrothermal atomic absorption spectrometric determination of aluminium in serum with a new technique for protein precipitation. Clin. Chem. *30*, 1216–1218.
162. Van der Voet, G.B., de Haas, E.J.M. and de Wolff, F.A. (1985) Monitoring of aluminium in whole blood, plasma, serum and water by a simple procedure using flameless atomic absorption spectrophotometry. J. Anal. Toxicol. *9*, 97–100.
163. Andersen, J.R. and Reimert, S. (1986) Determination of aluminium in human tissues and body fluids by Zeeman-corrected atomic absorption spectrometry. Analyst *111*, 657–660.
164. Fagioli, R., Locatelli, C. and Gilli, P. (1987) Determination of aluminium in serum by atomic absorption spectrometry with the L'vov platform at different resonance lines. Analyst *112*, 1229–1232.
165. Suzuki, Y., Imai, S. and Kamiki, T. (1989) Fluorimetric determination of aluminium in serum. Analyst *114*, 839–842.
166. Mauras, Y. and Allain, P. (1985) Automatic determination of aluminium in biological samples by inductively coupled plasma optical emission spectrometry. Anal. Chem. *57* 1706–1709.
167. De Wolff, F.A. and van der Voet, G.B. (1986) Biological monitoring of aluminium in renal patients. Clin. Chim. Acta *160*, 183–188.
168. Savory, J., Berlin, A., Courtoux, C., Yeoman, B. et al.. (1983) Summary report of an international workshop on "The role of biological monitoring in the prevention of aluminium toxicity in man: aluminium analysis in biological fluids". Ann. Clin. Lab. Sci. *13*, 444–451.
169. Taylor, A., Briggs, R.J. and Cevik, C. (1994) Findings of an external quality assessment scheme for determining aluminium in dialysis fluids and water. Clin. Chem. *40*, 1517–1521.
170. Cumming, A..D., Simpson, G., Bell, D., Cowie, J. et al.. (1982) Acute aluminium intoxication in patients on continuous ambulatory peritoneal dialysis. Lancet *1*, 103–104.
171. Sager, M. (1986) Spurenanalytik des Thalliums. Georg Thieme, Stuttgart & New York.
172. Kemper, F.H. & Bertram, H.P. (1991) Thallium. In: Metals and their compounds in the environment. Occurrence, analysis and biological relevance (Merian, E., ed.). VCH-Verlagsgesellschaft, Weinheim, New York, Basel, Cambridge.
173. Schaller, K.H., Manke, G., Raithel, H.J., Bühlmeyer, G., Schmidt, M. & Valentin, H. (1980) Investigation of thallium-exposed workers in cement factories. Int. Arch. Occup. Environ. Health *47*, 223–231.
174. Brandenberger, H. & Bader, H. (1963) Flammenemissions-Spektrophotometrie aus ätherischer Lösung. Die Erfassung von Gallium, Indium und Thallium in der Spurenanalyse. Helv. Chim. Acta *47*, 353–358.
175. Sighinolfi, G.P. (1973) Determination of thallium in geochemical reference samples by flameless atomic absorption spectroscopy *12*, 136–138.
176. Kopf-Maier, P., Janiak, C. & Schumann, H. (1988) Antitumor properties of organo metallic metallocene complexes of tin and germanium. J. Cancer. Res. Clin. Oncol. *114*, 502–506.

177. McMaster, M.L., Greco, F.A., Johnson, D.H. & Hainsworth, J.D. (1990) An evaluation of the combination 5-fluorouracil and spirogermanium in the treatment of advanced colerectal carcinoma. Invest. New Drugs *8*, 87–92.
178. Nagata, N., Yoneyama, T., Yanagida, K., Ushio, K., Yanagihara, S., Matsubara, O. & Eishi, Y. (1985) Accumulation of germanium on the tissues of a long-term user of germanium preparation who died of acute renal failure. J. Toxicol. Sc. *10*, 333–341.
179. Matsusaka, T., Fujii, M., Nakano, T., Terai, T., Kurata, A., Imaizumi,M. & Abe, H. (1988) Germanium-induced nephropathy: Report of two cases and review of the literature. Clin. Nephrology *30*, 341–345.
180. Wirth, W. & Gloxhuber, Ch. (1981) in: Toxikologie, 3rd Ed., pp. 120–121. Georg Thieme, Stuttgart & New York.
181. Analytical Methods for Atomic Absorption Spectroscopy using the MHS Mercury/Hydride System (1978). Bodenseewerk Perkin Elmer & Co., Ueberlingen.
182. Lüssi-Schlatter, B. & Brandenberger, H. (1976) Trace detection of arsine, germanium hydride, stibine, and tin hydride by gas chromatography with mass specific detection (GC-MD). In: Advances in Mass Spectrometry in Biochemistry and Medicine, Vol. II (Frigerio, A., ed.), pp. 231–248. Spectrum Publications, Inc., New York.
183. Schleich, C. & Henze, G. (1990) Trace analysis of germanium, Part 2: Polarographic behavior and determination by absorptive stripping voltametry. Z. Anal. Chem. *338*, 145–148.
184. Schleich, C. & Henze, G. (1990) Trace analysis of germanium, Part 1: Some new aspects for the determination by extraction-spectrophotometry and atomic spectrometry. Fresenius Z. Anal. Chem. *338*, 140–144.
185. Kisser, W. (1977) Biochemische Methoden zum Nachweis der Bleivergiftung. Arch. Toxicol. *37*, 173–193.
186. Alessio, L., Castoldi, M.R., Odone, P. and Franchini, I (1981) Behaviour of indicators of exposure and effect after cessation of occupational exposure to lead. Brit. J. Industr. Med. *38*, 262–267.
187. Ho, Y.W. (1990) Micro assay for urinary δ-aminolevulinic acid and porphobilinogen by high-performance liquid chromatography with pre-colum derivatization. J. Chromatog. *527*, 134–139.
188. Berlin, A. and Schaller, K.H. (1974) European standardized method for determination of δ-aminolevulinic acid dehydratase activity in blood. Z. klin. Chem. klin. Biochem. *12*, 389–390.
189. Gorodetsky, R., Fuks, Z., Peretz, T. and Ginsburg, H. (1985) Direct fluorometric determination of erythrocyte free and zinc protoporphyrins in health and disease. Clin. Biochem. *18*, 362–368.
190. Louro, O. and Tutor, J.C. (1994) Hematofluorometric determination of erythrocyte zinc protoporphyrin: oxygenation and derivatization of hemoglobin compared. Clin. Chem *40*, 369–372.
191. Scoble, H.A., McKeag, M., Brown, P.R. and Kavarnos, G.J. (1981) The rapid determination of erythrocyte porphyrins using reversed-phase high performance liquid chromatography. Clin. Chim. Acta *113*, 253–265.
192. Ho, J., Guthrie, R. and Tieckelmann, H. (1987) Quantitative determination of porphyrins, their precursors and zinc protoporphyrin in whole blood and dried blood by high-performance liquid chromatography with fluorimetric detection. J. Chromatog. *417*, 269–276.
193. Morrison, G.H. & Freiser, H. (1957) Solvent Extraction in Analytical Chemistry, p. 214. John Wiley, New York & London.
194. Rice, E.W., Fletcher, D.C. & Stumpff, A. (1970) Lead in blood and urine. In: Standard Methods of Clinical Chemistry *5*, 121–129. Academic Press, New York & London.
195. Ostapezuk, P. (1992) Direct determination of cadmium and lead in whole blood by potentiometric stripping analysis. Clin. Chem. *38*, 1995–2001.
196. Berman, E. (1984) The determination of lead in blood and urine by atomic absorption spectrophotometry. Atomic Absorption Newsletter *3*, 111–114.
197. Sprague, S. & Slavin, W. (1966) A simple method for the determination of lead in blood. Atomic Absorption Newsletter *5*, 9–10.
198. Mitchell, D.G., Ryan, F.J. & Aldous, K.M. (1972) The precise determination of lead in whole blood by solvent extraction – atomic absorption spectrometry. Atomic Absorption Newsletter *11*, 120–121.
199. Kahn, H.L., Peterson, G.E. & Schallis, J.E. (1968) Atomic absorption microsampling with the "sampling boat" technique. Atomic Absorption Newsletter *7*, 35–37.

200. Cernick, A. A. (1973) Some observations on the filter paper punched disk method for the determination of lead in capillary blood. Atomic Absorption Newsletter *12*, 42–44.
201. Fernandes, F. J. (1975) Micromethod for lead determination in whole blood by atomic absorption with use of a graphite furnace. Clin. Chem. *21*, 558–561.
202. Fernandez, F. J. & Hilligoss, D. (1982) An improved graphite furnace method for the determination of lead in blood using matrix modification and the L'vov platform. Atomic Spectroscopy *3*, 130–131.
203. Pruszkowska, E., Carnrick, G. R. & Slavin, W. S. (1983) Blood lead determination with the platform furnace technique. Atomic Spectroscopy *4*, 59–61.
204. Vela, N. P., Olson, L. K. and Caruso, J. A. (1993) Elemental speciation with plasma mass spectrometry. Anal. Chem. *65*, 55A–597A.
205. Polson, C. J. & Tattersall, R. N. (1969) Clinical Toxicology, 2nd ed., pp. 181–210. Pitman, London.
206. Cullen, W. R. & Reimer, K. J. (1989) Arsenic speciation in the environment. Chem. Rev. *89*, 713–764.
207. Irvin, T. R. & Irgolic, K. J. (1988) Arsenobetaine and arseno-choline; two marine arsenic compounds without embryo-toxicity. Appl. Organomet. Chem. *2*, 509–514.
208. Andreae, M. O. (1986) Organoarsenic compounds in the environment. In: Organometallic Compounds in the Environment (Craig, P. J., ed.), pp. 198–228. John Wiley, New York.
209. Le, X.-C., Cullen, W. R. & Reimer, K. J. (1994) Human urinary arsenic excretion after one-time ingestion of seaweed, crab and shrimp. Clin. Chem. *40*, 617–624.
210. Hamilton, E. I., Minski, M. J. & Cleary, J. J. (1972/73) Concentration and distribution of some stable elements in healthy human tissues from the United Kingdom; environmental study. Sci. Total Environ. *1*, 341–374.
211. Browning, E. (1961) Toxicology of Industrial Metals. Butterworth, London.
212. Vahter, M. (1994) What are the chemical forms of arsenic in urine, and what can they tell us about exposure? Clin. Chem. *40*, 679–680.
213. Buchet, J. P. & Lauwerys, R. (1983) Evaluation of exposure to inorganic arsenic in man. In: Analytical Techniques for Heavy Metals in Biological Fluids (Facchetti, S., ed.), pp. 75–90. Elsevier, Amsterdam.
214. Vahter, M. (1988) Environmental and occupational exposure to inorganic arsenic. Acta Pharmacol. Toxicol. *59*, 31–44.
215. Smith, H. (1967) The distribution of antimony, arsenic, copper and zinc in human tissue. J. Forensic Sci. Soc. *7*, 97–102.
216. Smith, H. (1964) The interpretation of the arsenic content of human hair. J. Forensic Sci. Soc. *4*, 192–199.
217. Le, X.-C., Cullen, W. R., Reimer, K. J. (1992) Decomposition of organoarsenic compounds by using a microwave oven and subsequent flow injection hydride generation atomic absorption spectrometry. Appl. Organomet. Chem. *6*, 161–171.
218. Fresenius, W. & Schneider, W. (1964) Bestimmung von geringen Mengen Arsen mit Silberdiäthyldithiocarbamid in der Mineralwasseranalyse. Z. Anal. Chem. *203*, 417–422.
219. Smales, A. A. & Pate, B. D. (1952) The determination of submicrogram quantities of arsenic by radioactivation III: The determination of arsenic in biological material. Analyst *77*, 196–202.
220. Xiao-quan, S., Zhe-ming, N. & Li, Z. (1984) Use of arsenic resonance line of 197.2 nm and matrix modification for determination of arsenic in environmental samples by graphite furnace atomic absorption spectrometry using palladium as a matrix modifier. Atomic Spectroscopy *5*, 1–4.
221. Fernandez, F. J. (1973) Atomic absorption determination of gaseous hydrides utilizing sodium borohydride reduction. At. Absorption Newsletter *12*, 93–97.
222. Le, X.-C., Cullen, W. R., Reimer, K. J. & Brindle, I. D. (1992) A new continous hydride generator for the determination of arsenic, antimony and tin by hydride generation atomic absorption spectrometry. Anal. Chim. Acta *258*, 307–315.
223. Lüssi, B. & Brandenberger, H. (1974) Trace detection of arsenic by gas chromatography with mass specific detection. Z. Klin. Biochem. *12*, 224 (Abstract).
224. Lüssi, B. & Brandenberger, H. (1980) Bestimmung von Arsen und Antimon (als Arsin bzw. Stibin) mittels Gaschromatographie mit Photoionisationsdetektion (GC-PID). Jahrbuch der Schweizer. Naturforschenden Gesellschaft, Wissenschaftlicher Teil *1980*, part 1, pp. 122–127. Birkhäuser, Basel.
225. Shibita, Y. & Morita, M. (1989) Specification of arsenic by reversed-phase high per-

formance liquid chromatography – inductively coupled plasma mass spectrometry. Anal. Sci. *5*, 107–109.

226. Le, X.-C., Cullen, W.R. & Reimer, K.J. (1994) Speciation of arsenic compounds by HPLC with hydride generation atomic absorption spectrometry and inductively coupled plasma mass spectrometric detection. Talanta *41*, 495–502.
227. McCallum, R.I. (1989) The industrial toxicology of antimony. J. Roy. Coll. Physicians of London *23*, 28–32.
228. Moeschlin, S. (1980) Klinik und Therapie der Vergiftungen, 6th Ed., pp. 180–181. Georg Thieme, Stuttgart & New York.
229. Gasser, R.A., Magill, A.J., Oster, C.N. et al. (1994) Pancreatitis induced by pentavalent antimonial agents during treatment of leishmaniasis. Clin. Infect. Dis. (US) *18*, 83–90.
230. Constantini, S., Giordano, R., Rizzica, M. & Benedetti, F. (1985) Applicability of anodic stripping voltammetry and graphite furnace atomic absorption spectrometry to the determination of antimony in biological matrices: Comparative study. Analyst *110*, 1355–1359.
231. Luke, C.L. (1953) Photometric determination of antimony in lead using the Rhodamine B method. Anal. Chem. *25*, 674.
232. Luke, C.L. & Campell, M.E. (1953) Determination of impurities in germanium and silicon. Anal. Chem. *25*, 1588–1593.
233. Burke, R.W. & Menis, O. (1966) Extraction-spectrophotometric determination of antimony as ternary complex. Anal. Chem. *38*, 1719–1722.
234. Smith, B.M. & Griffiths, M.B. (1982) Determination of lead and antimony in urine by atomic absorption spectroscopy with electrothermal atomisation. Analyst *107*, 253–259.
235. Chung, C.-H., Iwamoto, E., Yamamoto, M. & Yamamoto, Y. (1984) Selective determination of arsenic (III,V), antimony (III,V), selenium (IV, VI) and tellurium (IV, VI) by extraction and graphite furnace atomic absorption spectrometry. Spectrochim. Acta *39 B*, 459–466.
236. Fodor, P. & Barnes, R.M. (1983) Determination of some hydride-forming elements in urine by resin complexation and inductively coupled plasma atomic spectroscopy. Spectrochim. Acta *38 B*, 229–243.
237. Hutton, R.C. & Preston, B. (1983) A simple versatile hydride generation configuration for inductively coupled plasmas. Analyst *108*, 1409–1411.
238. Nakahara, T. & Kikui, N. (1985) Determination of trace concentrations of antimony by introduction of stibine into inductively coupled plasma for atomic emission spectrometry. Anal. Chim. Acta *172*, 127–138.
239. Bencze, K. & Barchet, R. (1991) Antimony. In: Analysen im biologischen Material 10 (Henschler, D., ed.). VCH, Weinheim.
240. Seeger, R. & Neumann, H.G. (1988) Giftlexikon. Ein Handbuch für Aerzte, Apotheker und Naturwissenschaftler. Deutscher Apothekerverlag, Stuttgart.
241. Berman, E. (1980) Toxic Metals and their Analysis. Heyden, London, Philadelphia, Rheine.
242. Baselt, R.C. (1982) Disposition of Toxic Drugs and Chemicals in Man, 2d edition. Biomedical Publications, Davis, Cal.
243. Madaus, S., Schulte-Frohlinde, E., Scherer, C., Kammereit, A. et al (1992) Comparison of plasma bismuth levels after oral dosing with basic bismuth carbonate or tripotassium dicitrato bismuthate. Aliment. Pharmacol. Ther. *6*, 241–249.
244. Froomes, R.A., Wan, A., Harrison, P.M. and Mclean, A.J. (1988) Improved assay for bismuth in biological samples by atomic absorption spectrophotometry with hydride generation. Clin. Chem. *34*, 382–384.
245. Dean, S., Tscherwonyl, P.J. and Riley, W.J. (1992) Elimination of matrix effects in electrothermal atomic absorption spectrophotometric determination of bismuth in serum and urine. Clin. Chem. *38*, 119–122.
246. Mauras, A., Premel-Cabic, A., Berre, S. and Allain, P. (1993) Simultaneous determination of lead, bismuth and thallium in plasma and urine by inductively coupled plasma mass spectrometry. Clin. Chim. Acta *218*, 201–205.
247. Vanhoe, H., Versieck, J., Vanballenberghe, L. and Dams, R. (1993) Bismuth in human serum: reference interval and concentrations after intake of a therapeutic dose of colloidal bismuth subcitrate. Clin. Chim. Acta *219*, 79–91.

Part 3
Toxicological Drug Analysis

3.1 Barbiturates

H. Brandenberger

3.1.1 Introduction

Barbiturates are a chemically well-defined class of hypnotics. Their structure is that off 2,4,6-trioxo-pyrimidines. They can be described as condensation products of urea, N-substituted ureas or thiourea with di-substituted malonic acids (fig. 1). The basic compound, malonyl urea or barbituric acid ($R_1 = R_2 = H$), synthesized in 1863 by A. v. Bayer (1), has no hypnotic effects. Pharmaceutical activity is found in barbituric acid derivatives with alkyl- or aryl-substituents in position 5 of the pyrimidine ring. The first active derivative, 5,5-diethyl barbituric acid, was synthesized shortly after 1900 by E. Fischer (2), and its therapeutic potential shown by von Mering (3). This first commercially important hypnotic has been marketed under many names such as Barbital, Barbitone, Dormonal, Malonal, Veronal and others. Occasionally, this compound is still used today. It has therefore filled a market position for nearly a century. This is amazingly long for a pharmaceutical.

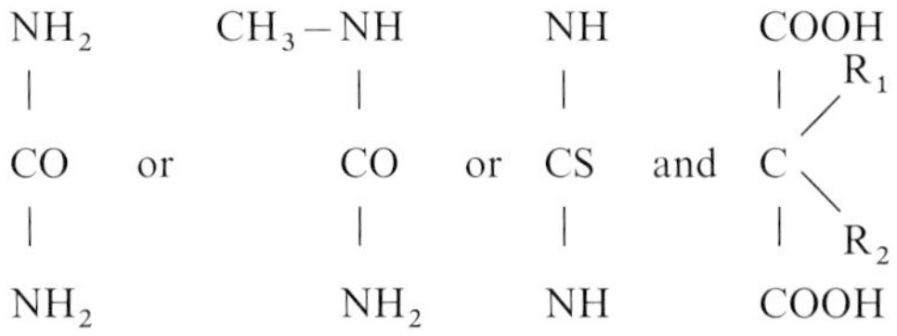

Figure 1. Structural Building Blocks of Barbituric Acids.

Until now, about 2500 barbiturates have been synthesized, but less than 100 have been shown to possess pharmaceutical activity. Table 1 lists the more important ones which have found the way to the market. All physiologically active members possess 2 lipophilic side chains in position 5. One of them must be an open chain. Together, the 2 chains contain usually between 4 and 9 carbon atoms. For many years, barbiturates have been the leading soporifics, and their class name has often been used as a substitute for hypnotics. 20 or more years ago, it was not uncommon for a physician to request "an analysis for barbiturates", when he really wanted a search for all hypnotics and strong sedatives. So, when we received materials "to analyze for barbiturates", I usually contacted the physician to find out if he meant for us to look for a specific barbiturate, for all barbiturates, or for all hypnotics and sedatives. Quite often, the latter was the case.

Today, the importance of barbiturates has diminished. This is largely due to the development of new pharmaceuticals with a benzodiazepine structure. Some members of this class possess a strong hypnotic action and have taken over the main markets. If we nevertheless devote this first chapter on drugs to barbiturates, it is because they are well-suited to illustrate many analytical possibilities for identification and quantification. This chapter is therefore not only a review of barbiturates, but also a survey of methods which are useful in a search for drugs in body fluids and tissues.

Table 1. 5-Substituents of Barbiturates with Pharmaceutical Activity

A) Normal di-substituted barbituric acids		
1. Ethyl-	Ethyl-	Barbital, Veronal, Malonal
2. Isopropyl-	Ethyl-	Probarbital, Ipral
3. n-Butyl-	Ethyl-	Butethal, Neonal, Butob.
4. 1-Methyl-propyl-	Ethyl-	Butabarbital, Noctinal
5. Isopentyl-	Ethyl-	Amobarbital, Amytal, Amal
6. 1-Methyl-butyl-	Ethyl-	Pentobarbiton, Nembutal
7. 1-Ethyl-butyl-	Ethyl-	Tetrabarbital, Butysal
8. n-Hexyl-	Ethyl-	Hexethal, Ortal, Hebaral
9. 1-Piperidyl-	Ethyl-	Eldoral, Ethylpiperidylb.
10. Propyl-	Propyl-	Propylbarbital, Proponal
11. 2-Furanyl-methyl-	Isopropyl-	Dormovit- Furfurylbarb.
12. Phenyl-	Methyl-	Heptobarbital, Rutonal
13. Phenyl-	Ethyl-	Phenobarbital, Luminal
14. Cyclopenten-2-yl-	Ethyl-	Pentenal, Cyclopentylb.
15. Cyclohexen-1-yl-	Ethyl-	Cyclobarbital, Phanodorm
16. Cyclohepten-1-yl-	Ethyl-	Heptabarbital, Medomin
17. 1-Methyl-buten-1-yl-	Ethyl-	Vinbarbital, Delvinal
18. Ethyl-	Allyl-	Ethallobarbital, Dormin
19. Isopropyl-	Allyl-	Alurate, Numal, Aprob.
20. 1-Methyl-propyl-	Allyl-	Talbutal, Lotusate
21. n-Butyl-	Allyl-	Idobutal, Butylallylbarb.
22. Isobutyl-	Allyl-	Butalbital, Sandoptal
23. 1-Methyl-butyl-	Allyl-	Secobarbital, Seconal
24. 1-Methyl-butyl-	Vinyl-	Vinylbital, Butylvinal
25. 2,2-Dimethyl-propyl-	Allyl-	Nealbarbital, Nevental
26. 2-Hydroxy-propyl-	Allyl-	Proxibarbital, Axeen
27. Phenyl-	Allyl-	Phenallymal, Alphenal
28. Cyclopenten-2-yl-	Allyl-	Cyclopal, Cyclopentobarb.
29. Allyl-	Allyl-	Allobarbital, Dial, Diadol
30. 1-Methyl-2-pentinyl-	Allyl-	Methohexital, Brevital
31. Isopropyl-	Bromoallyl-	Propallylonal, Noctal
32. 1-Methyl-propyl-	Bromoallyl-	Butallylonal, Pernoston
33. 1-Methyl-butyl-	Bromoallyl-	Recton, Recitidon, Sigmodal
34. Allyl-	Bromoallyl-	Brallobarbital, Vesperon
B) N-Methyl substituted barbituric acids		
35. Ethyl-	Ethyl-	Metharbital, Gemonil
36. Phenyl-	Ethyl-	Mephobarbital, Prominal
37. Cyclohexen-1-yl-	Methyl-	Hexobarbital, Evipan
38. Isopropyl-	Allyl-	Enallylpropymal, Narconumal
39. 1-Methyl-pentyn-2-yl	Allyl-	Brevital, Methohexital
40. Isopropyl-	Bromoallyl-	Eunarcon, Narcotal, Narcobar.
C) Thiobarbituric acids		
41. Ethyl-	Ethyl-	Thiobarbital, Ibition
42. 1-Methyl-propyl-	Ethyl-	Thiobutabarbital, Inactin
43. Isopentyl-	Ethyl-	Thioethamyl, Thioethymyl
44. 1-Methyl-butyl-	Ethyl-	Thiopental, Pentothal
45. Butyl-thio-methyl-	Ethyl-	Thionarcon
46. Methyl-thio-ethyl-	Methylbutyl-	Methitural, Thiogenal
47. 2-Methyl-allyl-	Ethyl-	Methallatal, Mosidal
48. Isobutyl-	Allyl-	Buthalital, Baytinal
49. 1-Methyl-butyl-	Allyl-	Thioamylal, Surital
50. Cyclohexen-2-yl-	Allyl-	Thialbarbital, Kemithal

3.1.2 Sub-Classes of Barbituric Acids

Three sub-classes of barbiturates must be distinguished. They possess not only different structures and analytical properties (Table 1), but also differ in respect to metabolism and pharmaceutical activity.

The largest sub-class consists of the condensation products of urea with di-substituted malonic acids. The resulting derivatives carry, in 5-position of the barbiturate ring, saturated alkyl, unsaturated alkenyl or a phenyl side chain(s). This group makes up the bulk of barbiturates on the market. A small number of barbiturates are composed of the building blocks methyl urea and malonic acid derivatives. Such N-substituted methyl-barbituric acids metabolize faster and are usually shorter-acting than their unmethylated analogs. Not included in Table 1, but also members of this sub-group, are the N-substituted barbiturates with larger substituents in position 1. They are listed separately in Fig. 2. The third group of barbiturates are the condensation products of thiourea with malonic acid derivatives. They have higher lipid solubility than the sulfur-free analogs and, as a consequence, a more rapid onset of action. Due to their extensive metabolism, they possess also a shorter duration of action.

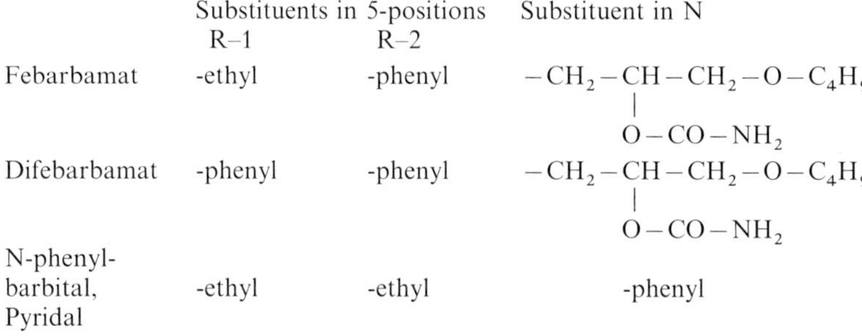

	Substituents in 5-positions		Substituent in N
	R–1	R–2	
Febarbamat	-ethyl	-phenyl	$-CH_2-CH(O-CO-NH_2)-CH_2-O-C_4H_9$
Difebarbamat	-phenyl	-phenyl	$-CH_2-CH(O-CO-NH_2)-CH_2-O-C_4H_9$
N-phenyl-barbital, Pyridal	-ethyl	-ethyl	-phenyl

Figure 2. Barbituric Acids with Large Substituents in N.

For each compound, Table 1 lists the two side chains in position 5 of the ring and 1 to 3 of the most common commercial names. It was difficult to make a choice among all the names of some of the market runners, especially since they vary from country to country, but the available space imposed a limitation. Within the main groups, the saturated barbiturates are mentioned first, followed by the phenyl-substituted substances, the compounds with unsaturated alicyclic and unsaturated aliphatic side chains, and finally the bromine-containing barbiturates. The entries are numbered to simplify the references in the text.

We considered it important to include quite a large number of barbiturates in our list, since the markets in the different continents and even different countries do not offer the same compounds. The number of barbituric acids sold in the USA, for example, is much smaller than in European countries like Germany, France, Italy or Switzerland.

3.1.3 The Pharmaceutical Action of Barbituric Acids

Barbiturates are not only used as hypnotics, but also as sedatives, antiepileptic agents, narcotics and injection anesthetics. As a somewhat too-generalized rule of thumb, it can be said that very low dosages lead to sedation, while higher dosages will have hypnotic-, and very high ones narcotic effect. This rule, however, should not be extended to the ultrashort acting compounds.

Today, the long acting barbiturates such as Barbital or Luminal (compounds 1 and 13 in Table 1), but also some others (compounds 5, 29 and 35), are commonly used as sedating agents or antiepileptics, less often as hypnotics. The intermediate and fast acting barbituric acids (the newer literature usually lists them together as fast acting barbiturates) are typical hypnotics. The ultra-fast acting substances, mostly N-methyl-barbiturates (compounds 37 to 40) and thiobarbiturates (compounds 42 to 45 and 48 to 50), are used as anesthetics (intravenous injection), just like a few compounds without thio-group but with a bromine atom in one of the side chains (i.e. compound 31). Among the anticonvulsants (compounds 12, 13, 36 and 40), the phenyl-substituted components are dominant. Some of the acids have also found other uses, i.e. as a thyroid inhibitor (compound 41) or an antiemetic (compound 47).

3.1.4 The Metabolic Fate of Barbituric Acids

The duration of action of the different barbiturates depends to a large extent on their metabolic breakdown in the body and the elimination pathways. At least one of the compounds, diethyl-barbituric acid (number 1), is not metabolized at all, but slowly excreted unchanged through the kidneys, hence its long duration of action.

Most other barbiturates undergo metabolic changes which are accompanied by a decrease of lipid and an increase of water solubility. Five metabolic routes must be considered:

1) Desulfuration of thio-barbituric acids.
2) N-dealkylation of N-substituted compounds.
3) Oxidation of side chains in position 5 with formation of hydroxyl, keto and carboxyl groups.
4) Coupling of the hydroxy metabolites, mainly with glucuronic acid.
5) Fission of the heterocyclic ring system.

Desulfuration and N-dealkylation are usually fast reactions. They lead to the corresponding compounds listed in part A of Table 1. If we have to search for a N-methyl-barbiturate or a thio-barbituric acid in blood or urine, we must extend our investigation also to the structurally analogous members without N-methyl- or thio group.

Apart from compound 1, side chain oxidations have been demonstrated for most barbiturates. They proceed relatively slowly; the basic compounds and their oxidation products may be found side by side in urine, in varying amounts, depending on the time elapsed between intake and urine collection, as well as on the dosage.

With therapeutic dosages, a much larger percentage of drug is oxidized than with elevated toxic doses. The hydroxy-metabolites undergo coupling reactions. They do not seem to represent a main metabolic pathway, since appreciable concentrations of hydroxy derivatives can be found also in non-hydrolyzed urine samples. For quantification, however, enzymatic hydrolysis must preceed urine extractions.

Usually, it is the larger of the substitutens in position 5 which is hydroxylated, for example, in compounds 12 and 13 the phenyl group, in compounds 2 to 8 the longer of the side chains in position 5, in compounds 14 to 16, 28, 37, 39 and 50 the alicyclic rings. Hydroxy-derivatives can be further oxidized to stable ketones, or, if hydroxylation occurred at the end position of an aliphatic side chain, to the corresponding carboxylic acids.

Fission of the heterocyclic ring is a side metabolic pathway. It probably occurs between the ring positions 1 and 6 and leads to substituted open-chain malonylureas. Its extent has not been very well-defined. This may be due to the fact that the open-chain metabolites are more difficult to detect and quantify than the barbiturates with an intact pyrimidine ring.

As is evident from Table 1, a large number of barbiturates has been put on the market. Some of them yield several metabolites. Since their detection and identification must form an integral part of every toxicological search for barbiturates, such investigations become more complex (but of course also more interesting).

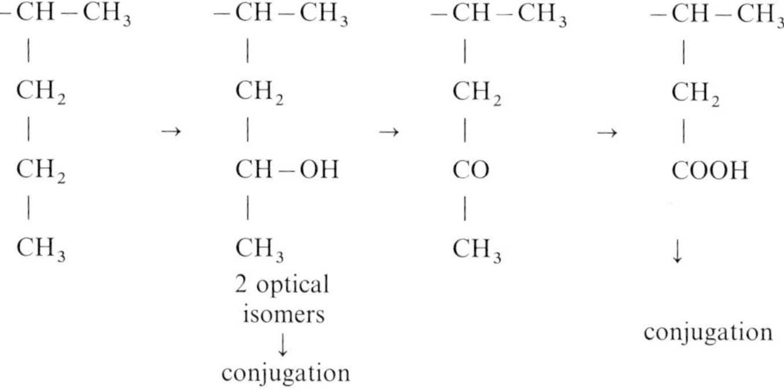

Figure 3. Metabolic Transformation of the 5-(1-methyl-butyl) Side Chain of Pentobarbital (Nembutal, compound 6).

Figure 3 gives an example for a metabolic transformation of a barbituric acid. We refrain from listing all barbiturate metabolites found so far and refer to "Clarke's Isolation and Identification of Drugs" (4), a handbook which should be on the desk of every toxicologist. It lists the metabolites reported so far. However, we must keep in mind, that with our modern and highly-sensitive analytical methods, it may still be possible to detect additional metabolites or to correct an assigned structure. For such a task, we cannot rely only on chromatographic retention data and compilations of spectroscopic tables. We must also be able to deduce a structure solely on the basis of the analytical information, without the help of library search programs.

3.1.5 Barbiturate Dosages, Body Levels and Toxicity

Luckily for the analyst, barbiturates are dosed fairly high. For therapeutic reasons, 200 mg/day seems to be an average. For some members of the group, dosage may be as low as 50 mg or as high as 500 mg/day. The minimal lethal doses (MLD) are estimated to be around 2 g for most barbiturates and as low as 1 g for some ultra-fast acting members used as anesthetics (5). Toxic- and in some cases even lethal doses therefore exceed therapeutic doses only by a factor of 10. This is one of the main disadvantages of this compound class.

Another disadvantage is said to be the high abuse potential. In fact, barbiturates have largely been illegally abused as narcotics, which has led to their inclusion into the "International List of Narcotic Drugs of Abuse". However, many of the newer sedatives and hypnotics, including the modern benzodiazepines, are now being abused exactly in the same way or even more so.

Due to the relatively high dosage, the concentrations of barbiturates in the body are substantial. Usually, they lie consistently higher in liver than blood, but blood levels are still sufficiently elevated to permit detection and quantification by a number of methods. As a rule of thumb, we may safely assume, that for short- and intermediate acting barbiturates, blood levels between 1 and 10 µg/ml can be found after therapeutic-, and over 10 µg/ml after toxic dosages. For long acting compounds, blood levels may reach 10 to 40 µg/ml after therapeutic-, and between 40 to 100 µg/ml after toxic doses (6, 7). For the ultra-short acting barbiturates, blood concentrations are much lower, but the corresponding desulfurated or N-dealkylated metabolites can also be found.

In intoxications, we like to analyze, side by side, stomach content, blood and urine. The comparision of the 3 concentrations can give a hint as to the time of poisoning. For clinical investigations, it can help the physician to select the appropiate medical treatment. If resorption is advanced (none or only little drug left in the stomach) and elimination efficient (high content in the urine), the patient may come out of coma without much medical help, aside from control and support of respiration. But if, with identical blood concentration, the stomach still contains a lot of the drug and the urine only a low level, the situation of the patient may worsen without appropiate medical care such as dialysis. The data from the analysis of all 3 body fluids are a sounder foundation for choosing a therapy than only isolated blood level values.

3.1.6 Preliminary Screening Tests

3.1.6.1 General remarks

Fast preliminary tests for the presence of members of a specific compound class or a single compound have always been and will continue to be in great demand. Such tests, however, are seldom 100% reliable. Nevertheless, screening of a body fluid (stomach content, stomach wash, urine or serum) often yields valuable indications for planning further analytical procedures.

In the case of barbituric acids, two possibilities have been or are being used for preliminary testing, a color reaction and immunochemical assays.

3.1.6.2 Color reactions: the Zwikker test

In the past, this test (8) has been extensively used, even though it lacks specificity and shows rather poor sensitivity. It relies on the fact that cobalt ions transform barbiturates into deep violet-blue complexes. Usually, this color persists for minutes. Many other compounds with related structural features also yield blue complexes with cobalt ions, but these colors are usually more transient.

When I entered the field of toxicological analysis in 1961 and even for many years afterwards, it was not uncommon for a physican to request that "only a Zwikker test" be carried out in a search for barbiturates. I could not comply with such wishes. Although the test can be carried out rapidly and with simple means (evaporation or extraction of a small urine sample, addition of a methanolic solution of cobalt acetate, alkalization with a methanolic solution of lithium hydroxide or with an organic base such as pyridine), it yields too many false positive reactions and even some false negative results. Furthermore, a reliable and not much more time consuming search for barbiturates is possible using UV-screening, as described later.

Today, the Zwikker reagent has lost its importance. It is still used as a spray reagent in thin-layer chromatography, but even for this purpose, more sensitive reagents are available.

3.1.6.3 Immunochemical assays

Immunochemical assays have become very important for fast preliminary screening. They have replaced most other screening tests, i.e. many color reactions. However, they have not yet been able to supplant the unequivocal physical identification techniques, as many colleagues had expected in the first years after their introduction. Immunochemical assays are indicated, when a problem cannot be handled by chemical techniques in comparable time and with comparable efforts (9). They can especially be recommended for a search of a specific compound or a compound class in a large number of samples. But only assays which yield a very low percentage of false negative and also a low percentage of false positive results should be used. All positive results must be verified by a different (non-immunochemical) procedure, unless they have no forensic or therapeutic consequences.

Many immunochemical techniques are available today. For analysing drugs, the most popular assay methods are:

- RIA (radioimmunoassays using β-radiation from ^{3}H or γ-radiation from ^{125}J),
- EIA (solid phase enzyme immunoassay),
- EMIT (enzyme multiple immunoassay technique),
- FPIA (fluorescence-polarization immunoassay).

The first 2 of these techniques require a separation step and are therefore called heterogenous immunotests. The other 2 do not require a separation and are called homogenous tests. The sensitivities of the tests, as well as their application ranges differ considerably (different detection techniques). The specificities, however, are quite similar, since the same or similar antibodies have been used to prepare the commercially available test kits.

Immunochemical assays for barbituric acids are possible with all the listed techniques. But since they have been conceived for the US market with its limited selection of barbiturates, they do not respond to some additional members of the class which are available outside the US. Especially in forensic work, it must also be considered that hypnotic substances may originate from sources other than a pharmacy or a doctor's office. An example from our daily service work can serve as an illustration:

Late at night, we received a call from a hospital. A young woman had been found in coma and poisoning was suspected. We asked the hospital to send stomach wash, blood and urine of the patient. It took us less than a half-hour, until we could inform the physician that the young woman had taken an overdose of a barbiturate, and in another half-hour we could specify that 5,5-diethyl-barbituric acid (Barbital) had been ingested, that only little was left in the stomach, and that the high urine concentration indicated a good elimination. The patient woke up the next morning. No medical treatment except the control and support of respiration and an infusion had been necessary. The woman, a laboratory technician, used to work with a barbital buffer (for electrolyses) and had swallowed several grams in a suicide attempt. The so-called "barbiturate immunotest" would have been unable to reveal the drug and would have misled the physician.

Due to the lack of specificity of the immunotests for barbituric acids, and also in consideration of the fact that highly reliable chemical detection methods for barbiturates are available, which can be extended to the identification and quantification of the specific compounds, we can not recommend the use of such tests in forensic work or clinical diagnostics (9).

3.1.7 Extractions

3.1.7.1 General remarks on liquid–liquid extraction

There are almost as many approaches to the extraction of drugs from body fluids or tissues as chemical-toxicological laboratories. Some investigators aim to simplify the procedure as much as possible by using one and the same operation for extracting "all drugs". They work at a slightly alkaline pH (8.5–9) and use a solvent with high polarity, such as ethyl acetate. Under such conditions, alkaline, amphoteric, neutral and even weakly acidic compounds pass into the organic phase. But such extracts often contain substantial concentrations of ballast materials like plasticizers and, due to the high solvent polarity, endogeneous substances. The simultaneous presence of compounds with different acidity and of large concentrations of contaminants renders subsequent identification and quantification more difficult. It narrows the choice of methods available for this task and excludes in particular spectrophotometric techniques, i.e. UV- and fluorescence spectrometry.

Another extreme is the use of different biological samples to extract acid and alkaline drugs separately. We do not think that is justified. An analyst has the obligation to use the submitted biological materials as efficiently as possible. Samples which are freed from only one type of drug should be used further and not discarded. Furthermore, a second extraction of a specimen yields a much cleaner extract than that of a fresh biological sample, since the neutral ballast compounds (i.e. most plasticizers) have already been removed, and less cross-contamination from only slightly acidic-, or alkaline drugs, must be expected.

3.1.7.2 Extraction procedure in a general search for drugs

A general search for a possible presence of unknown drugs or poisons is usually called by forensic toxicologists a "search for the general unknown". We recommend starting such a task by a systematic series of extractions at strictly controlled pH-values. Thereby, the extractable compounds are accumulated in separate fractions according to their acidity (10):

- strong acids } which must not al-
- weak acids } ways be subdivided
- neutral compounds
- alkaline compounds and
- amphoteric components.

Table 2 (p. 358) illustrates our methodology. Generally, the acid extraction step will precede the alkaline extraction, but for some analytical requests, it may be better to invert this order. While extractions with organic solvents always cause a dilution, re-extractions of acids and bases into aqueous phases can serve as concentration steps (less water phase than solvent, only 1 or 2 extraction steps).

Unfortunately, extraction is not always given the consideration it deserves. It is assigned to unskilled personnel with insufficient knowledge of the requirements of the work. Correct extraction, however, is a prerequisite for success, perhaps the most important part of an analysis. It should not be carried out by untrained technicians, unless they are constantly supervised. The responsible person must not only be able to follow a given scheme, but also to adapt it and to be aware of the implications of such adaptations.

A good way to control an extraction procedure is to follow it step by step with a so-called "UV-screening", as will be described in 3.1.8. The additional time for such a UV-control is well-invested, since it can serve as a guide for the further analytical procedure and facilitate or shorten subsequent investigations.

3.1.7.3 Extraction of barbituric acids

To extract barbituric acids, we recommend the scheme in Table 2. Acids and neutral components are extracted into ether, preferably isobutyl methyl ether. The strong acids are eliminated by back-extraction into phosphate buffer of pH 7.3. Then, the weak acids are concentrated by back-extraction into 0.5 mol/l KOH or NaOH. Bar-

Table 2. Extraction Scheme for a General Search for Drugs

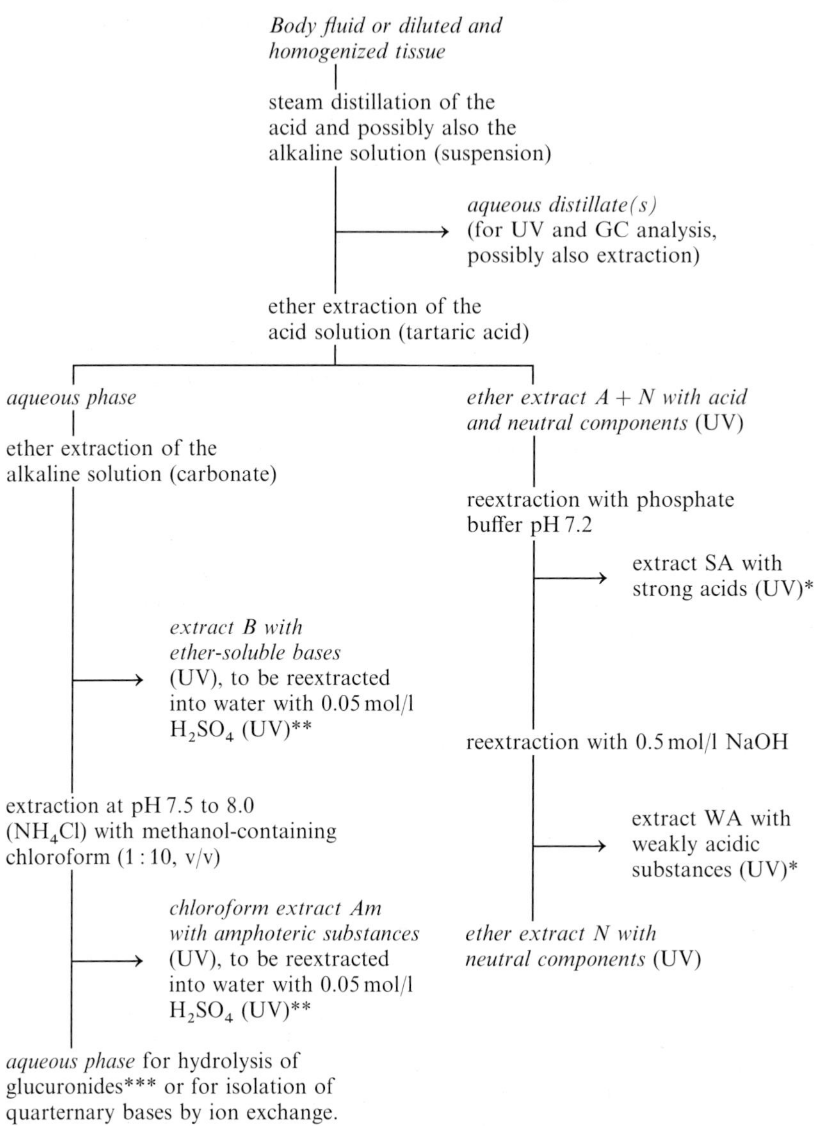

* after recording of UV-spectra (at 3 pH-values) acid extraction into ether (UV).
** after recording of UV-spectra (at 3 pH-values) alkaline extraction into organic phase.
*** with new extractions after hydrolysis.

biturates will pass into this aqueous phase, free from stronger acids (including most fatty acids) and neutral contaminants. The extract is acidified, and the barbiturates are re-extracted into ether. The organic phase is then dried and evaporated.

At the strong alkaline pH used for the back-extraction of the barbiturates into water, thiobarbituric acids are decomposed to the corresponding barbituric acids.

For their quantitative determination, as is for example requested for monitoring the blood pentothal level, direct extraction must be chosen, preferably with a less polar solvent than ether.

3.1.8 Ultraviolet-Spectrophotometry

3.1.8.1 UV-spectrophotometric control of extraction

The barbiturates are a perfect example to show the information which can be obtained with a control of extraction procedures by UV-spectrophotometry. Not only the organic extracts are analyzed, but also the re-extracted aqueous phases. Their spectra are recorded at three different pH values, as illustrated in Table 3. For a

Table 3. Procedures for Recording UV Absorption Spectra of Aqueous Extracts at 3 Different pH-Values

General outline:

The first spectrum is recorded with 2.5 ml of extract in the sample cell and 2.5 ml of extraction solvent in the reference cell. The pH shifts prior to the registration of the second and third spectra are effected by addition of 0.5 ml of the solutions listed below to the sample- and reference cells and mixing (parafilm covers).

Extracts:

SA = Extract of strong acids, in 0.1 mol/l phosphate pH 7.2.
WA = Extract of weak acids, in 0.1 mol/l NaOH.
B = Extract of bases, in 0.05 mol/l H_2SO_4.
Am = Extract of amphoteric substances, in 0.05 mol/l H_2SO_4.

Additions for the pH-shifts (0.5 ml each):

Extract	SA	WA	B	*Am*
first addition	0.5 mol/l NaOH	32 g/l NH_4Cl	1.0 mol/l NaOH	1.0 mol/l NaOH
second addition	0.5 mol/l H_2SO_4	0.5 mol/l H_2SO_4	0.65 mol/l NaH_2PO_4	0.65 mol/l NaH_2PO_4

complete extraction according to Table 2, up to twenty spectra will have to be recorded. At first sight, this may appear a disproportionate effort. However, the recordings are carried out with self-scanning spectrometers and are conceived so that every pH change is made directly in the UV absorption cell without sample transfer. An experienced person can perform all extraction and UV measurements within 2 hours. Should emulsions occur, the separatory funnels are replaced by centrifuge tubes and the layers separated, after centrifugation, with capillary pipets.

3.1.8.2 The UV-spectra of barbiturates

Barbiturates show up in the aqueous extract of the weakly acidic components. They exhibit strong absorption bands in alkaline-, but none in acid solution. The figures 4 to 6, copied from a previous article (10), show the spectra which are typical for

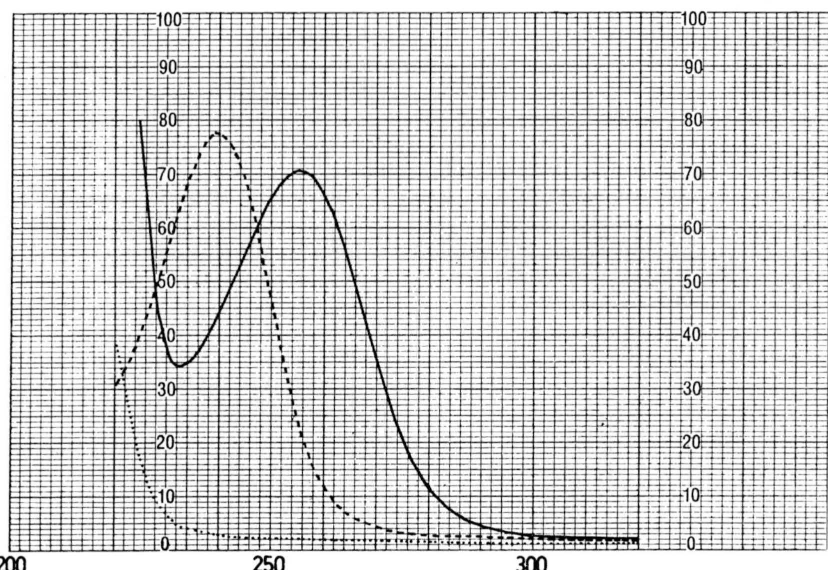

Figure 4. Typical UV-absorption pattern of only in 5-position disubstituted barbiturates, in water.
········ acid ----- pH 10 ——— pH 13 Abscissa: wavelength (nm).

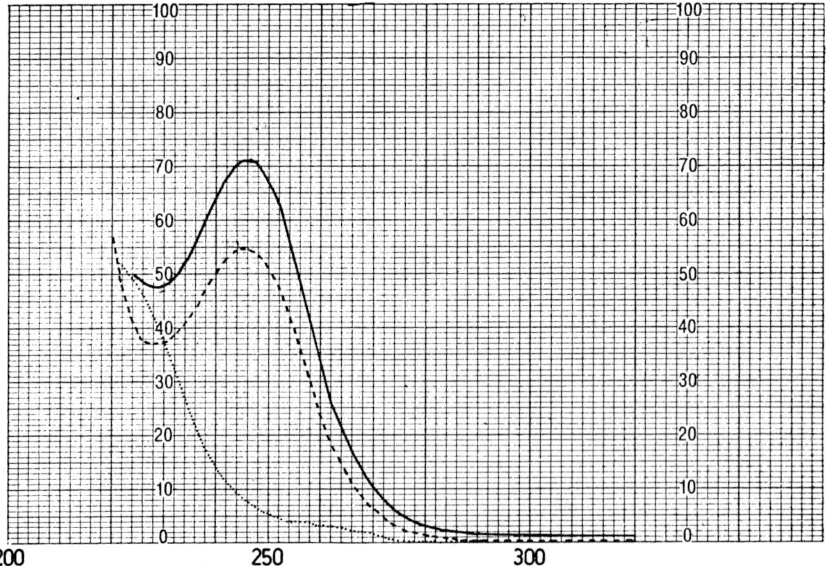

Figure 5. Typical UV-absorption pattern of *N*-methyl barbiturates, in water.
········ acid ----- pH 10 ——— pH 13 Abscissa: wavelength (nm).

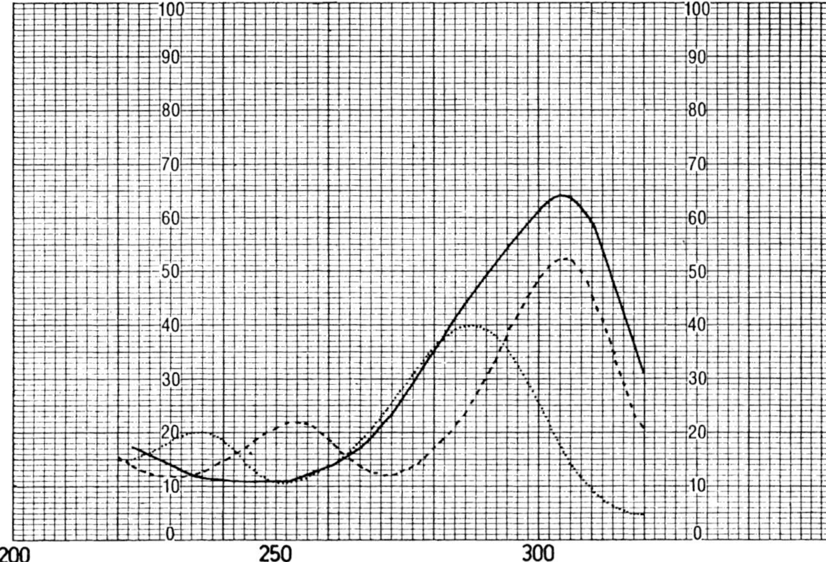

Figure 6. Typical UV-absorption pattern of thiobarbiturates, in water.
········ acid ----- pH 10 ——— pH 13

the members of the three sub-groups. Two ionization steps are possible for the barbiturates of group A (without substituted N-atoms) and only one for N-alkyl barbituric acids. The former yield therefore two different absorption bands (maxima near 240 nm in the region of pH 10, and near 255 above pH 12), the latter only one band near 245 nm. The UV maxima of thiobarbituric acids are of course shifted to higher wavelengths, to the region of 300 nm. With the extraction procedure we recommend, these maxima can usually be detected even after relatively low dosages.

In organic phase (ether extracts), the barbiturates without thio groups possess no UV-maxima, except for the sequence of much weaker aromatic bands with center near 255 nm exhibited by the phenyl-substituted compounds 12, 13, 27 and 36, if present at higher concentrations.

3.1.8.3 Quantification of barbiturates by UV-spectrometry

After the identity of a barbiturate has been determined (with the help of methods described later), it is possible to estimate its concentration on the basis of the spectra already recorded during the extraction process. We recommend using the absorption differences at alkaline and acid pH. The so-called ΔA-values, determined in our laboratory, are listed in Table 5 of reference (11). To use, at the corresponding maxima, the differences between the readings at alkaline and acid pH, rather than absolute absorption values, usually eliminates falsifications by the presence of UV-absorbing impurities. As a control, we recommend making the calculations with both ΔA-values, those at pH 10 and at pH 13. The UV-spectra of barbiturate metabolites formed by side chain oxidations are very similar to those of the parent compounds. The results obtained from the ΔA-values are therefore an estimate of the sum of a barbiturate and its metabolites with oxidized side chain.

3.1.9 Chromatographic Separations

Thin layer- (DC), liquid- (HPLC) and gas chromatography (GC), recently also capillary electrophoresis (CE), have successfully been used to identify single barbiturates as well as their metabolites.

For the DC Rf-values in 10 different solvent systems, we refer to a compilation edited jointly by the DFG (Deutsche Forschungsgemeinschaft) and TIAFT (The International Association of Forensic Toxicologists) (12). Valuable information can also be found in Clarkes Handbook of Toxicology (4). As visualization methods, we recommend short wave UV-light after alkalization of the plates, mercurous nitrate or mercuric-diphenylcarbazone and (less sensitive) the Zwikker reagent for revealing all barbiturates, acidified potassium permanganate for detecting the unsaturated compounds, and for the bromo-barbiturates a fluorescein spray after treatment with alkali (13).

HPLC has frequently been used to separate and identify barbiturates. It is very well possible that CE will replace it in the future, at least in part, due to its better resolution.

Our preferred method for separating barbituric acids is GC. The volatility of the barbiturates is sufficiently high, so that no derivatization is needed. The ease with which GC can be coupled with MS or with FTIR permits identifying not only the compounds on the market, but also their metabolites. In a previous article (14), we have listed the retention data of 24 barbituric acids on a slightly acidic column (Apiezon L). Newer data, obtained with different columns, are presented in Clarkes Handbook of Toxicology (4) and in a second joint compilation by the DFG and TIAFT (15).

3.1.10 Infrared Spectrophotometry

While UV is a good group (as well as sub-group) detection method for barbiturates, IR-spectrometry permits individual compound identifications. The thio-group can be recognized by 2 intensive bands at 1170 and 1540 cm^{-1} (8.55 and 6.5 μ), phenyl substituents in ring position 5 by bands at 1490 cm^{-1} (6.7 μ) and between 690 and 715 cm^{-1} (14.5 and 14 μ), allyl groups by absorption bands at 1000 and 1540 cm^{-1} (10 and 6.1 μ), N-methyl substituents by 2 bands near 1190 cm^{-1} (8.35 μ). In the fingerprint region, all barbiturates can be distinguished from each other, as is evident from our previously published tables containing the IR absorption bands of 24 barbiturates (10, 16, 17).

Before the time of GC and GC-MS combination, we used fractional vacuum sublimation of the weakly acid extracts for isolating barbituric acids in a clean state, prior to IR-analysis. After the availability of GC, we collected the bulk of the barbituric acid peaks at the outlet of the GC column with the help of an effluent split. The condensate in the trap (a simple capillary without additional cooling) was rinsed with chloroform directly onto KBr, before preparation of the micro-pellet. With a good micro-accessory, identifications are possible with less than 1 μg of sublimate, respectively condensate.

Today, such off-line methods can be replaced by on-line combinations of GC with FTIR. This eliminates handling of the separated barbiturate fractions. It is much faster and permits characterization of all fractions in a single GC-run. Three different techniques are so far available for an on-line combination of GC with FTIR (18, 19):

- the light pipe approach which furnishes, in real-time, relatively simple gas phase IR spectra,
- the matrix-isolation trapping technique, which yields, in a quasi on-line approach, matrix-spectra with much more fine structures, although not in real-time, and
- the direct deposition quasi on-line method, which yields – without the need for a surrounding matrix – normal solid phase IR spectra (comparable to the spectra in normal IR-libraries.

The deposition methods, especially matrix isolation, have better sensitivities (20), but the instrumentation is considerably more expensive and also more difficult to handle than a light pipe combination. The dimensions of the presently used IR light pipes (gold plated pyrex tubes with, typically, 1 mm inner diameter and 150 or 200 mm length) may not be ideal detector cells for narrow bore capillaries, but they are compatible with 0.3 or 0.5 mm capillary columns, which have become popular in chemical toxicology, since they require neither injection split nor make-up gas. The FTIR must of course be equipped with a nitrogen-cooled MCT detector (medium band, covering the spectral range from 750 to 4000 cm^{-1}). He, Ar or N_2 can be used as GC carrier gases. Since the sensitivity can be problematic, we recommend using thick-film capillaries permitting work with high sample loads.

An interesting aspect of light pipe GC-FTIR is the possibility to follow a chromatogram simultaneously is several ways:

- by checking the 3-dimensional spectral plot (IR-spectra versus elution time) displayed on the screen and stored in the computer,
- by recording a signal which can be considered as representative for the total IR absorption of the eluting fraction (i.e. a Gram-Schmidt plot), and/or
- by recording, in parallel, several so-called chemigrams, the absorption intensities of pre-selected spectral windows corresponding to specific absorption bands, respectively to specific functional groups.

In selecting such windows, due consideration must be given to the fact that the position of IR absorption bands in the gas phase can vary substantially from those in the solid or liquid phases. This holds particularly for functional groups capable of undergoing intermolecular interactions such as hydrogen bonding. For barbiturates, we have found a shift to higher wave numbers mainly for the N–H stretching vibration and the amide I band, but not for the C–H stretch (21). The knowledge of gas phase spectra is important for an optimal selection of spectral windows in recording chemigrams. With correctly adjusted narrow windows, much lower sensitivities can be obtained than with more generously set wider windows.

Compounds with reasonably strong absorbing functional groups such as barbituric acids can be detected and characterized by GC-FTIR, if present in the 10 to 100 ng range. For exceptionally strong absorbers and for revealing known compounds, the detection limits may even be lower. But they are still far inferior to those of GC-MS and GC with mass specific detection. However, in the field of barbiturates (as well as other pharmaceuticals which are dosed fairly high), GC-FTIR is capable of detecting, in extracts of body fluids, the unaltered drugs and occasionally also their main metabolites (21, 22). It should be used, if GC-MS is not available, and as a complementary technique for structural investigations.

Figure 7 (from 21) shows five chemigrams, recorded during a separation of seven barbiturates with a 0.53 mm DB-17 capillary. All seven are revealed by the amide I chemigram. The chemigram of the CH-stretch misses (the first eluting) 5,5-diallyl-

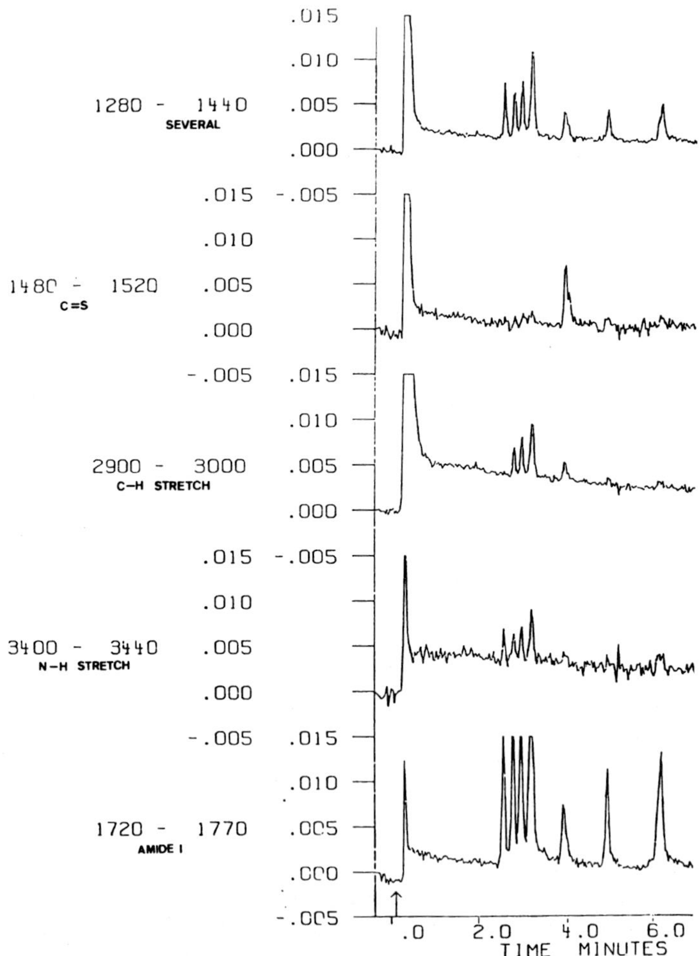

Figure 7. GC-Separation of Barbituric Acids with Recording of 5 Chemigrams by FTIR.

barbituric acid. The C = S band chemigram shows only thiopental (eluting in position 5 after 4 minutes). Figure 8 finally (also from 21) shows the IR spectrum of butalbital, isolated from urine. Additional peaks could be identified by GC-MS and GC-FTIR as metabolites of propyphenazone. This showed that the investigated person had driven under the influence of the pharmaceutical Optalidon. On-line combinations of GC with FTIR and MS (23) will certainly become very useful in metabolic studies.

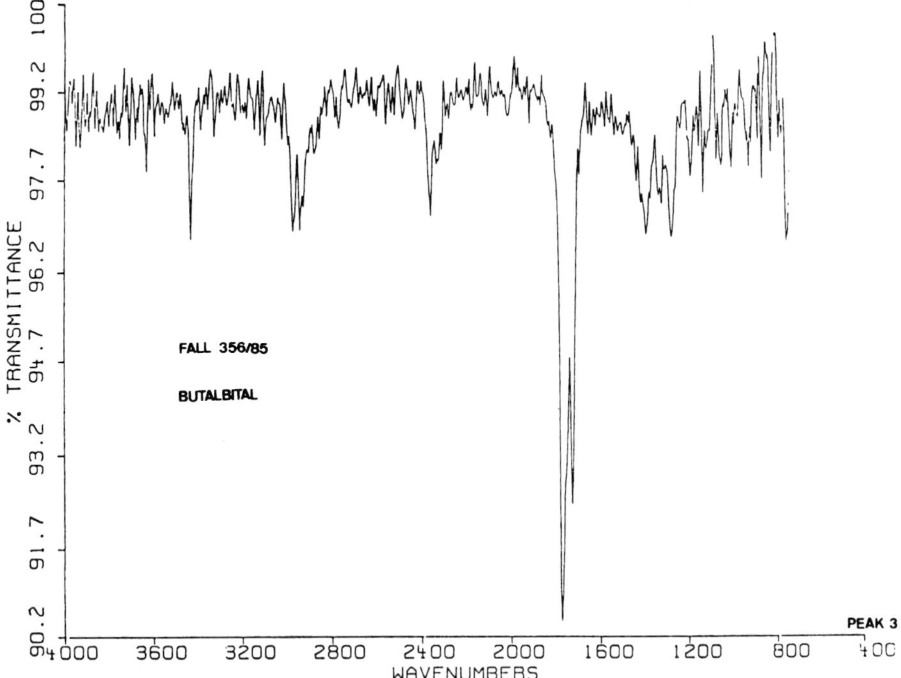

Figure 8. Identification of Butalbital by GC-FTIR in a Case Study.

3.1.11 Mass Spectrometry

3.1.11.1 Positive ion mass spectrometry of barbituric acids

The first application of electron impact (EI) mass spectrometry for the identification of barbiturates stems from Grützmacher and Arnold (24, 25). It was made before the introduction of the on-line GC-MS combination technique. Barbituric acids were separated from biological fluids by liquid extraction, isolated by thin-layer chromatography and analyzed by direct injection MS. Other possibilities, which we have used in the early times of MS, are direct inlet analysis of sublimates or GC-condensates isolated from weakly acid extracts.

The EI fragmentation scheme has been well described by Grützmacher and Arnold (24). Ionization by EI yields a large number of fragments ions; the molecular mass is seldom visible. Spectral interpretation is not easy, since the fragmentation is complicated by rearrangements. Ions in the low mass range contribute to the structural identification of the substituents in position 5.

In contrast to EI-MS, mass spectrometry by chemical ionization (CI) always furnishes molecular mass information (26). The intensive quasi-molecular ion $[M + H^+]$ is base ion; only scant information is supplied in the low mass range. A combination of EI- with CI-MS is a much sounder approach for identifying barbituric acids than the use of just one of the two methods.

3.1.11.2 Negative ion mass spectrometry of barbituric acids

Negative ion MS by CI, carried out at the high source pressure used for positive ion CI (near 1 torr), yields mainly the quasi-molecular ions $[M - H]^-$, that is molecular mass information, just like positive ion CI-MS.

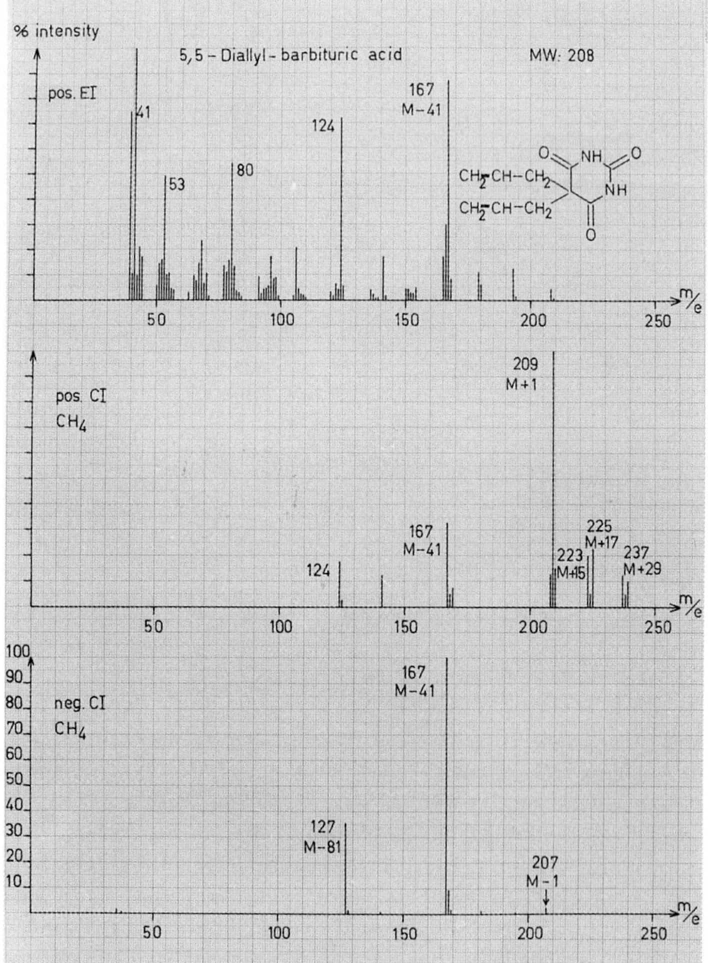

Figure 9. Positive EI, Positive CI and Low Pressure Negative CI Mass Spectra of 5,5-Diallyl-barbituric Acid.

In chapter 1.7.7.2 of this book, we have decribed that negative ion CI-MS (unlike positive ion CI-MS) works also at low reagent gas pressure, near 10^{-2} torr. Lower source pressure produces a harder fragmentation; the molecular and/or quasi-molecular ions formed by electron attachment and by interactions with negative reagent ions (mainly by negative charge exchange) show analytically useful fragmentations. In the case of barbiturates, the most striking degradations are the losses of the intact substituents in position 5 of the ring. The masses of these side chains are given by the differences between molecular ion and masses of two high mass fragment ions (27–30). In negative ion fragmentation, rearrangements play only a minor role (31). With barbiturates, a "negative McLafferty rearrangement" can be observed during the loss of the second side chain in 5-position, as is shown in the negative CI spectrum of 5,5-diallyl-barbituric acid in Figure 9.

A second important feature of negative ion MS in the analysis of barbiturates is the fact that it can be several factors of 10 more sensitive than positive EI-MS. This holds specifically for unsaturated compounds (32).

3.1.11.3 "Dual mass spectrometry" of barbituric acids

Dual MS is a combination of positive EI-MS with low-pressure negative CI-MS, that we have developed and used in our daily service work since 1980 (26, 31–33). The conditions which permit such a quasi-simultaneous combination of two different ionization techniques and its usefulness in stuctural elucidations are discussed in chapter 1.7.7.2. The barbiturates are an instructive example. Negative ion MS yields information on the molecular mass and the brutto composition of both substituents, positive EI-MS reveals their structures. With the information from the two different ionization modes, it is always possible to identify a barbituric acid or barbituric acid metabolite. Figure 10 (mass spectra of barbital and phenobarbital) shows the presentation of the "dual" mass spectra. For examples of identification of barbiturates and metabolites, we refer to the already quoted literature which we has been summarized and discussed in (27).

3.1.12 Concluding Remarks

In many countries, barbiturates are regarded as outdated compounds. But experience has shown, that once well-established "old" pharmaceuticals often return, either in their original composition or in a new form or combination. Every toxicological laboratory must therefore be able to detect and determine them.

Barbituric acids and their metabolites can easily be isolated from biological materials and identified by many instrumental techniques. We recommend recording the UV spectra of acid extracts at different pH values to verify their presence and to indicate, at the same time, the sub-group to which they belong. By far the most powerful approach for identifying the individual compounds and determining the structure of barbiturate metabolites is dual-MS in on-line combination with GC. It yields an optimum of structural information with a sensitivity superior to that of other detection techniques.

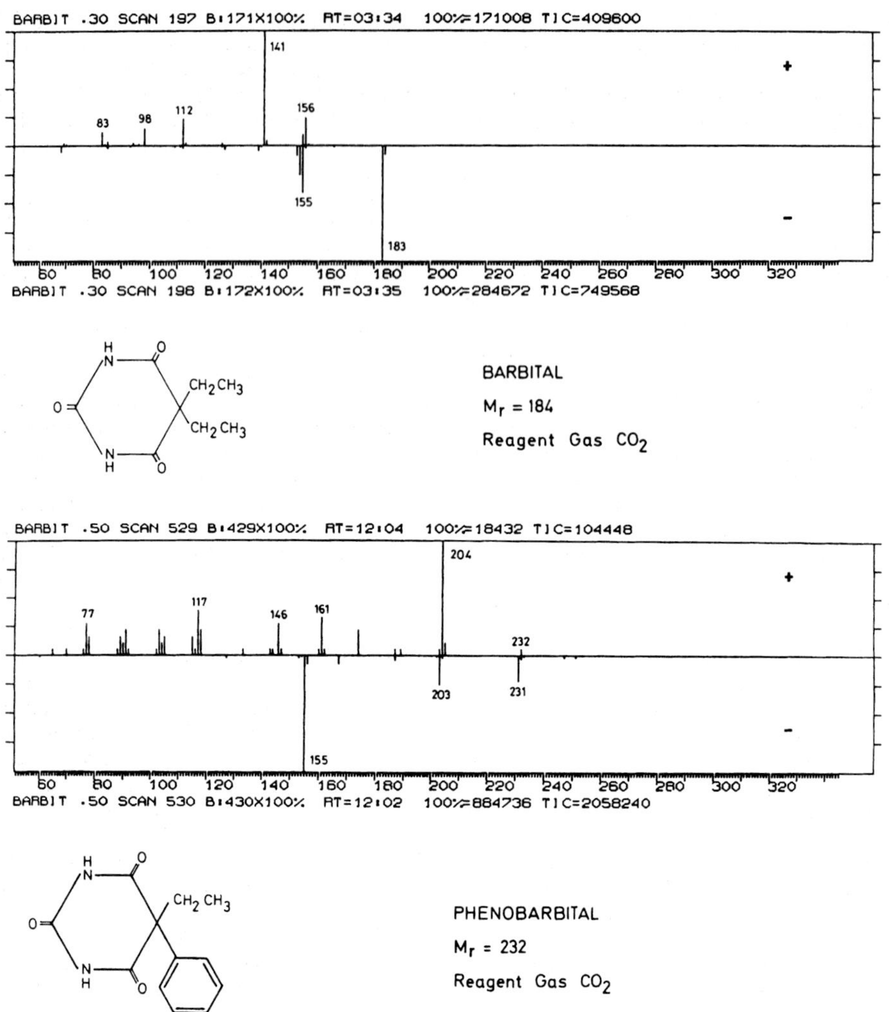

Figure 10. Dual Mass Spectra of Barbital and Phenobarbital.

References

1. v. Baeyer, A. (1863 & 1864) Untersuchungen über die Harnsäuregruppe (first and second communication). Ann. der Chem. & Pharm. *127*, 199–236 and *130*, 129–175.
2. Fischer, E. & Dilthey, A. (1904) Über C-Dialkylbarbitursäuren und über die Ureide der Dialkylessigsäuren. Ann. der Chem. & Pharm. *335*, 334–368.
3. Fischer, E. & v. Mering, J. (1903) Über eine neue Klasse von Schlafmitteln. In: Therapie der Gegenwart, Heft 3. Kurzreferat: Veronal. Therapeutische Monatsh. *17*, 208.
4. Clarke's Isolation and Identification of Drugs in Pharmaceuticals, Body Fluids and Post Mortem Materials, 2nd ed. (1986) (Moffat, A.C., Jackson, J.V., Moss, M.S. & Widdop, B., eds.). The Pharmaceutical Press, London.
5. Brandenberger, H. (1988) Wichtige exogene Gifte und ihre differentialdiagnostische Erkennung. In: Differentialdiagnose Innerer Krankheiten, 16th ed. (Siegenthaler, W., ed.), pp. 39.1–39.18. Georg Thieme, Stuttgart & New York.

6. Uges, D.R.A. (1994) Blood Level Data. In: Part 4 of this volume, where the additional literature references can be found.
7. See entries to the different barbituric acids in Clarke's Isolation and Identification of Drugs (reference 4).
8. Stevens, H.M. (1986) Colour Tests. In: Clarke's Isolation and Identification of Drugs (reference 4), p. 135.
9. Brandenberger, H. & Maes, R.A.A. (1988) Einsatz von immunochemischen Testen in der Suchtmittelanalytik. Empfehlungen zur klinisch-toxikologischen Analytik, Folge 1. Edited by the Senate Commission for Clinical-toxicological Analysis of the German Research Society. VCH, Weinheim.
10. Brandenberger, H. (1974) Toxicology. In: Clinical Biochemistry, Principles and Methods (Curtius, H.Ch. & Roth, M., eds.) Vol. II, pp. 1447–1453. Walter de Gruyter, Berlin & New York.
11. Brandenberger, H. (1974) in reference 10, pp. 1456–1457.
12. DFG (Deutsche Forschungsgemeinschaft) & TIAFT (The International Association of Forensic Toxicologists) (1992). Thin-Layer Chromatographic R_f Values of Toxicologically Revelant Substances on Standardized Systems, 2nd ed. (de Zeeuw, R., Franke, J.P., Degel, F., Machbert, G., Schütz, H. & Wijsbeek, J., eds.). Report XVII of the DFG Commission for Clinical-Toxicological Analysis. VCH, Weinheim.
13. Moffat, A.C. (1986) Thin layer chromatography. In: Clarke's Isolation and Identification of Drugs (reference 4), pp. 160–177.
14. Brandenberger, H. (1974) in reference 10, pp. 1454–1455 (Table 4).
15. DFG (Deutsche Forschungsgemeinschaft) & TIAFT (The International Association of Forensic Toxicologists) (1992). Gas Chromatographic Retention Indices of Toxicologically Relevant Substances on Packed or Capillary Columns with Dimethylsilicone Stationary Phases. 3rd ed. (de Zeeuw, R., Franke, J.P., Maurer, H.H. & Pfleger, K., eds.). Report XVIII of the DFG Comission for Clinical-Toxicologial Analysis, VCH, Weinheim.
16. Brandenberger, H. (1966) Methodische Entwicklung in der chemischen Toxikologie. Deutsche Zeitschr. für Ger. Medizin *57*, 300–310.
17. Brandenberger, H. (1967) Moderne chemische Methoden im Dienste der praktischen Medizin. Praxis *56*, 1302–1307.
18. Griffiths, P.R. & de Haseth, J.A. (1986) Fourier Transform Infrared Spectrometry. Wiley-Interscience, New York.
19. White, R. (1990) Chromatography/Fourier Transform Infrared Spectroscopy and its Applications. Marcel Dekker, New York.
20. Henry, E.H., Giorgetti, A., Haefner, M., Griffiths, P. & Gurka, D.F. (1987) Optimizing the optical configuration for lightpipe gas chromatography/Fourier transform infrared spectrometry. Anal. Chem. *59*, 2356–2361.
21. Brandenberger, H. & Wittenwiler, P. (1986) The combination of on-line gas chromatography with Fourier transform infrared spectrophotometry for chemical toxicology. In: Proceedings of 23rd International Meeting of "The International Association of Forensic Toxicologists" and the 2nd World Congress on new Compounds in Biological and Chemical Warfare, Ghent, Belgium (Heyndrickx, A., ed.), pp. 231–239. Department of Toxicology, State University of Ghent.
22. Brandenberger, H. (1987) The role of FTIR in chemical toxicology. Increased sensitivity of the on-line combination with GC by Derivatization. Proceedings of the 24th International Meeting of "The International Association of Forensic Toxicologists", Banff, Canada (Jones, G.R. & Singer, P.P., eds.), pp. 330–339. University of Alberta Printing Services, Alberta.
23. Wilkins, Ch.L. (1987) Linked gas chromatography-infrared-mass spectrometry. Anal. Chem. *59*, 571A–581A.
24. Grützmacher, H.F. & Arnold, W. (1966) Massenspektren von Barbitursäurederivaten. Tetrahedron Lett. *13*, 1365–1374.
25. Arnold, W. & Grützmacher, H.F. (1964) Mass spectrometric detection of drug metabolites (barbiturates, Noludar, Pyramidon) in forensic analysis. Z. anal. Chem. *247*, 179–188.
26. Fales, H.M., Milne, G.W.A. & Axenrod, T. (1970) Identification of barbiturates by chemical ionization mass spectrometry. Anal. Chem. *42*, 1432–1435.
27. Brandenberger, H. (1987) Low pressure negative chemical ionization mass spectrometry and dual mass spectrometry: a review of the work since 1977. In: TIAFT 22, Reports on Forensic Toxicology (Proceedings of the International Meeting of The International Association of Forensic Toxicologists, August 1985, Rigi Kaltbad, Switzerland (Brandenberger, H. & Brandenberger, R.,

eds.). CDC-Print (1987) available from the editors, CH-6356 Rigi Kaltbad.
28. Brandenberger, H. & Ryhage, R. (1977) Possibilities of negative ion mass spectrometry in forensic and toxicological analyses. Mass Fragments *1*, 1–4 (a LKB publication, Stockholm, Sweden) and Blood Drugs and other analytical challenges (Reid, E., ed.), pp. 173–184. Ellis Horwood, Chichester, U.K., distributed by John Wiley, Chichester, New York, London, Brisbane & Toronto.
29. Brandenberger, H. & Ryhage, R. (1978) New developments in the field of mass spectrometry and GC-MS and their applications to toxicological analysis. In: Recent Developments in Mass Spectrometry in Biochemistry and Medicine, Vol. I (Frigerio, A., ed.), pp. 327–341. Plenum Publ. Corp., New York.
30. Jones, V. & Whitehouse, M. J. (1981) Anion mass spectrometry of barbiturates. Biomedical Mass Spectrometry *8*, 231–236.
31. Brandenberger, H. (1983) Negative ion mass spectrometry by low pressure chemical ionization: inherent possibilities for structural analysis. Intern. Journ. of Mass Spectrometry and Ion Physics *47*, 213–216.
32. Brandenberger, H. (1979) Present state and future trends of negative ion mass spectrometry in toxicology, forensic and environmental chemistry. In: Recent Developments in Mass Spectrometry in Biochemistry and Medicine, Vol. II (Frigerio, A., ed.), pp. 227–255. Plenum Publ. Corp., New York.
33. Brandenberger, H. (1982) Negative ion mass spectrometry by low pressure chemical ionization and its combination with conventional positive ion mass spectrometry by electron impact. Abstracts of the 30th Annual Conference on Mass Spectrometry and allied topics, Honolulu, USA, pp. 468–469. Edited by the American Society for Mass Spectrometry.

3.2 Benzodiazepines

H. Schütz

3.2.1 Introduction

The story of the benzodiazepines starts in the mid-fifties (1–5). In May 1957 the first compound (chlordiazepoxide) revealed promising pharmacodynamic properties in several pharmacological screening tests in animals. This exciting new class, the benzodiazepines, was patented in July of 1959, and some months later chlordiazepoxide was marketed under the well known trade name Librium. Further activities resulted in the introduction of diazepam (1963) and oxazepam (1965). Within a short time over 3000 benzodiazepine derivatives were synthesized and tested for their pharmacological action. It was soon obvious that the substitution patterns (see formula in 3.2.1.2) have an important influence on the pharmacokinetic and pharmacodynamic activities. Significant positions are R^1, R^3, R^7 and $R^{2'}$ on the 1,4-(5-phenyl)-benzodiazepine ring. The knowledge of the structure-activity relationships (6) made it possible to predict the pharmacokinetic properties. Many variations were carried out and sophisticated derivatives were synthesized. Of particular interest are in the last years compounds in which an additional heterocyclic ring joins the positions 1 and 2 of the benzodiazepine ring system. This is exemplified by the triazolobenzodiazepines (e.g. alprazolam and triazolam) and the imidazolobenzodiazepine midazolam which is clinically used as a potent anesthetic agent.

At present over 40 1,4-benzodiazepines are on the market. Many of them are metabolites (e.g. nordazepam, oxazepam and temazepam are biotransformation products of diazepam). 1,5-Benzodiazepines (e.g. clobazam) are only of minor importance. The intensive present discussions on the addiction potential of benzodiazepines and/or the critical cost situation in the health sector may be the cause for a certain stagnation of new developments. On the other hand, benzodiazepines are now as before a highly valuable class of drugs with indispensable pharmacodynamic properties.

The structures of 44 benzodiazepines of national or international interest are listed in table 1. Most of the compounds may be derived from the basic structures A, B, C and D which are also presented below.

Within the scope of this contribution, only parent compounds are treated but some of them are also metabolites. The complete data of 115 benzodiazepines, including metabolites and important hydrolysis products (aminobenzophenones and benzoylpridine derivatives), were collected by Schütz (7–10).

Table 1. Structures of Benzodiazepines for Therapeutic Use

No.	Substance	Type	Substituents of Type A and B R^1	R^2(B)	R^3(A)	R^7	R$^{2'}$	Type C R^1	R$^{2'}$	Substituents of Type D R^2	R^3	R^7	R^{10}	R$^{2'}$
1	Adinazolam	C	–	–	–	–	–	–[a]	–H	–	–	–	–	–
2	Alprazolam	C	–	–	–	–	–	$-CH_3$	–H	–	–	–	–	–
3	Bromazepam	–	see structure no. 1											
4	Brotizolam	–	see structure no. 2											
5	Camazepam	A	$-CH_3$	–	–[b]	–Cl	–H	–	–	–	–	–	–	–
6	Chlordiazepoxide	–	see structure no. 3											
7	Clobazam	–	see structure no. 4											
8	Clonazepam	A	–H	–	–H	$-NO_2$	–Cl	–	–	–	–	–	–	–
9	Clorazepate	A	–H	–	–COOK	–Cl	–H	–	–	–	–	–	–	–
10	Clotiazepam	–	see structure no. 5											
11	Cloxazolam	D	–	–	–	–	–	–	–	–H	–H	–H	–Cl	–Cl
12	Delorazepam	A	–H	–	–H	–Cl	–Cl	–	–	–	–	–	–	–
13	Demoxepam	–	see structure no. 6											
14	Diazepam	A	$-CH_3$	–	–H	–Cl	–H	–	–	–	–	–	–	–
15	Estazolam	C	–	–	–	–	–	–H	–H	–	–	–	–	–
16	Ethyl-loflazepate	A	–H	–	–[c]	–Cl	–F	–	–	–	–	–	–	–
17	Etizolam	–	see structure no. 7											
18	Fludiazepam	A	$-CH_3$	–	–H	–Cl	–F	–	–	–	–	–	–	–
19	Flumazenil	–	see structure no. 8											
20	Flunitrazepam	A	$-CH_3$	–	–H	$-NO_2$	–F	–	–	–	–	–	–	–
21	Flurazepam	A	–[d]	–	–H	–Cl	–F	–	–	–	–	–	–	–
22	Flutazolam	D	–	–	–	–	–	–	–	–H	–H	–[e]	–Cl	–F
23	Halazepam	A	–[f]	–	–H	–Cl	–H	–	–	–	–	–	–	–
24	Haloxazolam	D	–	–	–	–	–	–	–	–H	–H	–H	–Br	–F
25	Ketazolam	–	see structure no. 9											

26	Loprazolam	–	see structure no. 10											
27	Lorazepam	A	$-H$	–	$-OH$	$-Cl$	$-Cl$	–	–	–	–	–	–	–
28	Lormetazepam	A	$-CH_3$	–	$-OH$	$-Cl$	$-Cl$	–	–	–	–	–	–	–
29	Medazepam	B	$-CH_3$	$-H, -H$	$-H$	$-Cl$	$-H$	–	–	–	–	–	–	–
30	Metaclazepam	B	$-CH_3$	$-H, -^{g}$	$-H$	$-Br$	$-Cl$	–	–	–	–	–	–	–
31	Mexazolam	D	–	–	–	–	–	–	–	$-H$	$-CH_3$	$-H$	$-Cl$	$-Cl$
32	Midazolam	–	see structure no. 11											
33	Nimetazepam	A	$-CH_3$	–	$-H$	$-NO_2$	$-H$	–	–	–	–	–	–	–
34	Nitrazepam	A	$-H$	–	$-H$	$-NO_2$	$-H$	–	–	–	–	–	–	–
35	Nordazepam	A	$-H$	–	$-H$	$-Cl$	$-H$	–	–	–	–	–	–	–
36	Oxazepam	A	$-H$	–	$-OH$	$-Cl$	$-H$	–	–	–	–	–	–	–
37	Oxazolam	D	–	–	–	–	–	–	–	$-CH_3$	$-H$	$-H$	$-Cl$	$-H$
38	Pinazepam	A	$-^{h}$	–	$-H$	$-Cl$	$-H$	–	–	–	–	–	–	–
39	Prazepam	A	$-^{i}$	–	$-H$	$-Cl$	$-H$	–	–	–	–	–	–	–
40	Quazepam	B	$-^{f}$	$=S$	$-H$	$-Cl$	$-F$	–	–	–	–	–	–	–
41	Temazepam	A	$-CH_3$	–	$-OH$	$-Cl$	$-H$	–	–	–	–	–	–	–
42	Tetrazepam	–	see structure no. 12											
43	Tofisopam	–	see structure no. 13											
44	Triazolam	C	–	–	–	–	–	$-CH_3$	$-Cl$	–	–	–	–	–

[a] $-CH_2-N(CH_3)_2$
[b] $-O-CO-N(CH_3)_2$
[c] $-CO-O-C_2H_5$
[d] $-(CH_2)_2-N(C_2H_5)_2$
[e] $-(CH_2)_2-OH$
[f] $-CH_2-CF_3$
[g] $-CH_2-O-CH_3$
[h] $-CH_2-C \equiv CH$
[i] $-CH_2-\triangleleft$

3.2.2 Structures of the Benzodiazepines

The basic structures of most of the compounds listed in table 1 belong to one of the following 4 types (A to D)

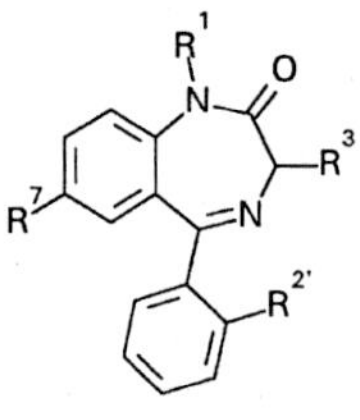

Type A

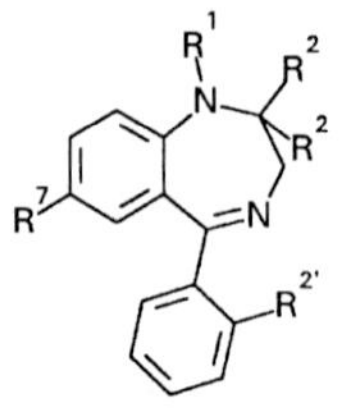

Type B

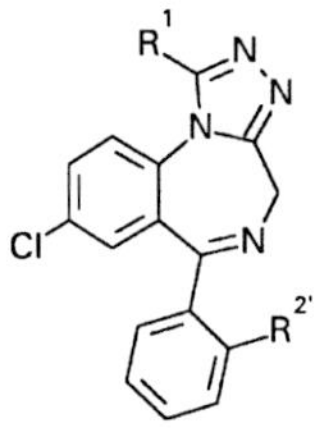

Type C

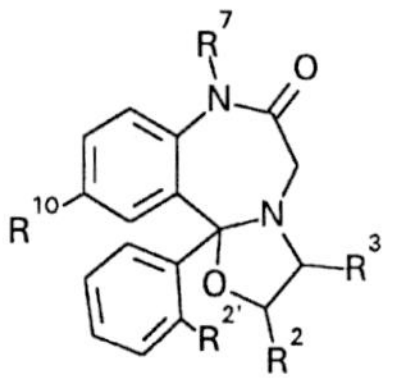

Type D

The structures of the following compounds are not covered by the general formulas A to D.

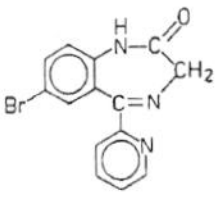

1 Bromazepam

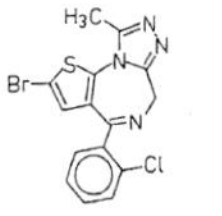

2 Brotizolam

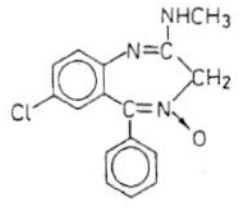

3 Chlordiazepoxide

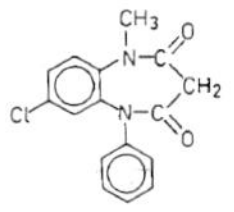

4 Clobazam

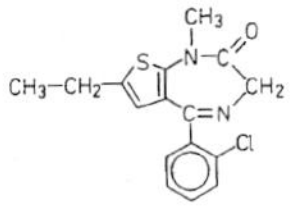

5 Clotiazepam

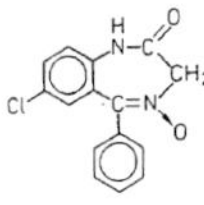

6 Demoxepam

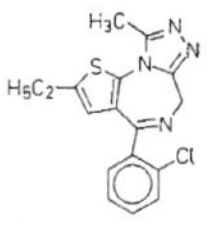

7 Etizolam

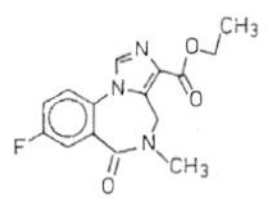

8 Flumazenil

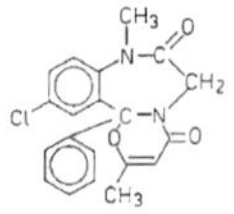

9 Ketazolam

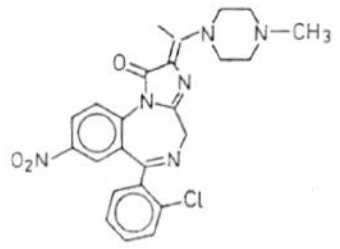

10 Loprazolam

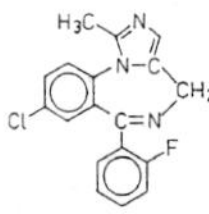

11 Midazolam

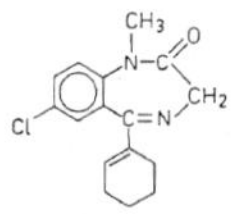

12 Tetrazepam

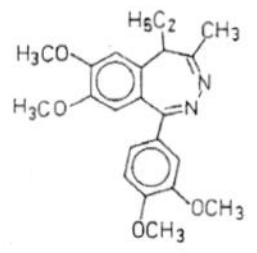

13 Tofisopam

3.2.3 Pharmacodynamic Properties of the Benzodiazepines

All benzodiazepine derivatives in clinical use possess **anxiolytic**, **hypnotic**, **anticonvulsant** and **muscle relaxant** properties within their pharmacodynamic spectrum. Only the predominance of certain components in the spectrum of action seems to vary, e.g. clonazepam is only used for the treatment of epilepsy and tetrazepam as central acting muscle relaxant. The popularity of the benzodiazepines is mainly caused by their wide therapeutic index, minimal serious adverse reactions, and the absense of undesirable autonomic nervous side effects especially when compared with formerly used psychotropic agents, e.g. meprobamate or phenobarbital (11).

Poser et al. (12, 13) published a list of equivalents of some benzodiazepines in relation to 10 mg diazepam (table 2).

Table 2. Equivalents of Benzodiazepines according to Poser et al. (12)

Substance	Equivalent Dose Relating to 10 mg Diazepam
Bromazepam	3 mg
Chlordiazepoxide	20 mg
Clobazam	20 mg
Clonazepam	2 mg
Clorazepate	20 mg
Clotiazepam	10 mg
Flunitrazepam	2 mg
Flurazepam	30 mg
Lorazepam	2 mg
Lormetazepam	1 mg
Medazepam	20 mg
Nitrazepam	5 mg
Oxazepam	40 mg
Prazepam	20 mg
Triazolam	1 mg

The action of benzodiazepines is based on specific benzodiazepine receptors mainly located in the limbic system and mediated by GABA and cyclic nucleotides (14–18). There is also evidence for naturally existing benzodiazepines in low concentrations (19–21), which may cumulate in subjects with hepatic impairment (hepatic encephalopathy) (22).

3.2.4 Pharmacokinetics of the Benzodiazepines

Biotransformation

Most benzodiazepines are extensively metabolized by phase I- (predominantly desalkylation, aliphatic and aromatic hydroxylation, reduction, acetylation) and phase II-reactions (formation of conjugates). In most cases the metabolites formed by phase I-reactions are also biologically active, whereas the conjugates possess no remarkable actions. Also see Beyer (23) and Pfeifer (24) for biotransformation of

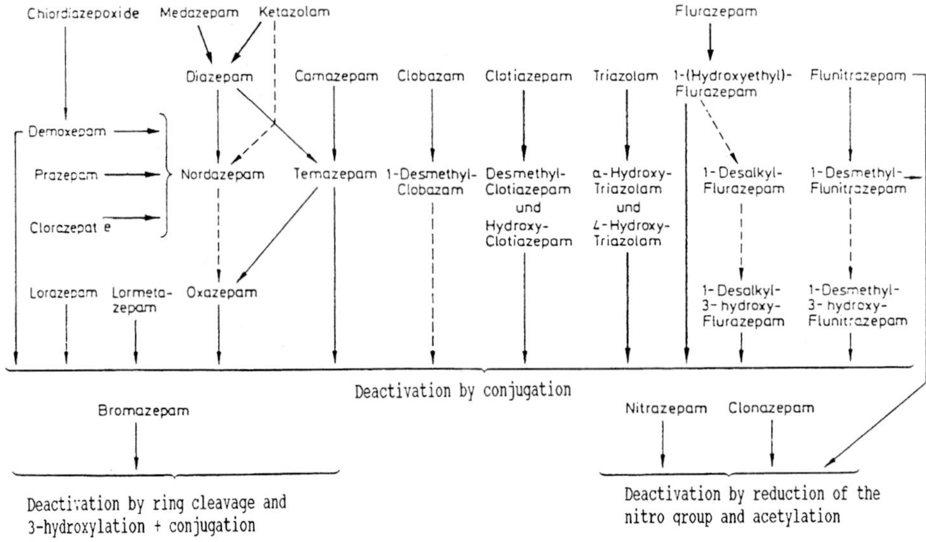

Figure 1. Simplified Survey of the Biotransformation of Numerous Benzodiazepines of Clinical Importance [25].

benzodiazepines and other drugs. Figure 1 describes a synoptical biotransformation scheme for many benzodiazepines with clinical importance (according to (25), modified).

As examples, the biotransformation of flunitrazepam, an important 7-nitrosubstituted benzodiazepine is presented in figure 2 and the metabolic pathways of triazolam (a triazolo benzodiazepine) are described in figure 3.

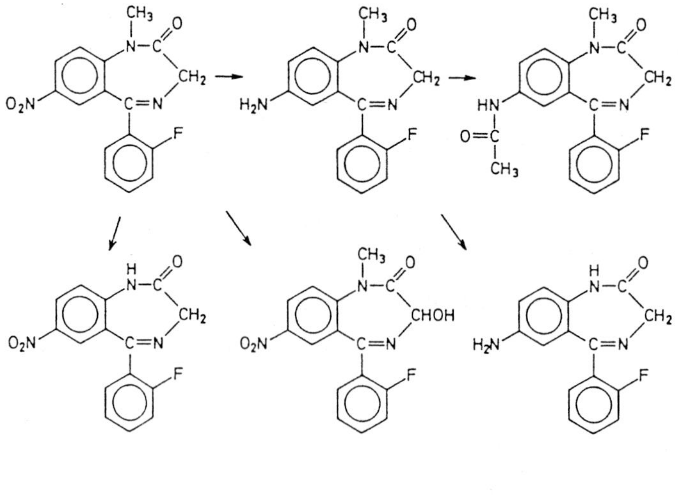

Figure 2. Biotransformation of Flunitrazepam (Kaplan et al. [26])

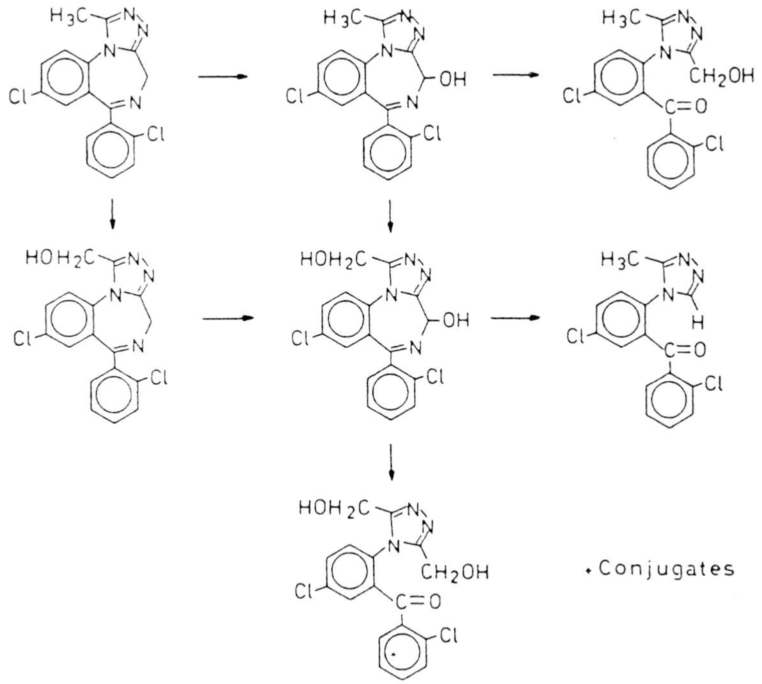

Figure 3. Biotransformation of Triazolam (Kitagawa et al. [27, 28], Eberts et al. [29])

Kinetics

Knowledge of the pharmacokinetics of benzodiazepines is the base for their therapeutic use, since kinetic data, e.g. the elimination half-life, have an important influence on the duration of their action. Table 3 presents elimination half-lifes, affinity values for ligand bindings and plasmaprotein-binding data according to Müller (30) and Pöldinger and Wider (31). A differentiation of benzodiazepines on the base of elimination half-lifes and the pharmacodynamic activities of their major metabolites is compiled in table 4 according to Greenblatt et al. (32). Many other pharmacokinetic data from literature were collected by Klotz et al. (11), Müller (30) and Schütz (7, 10).

Table 3. List of Elimination Half-Lifes, Affinity-Values (Müller (30)) and Plasmaprotein-Binding (Pöldinger and Wider (31)) of Benzodiazepines with Therapeutic Importance

Substance	Elimination Half-Time (hr)	IC_{50}* (nmol/l)	Plasmaprotein-Binding (%)
Alprazolam	10–18	20	70–80
Broazepam	10–24	18	45
Brotizolam	4– 8	1	90
Chlordiazepoxide	10–18	350	89–94
Clobazam	10–30	130	87–90
Clonazepam	24–56	2	
Clorazepate	2– 3	59	95–98

Table 3. (continued)

Substance	Elimination Half-Time (hr)	IC_{50}* (nmol/l)	Plasmaprotein-Binding (%)
Clotiazepam	3–15	2	99–100
Diazepam	30–45	8	96–98
Flunitrazepam	10–25	4	80
Flurazepam	2	15	15
Ketazolam	1– 2	1300	80–93
Loprazolam	7–10	5	
Lorazepam	10–18	4	85–94
Lormetazepam	9–15	4	>85
Medazepam	2	870	99
Midazolam	1– 3	5	95
Nitrazepam	20–50	10	85–88
Nordazepam	50–80	9	96–98
Prazepam	1– 3	110	97
Temazepam	6–16	16	76
Tetrazepam	12	34	
Triazolam	2– 4	4	89

* half-maximum value of inhibition concentration for specific ligand-binding (low values = high affinity)

Table 4. Pharmacokinetic Properties of Important Benzodiazepines and Metabolites according to Greenblatt et al. (32)

Substance		Active Metabolite	Elimination Half-Life
I. Benzodiazepines with long terminal Half-Life and long acting Metabolites			
Chlordiazepoxide		Nordazepam (+ others)	36–200 hr
Clorazepate*		Nordazepam	36–200 hr
Diazepam	20–40 hr	Nordazepam	36–200 hr
Flurazepam*		Desalkylflurazepam	40–250 hr
Medazepam*		Diazepam (+ others)	36–200 hr
Prazepam*		Nordazepam	36–200 hr
II. Benzodiazepines with medium to short terminal Half-Life and active Metabolites			
Bromazepam	10–20 hr	3-Hydroxybromazapem	
Estazolam	10–30 hr	Hydroxy-Metabolites	
Flunitrazepam	ca. 15 hr	7-Amino-Derivative	ca. 25 hr
III. Benzodiazepines with medium to short terminal Half-Life without active Metabolites			
Lorazepam	10–20 hr		
Lormetazepam	8–14 hr		
Nitrazepam	15–38 hr		
Oxazepam	4–15 hr		
Temazepam	8–14 hr		
IV. Benzodiazepines with ultrashort Half-Life and active Metabolites			
Brotizolam	4– 9 hr	4-Hydroxybrotizolam	4–9 hr
		α-Hydroxybrotizolam	4–9 hr
Midazolam	2– 4 hr	4-Hydroxymidazolam	ca. 1 hr
		α-Hydroxymidazolam	
Triazolam	2– 5 hr	4-Hydroxytriazolam	2–5 hr
		α-Hydroxytriazolam	2–5 hr

* May be regarded as prodrug

3.2.5 Benzodiazepine Doses, Serum Levels and Toxicity

Overdosage with benzodiazepines generally results in drowsiness, ataxia, muscular weakness and deep coma. It is said, that lethal monointoxications with classical benzodiazepines (e.g. chlordiazepoxide, diazepam, oxazepam) are not possible. However, they must be taken into consideration when benzodiazepines with extremely low elimination half-lives are used.

The concentrations presented in table 5 were published by Uges et al. (33) (see also part 4 of this book). They may serve as orientating values. Also see table 2 for equivalent doses of many benzodiazepines relating to 10 mg diazepam.

Table 5. Orientating values of Serum Concentrations of Benzodiazepines (Therapeutic and Toxic Range) according to Uges et al. (33)

Substance	Therapeutic Range (mg/l Serum)	Toxic Symptoms Observed (mg/l Serum)
Bromazepam	0.08–0.17	> 0.25–0.5
Camazepam	0.1–0.6	> 2
Chlordiazepoxide	0.7–2	> 3.5–10
(Demoxepam)		> 0.3–2.8
Clobazam	0.1–0.4	
(Norclobazam)	2–4	
Clonazepam	0.03–0.06	> 0.1
Clorazepate		
(Nordazepam)	0.25–0.75	> 2
Diazepam		> 1.5–3
(anxiolytic)	0.124–0.25	
(anticonvulsive)	0.25–0.50	
(eclampsia)	1–1.5	
Flunitrazepam	0.005–0.015	> 0.05
Flurazepam	0.005–0.01	> 0.15
(N-1-Desalkylfl.)	0.04–0.15	> 0.2
Ketazolam		
(Nordazepam)	0.2–0.8	> 2
Loprazolam	0.005–0.01	
Lorazepam	0.02–0.25	> 0.3–0.5
Lormetazepam	0.002–0.01	
Medazepam	0.01–0.15	> 0.6
(Nordazepam)	0.2–0.8	> 2
Midazolam	0.08–0.25	
Nitrazepam		
(anxiolytic)	0.03–0.05	> 0.2–0.5
(antiepileptic)	0.05–0.12	> 0.2–0.5
Nordazepam	0.2–0.8	> 2
Oxazepam	1–2	> 3–5
Prazepam	0.05–0.2	> 1
Temazepam	0.3–0.8	> 1
Triazolam	0.002–0.02	

Baselt and Cravey (34) collected the following blood concentrations (mg/l) in connection with fatal intoxications:

Alprazolam	0.122–0.39
Chlordiazepoxide	20–26
Diazepam	5–19
Flurazepam	3.2
Nitrazepam	1.2–9
Temazepam	0.9–14

The **interaction** of benzodiazepines with other centrally acting compounds (e. g. alcohol, tricyclic antidepressants, phenothiazines, barbiturates) may lead to fatal intoxications.

3.2.6 Screening of Benzodiazepines

3.2.6.1 Immunochemical methods

The screening of benzodiazepines is mainly based on immunochemical and chromatographic methods where thin-layer chromatography plays now as before an important role. The principle of immunochemical methods is the antigen-antibody reaction and the predominantly used tests are enzymimmunoassays (EIA, e. g. EMIT, ETS) and fluorescent-polarization immunoassays (FPIA, e. g. ADx, TDxFLx). These tests have proven their usefulness as screening procedures, but many pitfalls must be taken into account when they are uncritically applied, as false negative results may be caused by conjugate formation, poor cross-reactivities and/or low concentrations. Typical cases and modifications which enables one to obtain valid results are described by Schütz et al. (35, 36). Also see Brandenberger et al. (37) for recommendations concerning immunochemical tests.

As an example, actual **cross reactivities** of the FPIA TDxFLx are presented in table 6 (urine) and table 7 (serum).

Table 6. Cross-Reactivities of Important Benzodiazepines and Metabolites (TDx/TDxFLx Benzodiazepine (Urine))

Substance	Added (ng/ml)	Found (ng/ml)	Cross-Reactivity (%)
Alprazolam	2,400	1,451.1	60.5
	1,200	983.0	81.9
	800	768.7	96.1
	400	436.8	109.2
	200	232.9	116.5
Bromazepam	2,400	337.4	14.1
	1,200	240.9	20.1
	800	192.8	24.1
	400	127.6	31.9
	200	82.6	41.3

Table 6. (continued)

Substance	Added (ng/ml)	Found (ng/ml)	Cross-Reactivity (%)
Chlordiazepoxide	2,400	160.9	6.7
	1,200	115.2	9.6
	800	92.6	11.6
	400	65.1	16.3
	200	44.7	22.4
Clobazam	10,000	807.2	8.1
	1,000	272.3	27.2
Clonazepam	2,400	368.7	15.4
	1,200	283.1	23.6
	800	201.5	25.2
	400	149.7	37.4
	200	94.6	47.3
Demoxepam	2,400	301.4	12.6
	1,200	195.2	16.3
	800	160.2	20.0
	400	103.8	26.0
	200	66.6	33.3
Desalkylflurazepam	2,400	882.9	36.8
	1,200	558.9	46.6
	800	423.8	53.0
	400	226.4	56.6
	200	118.3	59.2
Diazepam	2,400	HI	–
	1,200	1,722.2	143.5
	800	1,102.1	137.8
	400	535.6	133.9
	200	246.2	123.1
Estazolam	1,000	881.5	88.2
	100	132.4	132.4
	10	ND*	–
Flunitrazepam	2,400	749.4	31.2
	1,200	421.0	35.1
	800	322.2	40.3
	400	209.3	52.3
	200	140.2	70.1
Flurazepam	2,400	656.9	27.4
	1,200	468.7	39.1
	800	386.8	48.4
	400	241.4	60.4
	200	149.9	75.0
1-N-Hydroxy-ethyl-flurazepam	2,400	1,246.2	51.9
	1,200	769.2	64.1
	800	573.8	71.7
	400	335.9	84.0
	200	179.6	89.8
Lorazepam	2,400	410.1	17.1
	1,200	289.0	24.1
	800	232.2	29.0
	400	153.3	38.3
	200	100.2	50.1

Table 6. (continued)

Substance	Added (ng/ml)	Found (ng/ml)	Cross-Reactivity (%)
Medazepam	2,400	1,133.2	47.2
	1,200	811.6	67.6
	800	637.9	79.7
	400	368.5	92.1
	200	197.5	98.8
Midazolamhydro-chloride	2,400	1,070.3	44.6
	1,200	743.9	62.0
	800	570.3	71.3
	400	351.4	87.9
	200	196.9	98.5
Nimetazepam	100,000	HI	–
	10,000	1,846.0	18.5
	1,000	419.0	41.9
	100	90.0	90.0
	10	ND*	–
Nitrazepam	2,400	736.9	30.7
	1,200	503.9	42.0
	800	391.7	49.0
	400	263.6	65.9
	200	178.2	89.1
Norchlordiazepoxide	2,400	177.9	7.4
	1,200	122.3	10.2
	800	95.6	12.0
	400	67.9	17.0
	200	44.4	22.2
Oxazepam	2,400	868.1	36.2
	1,200	557.7	46.5
	800	430.2	53.8
	400	261.3	65.3
	200	152.1	76.1
Prazepam	2,400	1,684.2	70.2
	1,200	1,037.4	86.5
	800	797.4	99.7
	400	458.6	114.7
	200	237.2	118.6
Temazepam	2,400	1,142.2	47.6
	1,200	688.2	57.4
	800	504.2	63.0
	400	270.9	67.7
	200	146.5	73.3
Triazolam	2,400	550.6	22.9
	1,200	429.2	35.8
	800	363.5	45.4
	400	261.9	65.5
	200	165.3	82.7

* ND = not detected (concentration lower than assay-sensitivity (40.00 ng/ml)

Table 7. Cross-Reactivities of Important Benzodiazepines and Metabolites (TDx/TDxFLx Benzodiazepine (Serum))

Substance	Added (ng/ml)	Found (ng/ml)	Cross-Reactivity (%)
Alprazolam	700	423.54	60.5
	300	231.30	77.1
	75	58.89	78.5
	25	15.74	63.0
Bromazepam	700	79.74	11.4
	300	43.65	14.6
	75	ND*	–
Chlordiazepoxide	700	63.55	9.1
	300	39.16	13.1
	75	ND*	–
Norchlordiazepoxide	700	111.41	15.9
	300	57.82	19.3
	75	13.92	18.6
Demoxepam	700	105.07	15.0
	300	62.78	20.9
	75	21.33	28.4
Clobazam	1000	180.00	18.0
	100	62.70	62.7
Clonazepam	700	133.34	19.0
	300	87.37	29.1
	75	33.56	44.7
7-Amino-clonazepam	700	77.20	11.0
	300	48.93	16.3
	75	15.56	20.7
Diazepam	700	669.46	95.6
	300	288.41	96.1
	75	63.71	84.9
Estazolam	1000	91.28	9.1
Flunitrazepam	700	217.19	31.0
	300	137.12	45.7
	75	45.35	60.5
Norflunitrazepam	700	168.67	24.1
	300	109.19	36.4
	75	35.19	46.9
Flurazepam	700	171.18	24.5
	300	108.76	36.3
	75	36.10	48.1
Desalkylflurazepam	700	380.66	54.4
	300	187.24	62.4
	75	53.21	70.9
1-N-Hydroxy-ethyl-flurazepam	700	367.35	52.5
	300	176.57	58.9
	75	45.75	61.0
Halazepam	700	255.50	36.5
	300	136.39	45.5
	75	36.66	48.9
Lorazepam	700	152.47	21.8
	300	90.06	30.0
	75	30.00	40.0
Medazepam	700	272.82	39.0
	300	172.93	57.6
	75	51.32	68.4

Table 7. (continued)

Substance	Added (ng/ml)	Found (ng/ml)	Cross-Reactivity (%)
Midazolam	1000	410.96	41.1
	100	114.17	114.2
Nimetazepam	700	262.71	37.5
	300	156.46	52.2
	75	46.11	61.5
Nitrazepam	700	217.75	31.1
	300	124.89	41.6
	75	42.34	56.5
7-Amino-nitrazepam	700	91.50	13.1
	300	55.17	18.4
	75	13.64	18.2
Oxazepam	700	282.70	40.4
	300	163.18	54.4
	75	57.05	76.1
Prazepam	700	392.40	56.1
	300	207.42	69.1
	75	50.03	66.7
	25	14.01	56.0
Temazepam	700	285.44	40.8
	300	184.65	61.6
	75	56.78	75.7
Triazolam	700	297.76	42.5
	300	169.49	56.5
	75	49.11	65.5
	25	12.39	49.6

* ND = not detected (concentration lower than assay-sensitivity (40.00 ng/ml)

3.2.6.2 TLC-methods

Thin-layer-chromatography (TLC) is the preferred method for screening benzodiazepines and their metabolites. The procedure involves hydrolysis to yield aminobenzophenone derivatives, which are then extracted, separated by TLC and photolytically dealkylated. The products are diazotized and coupled with azo-dyes (e.g. the Bratton-Marshall reagent) (fig. 4). See Schütz (8) and Rochholz et al. (38) for additional details.

Experimental

Recommended reference substances for commonly used benzodiazepines and their metabolites (hR_f-values in toluene):

ABFB = 2-amino-5-bromo-2′-fluorobenzophenone (e.g. from haloxazolam) ($hR_f = 32$)
ABP = (2-amino-5-bromophenyl)(2-pyridyl)methanone (e.g. from bromazepam) ($hR_f = 2$)
ACB = 2-amino-5-chlorobenzophenone (e.g. from oxazepam) ($hR_f = 27$)
ACFB = 2-amino-5-chloro-2′-fluorobenzophenone (e.g. from desalkylflurazepam) ($hR_f = 31$)

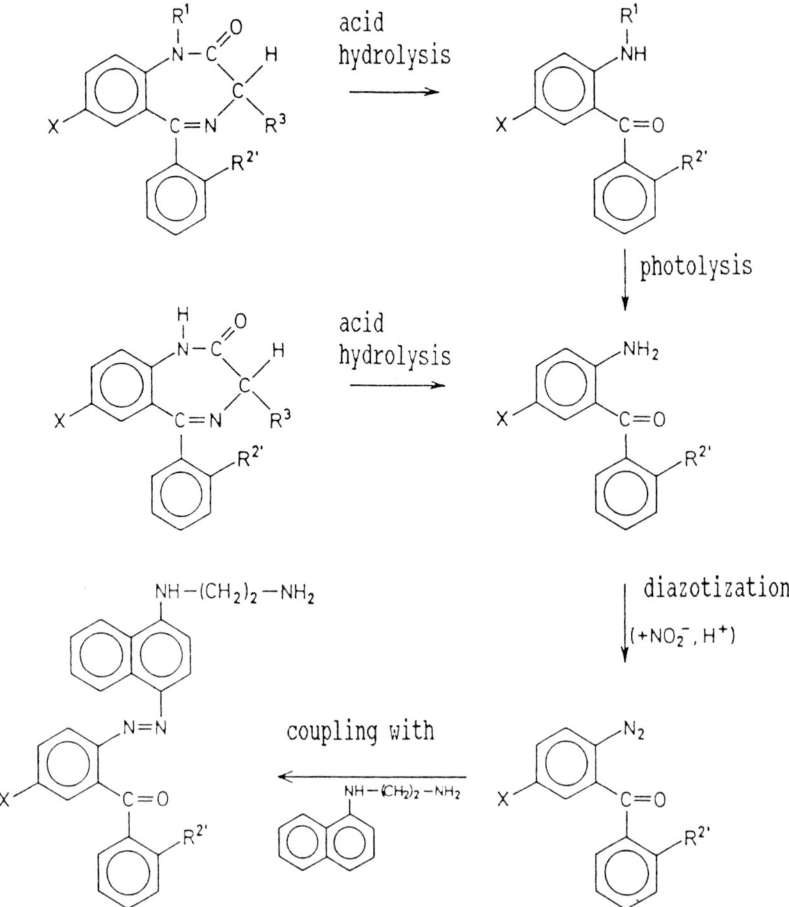

Figure 4. TLC-Screening of 1,4-Benzodiazepines via Azo Dyes (Bratton Marshall Detection)

ADB = 2-amino-2′,5-dichlorobenzophenone (e.g. from lorazepam) ($hR_f = 33$)
ANB = 2-amino-5-nitrobenzophenone (e.g. from nitrazepam) ($hR_f = 15$)
ANCB = 2-amino-2′-chloro-5-nitrobenzophenone (e.g. from clonazepam) ($hR_f = 16$)
ANFB = 2-amino-2′-fluoro-5-nitrobenzophenone (e.g. from 1-desmethylflunitrazepam) ($hR_f = 16$)
CCB = 5-chloro-2-[(cyclopropylmethyl)amino]benzophenone (e.g. from prazepam) ($hR_f = 68$)
CFMB = 5-chloro-2′-fluoro-2-(methylamino)benzophenone (e.g. from fludiazepam) ($hR_f = 55$)
CFTB = 5-chloro-2′-fluoro-2-(2,2,2-trifluoroethylamino)benzophenone (e.g. from quazepam) ($hR_f = 74$)
CPB = 5-chloro-2-(2-propinylamino)benzophenone (e.g. from pinazepam) ($hR_f = 57$)

MACB = 5-chloro-2-(methylamino)benzophenone (e.g. from diazepam) ($hR_f = 52$)
MDB = 2′,5-dichloro-2-(methylamino)benzophenone (e.g. from lormetazepam) ($hR_f = 57$)
MNB = 2-(methylamino)-5-nitrobenzophenone (e.g. from nimetazepam) ($hR_f = 29$)
MNFB = 2′-fluoro-2-(methylamino)-5-nitrobenzophenone (e.g. from flunitrazepam) ($hR_f = 29$)
TCB = 5-chloro-2-(2,2,2-trifluoroethylamino)benzophenone (e.g. from halazepam) ($hR_f = 74$).

Spray Solution (Bratton-Marshall Reagent)
Dissolve 1 g of N-(1-naphthyl)ethylenediamine in a mixture of 50 ml of dimethylformamide and 50 ml of 4 M hydrochloric acid, with warming if necessary. Filter the cooled solution if it is not clear. A slight violet colour does not affect its use. If kept in the refrigerator the solution is stable for about a year.

Standard Solution for TLC
Dissolve 1 mg each of the reference substances in 2 ml of methanol. Note that for screening not all the reference substances are absolutely necessary, but ANB, ACB, MACB and CCB should all be used for comparison purposes and for calculation of the corrected R_f-value. If stored in glass bottles in the refrigerator (4 °C) and protected from light, the solution is stable for several months.

To avoid interference in the TLC, no other substances should be present that give a color with the Bratton-Marshall reagent.

Hydrolysis
Place 100 ml of the urine sample in a 500-ml Erlenmeyer flask and add 50 ml of concentrated hydrochloric acid. Heat the mixture for 30 min under a reflux condenser, in a boiling waterbath, and if necessary rinse the condensate from the condenser into the flask with a little concentrated hydrochloric acid.

Neutralization and Extraction
After the hydrolysis, cool the solution to room temperature, and then, with further cooling, adjust the pH to between 8 and 9 (universal indicator paper) by addition of 8 M sodium hydroxide (about 5 ml or so will be needed). Wear safety goggles during this operation, which should be conducted under an efficient fumehood on account of the very unpleasant smell. Extract the aminobenzophenone derivatives with about 200 ml of diethyl ether. Note that the acid hydrolysis of bromazepam and its metabolites does not yield benzophenone derivatives, but benzoylpyridine derivatives. As the latter are also primary aromatic amines, they can be detected with Bratton-Marshall reagent as well. To increase the yield, the extraction can be repeated with 100 ml of diethyl ether, at pH 11. Reduce the combined extracts to a volume of about 3 ml in a rotary evaporator, transfer this concentrate to a glass-stoppered centrifuge tube and carefully evaporate it to dryness (at about 30–40 °C, it is not necessary to use reduced pressure). Cool the residue to 4 °C and reserve it for analysis; for this dissolve it in 0.1 ml of methanol.

Thin-Layer-Chromatography
Use 20 × 20 cm silica gel 60 F254 TLC plates (Merck), layer thickness 0.25 mm. Apply the sample and standard spots 1.5 cm from the lower edge of the plate, with 2 µl capillaries. For each sample use three capillary-loads overlapped to give an approximately straight line of sample. To avoid any cross-contamination apply the test solutions before the standards. Run the chromatogram until the solvent front has travelled 10 cm. Use the ascending method, without chamber saturation. No special activation of the plates is needed, and would not improve the results anyway. Use about 100 ml of toluene as the mobile phase.

Photolytic Dealkylation
After the development of the chromatogram (which takes 30–40 min), leave the plate to drip in the development tank for a short time, then dry it in a cold air-stream under the fume-hood. Expose the dried plate to a suitable ultraviolet source (e.g. a sunlamp) at a distance of 30–40 cm for about 20 min. For rapid analysis a 6-min exposure is sufficient. Immediately cool the plate to room temperature, or the yield in the diazotization step will be impaired.

Note that the dealkylation step is only necessary when testing for diazepam, camazepam, temazepam, ketazolam, prazepam, flurazepam, flunitrazepam, lormetazepam, fludiazepam, nimetazepam, pinazepam, quazepam and halazepam.

Diazotization
Place the cool dry plate in an empty chromatographic tank, on the bottom of which is a small beaker (20–50 ml) containing 10 ml of 20 % sodium nitrite solution. Pipette 5 ml of 25 % v/v hydrochloric acid into the beaker as fast as possible, to liberate nitrogen oxides, and seal the tank with its lid. Leave the plate in the tank for 3–5 min, which is sufficient time for diazotization of primary aromatic amine groups. Remove the lid, and when most of the nitrous gases have dispersed take out the plate and leave it under the fumehood for 20–30 min in a stream of cold air (e.g. from a fan heater set at "cold").

For rapid work it is sufficient to air the plate for only 5 min to remove the nitrous gases and then to spray it gently with a 1 % aqueous solution of urea. Alternatively, put the plate in a vacuum desiccator and draw vacuum for about 2 min to remove excess nitrous gases. Finally, spray the plate thinly and uniformly (meander pattern) with Bratton-Marshall reagent at 4 °C to couple the diazonium salts. The result: violet and redviolet colors.

N. B. As nitrous gases are poisonous, work in a fume-hood and wear gloves!

Evaluation
The hR_f-values are listed in table 8.

Sensitivity of the Screening-Procedure according to Bratton and Marshall:

Oxazepam, Diazepam and Bromazepam	0.05 mg/l
Lorazepam	0.10 mg/l
7-Amino-nitrazepam	0.15 mg/l

3.2.7 Separation of Benzodiazepines by Chromatographic Techniques

3.2.7.1 Thin layer chromatography (TLC)

Thin-layer chromatography is now as before a powerful tool for the screening of drugs and other substances, and the modern high performance thin-layer chromatography (HPTLC) may be regarded as a renaissance of this good practicable and effective method. One of the improvements was the introduction of the corrected R_f-value, as the reproducibility of the R_f-value is governed by many factors (e.g. activity of the sorbent, state of saturation of the development tank, running distance, amount of drug applied to the chromatogramm, geometry of the chamber, temperature, etc.). With the aim of eliminating some of these parameters, Galanos and Kapoulas (39) and de Zeeuw et al. (40) recommended a multiple correction graph which is based on a linear interpolation, carried out graphically or by calculation, using the following equation:

$$R_f^c(p) = \frac{\blacktriangle^c}{\blacktriangle} \{R_f(p) - R_f(t_n)\} + R_f^c(t_n)$$

wherein

$$\blacktriangle^c = R_f^c(t_n) - R_f^c(t_{n+1})$$

$$\blacktriangle = R_f(t_n) - R_f(t_{n+1})$$

$R_f^c(p)$: corrected R_f value of the unknown substance p

$R_f(t_n)$: measured R_f-value of the reference substance nearest to p (lower value, possible starting point = 0)

$R_f(p)$: measured R_f value of the unknown substance p

$R_f^c(t_n)$: corrected R_f value of the reference substance nearest to p (lower value, possible starting point = 0)

$R_f^c(t_{n+1})$: corrected R_f value of the other reference substance nearest to p (higher value, possible solvent front = 100)

See (40, 41, 74) for additional details concerning the corrected R_f-value.

Report VII of the *DFG Commission for Clinical-Toxicological Analysis (special issue of the TIAFT Bulletin)* (40) presents corrected R_f data of some 1,600 toxicologically relevant substances (drugs, illicit products, pesticides, metabolites, endogenous compounds) in 10 standardized TLC systems. The modern TLC-scanners permit a very precise determination of the uncorrected and corrected R_f-values by registration of the remission-distance curve. Interlaborative studies revealed, that 100 of 343 substances had exact the same corrected R_f-value. 94.5% of all compounds were observed within a search window of +/− 5 corrected R_f-values (98% within +/− 7 corrected R_f-values) (41).

Detection may be performed by visualization under UV-light (fluorescence quenching at 254 nm) or spraying with iodoplatinat or the reagent according to Dragendorff (42).

The hR_f^c-values of 44 benzodiazepins are listed in table 8.

Table 8. Chromatographic Data of Benzodiazepines

No.	Substance	RI	hR_f^c	
			A	B
1	Adinazolam	3060	9	51
2	Alprazolam	3050	7	40
3	Bromazepam	2660	13	47
4	Brotizolam	3145	15	53
5	Camazepam	2955	55	69
6	Chlordiazepoxide	2800	10	53
7	Clobazam	2690	53	70
8	Clonazepam	2885	35	56
9	Clorazepate	2460	34	57
10	Clotiazepam	2580	55	66
11	Cloxazolam	2405	39	63
12	Delorazepam	2650	35	57
13	Demoxepam	2530	15	42
14	Diazepam	2425	58	72
15	Estazolam	2955	7	44
16	Ethyl-loflazepate	2195	53	62
17	Etizolam	3090	11	50
18	Fludiazepam	2460	56	67
19	Flumazenil	2560	30	61
20	Flunitrazepam	2645	54	72
21	Flurazepam	2785	3	41
22	Flutazolam	2310	30	62
23	Halazepam	2335	59	70
24	Haloxazolam	2620	46	65
25	Ketazolam	2470	45	62
26	Loprazolam	nm	3	36
27	Lorazepam	2405	23	41
28	Lormetazepam	2700	46	60
29	Medazepam	2230	56	73
30	Metaclazepam	2690	47	71
31	Mexazolam	2670	59	68
32	Midazolam	2620	13	53
33	Nimetazepam	2530	53	71
34	Nitrazepam	2750	35	55
35	Nordazepam	2500	34	57
36	Oxazepam	2335	22	42
37	Oxazolam	2590	53	65
38	Pinazepam	2580	65	73
39	Prazepam	2640	64	72
40	Quazepam	2485	78	78
41	Temazepam	2630	51	65
42	Tetrazepam	2460	57	67
43	Tofisopam	3035	55	72
44	Triazolam	3090	5	41

3.2.7.2 Gas chromatography (GC)

Gas Chromatography is one of the most useful and frequently applied tools for the screening and identification of organic compounds (preferably in combination with spectroscopic methods, e.g., mass-spectrometry, and infrared spectroscopy in the FTIR mode). The primary measuring parameter is the retention time, which is governed by many variables, such as composition of the stationary phase, column length, flow of the carrier gas, oven temperatures and other variables. Therefore, the retention times are hardly appropriate to serve as reliable and intercomparable data. Moderate improvement can be seen in measurement of the relative retention time, but the most useful instrument is without any doubt the retention index (RI) developed and introduced by Kovats (43–45).

The retention index RI(A) can be calculated by using one of two essentially identical equations (46):

$$RI(A) = 100(y - x)\frac{\log\frac{t(A)}{t(X)}}{\log\frac{t(Y)}{t(X)}} + 100x$$

$$RI(A) = [RI(Y) - RI(X)]\frac{\log\frac{t(A)}{t(X)}}{\log\frac{t(Y)}{t(X)}} + RI(X)$$

wherein:

t(A) = net retention time of a substance A
t(X) = net retention time of the n-alkane C_xH_{2x+2} eluting immediately before A
t(Y) = net retention time of the n-alkane C_yH_{2y+2} eluting immediately after A
x = carbon number of the n-alkane C_xH_{2x+2}
y = carbon number of the n-alkane C_yH_{2y+2}
RI(A), RI(X), etc. = retention indices of substances A, X, etc.

The retention index RI can also be obtained by a simple graph (46). See Schütz and Wollrab (75) and Wollrab and Schütz (48) for additional details.

The interlaboratory standard deviation of the RI is in the order of 15–20 RI units (49, 50). According to *DFG* (46), a "search window" of $\pm$ 50–60 RI units should be taken into consideration when working under temperature-programmed conditions with an almost linear relationship between the carbon number of the n-alkanes and the retention time (51). The search window mentioned above will also take the temperature dependency of the RI into consideration. The quality of the column must be tested as described in (46) before starting the analysis.

The RI values of 44 benzodiazepines on OV-1/OV-101 (packed columns) are listed in table 8. Within the scope of this contribution only parent compound are treated but some of them are also metabolites. The complete data of 115 benzodiazepines, including metabolites and important hydrolysis products (aminobenzophenones and benzoylpyridine derivatives), were collected by Schütz (7, 10).

In conclusion, one can say that the introduction and use of the retention index has raised gas chromatography to a higher level with excellent interlaboratory reproducibility. Based on a compilation of Ardrey and Moffat (52), the DFG/TIAFT publications present data on about 1,600 substances in a third revised and enlarged edition (46), which is successfully used to solve screening problems all over the world.

3.2.7.3 High pressure liquid chromatography (HPLC)

High-pressure liquid chromatography (HPLC) is a good, practicable tool for the quantitative determination of benzodiazepines. Its value as a screening method is controversial, as the reproducibility of the retention data cannot be compared with that for gas chromatography. This is caused by restricted discrimination power and the difficulties connected with the standardization of the stationary phases, mobile phases, and certain apparatus parameters. Another problem is the establishment of a suitable reference system (comparable to n-alkanes in gas chromatography).

On the other hand, the combination of HPLC with diode-array detection (DAD) must be considered a highly effective screening method. Therefore, the principal absorption data are presented in table 9. Also see Brandenberger et al. (53) for additional details concerning the recommendation of HPLC.

Table 9. UV-Spectroscopic Data of Benzodiazepines

No.	Substance	UV-Maxima [nm]		
		EtOH*	HCl*	NaOH*
1	Adinazolam		263	
2	Alprazolam	222	264	258
3	Bromazepam	233/325	240/350	238/350
4	Brotizolam	240	254	241
5	Camazepam	230/255/315	227	232
6	Chlordiazepoxide	245/267	245/310	261
7	Clobazam	231/297	228/285	285
8	Clonazepam	218/250/312	270	366
9	Clorazepate	230	238/286	233/340
10	Clotiazepam	245/305/395	260/300/395	239/340
11	Cloxazolam	244	240/285/374	220/280
12	Delorazepam	320	238/286	227/345
13	Demoxepam	235/312	235	243/257
14	Diazepam	230/255	241/285/360	229
15	Estazolam	222	278	218
16	Ethyl-loflazepate	230/320	232	354
17	Etizolam	243	250/294/362	243
18	Fludiazepam	229	240/282	229
19	Flumazenil	246		
20	Flunitrazepam	221/253/311	275	400
21	Flurazepam	228/315	236/285/355	230
22	Flutazolam	245	241/368	245
23	Halazepam	225	233/285	248
24	Haloxazolam	247	242	
25	Ketazolam	242	242	242
26	Loprazolam	330	329	309

Table 9. (continued)

No.	Substance	UV-Maxima [nm] EtOH*	HCl*	NaOH*
27	Lorazepam	229/322	230	234/349
28	Lormetazepam	230/310	230	300
29	Medazepam	231/252/360	254	
30	Metaclazepam	370	251/460	370
31	Mexazolam	244	241/290/373	244
32	Midazolam	217	212	216
33	Nimetazepam	220/260	282	248/395
34	Nitrazepam	220/258/312	280	226/258/357
35	Nordazepam	228/325	237/282/370	340
36	Oxazepam	229/324	234/281	234/340
37	Oxazolam	247	237	247
38	Pinazepam	227	238/284/356	
39	Prazepam	228/256	240/287/360	230
40	Quazepam	286	273	
41	Temazepam	231/255/315	235/283/355	
42	Tetrazepam	227/225	240/283/345	310
43	Tofisopam	238/310	251/349	239/308
44	Triazolam	223	223	

* underlined numbers = main maxima

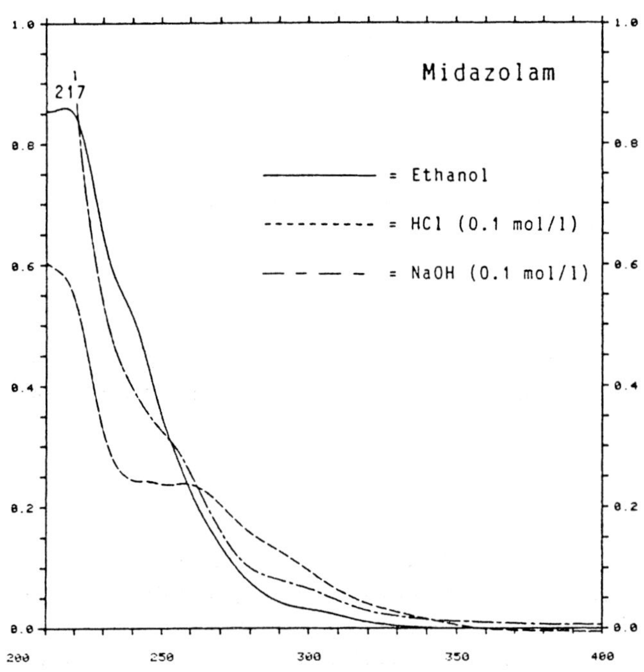

Figure 5. UV-Spectra of Midazolam

3.2.8 Detection of Benzodiazepines by Spectroscopic Techniques

3.2.8.1 Ultraviolet detection and quantification

Since the UV-spectra of many 1,4-benzodiazepines are not very characteristic (see fig. 5) with the exception of nitrosubstituted and some other benzodiazepines (see fig. 6), and since the concentrations in body fluids are very low (except cases of overdosage), the quantification of benzodiazepines in biological materials by UV-spectroscopy is not the method of choice.

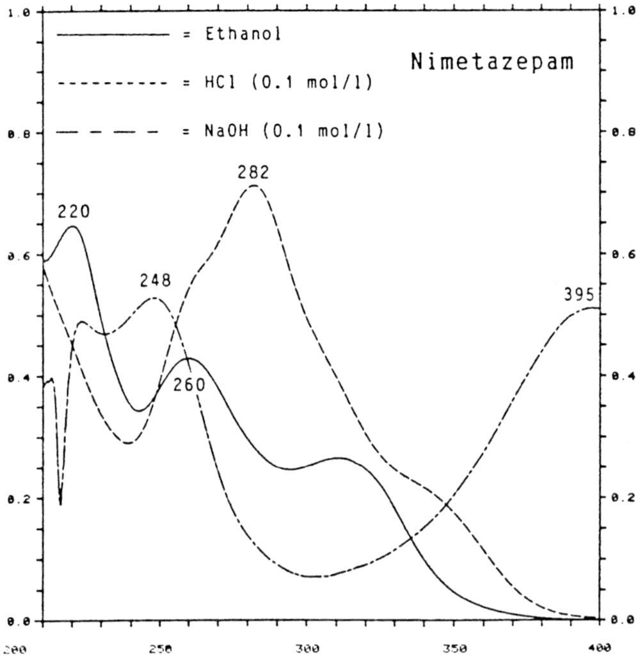

Figure 6. UV-Spectra of Nimetazepam

On the other hand, the quantification of pharmaceutical formulations by UV-spectroscopy is described in many pharmacopoeias. Some specific absorption values are compiled in table 10 (54).

3.2.8.2 Infrared spectroscopy

Infrared spectroscopy is still nowadays a highly effective tool for the identification of benzodiazepine derivatives. On the other hand, the concentrations are often very low and micro techniques must be applied.

The IR-spectra of 115 1,4- and 1,5-benzodiazepines, metabolites, rearrangement and hydrolysis products are published by Schütz (7, 10).

Table 10. UV-absorption Data (A_1^1-Values) of Selected Benzodiazepines (from Clarke (54))

Substance	Wavelength (nm)	Medium	A_1^1
Bromazepam	237	aqu. alk.	920
	233	methanol	1050
	320	methanol	61
Chlordiazepoxide	246	aqu. acid	1112
Clobazam	230	aqu. acid	1373
	289	aqu. acid	76
	286	aqu. alk.	193
Clonazepam	273	aqu. acid	645
	245	methanol	460
	309	methanol	360
Clorazepate	237	aqu. acid	747
Diazepam	242	aqu. acid	1020
Flunitrazepam	252	methanol	516
	308	methanol	332
Flurazepam	236	aqu. acid	620
	231	aqu. alk.	856
	312	aqu. alk.	53
Ketazolam	242	ethanol	492
Loprazolam	330	ethanol	884
Lorazepam	230	ethanol	1100
Lormetazepam	231	aqu. acid	1030
	311	aqu. acid	59
Medazepam	253	aqu. acid	860
Nitrazepam	280	meth. acid	910
Nordazepam	238	aqu. acid	1140
Prazepam	240	aqu. acid	1760
Temazepam	237	aqu. acid	980
	284	aqu. acid	283
	358	aqu. acid	68
	230	methanol	1090
	314	methanol	76
Tetrazepam	239	aqu. acid	940
	305	aqu. alk.	84

3.2.8.3 Mass spectrometry

A method for the identification and differentiation of the following benzodiazepines and their metabolites in urine after acid hydrolysis and acetylation was described by Maurer and Pfleger (55, 56): bromazepam, camazepam, chlordiazepoxide, clobazam, clonazepam, clorazepate, clotiazepam, cloxazolam, delorazepam, diazepam, ethyl loflazepate, flunitrazepam, flurazepam, halazepam, ketazolam, loprazolam, lorazepam, lormetazepam, medazepam, metaclazepam, midazolam, nitrazepam, nordazepam, oxazepam, oxazolam, prazepam, quazepam, temazepam and tetrazepam. The acetylated extract was analyzed by computerized gas chromatography/mass spectrometry. An on-line computer allowed rapid detection using ion chromatography with ions m/z 205, 211, 230, 241, 245, 249, 312 and 333. The identity of positive signals in the reconstructed ion chromatogram was confirmed by a comparison of the stored full mass spectra with reference spectra. The ion chromatograms, reference mass spectra and gas chromatographic retention indices (on OV-101) are documented.

Positive and negative ion mass spectrometry of benzophenones, the acid-hydrolysis products of 1,4-benzodiazepines was described by Suzuki et al. (57): positive electron impact (EI), positive chemical ionization (CI), and negative CI mass spectra of 14 benzophenones were investigated. In the positive EI mode, intense molecular peaks appeared for most compounds; some other peaks due to splitting at both sides of the carbonyl group also appeared. In the positive CI mode, $[M + 1]^+$ quasi-

Table 11. Mass-Spectrometric Data of Benzodiazepines

No.	Substance	Important m/z-Values							
1	Adinazolam	308	310	58	307	309	77	91	205
2	Alprazolam	308	279	204	273	77	310	307	102
3	Bromazepam	236	315	317	77	91	287	103	104
4	Brotizolam	394	392	245	316	318	369	118	174
5	Camazepam	78	72	58	57	77	271	256	371
6	Chlordiazepoxide	282	283	284	77	271	247	299	253
7	Clobazam	300	77	51	258	283	302	181	91
8	Clonazepam	280	314	315	234	289	287	75	76
9	Clorazepate	242	270	269	241	103	76	77	271
10	Clotiazepam	289	318	291	320	275	317	292	120
11	Cloxazolam	237	239	56	139	111	57	75	222
12	Delorazepam	304	269	280	281	282	306	305	303
13	Demoxepam	285	286	296	77	287	241	242	107
14	Diazepam	256	283	284	258	57	78	221	285
15	Estazolam	259	205	77	51	294	239	137	293
16	Ethyl-loflazepate	259	287	261	288	289	360	223	75
17	Etizolam	342	266	344	313	224	239	274	137
18	Fludiazepam	274	302	301	275	303	273	276	239
19	Flumazenil	229	257	302	230	201	158	94	132
20	Flunitrazepam	285	312	286	266	313	294	238	239
21	Flurazepam	86	58	99	87	56	71	84	387
22	Flutazolam	281	56	283	123	95	166	206	346
23	Halazepam	324	352	351	323	325	326	353	241
24	Haloxazolam	335	333	210	289	291	183	211	305
25	Ketazolam	256	283	84	284	69	257	285	325
26	Loprazolam	282	279	465	467	391	309	254	305
27	Lorazepam	291	239	75	274	293	111	138	275
28	Lormetazepam	304	75	306	57	50	102	333	152
29	Medazepam	207	242	244	165	243	270	271	57
30	Metaclazepam	349	347	351	321	136	233	285	394
31	Mexazolam	251	70	253	139	236	56	111	362
32	Midazolam	310	58	312	325	69	71	128	142
33	Nimetazepam	267	294	295	268	248	221	220	165
34	Nitrazepam	280	253	281	234	252	254	205	264
35	Nordazepam	242	270	269	241	77	243	103	271
36	Oxazepam	257	77	205	239	233	181	259	268
37	Oxazolam	253	251	70	77	105	252	202	283
38	Pinazepam	308	280	307	309	281	310	282	241
39	Prazepam	269	91	55	295	298	241	324	271
40	Quazepam	58	75	109	245	183	386	323	303
41	Temazepam	271	57	56	273	77	257	256	255
42	Tetrazepam	253	288	287	225	289	77	259	290
43	Tofisopam	326	382	341	327	383	310	354	342
44	Triazolam	342	313	238	315	344	102	75	105

molecular ions together with $[M + C_2H_5]^+$ peaks were observed for all compounds; some fragment peaks were common to those in the positive EI mode. In the negative CI mode the spectra were much simpler than those in the positive EI or CI mode. In the 1 Torr negative CI mode, some spectra showed only single molecular anions; in the 0.01 Torr negative CI mode, halogen or nitro peaks appeared in addition to the molecular anions.

Important *m/z*-values (EI/70 eV) of 44 benzodiazepine derivatives are listed in table 11. Also see Pfleger et al. (58) and Schütz (7, 10, 47) for full mass spectra.

3.2.9 Electroanalytical Methods

The polarographic behaviour of 12 therapeutically important 1,4-benzodiazepines (bromazepam, chlordiazepoxide, clonazepam, clorazepate, diazepam, flunitrazepam, flurazepam, lorazepam, medazepam, nitrazepam, oxazepam, prazepam) in *Britton-Robinson* universal buffers (pH 4.0 and pH 12.0) has been investigated by Smyth et al. (59). It is suggested that this procedure would be applicable to the analysis of unknown formulations or body fluids in forensic cases where the parent compound exists in relatively high concentrations compared with its metabolites.

Electrochemically active elements are the azomethine group (nearly all benzodiazepines), the N-oxide structure (e.g. chlordiazepoxide or demoxepam), the nitro group (e.g. clonazepam, flunitrazepam, nimetazepam, nitrazepam) and the pyridyl group (e.g. bromazepam).

Also see the following publications for the polarographic determination of benzodiazepines in body fluids (60):

Compound/Matrix	Method	Author(s)
1,4-Benzodiazepines + metab.	DPP (DME)	Brooks and de Silva (61)
Bromazepam + metab. in urine	DPP (DME)	de Silva et al. (62)
Bromazepam in blood	CRP (DME)	Sengün and Oelschläger (63)
Chlordiazepoxide + metab. in plasma	DPP (DME)	Hackman et al. (64)
Clonazepam + metab. in urine	DPP (DME)	de Silva et al. (65)
Diazepam + metab. in urine	DCP (DME)	Dugal et al. (66)
Flurazepam + metab. in plasma	DPP (DME)	Clifford et al. (67)
Flurazepam + metab. in blood plasma and urine	DPP (DME)	de Silva et al. (68)
Flurazepam + metab. in body fluids	DPP (DME)	Smyth and Groves (69)
Lorazepam + metab. in urine	DPP (DME)	de Silva et al. (70)
Medazepam + metab. in blood	CRP (DME)	Oelschläger and Geppert (71)
Nimetazepam in blood and serum	DCP (DME)	Kobiela (72)
Triazolam in plasma and serum	DPP	Oelschläger and Sengün (73)

References

1. Sternbach, L.H. (1971) 1,4-Benzodiazepines. Chemistry and some aspects of the structure-activity relationship. Angew. Chem. (Int. Ed.) *10*, 34–43.
2. Sternbach, L.H. (1978) The benzodiazepine story. Progr. Drug Res. *22*, 229–266.
3. Sternbach, L.H. (1979) The benzodiazepine story. J. Med. Chem. *22*, 1–7.
4. Randall, L.O. Discovery of Benzodiazepines. In: ref (17), pp. 15–22.
5. Schütz, H. (1982) Benzodiazepine – Entdekkung, Entwicklung und Zukunftsperspektiven. Pharmazie in unserer Zeit *11*, 161–176.
6. Sternbach, L.H., Randall, L.O. (1966) Some aspects of structure-activity relationship in psychotropic agents of the 1,4-benzodiazepine series. In: CNS Drugs, pp. 53–69, Symposium Regional Research Laboratory, Hyderabad, India (New Delhi: Council of Scientific and Industrial Research).
7. Schütz, H. (1982) Benzodiazepines – A Handbook (Vol. 1). Basic Data, Analytical Methods, Pharmacokinetics and Comprehensive Literature. Springer-Verlag, Berlin/Heidelberg/New York.
8. Schütz, H. (1986) Dünnschichtchromatographische Suchanalyse für 1,4-Benzodiazepine in Harn, Blut und Mageninhalt. Mitteilung der Senatskommission der Deutschen Forschungsgemeinschaft für Klinisch-toxikologische Analytik. VCH-Verlagsgesellschaft, Weinheim – Deerfield Beach – Basel.
9. Schütz, H. (1988) Modern Screening Strategies in Analytical Toxicology with Special Regard to Newer Benzodiazepines. Z. Rechtsmed. *100*, 19–37.
10. Schütz, H. (1989) Benzodiazepines II (A Handbook). Basic Data, Analytical Methods, Pharmacokinetics and Comprehensive Literature. Springer-Verlag, Berlin/Heidelberg/New York/London/Paris/Tokyo.
11. Klotz, U., Kangas, L., Kanto, J. (1980) Clinical Pharmacokinetics of Benzodiazepines. Gustav Fischer Verlag, Stuttgart/New York.
12. Poser, W., Poser, S. (1986) Abusus und Abhängigkeit von Benzodiazepinen. Internist *27*, 738–745.
13. Poser, W., Poser, S., Piesiur-Strehlow, B., Strehlow, U. (1985) Pharmakologische und klinische Daten zum Problem der Benzodiazepin-Abhängigkeit. In: Hippius, H. (Hrsg.) Buspiron-Workshop. Neue Entwicklungen der Pharmakotherapie der Angst. Edition Materia Medica, Socio-Medico, München.
14. Haefely, W.E. (1983) The biological basis of benzodiazepine actions. J. Psychoact. Drugs *15*, 19–39.
15. Haefely, W.E. (1987) Structure and function of the benzodiazepine receptor. Chimia *41*, 389–396.
16. Costa, E. (ed.) (1983) The Benzodiazepines – From molecular biology to clinical practice. Raven Press, New York.
17. Garattini, S., Mussini, E., Randall, L.O. (1973) The Benzodiazepines. Raven Press, New York.
18. Usdin, E., Skolnick, P., Tallman, jr., J.F., Greenblatt, D., Paul, S.M. (eds.) (1983) Pharmacology of the benzodiazepines. Verlag Chemie, Weinheim/Deerfield Beach.
19. De Blas, A.L., Sangameswaran, L. (1986) Demonstration and purification of an endogenous benzodiazepine from the mammalian brain with a monoclonal antibody to benzodiazepine, Life Sci. 1927–1936.
20. Müller, W.E. (1988) Sind Benzodiazepine 100% Natur? Dtsch. Apoth. Ztg. *128*, 672–674.
21. Wildmann, J., Möhler, H., Vetter, W., Ranalder, U., Schmidt, K., Maurer, R. (1987) Diazepam and N-desmethyldiazepam are found in rat brain and adrenal and may be of plant origin. J. Neurol. Transm. *70*, 383–398.
22. Köppel, C., Tenczer, J., Gerlach, J. (1991) Die Problematik der endogenen Benzodiazepine in der forensischen Toxikologie. GTFCh-Symposium 1991 Spurenanalyse. pp. 99–110 Verlag Dr. D. Helm, Heppenheim.
23. Beyer, K.-H. (1990) Biotransformation der Arzneimittel. Springer-Verlag, Berlin/Heidelberg/New York.
24. Pfeifer, S. Biotransformation von Arzneimitteln. Bd. 1–5 VEB Verlag Volk und Gesundheit, Berlin 1975–1983.
25. Kuschinsky, G., Lüllmann, H. (1984) Kurzes Lehrbuch der Pharmakologie und Toxikologie. Thieme, Stuttgart.
26. Kaplan, S.A., Alexander, K., Jack, M.L., Puglisi, C.V., de Silva, J.A.F., Lee, T.L., Weinfeld, R.E. (1974) Pharmacokinetic profiles of clonazepam in dog and humans and of flunitrazepam in dog. J. Pharm. Sci. *63*, 527–532.
27. Kitagawa, H., Esumi, Y., Kurosawa, S., Sekine, S., Yokoshima, T. (1979) Metabolism of 8-chloro-6-(o-chlorophenyl)-1-methyl-5H-S-triazolo(4,3-α)(1,4)benzodiazepine,

triazolam, a new central depressant. I. Absorption, distribution and excretion in rats, dogs and monkeys. Xenobiotica *9*, 415–428.

28. Kitagawa, H., Esumi, Y., Kurosawa, S., Sekine, S., Yokoshima, T. (1979) Metabolism of 8-chloro-6-(o-chlorophenyl)-1-methyl-5H-S-triazolo(4,3-α)(1,4)benzodiazepine, triazolam, a new central depressant. II. Identification and determination of metabolites in rats and dogs. Xenobiotica *9*, 429–439.
29. Eberts, F.S., Philopoulos, V., Vliek, R.W. (1979) Disposition of ^{14}C-triazolam, a short acting hypnotic in man. Pharmacologist *21*, 168.
30. Müller, W.E. (1988) Die Wirkung der Benzodiazepine auf neuronaler Ebene. In: Kubicki, St., Engfer, A. (Hrsg.) Schlaf- und Schlafmittelforschung. Neue Ergebnisse und therapeutische Konsequenzen. pp. 39–56 Vieweg Verlag, Braunschweig/Wiesbaden.
31. Pöldinger, W., Wider, F. (1985) Tranquilizer und Hypnotika. Gustav Fischer, Stuttgart.
32. Greenblatt, D.J., Shader, R.I., Divoll, M., Harmatz, J.S. (1981) Benzodiazepines: A summary of pharmacokinetic properties. Br. J. clin. Pharmac. *11*, 11S–16S.
33. Uges, D.R.A., von Clarmann, M., Geldmacher-von Mallinckrodt, M., van Heijst, A.N.P., Ibe, K., Oellerich, M., Schütz, H., Stamm, D., Wunsch, F. (1990) Orientierende Angaben zu therapeutischen und toxischen Konzentrationen von Arzneimitteln und Giften in Blut, Serum oder Urin. Mitteilung XV der Senatskommission der Deutschen Forschungsgemeinschaft für Klinisch-toxikologische Analytik. VCH-Verlagsgesellschaft, Weinheim.
34. Baselt, R.C., Cravey, R.H. (1989) Disposition of toxic drugs and chemicals in man. Third edition. Year Book Medical Publishers, Inc., Chicago/London/Boca Raton/Littleton.
35. Schütz, H. (1992) Screening von Drogen und Arzneimitteln mit Immunoassays. Ein Leitfaden für die Praxis. Abbott, Wiesbaden.
36. Schütz, H., Schneider, W.R., Schölermann, K. (1988) Verbessertes enzymimmunologisches Screeningverfahren für Benzodiazepine im Harn nach EXTRELUT®-Anreicherung. Ärztl. Lab. *34*, 130–136.
37. Brandenberger, H., Maes, R.A.A., Bäumler, J., Biro, G., Daldrup, T., Geldmacher-von Mallinckrodt, M., Gibitz, H.J., Machata, G., Schütz, H. (1988) Empfehlungen zur klinisch-toxikologischen Analytik. Folge 1: Einsatz von immunochemischen Testen in der Suchtmittelanalytik. Mitteilung X der Senatskommission der DFG für Klinisch-toxikologische Analytik. VCH-Verlagsgesellschaft, Weinheim.
38. Rochholz, G., Mayr, A., Schütz, H. (1993) TLC-Screening of Important Pharmaceutical Substances Using Fluorescence Detection. Fresenius' Z. Anal. Chem. *346*, 819–827.
39. Galanos, D.S., Kapoulas, V.M. (1964) The paper chromatographic identification of compounds using two reference compounds. J. Chromatogr. *13*, 128–138.
40. Zeeuw, de, R.A., Franke, J.P., Degel, F., Machbert, G., Schütz, H., Wijsbeek, J. (1992) Thin-Layer Chromatographic R_f-Values of Toxicologically Relevant Substances on Standardized Systems. Report XVII of the DFG Commission for Clinical-Toxicological Analysis/Special Issue of the TIAFT Bulletin. VCH-Verlagsgesellschaft, Weinheim.
41. Schütz, H. (1989) Der korrigierte R_f-Wert: Ermittlung und Bedeutung für die toxikologische Analytik. Pharmazie in unserer Zeit *18*, 161–168.
42. Müller, R.K. (ed.) (1991) Toxicological Analysis. Verlag Gesundheit, Berlin.
43. Kovats, E. (1958) Gaschromatographische Charakterisierung organischer Verbindungen. Teil 1: Retentionsindices aliphatischer Halogenide, Alkohole, Aldehyde und Ketone. Helv. Chim. Acta *41*, 1915–1932.
44. Kovats, E. (1961) Zusammenhänge zwischen Struktur und gaschromatographischen Daten organischer Verbindungen. Fresenius' Z. Anal. Chem. *181*, 351–364.
45. Kovats, E., Wehrli, A. (1959) Gaschromatographische Charakterisierung organischer Verbindungen: Teil 3: Berechnung der Retentionsindices aliphatischer, alicyclischer und aromatischer Verbindungen. Helv. Chim. Acta. *42*, 2709–2736.
46. Zeeuw, de, R.A., Franke, J.P., Maurer, H.H., Pfleger, K. (1992) Gas-Chromatographic Retention Indices of Toxicologically Relevant Substances on Packed or Capillary Columns with Dimethylsilicone Stationary Phases. Third, Revised and Enlarged Edition Report XVIII of the DFG-Commission for Clinical-Toxicological Analysis/Special Issue of the TIAFT Bulletin. VCH-Verlagsgesellschaft, Weinheim.
47. Schütz, H., Weiler, G. (1991) Optimierter Acht-Peak-Index für 83 Benzodiazepine und 29 Hydrolysenprodukte. pp. 577–583. In: Schütz, H., Kaatsch, H.-J., Thomsen, H. (Hrsg.) Medizinrecht – Psychopathologie –

Rechtsmedizin. Diesseits und jenseits der Grenzen von Recht und Medizin. Springer-Verlag, Berlin/Heidelberg/New York/London/Paris/Tokyo/Hong Kong/Barcelona.
48. Wollrab, A., Schütz, H. (1988) Der Retentions-Index in der Gaschromatographie. Pharmazie in unserer Zeit *17*, 65–70.
49. Moffat, A.C. (1975) Use of SE-30 stationary phase for the gas-liquid chromatography of drugs. J. Chromatogr. *113*, 69–95.
50. Berninger, H., Möller, M.R. (1977) Retentionsindices zur gaschromatographischen Identifizierung von Arzneimitteln. Arch. Tox. *37*, 295–305.
51. Peel, H.W., Perrigo, B.J. (1976) The practical gaschromatographic procedures for toxicological analysis. Can. Soc. Forens. Sci. *9*, 69–74.
52. Ardrey, R.E., Moffat, A. (1981) Gas-liquid chromatography retention indices of 1318 substances of toxicological interest on SE-30 or OV-1 stationary phase. J. Chromatogr. *220*, 195–222.
53. Brandenberger, H., Engelhardt, H., Haerdi, W., Bäumler, J., Daldrup, T., Geldmacher-von Mallinckrodt, M., Machata, G., Maes, R.A.A., Schütz, H. (1989) Empfehlungen zur klinisch-toxikologischen Analytik. Folge 3: Einsatz der Hochleistungsflüssigchromatographie in der klinisch-toxikologischen Analytik. Mitteilung XII der Senatskommission der DFG für Klinisch-toxikologische Analytik. VCH-Verlagsgesellschaft, Weinheim.
54. Clarke's Isolation and Identification of Drugs in Pharmaceuticals, Body fluids, and post-mortem Material. Second Edition, A.C. Moffat, J.V. Jackson, M.S. Moss, B. Widdop (eds.) (1986) The Pharmaceutical Press, London.
55. Maurer, H., Pfleger, K. (1981) Determination of 1,4-benzodiazepines and 1,5-benzodiazepines in urine using a computerized gas chromatographic-mass spectrometric technique. J. Chromatogr. *222*, 409–419.
56. Maurer, H., Pfleger, K. (1987) Identification and differentiation of benzodiazepines and their metabolites in urine by computerized gas-chromatography mass-spectrometry. J. Chromatogr. *422*, 85–101.
57. Suzuki, O., Hattori, H., Asano, M., Takahashi, T., Brandenberger, H. (1987) Positive and negative ion mass spectrometry of benzophenones, the acid-hydrolysis products of benzodiazepines. Z. Rechtsmed. *98*, 1–10.
58. Pfleger, K., Maurer, H.H., Weber, A. (1992) Mass spectral and GC data of drugs, poisons, pesticides, pollutants and their metabolites. Second edition. VCH-Verlagsgesellschaft, Weinheim/New York/Basel/Cambridge.
59. Smyth, W.F., Smyth, M.R., Groves, J.A., Tan, S.B. (1978) Polarographic methods for the identification of 1,4-benzodiazepines. Analyst *103*, 497–508.
60. Henze, G., Neeb, R. (1986) Elektrochemische Analytik. Springer-Verlag, Berlin/Heidelberg/New York/Tokyo.
61. Brooks, M.A., de Silva, J.A.F. (1975) Determination of 1,4-benzodiazepines in biological fluids by differential pulse polarography. Talanta *22*, 849–860.
62. De Silva, J.A.F., Bekersky, J., Brooks, M.A., Weinfeld, R.E., Glover, W., Puglisi, C.V. (1974) Determination of bromazepam in blood by electron-capture GLC and of its major urinary metabolites by differential pulse polarography. J. Pharm. Sci. *63*, 1440–1445.
63. Sengün, F.I., Oelschläger, H. (1975) Arzneimittelanalysen mittels Polarographie oder Oszillopolarographie. 18. Mitt.: Elektroanalytische Eigenschaften des Bromazepams und seine Bestimmung im Blut durch Kathodenstrahlpolarographie. Arch. Pharm. *308*, 720–723.
64. Hackman, M.R., Brooks, M.A., de Silva, J.A.F., Ma, T.S. (1974) Determination of chlordiazepoxide hydrochloride (librium) and its major metabolites by differential pulse polarography. Anal. Chem. *46*, 1075–1082.
65. De Silva, J.A.F., Puglisi, C.V., Munno, N. (1974) Determination of clonazepam and flunitrazepam in blood and urine by electron-capture GLC. J. Pharm. Sci. *63*, 520–526.
66. Dugal, R., Caille, G., Cooper, S.F. (1973) Urinary metabolites of diazepam in man: Chromatographic and polarographic methods. Union. Med. Can. *102*, 2491–2497.
67. Clifford, J.M., Smyth, M.R., Smyth, W.F. (1974) Assay of flurazepam in plasma by differential pulse polarography. Z. Anal. Chem. *272*, 198–201.
68. De Silva, J.A.F., Puglisi, C.V., Brooks, M.A., Hackman, M.R. (1974) Determination of flurazepam (dalmane) and its major metabolites in blood by electron-capture GLC and in urine by differential pulse polarography. J. Chromatogr. *99*, 461–483.
69. Smyth, W.F., Groves, J.A. (1981) The solvent-extraction of flurazepam and its major metabolites and their determination by polarography. Anal. Chim. Acta *123*, 175–186.
70. De Silva, J.A.F., Bekersky, J., Brooks, M.A. (1974) ECD-GLC determination of blood

levels of 7-chloro-1,3-dihydro-5-(2′-chlorophenyl)-2H-1,4-benzodiazepine-2-one (lorazepam) in humans and its urinary excretion as lorazepam determined by differential pulse polarography. J. Pharm. Sci. *63*, 1943–1945.

71. Oelschläger, H., Geppert, V. (1976) Drug analysis by polarographic methods. XIX: Determination of medazepam and its main metabolites by cathode ray polarography. Arch. Pharm. *309*, 1000–1006.
72. Kobiela-Krzyzanowska, A. (1977) Polarographic determination of nimetazepam in serum and whole blood. Pharmazie *32*, 693–695.
73. Oelschläger, H., Sengün, I. F. (1984) Analyses of drugs by polarography. 23. Determination by polarography (DPP and CRP) of triazolam in plasma and serum. Arch. Pharm. *317*, 69–73.
74. Schütz, H. (1991) Screening und Identifizierung klinisch-toxikologisch relevanter Giftstoffe. Pharmazie in unserer Zeit *20*, 271–277.
75. Schütz, H., Wollrab, A. (1988) Die Bedeutung des Retentionsindex für die toxikologische Analytik. Pharmazie in unserer Zeit *17*, 97–101.

3.3 Hypnotics and Sedatives Not Belonging to the Classes of Barbiturates and Benzodiazepines

H. Brandenberger

3.3.1 Introduction

Table 1 lists the most important substance classes which have been used or are still in use as hypnotics and sedatives. The order is chronological and the year is stated when the first member of the class has been introduced as a pharmaceutical. This does not hold for the class of carbamates. Since their members are hardly used as hypnotics (exceptions are Aponal, N-Oblivon and Valamin), but as sedatives, muscle relaxants or tranquilizers, we have put them at the end of the list.

The barbiturates, discussed in chapter 3.1, have been the leading class of hypnotics during the first three quarters of the nineteenth century. Today, a large part of the sedatives and hypnotics in daily use belongs to the class of benzodiazepines, treated in chapter 3.2. In some countries, barbiturates have almost vanished from the market, and so has a substantial part of the representatives of the other classes of hypnotics except for the benzodiazepines. However, some of the older hypnotics and sedatives are still in use and can be responsible for severe cases of intoxication. Especially in suicide and homocide cases, older hypnotics can be found fairly often. And since hypnotics and strong sedatives are among the most abused pharmaceuticals, it is advisable to extend a general search for compounds with hypnotic action to all classes listed in table 1.

Table 1. The Main Classes of Strong Sedatives and Hypnotics

1. Inorganic Bromide Salts (in use since 1826)
2. Alcohols and Aldehydes (introduction of Chloral Hydrate 1869)
3. Barbiturates (introduction of Barbital 1903)
4. Linear Ureides (introduction of Bromisoval 1905) and Amides
5. Piperidine-diones (introduction of Persedon 1949)
6. Quinazolinones (introduction of Methaqualone 1958)
7. Benzodiazepines (Sedatives since 1960, Hypnotics since 1965)
8. Carbamates (mostly used as Sedatives and Tranquilizers)

A differentiation between sedatives and hypnotics is often rather artificial (1). Furthermore, the borderlines to the narcotics and to the tranquilizers may sometimes also appear fluid. As a rough rule of thumb it can be said that some pharmaceuticals can lead to sedation if their dosage is only low, induce sleep at higher dosages, and even act as narcotics, if given in very high doses. It seems therefore justified to review sedatives and hypnotics together.

In this chapter, the pertinent members of the different chemical groups of hypnotics not already discussed in chapters 3.1 and 3.2 will be listed, and some data on their clinical use and toxicological importance given. The possibilities for detection and dosage in biological fluids will then be outlined. Several of the analytical methods already described in chapter 3.1 are useful also for the analysis of compounds treated in this chapter. In such cases, detailed descriptions are replaced by cross-references.

3.3.2 Inorganic Bromides

The oldest class of sedatives encompasses the inorganic bromides such as sodium bromide (also used as anticonvulsant, especially antiepileptic agent for children, and as medication against nervous heart conditions) and potassium bromide (often given to victims of whooping cough and vomiting during pregnancy). But also calcium, magnesium and lithium bromides have played roles as sedatives, hypnotics and anticonvulsants. Therapeutic dosages of these salts must be high. Up to 15 g per day have been given. Resorption is rapid and elimination only slow, and prolonged treatments carry the danger of accumulation with toxic effects. An analytical control of the therapy is therefore advisable.

Today, inorganic bromides still occur as constituents of mixed preparations. It should further be noted that bromide is often chosen as the anion for basic pharmaceuticals (i.e. phenothiazines). But in this form, its dosage is of course only low.

For the forensic chemist, it is important to realize that bromide salts can easily be obtained and used for self-medication or criminal purposes, since they are used as photgraphic chemicals, fumigants and for many other purposes. In this connection, it must also be kept in mind that bromides can be addicting and lead to a chronical use called "bromism".

Bromide concentrations in normal sera are usually not higher than 10 µg/ml and therapeutic treatments should not elevate the levels to much over 50 µg/ml, certainly not up to 100 µg/ml. At concentrations over 500 µg/ml, in some people already over 300 µg/ml, toxic effects are to be expected. Overdosages can lead to tremor, convulsions, paralysis of the central nervous system, coma and excitus.

Intoxications with inorganic bromides must be controlled by bromide analysis. This is discussed in chapter 2.8 on non-metallic toxic elements. We just want to point out that bromide concentrations from overdoses can usually be detected in urine with specific bromide electrodes. For more precise measurements and for lower concentrations in serum and urine, colorimetric methods have been described (2). Anion chromatography can also be used.

3.3.3 Alcohols and Aldehydes

3.3.3.1 General remarks

It is well known that alcohols possess hypnotic and narcotic properties. In the alcohols up to butanol, the narcotic effect dominates the hypnotic influence. It may take up to ten drinks, corresponding to an intake of 100 g of ethanol, to put an adult to sleep. But in higher alcohols such as tert-pentyl alcohol (tert-amyl alcohol, amylene hydrate), the hypnotic effect is more pronounced. This compound has therefore been used in doses between 2 and 4 g as a sleep inducing agent. However, the unpleasant odor and the action as local irritant have handicapped its acceptance. It has been replaced by compounds which are better tolerated. They carry substituents which strengthen the hypnotic effect, such as an ethynyl group, or 1 or several chlorine atoms, or both. With such hypnotics, dosages can be much lower, as a rule between 0.25 and 1 g.

Table 2. Volatile Sedatives and Hypnotics with Alcohol or Aldehyde Structure

$CH_3{-}CH_2{-}C(CH_3)(CH_3){-}OH$	tert-Pentyl alcohol, tert-Amyl alcohol, Ethyl dimethyl carbinol, Amylene Hydrate (Hypnotic)
$CH_3{-}CH_2{-}C(C{\equiv}CH)(CH_3){-}OH$	3-Methyl-pentyn-3-ol, Ethyl ethynyl methyl carbinol, Allotropal, Dormison, Meparfynol, Oblivon, Pentadorm, Somnesin (Hypnotic, Sedative)
$CH_3{-}CH_2{-}C(C{\equiv}CH)(CH{=}CHCl){-}OH$	*β*-Chlorovinyl ethyl ethynyl carbinol, Ethchlorvynol, Placidyl, Roeridorm (Hypnotic, Sedative)
$CCl_3{-}CH_2{-}OH$	Trichloroethanol, Trichloroethyl alcohol Active metabolite of Chloral Hydrate (Hypnotic, Sedative)
$CCl_3{-}C(CH_3)(CH_3){-}OH$	Trichloromethyl dimethyl carbinol, Trichlorobutanolum, Chlorobutanol, Chlorbutol (Sedative, vet. Sedative, Preservative)
$CCl_3{-}CH(OH)(OH)$	Trichloroacetaldehyde monohydrate, Chloral Hydrate, Chloral Durate (Hypnotic, Sedative)
$CH_3{-}CH(OH){-}CH_2{-}CHO$	3-Hydroxy-butanal, Aldol (Hypnoticum, Sedativum)
cyclic trimer: three CH_3-substituted C atoms alternating with three O atoms	Paraldehyde, Paral, Trimer of Acetaldehyde (Anticonvulsant, Hypnotic, Sedative)
thiazole ring (S, N) with $-CH_2{-}CH_2{-}OH$ and $-CH_3$	4-Methyl-5-thiazole ethanol (Hypnotic, Sedative)

Table 2 lists the most important members of this group of sedatives and hypnotics. Since the chemical formulas of the compounds are given, the official chemical names are in part omitted in favor of a good choice of trivial names.

3.3.3.2 Chloral hydrate

Particularly one of the compounds in table 2 deserves special attention. It is 2,2,2-trichloro-ethane-1,1-diol, a hydrated aldehyde, the well known chloral hydrate, synthezised by Liebig as far back 1832 and introduced as a hypnotic by Liebreich in 1869. This is the first organic hypnotic of pharmaceutical importance which has been put on the market, long before barbital, the first in the large series of barbiturates. And the amazing fact is that chloral hydrate has been used on and off during all these years and is still in use today, as a sedative and hypnotic for humans, especially children, and as a narcotic and anesthetic in the field of veterinary medicine.

A hypnotic dose for an adult is 0.5 to 1.0 g, for a child not over 0.25 g. The drug is mostly precribed as a syrup which may contain 1 g of chloral hydrate in 15 ml, or in form of gelatine capsules. It is used as sedative, hypnotic, tranquilizer and anticonvulsant, especially in pediatric medicine. Dosages over 1 g can lead to nausea with vomiting; higher doses can slow down the respiration and lead to unconsciousness. 10 g, according to some authors already 5 g (3), must be considered as minimal lethal dose for an adult. But in spite of the small security margin, the medicament has not been abandoned during the long period of more than 120 years since its introduction.

3.3.3.3 Analytical detection

The alcohols with sedative and hypnotic properties are neutral and volatile compounds. They can be separated from the biological fluids or tissues by steam distillation (4,5). This operation must precede the general extractions for drugs described in 3.1.7. By using super-heated steam, dilution of the distillation residue may be prevented, thus eliminating the need for a concentration step before extracting the non-volatiles. The steam distillate can be directly analysed by GC, provided that the detector used tolerates water. Alternatively, it can first be extracted with ether, and the GC analysis carried out with the organic extract. This allows a free choice of detector.

For the analysis of organic extracts, we recommend the use of a dual detection system, consisting of a flame ionization and an electron capture detector in parallel. This permits identification of the eluting compounds not only on the basis of retention properties, but also by the response ratios of the two different detector signals (4,5). We call this quotient (the electron capture divided by the flame ionization signal) the relative electron affinity or, short, REA-value. If it is adjusted to 10^1 for a compound with 1 Cl (like Roeridorm), it will lie in the range of 10^3 to 10^4 for compounds containing 3 Cl, such as chloroform, chloral hydrate or trichloroethanol. Molecules without electron attracting substituents, on the other hand, possess very low REA-values, usually (depending on the type of electron capture detector used) below 10^{-2}.

On non-polar or only moderately polar columns, the hypnotics in table 2 will emerge in the order amylene hydrate, chloral (since the hydrate is decomposed to the aldehyde), Oblivon, Roeridorm and trichloroethanol (4,5). In the cases of chloral, it is important to know that this aldehyde already undergoes dismutation in the stomach and, after resorption, an additional metabolic transformation to trichloroethanol, the major active metabolite, and trichloroacetic acid, the major urinary metabolite. In urine, both metabolites appear partly conjugated with glucuronic acid. This holds especially for trichloroethanol, which is derivatized to urochloralic acid. In the first 24 hours, an estimated 10 to 30% of an ingested dose of chloral hydrate is excreted as urochloralic acid and up to 10% as free trichloroethanol, while trichloroacetic acid elimination may take several days. In the presence of alcohol, excretion of trichloroethanol is retarded. For quantitative studies, enzymatic hydrolysis of urine is essential, preferably with a Helix pomatia extract containing both β-glucuronidase and aryl sulfatase. Such a hydrolytic step frees also the other members of this group of hypnotics which may have undergone metabolic conjugation.

Since the alcohols and aldehydes with hypnotic properties do not contain strongly UV-absorbing structural units, UV-spectrophotometry is not useful for their detection. We must recur to IR-absorption or MS, both used in combination with GC. The IR-technique does not tolerate the presence of water. If we want to work without organic extraction of the distillate, a GC-MS combination is certainly the best analytical approach. The excellent sensitivity of this technique permits direct injection of aqueous steam distillates. For trace analysis, we recommend the use of negative CI-MS in place of conventional positive EI-MS. For compounds possessing high electron affinity, such a switch-over can improve the detection limits by several factors of 10. The mass spectrometer can be used in the normal scanning mode or, for revealing pg quantities, as a selected ion detector, that is for monitoring one or several intensive fragment ions during the elution. In low pressure negative ion CI-MS, Cl-containing hypnotics release the chloride anions with masses 35 and 37. They are usually among the most intense peaks in the spectrum and can be recorded for a simultaneous trace detection of these compounds (6,7).

3.3.3.4 Preliminary tests

The reader may also ask about preliminary tests for the presence of hypnotic alcohols, to be carried out directly with body fluids. Several have been in use. Trichloro-compounds can be tested for in stomach content or urine samples by the Fujiwara test (8–10). The presence of ethchlorvynol can be controlled in the same body fluids by the so-called diphenylamine test (8, 9, 11) or by checking for a first white and then yellow precipitate on addition of an ammoniacal solution of silver nitrate (9, 11). Such tests may be sufficiently sensitive to detect therapeutic concentrations, but they lack the specificity needed for a final proof.

In place of preliminary color tests, we recommend a simple head space sampling method, which can be used not only for investigating body fluids, but also for tissues:

A controlled amount of the biological specimens (i.e. 1 g) is added to small serum vials (i.e. 10 ml) which are capped with silicone membranes. The vials are heated in a water bath, aliquots of the gas phase (i.e. 1 ml) are removed with a syringe and injected into the column of a GC-MS combination. The sensitivity of the detection system is capable of revealing even low therapeutic concentrations of the more volatile compounds and elevated therapeutic or toxic concentrations of the less volatile components such as trichloroethanol or trichloroacetic acid. The head space gas sampling for GC-analysis can also be used for quantifications, using internal or external calibration.

An even faster and less cumbersome method is direct injection of urine, serum or even full blood into the GC-column coupled to a flame ionization detector or, for higher sensitivity and better identification power, to a mass spectrometer. We have used this simple approach often for identifying chloral and its metabolites in intoxications (6, 7, 12). It is hard on the column life, but it requires only one single µl of biological material.

It must be added that the volatile members of this group of hypnotics can often be detected by their odor in the exhaled air of a patient. This holds especially for amylene hydrate and chloral hydrate. A physician may not be able to identify the origin of such an odor, but he should always report his observation to the laboratory investigator. In a search for an unknown hypnotic, such communications can greatly help to shorten the time of an analytical investigation.

3.3.3.5 Derivatives of chloral hydrate

Since chloral hydrate is still used today, it may be justified to mention in this context also the chloral adducts and derivatives which have reached the market. Table 3

Table 3. Chloral Hydrate Adducts and Derivatives

Chloral Betaine	Beta-Chlor Somilan	Inner salt between betaine hydrate and chloral hydrate (Sedative,Hypnotic)
Chloralantipyrine	Hypnal	1 : 1 mixture of phenazone with chloral hydrate (Hypnotic, Analgetic)
Dichloralantipyrine Dichloralphenazone	Bihypnal Bonadorm Dormwell Welldorm	1 : 2 mixture of phenazone with chloral hydrate (Sedative, Hypnotic)
Triclofos Trichloroethylphosphate	Trichloryl Tricloryl Triclos	2,2,2-Trichloroethanol dihydrogen phosphate (Na-salt) (Hypnotic, Sedative)
Carbocloral Chloral-urethane	Ural Uraline Uralium	2,2,2-Trichloro-1-hydroxyethyl carbamic acid ester (Hypnotic)
Chloral Formamide	Chloralamide Chloramide	Monoamide between chloral hydrate and formamide (Sedative, Hypnotic)
Chlorhexadol	Chloralodol Mecoral Medodorm	Monoether between chloral hydrate and 2-methyl-pentane-2,4-diol (Hypnotic)
Pentaerythritol Chloral	Periclor Petrichloral	Pentaerythritol etherified with 4 molecules chloral hydrate (Hypnotic, Sedative)
Alpha-Chloralose Glucochloral	Alphakil Chlorosane Dorcalm Somio	Double ether between glucose and trichloroacetaldehyde (Sedative, Hypnotic and veteriary Anesthetic)

lists them. Some of the compounds are still used, especially in veterinary medicine. Not all of them are as easy to detect as the free chloral hydrate. They may therefore be used for criminal intoxications.

The simple adducts or inner salts such as the combinations of phenazone with chloral hydrate dissociate in the body, and the constituents of the mixture can be determined separately. The esters can be hydrolysed and chloral as well as its metabolites measured by GC. With the ethers and amides, this is not possible. But by treating the biological specimens with alkali, the chloral moiety in these molecules is decomposed, and the chloroform released can be detected and measured with GC. If the decomposition is carried out directly in small serum flasks and the chloroform transferred by a syringe from the head space of the sealed reaction vessel to the GC column, analysis of these chloral derivatives becomes also quite simple.

3.3.4 Linear Ureides and Amides

3.3.4.1 Bromoureides

Table 4. Active Compounds with Linear Ureide Structure

$(CH_3-CH_2)_2C(Br)-CO-NH-CO-NH_2$	Bromodiethylacetylurea, Bromodiethylacetylcarbamide, Adalin, Bromadal, Carbromal, Diacid, Nyctal, Uradal (Sedative, Hypnotic)
$(CH_3-CH_2)_2C(Br)-CO-NH-CO-NH-CO-CH_3$	N-Acetyl-N-bromodiethylacetyl-urea, Acetyl Adalin, Abasin, Acecarbromal, Carbased, Sedamyl (Sedative, Hypnotic)
$(CH_3)_2CH-CH(Br)-CO-NH-CO-NH_2$	2-Bromo-isovalerylurea, Alluval, Bromisoval, Bromural, Brovaleryl, Bromvaletone, Bromvalurea, Dormigene, Isobromyl, Somnurol, Uvaleral (Sedative, Hypnotic)
$(CH_3)_2CH-CH(CH_2-CH=CH_2)-CO-NH-CO-NH_2$	Allylisopropylacetylurea, Apronal, Apronalide, Sedormid (Sedative, Hypnotic)
$(CH_3-CH_2)(CH_3)CH-CH(CH_2-CH_3)-CO-NH-CO-NH_2$	2-Ethyl-3-methylvalerylurea, Capuride, Pacinox (Hypnotic)
$CH_3-CH_2-C(=CH-CH_3)-CO-NH-CO-NH_2$	2-Ethylcrotonoyl-urea, Ectylurea, Astyn, Cronil, Ektyl, Levanil, Neuroprocin, Nostal, Nostyn, Pacetyn (Sedative, Hypnotic)
$C_6H_5-CH_2-CO-NH-CO-NH_2$	Phenylacetylurea, Phenacemide, Phenurone, Epiclase, Phacetur (Anticonvulsant)

The linear ureides with a bromine atom (the first three entries in table 4) are a well-defined old class of hypnotics which has been widely used, Bromoisoval since 1905, Bromadal since 1910. At low dosages, the compounds are sedatives, at higher dosages hypnotics. Furthermore, they also have been used as anticonvulsants (anti-epileptics) and as agents against psychic stress or fear (tranquilizers). For hypnotic action, doses between 0.5 and 1 g are needed. This is quite high, since a minimal lethal dosis is close to 10 g (3).

Long-term use of bromoureides may lead to symptoms resembling bromism. The substances are as addicting as the barbiturates. The following side effects have been recorded after prolonged use: stomach bleedings, affection of blood clotting mechanism and depression of the central nervous system. Overdosages can lead to coma and respiratory arrest, just as in the case of barbiturates. In most countries, barbituric acids have been placed under narcotic control a few years before attention has been given to bromoureides. During these years, bromoureides became widely abused hypnotics, responsible for a great number of intoxications, not few of them lethal. Addiction to these drugs became quite common. They also had to be submitted to medical control.

Today, bromoureides are seldom taken alone. However, they still exist as composites of mixed preparations, and it seems therefore advisable to check for their presence in every general search for hypnotics and sedatives in body fluids.

Bromoureides are readily absorbed and extensively metabolized. Bromodiethylacetylurea or 2-bromo-2-ethyl-butyrylurea, as an example, is hydrolyzed to 2-bromo-2-ethyl-butyrylamide, which is in part oxidized to 2-bromo-2-ethyl-3-hydroxy-butyrylamide. The basic compound as well as these metabolites lose bromide, which is the main metabolite excreted in urine, together with 2-ethyl-2-hydroxy-butyric acid and 2-ethyl-butyrylurea. After therapeutic dosages, only a small percentage of the unchanged drug is excreted. But in intoxication cases, considerable quantities of not metabolized bromoureides could be recovered from urine (13).

3.3.4.2 Ureides without bromine

The bromine-free linear ureides are of newer date. They have not often been reported as main cause of intoxications. However, such information can be misleading, since toxicological laboratories do not always search for bromine-free ureides or may miss them. Of the compounds reported in table 4, apronal (Sedormid) is probably best known as hypnotic. The next two have rather been used as sedatives, and phenurone, the last on the list, as anticonvulsant.

3.3.4.3 Amides

Table 5 lists some of the amides which have hypnotic and/or sedative properties. The first two entries contain bromine. They can be used just like the bromoureides as hypnotics or sedatives. The first compound on the list is actually a metabolite of carbromal, the second is structurally related to bromisoval. These bromo-amides are habit-forming like their related ureides.

In the bromine-free amides quoted in table 5, the hypnotic properties are less pronounced. They are used as sedatives.

Table 5. Amides used as Sedatives or Hypnotics

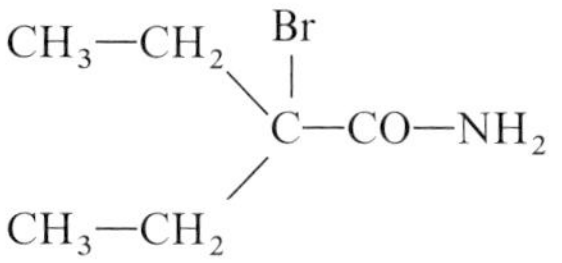

Structure	Name
$CH_3{-}CH_2$, Br C—CO—NH_2 $CH_3{-}CH_2$	2-Bromo-2,2-diethyl-acetamide, Diethylbromoacetamide, Carbromide, Neuronal, Metabolite of Carbromal (Sedative, Hypnotic)
CH_3 Br CH—C—CO—NH_2 CH_3 $CH_2{-}CH_3$	2-Bromo-2-ethyl-2-isopropyl-acetamide, 2-Bromo-2-ethyl-isovaleramide, Ibrotamide, Vagopropal, Vagoprol (Sedative, Hypnotic)
CH_3 $CH_2{-}CH_3$ CH—CH_2—CO—N CH_3 $CH_2{-}CH_3$	N,N-Diethyl-isovaleramide, N,N-Diethyl-3-methyl-butyramide, Isovaleryl Diethylamide, Valyl (Sedative)
CH_3O $CH_2{-}CH_3$ CH_3O—⟨O⟩—CO—NH—CH_2—CO—N CH_3O $CH_2{-}CH_3$	3,4,5-Trimethoxy-benzamido-N,N-diethylacetamide, (Trimethoxybenzoyl)-glycine-diethylamide, Tricetamide, Trimeglamide (Sedative)
CH_3O CH_3O—⟨O⟩—CO—N⟨ ⟩O CH_3O	3,4,5-Trimethoxybenzoyl-morpholine, Opaline, Trimethozine, Trioxazine (Sedative)
CH_3—CH—CH_2—CO—NH—CH_2—CH—$(CH_2)_3$—CH_3 OH $CH_2{-}CH_3$	N-(2-ethylhexyl)-3-hydroxy-butyramide, Butoctamide, composite of Listomine (Sedative, Hypnotic)
N⟨O⟩—CO—NH—CH—CH_2—CH_2—N⟨ ⟩N—⟨O⟩—F CH_3	(p-Fluorophenyl)-1-piperazinyl-1-methyl-propyl-nicotinamide, Niaprazine, Nopron (Sedative, Hypnotic)

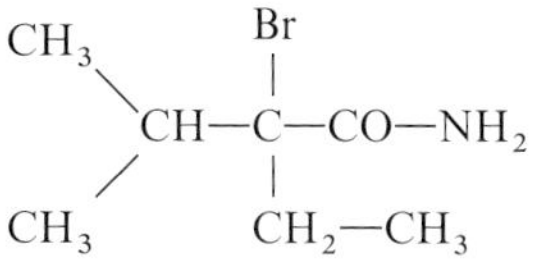

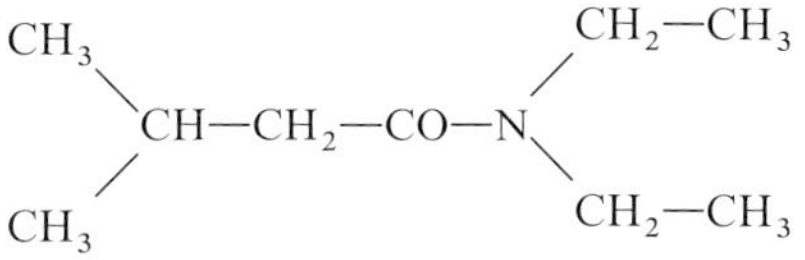

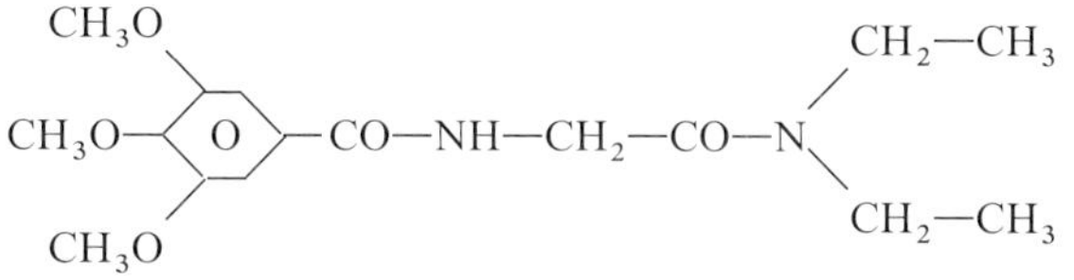

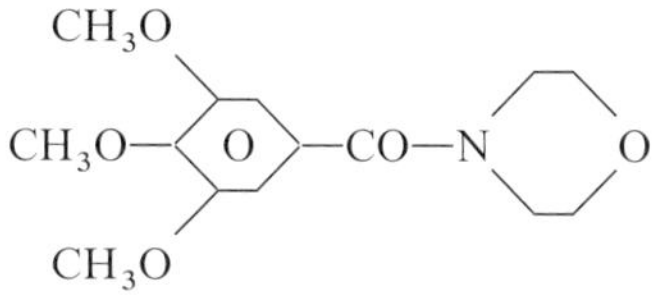

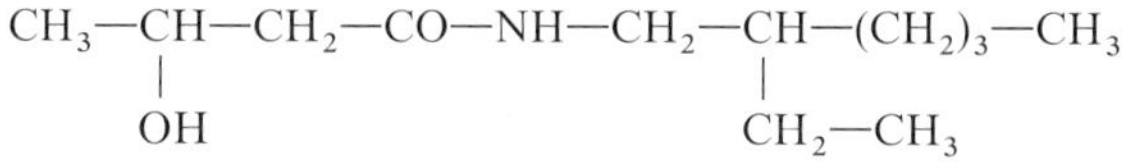

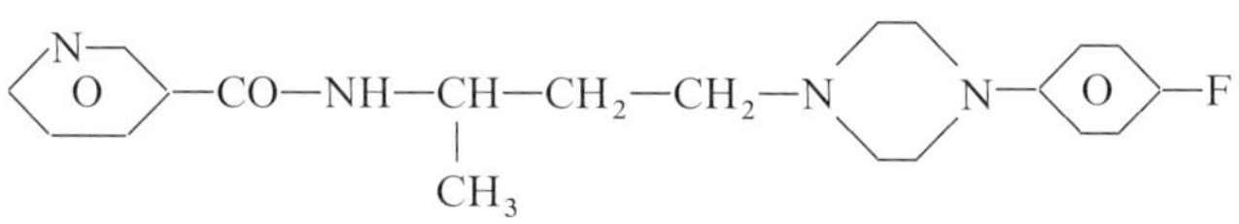

3.3.4.4 Analytical detection

All the ureides and amides listed in tables 4 and 5 are neutral compounds and fairly well soluble in organic solvents like ether. They will pass into the organic phase in the first extraction step used, regardless of the pH of the aqueous fluid or suspension. We can therefore choose whether we want to separate them from the biological phase together with the acidic or together with the basic components. We usually have had better luck by starting our extractions at acid pH. After this first organic extract has been freed from the acid compounds by back-extracting them into basic aqueous solution, several different analytical pathways can be selected for an identification of the remaining neutral components:

- The organic phase can be taken to dryness and submitted to a fractional vacuum sublimation. Most of the ureides and also some of the amides sublime easily and can be identified by melting point and mixed melting point analysis. Thanks to the high dosages, this class of compounds is ideally suited for such an "old fashioned" and slow identification method, which is however straightforward and does not involve much manual labor. The melting points can be found in the Merck Index (14), the handbook of Clarke (15), and in other compilations on the desk of every toxicological chemist.
- The organic extract with the neutral components can be submitted to TLC. After migration, the plates or sheets are treated with mercurous nitrate or mercuric nitrate and diphenylcarbazone for the detection of all ureides and some amides, with Nessler reagent, which is especially suited for revealing the aliphatic ureides and amides (9, 16), and with N,N-dimethylamino-phenylenediamine, a good reagent for bromine and chlorine containing compounds (17, 18). The Rf-values in a number of flow systems have been reported (15, 18–20).
- The neutral extracts can also be analyzed by GC. Retention data on different columns are available (15, 21). By using a dual detection system consisting of flame ionization and electron capture detector in parallel, it is again possible to pick out at first sight the bromine-containing compounds on the basis of their high REA-values (near 10^2) (4, 5). For the nitrogen-containing compounds, thermionic detection, if possible in parallel with flame ionization, is recommended. The best approach is of course again GC-MS: excellent sensitivity and identification power, independence from retention data. Positive ion MS (EI supplemented by CI) is well suited for identifying the sedatives and hypnotics listed in tables 4 and 5 (22). In a search for bromine-containing ureides and amides, negative CI-MS is the most sensitive approach (23). By monitoring the 2 isotopic bromine ions with masses 79 and 81, Br-containing compounds can be detected on a pg- or sub-pg-level (24, 25). With a mass spectrometer which can record simultaneously positive and negative ions (7, 26), such a recording can be complemented by incorporating substance-typical positive ions into the analytical scheme (27).
- HPLC can of course also be used for revealing the presence of hypnotics or sedatives with ureide or amide structure. But since most substances do not possess UV absorption maxima over 200 nm or fluorescence, the sensitivity of HPLC is only limited and compound identification cumbersome.

It should be noted that the bromoureides and bromoamides must be searched for primarily in stomach content and blood and not in urine. Urine samples, on the other hand, should also be tested for bromide anions. For quantifications, the total bromide concentration of all the specimens under investigation can be measured after decomposition of the organic material.

3.3.4.5 Preliminary tests

Immunochemical assays have not been developed for these old classes of drugs. The available preliminary tests are based on precipitation and color reactions. We recommend the following procedure for checking for organo-bromine compounds (8, 28) in the neutral organic extract, but not directly in biological fluids:

A few drops of the extract are taken to dryness in presence of sodium hydroxide and a drop of fluorescein solution. Glacial acetic acid and hydrogen peroxide are added. In presence of organo-bromine compounds, the red color of eosine appears during slow evaporation to dryness on a water bath.

Since one of the main metabolic pathways of the bromoureides and bromoamides is the release of bromine, overdoses can be detected in urine samples, exactly as in an abuse of inorganic bromides, by measuring the bromide concentration (chapter 2.8).

3.3.5 Piperidine-diones and Tetrahydropyridine-diones

3.3.5.1 General remarks

The hypnotics with piperidine-dione and tetrahydropyridine-dione structure have made a swift appearance on the market in the 1950s and 1960s. But in contrast to the aforementioned substance classes, they did not last very long. Their fast ascent was followed, in less than 20 years, by an even faster descent; and it does not seem very likely that they will make a reappearance as hypnotics. Too many side effects have been connected with their chronic use, and their abuse-potential has been shown to be of a considerable dimension.

Table 6. Hypnotic/Sedatives with Piperidine-dione Structure

2,4-Dioxo-3,3-diethyl-tetrahydro-pyridine	Pyrithyldione, Persedon, Presidon, Benedorm, Tetridin	introduced in 1949
2,4-Dioxo-3,3-diethyl-5-methyl-piperidine	Methyprylone, Noludar, Noctan. Dimerin	introduced in 1955
2,4-Dioxo-3,3-diethyl-piperidine	Dihyprylone, Sedulon, Tusseval	Sedative and Antitussive
2,6-Dioxo-3-ethyl-3-phenyl-piperidine, α-Ethyl-α-phenyl-glutarimide	Glutethimide, Doriden, Dorimide, Elrodorm	introduced in 1954 abandoned
2,6-Dioxo-3-phthal-imido-piperidine, 3-Phthalimido-glutamic acid imide, N-phthaloyl-glutamic acid imide	Thalidomide, Contergan, Distaval, Kevadon, Neurosedyn, Pantosediv, Sedalis, Softenon, Talimol	introduced in 1957, withdrawn around 1961

Table 6 list the few members of the group which have been manufactured and used as hypnotics or sedatives on a large scale. The commercial names are listed, as well as the year the specific compound has reached the market. The third entry on the list, dihyprylone, has been prescribed as an antitussive and not as a hypnotic. In figure 1, we compare the chemical structures of piperidine-diones with those of barbiturates (pyrimidine-triones). The striking resemblance has probably been the starting point for the interest of the chemical designers of hypnotics and sedatives in piperidine- and tetrahydropyridine-diones.

```
            O   Pyrithyldione                        O   Barbital
            ||                                       ||
      NH—C      CH2CH3                         NH—C      CH2CH3
     /      \  /                              /      \  /
  CH         C                             O=C         C
   \\       /  \                              \      /  \
      CH—C      CH2CH3                         NH—C      CH2CH3
            ||                                       ||
            O                                        O

            O   Methyprylone                         O   Metharbital
            ||                                       ||
      NH—C      CH2CH3                         NH—C      CH2CH3
     /      \  /                              /      \  /
  CH2        C                             O=C         C
     \      /  \                              \      /  \
      CH—C      CH2CH3                          N — C    CH2CH3
     /      ||                                 /     ||
  CH3       O                               CH3      O

            O   Glutethimide                         O   Phenobarbital
            ||                                       ||
      NH ——C      CH2CH3                       NH—C      CH2CH3
     /       \  /                             /      \  /
  O=C          C                           O=C         C
     \       /  \                             \      /  \
      CH2—CH2    C6H5                          NH—C      C6H5
                                                     ||
                                                     O
```

Figure 1. Similarities between the Structures of Piperidine-diones and Barbiturates.

3.3.5.2 Advantages and disadvantages

The main advantage of this class of hypnotics over the older barbiturates and bromoureides was considered to be the wider span between hypnotic and minimal lethal doses. For pyrithyl-dione, the first member of the class, between 200 to 400 mg are given to induce sleep. The minimal lethal dosis is estimated to be near 10 g (3). This is a safer margin than that for barbital (therapeutic dose up to 600 mg, estimated minimal lethal dose 2 g). Similar figures can be quoted for the structurally related hypnotics methyprylone and phenobarbital.

However, this advantage was soon overshadowed by some side effects attributed to members of the piperidine-dione class. Thalidomide, if taken during pregnancy over longer periods, was found to be the cause of a great number of very severe fetal abnormalities (worldwide over 10000), and it had to be withdrawn from the market only a few years after its introduction. It is understandable that this disaster has also affected the sales of glutethimide, the second hypnotic with 2,6-dioxo-piperidine structure. In the early 60s, this compound has caused a considerable number of lethal intoxications. To induce sleep, it must be given in somewhat higher doses than the 2,4-dioxo-piperidines and tetrahydropyridines (up to 600 mg), but the minimum lethal dose may be as low as 5 g (3). The compound is not sold any longer. Pyrithyldione and methyprylon have had a longer market life. But the suspicion arose, that a long-term use of pyrithyldione may lead to undesirable side effects such as decrease of blood platelets, damage to the spinal marrow and affection of blood clotting mechanism. Since methyprylone metabolizes in part to pyrithyldione, the sale of both 2,4-dioxo-piperidines has also been stopped in most countries. Nevertheless, both compounds are still encountered occasionally in forensic investigations, be it that they stem from an old stock, or that they originate from a country where they are still available on the market.

All piperidine-diones are, like the barbiturates, habit forming. Not many years after their appearance they became drugs of abuse and had to be added to the growing list of controlled drugs. In Germany, methyprylone has been available for awhile in liquid form. This has been used for criminal purposes. Especially in the night scene of the city of Hamburg, but later also elsewhere, the so-called knock-out drops were added to alcoholic drinks in order to put a "client" fast asleep and rob him (29). The trick was extremely effective, since ethyl alcohol strongly intensifies the hypnotic action of the drugs and speeds up the onset of hypnosis. An additional factor for the "success" of the undertaking is the instability of methyprylon solutions. Upon air contact, the drug slowly oxidizes to methyl-persedon, which is a stronger and faster acting hypnotic than methyprylone.

3.3.5.3 Metabolism

No metabolites with an intact ring system have been found for pyrithyldione. The compound passes into the urine.

Methyprylone, on the other hand, is characterized by an extensive metabolism which has been well-studied (30). It is illustrated in figure 2. The pyrimidine ring is first dehydrogenated to a tetrahydropyridine ring, yielding 2,4-dioxo-3,3-diethyl-5-methyl-tetrahydropyridine or 5-methyl-persedon. The methyl group is then successively oxidized to a hydroxymethyl-, an aldehyde- and finally a carboxyl group. Decarboxylation converts 5-carboxy-persedon to persedon. An alternate metabolic pathway is oxidation to 2,4,6-trioxo-3,3-diethyl-piperidine. Thc pharmacologically most active compound among all these metabolites is 5-methyl-persedon.

Methyprylone, Noludar
(2,4-dioxo-3,3-diethyl-
5-methyl-piperidine)

C_2H_5 — C(— C_2H_5) — CO — CH-CH_3 — CH_2 — NH — OC (ring)

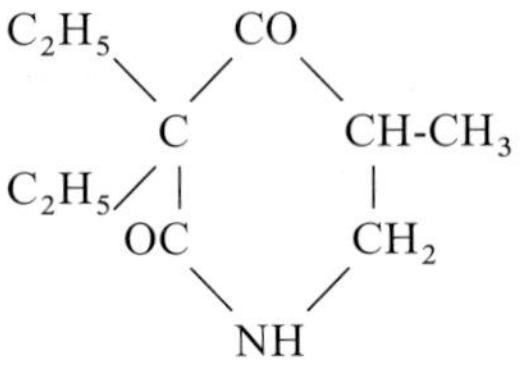

↓ dehydrogenation

5-Methyl-pyrithyldione,
5-Methyl-persedon
(2,4-dioxo-3,3-diethyl-5-
methyl-tetrahydropyridine)

C_2H_5 — C(— C_2H_5) — CO — C-CH_3 ‖ CH — NH — OC (ring)

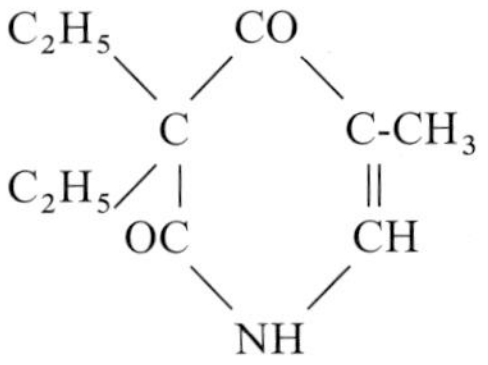

↓ oxidation

5-Hydroxymethyl-pyrithyldione
(2,4-dioxo-3,3-diethyl-5-hyd-
roxymethyl-tetrahydropyridine)

C_2H_5 — C(— C_2H_5) — CO — C-CH_2OH ‖ CH — NH — OC (ring)

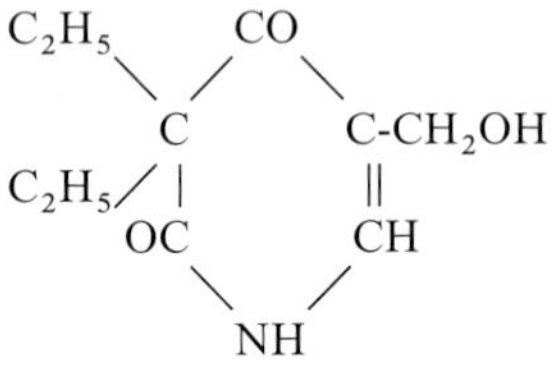

↓ oxidation (with alde-
hyde as intermediate)

5-Carboxy-pyrithyldione
(2,4-dioxo-3,3-diethyl-5-
carboxy-tetrahydropyridine)

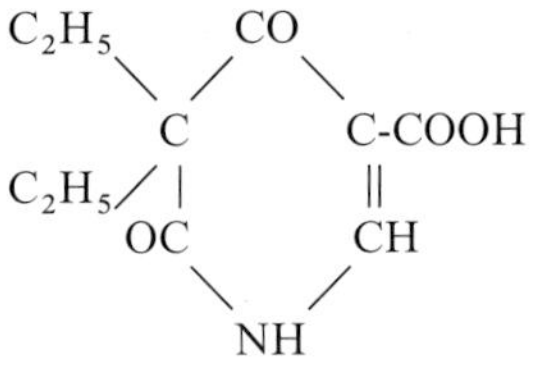

C_2H_5 — C(— C_2H_5) — CO — C-COOH ‖ CH — NH — OC (ring)

↓ decarboxylation

Pyrithyldione, Persedon
(2,4-dioxo-3,3-diethyl-
tetrahydro-pyridine)

C_2H_5 — C(— C_2H_5) — CO — CH ‖ CH — NH — OC (ring)

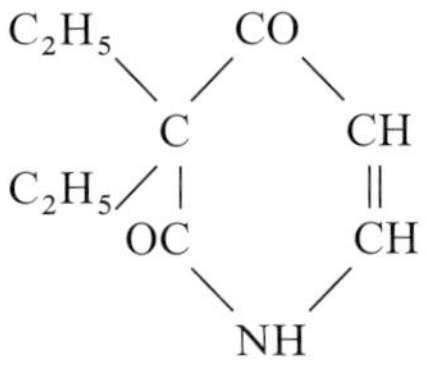

Another metabolic reaction
oxidizes methyprylone to

2,4,6-trioxo-3,3-diethyl-
5-methyl-piperidine

C_2H_5 — C(— C_2H_5) — CO — CH-CH_3 — CO — NH — OC (ring)

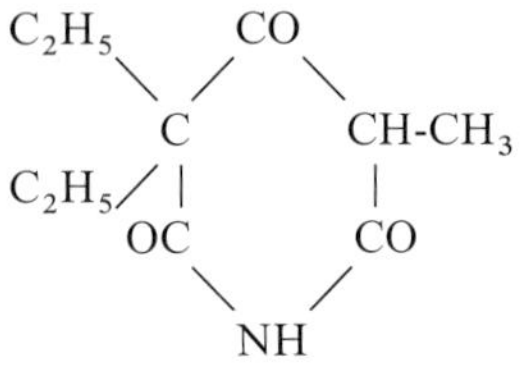

Figure 2. Metabolic Transformation of Methyprylone (Noludar)

3.3.5.4 Detection and dosage of piperidine-2,4-diones and tetrahydropyridine-2,4-diones

The compounds are weak acids, just like the barbiturates. In alkaline aqueous solution, they exist in their enolic forms. In our systematic extraction procedure, described in chapter 3.1.7, they can be found in the extract of weak acids, often fairly free from neutral contaminents (fat constituents and plasticizers) and from stronger acidic components. Two metabolites of methyprylone, however, the carboxy- and hydroxymethyl-derivatives, pass in part also into the extract containing the stronger acids. If a quantitative search should include these metabolites, both acid extracts must be analyzed, or, as an alternative, the separation into strong and weak acids omitted by back-extracting the inital organic extract directly with a strongly alkaline aqueous solution.

UV-analysis of these aqueous extracts at different pH-values, carried out according to chapter 3.1.8, reveals the presence of all 2,4-dioxo-tetrahydropyridines, but not that of 2,4-dioxo-piperidines, which do not possess intensive UV-maxima over 200 nm. The maxima of the tetrahydropyridines in diethyl ether and in aqueous solution at three different pH-values are reported in table 7. In addition, the fluorescence spectra are also very helpful for detecting traces of dioxo-tetrahydropyridines. In alkaline solution, the compounds show a strong blue emission. Its maxima, in alcoholic as well as in aqueous alkali, are also included in table 7.

Table 7. UV-Absorption and FS-Emission (Activation at 365 nm): Maxima of 2,4-Dioxo-3,3-diethyl-tetrahydropyridins (Persedon and Metabolites of Noludar, from ref. 5)

Substituent in position 5	UV-Maxima (nm) in				FS-Maxima (nm) in	
	ethyl ether	water pH 2	water pH 10	water pH 12	aqueous KOH	alcoholic KOH
H (Persedon)	303	306	306	366	474	466
CH_3	305	315	315	376	492	498
CH_2OH	305	307	307	367	494	490
CHO	315	335	366	350	not measured	
COOH	315	313	313	356	489	478

The intensive fluorescence can be used as a screening test: A drop of the extract is mixed with a drop of aqueous or alcoholic KOH under the light of a long wavelength UV-lamp. A blue fluorescence which disappears upon neutralization indicates the presence of 2,4-dioxo-tetrahydropyridines.

The 2,4-dioxo-piperidine and -tetrahydropyridine derivatives can be separated by GC (31, 32). Their retention times on various columns have been reported (15, 21). From a column with medium polarity like XF-1150, they elute in the order Noludar, Sedulon, methyl-persedon and Persedon (4, 5). For detection, we recommend a N-specific thermionic detector or a parallel use of flame ionization and electron capture. The tetrahydropyridines produce, in contrast to the piperidine derivatives, high electron capture signals (4, 5), permitting differentiation between the two types of compounds. Mass spectrometric detection is of course also a good choice, but not absolutely necessary, since this class of hypnotics is fairly easy to identify also by simpler means.

HPLC can be used as an alternative to GC-separation. It will reveal the tetrahydropyridines by UV-detection, in organic or acid and neutral aqueous solution at 310, and in alkaline solution at 360 nm. In alkaline solution, recording of the fluorescence emission in the region of 480 nm is also possible. However, it should be kept in mind, that the stability of the compounds in alkaline solution is limited. Since the piperidines do not show strong UV- or fluorescence spectra, they are easily overlooked. HPLC is therefore not the best method for their detection.

Pyrithyldione and the metabolites of methyprylone can be detected in TLC by spraying the developed sheets with an alcoholic solution of KOH. For revealing piperidine derivatives, they must first be oxidized, preferably by spraying with H_2O_2. The use of TLC sheets without fluorescent indicator is indicated, the blue fluorescent spots are then better visible.

3.3.5.5 Detection of glutethimide and its metabolites

Glutethimide (or 2-ethyl-2-phenyl-glutarimide or 2,6-dioxo-3-ethyl-3-phenyl-piperidine) metabolizes to different hydroxylated compounds. The most abundant ones are probably the 4-hydroxy-, the 3-hydroxyethyl- and the 3-(4-hydroxyphenyl)-derivatives, which are all in part converted to glucuronides. In addition, 2-phenylglutarimide and 2-ethyl-2-phenylglutaconimide could also be found, as well as other metabolites without or with one or two hydroxyls. In our extraction scheme (chapter 3.1.7), these compounds accumulate in the neutral and in the weak acid extracts.

Glutethimide and derivatives show a series of weak UV-maxima between 250 and 270 nm (for glutethimide 252, 258 and 264 nm), like other aromatic compounds. Due to their low intensity, they may escape detection in extracts.

TLC has been extensively used for identifying glutethimide and its metabolites. Mercurous nitrate, diphenylcarbazone reagent with mercuric chloride, or iodoplatinate solution can be used as spray reagents. Dragendorff reagent responds only weakly.

For quantifications, GC has been recommended (33, 34). From a moderately polar column, the 2,6-dioxo-piperidines elute after the 2,4-dioxo-piperidines (4, 5). For identifying the many metabolites, GC-MS is of course the method of choice. It has served to identify glutethimide and 6 of its metabolites from body fluids by selected ion monitoring (35).

3.3.6 Quinazolinones

3.3.6.1 General remarks

This class of hypnotics contains only two members. Both are derivatives of 2-methyl-3-phenyl-quinazolin-4(3H)-one. Their structures are given in figure 3. The only difference is the substituent in position 3. In methaqualone it is a 2-methyl-phenyl, in mecloqualone a 2-chlorophenyl group. Both compounds are strong hypnotics, but the first one has found much larger acceptance on the market, under a great number of names such as Dormigen, Fadormir, Mequelon, Mequin, Metadorm, Noctilene,

Normi-Nox, Optimil, Parest, Parminal, Paxidorm, Revonal, Riporest, Sleepinal, Somnium, Sopor, Toquilone, Torinal and Tualone. It has also been sold as a mixture with the antihistamine diphenhydramine under market names such as Mandrax or Toquilone compositum, as well as in other mixtures. Mecloqualone has been marketed under the name Nubarene.

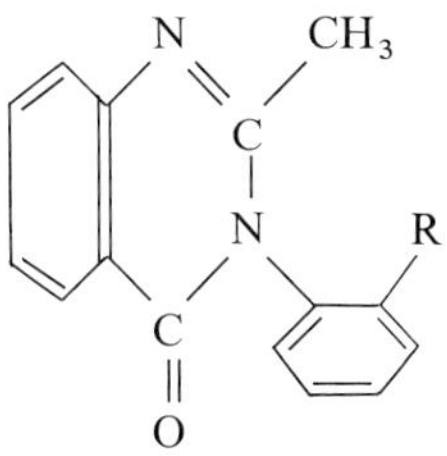

$R = CH_3$: Methaqualone (Normi-Nox)
2-methyl-3-o-tolyl-
quinazolin-4(3H)-one

$R = Cl$: Mecloqualone (Nubarene)
2-methyl-3-o-chlorophenyl-
quinazolin-4(3H)-one

Figure 3. Structure of the 2 Hypnotics with Quinazolinone-Structure

The two compounds are given as hypnotics in doses up to 300 mg daily. The estimated minimum lethal dose for a non-tolerant adult is 5 g. After oral administration, methaqualone is well absorbed and distributed in the body. It is metabolically hydroxylated, especially on the two methyl groups (to 2-hydroxy-methyl- and 3-(2-hydroxymethyl-phenyl)-derivatives), but also in 6-position of the quinazolinone ring and in the 3- and 4-positions of the o-tolyl ring (36). The hydroxy compounds are mostly excreted in conjugated forms.

The plasma half life of methaqualone is long; figures from one to three days have been reported. In long term use, drug accumulation with toxic effects can therefore easily occur. In particular, neurotoxic symptoms have been reported. Methaqualone is especially dangerous if taken together with ethanol. In this combination, it has been the cause of a great number of severe intoxications, many of them lethal. The drug is highly addictive and has become an illicit drug of abuse.

Methaqualone may not be prescribed any longer by physicians, but it would be wrong to assume that it has disappeared. It is still around, and a toxicological chemist may be confronted with the substance and its metabolites.

3.3.6.2 Analysis for 3-phenyl-quinazolin-4(3H)-ones

Since methaqualone has gained a wide negative reputation as a narcotic drug, immunoassays have been developed for its direct detection in biological fluids (blood, serum, urine). But chemical analysis is also simple and gives very reliable results. If one does not have to check a great number of samples, the incorporation of a preliminary immunotest into the analytical scheme does not seem to be necessary (37).

Both members of this group of hypnotics are only weakly basic compounds and may be found in our extraction scheme (see chapter 3.1.7) in the neutral and in the basic extracts, if we start with an acid extraction. If it seems possible that methaqualone may be the cause of an intoxication, it is therefore advisable to let the basic extraction precede the acid one. From the first extract, containing the bases as well as the neutral components, methaqualone and mecloqualone will then quantitatively back-extract from organic solution into acid aqueous phase.

Both methaqualone and mecloqualone have an intensive and practically identical UV-absorption. In acid solution, they show an intense UV-maximum at 234 nm and a less intense peak at 269 nm with shoulders around 284 and 296 nm. In alkaline solution, a shift to 227 nm (with intensification) and 264 nm (shoulder around 275 nm) occurs, and additional maxima appear at 306 and 316 nm. Almost the identical spectrum is obtained in organic solution. Since the absorption intensity is high, it is possible to pick out a presence of methaqualone (or mecloqualone) by UV-screening of extracts as described in chapter 3.1.8. The UV-spectra can also serve as basis for quantification.

For the identification of the single compounds, TLC, GC and HPLC can be used:

TLC R_f-values in different solvent systems are reported (15, 19, 20). To reveal the two hypnotics as well as their metabolites on thin layer sheets, detection under UV-light and general alkaloid reagents such as a Dragendorff spray or acidified iodoplatinate solution (38) can be used.

In GC, the two hypnotics and their metabolites except the N-oxide elute without decomposition from the conventional GC-columns. They can be detected by a thermionic or a simple flame ionization detector. The retention data are reported for both drugs and five metabolites of methaqualone (15, 21, 36, 39). Especially for identifying metabolites, GC-MS can be of great help. The EI-spectra of methaqualone and some of its metabolites can be found in the literature (22, 36). In addition to the positive EI-spectrum, the negative chemical ionization spectrum of methaqualon, recorded with various ionization gases at different source pressure levels, has also been described (6). For structure elucidation, we recommend a simultaneous recording of positive EI- and negative CI-spectra (7, 40, 41).

Since methaqualone is dosed fairly high, GC-FTIR instrumentation can also be used in place of GC-MS (42). On-line combination of GC with FTIR as well as MS should certainly be able to solve any open structural problems.

Thanks to the intensive UV absorption of the quinazolinones, HPLC with UV-detection is also a possible analytical tool, preferably with a scanning UV-detector. But since the spectra of the 2 basic compounds and their main metabolites are almost identical, the identification of the peaks rests on the availability of retention data.

Methaqualone and probably also mecloqualone have a strong tendency to bind to protein. The hydroxy-metabolites are mostly excreted in conjugated form. It is therefore important to submit biological samples, before extraction, to enzymatic treatments, either with a mixture of β-glucuronidase and aryl-sulfatase (for urine samples) or with the proteolytic enzyme subtilisin. In an extraction of methaqualone from ground liver, carried out by extractive dialysis (43) against ethyl ether, we could recover 12 mg of the drug without addition of subtilisin, 17 mg if dialysis was carried out in presence of subtilisin, and 20 mg after overnight incubation of the tissue with the proteolytic enzyme.

3.3.7 Hypnotics, Sedatives and Tranquilizers with Urethane Structure

3.3.7.1 General remarks

The best known members of this large class of pharmaceuticals are listed in table 8. Only three of them can be called hypnotics, the tert-amyl carbamate Aponal, Mepentamate (N-Oblivon or Oblivon C) and Ethinamate (Valamin). Among the tranquilizers, Meprobamate (Miltown) is best known. Today, the interest in this compound class has diminished and we shall treat it therefore rather briefly.

Table 8. Some Urethanes with Pharmacological Action

Structure	Name	Action
$CH_3-CH_2-O-CO-NH_2$	Ethyl carbamate, Ethyl urethane, Urethan	Antineoplastic
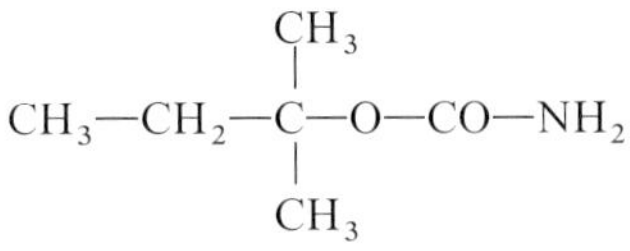 CH_3 $CH_3-CH_2-C-O-CO-NH_2$ CH_3	1,1-Dimethyl-propyl carbamate, tert-butyl carbamate, Aponal	Hypnotic
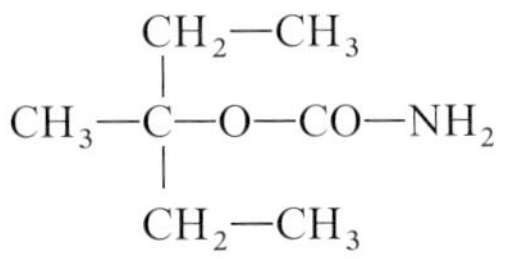 CH_2-CH_3 $CH_3-C-O-CO-NH_2$ CH_2-CH_3	1-Ethyl-1-methyl-propyl carbamate, Emylcamate	Tranquilizer, Muscle Relaxant
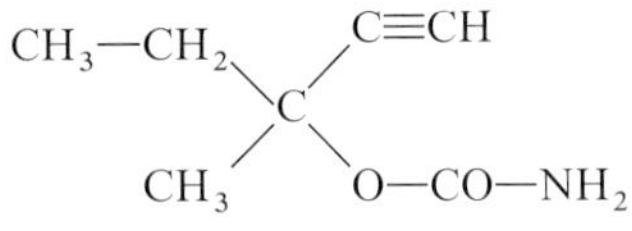 CH_3-CH_2 $C{\equiv}CH$ C CH_3 $O-CO-NH_2$	Methylpentynol carbamate, Mepentamate, N-Oblivon, Oblivon C	Sedative, Hypnotic
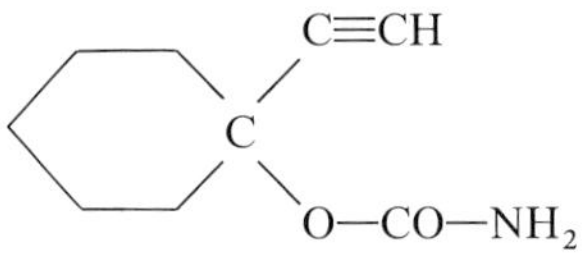 (cyclohexane ring) C with $C{\equiv}CH$ and $O-CO-NH_2$	Ethynyl-cyclo-hexyl carbamate, Ethinamate, Valamin, Valmid	Sedative, Hypnotic
$C_6H_5-(CH_2)_3-O-CO-NH_2$	3-Phenylpropyl carbamate, Phenprobamate, Proformiphen, Gamaquil	Tranquilizer, Anxiolytic, Muscle Relaxant
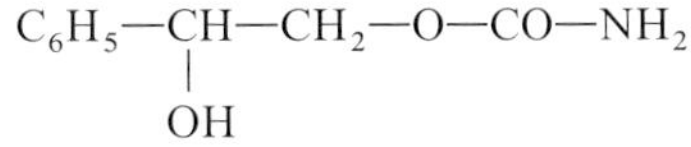 $C_6H_5-CH-CH_2-O-CO-NH_2$ OH	2-Hydroxy-2-phenyl-ethyl carbamate, Styramate, Sinaxar	Muscle Relaxant

Table 8. (continued)

Structure	Name	Use
$C_6H_5-C(OH)(CH_2-CH_3)-CH_2-O-CO-NH_2$	2-Hydroxy-2-phenyl-butyl carbamate, Hydroxyphenamate, Oxyphenamate, Listica	Tranquilizer, Anxiolytic
$C_6H_5-C(C{\equiv}CH)-O-CO-NH_2$	Phenylethynyl-carbinol carbamate, Carfinate, Equilium	Sedative, Hypnotic
$C_6H_4(O-CH_3)-O-CH_2-CH(OH)-O-CO-NH_2$	Guaiphenesin carbamate, Methocarbamol, Neuraxin	Muscle Relaxant
$(CH_3-(CH_2)_2)(CH_3)C(CH_2-O-CO-NH_2)_2$	Meprobamate, Meprotanum, Equanil, Miltown	Tranquilizer
$(CH_3-CH_2-CH(CH_3))(CH_3)C(CH_2-O-CO-NH_2)_2$	Mebutamate, Mebutina, Sigmafon	Tranquilizer
$(CH_3-(CH_2)_2)(CH_3)C(CH_2-O-CO-NH_2)(CH_2-O-CO-NH-CH_2-CH_2-CH_3)$	N-Propyl-meprobamate, Tybamate, Tybatran	Tranquilizer
$(CH_3-(CH_2)_2)(CH_3)C(CH_2-O-CO-NH_2)(CH_2-O-CO-NH-CH-(CH_3)_2)$	N-Isopropyl-meprobamate, Carisoprodol, Sanoma, Soma, Somadril, Rela	Muscle Relaxant

3.3.7.2 Hypnotics with carbamate structure

The name Aponal has not only been given to tert-amyl carbamate, but also to the antidepressant Doxepin. This may cause some confusion. The hypnotic Aponal is the carbamate derivative of the tert-amyl alcohol mentioned in chapter 3.3.2. Oblivon C or N-Oblivon is the carbamate of the hypnotic Oblivon, also mentioned in chapter 3.3.2 and listed in table 2. In both cases, the related compounds have not only similar structures, but also similar pharmacological properties and toxicity.

Not only the alcohols, but also the carbamates are quite volatile and can be analyzed by GC with a head space sampling approach. In our extraction scheme, the compounds accumulate in the neutral extract. TLC-methods for the analysis of this extract have been described, but today, GC or GC-MS is of course the method of choice.

The alcohol which would correspond to the carbamate Ethinamate (Valamin) has not appeared as a hypnotic. That may be the reason why Ethinamate has found the most interest among the carbamates with hypnotic activity. It is given in doses between 0.25 and 1 g. The estimated minimum lethal dose is 15 g. Ethinamate is rapidly absorbed and to a large part metabolized to hydroxy-derivatives. 4-Hydroxy-ethinamate is the main metabolite excreted in urine, but 2- and 3-hydroxy-ethinamate have also been observed, all 3 to a large part as glucuronic acid conjugates. A hydrolytic step should therefore precede the analytical investigation of urine. In our extraction scheme, Ethinamate and metabolites will pass into the neutral extract. TLC has been used for analysis, but GC or GC-MS is preferable.

3.3.7.3 Main tranquilizers with carbamate structure

Meprobamate respectively 2-methyl-2-propyl-1,3-propanediol dicarbamate was the first and most important member of this class of tranquilizers. It has gained world-wide acceptance before the benzodiazepines took hold. We probably could call Meprobamate the "Valium" of the late 50s and early 60s. Table 8 lists a few more tranquilizers with structural similarity to Meprobamate. They have been put on the market somewhat later and have gained less acceptance. It should be mentioned that Tybamate (a tranquilizer) and Carisoprodol (a muscle relaxant) metabolize in part to Meprobamate. The main metabolic pathway of the compounds is however hydroxylation. Meprobamate is excreted in part as 2-hydroxypropyl-derivative.

Daily doses for these compounds lie between 1 and 2.5 g. The chemical toxicologist should have no difficulty with identification. In TLC, they can be revealed by the Dragendorff spray, acidified iodoplatinate, Marquis reagent, ninhydrin spray or furfuryl-aldehyde reagent (38). GC can be used for detecting all 3 components, GC-MS is recommended.

3.3.8 Hypnotics with Different Structures

It would be wrong to assume that with the discussion of the classes of hypnotics in the chapters 3.1, 3.2 and 3.3, the subject would be exhausted. First of all, our tables do not always list all members of the different hypnotic classes. Secondly, a few substances which do not fit into the aforementioned classes have also made shorter or longer appearances as hypnotics. Some are listed in table 9.

In this place, we have to mention first the thiazole derivative 5-chloroethyl-4-methyl-thiazol or Clomethiazol (Hemineurin or Heminevrin). It is structurally related to the alcohol listed already in table 2. While the alcohol has never gained success as a hypnotic, the chlorinated compound has become a market runner in Great Britain. It is widely used as a sedative in treating alcoholism, but also as a sleep-

Table 9. Hypnotics with Different Structures

Structure	Name
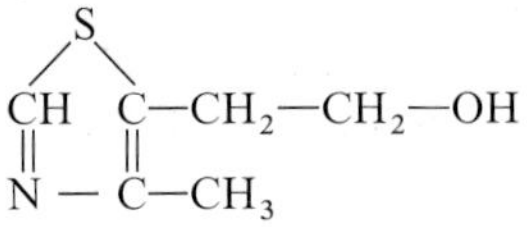 S CH C—CH_2—CH_2—OH ‖ ‖ N — C—CH_3	5-(2-Hydroxyethyl)-4-methyl-thiazole, 4-methyl-5-thiazole ethanol (Hypnotic, Sedative)
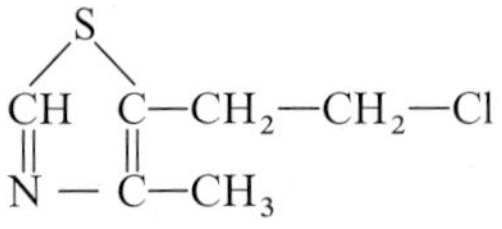 S CH C—CH_2—CH_2—Cl ‖ ‖ N — C—CH_3	5-(2-Chloroethyl)-4-methyl-thiazole, chloroethiazole, chlormethiazole, Clomethiazole, Distraneurin, Hemineurin, Heminevrin (Hypnotic, Sedative, Anticonvulsant)
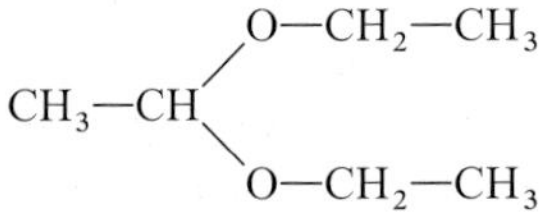 O—CH_2—CH_3 CH_3—CH O—CH_2—CH_3	1,1-Diethoxy-ethane, acetaldehyde-diethyl acetal, diethyl acetal, acetal (Hypnotic)
$CHBr_3$	Bromoform, tribromomethane (Hypnotic, Sedative, Antitussive)
HBr	Hydrobromic acid (Sedative)
C_6H_5—CO—CH_3	Phenyl-methyl-ketone, Acetophenone (Hypnotic)
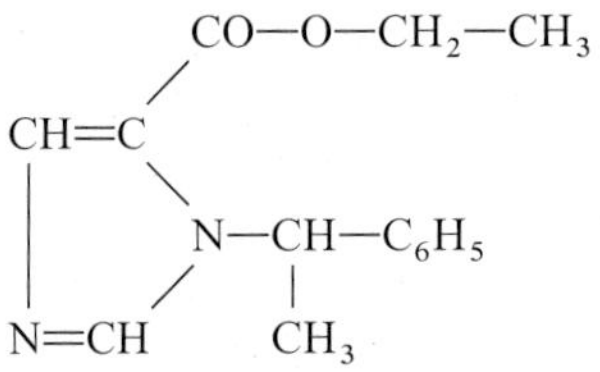 CO—O—CH_2—CH_3 CH=C N—CH—C_6H_5 N=CH CH_3	1-(1-Phenylethyl)-1H-imidazole-5-carboxylic acid ethyl ester, Etomidate, Amidate, Hypnomidate (Hypnotic)
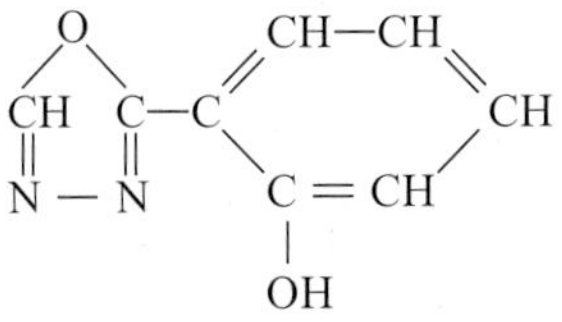 O CH—CH CH C—C CH ‖ ‖ N — N C = CH OH	2-(1,3,4-Oxadiazol-2-yl)phenol, 2-(o-hydroxyphenyl)-1,3,4-oxadiazole, Fenadiazole, Hypnazole (Hypnotic)
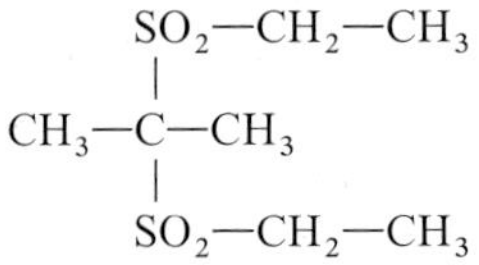 SO_2—CH_2—CH_3 CH_3—C—CH_3 SO_2—CH_2—CH_3	2,2-Bis(ethylsulfonyl)propane, Sulfonmethane, Sulfonal (Hypnotic)
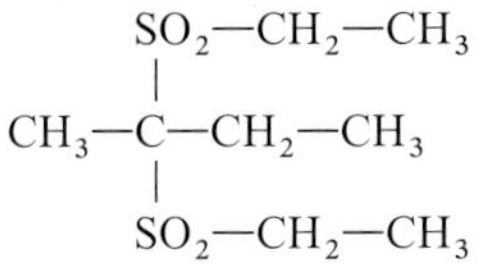 SO_2—CH_2—CH_3 CH_3—C—CH_2—CH_3 SO_2—CH_2—CH_3	2,2-Bis(ethylsulfonyl)butane, Methylsulfonal, Ethylsulfonal, Trional (Hypnotic)
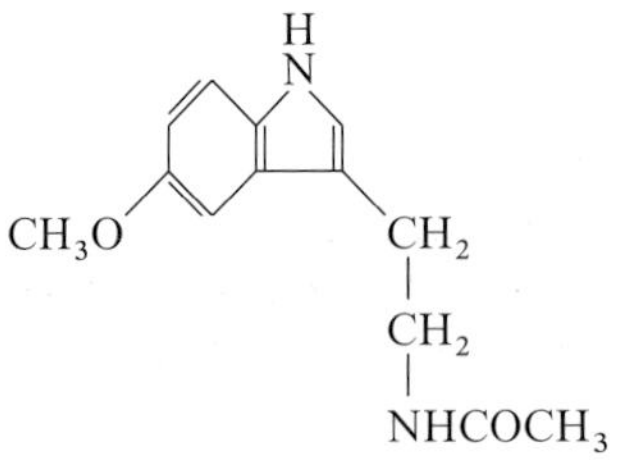 H N CH_3O CH_2 CH_2 $NHCOCH_3$	N-Acetyl-5-methoxy-tryptamine, Melatonin (synthetically available hormone of the pinary gland) (hypnotic, i.e. for preventing jet lag; also in use as contraceptive, antineoplastic, antiparkinsonian and for delaying aging)

inducing agent. It is sold in 200 mg capsules or in form of a syrup (50 mg per ml). Up to 12 capsules have been prescribed for alcohol withdrawal. Fatalities have occurred due to overdosages and the simultaneous action of ethanol.

Clomethiazole is volatile and can be extracted from the biological matrix by distillation with super-heated steam. The UV-spectrum of the distillate may discern the maximum of the molecule, in alkaline solution at 258 nm, in acid solution at 250 nm. Identification can be carried out by GC or GC-MS.

Other compounds used as hypnotics or sedatives are tribromomethane (bromoform), hydrobromic acid, phenyl-methyl-ketone (acetophenone), the imidazole derivative Etomidate, the oxadiazole derivative Fenodiazole (Hypnazole) and the sulfones Sulfonal (Sulfomethane) and Ethylsulfonal (Trional) (table 9).

A group of hypnotic agents not listed in our tables are the sleep-inducing composites in some plant extracts. The active ingredients of hops could be identified as Humulon and Lupulon. The active agents from valerian have been isolated and put on the market as Valmane, a mixture of the iroide drugs Valtratum, Dihydrovaltratum and Acevaltratum (44). In this connection, we should also mention menthyl valerate (Validol) and the esters of the plant constituent borneol with isovaleric or bromoisovaleric acid. All have been used as sedatives or hypnotics.

Already many years ago, it has been described that some endogenous human peptides are sleep-inducing. But the attempts to exploit this possibility commercially have so far not met with success. The chemical toxicologist should be grateful for that, since isolation of such peptides from biological fluids would be an extremely difficult task.

Since 1995, the pineal gland hormone Melatonin (N-acetyl-5-methoxytryptamine or 3-acetylaminoethyl-5-methoxyindole, last entry in table 9), is used increasingly as a sleeping agent, especially in the US (45–47). The human body produces the hormone only in the absence of light. Nighttime blood levels are sizable during childhood (up to 125 pg/ml at age 6 when secretion is highest), decline sharply during puberty and fade out in old age. Melatonin has been known for years as a hormone which keeps us synchronized with the rhythm of the day, and animals with seasonal cycles (46–48). Since the synthetic product is readily available without prescription (in the US it is sold inexpensively in health-food and drug stores), it is understandable that it has become popular for easing insomnia, especially in night workers and intercontinental travelers (as an antidote to jet lag). However, the main reasons why Melatonin has become a market runner may not be its quite well-documented usefulness for combating sleep disturbances in persons with an abnormal daily rhythm and in old people, but more recent claims, based in part only on test-tube or animal trials, that the hormone is capable of

- protecting cells from free-radical damage (47–50),
- boosting the immune system (47, 51),
- preventing cancer (47, 51, 52) and
- slowing the calendar of aging, thus extending life (53, 54).

The last claim may be the main driving force for millions of people to take Melatonin, in daily doses of 1 mg or much higher. The abuse potential is therefore considerable. No adverse reactions have been confirmed so far, but toxicologists must nevertheless be prepared for analytical testing.

Melatonin can be extracted from body fluids by fairly polar solvents such as ether, chloroform or methylene chloride. Like other indoles, it possesses 2 UV-maxima, one lies at 223 nm ($\varepsilon = 27500$), the weaker one at 278 nm ($\varepsilon = 6300$). The EI-mass spectrum shows a weak molecular ion at m/z 232, together with much more intensive fragment ions at m/z 173, 160 (base ion, by β-fragmentation of the side chain), 144, 117 and 43 (acetyl). Melatonin has been detected and measured by GC-MS, HPLC-MS and with an immunotest (55, 56). For trace analysis by chromatography with mass specific detection, we recommend recording mass 160 as tracer ion, if possible supplemented by the masses 173 and 144. The main urinary metabolite seems to be 6-hydroxy-melatonin (51).

As we have stated in the introduction, the borderline between narcotics and hypnotics is a fluid one. Narcotic drugs such as opiates, Pejote or Kat have often been used in the past as sleeping agents. In default of commercial hypnotics, this may still be the case today. Since most of the efficient hypnotics on the market have been submitted to medical control or even added to the list of narcotic drugs, some users are also trying to replace them by antihistamines, which often possess a hypnotic component (see chapter 3.9).

If some people try to substitute hypnotics by narcotics, the contrary is also true. Drug addicts without access to heroin or other opiates like to swallow or inject fast acting hypnotics. Earlier, they used the preparation Vesparax (containing allo- and brallobarbital together with the tranquilizer Hydroxyzine), chloral hydrate or Methaqualone. Today, they prefer the benzodiazepine Rohypnol or the annulated benzodiazepines Dormicum or Halcion (see chapter 3.2).

3.3.9 Concluding Remarks

Hypnotics and sedatives are responsible for a large percentage of intoxications. They are prescribed much too freely and much too often. Very frequently, dosages are too high. In many lethal intoxications that we had to investigate, it was not possible to determine with certainty, if death had occurred accidentally (i.e. by non-intentional overdosing) or through intentional suicide. It has been reported that in some countries up to 10% of the population are using hypnotics regularly. They have become addicted and will sooner or later be victims of chronic intoxications.

In this chapter, we have given a review of the hypnotics and sedatives which are structurally different from barbiturates and benzodiazepines. It might be argued that today, the knowledge of benzodiazepines is sufficient, since all the other hypnotic classes are outdated. We do not agree. The pharmaceutical industry has always been searching for a hypnotic which does

- not change the physiological rhythm of sleep,
- not lead to narcosis,
- not involve a hang-over,
- not lead to tolerance and
- not lead to addiction.

We do not believe that such a drug will ever exist. All hypnotics and strong sedatives synthesized so far have shown similar disadvantages. Some produce a hang-over if taken at the wrong time. Most lead to tolerance upon chronical use. For a certain category of people, all hypnotics will sooner or later become drugs of abuse, just like narcotics. And an abuse of the most potent hypnotics for suicides and for criminal intentions will always be found.

The new benzodiazepines have been praised as almost ideal hypnotics, as sleep-inducing and not sleep-enforcing, as non-addicting and not leading to tolerance. But after they became "everyday drugs" and the drugs of "everyone", we had to learn, that the disadvantages stated above are also associated with these new sedatives and hypnotics. This holds especially for their addicting properties. Disappointment will extend also to this new class of drugs, and the time will come when – in want of a new pharmaceutical (a new hope) – members of an older class of hypnotics will reappear on the market, either in a new form or a new combination. That is why a chemical toxicologist must be able to detect not only today's pharmaceuticals, but even the drugs of the past. In addition, he must also be prepared to deal with completely unexpected new active compounds which may hit the market. This calls for a versatile and flexible methodology and for the capability to use the analytical data, especially the spectra, for structural elucidations.

References

1. Auterhoff, H. (1978) Lehrbuch der Pharmazeutischen Chemie, 9th ed., p. 381. Wissensch. Verlagsgesellschaft, Stuttgart.
2. Yeoman, W. B. (1986) Metals and Anions. In: Clarke's Isolation and Identification of Drugs, 2nd ed. (Moffat, A. C., Jackson, J. V., Moos M. S. & Widdop, B., eds.), pp. 55–69. The Pharmaceutical Press, London.
3. Brandenberger, H. (1988) Wichtige exogene Vergiftungen und ihre differentialdiagnostische Erkennung. In: Differentialdiagnose Innerer Krankheiten (Siegenthaler, W., ed.), pp. 39.1–39.18. Georg Thieme, Stuttgart & New York.
4. Brandenberger, H. (1970) Die Gaschromatographie in der toxikologischen Analytik: Verbesserung von Selektivität und Empfindlichkeit durch Einsatz multipler Detektoren. Pharmaceutica Acta Helvetiae *45*, 394–413.
5. Brandenberger, H. (1974) Toxicology. In: Clinical Biochemistry, Principles and Methods (Curtius, H. Ch. & Roth, M., eds.), pp. 1425–1467. Walter de Gruyter, Berlin & New York.
6. Brandenberger, H. (1979) Present state and future trends of negative ion mass spectrometry in toxicology, forensic and environmental chemistry. In: Recent Developments in Mass Spectrometry in Biochemistry and Medicine, Vol. 2 (Frigerio, A., ed.), pp. 227–255. Plenum Publishing Corp., New York.
7. Brandenberger, H. (1987) Low pressure negative chemical ionization mass spectrometry and dual mass spectrometry: A review of the work since 1977. In: Reports on Forensic Toxicology, Proceedings of the 22nd International Meeting of TIAFT, 1985 in Rigi Kaltbad, Switzerland (Brandenberger, H. & R., eds.), pp. 437–453. Publ. by Branson Research, CH-6356 Rigi Kaltbad.
8. Widdop, B. (1986) Hospital Toxicology and Drug Abuse Screening. In: Clarke's Isolation and Identification of Drugs, 2nd ed. (Moffat, A. C., Jackson, J. V., Moos, M. S. & Widdop, B., eds.), pp. 1–34. The Pharmaceutical Press, London, England.
9. Stevens, H. M. (1986) Colour Tests. In Clarke's Isolation and Identification of Drugs, 2nd ed. (Moffat, A. C., Jackson, J. V., Moos, M. S. & Widdop, B., eds.), pp. 128–147. The Pharmaceutical Press, London.
10. McBay, A. J., Boling, V. R. & Reynolds, P. C. (1980) Spectrometric determination of trichloroethanol in chloral hydrate poisoning. J. analyt. Toxicol. *4*, 99–101.
11. Bridges, R. R. & Jennison, T. A. (1984) Analysis of ethchlorvynol (Placidyl). Evaluation

of a comparison performed in a clinical laboratory. J. analyt. Toxicol. *8*, 263–268.

12. Brandenberger, H. (1972) The search for drugs of abuse by gas chromatography with mass specific detection systems. In: Proceedings of the International Symposium on Gas Chromatography-Mass Spectrometry, Isle of Elba, Italy (Frigerio, A., ed.), pp. 37–66. Tamburini Editore, Milano.
13. Gibitz, H. & Homma, H. (1959) Über einen Fall von tödlicher Bromuralvergiftung. Arch. f. Toxikol. *17*, 295–298.
14. The Merck Index, 12th ed. (1996), Merck & Co., Inc., Rahway, N.J.
15. Clarke's Isolation and Identification of Drugs, 2nd ed. (1986) (Moffat, A.C., Jackson, J.V., Moss, M.S. & Widdop, B., eds.). The Pharmaceutical Press, London.
16. Feigl, F. (1954) Spot Tests, Vol. II, Organic Applications, 4th ed. Elsevier, Amsterdam, London & New York.
17. Bäumler, J. & Rippstein, S. (1961) Dünnschichtchromatographischer Nachweis von Insektiziden. Helv. Chim. Acta *44*, 1162–1164.
18. Bäumler, J. & Rippstein, S. (1963) Über den Nachweis von Carbromal (Adalin) und dessen Metaboliten. Archiv der Pharmazie *296*, 301–306.
19. Machata, G. & Schütz, H. (1989) Empfehlungen zur Dünnschichtchromatography (Folge 5 der Empfehlungen zur klinisch-toxikologischen Analytik der DFG. Report XVI of the DFG Commission for Clinical-Toxicological Analysis), VCH Verlagsgesellschaft, Weinheim.
20. Zeeuw, de, R.A., Franke, J.P., Degel, F., Machbert, G., Schütz, H. & Wijsbeek, J. (1992) Thin-Layer Chromatographic R_f Values of Toxicologically Relevant Substances on Standardized Systems, 2nd ed. (Report XVII of the DFG Commission for Clinical-Toxicological Analysis, Special Issue of the TIAFT Bulletin). VCH Verlagsgesellschaft, Weinheim.
21. Zeeuw, de, R.A., Franke, J.F., Maurer, H.H. & Pfleger K. (1992) Gas Chromatographic Retention Indices of Toxicologically Relevant Substances on Packed or Capillary Columns with Dimethylsilicone Stationary Phases, 3rd ed. (Report XVIII of the DFG Commission for Clinical-Toxicological Analysis. Special Issue of the TIAFT Bulletin). VCH Verlagsgesellschaft, Weinheim.
22. Pfleger, K., Maurer, H. & Weber, A. (1992) Mass Spectral and GC Data of Drugs, Poisons, Pesticides, Pollutants and their Metabolites, Part I, II & III. 2nd ed. VCH Verlagsgesellschaft, Weinheim.
23. Frangi-Schnyder, D. & Brandenberger, H. (1978) Anion-CI-Mass Spectrometry of the Main Classes of Hypnotics. Fresenius Z. Anal. Chem. *290*, 153–154.
24. Brandenberger, H. & Ryhage, R. (1978) Possibilities of negative ion mass spectrometry in forensic and toxicological analysis. In: Blood Drugs and other Analytical Challenges, Vol. 7 of Methodological Surveys in Biochemistry (Reid, E., ed.), pp. 173–184. Ellis Horwood Ltd., Chichester, a division of John Wiley.
25. Brandenberger, H. (1978) Negative ion mass spectrometry; a new tool for the forensic toxicologist. In: Instrumental Applications in Forensic Drug Chemistry, Proceedings of the International Symposium of the US Drug Enforcement Administration in Arlington, Virginia (Klein, M., Kruegel, A.V. & Sobol, S.P., eds.), pp. 48–59. US Government Printing Office, Washington, D.C.
26. Brandenberger, H. (1982) Negative ion MS by low pressure chemical ionization and its combination with conventional positive ion MS. In: Abstracts of the 30th Annual Conference on Mass Spectrometry and Allied Topics in Honolulu, pp. 468–469. American Society for Mass Spectrometry.
27. Yamada, M. & Brandenberger, H. (1986) Analysis of primary bioactive amines by positive and negative ion mass spectrometry of their isothiocyanates. In: Proceedings of the 23rd Meeting of the International Association of Forensic Toxicologists and the 2nd World Congress on New Compounds in Biological and Chemical Warfare (B. Heyndrickx, ed.), pp. 165–171. Publ. by the Dep. of Toxicology of the State University of Ghent.
28. Curry, A. (1969) Poison Detection in Human Organs, 2nd ed., pp. 113–114, Charles C Thomas, Springfield, Ill. and Fort Lauderdale, Fl.
29. Arnold, W. (1966) Unbemerkte Beimischung von Schlafmitteln in alkoholischen Getränken, Kriminalistik *20*, 363–364.
30. Pribilla, O. (1959) Studien zur Toxikologie der Schlafmittel aus der Tetrahydropyridin- und piperi-dion-Reihe. Arch. f. Toxicol. *18*, 1–86.
31. Bridges, R.R. & Peat, M.A. (1979) Gas-liquid chromatographic analysis of methyprylon and its major metabolite (2,4-dioxo-3,3-diethyl-5-methyl-1,2,3,4-tetrahydropyridine) in an overdose case. J. analyt. Toxicol. *3*, 21–25.

32. Van Boven, M. & Sunshine, J. (1979) Capillary gas chromatography for drug analysis. J. analyt. Toxicol. *3*, 174–176.
33. Kadar, D. & Kalow, W. (1972) A method for measuring glutethimide (Doriden) in human serum after intake of therapeutic doses. J. of Chromat. *72*, 21–27.
34. Hansen, A. R. & Fischer, L. J. (1974) Gaschromatographic simultaneous analysis for Glutethimide and an active hydroxylated metabolite in tissues, plasma and urine. Clin. Chem. *20*, 236–242.
35. Kennedy, K. A. et al. (1978) A selected ion monitoring method for Glutethimide and six metabolites. Application to blood and urine from humans intoxicated with Glutethimide. Biomedical Mass Spectrometry *5*, 679–685.
36. Kazyak, L., et al. (1977) Methaqualone metabolites in human urine after therapeutic doses. Clin. Chem. *23*, 2001–2006.
37. Brandenberger, H. & Maes R. A. A. (1988) Empfehlung zur klinisch-toxikologischen Analytik. Folge 1: Einsatz von immunochemischen Testen in der Suchtmittelanalytik. Mitteilung X der Senatskommission für Klinisch-toxikologische Analytik der Deutschen Forschungsgemeinschaft (DFG). VCH Verlagsgesellschaft, Weinheim.
38. Anfärbereagenzien für Dünnschicht- und Papier-Chromatographie (1980), published by E. Merck, Darmstadt.
39. Peat, M. A. & Finkle, B. S. (1980) Determination of Methaqualone and its major metabolites in plasma and saliva after single oral doses. J. analyt. Toxicol. *4*, 114–118.
40. Brandenberger, H. (1983) International Journal for Mass Spectrometry and Ion Physics *47*, 213–216, and in: Mass Spectrometry Advances 1982 (Schmid, E. R., Varmuza, K. & Fogy, I., eds.), pp. 213–216. Elsevier, Amsterdam, Oxford & New York.
41. Brandenberger, H. (1984) Negativionen-Massenspektrometrie. Fesenius Z. Anal. Chem. *317*, 634–635.
42. Brandenberger, H. & Wittenwiler, P. (1986) The combination of on-line gas chromatography with Fourier transform infrared spectrophotometry for chemical toxicology. In: Proceedings of the 23rd Meeting of the International Association of Forensic Toxicologists and the 2nd World Congress on New Compounds in Biological and Chemical Warfare in Ghent (Heyndrickx, B., ed.), pp. 231–239. Publ. by the Dep. of Toxicology of the State University of Ghent.
43. Brandenberger, H. & Bucher, M. (1980) Extractive Dialysis. In: Forensic Toxicology, Proceedings of the European Meeting of TIAFT 1979 in Glasgow (Oliver, J., ed.), pp. 104–108. Croom Helm, London.
44. Sticher, O. (1977) The analysis of iridoid drugs. Pharm. Acta Helv. *52*, 20–32.
45. Sahelian, R. (1996) Melatonin: Nature's sleeping pill. Happier Press, USA.
46. Reiter, R. J. & Robinson, J. (1996) Melatonin: Your body's natural wonder drug. Bantan Books, New York.
47. Yu, Hing-Sing & Reiter, R. J. (eds.) (1993) Melatonin, Biosynthesis, Physiological Effects and Clinical Applications. CRC Press, Boca Raton.
48. Wetterberg, L. (1995) Seasonal affective disorder, Melatonin and light, in: The Pineal Gland and its Hormones, Fundamentals and Clinical Perspectives (Fraschini, F., Reiter, R. J. & Stankov, eds.), pp. 199–208. Plenum Press, New York & London.
49. Reiter, R. J., Tan, D. X., Poeggeler, B., Menendez-Pelaez, A., Chen, L. D. & Saarela, S. (1994) Melatonin as a free radical scavenger: Implications for aging and age related diseases. Ann. N. Y. Acad. Sci. *719*, 1–12.
50. Reiter, R. J. (1996) Cellular and Molecular Actions of Melatonin as an Antioxidant. 2nd Locarno Meeting on Neuroendocrinoimmunology: Therapeutic Potential of Melatonin, Abstract 18.
51. Conti, A., Haran-Ghera, N. & Maestroni, G. J. M. (1992) Role of pineal Melatonin and melatonin-induced-immunoopioids in murine leukemogöenesis. Med. Oncol. & Tumor Pharm. *9*, 87–92.
52. Conti, A. & Maestroni, G. J. M. (1995) The clinical neuroimmunotherapeutic role of Melatonin in oncology. J. Pineal Res. *19*, 103–110.
53. LeVert, S. (1995) Melatonin: The anti-aging hormone. Avon Books, New York.
54. Pierpaoli, W. & Regelson, W. (1995) The Melatonin Miracle: Nature's Age-Reversing, Disease-Fighting, Sex-Enhancing Hormone. Simon & Schuster, New York.
55. Skene, D. J., Leone, R. M., Young, R. E. & Silman, R. E. (1983) The assessment of a plasma melatonin assay using gas chromatography negative chemical ionization mass spectrometry. Biomed. Mass Spectrom. *10*, 655–659.
56. Dubbels, R., Reiter, R. J., Klenke, E., Goebel, A., et al. (1955) Melatonin in edible plants identified by radioimmunoassay and by high performance liquid chromatography-mass spectrometry, J. Pineal Res. *18*, 28–31.

3.4 Anticonvulsants Not Belonging to the Classes of Barbiturates and Benzodiazepines

W. R. Külpmann

3.4.1. General Introduction

During many years the search for new antiepileptic drugs was guided by the conception, that as a prerequisite of the antiepileptic activity of a substance a certain structure within the molecule is necessary:

```
—C—N—C—C—
 ‖  ‖
 O  O
```

This conception proved to be helpful for developing new, more effective and less toxic antiepileptic drugs. On the other hand it is well understandable that the antiepileptic properties of valproic acid could only be detected by chance as long as one followed strictly the conception of a characteristic "antiepileptic" molecular strucutre. (Valproic acid does not contain any carbonyl function nor a nitrogen atom). The antiepileptic activity of valproic acid was detected, when it was used as a solvent for new substances under investigation for seizure control. As a consequence, the old conception of an "antiepileptic" molecule structure is now mostly overruled by the idea that all substances, which cause a change of synthesis/release of neutrotransmitters or an alteration of other cellular metabolic processes may exert antiepileptic activities. This new guideline helped to detect new anticonvulsants drugs, which are included in figure 1.

3.4.2 Carbamazepine

3.4.2.1 Introduction

Carbamazepine is an iminostilbene derivative (5-carbamoyl-5H-dibenz[b,f]azepine) with a relative molecular mass of 236.3. It is nearly insoluble in water, but soluble in ethanol, acetone, and propylene glycol. Carbamazepine was introduced into clinical use in the early 1960s in Europe and in 1974 in the US and is now applied for the treatment of generalized tonic-clonic seizures, partial seizures with complex symptomatology and mixed seizure patterns.

I. IMINOSTILBENES

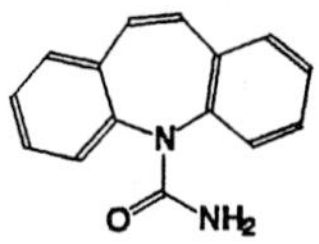

CARBAMAZEPINE

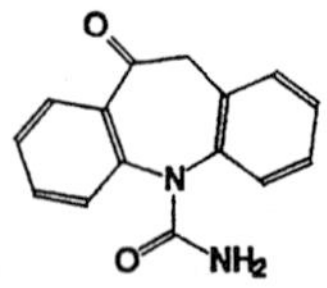

OXCARBAZEPINE

II. SUCCINIMIDES

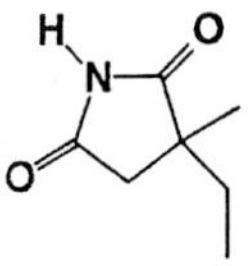

ETHOSUXIMIDE

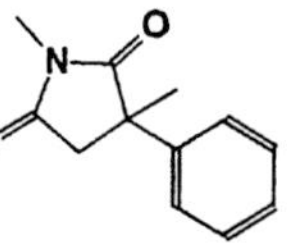

METHSUXIMIDE

III. HYDANTOINS

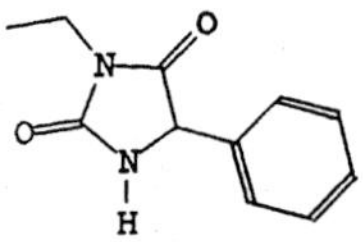

ETHOTOIN

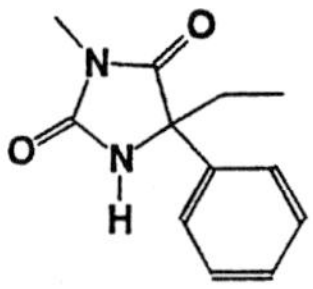

MEPHENYTOIN

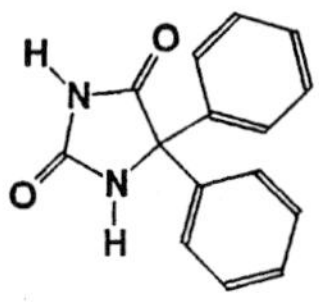

PHENYTOIN

IV. OXAZOLIDINEDIONE

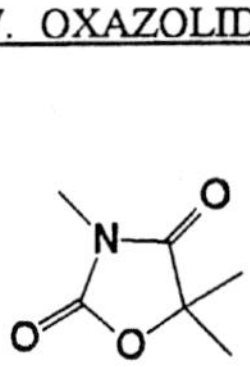

TRIMETHADION

Figure 1. Structural Formula of Antiepileptic Drugs.

3.4.2.2 Methods

3.4.2.2.1 Quantiative methods

Carbamazepine in serum is usually determined either by an immunoassay or by a chromatographic technique. According to the data from an external quality assessment scheme (1, 2) the use of immunoassays increased by 25% within about three years, whereas the frequency of chromatographic determinations remained nearly constant. Among the immunoassays-enzyme multiplied immunoassay technique and fluorescence polarization immunoassay (FPIA) were preferred. The use of the latter increased by about 60% within three years. High performance liquid chromatography (HPLC) becomes more and more popular as compared to gas-liquid chromatography. Carbamazepine-10,11-epoxide is predominantly determined by HPLC.

Immunoassays
Both, enzyme multiplied immunoassay technique (EMIT) (3) and fluorescence polarization immunoassay technique (FPIA) (4) are tests which are commercially available. EMIT tests (principle see 3.4.2.2) may be run on dedicated analyzers, but can be adapted to many analyzers of routine clinical chemistry, which may lead to a reduction of sample volume, of reagents and of costs (5). Tittle et al. (6) describe the use of EMIT reagents for carbamazepine determination using immunoassay plates and a plate-reading spectrophotometric instrument. The applicability of FPIA, on the other hand, is more strictly confined to a dedicated, closed system.

Nephelometric inhibition immunoassay (NIIA) is run on the Beckman immunochemistry system, which was primarily developed for the determination of individual proteins. Carbamazepine bound to a carrier protein and carbamazepine specific antibodies are mixed in amounts to yield maximal precipitation. Depending on the concentration of carbamazepine in the sample, precipitation and hence intensity of scatter light is reduced.

In substrate labeled fluorescence immunoassay (SLFIA), a substrate, which is labeled by a fluorogenic compound, is attached to the drug (7). The labeled carbamazepine and the (unlabeled) carbamazepine of the sample compete for the antibody binding sites. On binding to the antibody, labeled substrate cannot be attacked by a substrate specific enzyme to yield a fluorescent product. Intensity of fluorescence is proportional to carbamazepine concentration.

Recently a mechanized homogeneous liposome immunoassay was presented (8). Carbamazepine-phospholipid conjugate was entrapped in artificial liposomes, as well as glucose-6-phosphate dehydrogenase. In the first step, sample, antibody solution and glucose-6-phosphate are mixed. Then the liposomes and guinea pig complement are added. The immunolysis of liposomes caused by activation of complement is proportional to unbound antibody concentration, that is indirectly proportional to carbamazepine concentration of the sample. Immunolysis is determined from NAD^+-reduction caused by oxidation of glucose-6-phosphate, which is catalyzed by liberated glucose-6-phosphate dehydrogenase.

For emergency use and small batches in outpatient clinics, the dry-phase apoenzyme reactivation immunoassay system (ARIS) was developed (9, 10), which is measured by a dedicated reflectance photometer (Seralyzer). Carbamazepine of the sample and carbamazepine labeled with flavin adenine dinucleotide (FAD) compete for a limited number of binding sites on the carbamazepine-specific antibody. Labeled carbamazepine bound to the antibody is not able to reactivate the apoglucose oxidase. The higher the concentration of carbamazepine in the sample, the less the FAD binding, and the better the reactivation of the enzyme. Enzyme activity is determined by measuring oxidation of glucose, which is catalyzed by (activated) glucose oxidase and peroxidase. As indicator, 3,3′, 5,5′-tetramethylbenzidine is used.

For the same purpose a non-instrumented immunochromatographic assay (Aculevel[R]) is available, which needs 12 μl of whole blood (11, 12). The sample is drawn into a calibrated pipette and mixed with a reagent containing glucose oxidase and horseradish peroxidase conjugated carbamazepine. A plastic cassette with chromatographic paper coated with carbamazepine specific monoclonal antibodies is put into the mixture, the liquid migrates up and labeled and unlabeled carbamazepine is bound to the antibodies. After about 15 min the cassette is removed from the mixture and placed into a developing solution, which shows how far the labeled (and unlabeled) carbamazepine migrated. The distance is proportional to the carbamazepine concentration of the sample; the concentration is read from a conversion table.

Chromatographic methods

Gas-liquid chromatography (GC) and high-performance liquid chromatography (HPLC) are used for the quantitative determination of the antiepileptic drugs. These techniques are versatile and are used now primarily for newly introduced drugs, in case of small workload and for evaluation of e.g. immunoassays. Methods are to be preferred, which allow the simultaneous measurement of at least the mainly used antiepileptic drugs in the same extract. Comedication is not rare and it saves time, if the instruments can be run for all drugs with the same settings, same column and same detector, and the prepurification of the sample is equally well suited for all drugs. As the antiepileptic drugs differ considerably in their physico-chemical properties, the procedures may be not optimal for each individual drug. For the determination of one drug, a dedicated method may be more reliable than the "group" assay. HPLC is preferred to GC for the measurement of the (water soluble) metabolites of the drugs.

Gas-Liquid chromatography

This technique was mainly used when therapeutic drug monitoring started (13), as carbamazepine could now be determined specifically, separating the metabolites from the drug. At first, carbamazepine is extracted at neutral pH by organic solvents. The dried extract is redissolved in about 100 μl and an aliquot is injected into the gas chromatograph. Various internal standards were recommended depending on the drugs that are concomitantly determined. As carbamazepine is susceptible to decomposition at high temperature, it was tried to form more stable derivatives (13, 14). On the other hand, stationary phases (SP 2510 DA, Dexil-300) became available that allowed to chromatograph the underivatised compound (15, 16). Methods using gas chromatography/mass spectrometry (GC/MS) were introduced, which are still more specific than conventional GC using flame ionization or nitrogen selective detectors. GC/MS may be especially useful for the determination of carbamazepine metabolites (17).

High performance liquid chromatography (HPLC)
The samples are prepared for HPLC analysis

1) by extraction of the drugs with organic solution (e.g. chlorinated solvent, diethylether, ethylacetate),
2) by absorption of the drugs on to a solid material (e.g. reversed phase silica gel, divinylbenzene Kieselguhr), from which a clean eluate is obtained,
3) by protein precipitation using a solvent miscible with water (e.g. acetonitrile) if a reversed phase column is used for HPLC.

Procedure 3) may be the fastest, however, all water soluble impurities are injected onto the HPLC column; procedure 2) can be performed on-line.

For the chromatographic analysis, reversed phase silica gel is preferred, as the separation is independent from the water content of the sample and a gradient elution is possible. As mobile phase a mixture of acetonitrile and perchloric acid may be used. Signals are recorded by a UV detector.

A procedure for the simultaneous determination of carbamazepine, ethosuximide, phenobarbital, phenytoin and primidone (and some of their metabolites) was published by Juergens (18), who cites more than a hundred papers on HPLC determinations of antiepileptic drugs. Coefficient of variation between days ranges from 1.6% to 2.4% for carbamazepine. Methods for the exclusive determination of carbamazepine and its metabolites in plasma and saliva were published by other authors (19, 20, 21).

3.4.2.2.2 Qualitative methods

Carbamazepine can be detected by thin-layer chromatography (TLC). After extraction by organic solvent from aqueous alkaline solution, the extract is applied to TLC and developed by use of methanol/ammonia (100/1.5 v/v). The drug is detected by potassium permanganate (22).

3.4.2.3 Quality assessment

Precision: According to an external quality assessment scheme (2), interlaboratory coefficient of variation was between 8% and 10% for several immunoassay techniques (FPIA, NIIA, EMIT) and HPLC, but about 12% for gas chromatographic techniques including derivatisation (GLC-D).

Accuracy: Smallest deviations (below 4%) from the spike value were obtained with EMIT, FIA and HPLC, highest bias was observed in case of GLC-D and FPIA (2). The inaccuracy of FPIA was attributable to the use of calf serum, whereas in case of human sera, accuracy of FPIA was satisfactory. It must be added that the antibodies of immunoassays exhibit cross-reactivity with carbamazepine-10, 11-epoxide, which is, however, usually negligible (see 3.4.2.5).

3.4.2.4 Pharmacokinetics

Absorption
As an intravenous preparation of carbamazepine is not available, bioavailability can only be roughly estimated to be 80% for tablets, but decreasing at high doses. Peak serum levels are reached after 6h to 18h after a single dose, but 1h to 5h in case of chronic treatment due to enhanced elimination.

Distribution
75% of total carbamazepine in serum is bound to albumin and α_1-acid glycoprotein. Due to high tissue binding, volume of distribution is rather high. Carbamazepine crosses the placenta and is excreted into breast milk. The concentration of the drug in tears and saliva closely approximates its unbound fraction in serum.

Elimination
Only a small fraction of carbamazepine is excreted unchanged by the kidneys, more than 98% is metabolized in the liver to form carbamazepine-10,11-epoxide, and further transdiol and hydroxymethyl acridane derivatives. The percentage of these derivatives is increased in chronic treatment due to auto-induction and furthermore in case of concomitant administration of enzyme inducing drugs e.g., phenobarbital, phenytoin and primidone. The reduction of the epoxide is inhibited by valproate and the fraction of this presumably active metabolite is elevated. Body clearance is enhanced and half life decreased in chronic treatment, which complicates the dosage during initiation of therapy to achieve the therapeutic interval. Although half life of carbamazepine-10, 11-epoxide is still shorter, in practice its concentration parallels total carbamazepine concentration limited by its rate of formation.

Influence on pharmacokinetics
In children and during pregnancy, half life of carbamazepine is shortened and higher dosages may be needed.

Carbamazepine is predominantly cleared by the liver, although it is not highly extracted. Therefore in case of hepatic diseases with reduced enzyme activity, carbamazepine concentration is elevated. Its unbound (active) fraction is increased, if protein concentration is decreased, which happens in uraemia and in diseases of the liver. For these states, dosage adjustment guided by drug monitoring is necessary.

Enzyme inducing drugs, such as phenobarbital, phenytoin and primidon merease the formation of the epoxide and lower carbamazepine concentration; valproate inhibits the epoxide metabolism. Enzyme inhibitors (e.g. cimetidine, erythromycin, propoxyphene) may require a reduction of the dose. Carbamazepine is enzyme inducing itself and increases the metabolism of e.g. clonazepam, ethosuximide, valproic acid and warfarin.

3.4.2.5 Medical interpretation

Therapeutic range of carbamazepine and the potentally toxic or lethal concentrations are given in table 1 (23, 24, 25). Ingestion of 4 to 60 g were reported to be fatal (26, 27). Serum concentration and clinical status seem to correlate quite well, although it is said that the evolution of the intoxication correlates more closely with the course of the epoxide level than with the concentration of carbamazepine itself.

Table 1. Pharmacokinetic Summary Table: Carbamazepine (23, 24, 25)

Carbamazepine	
Therapeutic range:	17 µmol/l–42 µmol/l (4 mg/l–10 mg/l)
Potentially toxic concentration:	> 42 µmol/l (> 10 mg/l)
Potentially lethal concentration:	> 85 µmol/l (> 20 mg/l)
Plasma half life:	36 h (single dose) 15 h (chronic treatment) (Carbamazepine-10,11-epoxide: 6 h)
Volume of distribution:	1.4 l/kg
Plasma protein binding:	75 % (Carbamazepine-10,11-epoxide: 50 %)
Elimination:	< 1 % excreted unchanged renally > 99 % metabolised to carbamazepine-10,11-epoxide by liver and further on to 10,11-dihydroxy-carbamazepine, which is esterified with glucuronic acid
Active metabolites:	Carbamazepine-10,11-epoxide
Sampling:	Before next dose (trough value)

For the treatment of overdoses, haemodialysis is not very promising, as most of the drug is tightly bound to proteins and volume of distribution is rather high. Charcoal, given orally, however, inhibits the absorption of carbamazepine effectively. In threatening intoxication, haemoperfusion may be used (28), although it is associated with several complications (e.g. thrombocytopenia). The anticonvulsant drug has antiarrhythmic, antidiuretic, sedative, anticholinergic, antidepressant, and muscle relaxant properties. In overdose, usually respiratory depression and coma are observed, but sometimes excitation and aggression. Determination of carbamazepine in saliva or tears is not widely used (29, 30), but reflects roughly the unbound fraction.

3.4.3 Ethosuximide

3.4.3.1 Introduction

Ethosuximide is a derivative of succinimide: 2-ethyl-2-methylsuccinimide with a relative molecular mass of 141.2. It is scarcely soluble in water (1 g in 4.5 l water), but well soluble in ethanol, diethylether and chloroform (1 g in less than 1 l). The drug was introduced into clinical use in 1958 and is now used especially for the treatment of absence epilepsy.

3.4.3.2 Methods

3.4.3.2.1 Quantitative methods

Determinations of ethosuximide in serum are performed either by immunoassays or by chromatographic techniques. During the last years, immunoassay determinations became most popular, especially due to an increased use of fluorescence polarization immunoassay, whereas the use of gas-liquid chromatography decreased. The mean frequency of methods used for measurement of ethosuximide in serum was discussed according to an external quality assessment scheme (1, 2).

Determination of ethosuximide in saliva or tears is not performed routinely.

Immunoassays

Most immunoassays are commercially available tests. The enzyme multiplied immunoassay technique (EMIT) (3) for the determination of ethosuximide in serum is based on the following principle: The supplied reagent contains ethosuximide, which is labeled with an enzyme (glucose-6-phosphate dehydrogenase). When the enzyme-labeled drug is bound to an antibody against ethosuximide, the activity of the enzyme is reduced. Ethosuximide in the sample competes with the enzyme-labeled ethosuximide for the antibody, and enzyme activity correlates with the concentration of the drug in the sample. A separation step is not necessary. This allows the application of the test to various fully mechanized clinical chemistry analyzers.

Ethosuximide in serum can be determined by fluorescence polarization immunoassay (4) as well: The drug in the sample competes with the fluoresceine labeled ethosuximide for the antibody. At low concentration of the drug in the sample, many of the labeled antigens are bound and polarization is high, and vice versa at higher drug concentrations. Ethosuximide concentration of the sample correlates indirectly to the degree of polarization. A separation step is not necessary. The tests are run on dedicated analyzers.

Chromatographic techniques

Gas-liquid chromatography

Drug monitoring of ethosuximide was performed at first mainly by gas-liquid chromatography (31–36). The extraction of the sample is performed by organic solvents at low pH. The organic phase is either injected directly or after evaporation and redissolution. As favourite internal standard, 2,2,3-trimethylsuccinimide is used. Some authors recommend derivatisation of the analyte with butyl iodide (31), tetrabutyl ammonium hydroxide (32), or pentafluorobenzoyl chloride (35). For chromatographic separation OV-101, OV-17 and OV-225 as well as SP 1000 are used as stationary phases. Most measurements are performed by flame ionisation detector, but nitrogen-phosphorus (32) or electron-capture detectors (35) are used, too. Precise determinations (coefficient of variation below 3%) are obtained, when the extract is injected directly into the gas chromatograph, whereas methods encompassing an evaporation step yield less precise results, probably due to the volatility of the drug.

For the determination of ethosuximide in saliva (and serum), gas chromatography-mass spectrometry was proposed (37).

High performance liquid chromatography (HPLC)
Ethosuximide is extracted by organic solvents or by use of Bond-Elut columns (from Analytichem, Harbor City, USA). The extract is directly applied to chromatograph and analyzed by use of C-18 reversed phase with a mixture of acetonitrile-methanol-phosphate buffer as eluent. Alternatively, samples are deproteinized with acetonitrile and the supernatant is analyzed. Signals are recorded by a UV detector at the wavelength of 195 nm (38–40). Coefficients of variation for imprecision between days range from 1.0% to 5.6%.

3.4.3.2.2 Qualitative methods

Ethosuximide can be detected in urine, gastric juice or plasma by thin-layer chromatography (41). Ethosuximide is extracted at low pH, the extract is applied to thin-layer sheets and developed by use of methanol/ammonia (100: 1.5 v/v). Detection is performed by iodine vapour (42).

For detection of ethosuximide in pharmaceuticals two tests can be used:

a) At high pH, ethosuximide will show a green fluorescence after heating in presence of resorcinol (43).
b) In ethanol an absorption peak at 248 nm, and at pH 12 an absorption peak at 218 nm (which is not seen at low pH) are indicative of ethosuximide (42).

3.4.3.3 Quality assessment

Precision: In a study on the performance of the techniques for measurement of therapeutic drugs in serum (2), outliers were rare and equally frequent for the different immunoassays and chromatographic techniques, but more frequent for gas-liquid chromatography after derivatisation. Interlaboratory imprecision was not significantly different for the various methods, and coefficients of variation ranged from 7% to 10%. Accuracy: The smallest bias as compared to the spike value was obtained with chromatographic techniques (gas-liquid chromatography without derivatisation, high performance liquid chromatography), the highest with measurements by gas-liquid chromatography after derivatisation (2).

3.4.3.4 Pharmacokinetics

Absorption
Peak serum levels are reached 3 to 7h after (oral) dosing depending on formulation: Syrup is faster absorbed than capsules. Independent from formulation, absorption is said to be nearly complete, and biovailability is about 100% as the drug does not undergo a significant first-pass effect.

Distribution
Ethosuximide is not bound to proteins and does not distribute into fat. That is why the volume of distribution is rather low (0.7 l/kg) and concentrations in plasma, saliva and cerebrospinal fluid are alike. Ethosuximide crosses the placenta and is excreted into breast milk.

Elimination

About 75% of ethosuximide is metabolised in the liver prior to renal excretion, about 25% is excreted unchanged. In the liver, the aliphatic side chains are hydroxylated and afterwards conjugated with glucuronic acid. About 40% of a dose is metabolized to 2-(1-hydroxyethyl-)2-methylsuccinimide, but none of the metabolites is considered to be pharmacologically active. Apparent body clearance was estimated to be 10 $\pm$ 4 ml/h per kg body weight.

Influences on pharmacokinetics

Some patients exhibit non-linearity following normal doses, i.e. the serum concentration is difficult to predict from dosage, and drug monitoring is mandatory in ethosuximide therapy. The reasons for non-linearity are not clearly understood. Besides body weight, the following factors infuence serum concentration:

Sex: In female patients ethosuximide concentration is higher than in male obtaining the same dosage.

Age: Young patients need a higher dose to reach the same serum concentration as adults.

In pregnancy clearance may be increased. One should bear in mind that in case of liver disease, metabolic clearance of ethosuximide will be reduced and dosage has to be adjusted. Interaction by other drugs is rare. Valproic acid may cause a rise of ethosuximide concentration, presumably due to its capacity-limited metabolism. Dialysed epileptics treated with ethosuximide will need supplemental dosing.

3.4.3.5 Medical interpretation

The therapeutic range of ethosuximide concentration in serum is considered to be between 280 and 710 µmol/l (table 2) (24, 44, 44a, 44b). However, some patients may require higher levels. Potentially toxic serum concentrations are between 710 and 1420 µmol/l (24), although Least (45) reported on a patient with a ethosuximide

Table 2. Pharmacokinetic Summary Table: Ethosuximide (24, 44a, 44b)

Ethosuximide
Therapeutic range: 280 µmol/l–710 µmol/l
(40 mg/l–100 mg/l)

Potentially toxic concentration: between 710 µmol/l and 1420 µmol/l
(100 mg/l and 200 mg/l)

Potentially lethal concentration: > 1780 µmol/l (> 250 mg/l)
Plasma half life: adults: 45 h
children: 30 h

Volume of distribution: 0.7 l/kg
Plasma protein binding: 0%
Elimination: 25% excreted unchanged renally
75% metabolised by liver
Active metabolites: none
Sampling: within dosage interval, preferable in a fixed relationship to dosage

concentration of 1190 µmol/l and no signs of intoxication. Toxic effects are not closely related with serum levels. Concentrations higher than 1780 µmol/l are considered to be potentially lethal. The estimated minimum lethal dose is 5 g (42). In case of severe poisoning, charcoal haemoperfusion may be effective, although conclusive data were not reported, as well as for application of haemodialysis, peritoneal dialysis and exchange transfusion. Forced diuresis is not recommended, because of the poor renal excretion of the drug.

The most common adverse effects are nausea and vomiting. Less often, abdominal discomfort, anorexia, drowsiness, fatigue, dizziness, hiccups and headache are reported.

Determination of ethosuximide in saliva or tears has not become popular and does not add any information as total and free fraction are identical (46).

3.4.4 Phenytoin

3.4.4.1 Introduction

Phenytoin (5,5-diphenylhydantoin or 5,5-diphenyl-2,4-imidazolidinedione) has a relative molecular mass of 252.3. It is almost insoluble in water, fairly soluble in ethanol, but less soluble in diethylether or chloroform. Its antiepileptic properties were detected in the late 1930's. It is used for the treatment of primary or secondary generalized tonic-clonic seizures, simple and complex partial seizures, mixed seizure types as well as tonic-clonic status epilepticus.

3.4.4.2 Methods

3.4.4.2.1 Quantitative methods

The favourite techniques for the determination of phenytoin in serum seem to be immunoassay (about two third) and chromatography (about one third) (1, 2). In recent years the use of immunoassays has further increased, whereas the number of laboratories using chromatographic techniques did not significantly change. Among the immunoassays, enzyme multiplied immunoassay (EMIT) and fluorescence polarization immunoassay (FPIA) were preferred and their use increased in an external quality assessment by 14 % and 62 % respectively within about 3 years. Nephelometric and turbidimetric immunoassays increased their share among the immunoassays from 3 % to 6 % within three years, whereas radioimmmunoassay is no longer used. Regarding chromatographic techniques, about half of the analysts gave up gas-liquid chromatography in favour of high performance liquid chromatography (HPLC).

Immunoassays

The principle of the tests is described in 3.4.2.2 and 3.4.3.2. As compared to FPIA, EMIT can be adapted to analyzers usually already available in the laboratory and can be scaled down to allow the analysis of small sample volumes. EMIT and FPIA were thoroughly evaluated (16, 47, 48, 49). Reports on the reliability of phenytoin determinations by SLFIA (50) were also published. Furthermore, a mechanized homogeneous liposome immunoassay system was developed (8). For outpatient clinics, the apoenzyme reactivation immunoassay system (ARIS) (51) and a non-instrumental immuno-chromatographic assay (52) are available. For continuous, reversible measurement of phenytoin, a fiber optic immunochemical sensor was developed (53).

Chromatographic methods
General remarks: see 3.4.2.2
Gas-liquid chromatography
Gas liquid chromatography was the favourite technique in the beginning of routine therapeutic drug monitoring (53 a). Usually phenytoin was extracted from the sample by organic solvents or by collection on solid materials. The organic phase was separated and evaporated, the dry residue redissolved and injected into the gas chromatograph.

Phenytoin can be analysed without derivatisation when using inert polar stationary phases (SP 2510 DA; SP 2250 DA) (16, 54). Specificity and detectability (in order to determine free phenytoin concentration) are considerably improved by use of fused silica capillary columns coated with phenylmethylsilicone and a nitrogen selective detector (47). Other authors recommend derivatisation, e.g. methylation prior to chromatography (53 a, 55). However, data from external quality assessment are not in favour of these procedures, although these results do not take into account the reliability of the method in the hands of the expert. The use of gas chromatography/mass spectrometry is restricted to studies of pharmacokinetics and metabolism, where utmost specificity and detectability are needed (56, 57).

High performance liquid chromatography
Usually, the group assays will encompass phenytoin (18, 29, 30, 58–63), although not all methods are sufficiently sensitive to determine free phenytoin concentration. On the other hand, procedures for the exclusive measurement of phenytoin to improve analytical reliability are not needed. In the beginning, drugs were extracted by organic solvents and HPLC was performed on silica gel or an ion-exchange column. Since the availability of reversed phase colums, methods became popular which use acetonitrile to precipitate proteins. Furthermore, Extrelut columns have been successfully applied to the pre-purification of the sample (64).

3.4.4.2.2 Qualitative methods

Thin-layer chromatography (TLC): Phenytoin is extracted from acidified liquids by diethylether. The extract is submitted to TLC, which is performed by use of acetone/chloroform (1:9 v/v). The drug can be detected by mercurous nitrate (65).

3.4.4.3 Quality assessment

Precision: Interlaboratory coefficient of variation was lowest (about 6 %) for FPIA and turbidimetric immunoassay, highest (10 % and 8 %) for nephelometric immunoassay (NIIA) and gas-liquid chromatography after derivatisation (2).

Accuracy: In the mean, deviation from the spike value was low (below 4 %) for all methods in an external quality assessment scheme (2). Best accuracy was observed when EMIT and FPIA were used; gas-liquid chromatography after derivatisation, nephelometric inhibition immunoassay and turbidimetric immunoassay were the least accurate methods. Nevertheless, the individual laboratory may obtain deviating performances depending on its engagement and skill.

3.4.4.4 Pharmacokinetics

Absorption

Phenytoin given orally is nearly completely absorbed, mainly in the duodenum. Absorption rate is dependent on formulation and rate of dissolution in water. The absorbed drug is not well extracted by the liver and first-pass effect is small.

Distribution

Phenytoin distributes by passive diffusion into body fluids. It is bound to tissue and serum proteins. In serum, most of phenytoin is bound to albumin, in the mean 90 %. However, the bound fraction may vary between 69 % and 95 %, depending on albumin concentration or competitors for albumin binding, such as bilirubin or other drugs, and in case of diseases of the liver or the kidneys (under these circumstances concentration of total phenytoin in serum is misleading and determination of the free fraction is preferred).

Phenytoin crosses placenta and is excreted to a low extent into breast milk. The concentration of the drug in cerebrospinal fluid and saliva resembles the free fraction in serum.

Elimination

Excretion of the unchanged drug by the kidneys is negligible. Phenytoin is mainly metabolised in the liver to form 5-(4′-hydroxyphenyl)-5 phenylhydantoin, which is esterified with glucuronic acid and eliminated in the urine. The rate of elimination is concentration-dependent: Half-life increases with increasing concentration. This effect is due to the metabolising enzyme system, which is saturable already within the therapeutic range of the drug. As a consequence, a small dose increment may lead to a greater than proportional increase of serum concentration and – taking into account intra-individual variability – the sequals of dose adjustments on serum concentration are difficult to predict. Therefore half-life is an inappropriate parameter for the characterization of the pharmacokinetics of phenytoin as its value varies with concentration. Furthermore, half-life is dependent on age and is influenced by drugs given concomitantly, which induce or inhibit the relevant enzyme system.

Influence on pharmacokinetics

In children and during pregnancy, phenytoin clearance is increased and may require shorter dosing intervals to avoid large fluctuations of serum concentration. In severe renal diseases as well as in hepatic diseases (which reduce the activity of the relevant enzymes or lead to hypalbuminaemia), the free fraction of the drug (representing the active drug) may be elevated. Under these circumstances, drug monitoring should rely on the concentration of unbound phenytoin. Although total concentration may be within the normal range, the determination of the free fraction measures toxicity. Haemodialysed patients will not need supplemental dosing as only minor amounts of the drug are eliminated.

Antacids reduce absorption, enzyme inducing agents such as carbamazepine, phenobarbital or ethanol enhance metabolism. In any case, a higher dosage of phenytoin may be necessary. Enzyme inhibiting drugs e. g. valproic acid, disulfiram, isoniazid, bishydroxycoumarin, propoxyphene or cimetidine as well as displacement from albumin by salicylates or valproic acid may require a reduction of the dosage.

3.4.4.5 Medical interpretation

Thepareutic range of phenytoin and its potentially toxic or lethal concentrations are given in table 3 (24, 65a, 65b). The ingestion of 7.5 g phenytoin was fatal, whereas another patient, after intake of 1.5. g, survived. In general, concentration of free phenytoin is more meaningful than concentration of the total drug.

Table 3. Pharmacokinetic Summary Table: Phenytoin (24, 65a, 65b)

Phenytoin

Therapeutic range (anticonvulsive):	Phenytoin, total:	20 µmol/l–79 µmol/l (5 mg/l–20 mg/l)
	Phenytoin, unbound:	2 µmol/l–8 µmol/l (0.5 mg/l–2.0 mg/l)

Potentially toxic concentration:	> 79 µmol/l–99 µmol/l (> 20 mg/l–25 mg/l)
Potentially lethal concentration:	> 198 µmol/l (> 50 mg/l)
Plasma half life:	20 h
Volume of distribution::	0.6 l/kg
Plasma protein binding:	90 %
Elimination:	2 % excreted unchanged by the kidneys 5 % excreted with faeces 93 % metabolised by the liver to 4-hydroxyphenytoin and 4-hydroxyphenytoin glucuronide
Active metabolites:	not known
Sampling:	before next dose (trough value)

In case of threatening intoxication, intestinal absorption can be inhibited by activated charcoal. Haemodialysis is not effective for removing the drug, haemoperfusion may be useful, but the benefit-risk ratio for this treatment has to be considered deliberately.

Phenytoin stabilizes membranes in the brain (and hence suppresses seizures) and in the heart (and hence suppresses arrhythmias). In case of intoxication, nystagmus and ataxia are observed. Severely poisoned patients are confused and disorientated, but usually not in a totally unresponsive state. Death is rare and due to cardiac arrest or ventricular fibrillation.

Phenytoin concentration in saliva can be determined easily and is considered as an appropriate estimation of the free fraction in serum.

3.4.5 Valproic Acid

3.4.5.1 Introduction

Valproic acid (n-dipropylacetic acid) is a branched-chain carboxylic acid with a relative molecular mass of 144.2 and was primarily used as an organic solvent. It is only poorly soluble in water, but well soluble in organic solvents. Its antiepileptic activity was discovered in 1963 by Meunier. It is used for a broad range of clinical seizure disorders, especially for absence epilepsy, generalized tonic-clonic and myoclonic seizures.

3.4.5.2 Methods

3.4.5.2.1 Quantitative methods

Valproic acid in serum is determined by immunoassays or by chromatographic techniques. During the last years, immunoassays became more popular, especially measurements by fluorescence polarization immunoassays, whereas the number of analyses by gas-liquid chromatography (and high performance liquid chromatography) decreased.

Unbound valproate can be determined after ultrafiltration. Errors arise from adsorption to the membrane, from insufficient control of temperature and pH or increase of free fatty acids during storage (66–68).

Immunoassays

Valproate in serum can be determined by enzyme multiplied immunoassay technique (EMIT) (3) as described (see 3.4.2.2), by use of enzyme-labeled valproate and by fluorescence polarization immunoassay (4), for which fluoresceine labeled valproate is supplied (see 3.4.2.2). Furthermore, substrate-labeled fluorescence immunoassay (69) and nephelometric inhibition immunoassay are available. A mechanized fluoroimmunoassay for valproic acid by flow injection analysis using high performance liquid chromatography devices was presented (70).

Chromatographic techniques

Gas-liquid chromatography

During many years, gas-liquid chromatography was by far the most popular method for the determination of total valproic acid in serum (71–73) and various procedures for this purpose were published. They allow the simultaneous determination of ethosuximide (36, 74). The most advantageous technique is to acidify the sample, add an appropriate organic solvent and vortex. An aliquot of the organic phase is injected directly into the gas-chromatograph without derivatisation. This procedure can be adapted to head space analysis (75). Most often sulfuric acid is used for acidifying, and chloroform, n-hexan or pentane for extraction. With SP 1000, Carbowax 20 M or Carbowax 6000, adequate separation is obtained in gas-liquid chromatography. Coefficients of variation are usually below 5% for imprecision between days.

For the determination of free valproic acid or its various metabolites, capillary gas-liquid chromatography – mass spectrometry is needed after derivatisation of the compounds (76–78).

High performance liquid chromatography (HPLC)

Determination of valproic acid by HPLC is hampered by its volatility, which impedes extraction and evaporation for enrichment, and by its low detectability in the ultraviolet region of the spectrum. To overcome these difficulties, van der Horst et al. (79) and Wolf et al. (80) propose to derivatize valproic acid by use of 4-bromomethyl-7-methoxy-coumarin in order to obtain a fluorescent compound, which can be detected sensitively by a fluorescence detector. Extraction is not necessary. However, at least Wolf and coworkers (80) observed double peaks for all acids after derivatisation (e.g. valproic acid, nonanoic acid (internal standard). Metabolites of valproic acid can be determined probably in case of accumulation, they are, however, not distinctly separated from the mother compound with the proposed HPLC system.

Kushida et al. (81) present a method for the simultaneous measurement of valproic acid and other antiepileptic drugs. To improve detectability by ultraviolet detection, samples are extracted. Internal standardisation is used to compensate for the extraction losses of valproic acid (about 10%).

On the whole, HPLC does not seem to offer any evident advantages as compared to immunoassays or gas-liquid chromatography in valproic acid determination.

Other techniques

Valproate can be determined by isotachophoresis (82), and a valproate-selective electrode was developed for quantitative analysis of the drug in pharmaceutical preparations (83).

3.4.5.3 Quality assessment

Precision: In an evaluation of results of an external quality assessment scheme (2), best precision was obtained with fluorescene polarization immunoassay, least precise with measurements by high performance liquid chromatography and gas-liquid chromatography after derivatisation. Coefficients of variation for interlaboratory imprecision ranged from 6% to 10% for the different methods. Accuracy: There was a negative bias for all methods as compared to the spike value of the control materials: – 8% for measurements by fluorescence polarization immunoassay, – 5% to – 4% for chromatographic techniques, and – 2% for EMIT (2). It is supposed that the spiked value is lower than declared due to hygroscopy of valproic acid, which causes weighing errors.

3.4.5.4 Pharmacokinetics

Absorption

Valproic acid is rapidly absorbed from an empty stomach, and serum peak concentrations are reached 1 to 4h (depending on formulation) after ingestion. Absorption is fastest for syrups, less quick for capsules and slowest for enteric coated tablets. Absorption is delayed when the drug is taken after eating; nevertheless it is always about 100%.

Distribution

Valproate is highly bound to serum proteins (90% to 95%) (24, 44, 84, 85), but not to tissue proteins, and the volume of distribution is rather low (table 4). The free fraction is dependent on valproate concentration and increases already at higher concentrations of total, which are still within the therapeutic range. I.e. there is a saturable protein binding, which is effective within the therapeutic range. Concentration of valproate in cerebrospinal fluid resembles its free concentration in serum. Very small amounts are excreted into breast milk.

Table 4. Pharmacokinetic Summary Table: Valproate (24, 84, 85)

	Valproic Acid
Therapeutic range:	350 µmol/l–700 µmol/l (50 mg/l–100 mg/l)
Potentially toxic concentration:	> 1400 µmol/l (> 200 mg/l)
Potentially lethal concentration:	no data
Plasma half life:	14 h
Volume of distribution:	0.13 l/kg
Plasma protein binding:	93% (concentration dependent)
Elimination:	2% excreted unchanged renally 98% metabolised by liver
Active metabolites:	several
Sampling:	prior to a (morning) dose (trough value) 2 h after ingestion of the capsule or syrup (peak value)

Elimination

About 95% of valproate is eliminated by hepatic metabolism, less than 5% are excreted unchanged in urine. More than 10 metabolites have been identified, e.g. 2-propyl-pentanoic acid and 2-propyl-3-keto-pentanoic acid; 4-en-valproic acid was found only in minor amounts, however, hepatotoxicity and teratogenic potential are attributed to this metabolite. 40% of a dose is excreted as glucuronides. Some metabolites possess antiepileptic activities, but as their concentration in brain is rather low, they will not contribute substantially to the antiepileptic effects. Total body clearance (clear) is higher in children (15 ml/h per kg body weight) than in adults (8 ml/h per kg body weight) and half-life ($T_{1/2}$) is shorter, because volume of distribution (V_d) is independent of age ($T_{1/2} = \ln 2 \cdot V_d/\text{Clear}$). Increase of the free fraction due to an increase of valproate serum concentration will lead to an increase of both clearance and volume of distribution. Thus half-life will be unchanged.

Influence on pharmacokinetics
In case of hepatic disease, elimination half-life is prolonged, but not in renal disease. Administration of enzyme inducing drugs such as carbamazepine, phenobarbital or phenytoin increases clearance of valproate and shortens half-life. Salicylates displace valproate from proteins and lead to an increase of free fraction and clearance.

3.4.5.5 Medical Interpretation

It is generally accepted that most patients will benefit from serum concentrations of valproate between 350 and 700 μmol/l (table 4) (24, 84, 85). Toxic symptoms can appear at concentrations between 840 and 1050 μmol/l (85); according to Schulz et al. (24) toxicity is common above 1400 μmol/l. In a case of fatal poisoning, valproate concentration was as high as 13270 μmol/l (86). Unconsciousness was observed when > 200 mg/kg body weight were administered, although after a dose of 25 g the patient recovered. Due to the widespread use, poisoning is not rare (86). Poisoned patients were treated by gastric lavage and application of activated charcoal. No systematic studies are available to prove the usefulness of forced diuresis, haemodialysis or haemoperfusion. Due to minimal excretion and high degree of protein binding, these means seem not very promising for treatment of valproate poisoning. The following adverse effects were reported: nausea and vomiting, hepatotoxicity, pancreatitis, coagulopathy, tremor and incoordination. Hepatic necrosis (Reye-like syndrome) or acute pancreatitis may occur unpredictably at concentrations within the therapeutic range. Sixty reports of fatal hepatic necrosis due to valproate were published (87), mainly in children and adolescents during the first 6 months of treatment. The aspartate aminotransferase activity may, on elevation, signal the need to discontinue valproic acid therapy.

Measurement of the free fraction is recommended in case of hypo- or hyperproteinaemia, and if displacement from protein through drug interaction is suspected. Due to its low distribution into saliva (0.5% to 4.5%), determination of valproic acid is not recommended in this material.

3.4.6 Other Antiepileptic Drugs

In recent years some new drugs were developed for seizure control. They are mostly determined by high performance liquid chromatography, whereas for most of the antiepileptic drugs known since a long time, gas chromatographic methods are available (table 5).

Table 5. Other Antiepileptic Drugs: Methods of Determination

Drug	Method	Reference
Acetazolamide	GC, HPLC	88, 89
Bromide	AS	90
Ethotoin	HPLC, GC-MS	91, 92
Felbamate	HPLC	59, 61
Flunarizine	GC, HPLC	93, 94
Gabapentin	HPLC	95
Lamotrigine	HPLC	96, 97
Mephenytoin	GC, HPLC	98, 99
Methsuximide	GC-MS	100
N-Desmethylmethsuximide	EMIT, FPIA, GC, HPLC	101, 35, 38
Oxcarbazepine	HPLC	102, 103
Paraldehyde	GC	104
Progabide	GC, HPLC	105, 106
Stiripentol	HPLC	107
Trimethadione	GC, HPLC	108, 109
Vigabatrin	HPLC	110, 111
Zonisamide	HPLC, EIA	58, 112

AS: Absorption spectrometry after reaction according to Kisser (90)
EMIT: Enzyme multiplied immunoassay technique
FPIA: Fluorescence polarization immunoassay
GC: Gas-liquid chromatography
GC-MS: Gas-liquid chromatography-mass spectrometry
HPLC: High performance liquid chromatography
EIA: Immunoassay

References

1. Wilson, J. F., Tsanaclis, L. M., Williams, J., Tedstone, J. E. & Richens, A. (1989) Evaluation of assay techniques for the measurement of antiepileptic drugs in serum: A study based on external quality assurance measurements. Ther. Drug Monit. *11*, 185–195.
2. Wilson, J. F., Tsanaclis, L. M., Perrett, J. E., Williams, J., Wicks, J. F. C. & Richens, A. (1992) Performance of techniques for measurement of therapeutic drugs in serum. A comparison based on external quality assessment data. Ther. Drug Monit. *14*, 98–106.
3. Rubenstein, K. E., Schneider, R. S. & Ulman, E. F. (1972) Homogeneous enzyme immunoassay. A new immunochemical technique. Biochem. Biophys. Res. Commun. *47*, 846–851.
4. Jolley, M. E. (1981) Fluorescence polarization immunoassay for the determination of therapeutic drug levels in human plasma. J. Anal. Toxicol. *5*, 236–240.
5. Moore, F. M. L. & Simpson, D. (1989) Cost effective assays for use in monitoring carbamazepine, phenobarbital, and phenytoin in serum. Clin. Chem. *35*, 1782–1784.
6. Tittle, T. V. & Schaumann, B. A. (1992) A micro-enzymemultiplied immunoassay technique plate assay for antiepileptic drugs. Ther. Drug Monit. *14*, 159–163.
7. Nakamura, R. M. (1983) Advances in analytical fluorescence immunoassays: Methods and clinical application. In: Clinical Laboratory Assays, edited by R. M. Nakamura, W. R. Dito, E. S. Tucker III, pp. 33–60. Masson Publishing USA, Inc.
8. Kubotsu, K., Goto, S., Fujita, M., Tuchiya, H., Kida, M., Takano, S., Matsuura, S. & Sakurabayashi, I. (1992) Automated homogeneous liposome immunoassay systems for anticonvulsant drugs. Clin. Chem. *38*, 808–812.
9. Croci, D., Nespolo, A. & Tarenghi, G. (1988) A dry-reagent strip for quantifying carbamazepine evaluated. Clin. Chem. *34*, 388–392.

10. Miles, M.V., Tennison, M.B., Greenwood, R.S., Benoit, S.E., Thorn, M.D., Messenheimer, J.A. & Ehle, A.L. (1990) Evaluation of the Ames Seralyzer for the determination of carbamazepine, phenobarbital and phenytoin concentrations in saliva. Ther. Drug Monit. *12*, 501–510.
11. Zuk, R.F., Ginsberg, V.K., Houts, T., Rabbie, J., Merrick, H., Ullman, E.F., Fischer, M.M., Sizto, C.C., Stiso, S.N. & Litman, D.J. (1985) Enzyme immunochromatography-a quantitative immunoassay requiring no instrumentation. Clin. Chem. *31*, 1144–1150.
12. Cochran, E.B., Massey, K.L., Phelps, S.J., Cramer, J.A., Toftness, B.R., Denio, L.S. & Drake, M.E. (1990) Comparison of a noninstrumented immunoassay for carbamazepine to high performance liquid chromatography and fluorescence polarization immunoassay. Epilepsia *37*, 480–484.
13. Kupferberg, H.J. (1972) Gas liquid chromatographic determination of carbamazepine in plasma. J. Pharm. Sci. *61*, 284–286.
14. Bathia, H.M. (1986) Improved gas-chromatographic determination of carbamazepine. Clin. Chem. *32*, 563–564.
15. Külpmann, W.R. (1980) Eine gas-chromatographische Methode zur Bestimmung von Carbamazepin, Phenobarbital, Phenytoin und Primidon im gleichen Serumextrakt. J. Clin. Chem. Clin. Biochem. *18*, 227–232.
16. Külpmann, W.R. & Oellerich, M. (1981) Monitoring of therapeutic serum concentrations of antiepileptic drugs by a newly developed gas chromatographic procedure and enzyme immunoassay (EMIT): A comparative study. J. Clin. Chem. Clin. Biochem. *19*, 249–258.
17. Lynn, R.K., Smith, R.G., Thompson, R.M., Deiner, M.L., Griffin, D. & Gerber, N. (1978) Characterization of glucuronide metabolites of carbamazepine in human urine by gas chromatography and mass spectrometry. Drug Metab. Dispos. *6*, 494–501.
18. Juergens, U. (1987) HPLC analysis of antiepileptic drugs in blood samples: microbore separation of fourteen compounds. J. Liquid Chromatogr. *10*, 507–532.
19. Shibukawa, A., Nishimura, N., Nomura, K., Kuroda, Y. & Nakagawa, T. (1990) Simultaneous determination of the free and total carbamazepine concentrations in human plasma by high-performance frontal analysis using an internal-surface reversed-phase silica support. Chem. Pharm. Bull. *38*, 443–447.
20. Riad, L.E. & Sawchuk, R.J. (1988) Simultaneous determination of carbamazepine and its epoxide and transdiol metabolites in plasma by microbore liquid chromatography. Clin. Chem. *34*, 1863–1866.
21. Hartley, R., Lucock, M., Cookman, J.R., Becker, M., Smith, I.J., Smithells, R.W. & Forsythe, W.I. (1986) High-performance liquid chromatographic determination of carbamazepine and carbamazepine-10,11-epoxide in plasma and saliva a following solid-phase sample extraction. J. Chromatogr. *380*, 347–356.
22. Clarke, E.G.C. (1974) Isolation and identification of drugs. Vol. 1 p. 338. Pharmaceutical Press, London.
23. Bircher, J. & Lotterer, E. (1988) Klinisch-Pharmakologische Datensammlung, S. 177–178. G. Fischer Verlag, Stuttgart, New York.
24. Schulz, M. & Schmoldt, A. (1991) Therapeutische und toxische Plasmakonzentrationen sowie Eliminationshalbwertzeiten gebräuchlicher Arzneistoffe. Pharm. Ztg. Wiss. *136*, 87–94.
25. Uges, D.R.A. (1990) Orientierende Angaben zu therapeutischen und toxischen Konzentrationen von Arzneimitteln und Giften in Blut, Serum oder Urin. Mitteilung XV der Senatskommission für Klinisch-toxikologische Analytik, S. 16. VCH Verlag, Weinheim.
26. Berry, D.J., Wisemann, H.M. & Volans, G.N. (1983) A survey of non-barbiturate anticonvulsant drug overdose reposted to the Poison Information Service (UK). Human Toxicol. *2*, 357–360.
27. Lehrmann, S.N. & Baumann, M.L. (1984) Carbamazepine overdose. Am. J. Dis. Child *135*, 768–769.
28. Nilson, C., Sterner, G. & Idvall, J. (1984) Charcoal hemoperfusion for treatment of serious carbamazepine poisoning. Acta Med. Scand. *216*, 137–140.
29. Herkens, G.K., McKinnon, G.E. & Eadie, M.J. (1989) Simultaneous quantitation of salivary carbamazepine, carbamazepine-10, 11-epoxide, phenytoin and phenobarbitoneby high-performance liquid chromatography. J. Chromatogr. *496*, 147–154.
30. Schramm, W., Annesley, T.M., Siegel, G.J., Sackellares, J.C. & Smith, R.H. (1991) Measurement of phenytoin and carbamazepine in an ultrafiltrate of saliva. Ther. Drug Monit. *13*, 452–460.

31. Least, C.J.Jr., Johnson, G.F. & Solomon, H.M. (1975) A quantitative gas-chromatographic determination of ethosuximide based on N-butylation. Clin. Chim. Acta *60*, 285–292.
32. Menyharth, P., Lehane, D.P. & Levy, A.L. (1977) Rapid gas-chromatographic method for the determination of ethosuximide in serum. Clin. Chem. *23*, 1795–1796.
33. Sun, L. & Szafir, J. (1977) Comparison of enzyme immunoassay and gas chromatography for determination of carbamazepine and ethosuximide in human serum. Clin. Chem. *23*, 1753–1756.
34. Berry, D.J. & Clarke, L.A. (1978) Gas chromatographic analysis of ethosuximide (2-ethyl-2-methylsuccinimide) in plasma at therapeutic concentrations. J. Chromatogr. *150*, 537–541.
35. Wallace, J.E., Schwertner, H.A., Hamilton, H.E. & Shimek, E.L.Jr. (1979) Electron-capture gas-liquid chromatographic determination of ethosuximide and desmethylmethsuximide in plasma or serum. Clin. Chem. *25*, 252–255.
36. Külpmann, W.R. (1980) Eine gas-chromatographische Methode zur gleichzeitigen Bestimmung von Ethosuximid und Valproinat im Serum. J. Clin. Chem. Clin. Biochem. *18*, 339–344.
37. Horning, M.G., Brown, L., Nowlin, J., Lertratanangkoon, K., Kellaway, P. & Zion, T.E. (1977) Use of saliva in therapeutic drug monitoring. Clin. Chem. *23*, 157–164.
38. Szabo, G.K. Browne, T.R. (1982) Improved isocratic liquid-chromatographic simultaneous measurement of phenytoin, phenobarbital, primidone, carbamazepine, ethosuximide, and N-desmethylmetsuximide in serum. Clin. Chem. *28*, 100–104.
39. Kabra, P.M., Nelson, M.A. & Marton, L.J. (1983) Simultaneous very fast liquid-chromatographic analysis of ethosuximide, primidone, phenobarbital, phenytoin, and carbamazepine in serum. Clin. Chem. *29*, 473–476.
40. Wad, N. (1985) Rapid extraction method for ethosuximide and other antiepileptics in serum for determination by high-performance liquid chromatography. J. Chromatogr. *338*, 460–462.
41. Moffat, A.C., Franke, J.-P., Stead, A.H., Gill, R., Finkle, B.S., Möller, M.R., Müller, R.K., Wunsch, F. & de Zeeuw, R.A. (1987) Thin-layer chromatographic R_F values of toxicologically relevant substances on standardized systems. Report VII of the DFG commission for clinical-toxicological analysis, special issue of the TIAFT bulletin, p. 50. VCH Verlag, Weinheim.
42. Clarke, E.G.C. (1974) Isolation and identification of drugs. Vol. 1, p. 338. Pharmaceutical Press, London.
43. British Pharmacopeia (1969) Her Majesty's stationary office, p. 532. London.
44. Brodie, M.J. & Hallworth, M.J. (1989) Therapeutic monitoring of sodium valproate and ethosuximide. Hospital Update *15*, 351–364.
44a. Bircher, J. & Lotterer, E. (1988) Klinisch-Pharmakologische Datensammlung, S. 177–178. G. Fischer-Verlag, Stuttgart, New York.
44b. Uges, D.R.A. (1990) Orientierende Angaben zu therapeutischen und toxischen Konzentrationen von Arzneimitteln und Giften in Blut, Serum oder Urin. Mitteilung XV der Senatskommission für Klinisch-toxikologische Analytik, S. 20. VCH Verlag, Weinheim.
45. Least, C.J., Johnson, G.F. & Solomon, H.M. (1983) Ethosuximide. Type C procedure. Pages 160–162. In: Methodology for analytical toxicology (ed. Sunshine, J.) CRC Press, Boca Raton, Florida.
46. Piredda, S. & Monaco, F. (1981) Ethosuximide in tears, saliva and cerebrospinal fluid. Ther. Drug Monit. *3*, 321–323.
47. Külpmann, W.R., Gey, S., Beneking, M., Kohl, B. & Oellerich, M. (1984) Dermination of total and free phenytoin in serum by non-isotopic immunoassays and gas chromatography. J. Clin. Chem. Clin. Biochem. *22*, 773–779.
48. Booker, H.E. & Darcey, B.A. (1975) Enzymatic immunoassays vs gas/liquid chromatography for determination of phenobarbital and diphenylhydantoin in serum. Clin. Chem. *21*, 1766–1768.
49. Lu-Steffes, M., Pittluck, G.W., Jolley, M.E., Panas, H.N., Olive, D.L., Wang, C.H., Nystrom, D.D., Keegan, C.L., Davis, T.P. & Stroupe, S.D. (1982) Fluorescence polarization immunoassay IV. Determination of phenytoin and phenobarbital in human serum and plasma. Clin. Chem. *28*, 2278–2282.
50. Wong, R.C., Burd, J.F., Carrico, R.J., Buckler, R.T., Thoma, J. & Boguslaski, R.C. (1979) Substrate-labeled fluorescent immunoassay for phenytoin in human serum. Clin. Chem. *25*, 686–691.

51. Sommer, R., Nelson, C. & Greenquist, A. (1986) Dry-reagent strips for measuring phenytoin in serum. Clin. Chem. *32*, 1770–1774.
52. Morris, R.G. & Schapel, G.J. (1988) Phenytoin and phenobarbital assayed by the ACCULEVEL method compared with EMIT in an outpatient clinic setting. Ther. Drug Monit. *10*, 469–473.
53. Anderson, F.P. & Miller, W.G. (1988) Fiber optic immunochemical sensor for continuous, reversible measurement of phenytoin. Clin. Chem. *34*, 1417–1421.
53a. Kupferberg, H.J. (1970) Quantitative estimation of diphenylhydantoin, primidone and phenobarbital in plasma by gas-liquid chromatography. Clin. Chim. Acta *29*, 283–288.
54. Thoma, J.J., Ewald, T. & McCoy, M. (1978) Simultaneous analysis of underivatized phenobarbital, carbamazepine, primidone, and phenytoin by isothermal gas-liquid chromatography. J. Anal. Toxicol. *2*, 219–225.
55. Hill, R.E. & Latham, A.N. (1977) Simultaneous determination of anticonvulsant drugs by gas-liquid chromatography. J. Chromatogr. *131*, 341–346.
56. Hoppel, C., Garle, M. & Elander, M. (1976) Mass fragmentographic determination of diphenylhydantoin and its main metabolite, 5-(4-hydroxylphenyl)-5-phenylhydantoin, in human plasma. J. Chromatogr. *116*, 53–61.
57. Rane, A., Garle, M., Borga, O. & Sjoqvist, F. (1974) Plasma disappearance of transplacentally transferred diphenylhydantoin in the newborn studies by mass fragmentography. Clin. Pharmacol. Ther. *15*, 39–45.
58. Juergens, U. (1987 Simultaneous determination of zonisamide and nine other antiepileptic drugs and metabolites in serum. A comparison of microbore and conventional high-performance liquid chromatography. J. Chromatogr. *385*, 233–240.
59. Remmel, R.P., Miller, S.A. & Graves, N.M. (1990) Simultaneous assay of felbamate plus carbamazepine, phenytoin, and their metabolites by liquid chromatography with mobile phase optimization. Ther. Drug Monit. *12*, 90–96.
60. Asberg, A. & Haffner, F. (1987) Analysis of serum concentration of phenobarbital, phenytoin, carbamazepine and carbamazepine-10,11-epoxide by solvent-recycled liquid chromatography. Scand. J. Clin. Lab. Invest. *47*, 389–392.
61. Remmel, R.P., Miller, S.A. & Graves, N.M. (1990) Simultaneous assay of felbamate plus carbamazepine, phenytoin, and their metabolites by liquid chromatography with mobile phase optimization. Ther. Drug Monit. *12*, 90–96.
62. Mira, A.S., El-Sayed, Y. & Islam, S.I. (1987) Simultaneous measurement of phenobarbital, carbamazepine, phenytoin and 5-(4-hydroxyphenyl)-5-phenylhydantoin in serum by high-performance liquid chromatography. Analyst. *112*, 57–60.
63. Berg, J.D. & Buckley, B.M. (1987) Rapid measurement of anticonvulsant drug concentrations in the out-patient clinic, using HPLC with direct injection of plasma. Ann. Clin. Biochem. *24*, 488–493.
64. Helmsing, P.J., Van der Woude, J. & Van Eupen, O.M. (1978) A micromethod for simultaneous estimation of blood levels of some commonly used antiepileptic drugs. Clin. Chim. Acta, *89*, 301–309.
65. Clarke, E.G.C. (1974) Isolation and identification of drugs. Vol. 1, p. 496. Pharmaceutical Press, London.
65a. Bircher, J. & Lotterer, E. (1988) Klinisch-Pharmakologische Datensammlung. S. 326–328. G. Fischer-Verlag, Stuttgart, New York.
65b. Uges, D.R.A. (1990) Orientierende Angaben zu therapeutischen und toxischen Konzentration von Arzneimitteln und Giften in Blut, Serum oder Urin. Mitteilung XV der Senatskommission für Klinisch-toxikologische Analytik. S. 26. VCH-Verlag, Weinheim.
66. Barré, J., Chamouard, J.M., Houin, G. & Tillement, J.P. (1985) Equilibrium dialysis, ultrafiltration, and ultracentrifugation compared determining the plasma-protein-binding characteristics of valproic acid. Clin. Chem. *31*, 60–64.
67. Kara, M.H. & Tweeddale, M.G. (1986) Determination of free valproic acid levels by ultrafiltration. Ther. Drug Monit. 8, 245–246.
68. Tarasidis, C.G., Garnett, W.R., Kline, B.J. & Pellock, J.M. (1986) Influence of tube type, storage time, and temperature on the total and free concentration of valproic acid. Ther. Drug Monit. 8, 373–376.
69. Feinstein, H., Hovav, H., Fridlender, B., Inbar, D. & Buckler, R.T. (1982) Substrate-labeled fluorescent immunoassay (SLFIA) for valproic acid. Clin. Chem. *28*, 1665–1666.
70. Allain, P., Turcant, A. & Prémel-Cabic, A. (1989) Automated fluoroimmunoassay of

theophylline and valproic acid by flow-injection analysis with use of HPLC instruments. Clin. Chem. *35*, 469–470.

71. Rege, A.B., Lertora, J.J.L., White, L.E. & George, W.J. (1984) Rapid analysis of valproic acid by gas chromatography. J. Chromatogr. *309*, 397–402.
72. Tosoni, S., Signorini, C. & Albertini, A. (1983) Gas-chromatographic determination of valproic acid in serum without derivatization. Clin. Chem. *29*, 990.
73. Lin, W. & Kelly, A.R. (1985) Determination of valproic acid in plasma or serum by solid-phase column extraction and gas-liquid chromatography. Ther. Drug Monit. *7*, 336–344.
74. Bousquet, E., Puglisi, G. & Santagati, N.A. (1986) Improved procedure for simultaneous determination of sodium valproate and ethosuximide in serum by gas-liquid chromatography. Farmaco *41*, 404–410.
75. Degel, F., Heidrich, R., Schmid, R.D. & Weidemann, G. (1984) Quantitative determination of valproic acid by means of gas chromatographic headspace analysis. Clin. Chim. Acta *139*, 29–36.
76. Abbott, F.S. Kassam, J., Acheampong, A., Ferguson, S., Panesar, S., Burton, R., Farrell, K. & Orr, J. (1986) Capillary gas chromatography-mass spectrometry of valproic acid metabolites in serum and urine using tert.-butyldimethylsilyl derivatives. J. Chromatogr. *375*, 285–298.
77. Tatsuhara, T., Muro, H., Matsuda, Y. & Imai, Y. (1987) Determination of valproic acid and its metabolites by gas chromatography-mass spectrometry with selected ion monitoring. J. Chromatogr. *399*, 183–195.
78. Kassahun, K., Farrell, K., Zheng, J. & Abbott, F. (1990) Metabolic profiling of valproic acid in patients using negative-ion chemical ionization gas chromatography-mass spectrometry. J. Chromatogr. *527*, 327–341.
79. Van der Horst, F.A.L., Eikelboom, G.G. & Holthuis, J.J.M. (1988) High-performance liquid chromatographic determination of valproic acid in plasma using a micelle-mediated pre-column derivatization. J. Chromatogr. *456*, 191–199.
80. Wolf, J.H., Veenma-van der Duin, L. & Korf, J. (1989) Automated analysis procedure for valproic acid in blood, serum and brain dialysate by high-performance liquid chromatography with bromomethylmethoxycoumarin as fluorescent label. J. Chromatogr. *487*, 496–502.
81. Kushida, K. & Ishizaki, T. (1985) Concurrent determination of valproic acid with other antiepileptic drugs by high-performance liquid chromatography. J. Chromatogr. *338*, 131–139.
82. Mikkers, F., Verheggen, T. & Everaerts, F. (1980) Direct determination of valproate in serum by isotachophoresis. Comparison with a gas chromatographic method. J. Chromatogr. *182*, 496–500.
83. Suzuki, H., Akimoto, K. & Nakagawa, H. (1991) Quantitative analysis of sodium valproate in pharmaceutical preparations by a valproate selective electrode. Chem. Pharm. Bull. *39*, 133–136.
84. Bircher, J. & Lotterer, E. (1988) Klinisch-Pharmakologische Datensammlung, S. 406–407. G. Fischer-Verlag, Stuttgart, New York.
85. Uges, D.R.A. (1990) Orientierende Angaben zu therapeutischen und toxischen Konzentrationen von Arzneimitteln und Giften in Blut, Serum oder Urin. Mitteilung XV der Senatskommission für Klinisch-toxikologische Analytik, S. 30. VCH Verlag, Weinheim.
86. Garnier, R., Boudignat, O. & Fournier, P.E. (1982) Valproate poisoning. Lancet *2*, 97.
87. Ellenhorn, M.J. & Barceloux, D.G. (1988) Medical Toxicology. p. 264. Elsevier, New York, Amsterdam, London.
88. Wallace, S.M., Shah, V.B. & Riegelman, S. (1977) GLC analysis of acetazolamide in blood, plasma, and saliva following oral administration to normal subjects. J. Pharm. Sci. *66*, 527–530.
89. Chapron, D.J. & White, L.B. (1984) Determination of acetazolamide in biological fluids by revese-phase high performance liquid chromatography. J. Pharmaceutical Sci. *73*, 985–989.
90. Kisser, W. (1967) Über den Nachweis und die quantitative Bestimmung bromierter Harnstoffderivate in der Toxikologie. Arch. Toxikol. *22*, 404–409.
91. Inotsume, N., Higashi, A., Kinoshita, E., Matsuoka, T. & Nakano, M. (1986) Rapid and sensitive determination of ethotoin as well as carbamazepine, phenobarbital, phenytoin and primidone in human serum. J. Chromatogr. Biomed. Appl. *383*, 166–171.
92. Inotsume, N., Fujii, J., Honda, M., Nakano, M., Higashi, A. & Matsuda, I. (1988) Stereoselective analysis of the enantiomers of ethotoin in human serum using chiral stationary phase liquid chromatography and gas chromatography-mass spectro-

metry. J. Chromatogr. Biomed. Appl. *428*, 402–407.

93. Woestenborgs, R., Michielsen, L., Lorreyne, W. & Heykants, J. (1982) Sensitive gas chromatographic method for the determination of cinnarizine and flunarizine in biological samples. J. Chromatogr. *232*, 85–91.
94. Binnie, C.D. (1989) Potential antiepileptic drugs. In: Antiepileptic drugs (Levy, R.H. Dreifuss, F.E., Mattson, R.H., Meldrum, B.S. & Penry, J.K., eds.) 3rd edition, p. 971. Raven Press, New York.
95. Hengy, H. & Kölle, E.-U. (1985) Determination of gabapentin in plasma and urine by high-performance liquid chromatography and precolumn labelling for ultraviolet detection. J. Chromatogr. *341*, 473–478.
96. Cociglio, M., Alric, R. & Bouvier, O. (1991) Performance analysis of a reversed-phase liquid chromatographic assay of lamotrigine in plasma using solvent-demixing extraction. J. Chromatogr. Biomed. Appl. *572*, 269–276.
97. Fazio, A., Artesi, C., Russo, M., Trio, R., Oteri, G. & Pisani, F. (1992) A liquid chromatographic assay using a high-speed column for the determination of lamotrigine, a new antiepileptic drug, in human plasma. Ther. Drug Monit. *14*, 509–512.
98. Yonekawa, W.D. & Kupferberg, H.J. (1979) Measurement of mephenytoin (3-methyl-5-ethyl-5-phenylhydantoin) and its demethylated metabolite by selective ion monitoring. J. Chromatogr. *163*, 161–167.
99. Kupfer, A., Carr, J.R. & Branch, R. (1982) Analysis of hydroxylated and demethylated metabolites of mephenytoin in man and laboratory animals using gas-liquid chromatography and high performance liquid chromatography. J. Chromatogr. *232*, 93–100.
100. Porter, R.J., Penry, J.K., Lacy, J.R., Newmark, M.E. & Kupferberg, H.J. (1979) Plasma concentrations of phensuximide, methsuximide, and their metabolites in relation of clinical efficacy. Neurology *29*, 1509–1513.
101. Miles, M.V., Howlett, C.M., Tennison, M.B., Greenwood, R.S. & Cross, R.E. (1989) Determination of N-desmethylmethsuximide serum concentrations using enzyme-multiplied and fluorescence polarization immunoassays. Ther. Drug Monit. *11*, 337–342.
102. Menge, G.P., Dubois, J.P. & Bauer, G. (1987) Simultaneous determination of carbamazepine, oxcarbazepine and their main metabolites in plasma by liquid chromatography. J. Chromatogr. Biomed. Appl. *414*, 477–483.
103. Flesch, G., Francotte, E., Hell, F. & Degen, P.H. (1992) Determination of the R-(−) and S-(+) enantiomers of the monohydroxylated metabolite of oxcarbazepine in human plasma by enantioselective high-performance liquid chromatography. J. Chromatogr. Biomed. Appl. *581*, 147–151.
104. Lockman, L.A. (1989) Other antiepileptic drugs. In: Antiepileptic drugs (Levy, R.H., Dreifuss, F.E., Mattson, R.H., Meldrum, B.S. & Penry, J.K., eds.) 3rd edition, p. 881. Raven Press, New York.
105. Gillet, G., Fraisse-André, J., Lee, C.R., Dring, D. & Morselli, P.L. (1982) Gas chromatographic method for the determination of progabide (SL 76002) in biological fluids. J. Chromatogr. Biomed. Applic. *230*, 154–161.
106. Ascalone, V., Catalani, B. & DalBo, L. (1985) Determination of progabide and its acid metabolite in biological fluids by HPLC on silica column and UV detector. J. Chromatogr. Biomed. Applic. *344*, 231–239.
107. Lin, H.S. & Levy, R.H. (1983) Pharmacokinetic profile of a new anticonvulsant, stiripentol, in the rhesus monkey. Epilepsia *24*, 692–702.
108. Tanaka, E. & Misawa, E. (1987) Simultaneous determination of serum trimethadione and its metabolite by gas chromatography. J. Chromatogr. *413*, 376–378.
109. Tanaka, E., Hagino, S., Yoshida, I. & Kuroiwa, Y. (1984) Simultaneous determination of trimethadione and its metabolite in rat and human serum by high-performance liquid chromatography. J. Chromatogr. *308*, 393–397.
110. Tsanaclis, L.M., Wicks, J., Williams, J. & Richens, A. (1991) Determination of vigabatrin in plasma by reversed-phase high-performance liquid chromatography. Ther. Drug Monit. *13*, 251–253.
111. Löscher, W., Fassbender, C.P., Gram, L., Gramer, M., Hörstermann, D., Zahner, B. & Stefan, H. (1993) Determination of GABA and vigabatrin in human plasma by a rapid and simple HPLC method: Correlation between clinical response to vigabatrin and increase in plasma GABA. Epilepsy Res. *14*, 245–255.
112. Kaibe, K., Nishimura, S., Ishii, H., Sunahara, N., Naruto, S. & Kurooka, S. (1990) Competitive binding enzyme immunoassay for zonisamide, a new antiepileptic drug, with selected paired-enzyme labeled antigen and antibody. Clin. Chem. *36*, 24–27.

3.5 Neuroleptics

R. Whelpton

3.5.1 Introduction

Neuroleptic drugs, also known as anti-psychotics or major tranquillizers, are primarily used in the treatment of psychotic disorders, chiefly schizophrenia and paranoia. They are not disease specific and provide benefit for a range of syndromes; thus, they are used in mania, delirium, agitated depression and behavioral disturbances. They may be useful in severe anxiety and some have anti-depressive properties. Droperidol, which is short acting and has marked sedative properties, is used in anaesthesia. Benperidol has been used to control deviant anti-social sexual behaviour. As anti-emetics they are used to control vomiting and nausea, for example following cancer chemotherapy, and may be given in combination with opioid analgesics, which they potentiate. They are used in Huntington's chorea and Meniere's disease. The term "major tranquillizers" is misleading with regard to their use in schizophrenia for which their tranquillizing effect is of secondary importance. However, agents in this class are marketed as tranquillizers for veterinary use and, as they pose toxicological problems, are considered here.

Most neuroleptics belong to four chemical groups: the phenothiazines, thioxanthenes, butyrophenones und diphenylbutylpiperidines. In addition there are a number of miscellaneous compounds such as clozapine and sulpiride. Their anti-psychotic effects are thought to be due to antagonism of dopamine receptors, probably D_2. Blockade of dopamine receptors also explains their anti-emetic properties and the elevation of prolactin levels that frequently accompanies therapy. Chlorpromazine, introduced in the early 1950's, is the class prototype. In addition to dopamine, several other receptors are affected, giving rise to anti-muscarinic, anti-histaminic and anti-5HT effects. Blockade of alpha-receptors contributes to the postural hypotension that may be seen with some members of the class.

Therapeutically useful phenothiazines have 10-substituents, and several also have 2-substituents, giving rise to a large number of compounds. Neuroleptic phenothiazines may be subdivided into three groups on the basis of the 10-substituent; aminopropyl, piperidine and piperazine, but all have a common, $-C-C-C-N-$, structure which is required for neuroleptic activity. Compounds with only two carbons in the bridge are not neuroleptic but have marked anti-histaminic (promethazine) or anti-muscarinic (ethopropazine) properties, and although methods for their analyses may be similar, are discussed elsewhere. Those with aminopropyl side chains (chlorpromazine) are sedative, with some antihistaminic and anticholinergic activity while the piperidine-substituted compounds (thioridazine, pipothiazine) are sedative with marked anti-cholinergic effects. The piperazine derivatives (trifluoperazine, fluphenazine) are the most potent and have the least sedative properties of the class.

They have little anti-cholinergic activity but the incidence of parkinsonian symptoms is high.

Although there are fewer thioxanthenes in use, they have similar 2- and 10-substituents to those found in the phenothiazines and have similar pharmacologies, although they tend to be less potent than their phenothiazine counterparts. Because of the double bond, thioxanthenes exist as cis(Z)- and trans(E)-isomers (fig. 1). The cis-isomers are approximately twenty times more active as neuroleptics, and some preparations are available containing predominantly the active isomer, e.g. zuclopenthixol.

The butyrophenones and diphenylbutylpiperidines are chemically similar; several butyrophenones can be considered as substituted piperidines, although some, azaperone and fluanisone, are piperazine derivatives. Some substituents are found in both series so the structures may be more easily remembered by pairing, e.g. benperidol with pimozide; spiroperidol with fluspirilene. Haloperidol is less sedative than chlorpromazine and has less anti-muscarinic properties but extrapyramidal symptoms are more frequent. Sulpiride, pimozide and clozapine have been classed as atypical neuroleptics, since they are less prone to cause parkinsonian symptoms. Clozapine was withdrawn because of a high incidence of agranulocytosis but has been re-introduced as it may be of use when other neuroleptics fail. It should only be used when regular blood monitoring is possible and administration stopped at the first sign of leucopenia.

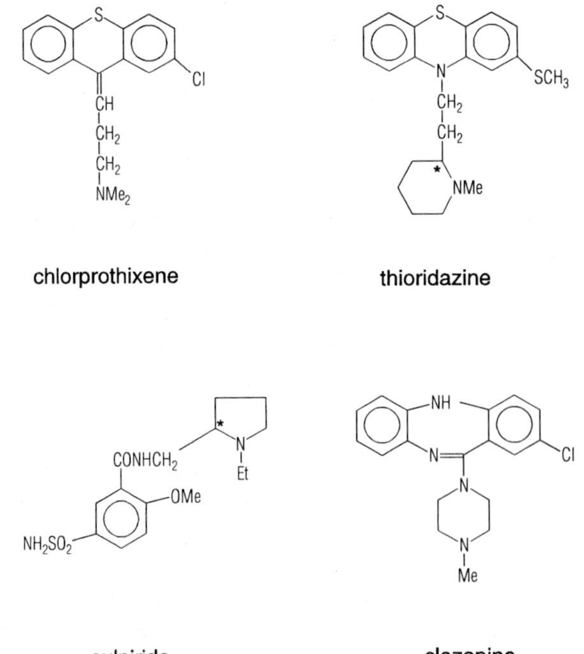

Figure 1. Chemical Structures of Some Neuroleptics.
Sulpiride and Clopazine have been referred to as "atypical neuroleptics". The asterisk indicates the position of a chiral carbon atom.

3.5.2 Pharmacokinetics

Owning to the lipophilic nature of this group of compounds, they are generally well absorbed from the gastrointestinal tract, but the systemic availability may be reduced by extensive first-pass metabolism. The oral bioavailability of chlorpromazine is approximately 30 % which should be taken into account if other routes of administration are proposed. First-pass metabolism of haloperidol is about 40 %.

The drugs are bound to plasma proteins (albumin and α_1-acid glycoprotein) such that the binding is > 90 % and, for some drugs, is in excess of 99 %. Thus CSF concentrations are considerably less than those in plasma. Extensive tissue binding is reflected in the high apparent volumes of distribution (> 20 L/kg) documented for a large proportion of this class of compounds, although there are some exceptions (e.g. sulpiride ca. 2 L/kg). Despite high hepatic clearances, the high apparent volumes of distributions lead to elimination half-times of 10–30 hours. The very lipophilic compounds such as pimozide and penfluridol have elimination half-times of around 100–200 hours, whereas sulpiride, which is relatively hydrophillic by comparison, has a half-time of approximately 10 hours, and is largely excreted in the urine unchanged. Not surprisingly, less than 1 % of a dose of the lipophilic compounds may be recoverable from urine, and the urine concentrations of unchanged drug may be less than the concentrations in plasma.

Intramuscular injections of long acting neuroleptics overcome some of the problems of compliance and erratic absorption/availability. Apart from fluspirilene, the drugs are administered as ester pro-drugs in an oily base. By using different esters the apparent elimination half-times can be controlled. For example the half-time of fluphenazine following injection of the enanthate is 3.5 days, whereas after the decanoate, the apparent elimination half-time is 7–10 days. These values can be compared with the half-time of 15–20 hours following oral or intramuscular administration of the non-esterified molecule. As esters were undetectable in plasma, it appears that it is the release from the injection site that is rate limiting (1). The drugs must contain an alcoholic group and be relatively potent so that the size of the dose is not prohibitive. Several fatty acids have been used and the range of preparations includes: fluphenazine, flupenthixol, haloperidol and zuclopenthixol as decanoate esters, pipothiazine palmitate and undeconoate, and fluphenazine enanthate. It is recommended that they are injected at 2–4 week intervals. Zuclopenthixol acetate is available for short term management.

Metabolism

Chlorpromazine metabolism has been extensively studied. The important reactions were elucidated by Fishman and Goldberg in the early 1960's, using paper chromatography and group specific spray reagents (fig. 2). Their findings have been confirmed by others using alternative techniques but these studies have added little to our understanding of chlorpromazine metabolism. Chlorpromazine sulphoxide was isolated and quantified as a plasma metabolite by Salzman and Brodie (2) in 1956. Demethylation gives the mono (nor_1) and didesmethyl (nor_2) metabolites. Chlorpromazine N-oxide was isolated and identified, partly, from its decomposition to parent drug and the allyl derivative via Cope elimination (3). Ring hydroxylation occurs at position 7 and, to a limited extent in man, position 3 (4). A large proportion of the phenolic metabolites are present as conjugates that can be hydrolysed by β-glucuronidase/aryl sulphatase. The presence of small quantities of 3-hydroxypromazine showed that de-halogenation occurred, a fact that has been confirmed using GC-MS (5). Deamination gives 2-chlorophenothiazine-10-propionic acid and, by complete loss of the side chain, 2-chlorophenothiazine. It has been calculated that by combining the known metabolic pathways there are over 150 possible chlorpromazine metabolites.

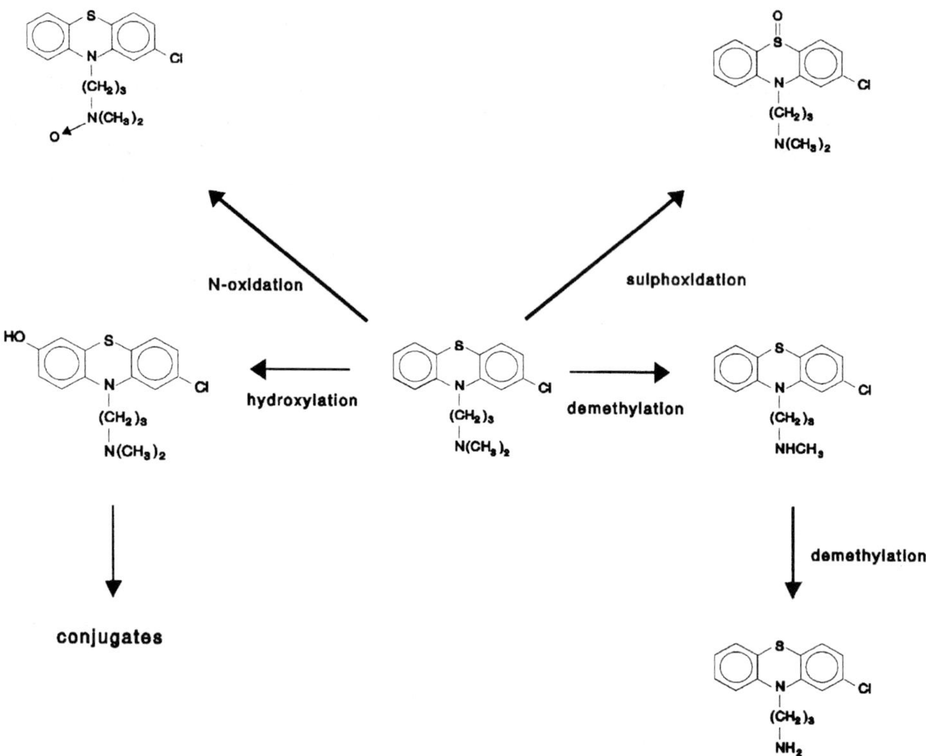

Figure 2. Major Metabolic Pathways of Chlorpromazine.

Other phenothiazines follow similar metabolic pathways, but, naturally, those with more complicated substituents have even more metabolites. The 2-thiomethyl group in thioridazine is oxidized to the sulphoxide (mesoridazine) and the sulphone (sulphoridazine) (6), both of which have neuroleptic activity. A large "phenolic" fraction was isolated but not investigated further (6). Papadopoulus et al. (7) identified 7- and 3-hydroxylated thioridazine metabolites in man and cyclic amides formed by oxidation of the piperidine ring (8). Whelpton et al. (unpublished) isolated similar metabolites from rat faecal extracts: 4 major spots on thin layer plates, giving navy-blue free radical colours. Two were amines and two appeared to be amides. They were tentatively identified as 7-hydroxythioridazine, 7-hydroxydesmethylthioridazine and the corresponding piperidones. The 2-O-methyl group in levomepromazine is demethylated. This phenothiazine is also different in that 3-hydroxylation is far more pronounced (9). The side chain carbonyl in propiomazine is reduced to the alcohol and subsequent loss of water gives the 2-propenyl metabolite (10).

The metabolism of drugs with a piperazine side chain is yet more complicated. The alkyl and hydroxyethyl groups are lost in compounds such as fluphenazine, perphenazine and trifluoperazine (11). Furthermore, piperazine side chains undergo ring opening to give ethylene diamine derivatives and total loss of the ring carbons to produce primary amine metabolites identical with those formed by didesmethylation of the dimethylamino-substituted parent compounds (12).

The known and postulated metabolic pathways of haloperidol are shown in figure 3 (13). Oxidative dealkylation gives 1-(4-fluorobenzoyl)propionic acid and its glycine conjugate and 4-fluorophenylacetic acid and its glycine conjugate, 4-fluorophenylaceturic acid (14), which are pharmacologically inactive. Haloperidol can be conjugated, probably with glucuronate or sulphate, without prior metabolism. Reduction gives the secondary alcohol, often referred to as "reduced haloperidol", RHAL (15). This compound may be important in that its has some pharmacological activity (about 25 % of the parent) and may be oxidized back to haloperidol. Similar metabolic pathways occur with other butyrophenones, e.g. bromperidol (16).

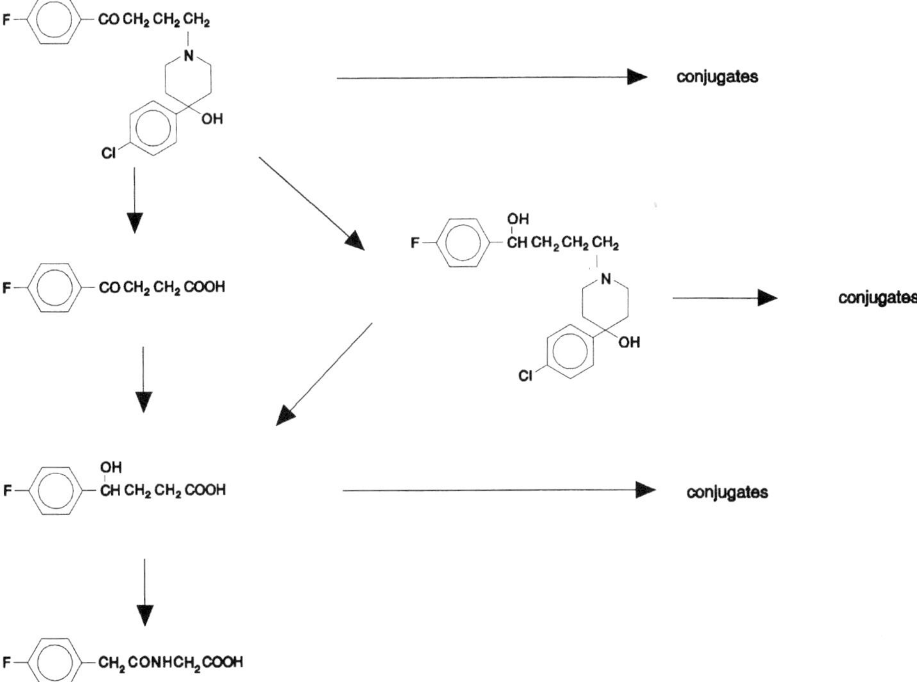

Figure 3. Major Metabolic Pathways of Haloperidol (after Miyazaki et al (13)).

Oxidation of sulpiride (fig. 1) at the 2-position in the pyrrolidine ring gives the pyrrolidinone (17). Pimozide may undergo several oxidative N-dealkylation reactions to give benzimidazolin-2-one, 1-(4-piperidyl)-benzimidazolin-2-one, 4-bis(4-

fluorophenyl)butyric acid, ß-oxidation of which gives 2-bis (4-fluorophenyl)acetic acid (18, 19). Metabolites of clozapine include the N-oxide and N-desmethylclozapine (20, 21). Demethylation of loxapine gives the anti-depressive, amoxapine.

Metabolism may lead to stereoisomeric products. For example, 5-sulphoxide metabolites of 2-substituted phenothiazines and thioxanthenes are chiral. This has been largely ignored although it is over 30 years since Salzman and Brodie (2) subjected chlorpromazine sulphoxide to polarimetry to see if it was optically active. 5-Sulphoxides of symmetrical phenothiazines are not chiral but it should be remembered that metabolism, e.g. ring hydroxylation, may produce asymmetry. When the molecule already contains an asymmetric centre this will lead to diastereoisomers that may separate on achiral phases. Thioridazine 5-sulphoxides are easily resolved on silica thin-layer plates and HPLC columns (fig. 4). As these diastereomers are usually present in approximately equal quantities, the isolation and quantification of only one will underestimate the true amount. The 2-sulphoxide (mesoridazine) and N-oxide metabolites of thioridazine are also chiral so that, if all three positions (2-, 5- and N-) are oxidized, the resulting tri-oxide exists as 16 stereoisomers (22).

Reduction of the carbonyl groups in propiomazine and haloperidol for example will introduce an asymmetric carbon although again the significance of this appears to have been ignored. The configuration of reduced haloperidol could influence its pharmacological activity and its reversion to parent compound.

3.5.3 Analysis

Reference compounds and internal standards

Pharmaceutical companies and colleagues may provide reference compounds if they are not available from commercial sources. Sometimes suitable reference compounds are not available or are required at short notice. Under these circumstances analysts may be able to make their own, purifying the small amounts that are needed by TLC, for example. Oxidation of phenothiazines and thioxanthenes with H_2O_2 will give the 5-sulphoxides, although the 2-thiomethyl group in thioridazine will also be oxidized, and there may be some N-oxidation. Nitrous acid is selective for 5-oxidation (23), and is to be recommended. Amine oxides are best prepared using 3-chloroperbenzoic acid (24). Aromatic hydroxylation can be accomplished using Fenton's Reagent, $FeCl_2$, as described Coccia and Westerfield (25). Treating N-oxides with ferrous salts may result in demethylation as well as reduction to the parent amine, although sometimes small amounts of the secondary amine may be obtained from the oxidation reaction directly (26). The products can be identified from their behaviour with reducing agents (sulphoxides are reduced by Zn/HCl or $TiCl_3$, N-oxides by metabisulphite) and group specific spray reagent such as nitroprusside/acetaldehyde (secondary amines) ninhydrin (primary amines) and silver nitrate (phenols). The free radical colours (below) give some indication of ring substitution while characteristic UV spectra confirm 5-sulphoxides.

Suitable analogues can be used as internal standards although these may not always be appropriate. Substituting a trifluoromethyl for chlorine (e.g. trifluoperazine for prochloperazine shifts the heptane-pH extraction curve by approximately

2 pH units. Drugs which may be present in the sample must be avoided but so too should compounds that might be present as metabolites, e.g. promazine is unsuitable as an internal standard for chlorpromazine as it is a known metabolite. It may be better to use a compound which cannot be present, e.g. a homologue prepared in the laboratory. The reason for inclusion of an internal standard should be considered. For example, thioridazine was used as internal standard in the GC-NPD assay of fluphenazine to correct for changes in detector response which occurred on repeat injections of the trisilyl derivative (27). The peak-height response ratio was shown to be constant with time and when the rubidium bead temperature or hydrogen flow rate were varied. It may not be necessary to use an internal standard or the improvement in precision may be marginal, particularly when solid phase extraction is used.

Problems of analysis

Adsorption

Being lipophilic bases, the compounds under discussion may be lost by adsorption onto glass or certain plastics. It is not possible to accurately predict which compounds will be affected, but the more hydrophobic the drug the more likely it will escape the aqueous environment. Trifluoperazine hydrochloride was rapidly lost from water solutions, not only onto glass, but also into plastics such as PTFE (28). Silanizing the glassware reduced the adsorption (table 1) but increased the formation of emulsions with plasma, so that smaller portions had to be taken and there was no overall gain in recovery. The loss from plasma was not a great problem, presumably due to plasma protein binding competing for glass binding sites, or deactivation of the glass by endogenous materials in the plasma. Sometimes a second drug may be introduced to compete for binding sites, this is particularly useful when a drug with a low detector response can be found, e.g. the addition of promazine in the GC-ECD analysis of chlorpromazine (29). Organic bases such as triethylamine or ethanolamine have been used to mask the binding sites.

Table 1. Loss of Trifluoperazine from Aqueous Solution to Different Materials

	Percent [^{3}H]-trifluoperazine remaining at:				
Material	0.1 h	0.25 h	0.50 h	1.0 h	18 h
Soda glass	72.7	67.8	52.5	41.6	15.5
Pyrex glass	78.2	70.1	60.3	52.9	24.0
Silanized Pyrex	87.4	84.7	76.4	75.3	73.6
PTFE	84.9	67.9	58.3	47.3	24.8
Polycarbonate	84.1	72.3	63.9	54.9	33.9
Polypropylene	80.0	71.4	58.2	45.9	16.1
Polystyrene	79.9	67.8	55.8	44.4	25.4

Adsorbed phenothiazines may survive routine laboratory glass washing procedures and contaminate subsequent analyses (30). In the past we have had to resort to soaking glassware in "chromic acid" but now prefer to use disposable glassware.

Metabolite reversion

Curry (31) noted that the precision of replicate chlorpromazine analyses of experimental samples was worse than that obtained from standard solutions, and reanalysis of stored "real" samples would sometimes give higher values than the original assay whereas this was not the case with stored standards. Amongst the possible explanations was metabolite reversion to parent compound.

The most probable candidates were the sulphoxide and N-oxide metabolites. Solutions of chlorpromazine, chlorpromazine sulphoxide and chlorpromazine-N-oxide in human plasma were shown to be stable for up to 7 days at 4 °C and up to 9 weeks (the duration of the experiment) at − 18 °C and during a series of freeze-thaw cycles (32). Reduction of the N-oxide to chlorpromazine at week 2 was shown to be due to the labile nature of amine oxides in alkaline solutions. Addition of metabisulphite or ascorbic acid reduced the N-oxide, but not the sulphoxide, to parent drug. In a subsequent study the N-oxide was only reduced in alkaline plasma but not in protein free buffer solutions (33). As there is no reason to suppose that other amine oxides do not undergo similar reduction, the use of reducing agents as stabilisers, eg. addition of ascorbate to urine samples as described by Leonnechen (34), is questionable. Metabolites may be unstable under chromatographic conditions or during derivatization. N-oxides are reduced to parent compound on GLC columns, but, as there are likely to be Cope elimination products, these should alert the analyst to a problem. Promethazine sulphoxide has been reported to undergo reduction on GC columns (35). Phenothiazine sulphoxides may be reduced by derivatizing agents such as trifluoroacetic anhydride (36).

Solvents

Drugs or their metabolites can react with solvents or impurities/additives in solvents. Primary amines will form imines with aldehydes and ketones. In the original GC-ECD assay (37), didesmethylchlorpromazine quantitatively reacted with iso-valeraldehyde, an impurity in iso-amyl alcohol (IAA), the imine from which was completely resolved from monodesmethylchlorpromazine on OV−17 GC columns. By selecting the appropriate aldehyde, a product with a suitable retention time could be selected (38). The suggestion that a small quantity of the aldehyde be added to the extraction solvent is a sensible one (39). A similar interaction has been observed by Beckett, et al. (40), when acetaldehyde, an impurity in the diethyl ether extraction solvent, reacted with primary hydroxylamine metabolites to form an "acetaldehyde nitrone".

Solvents may contain additives as stabilizers, and different manufacturers use different ones. For example, chloroform frequently contains ethanol, and dichloromethane may contain ethanol or pentene as stabilizer. Diethyl ether has to be stabilized with an anti-oxidant to prevent peroxide formation; pyrogallol, hydroquinol and butylated hydroxytoluene (BHT) may be used and these will interfere with electrochemical detection unless they are removed. The ether must freshly distilled before use, as storage of distilled ether is not only dangerous but the accumulated peroxides may oxidize the analyte. Methyl tert-butyl ether (MTBE, b.pt. 51 °C), which is supplied without stabilizer, is recommended as an alternative.

Plasticisers
Often a spurious peak or sets of spurious peaks on a chromatogram may be put down to plasticisers when the sample has been in contact with plastic material at some stage, but only rarely is this documented and even more rarely is the identity of the plasticiser confirmed. The classic example of a plasticiser interaction is that described by Borga et al. (41), when plasma imipramine concentrations were reduced due to displacement from plasma proteins with a subsequent increase in red cell: plasma ratio. Although not a neuroleptic, the observation highlights a potential problem. We have observed peaks during GC-ECD determinations of chlorpromazine that arose from the wadding of certain screw capped extraction tubes, and more recently, peaks in HPLC-ED chromatograms that were associated with the use of Eppendorff Combi-tips (unpublished).

Sample preparation
Details of sample preparation prior to analysis can be found in the references cited below and some indication of the methods adopted is given in the following sections, however a few generals points can be made.

Liquid-liquid extraction
Most of the drugs are readily extracted from alkaline plasma into relatively non-polar solvents, and inspection of the literature shows that n-heptane, n-hexane and n-pentane are frequently used, usually with a small proportion of iso-amyl (iso-pentyl) or iso-propyl alcohol (IPA) to reduce adsorptive losses. The benzamides, being more polar, require a more polar extraction solvent and chloroform has been used. More polar solvents may be required if metabolites are to be quantified, and diethyl ether, dichloromethane and ethyl acetate have been used.

For gas-chromatographic analyses, particularly with electron-capture detection (ECD), an acid backwash is needed, followed by re-extraction into organic solvent after pH adjustment. Where the partition characteristics allow, extraction into a small volume of organic phase to avoid the need for evaporation is good practice. Samples that are to be assayed by GC with nitrogen detection (NPC) or by HPLC may not need the acid backwash. It is also worth remembering that if the acid phase can be injected onto an HPLC column, samples can be concentrated by extraction to a small volume of acid.

Solid-phase extraction
Solid-phase extraction (SPE) methods have been described, including one developed for chlorpromazine and 13 of its metabolites (42). The C-8 columns (100 mg, 1 mL) were prepared by sequentially washing with reservoir volumes of 0.2 M phosphoric acid, 0.25% sodium carbonate, methanol and, finally, water. Plasma (1 mL) was acidified with H_3PO_4 (3 M, 0.05 mL) and applied to the columns, which were sequentially washed with water, methanol (0.5 mL) and Na_2CO_3 (25%). The analytes were eluted with methanol (2×150 µL). The mechanism appears to be mild ion exchange, using residual silanol groups on the SPE phase, which allows the columns to be washed with methanol before pH adjustment prior to elution. The use of acidified plasma may be useful for situations in which the use of alkali is best avoided. It is likely that more SPE methods will be developed particularly as the use of solvents becomes restricted for reasons of safety.

Enzyme hydrolysis

As the drugs are extensively bound to plasma and tissue proteins, protein precipitation is not recommended. Digestion with proteolytic enzyme has been described for chlorpromazine and its metabolites in post-mortem tissues (43) and similarly for thioridazine (44). Hydrolysis of glucuronide and sulphate conjugates with β-glucuronidase/aryl sulphatase is preferable to acid hydrolysis, particularly as phenolic metabolites are more readily oxidized in acid.

Colour reactions

Colour reactions that are applicable to this group of compounds are well documented in Clarke (45). Colour is a useful way of detecting ring substitution. Phenothiazines are readily oxidized to free radicals, the colours of which are characteristic of the ring substituents. Although the descriptive colours may depend on the whim of the author, stabilization of the radicals in 50% (v/v) H_2SO_4 allows determination of the unequivocal λ_{max} values (table 2). The sulphoxides may need warming before the colour is apparent. Generally electron withdrawing groups shift the colours from green-blue (650–600 nm) to red (530 nm) and salmon-orange (500 nm). 3-Hydroxylation increases the wavelength more than 7-hydroxylation, the two combining to give an even larger shift than each alone. Methyl esters, e.g. 7-methoxychlorpromazine, give the same colours as the phenols whereas the O-acetyl derivatives give the free radical colour of the parent drug (Whelpton, unpublished). Compounds with 2-electron withdrawing groups tend to be more stable, e.g. to photooxidation, than those without or with electron donating groups. The phenolic metabolites are less stable than the parent drugs.

Table 2. Free Radical Colours in 50% H_2SO_4

Compound	2-subs	Colour	(nm)
promazine	H	–	514
2-methylpromazine	CH_3	–	528
acepromazine	$COCH_3$	–	514
propiomazine	COC_2H_5	–	520
methotrimeprazine	OCH_3	–	568
thioridazine	SCH_3	turquoise	640
sulphoridazine	$SOCH_3$	salmon	510
mesoridazine	SO_2CH_3	pink	525
chlorpromazine	Cl	pink	532
7-OHCPZ	Cl	lavender	562
3-OHCPZ	Cl	dark blue	580
6-OHCPZ	Cl	red-pink	522
8-OHCPZ	Cl	blue	587
9-OHCPZ	Cl	pink	529
3,7-diOHCPZ	Cl	aqua	620
fluphenazine	CF_3	salmon	501
triflupromazine	CF_3	–	501
7-OHFPZ	CF_3	pink	559

Thin-layer chromatography

The use of paper chromatography in the early 1960's to elucidate chlorpromazine metabolism is cited above, but now thin-layer methods are used. Papers detailing metabolic studies are a useful source of thin-layer eluents which may be applicable to a particular problem. Zingales (46) detailed the chromatographic properties of 45 psychotropic drugs in 5 solvent systems. Unmodified phase silica is usually used, but Hushoff and Perrin (47) have applied a reversed-phase technique.

Generally, thin-layer chromatography is used for qualitative or semi-quantitative screening, however quantitative methods have been published. Kaul and his group determined 11 chlorpromazine metabolites after dansylation and chromatography on silica gel. The fluorescent spots were removed and eluted with dioxane prior to quantification (48). The problem of quantifying the tertiary amine parent compound was overcome by quaternization with 9-bromomethylacridine. The product was separated by TLC and the plate exposed to UV-light to yield fluorescent products (49). Compounds may be quantified in situ using densitometric scanning, e. g. as applied to trifluoperazine (50) and perazine (51). Alternatively, the plate or its contents may be modified prior to scanning. Davis and Fenimore (52) quantified fluphenazine with a sensitivity of 0.1 μg/L, although the calibration was run between 0.5–5 μg/L. The coefficient of variation at 0.5 μg/L was 8 %. Plasma (4 mL) with trifluoperazine added as internal standard was extracted with heptane containing 0.5% v/v IAA. After back extraction into HCl, the pH was adjusted with sodium carbonate and the drug extracted with pentane. Two-dimensional chromatography was performed on 10 × 10 cm HPTLC plates. The first development in toluene-acetone (60 + 40) was to remove interfering substances. Fluphenazine (R_f 0.2) and internal standard (R_f 0.5) were separated with toluene-acetone-NH_4OH (60 + 40 + 2). After the plates had been exposed to UV light, the compounds were quantified by fluorescence. In 1984, a similar approach was applied to the determination of chlorpromazine and thioridazine (53). The plates were exposed to nitrous acid vapour and the spots quantified by scanning reflective densitometry at 365 nm (CPZ) or 375 nm (TDZ). The sensitivity of the method was 10 μg/L. Thiothixene was quantified on HPTLC plates after oxidation with ozone, giving a limit of detection of ca. 0.1 μg/L (54). This paper contains a useful table of the R_f-values of 54 drugs and metabolites. Plates with fluorescent indicator have been used to quantify azaperone, azaperol propiomazine and carazolol in animal tissues. Again two dimensional development was used. Detection was at 254 nm or 366 nm in the case of propiomazine, which could be detected down to 25 ng/g of tissue (20 g sample) (55). Sulpiride and other benzamides have been quantified by reflective densitometry at 293 nm (56). Silica plates, without indicator were developed in acetone-butan–1-ol-water-ammonia (66 + 30 + 3 + 1). The limit of detection was 2 mg/L.

Gas Chromatography

Flame ionisation detection

Neuroleptics tend to be relatively potent, lipophilic compounds. Thus, the combination of low dose and high apparent volumes of distribution leads to low plasma concentrations during routine therapy. Therefore, flame ionisation is only useful for plasma determinations of a limited number of drugs, metabolite studies on urine, or, possibly, in cases of acute overdose. Driscoll and his colleagues (57) demonstrated the use of GC-FID for the analysis of chlorpromazine and some of its metabolites in urine in 1964. The column was 5% SE-30 and the desmethyl metabolites were resolved as their acetamides. The CPZNO and CPZNOSO were selectively extracted and assayed by UV (see problems of NO, above) and the sulphoxides were reduced to the corresponding sulphides prior to assay. Kelsey and Moscatelli (58) used OV-3 and OV-17 columns for their study of phenolic and non-phenolic urinary metabolites of chlorpromazine. They derivatized with trimethylsilylimidazole but unfortunately extracted and injected CPZNO with no consideration of the fact that is would undergo Cope elimination. The peak labelled CPZNO is probably the decomposition product, 10-allyl-2-chlorophenathiazine.

Plasma concentrations of thioridazine may be high enough for detection by FID. Curry and Mould (59) were the first to describe the use of 3% OV-17 columns for thioridazine. However, when desmethylthioridazine became available it was apparent that many gas-chromatographic stationary phases, including various loadings of OV-17, OV-1, OV-210 & OV-225, would not resolve this metabolite from the parent compound (Watkins, Curry and Whelpton, unpublished). This problem was ignored by Dinovo et al. (60), but confirmed by Kilts et al. (61), who failed to find a suitable GC stationary phase. Packed column GC is not to be recommended as a method for TDZ, unless steps are taken to separate the desmethyl metabolites or derivatize them prior to chromatography (36).

Electron capture detection

With the introduction of the ^{63}Ni electron capture detector and improved silicone stationary phases, e.g. OV-17, it was possible to monitor chlorpromazine plasma concentrations during routine therapy. The first plasma assay for chlorpromazine and its desmethyl and sulphoxide metabolites was described by Curry (37) in 1968. The compounds were extracted from alkaline plasma using 1.5% IAA in heptane, back extracted into acid and, after pH adjustment, extracted into a small volume of 15% IAA in toluene. The method did not use an internal standard but the coefficient of variation at 210 μg/L was approximately 6%. The limit of detection was 10 μg/L for a 1 mL sample. Several similar methods followed, but most were only minor modifications of the original. Davis el al (29) made several modifications, including the use of OV-225 as suggested previously (62), promazine as a "carrier" and nickel columns, which reduced the limit of detection to 1 μg/L for a 1 mL sample.

Because of its 2-chloro-substituent perphenazine is also electron capturing, Hansen and Larsen (63, 64) have quantified blood and plasma samples using GC-ECD. Five millilitres of blood were required to give a limit of detection of 0.2 μg/L. The compound was derivatized with acetic anhydride prior to chromatography.

In 1971, Marcucci et al. (65), published a GC-ECD method for haloperidol and trifluperidol using the native electron-capturing properties of these compounds, but, with a sensitivity of 1–2 ng injected, was of limited use. Recently, Tyndale and Inaba (66) described the use of microbore capillary columns (DB-17, 15 m × 0.53 mm) or packed columns (3% OV-17) for GC-ECD determination of haloperidol and its reduced metabolite. Flurazepam was the internal standard. For the packed column, the compounds were acylated with TFAA, which presumably imparted electron capturing properties as well as improving the chromatography.

Derivatization of fluphenazine and perphenazine with HFBA has been described (67) but never appears to have been applied; this group subsequently used the native electron capturing properties to assay perphenazine (63, 64). Our attempts at derivatizing fluphenazine for electron capture detection were remarkably unsuccessful, TFA and HFBA derivatives were never isolated, but the pentafluorobenzoyl ester (from the appropriate imidazole) was isolated on thin layer plates. However, the product was thermally labile, as determined by the fragmentation of radiolabelled material, and decomposed on GC-columns under all the conditions tried (Curry and Whelpton, unpublished). Derivatization with PFPI has been successfully applied to pipothiazine. Using prochlorperazine as internal standard, Cooper and Lapierre (68) quantified the long acting drug down to 10 µg/L, although the standard deviation was probably large: at 50 µg/L the CV was 9.4%. Cooper and Kelly (69) determined loxapine, amoxapine and their metabolites in serum and urine, after derivatization with TFAA and N-trimethylsilyldiethylamine using ECD and FID.

Nitrogen-phosphorus detection

The applicability of gas chromatography was extended with the availability of the nitrogen selective detector (NPD) which, with its greater linear range and stability, could be considered to be an improvement over the ECD. We found that the NPD sensitivity to chlorpromazine (considered typical of the electron capturing phenothiazines) was very similar to that achieved with the ECD, and consequently switched CPZ assays to NPD without modification and extended it to include (non-electron capturing) promazine (70). Bailey and Guba (71) used NPD for chlorpromazine, but derivatized the desmethyl metabolites with TFAA, which necessitated a second extract and run as the process reduced the CPZSO to parent compound. Ninci et al. (72) compared the use of liquid-liquid and liquid-solid phase extractions for chlorpromazine and the desmethyl metabolites prior to GC-NPD. The primary and secondary amines were derivatized with TFAA, but the problem of sulphoxide reduction was not addressed. Some thermal decomposition in the injection port was noted.

Fluphenazine, 7-hydroxyfluphenazine and fluphenazine sulphoxide have been detected in urine using a nitrogen detector based on a heated rubidium silicate bead (27). The compounds were chromatographed as their TMS ethers rather than the acetate esters, as the acetyl derivatives of the phenolic and sulphoxide metabolites were not completely resolved. Use of TMS derivatives resulted in a gradual loss of detector sensitivity which was accounted for by the use of thioridazine as internal standard. Periodically increasing the bead temperature for 2 minutes restored the sensitivity. In 1978 Franklin et al. (73) published a plasma fluphenazine assay based on silylation and nitrogen detection using the same type of detector as Whelpton and Curry. The sensitivity was 2 µg/L. The same year, Dekirmenjian et al. (74) published their method for several neuroleptics, including acetophenazine, fluphenazine, haloperidol, perhenazine and piperacetazine. This group chose acetic anhydride as the derivatizing agent, partly, one assumes, because their nitrogen detector was not suitable for TMS derivatives. Javaid et al. (75) applied the method to FPZ in plasma from 20 patients, and although a sensitivity of 0.2 µg/L (intra-assay CV = 20%) was achieved using 5 millilitres of plasma, they only measured FPZ concentrations for up to 4 days following intramuscular injection of fluphenazine decanoate.

Nitrogen detection for haloperidol was developing in parallel. Bianchetti and Morselli (76) extracted haloperidol with diethyl ether, using azaperone as internal standard, and chromatographed it without derivatization. The sensitivity was 1 μg/L for a 2 mL sample with a precision of 6% at 2.5 μg/L. Using 5 mL of plasma and chlorohaloperidol as internal standard, Franklin achieved a sensitivity of "0.5–1 μg/L" but the within-batch precision at 5 μg/L was 6.9% and between-batch 11.9% at 12.32 μg/L. The method of Abernethy et al. (77) was little different from that of Franklin, using similar extraction methods and having similar detection limits and precision, although they were able to show that reduced haloperidol, which by then had been identified as a metabolite, did not interfere. The method was applied to pharmacokinetic studies in dogs.

Flupenthixol has been assayed by GC-NPD after acetylation (78). The method did not differentiate between the cis- and trans-isomers, so the authors divided the results by two on the assumption that the cis/trans ratio remained one. The sensitivity was 0.5 μg/L, with coefficients of variation of 9.3% (1 μg/L, intra-assay) and 9.4% (2 μg/L, inter-assay). Clozapine (79) as been determined with a sensitivity of 1–2 μg/L. A single diethyl ether extraction was performed on 1 mL of serum using acetylated maprotiline as internal standard. A DB-5 wide bore capillary column was used with a helium carrier gas flow rate of 6 mL/min. The structurally similar loxapine has been determined by GC-NPD (80); the limit of detection being 2 μg/L.

Mass spectrometric detection

GC coupled to mass spectrometry (GC-MS) may be used to identify an unknown species from its fragmentation pattern and/or to quantify concentrations of drugs. The use of selective ion monitoring allows low levels of analyte to be detected, not because the method is particulary sensitive, but since the high degree of selectivity allows specific detection in the presence of other material. The method may be used when others are not applicable or as a "yard-stick" by which alternative assays may be judged.

Hammar et al. (81) used mass fragmentography, as they called it, to detect chlorpromazine and some of its metabolites in plasma. In 1976, Alfredsson (82) used GC-MS to determine chlorpromazine, nor$_1$chlorpromazine and 7-hydroxychlorpromazine in plasma and CSF. Using [^{2}H]-labelled materials as internal standards they were able to quantify as little as 1 μg/L and show that the amount of chlorpromazine in CSF was 3% that in plasma, which suggests that the "free fraction" in plasma is in equilibrium with CSF. A similar approach for the determination of plasma trifluoperazine was adopted by Whelpton et al. (83), who used [^{2}H-methyl]-trifluorperazine as internal standard; the limit of detection was 0.2 μg/L. Midha's group has prepared internal standards based on [^{2}H]-labelled piperazine for CG-MS of trifluoperazine (84) and fluphenazine (85). For chlorpromazine they used prochlorperazine as internal standard (86).

Two GC-MS methods for haloperidol were published in 1979. In one the EI and CI (isobutane) spectra of haloperidol and its chlorinated analogue were compared (87). Chromatography was performed with either SE-30 or OV-17 wall-coated open-tubular columns. The worse than expected sensitivity of the method was attributed to poor peak shapes and adsorptive losses. The second method also used chemical ionization but this time methane or methane/ammonia were used as reagent gas and trifluperidol was used as internal standard (88). An "improved" assay (89) using d_4-haloperidol as internal standard and thioridazine as a "chaser" to reduce adsorpiton to the SP-2100 packed column, was presented in 1985. The limit of detection was not given, but calibration samples were prepared between 1 and 20 µg/L, and the inter-assay coefficient of variation at 2 µg/L was 10.1 %. Using fused silica bonded phase capillary columns, Häring et al. (90) developed a GC-MS method capable of assaying 0.2 ng in a 2 mL sample. To achieve this sensitivity, haloperidol and the d_4 internal standard were trifluoroacetylated and the mass spectrometer run in the negative ion chemical-ionization mode using ammonia as reagent gas.

Clozapine and its desmethyl metabolite have been determined in patient plasma using a fused silica capillary column and electron impact selected ion monitoring (20). The N-propyl homologue was used as internal standard. The sensitivities were 1 and 5 µg/L for parent drug and metabolite, respectively.

Liquid chromatography

Conditions

Early methods for neuroleptics were generallly based on straight phase or ion-pair methods. In 1974, Muusze and Huber (91) described a method for thioridazine and several metabolites based on a rather complex mixture of trimethylpentane, 2-aminopropane, acetonitrile and ethanol. Knox and Jurand (92) described an ion-pair technique for several phenothiazines and related compounds. Perchlorate buffer, supported on silica, has been used to chromatograph fluphenazine, using dichloroethane-butanol as mobile phase (93). Since then (table 3) most methods have used modified silica, although unmodified silica is used by some groups. Smith (94) used C-2 silica for determination of chlorpromazine and some of its related compounds in tablets, but it was Curry (95–98) and Midha (99) who popularized nitrile columns for these compounds. Their eluents contained a high volume of organic modifier, typically 90 % v/v methanol or acetonitrile with ammonium acetate or acetate buffer. Curry's group demonstrated the chromatography of over 20 drugs and some of their metabolites, the list included: acetophenazine, butaperazine, chlorpromazine, fluphenazine, haloperidol, mesoridazine, promazine, thioridazine, thiothixene, trifluoperazine, triflupromazine, and trimeprazine. The applicability of the methods to biological samples was exemplified. Other workers have chosen nitrile phases, but C-18 is also popular. To prevent tailing, organic bases, e.g. triethylamine and dibutylamine, may be added and/or "base-deactivated" columns used. Some workers have chosen polymer columns (table 3).

An alternative way of improving peak shape is the use of alkaline eluents, for example Heyes and Salmon (100) used methanol-acetonitrile-ammonium carbonate (1 + 1 + 1) with a modified silica column to separate fluphenazine and its esters, but later used methanol-ammonia (9 + 1) to separate fluphenazine and its N- and S-oxides on an unmodified silica column (101). Silica columns can provide good peak shapes and excellent separations when the correct mobile phases are used. An advantage is that TLC methods translate well. Flanagan and Jane (102) have described a "non-aqueous" solvent system based on methanol with ammonium perchlorate that they have applied to numerous basic drugs. Alternatively, 10 % v/v buffer (ammonium nitrate or acetate) in organic modifier can be used. As the

Table 3. Summary of HPLC Methods for Neuroleptic Drugs

Compound	Column	Detector	Extraction	Sensitivity	Ref
Benperidol	C-8	ED PG 0.2/0.65 V	SPE-C8	0.2 μg/L	108
Benperidol	C-18	UV-254	hex/IPA	0.5 μg/L (4 mL)	109
Bromperidol	TSK gel	UV-245	hept + ColSwi	0.3 μg/L	110
Chlorpromazine, metab	C-2	UV-254	NA (tabs)	–	94
Chlorpromazine	CN	UV-254	hept/IPA	1 μg/L	111
Chlorpromazine, metab	SiO_2	UV-254	subtilisin	2 μg/L	43
Chlorpromazine	CN	ED CG 0.9 V	pent/IPA	0.1 μg/L	99
Chlorprothixane, sulphoxide	CN	UV-229 ED GC	hept/IAA	5 μg/L	112
Chlorpro, TDZ, metab	C-18	ED PG 0.54/-0.15 V	THF	0.1 ng/mg	113
Clopenthixol, desalkyl-	SiO_2	UV-254	hex/$IPrNH_2$	0.5 and 2.5 μg/L	114
Clozapine	C-18	UV-254	hex	5 μg/L	115
Clozapine, desmethyl-N-oxide	C-18	UV-254	ether	5 μg/L	21
Clozapine, desmethyl-	C-18	UV-230	hex/IAA	15 and 30 μg/L	116
Clozapine	C-8	ED GC 0.7 V	hex	20 μg/L	117
Droperidol	C-18	UV-226/254	SPE-C18	?	118
Droperidol	C-18	UV-248	BuCl	4 μg/L (2 mL)	119
Fluphazine	perchlorate	UV-257	ether	? < 1 μg/L	93
Fluophazine, metab	C-18	RIA	hept/IAA	?	120
Fluphenazine	C-18	ED GC 0.85 V	IAA	300 fmol	121
Fluphenazine	CN	ED PG 0.5/0.75 V	EtAc/pent	10 ng/L	122
Haloperidol	C-18	UV-250	$CHCl_3$	2 μg/L (2 mL)	123
Haloperidol, reduced	CN	ED GC 0.9 V	hex/IAA	0.5 μg/L (4 mL)	124
Haloperidol	SiO_2	UV-244	hept/IAA	2 μg/L	125
Haloperidol	C-8	UV-254	hept/IAA	1 μg/L	126
Haloperidol, reduced	C-18	UV-196	hex/IAA	2 μg/L?	127
Haloperidol, reduced	CN	ED GC 0.5/0.9 V	SPE-CN	20 pg	128
Haloperidol, reduced	CN	ED PG 0.7/0.85 V	pent/IPA	50 ng/L	129
Haloperidol, reduced	C-18	UV-220	SPE	1 and 2.5 μg/L	130
Haloperidol, reduced	C-8	ED PG 0.4/0.9 V	pent/IPA	50 ng/L	131
Haloperidol	C-18	UV-248	ether/hept	0.5 nmol/L	132
Haloperidol, reduced	C-18	UV-220/246	hex/IAA	0.25 and 0.1 μg/L	104, 133
Levomepromazine, metab	C-18	UV-254	–	qual	134
Levomepromazine, CPZ	C-18	ED GC 0.85 V	hept/IAA	2 μg/L (plasma) 0.5 (Urine)	135
Levomepromazine, metab	C-18	UV-254	SPE C-2	15 nM	34
Pimozide	TSK-C18	Fl 210/320	hex/IAA	0.3 μg/L	136
Pipothiazine	SiO_2	Fl 263/470	hex/DCM	??	137
Prochlorperazine	CN	ED GC 0.85 V	ether/$CHCl_3$	0.2 μg/L (5 mL)	138
Prochlorperazine	CN	ED GC 0.85 V	$CHCl_3$	1 μg/L (2 mL)	139
Promazine	SiO_2	UV-240	–	–	140
Spiroperidol	C-18	UV-254	hept/IAA	1.5 ng	141
Sulpiride	C-18	Fl 299/342	$CHCl_3$	10 μg/L	142
Sulpiride, sultopiride	?	UV-226	$CHCl_3$	10 and 15 μg/L	143
Sulpiride	C-18	Fl 300/360	$CHCl_3$/EtAc	10 μg/L	144
Trimeprazine	CN	ED GC 0.9 V	pent/IPA	0.25 μg/L	145
Thioridazine, metab	SiO_2	UV-254	ether/hex	10–20 μg/L	146
Thioridazine, metab	CN	ED GC 0.9 V	ether/hex	0.1 ng	147
Thioridazine, metab	SiO_2	UV-254	hex/IAA	–	22
Thioridazine, metab	C-18	UV-254	SPE-C18	0.5 μmol/L?	148
Thiothixene	CN	UV-229	hex/IAA	0.5 μg/L	149
Thiothixene	SiO_2	UV-230	hex/ether	50 ng/L	150
Thiothixene	CN	ED PG 0.3/0.8 V	pent/IPA	0.2 μg/L	151

ED – electrochemical detection PG – porous graphite GC – glassy carbon analytes are lipophilic bases, relative retention can be affected by manipulation of pH. The substitution of acetonitrile for all or some of the methanol can improve peak shape and resolution, and has the further advantage of reducing back pressure. Lipophilic impurities elute early, and generally the more polar metabolites elute after the parent compound. A typical separation is shown in figure 4. Rather than inorganic bases, organic ones, e. g. ethanolamine (43, 103), may be used with native silica.

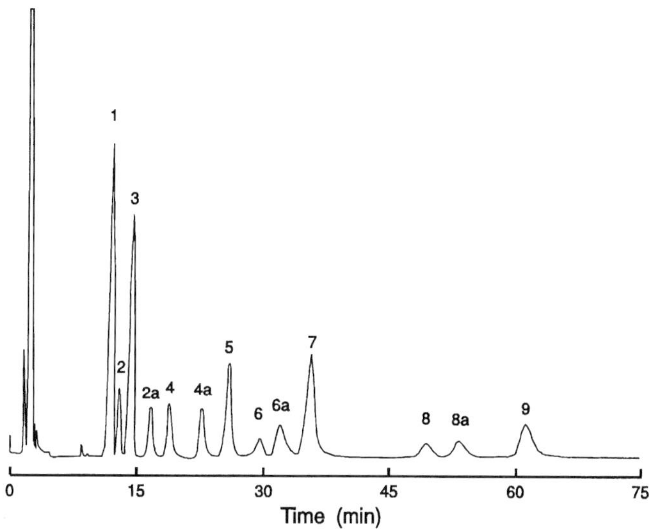

Figure 4. HPLC Separation of Thioridazine (3) and Some of Its Metabolites on a Silica Column (after Watkins et al[22]).
Thioridazine-2-sulphone (1), thioridazine-2-sulphone-5-suphoxide (2,2a), thioridazine-5-sulphoxide (4,4a), thioridazine-2-sulphoxide (5), thioridazine-2,5-disulphoxide (6,6a), desmethylthioridazine (7), desmethylthioridazine-5-sulphoxide (8,8a) and desmethylthioridazine-2-suphoxide (9). UV detection at 254 nm.

Detection

Phenothiazines have ultraviolet absorption maxima with extinction coefficients of approximately 30,000 M^{-1} cm^{-1} in the region 245–260 nm, which makes them suitable for UV detection. With optimal chromatography one would predict detection limits of a few µg/L. Similarly, haloperidol has a λ_{max} at 245 nm, $\varepsilon \approx 13{,}000$ M^{-1} cm^{-1}. The λ_{max} of reduced haloperidol is at 200 nm, with little absorbance at 250 nm, so the lower wavelength may be used if both compounds are to be assayed, or the higher one if only haloperidol is required (table 3). Vatassery et al. (104) used both wavelengths.

Phenothiazines may be oxidized to give fluorescent products (see TLC above), and this is particularly true of thioridazine which has been oxidized post-column with $KMnO_4$ and the excess permanganate removed by reaction with H_2O_2 in a second reaction coil. The products were measured by fluorimetry (cf. Ragland and Kinross-Wright (105)). The native fluorescence of some compounds allows quantification without derivatization or chemical modification; this is true of sulpiride and pimozide.

Many members of the group are amenable to electrochemical detection. A useful source on electrochemical behaviour is the paper by Jane et al. (106), which lists the relative ultraviolet absorbtion at 254 nm and electrochemical response at a glassy carbon electrode, operated at + 1.2 V relative to a silver/silver chloride reference electrode, of some 462 basic compounds. Most of the neuroleptics fall into their B or C categories, indicating that the ED : UV ratio is between 10 and 50. Oxidation of the heterocyclic sulphur in phenothiazines and thioxanthenes probably makes the major contribution to their electroactivity. This explains why the thioridazine N-demethylated metabolites and N-oxides show similar electrochemical responses to those of the parent compounds, whereas the 5-sulphoxides give little or no response in the oxidation mode. Sulphoxides may be detected using a two electrode system so that they are reduced at the first electrode (107) and detected along with the other compounds at the second electrode.

Radioimmunoassays and radioreceptor assays

The small sample sizes required by radioimmunoassay and radioreceptor assays and large sample throughput make these techniques appealing to forensic scientists and clinical chemists, both of whom may have large numbers of samples but often limited sample volumes. The limited specificity of the assays may be exploited when screening unknown samples to determine the class of drug present, and with suitable controls, antiserea raised to one drug may be used to quantify related compounds.

A radioimmunoassay for chlorpromazine, capable of quantifying 10 pg in 10 µL, was published in 1975 (152). The specificity of the antisera when tested against known chlorpromazine metabolites was good. The ID_{50} for 7-hydroxychlorpromazine was 1031 pg compared to 118 pg for the parent drug. The hapten was prepared by first coupling diazotized p-aminobenzoic acid to chlorpromazine and then conjugation to BSA using a mixed anhydride method. The authors did not determine the site where diazo coupling had taken place, but it is likely that it was either 3- or 7-. Coupling at 3- might explain the relative insensitivity of the assay to 7-hydroxychlorpromazine.

Radioimmunoassay of flupenthixol in plasma was described by Robinson and Risby (153). The cross-reactivity to the trans-isomer (77.5 %) and 7-carboxy-flupenthixol (77.5 %) which had been used to prepare the hapten was high. The sulphoxide (32.6 %) and N-desalkyl (27.9 %) metabolites, and the structurally similar drug, fluphenazine (59.6 %), also cross-reacted. Using a similar approach, Jørgensen (154) produced an antiserum that only gave 7 % cross-reaction with the trans isomer. The cross-reactivities to the metabolites and decanoate and palmitate esters were < 7 %. Cis(Z)-clopenthixol and fluphenazine cross-reacted to 14 and 17 % respectively. Comparison with GC-NPD showed that RIA over-estimated the concentration of cis(Z)-flupenthixol (78).

Wiles and Franklin (155) used an antiserum raised to a flupenthixol conjugate to determine plasma fluphenazine concentrations down to 50 ng/L. The lack of specificity of antisera was used by Goldstein and Vunakis (120) to detect low concentrations of fluphenazine, its metabolites and other phenothiazines after HPLC separation. Their antisera were raised to a fluphenazine conjugate coupled via the 10-side chain and initially [^{3}H]-CPZ was used as label. The relative sensitivites, given as the ratio of IC_{50} FPZ: IC_{50} test compound, showed that CPZNO (3.8) perphenazine (5.9), prochlorperazine (3.3) and trifluoperazine (0.83) were strongly bound by the antibody. Unlike those antisera raised by linking the hapten via the aromatic nucleus, the cross-reactivity to $\pm$ flupenthixol (sic) was slight. Lo et al. (156) preferred to extract fluphenazine from plasma with heptane: IAA to maintain a high degree of specificity. The limit of detection, using 1 mL plasma, was 20 ng/L.

A number of radioimmunoassays for haloperidol have been published, one of the first appearing in 1977 (157). As the hapten was joined via the carbonyl group, the antiserum had poor specificity with regard to the reduced metabolite and other metabolites, so a second antiserum was raised by conjugation via the tertiary alcohol (158). This antibody showed good selectivity for haloperidol in the presence of metabolites, but reacted with other butyrophenones such as bromperidol, moperone, trifluperidol, droperidol and spiperidone. More recently, an antibody has been raised in guinea-pigs using a hapten produced by linkage to the position normally carrying the fluorine atom, that cross-reacted to less than 1 % with known metabolites (159). Radioimmunoassays for bromperidol have been developed using antisera raised to haloperidol conjugates. One used the commercially available haloperidol kit (IRE, Fleurus, Belgium) to determine bromperidol after extraction (160), and a similar approach was adopted by Tischo et al. (161) who showed that apparent bromperidol concentrations were lower after ether extraction.

Mizuchi et al. (162) developed an immunossay for sultopride, and although the antibody reacted to < 1 % with known sultopride metabolites, the cross reactivity to sulpiride (4.5 %) allowed the antiserum to be used for the quantification of this drug also. The method was applied to a study of the regional distribution of these drugs in rat brain (163). The antibody raised to pimozide also bound cloimozide (100 %) and fluspiriline (42 %) (164).

Kamal Midha and his colleagues have devoted considerable time and effort, not only to developing immunoassays for parent phenothiazines (chlorpromazine (165), fluphenazine (166), perphenazine (167), trifluoperazine (168), prochlorperazine (169), thioridazine (170), but also metabolites, including chlorpromazine sulphoxide (171), 7-hydroxychlorpromazine (172), chlorpromazine N-oxide (173), 7-hydroxytrifluoperazine (174), trifluoperazine N-oxide (175) and sulphoridazine (176). This group has also developed immunoassays based on monoclonal antibodies, e.g. for fluphenazine (177). But interestingly, although the slope of the correlation line for the monoclonal antibody results versus an HPLC assay was closer to unity than when the a polyclonal antibody was used, the correlation was not so good.

A radioreceptor assay for neuroleptics was developed by Creese and Snyder (178) following their work on receptor binding. It is a saturation assay based on the displacement of [^{3}H]-haloperidol or [^{3}H]-spiroperidol from dopamine receptor preparations from rat striatum. The sensitivities (µg/L) varied with the compound being assayed: fluphenazine, 1.8; trifluoperazine, 2.2; haloperidol, 2.5; chlorpromazine, 8.6 and thioridazine, 30. The more potent the compound, the lower the limit of quantification, so those compounds which are expected to be present in low concentrations have the highest sensitivities, but even so the values are rather high compared with more recent methods. Other receptor preparations have been tried, for example, calf caudate (179) and porcine striata (180), and although it was claimed that the preparation gave improved sensitivity, this only appears to be the case for haloperidol. The lack of specificity was considered an advantage for clinical monitoring, as the assay detects "total neuroleptic" activity. This early promise has not been fully realized, probably in part due to the fact that the proportions of drug and metabolites in tissues are different from those in plasma and that receptor populations and affinities in human brain are different from those in animals.

3.5.4 Plasma Concentrations During Therapy and Overdose

Because of the nature of the illness, the complication of active metabolites and the fact that a combination of drugs is likely to be used, defining a therapeutic range or "window" for neuroleptics has not always been possible. Several reviews on plasma level monitoring (181–184), therapeutic effects and pharmacokinetics (185–187) of these drugs have appeared and the reader is advised to consult these for more detailed information. Similarly, attributing unwanted effects and fatalities to a given neuroleptic is likely to be complicated by the presence of other drugs. Neuroleptics will potentiate the sedative effects of other drugs. Based on animal data, thioridazine appears to have the lowest therapeutic index (20), the value for chlorpromazine is about 10 times higher and some of the potent agents may have values exceeding 1000.

To be effective, chlorpromazine plasma concentrations should probably be > 30 μg/L, and patients are unlikely to gain benefit from concentrations greater than 300–350 μg/L. Toxic effects may be associated with concentrations > 500 μg/L, but lethal concentrations may be in excess of 2000 μg/L. Bailey and Guba (71) reported concentrations between 500 and 1500 μg/L in serum from patients who survived 1.6–1.8 g. Post mortem concentrations in the range 3–35 mg/L have been reported but unfortunately these were determined by a nonselective method and would be elevated by the presence of metabolites (188).

Potent phenothiazines are effective at concentrations considerably lower than those required for chlorpromazine. A problem with early reports of fatalities is that the analytical methods may have been insufficient to determine blood concentrations. Thus fluphenazine (given as the decanoate ester) may be active at as little as 0.2 μg/L and it should not be necessary for concentrations to be above 2–3 μg/L. Following two fatalities, fluphenazine concentrations in liver were 5 and 23 mg/kg, respectively (188). For the slightly less potent perphenazine, the therapeutic window is a little higher, but possibly narrower, with 1.2–2.4 μg/L being suggested as optimum (182). Trifluoperazine concentrations determined in a patient receiving 30 mg/day were 4.2 μg/L, pre-dose, rising to 5.3 μg/L at 6 h (83). However, in a patient receiving 80 mg/day, the peak concentration was nearly 30 μg/L (189). Curry has suggested that the concentration should be somewhere between 2 and 30 μg/L (184).

Defining a therapeutic range for thioridazine is complicated by high concentrations of active metabolites; at steady-state conditions mesoridazine concentrations will be comparable to those of the parent drug. Concentrations > 100 μg/L of each are probably required for therapeutic efficacy. In eight fatal cases of poisoning, blood concentrations ranged from 300–8500 μg/L, with a similar range for mesoridazine (44).

The approximate therapeutic range for haloperidol is 5–15, or possibly 20 μg/L. Unwanted effects occur at about 50–100 μg/L. Fatalities have been associated with blood concentrations of 1 μg/L (188).

Thiothixene levels between 10 and 22.2 μg/L have been reported in samples from responding patients, taken 2–2.5 hours after the final dose (5–20mg) (190). The post-mortem blood concentration following ingestion of 250mg thiothixene (and an unknown quantity of doxepin) was 130 μg/L. The maximum level determined in a patient who ingested 800mg was 520 μg/L, falling to 47 μg/L after 12 hours (191). However, the patient was admitted only 20 minutes post administration, and the rapid fall in blood concentration suggests a marked redistributional phase.

In a report of a fatality involving chlorprothixene and benztropine ingestion, Poklis et al. (192) reported parent drug and sulphoxide metabolite levels as 100 and 600 μg/L, respectively.

References

1. Curry, S.H., Whelpton, R., de Schepper, P.J., Vranckx, S. & Schiff, A.A. (1979) Kinetics of fluphenazine after fluphenazine dihydrochloride, enanthate and decanoate administration to man. Br. J. Clin. Pharmacol. *7*, 325–331.
2. Salzman, N.P. & Brodie, B.B. (1956) Physiologicial disposition and fate of chlorpromazine and a method for its estimation on biological material. J. Pharmacol. Exp. Ther. *118*, 46–54.
3. Fishman, V., Heaton, A. & Goldenberg, H. (1962) Metabolism of chlorpromazine. III. Isolation and identification of chlorpromazine N-oxide. Proc. Soc. Exptl. Biol. Med. *109*, 548–552.
4. Fishman, V. & Goldenberg, H. (1965) Side-chain degradation and ring hydroxylation of phenothiazine tranquilizers. J. Pharmacol. Exp. Ther. *150*, 122–128.
5. Sgaragli, G., Ninci, R., Della Corte, L., Valoti, M., Nardini, M., Andreoli, V. & Moneti, G. (1986) Promazine. A major plasma metabolite of chlorpromazine in a population of chronic schizophrenics. Drug Metab. Dispos. *14*, 263–266.
6. Zehnder, K., Kalberer, F., Kreis, W. & Rutschmann, J. (1962) The metabolism of thioridazine and one of its pyrrolidine analogues in the rat. Biochem. Pharmacol. *11*, 535–550.
7. Papadopoulos, A.S., Crammer, J.L. & Cowan, D.A. (1985) Phenolic metabolites of thioridazine in man. Xenobiotica *15*, 309–316.
8. Papadopoulos, A.S. & Crammer, J.L. (1986) Sulphoxide metabolites of thioridazine in man. Xenobiotica *16*, 1097–1107.
9. Dahl. S.G., Kauffmann, E., Mompon, B. & Purcell, T. (1987) Nuclear magnetic resonance analysis of methotrimeprazine (levomepromazine) hydroxylation in humans. J. Pharm. Sci. *76*, 541–544.
10. Park, J., Shin, Y.O. & Choo, H.Y. (1989) Metabolism and pharmacokinetic studies of propionylpromazine in horses. J. Chromatogr. *489*, 313–321.
11. Gaertner, H.J., Breyer, U. & Liomin, G. (1974) Metabolism of trifluoperazine, fluphenazine, prochlorperazine and perphenazine in rats: in vitro and urinary metabolites. Biochem. Pharmacol. *23*, 303–311.
12. Breyer, U., Gaertner, H.J. & Prox, A. (1974) Formation of identical metabolites from piperazine- and dimethylamino-substituted phenothiazine drugs in man, rat and dog. Biochem. Pharmacol. *23*, 313–322.
13. Miyazaki, H., Matsunaga, Y., Nambu, K., Oh-e, Y., Yoshida, K. & Hashimoto, M. (1986) Disposition and metabolism of [14C]-haloperidol in rats. Arzneimittelforschung. *36*, 443–452.
14. Soudijn, W., Van Wijngaarden, I. & Allewijn, F. (1967) Distribution, excretion and metabolism of neuroleptics of the butyrophenone type. I. Excretion and metabolism of haloperidol and nine realted butyrophenone-derivatives in the Wistar rat. Eur. J. Pharmacol. *1*, 47–57.
15. Pape, B.E. (1981) Isolation and identification of a metabolite of haloperidol. J. Anal. Toxicol. *5*, 113–117.
16. Wong, F.A., Bateman, C.P., Shaw, C.J. & Patrick, J.E. (1983) Biotransformation of bromperidol in rat, dog, and man. Drug Metab. Dispos. *11*, 301–307.

17. Brennan, J.J., Imondi, A.R., Westmoreland, D.G. & Williamson, M.J. (1982) Isolation, identification, and synthesis of the major sulpiride metabolite in primates. J. Pharm. Sci. *71*, 1199–1203.
18. Soudijn, W. & Van Wijngaarden, I. (1969) The metabolism and excretion of the neuroleptic drug pimozide (R 6238) by the Wistar rat. Life Sci. *8*, 291–295.
19. Pinder, R.M., Brogden, R.N., Swayer, R., Speight, T.M., Spencer, R. & Avery, G.S. (1976) Pimozide: a review of its pharmacological properties and therapeutic uses in psychiatry. Drugs *12*, 1–40.
20. Bondesson, U. & Lindstrom, L.H. (1988) Determination of clozapine and its N-demethylated metabolite in plasma by the use of gas chromatography-mass spectrometry with single ion detection. Psychopharmacology Berl. *95*, 472–475.
21. Wang, Z., Lu, M., Xu, P & Zeng, Y. (1991) Determination of clozapine and its metabolites in serum and urine by reversed phase HPLC. Biomed. Chromatogr. *1*, 53–57.
22. Watkins, G.M., Whelpton, R., Buckley, D.G. & Curry, S.H. (1986) Chromatographic separation of thioridazine sulphoxide and N-oxide diastereoisomers: identification as metabolites in the rat. J. Pharm. Pharmacol. *38*, 506–509.
23. Juenge, E.C., Wells, C.E., Green, D.E., Forrest, I.S. & Shoolery, J.N. (1983) Synthesis, isolation, and characterization of two stereoisomeric ring sulfoxides of thioridazine. J. Pharm. Sci. *72*, 617–621.
24. Essien, E.E., Cowan, D.A. & Beckett, A.H. (1975) Metabolism of phenothiazines: identification of N-oxygenated products by gas chromatography and mass spectrometry. J. Pharm. Pharmacol. *27*, 334–342.
25. Coccia, P.F. & Westerfield, W.W. (1967) The metabolism of chlorpromazine by the liver microsomal enzyme systems. J. Pharmacol. Exp. Ther. *157*, 446–458.
26. Choo, H.Y., Shin, Y.O. & Park, J. (1990) Study of the metabolism of phenothiazines: determination of N-demethylated phenothiazines in urine. J. Anal. Toxicol. *14*, 116–119.
27. Whelpton, R. & Curry, S.H. (1976). Methods for the study of fluphenazine kinetics in man. J. Pharm. Pharmacol. *28*, 869–873.
28. Whelpton, R. (1984) Isotope-labelled materials as internal standards. In: Drug determination in therapeutic and forensic contexts (Reid, E.R. & Wilson, I.D., eds.) pp. 39–44, Plenum, New York.
29. Davis, C.M. Meyer, C.J. & Fenimore, D.C. (1977) Improved gas chromatographic analysis of chlorpromazine in blood serum. Clin. Chim. Acta *78*, 71–77.
30. Spirtes, M.A. (1972) Artifactual contamination of control serum in gas chromatographic analysis for chlorpromazine. Clin. Chem. *18*, 317–318.
31. Curry, S.H. (1974) Chlorpromazine analysis by gas chromatography with an electoncapture detector. In: The Phenothiazines and Structurally Related Drugs (Forrest, I.S., Carr, C.J. & Usdin, E., eds.) pp. 335–345, Raven Press, New York.
32. Whelpton, R. (1992) What happens when we freeze and thaw plasma? Acta Pharm. Suecica *15*, 458.
33. Hubbard, J.W., Cooper, J.K., Hawes, E.M., Jenden, D.J., May, P.R., Martin, M., McKay, G., Van Putten, T. & Midha, K.K. (1985) Therapeutic monitoring of chlorpromazine I: Pitfalls in plasma analysis. Ther. Drug Monit. *7*, 222–228.
34. Loennechen, T., Andersen, A., Hals, P.A. & Dahl, S.G. (1990) High-performance liquid chromatographic determination of levomepromazine (methotrimeprazine) and its main metabolites in serum and urine. Ther. Drug Monit. *12*, 574–581.
35. Patel, R.B. & Welling, R.G. (1981) On column reduction of promethazine metabolite during gas-chromatography. Clin. Chem. *27*, 1780–1781.
36. Ng, C.H. & Crammer, J.L. (1977) Measurement of thioridazine in blood and urine. Br. J. Clin. Pharmacol. *4*, 173–183.
37. Curry, S.H. (1968) Determination of nanogram quantities of chlorpromazine and some of its metabolites in plasma using gas-liquid chromatography with an electron capture detector. Anal. Chem. *40*, 1251–1255.
38. Whelpton, R. & Curry, S.H. (1976) Chromatographic properties of imines formed from dedimethylchlorpromazine and various aldehydes and ketones. J. Chromatogr. *121*, 88–92.
39. Curry, S.H. (1976) Gas-chromatographic methods for the study of chlorpromazine and some of its metabolites in human plasma. Psychopharmacol. Commun. *2*, 1–15.
40. Beckett, A.H., Navas, G.E. & Hutt, A.J. (1988) Metabolism of chlorpromazine and promazine in vitro: isolation and characte-

rization of N-oxidation products. Xenobiotica *18*, 61–74.

41. Borga, O., Piafsky, K.M. & Nilsen, O.G. (1977) Plasma protein binding of basic drugs. I. Selective displacement from alpha$_1$-acid glycoprotein by tris(2-butoxyethyl)phosphate. Clin. Pharmacol. Ther. *22*, 539–544.
42. Smith, C.S., Morgan, S.L., Greene, S.V. & Abramson, R.K. (1987) Solid-phase extraction and high-performance liquid chromatographic method for chlorpromazine and thirteen metabolites. J. Chromatogr. *423*, 207–216.
43. Allender, W.J., Archer, A.W. & Dawson, A.G. (1983) Extraction and analysis of chlorpromazine and its major metabolites in post mortem material by enzymic digestion and HPLC. J. Anal. Toxicol. *7*, 203–206.
44. Allender, W.J. (1985) High-pressure liquid chromatographic determination of thioridazine and its major metabolites in biological tissues and fluids. J. Chromatogr. Sci. *23*, 541–545.
45. Moffat, A.C. (ed) (1986) Clarke's Isolation and Identification of Drugs, 2nd Ed. Pharmaceutical Press, London.
46. Zingales, I. (1967) Systematic identification of psychotropic drugs by thin-layer chromatography. 1. J. Chromatogr. *31*, 405–419.
47. Hulshoff, A. & Perrin, J.H. (1976) A reversed-phase thin-layer chromotographic method for the determination of relative partition coefficients of very lipophilic compounds. J. Chromatogr. *120*, 65–80.
48. Kaul, P.N., Conway, M.W., Clark, M. & Huffine, J. (1970) Chlorpromazine metabolism I: quantitative fluorometric method for 11 chlorpromazine metabolites. J. Pharm. Sci. *59*, 1745–1749.
49. Lehr, R.E. & Kaul, P.N. (1975) Chlorpromazine metabolism IV: quaternization as key to determination of picomoles of chlorpromazine and other tertiary amine drugs. J. Pharm. Sci. *64*, 950–953.
50. Breyer, U. & Schmalzing, G. (1977) A thin-layer chromatographic method for the measurement of trifluoperazine and its metabolites in rat tissues. Drug Metab. Dispos. *5*, 97–115.
51. Kreese, M., Schley, J. & Muller-Oeblinghausen, B. (1980) Reliable routine method for the determination of perazine in serum by thin-layer chromatography with an internal standard. J. Chromatogr. *183* , 475–482.
52. Davis, C.M. & Fenimore, D.C. (1983) Determination of fluphenazine in plasma by high-performance thin-layer chromatography. J. Chromatogr. *272*, 157–165.
53. Davis, C.M. & Harrington, C.A. (1984) Quantitative determination of chlorpromazine and thioridazine by high-performance thin layer chromatography. J. Chromatogr. Sci. *22*, 71–74.
54. Davis, C.M. & Harrington, C.A. (1988) Quantitation of thiothixene in plasma by high-performance thin-layer chromatography and fluorometric detection. Ther. Drug Monit. *10*, 215–223.
55. Haagsma, N., Bathet, E.R. & Engelsma, J.W. (1988) Thin-layer chromatographic screening method for the tranquilizers azaperone, propiopromazine and carazolol in pig tissues. J. Chromatogr. *436*, 73–79.
56. Bressolle, F., Bres, J., Brun, S. & Rechencq, E. (1979) Quantitative determination of drugs by in situ spectrophotometry of chromatograms for pharmacokinetic studies. I. Sulpiride and other benzamides, vincamine, naftazone. J. Chromatogr. *174*, 421–433.
57. Driscoll, J. L., Matin, H. F. & Gudzinowicz, B. J. (1964) A gas chromatographic method for the quantitative analysis of some urinary metabolites of chlorpromazine. J. Gas Chromatogr. *2*, 109–114.
58. Kelsey, M.I. & Moscatelli, E. (1973) Gas-liquid chromatographic analysis of phenolic and non-phenolic chlorpromazine metabolites in the urine of schizophrenic patients. J. Chromatogr. *85*, 65–74.
59. Curry, S.H. & Mould, G.P. (1969) Gas chromatographic identification of thioridazine in plasma, and a method for routine assay of the drug. J. Pharm. Pharmacol. *21*, 674–677.
60. Dinovo, E.C., Gottschalk, L.A., Nandi, B.R. & Geddes, P.D. (1976) GLC analysis of thioridazine, mesoridazine and their metabolites. J. Pharm. Sci. *65*, 667–669.
61. Kilts, C.D., Patrick, K.S., Breese, G.R. & Mailman, R.B. (1982) Simultaneous determination of thioridazine and its S-oxidized and N-demethylated metabolites using high-performance liquid chromatography on radially compressed silica. J. Chromatogr. *231*, 377–391.
62. Whelpton, R. & Curry, S.H. (1975) Gas-chromatographic separation of chlorpromazine, diazepam, and N-desmethyldiazepam. J. Pharm. Pharmacol. *27*, 970–971.

63. Hansen, C.E. & Larsen, N.E. (1974) Perphenazine concentrations in human whole blood. A pilot study during anti-psychotic therapy using different administration forms. Psychopharmacology Berl. *37*, 31–36.
64. Hansen, L.B. & Larsen, N.E. (1977) Plasma concetration of perphenazine and its suphoxide metabolite during continuous oral treatment. Psychopharmacology Berl. *53*, 127–130.
65. Marcucci, F., Airoldi, L., Mussini, E. & Garrattini, S. (1971) A method for the gas chromatographic determination of butyrophenones. J. Chromatogr. *59*, 174–177.
66. Tyndale, R.F. & Inaba, T. (1990) Simultaneous determination of haloperidol and reduced haloperidol by gas chromatography using a megabore column with electron-capture detection: application to microsomal oxidation of reduced haloperidol. J. Chromatogr. *529*, 182–188.
67. Larsen, N.E. & Naestoft, J. (1973) Determination of perphenazine and fluphenazine in whole blood by gas chromatography. Med. Lab. Tech. *30*, 129–132.
68. Cooper, S.F. & Lapierre, Y.D. (1981) Gas-liquid chromatographic determination of pipotiazine in plasma of psychiatric patients. J. Chromatogr. *222*, 291–296.
69. Cooper, T.B. & Kelly, R.G. (1979) GLC analysis of loxapine, amoxapine and their metabolites in serum and urine. J. Pharm. Sci. *68*, 216–219.
70. Harris, J.P., Phillipson, O.T., Watkins, G.M. & Whelpton, R. (1983) Effects of chlorpromazine and promazine on the visual aftereffects of tilt and movement. Psychopharmacology Berlin. *79*, 49–57.
71. Bailey, D.N. & Guba, J.J. (1979) Gas-chromatographic analysis for chlorpromazine and some of its metabolites in human serum, with use of a nitrogen detector. Clin. Chem. *25*, 1211–1215.
72. Ninci, R., Giovannini, M.G., Della Corte, L. & Sgaragli, G. (1986) Isothermal gas chromatographic determination of nanogram amounts of chlorimipramine, chlorpromazine and their N-desmethyl metabolites in plasma using nitrogen-selective detection. J. Chromatogr. *381*, 315–322.
73. Franklin, M., Wiles, D.H. & Harvey, D.J. (1978) Sensitive gas chromatographic determination of fluphenazine in human plasma. Clin. Chem. *24*, 41–44.
74. Dekirmenjian, H., Javaid, J.I., Duslak, B. & Davis, J.M. (1978) Determination of antipsychotic drugs by gas-liquid chromatography with a nitrogen detector using a simple acetylation technique. J. Chromatogr. *160*, 291–296.
75. Javaid, J.I., Dekirmenjian, H., Liskervych, U., Lin, R.-L. & Davis, J.M. (1981) Fluphenazine determination in human plasma by a sensitive gas chromatographic method using nitrogen-phosphorus detection J. Chromatogr. *19*, 439–443.
76. Bianchetti, G. & Morselli, P.L. (1978) Rapid and sensitive method for determination of haloperidol in human samples using nitrogen-phosphorus selective detection. J. Chromatogr. *153*, 203–209.
77. Abernethy, D.R., Greenblatt, D.J., Ochs, H.R., Willis, C.R., Miller, D.D. & Shader, R.I. (1984) Haloperidol determination in serum and cerebrospinal fluid using gas-liquid chromatography with nitrogen-phosphorus detection: application to pharmacokinetic studies. J. Chromatogr. *307*, 194–199.
78. Balant Gorgia, A.E., Balant, L.P., Genet, C. & Eisele, R. (1985) Comparative determination of flupentixol in plasma by gas chromatography and radioimmunoassay in schizophrenic patients. Ther. Drug Monit. *7*, 229–235.
79. Richter, K. (1988) Determination of clozapine in human serum by capillary gas chromatography. J. Chromatogr. *434*, 465–468.
80. Vasiliades, J., Sahawneh, T.M. & Owens, C. (1979) Determination of therapeutic and toxic concentrations of doxepin and loxapine using gas-liquid chromatography with a nitrogen-sensitive detector, and gas chromatography-mass spectrometry of loxapine. J. Chromatogr. *164*, 457–470.
81. Hammar, C.-G., Holmstedt, B. & Ryhage, R. (1968) Mass fragmentography. Identification of chlorpromazine and its metabolites in human blood by a new method. Anal. Biochem. *25*, 532–548.
82. Alfredsson, G., Wode-Helgot, B. & Sedvall, G. (1976) A mass fragmentographic method for the determination of chlorpromazine and two of its active metabolites in human plasma and CSF. Psychopharmacology Berl. *48*, 123–131.
83. Whelpton, R., Curry, S.H. & Watkins, G.M. (1982) Analysis of plasma trifluoperazine by gas chromatography and selected ion monitoring. J. Chromatogr. *228*, 321–326.
84. Midha, K.K., Roscoe, R.M.H., Hall, K., Hawes, E.M., Cooper, J.K., McKay, G. &

Shetty, H.V. (1982) A gas-chromatographic mass spectrometric assay for plasma trifluoperazine concentrations following single oral doses. Biomed. Mass. Spectrom. *9*, 186–190.

85. McKay, G., Hall, K., Edom, R., Hawes, E.M. & Midha, K.K. (1983) Subnanogram determination of fluphenazine in human plasma by gas chromatography mass spectrometry. Biomed. Mass. Spectrom. *10*, 550–555.
86. McKay, G., Hall, K., Cooper, J.K., Hawes, E.M. & Midha, K.K. (1982) Gas-chromatographic-mass spectrometric procedure for the quantification of chlorpromazine in plasma and its comparison with a new high-performance liquid chromatographic assay with electrochemical detection. J. Chromatogr. *232*, 275–282.
87. Moulin, M.A., Camsonne, R., Davy, J.P., Poilpre, E., Morel, P., Debruyne, D. & Bigot, M.C. (1979) Gas chromatography-electron-impact and chemical-ionization mass spectrometry of haloperidol and its chlorinated homologue. J. Chromatogr. *178*, 324–329.
88. Hornbeck, C.L., Griffiths, J.C., Neborsky, R.J. & Faulkner, M.A. (1979) A gas chromatographic mass spectrometric chemical ionization assay for haloperidol with selected ion monitoring. Biomed. Mass. Spectom. *6*, 427–430.
89. Szczepanik van Leeuwen, P.A. (1985) Improved gas chromatographic-mass spectrometric assay for haloperidol utilizing ammonia chemical ionization and selected-ion monitoring. J. Chromatogr. *339*, 321–330.
90. Haring, N., Salama, Z., Todesko, L. & Jaeger, H. (1987) Gas chromatographic-mass spectrometric determination of haloperidol in plasma. Application to pharmacokinetics. Arzneimittelforschung. *37*, 1402–1404.
91. Muusze, R.G. & Huber, J.F.K. (1974) Determination of the psychotropic drug thioridazine and its metabolites in blood by means of high-performance liquid chromatography in combination with fluorometric reaction detection. J. Chromatogr. *12*, 779–789.
92. Knox, J.H. & Jurand, J. (1975) Separation of tricyclic psychosedative drugs by high speed ion-pair partition and liquid-solid chromatography. J. Chromatogr. *103*, 311–316.
93. Johansson, R., Borg, K.O. & Gabrielsson, M. (1976) Determination of fluphenazine in plasma by ion-pair partition chromatography. Acta Pharm. Suecica. *13*, 193–200.
94. Smith, D.J. (1981) The separation and determination of chlorpromazine and some of its related compounds by reversed-phase high performance liquid chromatography. J. Chromatogr. *19*, 65–71.
95. Curry, S.H. & Brown, E.A. (1981) HPLC assay of phenothiazines, thioxanthene and butyrophenone neuroleptics and antihistamines in plasma: I Separations and UV detection. IRCS Med. Sci. *9*, 166–167.
96. Curry, S.H., Brown, E.A. & Perrin, J.H. (1981) HPLC assay of phenothiazine, thioxanthene and butyrophenone neuroleptics and antihistamines in plasma: II Radial compression columns and electrochemical detection. IRCS Med. Sci. *9*, 168–169.
97. Curry, S.H., Brown, E.A. & Perrin, J.H. (1981) HPLC assay of phenothiazine, thioxanthene and butyrophenone neuroleptics and antihistamines in plasma: III Quantitative aspects and clinical data. IRCS Med. Sci. *9*, 170–171.
98. Curry, S.H., Brown, E.A., Hu, O.Y.-P. & Perrin, J.H. (1982) Liquid chromatographic assay of phenothiazine, thioxanthene and butyrophenone neuroleptics and antihistamines in blood and plasma with conventional and radial compression columns and UV and electrochemical detection. J. Chromatogr. *231*, 361–376.
99. Cooper, J.K., McKay, G. & Midha, K.K. (1983) Subnanogram quantitation of chlorpromazine in plasma by high-performance liquid chromatography with electrochemical detection. J. Pharm. Sci. *72*, 1259–1262.
100. Heyes, W.F. & Salmon, J.R. (1978) Some aspects of the high-performance liquid chromatography of fluphenazine and its esters. J. Chromatogr. *156*, 309–316.
101. Heyes, W.F., Salmon, J.R. & Marlow, W. (1980) High-performance liquid chromatographic separation of the N- and S-oxides of fluphenazine and fluphenazine decanoate. J. Chromatogr. *194*, 416–420.
102. Flanagan, R.J. & Jane, I. (1985) High-performance liquid chromatographic analysis of basic drugs on silica columns using non-aqueous ionic eluents. I. Factors influencing retention, peak shape and detector response. J. Chromatogr. *232*, 173–189.
103. Mehta, A.C. (1981) High-performance liquid chromatographic determination of chlorpromazine and thioridazine hydrochlorides in pharmaceutical formulations. Analyst *106*, 1119–1122.

104. Vatassery, G.T., Herzan, L.A. & Dysken, M.W. (1988) Simultaneous determination of very low concentrations of haloperidol and reduced haloperidol in human serum by a liquid chromatographic method (published erratum appears in J. Chromatogr. 1989 May 30; 490(2):492). J. Chromatogr. *433*, 312–317.
105. Ragland, J.B. & Kinross-Wright, V.J. (1964) Spectrofluorometric measurement of phenothiazines. Anal. Chem. *36*, 1356–1359.
106. Jane, I., McKinnon, A. & Flanagan, R.J. (1985) High-performance liquid chromatographic analysis of basic drug on silica columns using non-aquenous ionic eluents. II. Application of UV, fluorescence and electrochemical oxidation detection. J. Chromatogr. *232*, 191–225.
107. Hoffman, D.W., Edkins, R.D. & Shillcutt, S.D. (1988) Human metabolism of phenothiazines to sulfoxides determined by a new high performance liquid chromatography-electrochemical detection method. Biochem. Pharmacol . *37*, 1773–1777.
108. Furlanut, M., Perosa, A., Benetello, P. & Colombo, G. (1987) Electrochemical detection of benperidol in serum for drug monitoring in humans. Ther. Drug Monit. *9*, 343–346.
109. Kruger, R., Mengel, I. & Kuss, H.J. (1984) Determination of benperidol in human plasma by high-performance liquid chromatography. J. Chromatogr. *311*, 109–116.
110. Hikida, K., Inoue, Y., Miyazaki, T., Kojima, N. & Ohkura, Y. (1989) Determination of bromperidol in serum by automated column-switching high-performance liquid chromatography. J. Chromatogr. *495*, 227–234.
111. Midha, K.K., Cooper, J.K., McGilveray, I.J., Butterfield, A.G. & Hubbard, J.W. (1981) High-performance liquid chromatographic assay for nanogram determination of chlorpromazine and its comparison with a radioimmunoassay. J. Pharm. Sci. *70*, 1043–1046.
112. Brooks, M.A., DiDonato, G. & Blumenthal, H.P. (1985) Determination of chlorprothixene and its sulfoxide metabolite in plasma by high-performance liquid chromatography with ultraviolet and amperometric detection. J. Chromatogr. *337*, 351–362.
113. Svendsen, C.N. & Bird, E.D. (1986) HPLC with electrochemical detection to measure chlorpromazine, thioridazine and metabolites in human brain. Psychopharmacology Berl. *90*, 316–321.
114. Aaes Jorgensen, T. (1980) Specific high-performance liquid chromatographic method for the estimation of the cis(Z)- and trans(E)-isomers of clopenthixol and a N-dealkyl metabolite. J. Chromatogr. *183*, 239–245.
115. Haring, C., Humpel, C., Auer, B., Saria, A., Barnas, C., Fleischhacker, W. & Hinterhuber, H. (1988) Clozapine plasma levels determined by high-performance liquid chromatography with ultraviolet detection. J. Chromatogr. *428*, 160–166.
116. Lovdahl, M.J., Perry, P.J. & Miller. D.D. (1991) The assay of clozapine and N-desmethyclozepine in human plasma using high-performance liquid chromatography. Ther. Drug Monit. *13*, 69–72.
117. Humpel, C., Haring, C. & Saria, A. (1989) Rapid and sensitive determination of clozapine in human plasma using high-performance liquid chromatography and amperometric detection. J. Chromatogr. *491*, 235–239.
118. Wilhelm, D. & Kemper, A. (1990) High-performance liquid chromatographic procedure for the determination of clozapine, haloperidol, droperidol and several benzodiazepines in plasma. J. Chromatogr. *525*, 218–224.
119. Tan, S.T. & Boniface, P.J. (1990) High-performance liquid chromatography of droperidol in whole blood. J. Chromatogr. *532*, 181–186.
120. Goldstein, S.A. & Vunakis, H.V. (1981) Determination of fluphenazine and related phenothiazine drugs and metabolites by combined high-performance chromatography and radio-immunoassay. J. Pharmacol Exp. Ther. *217*, 36–43.
121. Hoffman, D.W., Edkins, R.D., Shillcutt, S.D. & Salama, A. (1987) New high-performance liquid chromatographic method for fluphenazine and metabolites in human plasma. J. Chromatogr. *414*, 504–509.
122. Cooper, J.K., Hawes, E.M., Hubbard, J.W., McKay, G. & Midha, K.K. (1989) An ultrasensitive method for the measurement of fluphenazine in plasma by high-performance liquid chromatography with coulometric detection. Ther. Drug Monit. *11*, 354–360.
123. Miyazaki, K., Arita, T., Oka, I., Koyama, T. & Yamashita, I. (1981) High-performance liquid chromatographic determina-

tion of haloperidol in plasma. J. Chromatogr. *223*, 449–453.
124. Korpi, E.R., Phelps, B.H., Granger, H., Chang, W.H., Linnoila, M., Meek, J.L. & Wyatt, R.J. (1983) Simultaneous determination of haloperidol and its reduced metabolite in serum and plasma by isocratic liquid chromatography with electrochemical detection. Clin. Chem. *29*, 624–628.
125. McBurney, A. & George, S. (1984) High-performance liquid chromatography of haloperidol in serum at the concentrations achieved during chronic therapy. J. Chromatogr. *308*, 387–392.
126. Dhar, A.K. & Kutt, H. (1984) Improved liquid-chromatographic determination of haloperidol in plasma. Clin. Chem. *30*, 1228–1230.
127. Miller, R.L. & DeVane, C.L. (1986) Measurement of haloperidol and reduced haloperidol in human plasma using reversed-phase high-performance liquid chromatography. J. Chromatogr. *374*, 405–408.
128. Eddington, N.D. & Young, D. (1988) Sensitive electrochemical high-performance liquid chromatography assay for the simultaneous determination of haloperidol and reduced haloperidol. J. Pharm. Sci. *77*, 541–543.
129. Midha, K.K., Copper, J.K., Hawes, E.M., Hubbard, J.W., Korchinski, E.D. & McKay, G. (1988) An ultrasensitive method for the measurement of haloperidol and reduced haloperidol in plasma by high-performance liquid chromatography with coulometric detection. Ther. Drug Monit. *10*, 177–183.
130. Cahard, C., Rop, P.P., Conquy, T. & Viala, A. (1990) High-performance liquid chromatographic analysis of haloperidol and hydroxyhaloperidol in plasma after solid-phase extraction. J. Chromatogr. *532*, 193–202.
131. Hariharan, M., Kindt, E.K., VanNoord, T. & Tandon, R. (1989) An improved sensitive assay for simultaneous determination of plasma haloperidol and reduced haloperidol levels by liquid chromatography using a coulometric detector. Ther. Drug Monit. *11*, 701–707.
132. Nilsson, L.B. (1988) Reversed-phase ion-pair liquid chromatographic method for the determination of low concentrations of haloperidol in plasma. J. Chromatogr. *431*, 113–122.
133. Vatassery, G.T., Herzan, L.A. & Dysken, M.W. (1990) Liquid chromatographic determination of reduced haloperidol and haloperidol concentrations in packed red blood cells from humans. J. Anal. Toxicol. *14*, 25–28.
134. Loennechen, T. & Dahl, S.G. (1990) High-performance liquid chromatography of levomepromazine (methotrimeprazine) and its main metabolites. J. Chromatogr. *503*, 205–215.
135. Murakami, K., Murakami, K., Ueno, T., Hijikata, J., Shirasawa, K & Muto, T. (1982) Simultaneous determination of chlorpromazine and levomepromazine in human plasma and urine by high-performance liquid chromatography using electrochemical detection. J. Chromatogr. *227*, 103–112.
136. Miyao, Y., Suzuki, A., Noda, K. & Noguchi, H. (1983) A sensitive assay method for pimozide in human plasma by high-performance liquid chromatography with fluorescence detection. J. Chromatogr. *275*, 443–449.
137. Ogden, D.A., McKechnie, A.A., Bolton, D.A., Rankin, R.B., Dobbins, S.E. & Clements, J.A. (1989) Determination of pipothiazine in human plasma by reversed-phase high-performance liquid chromatography. J. Pharm. Biomed. Anal. *7*, 1273–1280.
138. Sankey, M.G., Holt, J.E. & Kaye, C.M. (1982) A simple and sensitive H.P.L.C. method for the assay of prochlorperazine in plasma. Br. J. Clin. Pharmacol. *13*, 578–580.
139. Fowler, A., Taylor, W. & Bateman, D.N. (1986) Plasma prochlorperazine assay by high-performance liquid chromatography-electrochemistry. J. Chromatogr. *380*, 202–205.
140. Tebbett, I.R. (1986) Analysis of promazine in pharmaceutical preparations by high-performance liquid chromatography. J. Chromatogr. *356*, 227–229.
141. Parkinson, D. (1985) Sensitive analysis of butyrophenone neuroleptics by high-performance liquid chromatography with ultraviolet detection at 254 nm. J. Chromatogr. *341*, 465–472.
142. Alfredsson, G., Sedvall, G. & Wiesel, F.A. (1979) Quantitative analysis of sulpiride in body fluids by high-performance liquid chromatography with fluorescence detection. J. Chromatogr. *164*, 187–193.
143. Bressolle, F. & Bres, J. (1985) Determination of sulpiride and sultopride by high-performance liquid chromatography for phar-

macokinetic studies. J. Chromatogr. *341*, 391–399.
144. Nicolas, P., Fauvelle, F., Ennachachibi, A., Merdjan, H. & Petitjean, O. (1986) Improved determination of sulpiride in plasma by ion-pair liquid chromatography with fluorescence detection. J. Chromatogr. *381*, 393–400.
145. McKay, G., Cooper, J.K., Midha, K.K., Hall, K. & Hawes, E.M. (1982) Simple and sensitive high-performance liquid chromatographic procedure with electrochemical detection for the determination of plasma concentrations of trimeprazine following single oral doses. J. Chromatogr. *233*, 417–422.
146. Mailman, R.B., Pierce, J.P., Crofton, K.M., Petitto, J., DeHaven, D.L., Kilts, C.D. & Lewis, M.H. (1984) Thioridazine and the neuroleptic radioreceptor assay. Biol. Psychiatry *19*, 833–847.
147. Stoll, A.L., Baldessarini, R.J., Cohen, B.M. & Finklestein, S.P. (1984) Assay of plasma thioridazine and metabolites by high-performance liquid chromatography with amperometric detection. J. Chromatogr. *307*, 457–463.
148. Svensson, C., Nyberg, G., Soomagi, M. & Martensson, E. (1990) Determination of the serum concentrations of thioridazine and its main metabolites using a solid-phase extraction technique and high-performance liquid chromatography. J. Chromatogr. *529*, 229–236.
149. Narasimhachari, N., Dorey, R.C., Landa, B.L. & Friedel, R.O. (1984) Improved high-performance liquid chromatographic method for the quantitation of cis-thiothixene in plasma samples using trans-thiothixene as internal standard. J. Chromatogr. *311*, 424–429.
150. Dilger, C., Salama, Z. & Jaeger, H. (1988) Improved high-performance liquid chromatographic method for the determination of tiotixene in human serum. Arzneimittelforschung. *38*, 1522–1525.
151. Hariharan, M., VanNoord, T., Kindt, E.K. & Tandon, R. (1991) A simple, sensitive liquid chromatographic assay of cis-thiothixene in plasma with coulometric detection. Ther. Drug Monit. *13*, 79–85.
152. Kawashima, K., Dixon, R. & Spector, S. (1975) Development of radioimmunoassay for chlorpromazine. Eur. J. Pharmacol. *32*, 195–202.
153. Robinson, J.D. & Risby, D. (1977) Radioimmunoassay for flupenthixol in plasma. Clin. Chem. *23*, 2085–2088.
154. Jorgensen, A. (1978) A sensitive and specific radioimmunoassay for cis(Z)-flupenthixol in human serum. Life Sci. *23*, 1533–1542.
155. Wiles, D.H. & Franklin, M. (1978) Radioimmunoassay for fluphenazine in human plasma. Br. J. Clin. Pharmacol. *5*, 265–268.
156. Lo, E.S., Fein, M., Hunter, C., Suckow, R.F. & Cooper, T.B (1988) A highly sensitive and specific radioimmunoassay for quantitation of plasma fluphenazine. J. Pharm. Sci. *77*, 255–258.
157. Clark, B.R., Tower, B.B. & Rubin, R.T. (1977) Radioimmunoassay of haloperidol in human serum. Life Sci. *20*, 319–325.
158. Rubin, R.T., Tower, B.B., Hays, S.E. & Poland, R.E. (1982) Radioimmunoassay of haloperidol. Methods Enzymol. *84*, 532–542.
159. Terauchi, Y., Ishikawa, S., Oida, S., Nakao, M., Kagemoto, A., Oida, T., Utsui, Y. & Sekine, Y. (1990) Direct radioimmunoassay for haloperidol in human serum. J. Pharm. Sci. *79*, 432–436.
160. Van Den Eeckhout, E., Belpaire, F.M., Bogaert, M.G. & DeMoerloose, P. (1980) Radioimmunoassay of bromperidol and haloperidol in human and canine plasma. Eur. J. Drug Metab. Pharmacokinet. *5*, 45–48.
161. Tischio, J., Hetyei, N. & Patrick, J. (1984) Bromperidol radioimmunoassay: human plasma levels. J. Pharm. Sci. *73*, 546–548.
162. Mizuchi, A., Saruta, S., Kitagawa, N. & Miyachi, Y. (1981) Development of radioimmunoassay for sultopride and sulpiride. Arch. Int. Pharmacodyn. Ther. *254*, 317–326.
163. Mizuchi, A., Kitagawa, N. & Miyachi, Y. (1983) Regional distribution of sultopride and sulpiride in rat brain measured by radioimmunoassay. Psychopharmacology Berlin. *81*, 195–198.
164. Michiels, M., Hendriks, R. & Heykants, J. (1982) Radioimmunoassay of pimozide. Methods Enzymol *84*, 542–551.
165. Midha, K.K., Loo, J.C.K., Hubbard, J.W., Rowe, M.L. & McGilveray, I.J. (1979) Radioimmunoassay for chlorpromazine in plasma. Clin. Chem. *25*, 166–168.
166. Midha, K.K., Cooper, J.K. & Hubbard, J.W. (1981) Radioimmunoassay for fluphenazine in human plasma. Commun. Psychopharmacol. *4*, 107–114.
167. Midha, K.K., MacKonka, C., Cooper, J.K., Hubbard, J.W. & Yeung, P.F.K. (1981) Radioimmunoassay for perphenazine in human plasma. Br. J. Clin. Pharmacol. *12*, 85–88.

168. Midha, K.K., Hubbard, J.W., Cooper, J.K., Hawes, E.M., Fournier, S. & Yeung, P. (1981) Radioimmunoassay for trifluoperazine in human plasma. Br. J. Clin. Pharmacol. *12*, 189–193.
169. Midha, K.K., Hawes, E.M., Rauw, G., McVittie, J., McKay, G., Cooper, J.K. & Shetty, H.U. (1983) Radioimmunoassay for prochlorperazine in human plasma. Ther. Drug Monit. *5*, 117–121.
170. Chakraborty, B.S., Hawes, E.M., Rosenthaler, J. & Midha, K.K. (1987) A new radioimmunoassay of thioridazine and its comparison with a high performance liquid chromatographic method and another radioimmunoassay. Ther. Drug Monit. *9*, 227–235.
171. Yeung, P.K., Hubbard, J.W., Cooper, J.K. & Midha, K.K. (1983) A study of the kinetics of chlorpromazine sulfoxide by a specific radioimmunoassay after a single oral dose of chlorpromazine in healthy volunteers. J. Pharmacol. Exp. Ther. *226*, 833–838.
172. Yeung, P.K., McKay, G., Ramshaw, I.A., Hubbard, J.W. & Midha, K.K. (1985) A comparison of two radioimmunoassays for 7-hydroxychlorpromazine: rabbit polyclonal antibodies vs. mouse monoclonal antibodies. J. Pharmacol. Exp. Ther. *233*, 816–822.
173. Yeung, P.K., Hubbard, J.W., Korchinski, E.D. & Midha, K.K. (1987) Radioimmunoassay for the N-oxide metabolite of chlorpromazine in human plasma and its application to a pharmacokinetic study in healthy humans. J. Pharm. Sci. *76*, 803–808.
174. Aravagiri, M., Hawes, E.M. & Midha, K.K. (1985) Radioimmunoassay for the 7-hydroxy metabolite of trifluoperazine and its application to a kinetic study in human volunteers. J. Pharm. Sci. *74*, 1196–1202.
175. Aravagiri, M., Hawes, E.M. & Midha, K.K. (1986) Development and application of a specific radioimmunoassay for trifluoperazine N4'-oxide to a kinetic study in humans. J. Pharmacol. Exp. Ther. *237*, 615–622.
176. Chakraborty, B.S., Hawes, E.M. & Midha, K.K. (1988) Development of a radioimmunoassay procedure for sulphoridazine and its comparison with a high-performance liquid chromatographic method. Ther. Drug Monit. *10*, 205–214.
177. McKay, G., Steeves, T., Cooper, J.K., Hawes, E.M. & Midha, K.K. (1990) Development and application of a radioimmunoassay for fluphenazine based on monoclonal antibodies and its comparison with alternate assay methods. J. Pharm. Sci. *79*, 240–243.
178. Creese, I. & Snyder, S.H. (1977) A simple and sensitive radioreceptor assay for antischizophrenic drugs in blood. Nature *270*, 180–182.
179. Lader, S.R. (1980) A radioreceptor assay for neuroleptic drugs in plasma. J. Immunoassay *1*, 57–75.
180. Rao, M.L. (1986) Modification of the radioreceptor assay technique for estimation of serum neuroleptic drug levels leads to improved precision and sensitivity. Psychopharmacology Berlin *90*, 548–553.
181. Cooper, T.B. (1978) Plasma level monitoring of antipsychotic drugs. Clin. Pharmacokinet. *3*, 14–38.
182. Dahl, S.G. (1986) Plasma level monitoring of antipsychotic drugs. Clinical utility. Clin. Pharmacokinet. *11*, 36–61.
183. Balant Gorgia, A.E. & Balant, L. (1987) Antipsychotic drugs. Clinical pharmacokinetics of potential candidates for plasma concentration monitoring. Clin. Pharmacokinet. *13*, 65–90.
184. Curry, S.H. (1985) The strategy and value of neuroleptic drug monitoring. J. Clin. Psychopharmacol. *5*, 263–271.
185. Ko, G.N., Korpi, E.R. & Linnoila, M. (1985) On the clinical relevance and methods of quantification of plasma concentrations of neuroleptics. J. Clin. Psychopharmacol. *5*, 253–262.
186. Froemming, J.S., Lam, Y.W., Jann, M.W. & Davis, C.M. (1989) Pharmacokinetics of haloperidol. Clin. Pharmacokinet. *17*, 396–423.
187. Bressolle, F., Bres, J. & Mourad, G. (1989) Pharmacokinetics of sulpiride after intravenous administration in patients with impaired renal function. Clin. Pharmacokinet. *17*, 367–373.
188. Baselt, R.C. & Cravey, R.H. (1989) Disposition of Toxic Drugs and Chemicals in Man, 3rd. Ed. Year Book Medical, Chicago.
189. Curry, S.H., Stewart, R.B., Springer, P.K. & Pope, J.E. (1981) Plasma trifluoperazine concentrations during high dose therapy. Lancet *1*, 395–396.
190. Hobbs, D.C., Welch, W.M., Short, M.J., Moody, W.A. & van der Velde, C.D. (1974) Pharmacokinetics of thiothixene in man. Clin. Pharmacol. Ther. *16*, 473–478.

191. Kemal, M. & Imami, R.H. (1985) Acute thiothixene overdose. J. Anal. Toxicol. *9*, 94–95.

192. Poklis, A., Maginn, D. & Mackell, M.A. (1983) Chlorprothixene and chlorprothixene-sulfoxide in body fluids from a case of drug overdose. J. Anal. Toxicol. *7*, 29–32.

3.6 Antidepressants

R. A. Braithwaite

3.6.1 Introduction

The first tricyclic antidepressant to be introduced was imipramine (Tofranil) by Geigy in 1957. Following its success, the synthesis of a large number of chemically related compounds soon yielded many other effective antidepressants, the most well known of which was amitriptyline. The use of this group of drugs expanded rapidly during the 1960's and 1970's and became agents of choice in the treatment of depressive disorders. However, reports of toxicity, particularly in overdosage, soon appeared in the medical literature (1). At the present time tricyclic antidepressants, because of their proven efficacy and low cost, are widely prescribed, but remain one of the commonest causes of acute self-poisoning world-wide, and are responsible for a large number of overdose fatalities (2–6). Approximately 10–20 % of all patients who present to hospital with self-poisoning may have taken a tricyclic antidepressant, also a high proportion of these patients require intensive care (4). Tricyclic antidepressants are also associated with a relatively high incidence of adverse-effects in normal therapeutic usage (7).

A number of "second generation" heterocyclic antidepressants were developed during the 1970's such as trazodone, mianserin, maprotiline and amoxapine (8). Although structurally unrelated to the tricyclic group of antidepressants, some of these drugs show a similar pharmacological action and toxicity profile (8). In the last decade a "third generation" of specific serotonin re-uptake inhibitor antidepressants (SSRI's) has been introduced, the most well known member of this group being fluoxetine (Prozac) (9). This new group of antidepressants appears to be clinically effective, with a greatly reduced toxicity in overdose (10).

There is therefore an important requirement for the availability of specific and sensitive analytical techniques for the qualitative and quantitative measurement of antidepressant drugs and their active metabolites in body fluids and tissues. The main clinical applications for the measurement of these drugs is shown in table 1.

Table 1. Main Clinical Applications for the Measurement of Tricyclic and Newer Antidepressants

1. Emergency toxicology
2. Therapeutic drug monitoring
3. Investigation of adverse drug reactions
4. Forensic toxicology
5. Clinical trials of antidepressant action
6. Pharmacokinetic studies

3.6.2 Pharmacokinetics

Chemical structures of the tricyclic and some newer antidepressants

The tricyclic heterocyclic and newer selective serotonin reuptake inhibitor (SSRI) antidepressants form a chemically very heterogeneous group of compounds. The three ring so-called "tricyclic" dibenzodiazepine structure of imipramine and some congeners (e.g. clomipramine, lofepramine) and the dibenzocycloheptadiene structure of amitriptyline are shown in figure 1. Minor modification of the central ring structure is found with dothiepin and doxepin, or outer aromatic ring in the case of clomipramine. The three ring system incorporating a central seven membered ring is the common feature of all of the drugs in this group. All these drugs, apart from lofepramine, have short aliphatic tertiary or secondary amino side-chains. The N-desmethyl metabolites of amitriptyline and imipramine (nortriptyline and desipramine respectively) are pharmacologically active, and were also developed as effective antidepressants in their own right. There is therefore a need to have analytical techniques that are able to measure tertiary amine parent drugs and their active N-desmethyl metabolites. The most important "second generation" heterocyclic antidepressants and the four major "third generation" SSRI's are shown in figure 2.

Pharmacokinetics and metabolism

The pharmacokinetics and metabolism of the tricyclic antidepressants have been intensively studied over many years, particularly in the case of amitriptyline and imipramine and their respective metabolites, nortriptyline and desipramine (11). These compounds are generally well absorbed from the gastrointestinal tract, and because of their high lipid solubility, drug concentrations found in whole blood and plasma are much lower than in tissues. Drug distribution volumes are corres-

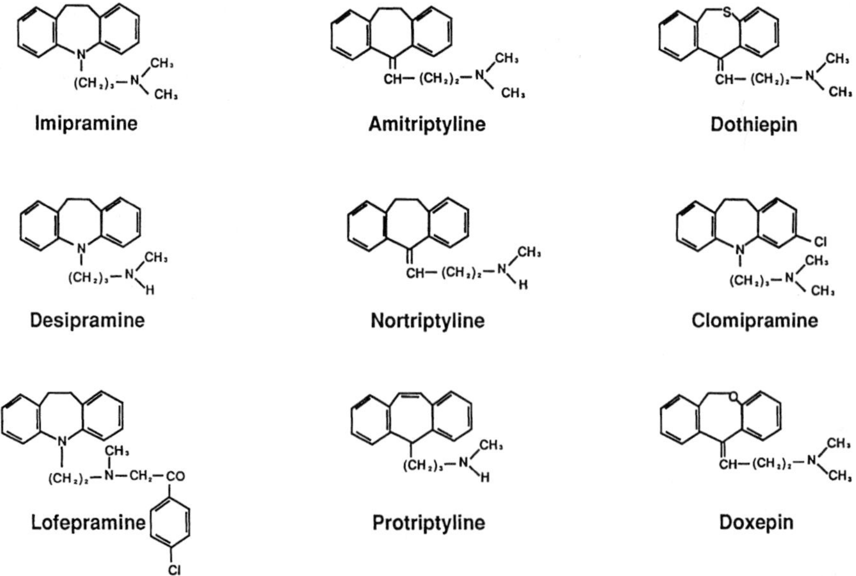

Figure 1. Chemical Structures of Well Known Tricyclic Antidepressants.

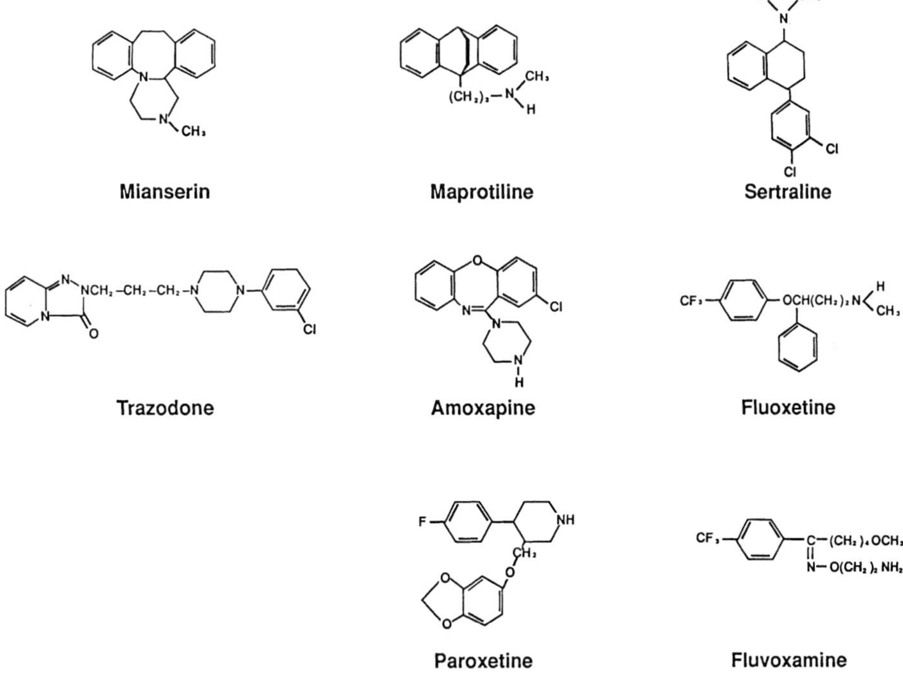

Figure 2. Chemical Structures of New Heterocyclic Antidepressants.

pondingly high and are of the order of 20–40 L/kg. Following oral ingestion of these drugs, they are extensively metabolised by the liver, mainly by N-demethylation and hydroxylation, with subsequent glucuronide conjugation and excretion in urine. The main metabolic pathways for imipramine and the formation of its two main active metabolites, desipramine and 2-hydroxydesipramine, are shown in figure 3. The plasma elimination half-lives of most tricyclic antidepressants are between 0.5–

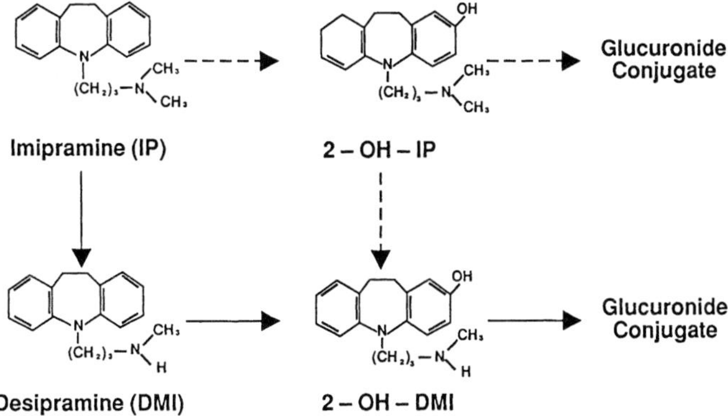

Figure 3. Metabolic Pathways of Imipramine. The Major Route of Metabolism is Shown in Bold Arrows.

2.0 days, although in some patients, particularly the elderly, half-lives in excess of 2 days may be observed (11). During repeated drug administration, plasma concentrations will rise to a plateau, the so-called "steady-state" concentration. The time to reach steady-state is a function of the plasma drug elimination half-life; 90% of the steady-state will be reached in 3.5 half-lives, and for most individuals this will be reached within 5–7 days. In those patients with slow or "defective" drug hydroxylation, steady-state concentrations of parent drug or active metabolites may not be reached for several weeks, with a consequent risk of delayed toxicity (11).

Early studies of nortriptyline and desipramine showed a 20–30 fold difference in steady-state plasma drug concentrations in patients receiving the same daily doses. Extensive work has been reported on the pharmacokinetic and pharmacogenetic mechanisms responsible for these large interindividual variations in plasma drug concentrations of nortriptyline and desipramine (12). Recent studies have investigated the genetic polymorphism in the activity of the cytochrome P-450 group of enzymes responsible for drug hydroxylation. Poor metabolisers of these drugs are homozygous for a defective CYP2D6 gene which has now been identified (13). These individuals are at special risk of accumulating extremely high plasma drug concentrations and consequently symptoms of toxicity (14).

3.6.3 Analysis

Introduction

There is an extensive literature concerning the qualitative and quantitative analysis of tricyclic and newer antidepressant drugs in biological fluids. Several hundred individual methods using a wide variety of analytical techniques (mainly chromatographic) have now been published. There are also a large number of reviews available (11, 15–20).

To some extent, immunoassay methods complement chromatographic techniques and there are clinical situations where certain analytical techniques are superior. As an example, it is sometimes difficult to resolve mixtures of some antidepressants and their metabolites using HPLC techniques when more than one antidepressant has been ingested. In contrast, gas chromatography, particularly capillary gas chromatography, is much better at resolving complex mixtures of antidepressants and is generally superior as a routine technique for drug identification, particularly when used with the accurate calculation of rentention indices (21).

Reference compounds and internal standards

Pure reference specimens of individual antidepressants are often available from individual manufacturers. Drugs are generally in the form of stable salts (e.g. hydrochloride) which are readily water soluble. Aqueous stock solutions in dilute acid are generally stable for many months when stored in a refrigerator. The acquisition of metabolites of antidepressants may present difficulties, particularly for older drugs such as imipramine and amitriptyline; the best source for metabolites is the manufacturer of the parent drug.

Calibration standards of mixtures of antidepressants can be prepared in bovine and equine serum and stored deep frozen (– 20 °C) for immediate use. Such standards are generally stable for many months, although stability problems have been found with prolonged storage of some drugs, e.g. imipramine and dothiepin.

The choice of an appropriate internal standard can be difficult, particularly where the identity of the antidepressant (and its metabolites) is uncertain. In most cases, such as in therapeutic drug monitoring or quantitative measurements in overdose, the identity of the drug is known from the history or preliminary qualitative investigations carried out on blood or urine. The choice of internal standard depends on the antidepressant being analysed, but the most common approach is to use another antidepressant drug that may be easily resolved by the analytical system being used (gas or liquid chromatography). In many cases it is useful to choose an antidepressant that is rarely prescribed and therefore unlikely to be found in patient specimens. Some authors have explored the use of more than one internal standard, e.g. a tertiary and secondary amine type compound, but this is not necessary for most analytical work.

Problems of analysis

Specimens collection is an important aspect of drug analysis that is often overlooked. Important factors to consider are: choice of specimens (serum, plasma, whole blood, urine), site of collection (post-mortem cases), time of sampling following dosage (therapeutic drug monitoring), specimen transport, storage and stability of drug and metabolites.

Urine is often used for drug screening, but qualitative and quantitative findings are unable to differentiate between therapeutic ingestion or overdosage. Further, in the case of acute poisoning, including fatalities, where urine has been collected very soon after ingestion, or death has been very rapid, very little drug may be found in urine, and blood should always be screened for the presence of ingested drug.

A number of early studies suggested that there was a problem with certain types of blood collection tubes, which may cause redistribution of drug between plasma and red-cells and spuriously low values in plasma. This problem has largely been resolved with the introduction of improved anticoagulated blood collection tubes and the issue has been reviewed by a number of authors (18, 20, 22). In particular, there appears to be very little difference between plasma or serum collected using the majority of evacuation type tubes (e.g. vacutainer). However, care should always be taken when any new blood collection tube system is introduced, particularly in the case of serum collection tubes where separating gells may be employed (22).

The timing of blood specimens is an important issue in therapeutic drug monitoring. The most appropriate collection time is 12–16 hours after a single night time dose or just before a dose when divided daily dosage regimes are used.

An aspect of specimen collection that is very often ignored is the stability of drug and metabolite during transport and storage of the specimens prior to analysis. In general, all of the tricyclic group of antidepressants are stable, even at room temperature for several days (23). Some newer antidepressants have been shown to be unstable, as in the case of nomifensine, a drug now withdrawn, but where analysis of plasma nomifensine was found to be very 'unreliable' due to the presence of high concentrations of an unstable N-glucuronide metabolite which was easily converted back to the parent drug (24).

Sample preparation

Specimen preparation and extraction of tricyclic antidepressants has recently been extensively reviewed by Wong (18) and Gupta (20). Most of the recently published methods on tricyclic antidepressant analysis have paid attention to the problems of recovery, since this is a critical issue in the quantitative analysis of these drugs. Various techniques have been used to resolve the problem of drug adsorption and poor recovery. In earlier studies silanised glassware was widely used, but this has now been replaced by the use of polypropylene tubes which are generally superior to the use of silanised glassware. In most liquid extraction methods the specimen is made alkaline, then extracted into a relatively non-polar organic solvent e.g. n-hexane; the drug is then back extracted into a small volume of acid and analysed directly by HPLC or re-extracted into a small volume of organic solvent for gas chromatographic analysis. More recently the use of solid phase extraction techniques has replaced liquid extraction and a number of commercial solid phase materials are suitable (20). Gupta (25) has also recently described a useful protein precipitation procedure to be used prior to solid phase extraction of drugs.

Thin-layer chromatography

Qualitative techniques. Thin layer chromatography following solid or liquid phase extraction of urine or stomach contents is still a widely used technique for drug screening that may be applied to the detection of tricyclic and other antidepressants. Many solvent systems have been described and many laboratories have their own in-house procedures (26). Visualisation of antidepressants can be carried out using a variety of spray reagents such as Dragendorff spray (platinic acid, potassium iodide, hydrochloric acid) and Mandelin's spray (Ammonium vanadate, concentrated sulphuric acid). Of these spray reagents, Mandelin's spray provides a relatively specific and sensitive means of detecting those tricyclic antidepressants that are related to imipramine or amitriptyline. A range of different colours may be observed which also change on heating. Some drugs e.g. amitriptyline and nortriptyline give very characteristic colours when viewed under UV light.

A commercial TLC drug screening procedure (TOXI-Lab) has also been developed which may be applied to the detection of these drugs in urine or stomach contents.

Quantitative techniques. A number of methods have been described for the quantitative analysis of tricyclic antidepressants by TLC (27–30). These methods require the use of a reliable scanning densitometer which may also be applied to the measurement of other drugs. Such methods are currently little used having largely been replaced by liquid chromatographic techniques.

Gas chromatography

Gas chromatography still remains an important technique for the qualitative identification of tricyclic and other antidepressant drugs, particularly with the use of capillary columns (21). Although gas chromatography was the method of choice for the quantitative analysis of tricyclic antidepressants during the 1970's and early 1980's, liquid chromatographic techniques are now the methods of choice for the quantitative analysis of all these drugs.

The earliest gas chromatographic method used flame ionisation detection (FID) but these methods are now only of historical interest, due to the limited sensitivity of detection. Electron-capture detection following derivative formation is a technique used in the early 1970's for the measurement of desipramine and nortriptyline (31). Such methods were only applicable to the direct measurement of secondary amine antidepressants, which can be derivatised with reagents such as trifluoroacetic anhydride or heptafluoro-butyric anhydride. Subsequent methods described procedures for the separate determination of tertiary and secondary amine type antidepressants using pre-column extraction and separation followed by derivative formation. There has recently been renewed interest in electron-capture detection for the specific analysis of "third generation" SSRI antidepressants such as fluvoxamine (32, 33).

Nitrogen specific detection without derivative formation using packed columns (e.g. OV 17, SP 2250) was a common approach to the analysis of tricyclic antidepressants during the 1970's and numerous methods were described for the analysis of most compounds and their N-desmethyl metabolites (34–39). These techniques may also be applied to the quantitative analysis of other antidepressants. One disadvantage of these techniques is that they cannot easily be applied to the analysis of polar hydroxylated metabolites.

The introduction of capillary columns has improved efficiency considerably, which is of great importance in the qualitative identification of antidepressants in drug screening procedures. Calculation of drug retention indices of antidepressants and metabolic association products is a useful approach (21).

Mass-spectrometry is a widely applied tool to drug identification, particularly when used as a specific detector following capillary gas chromatography. A number of quantitative methods using mass fragmentography have been published (40–42). Such methods are largely of historical interest, although in certain forensic cases it is important to provide mass spectrometric evidence of drug and metabolite identity. Bench-top gas chromatography systems are in common use in laboratories which now makes this technique a standard analytical approach to the "confirmation" of drug identity.

High performance liquid chromatography (HPLC)

The application of HPLC to the measurement of tricyclic antidepressants began in the mid 1970's with the use of straight phase and CN bonded silica phases. Rapid development in HPLC columns and separation techniques during the 1980's led to greatly improved performance in column efficiency and limits of detection. HPLC has now replaced gas chromatography as the technique of choice for the quantitative analysis of tricyclic and other antidepressant drugs, also their polar metabolites. Two recent extensive reviews concerning HPLC of tricyclic antidepressants have been published giving detailed information on column selection, mobile phase and detection wavelength (18, 20). A common approach is the use of C_8 or C_{18} modified silica columns with the use of mobile phases containing ion-pairing reagents (e.g. perchlorate) or bases (e.g. methylamine or triethylamine) to reduce peak tailing and improve resolution of closely related drugs (25). Early methods used UV detection at relatively high wavelengths (254 nm), however greatly improved detection limits can be successfully achieved at much lower wavelengths (210–215 nm). More recently McIntyre et al. (43) have published a dual ultraviolet wavelength HPLC method for the analysis of a wide range of antidepressants suitable for forensic and clinical specimens. Fluorescence detection has also been used for those antidepressants with suitable characteristics such as imipramine and its analogues (44). Electrochemical detection has also been successfully used for the detection of imipramine and its metabolites (45). HPLC has been successfully applied to the analysis of fluoxetine and norfluoxetine and may be applied to other antidepressants (46–50).

HPLC in combination with diode array detection also offers a powerful technique for both the qualitative and quantitative analysis of tricyclic and other antidepressants. More recently commercial automated HPLC systems have been developed by Bio-Rad Laboratories for both qualitative and quantitative analysis of a wide variety of drugs including antidepressants.

The determination of tricyclic antidepressants has recently been carried out using micellar electrokinetic capillary chromatography (51).

Immunoassays

Radioimmunoassays. The availability of specific antibodies to the tricyclic antidepressant ring structure was a promising development during the 1970's, which led to the publication of a number of sensitive methods (52–55). However, all of these methods demonstrated variable cross-reactivity between different tricyclic antidepressant drugs, also between tertiary amine parent drugs and their desmethyl metabolites. Although there were subsequent improvements in methodology, radioimmunoassay methods have fallen into disuse and are now only of historical interest. This has largely been due to the rapid development of non-isotopic immunoassay techniques for most drugs of clinical interest.

Non-isotopic immunoassays. During the 1980's there was an important and useful development of qualitative and subsequently quantitative methods using EMIT (Syva Corporation) and FPIA (Abbott Diagnostic) for the determination of tricyclic antidepressants. A range of different EMIT methods were developed for both therapeutic drug monitoring and diagnosis of tricyclic antidepressant overdosage. (19, 56–59). These methods are unable to differentiate between different tricyclic structures unless there is a prior solid phase separation procedure. Because of differences in cross reactivity profile, such methods can be unreliable for the accurate measurement of tricyclic antidepressant concentrations in overdosage, particularly where the identity of the particular antidepressant is unknown. EMIT serum assays have also been applied to analysis of urine (60, 61). One advantage of EMIT immunoassay methods is that they may be adapted to running on a large clinical laboratory analyser (19).

A quantitative assay for tricyclic antidepressants based on FPIA is also available from Abbott instruments (TDX and ADX). This assay is rapid to perform and is useful in the diagnosis of tricyclic antidepressant overdosage (62–63) but, similar to EMIT immunoassay methods, the assay exhibits different cross reactivity to different antidepressants (62–64). Most recently, new FPIA reagents have been described for the specific determination of amitriptyline and nortriptyline in the presence of each other (65).

The serious limitations in both EMIT and FPIA methods for the quantitative and qualitative analysis of tricyclic antidepressants have become of greater significance with the introduction of newer non-tricyclic antidepressants such as second generation "heterocyclics" and "third generation" SSRI drugs, all of which **cannot** be detected by these methods. Thus, immunoassay techniques are unreliable as simple laboratory methods for the diagnosis of "antidepressant" overdosage.

3.6.4 Drug Concentrations in Therapy and Overdose

Therapeutic drug monitoring

The relationship between plasma concentrations of tricyclic antidepressant drugs and their therapeutic effects has been the subject of intense investigation for more than 20 years and the subject is still somewhat controversial (11, 66). There is, however, better agreement on the relationship between plasma drug concentrations and the incidence of adverse reactions. A number of studies have shown that in both adults and children, high plasma concentrations are associated with an increased incidence of CNS associated toxicity (67–71). In a review of the central nervous system toxicity of tricyclic antidepressants by Preskorn and Jerkovich (71), the risks of toxicity were reported in 22–35% of patients whose tricyclic antidepressant plasma concentrations were above 300 µg/L, compared with only 0–8% of those patients with plasma levels below 300 µg/L. Moreover, studies showed that 56–86% of those patients with tricyclic antidepressant concentrations above 450 µg/L experienced toxicity, compared with only 0–7% of patients where levels were below 450 µg/L, representing a greater than 10 fold increased risk of toxicity. The early stages of tricyclic antidepressant induced toxicity can be mistaken for a worsening of depressive symptoms with a consequent risk of an increase in drug dosage (71). The accumulation of high plasma drug concentrations during therapy is related to poor or defective metabolism of these drugs, which is largely under genetic control. Pharmacogenetic studies have shown that patients who develop exceptionally high plasma drug concentrations are typically poor hydroxylators (12, 14).

There have been a number of published guidelines for the therapeutic monitoring of tricyclic antidepressants (11, 72–74). In general, the main indications for therapeutic drug monitoring of tricyclic antidepressants are shown in table 2. Monitoring is best established for antidepressants such as imipramine, desipramine, amitriptyline and nortriptyline.

For other tricyclic antidepressants and newer agents there is no consensus view, although some of the reasons for monitoring tricyclic antidepressants shown in table 2 will still apply.

Diagnosis and management of acute poisoning

Measurement of plasma concentrations of tricyclic and other antidepressants can be useful in the diagnosis of suspected drug overdosage. The situation is less clear regarding the prediction of severity of poisoning or in its management.

Earlier studies indicated that patients with plasma drug concentrations in excess of 1000 µg/L (1 mg/L) were associated with more severe symptoms of poisoning

(75–77). Subsequent studies have shown a more complex picture. A large multicentre study reported by Hulten et al. (78) indicated that the grade of coma on admission was the best predictor of outcome, although plasma drug concentrations in patients with complications were significantly higher than those without complications. A major study by Boehnert and Lovejoy (79) also showed that plasma drug concentrations were unable to predict the risk of seizures or ventricular arrhythmias.

Single or repeated measurements of plasma drug concentrations of tricyclic and other antidepressants are of little value in the management of poisoned patients where the diagnosis is already clearly established. Laboratory measurements are generally only of use in the investigation of acutely poisoned patients where the diagnosis is uncertain, such as unconscious patients or children (77). Measurement of the ratio of parent drug to main active metabolite can also have diagnostic value in differentiating between acute and chronic ingestion (80). A recent study by Amitai et al. (81) investigated the relative concentrations of parent drug, tertiary amine antidepressant and secondary N-desmethyl metabolites in plasma and red cells of a small series of poisoned patients. Higher concentrations of parent drug were found in plasma compared with red-cells; in the case of active metabolite, the concentration in red cells gave a better correlation with the QRS duration than the metabolite concentration in plasma (81). Chromatographic techniques are required for the analysis of individual concentrations of parent drug and active metabolite in the investigation of poisoning. These methods are complex and laborious and generally not available on an emergency basis from clinical laboratories. Immunoassays (e.g. Syva EMIT, Abbott TDX) can provide a rapid semi-quantitative analysis of 'total' tricyclic antidepressant concentrations suitable for rapid diagnostic purposes. But, they are unable to detect the presence of heterocyclic or newer SSRI antidepressants; conventional chromatographic techniques are required in the analysis of these drugs.

Post-mortem investigation of poisoning

The measurement of tricyclic antidepressant drugs in post-mortem blood (and tissues in some cases) is of great importance in establishing a cause of death (82–83). Most of the methods applied to serum or plasma can be applied to post-mortem blood with few problems. Interpretation of findings can however be difficult if the site of blood specimen collection is unknown. It is well established that for many drugs, including antidepressants, which have high volumes of distribution, there are great differences in the concentrations of drug and metabolite between different sites of post-mortem sampling (84). Exceptionally 'high' values can be found in heart blood or from other 'central' sites. The most consistent findings are obtained for specimens taken from a 'peripheral' region, e.g. femoral vein. It is a common finding that concentrations of antidepressants will **increase** in blood following death due to the phenomenon of post-mortem redistribution (84). Great care and experience is required in the interpretation of findings. The concentration of drug found in post-mortem blood is not equivalent to that in plasma or serum at the time of death.

Acknowledgement

I would like to thank Mrs. S. Samra for secretarial assistance in the preparation of this manuscript.

References

1. Noble, J. and Mathew, H. (1969) Acute poisoning by tricyclic antidepressants. Clinical features and management of 100 patients. Clin. Toxicol. *2*, 403–421.
2. Crome, P. (1986) Poisoning due to tricyclic antidepressant overdosage – clinical presentation and treatment. Med. Toxicol. *1*, 261–285.
3. Cassidy, S. and Henry, J. (1987) Fatal toxicity of antidepressant drugs in overdose. Brit. Med. J. *295*, 1921–1024.
4. Fromer, D.A., Kulig, K.W., Marx, J.A. and Rumack, B. (1987) Tricyclic antidepressant overdose – A review J. Amer. Med. Assoc. *257*, 521–526.
5. Dziukas, L.J. and Vohra, J. (1991) Tricyclic antidepressant poisoning. Med. J. Aust. *154*, 344–350.
6. Buckley, N.A., Dawson, A.H., Whyte, I.M. and Henry, D.A. (1994) Greater toxicity in overdose of dothiepin than of other tricyclic antidepressants. Lancet *343*, 159–162.
7. Henry, J.A. and Martin, A.J. (1987) The risk-benefit assessment of antidepressant drugs. Medical Toxicology *2*, 445–462.
8. Rudorfer, M.V. and Potter, W.Z. (1989) Antidepressants. A comparative review of the clinical pharmacology and therapeutic use of the 'newer' versus the 'older' drugs. Drugs *37*, 713–738.
9. Grimsley, S.R. and Jann, M.W. (1992) Paroxetine, sertraline and fluvoxamine: new selective serotonin re-uptake inhibitors. Clin. Pharm. *11*, 930–57.
10. Borys, D.J., Setzer, S.C., Ling, L.J., Reisdorf, J.J., Day, L.C., Krenzelok, E.P. (1992) Acute fluoxetine overdose: a report of 234 cases. Am. J. Emerg. Med. *10*, 115–120.
11. Braithwaite, R.A. (1985) The tricyclic antidepressants. In: Therapeutic Drug Monitoring (B. Widdop edit) pp. 211–239. Churchill Livingstone.
12. Sjöqvist, F. (1989) Pharmacogenetics of antidepressants. In: Clinical Pharmacology in Psychiatry Editors: Dahl, S.G. and Gram, L.F. Psychopharmacol. (Series 7). Springer-Verlag, Berlin, pp. 181–191.
13. Heim, M. and Meyer, U.A. (1990) Genotyping of poor metabolisers of debrisoquine by allele-specific amplifaction. Lancet *336*, 529–532.
14. Tache, U., Leinonen, E., Lillsunde, P., Seppala, T., Arvela, P., Pelkonen, O. and Ylitalo, P. (1992) Debrisoquine hydroxylation phenotypes of patients with high versus low to normal serum antidepressent concentration. J. Clin. Psychopharm. *12*, 262–267.
15. Scoggins, B.A., Maguire, K.P., Norman, T.R., Burrows, G.D. (1980) A Measurement of tricyclic antidepressants Part I. A review of methodology. Clin. Chem. *26*, 5–17.
16. Scoggins, B.A., Maguire, K.P., Norman, T.R., Burrows, G.D. (1980) Measurement of tricyclic antidepressants. Part II. Applications of methodology. Clin. Chem. *26*, 805–815.
17. Norman, T.R. and Maguire, K.P. (1985) Analysis of tricyclic antidepressants drugs in plasma and serum by chromatographic techniques. J. Chromatog. *340*, 173–97.
18. Wong, S.H.Y. (1988) Measurement of antidepressants by liquid chromatography: a review of current methodology Clin. Chem. *34*, 848–855.
19. Orsulak, P.J., Haven, M.C., Burton, M.E. and Akers, L.C. (1989) Issues in methodology and applications for therapeutic drug monitoring of antidepressant drugs. Clin. Chem. *35*, 1318–1325.
20. Gupta, R. (1992) Drug level monitoring: antidepressants. J. Chromatog. *576*, 183–211.
21. deZeeuw, R.A., Franke, J.P., Maurer, H.H. and Pfleger, K. (1992) Gas chromatographic retention indices of toxicologically relevant substances on packed or capillary columns with dimethylsilicone stationary phases. Third Edition, VCH, Weinheim.
22. Nyberg, G. and Martensson, E. (1986) Preparation of serum and plasma samples for determination of tricyclic antidepressants: effects of blood collection tubes and storage. Ther. Drug Monit. *8*, 478–482.
23. Burch, J.E., Raddats, M.A. and Thompson, S.G. (1979) Reliable routine method for the determination of plasma amitriptyline and nortriptyline by gas chromatography. J. Chromatog. *162*, 351–366.
24. Dawling, S. and Braithwaite, R.A. (1980) The stability of the antidepressive agent nomifensive in human plasma. J. Pharm. Pharmacol. *32*, 304–305.
25. Gupta, R.N. (1993) An improved solid phase extraction procedure for the determination of antidepressants in serum by column liquid chromatography. J. Liqu. Chromatog. *16*, 2751–2765.
26. de Zeeuw, R.A., Franke, J.P., Degel, F., Machbert, G., Schutz, H. and Wijsbeek, J. (1992) Thin-layer chromatographic Rf val-

ues of toxicologically relevant substances on standardized systems. Second Edition, VCH, Weinheim.

27. Haefelfinger, P. (1978) Determination of amitriptyline and nortriptyline in human plasma by quantitative thin-layer chromatography. J. Chromatog. *145*, 445–451.
28. Haefelfinger, P. (1980) Experience with quantitative thin-layer chromatography in the analysis of drugs in plasma. J. Liq. Chromatog. *3*, 797–808.
29. Breyer, U. and Villumsen, K. (1976) Measurement of plasma levels of tricyclic psychoactive drugs and their metabolites by UV reflectance photometry of thin layer chromatograms. Eur. J. of Clin. Pharmacol. *9*, 457–465.
30. Fenimore, D.C., Meyer, C.J., Davis, C.M., Hsu, F., Zlatkis, A. (1977) High performance thin-layer chromatographic determination of psychopharmacologie agents in blood serum. J. Chromatog. *142*, 399–409.
31. Borga, O., Garle, M. (1972) A gas chromatographic method for the quantitative determination of nortriptyline and some of its metabolites in human plasma and urine. J. Chromatog. *68*, 77–88.
32. Hurst, H.E., Jones, D.R., Jarboe, C.H. (1981) Determination of clovoxamine concentration in human plasma by electron-capture gas chromatography. Clin. Chem. *27*, 1210–1212.
33. Lantz, R.J., Farid, K.Z., Koons, J., Tenbarge, J.B., Bopp, R.J. (1993) Determination of fluoxetine and norfluoxetine in human plasma by capillary gas chromatography with electron capture detection. J. Chromatog. *614*, 175–179.
34. Gifford, L.A., Turner, P., Pare, C.M.B. (1975) Sensitive method for the routine determination of tricyclic antidepressants in plasma using a specific nitrogen detector. J. Chromatog. *105*, 107–113.
35. Bailey, D.N., Jatlow, P.I. (1976) Gas chromatographic analysis for therapeutic concentrations of amitriptyline and nortriptyline in plasma with use of a nitrogen detector. Clin. Chem. *22*, 777–781.
36. Dorrity, F., Linnoila, M., Habig, R. (1977) Therapeutic monitoring of tricyclic antidepressants in plasma by gas chromatography. Clin. Chem. *23*, 1326–1328.
37. Hucker, H.B. and Stauffer, S.C. (1977) Rapid, sensitive gas-liquid chromatographic method for determination of amitriptyline and nortriptyline in plasma using a nitrogen sensitive detector. J. Chromatog. *138*, 437–442.
38. Dawling, S. and Braithwaite, R.A. (1978) Simplified method of monitoring tricyclic antidepressant therapy using gas-liquid chromatography with nitrogen detection. J. Chromatog. *146*, 449–456.
39. Dhar, A.K. and Kutt, H. (1979) An improved gas-liquid chromatographic procedure for the determination of amitriptyline and nortriptyline levels in plasma using nitrogen – sensitive detection. Ther. Drug Monit. *1*, 209–216.
40. Wilson, J.M., Williamson, L.J., Raisys, V.A. (1977) Simultaneous measurement of secondary and tertiary tricyclic antidepressants by GC/MS chemical ionisation mass fragmentography. Clin. Chem. *23*, 1012–1017.
41. Jenkins, R.G. and Friedel, O. (1978) Analysis of tricyclic antidepressants in human plasma by GLC chemical ionisation mass spectrometry with selective ion monitoring. J. of Pharm. Sc. *67*, 17–23.
42. Garland, W.A., Muccino, R.R., Min, B.H., Cupano, J., Fann, W.E. (1979) A method for the determination of amitriptyline and its metabolites nortriptyline, 10-hydroxyamitriptyline and 10-hydroxynortriptyline in human plasma using stable isotope dilution and gas chromatography chemical ionisation mass spectrometry. Clin. Pharm. Ther. *25*, 844–856.
43. McIntyre, I.M., King, C.V., Skafidis, S. and Drummer, O.H. (1993) Dual ultraviolet wavelength high-performance liquid chromatographic method for the forensic or clinical analysis of seventeen antidepressants and some selected metabolites. J. Chromatog. Biomed. Appl. *621*, 215–223.
44. Sutfin, T.A. and Jusko, W.J. (1979) High performance liquid chromatographic assay for imipramine, desipramine and their 2-hydroxylated metabolites. J. of Pharm. Sci. *68*, 703–705.
45. Suckow, R.F. and Cooper, T.B. (1981) Simultaneous determination of imipramine, desipramine and their 2-hydroxy metabolites in plasma by ion-pair reversed phase high performance liquid chromatography with amperometric detection. J. Pharm. Sci. *70*, 257–261.
46. Thomare, P., Wang, K.G., Van der Meersch-Mougeot, V., Diquet, B. (1992) Sensitive micromethod for column liquid chromatographic determination of fluoxetine and norfluoxetine in human plasma J. Chromatog. *583*, 217–221.
47. El Maanni A., Combourieu, I., Bonini, M., Creppy, E.E. (1993) Fluoxetine, an antide-

pressant, and norfluoxetine, its metabolite determined by HPLC with a C_8 column and ultraviolet detection. Clin. Chem. *39*, 1749–1750.

48. Nichols, J.H., Charlson, J.R. and Lawson, G.M. (1994) Automated HPLC assay of fluoxetine and norfluoxetine in serum. Clin. Chem. *40*, 1312–1316.
49. El-Yazigi, A. and Raines, D.A. (1993) Concurrent liquid chromatographic measurement of fluoxetine, amitriptyline, imipramine, and their active metabolites norfluoxetine, nortriptyline and desipramine in plasma. Ther. Drug. Monit. *15*, 305–309.
50. Wang, P.T., Van Der Meersch-Mougeot, V. and Diquet, B. (1992) Sensitive micromethod for column liquid chromatographic determination of fluoxetine and norfluoxetine in human plasma. J. Chromatog. *583*, 217–221.
51. Lee, K.-J., Lee, J.J. and Moon, D.C. (1993) Determination of tricyclic antidepressants in human plasma by micellar electrokinetic capillary chromatography. J. Chromatog. Biomed. Appl. *616*, 135–143.
52. Lucek, R. and Dixon, R. (1977) Specific radioimmunoassay for amitriptyline and nortriptyline in plasma. Res. Commun. Chem. Path. Parm. *18*, 125–135.
53. Robinson, J.D., Risby, D., Riley, G., Aherne, G.W. (1978) A radioimmunoassay for the determination of combined amitriptyline and nortriptyline concentrations in microlitre samples of plasma. J. Pharmac. Exp. Ther. *205*, 499–502.
54. Brunswick, D.J., Needelman, B., Mendels, J. (1979) Specific radioimmunoassay of amitriptyline and nortriptyline. Brit. J. Clin. Pharmacol. *7*, 263–269.
55. Virtanen, R. (1980) Radioimmunoassay for tricyclic antidepressants. Scand. J. Clin. Lab. Invest. *40*, 191–197.
56. Schroeder, T.J., Tasset, J.J., Otten, E.J. and Hedges, J.R. (1986) Evaluation of Syva EMIT toxicological serum tricyclic antidepressant assay. J. Anal. Tox. *10*, 221–224.
57. Pankey, S., Collins, C., Jaklitsch, A., Izutsu, A., Hu, M., Piro, M. and Singh, P. (1986) Quantitative homogeneous enzyme immunoassays for amitriptyline, nortriptyline and desipramine. Clin. Chem. *32*, 768–772.
58. Benitez, J., Dahlqvist, R., Gustafsson, L.L., Magnusson, A., Sjoquist, F. (1986) Clinical pharmacological evaluation of an assay kit – intoxications with tricyclic antidepressants. Ther. Drug Monit. *8*, 102–105.
59. Dorey, R.C., Preskorn, S.H. and Wildener, P.K. (1988) Results compared for tricyclic antidepressants as assayed by liquid chromatography and enzyme immunoassay. Clin. Chem. *34*, 2348–2351.
60. Asslin, W.M. and Leslie, J.M. (1990) Use of EMIT TOX serum tricyclic antidepressant assay for the analysis of urine samples. J. Anal. Tox. *14*, 168–171.
61. Meenan, G.M., Barlotta, S. and Lehrer, M. (1990) Urinary tricyclic antidepressant screening: comparison of results obtained with Abbott FPIA reagents and Syva EIA reagents. J. Anal. Tox *14*, 273–276.
62. Poklis, A., Sogholan, D., Crooks, C.R., Saady, J.J. (1990) Evaluation of the Abbott ADX total serum tricyclic immunoassay. Clin. Toxicol. *28*, 235–248.
63. Rao, M.L., Staberock, U., Baumann, P., Hiemke, C., Deister, A., Cuendet, C., Amey, M., Hartter, S. and Kraemer, M. (1994) Monitoring tricyclic antidepressant concentrations in serum by fluorescence polarization immunoassay compared with gas chromatography and HPLC. Clin. Chem. *40*, 929–933.
64. Nebinger, P. and Koel, M. (1990) Specificity data of the tricyclic antidepressants assay by fluorescent polarisation immunoassay. J. Anal Tox. *14*, 219–221.
65. Adamczyk, M., Fishpaugh, J.R., Harrington, C.A., Hartter, D.E., Vanderbilt, A.S., Orsulak, P. and Akers, L. (1994) Immunoassay reagents for psychoactive drugs. Part 4. Quantitative determination of amitriptyline and nortriptyline by fluorescence polarisation immunoassay. Ther. Drug. Monit. *16*, 298–311.
66. Perry, P.J., Pfohl, B.M. and Holstand, S.G. (1987) The relationship between antidepressant response and tricyclic antidepressant plasma concentrations. A retrospecitve analysis of the literature using logistic regression analysis. Clin. Pharmacokin. *13*, 381–392.
67. Loo, H., Benyacoub, A.K., Rovei, V., Altramura, C.A., Vadrot, M., Morselli, P.L. (1980) Long-term monitoring of tricyclic antidepressant plasma concentrations. Brit. J. of Psych. *137*, 444–451.
68. Dawling, S. (1982) Monitoring of tricylic antidepressant therapy. Clin. Biochem. *15*, 56–61.
69. Meador-Woodruff, J.H., Abil, M., Wisner-Carson, R., Grunaus, L. (1988) Behavioural and congitive toxicity related to elevated plasma tricyclic antidepressant levels. J. Clin. Psychopharm. *8*, 28–32.

70. Preskorn, S.H., Jerkovish, G.S., Beber, J.H., Widener, P. (1989) Therapeutic drug monitoring of tricyclic antidepressants: a standard of care issue: Psychopharm. Bulletin *25*, 281–184.
71. Preskorn, S.H. and Jerkovich, G.S. (1990) Central nervous system toxicity of tricyclic antidepressants: phenomenology, course, risk factors and role of therapeutic drug monitoring. J. Clin. Psychopharm. *10*, 88–95.
72. Preskorn, S.H. (1988) Therapeutic drug monitoring of antidepressants. Clin. Chem. *34*, 822–828.
73. Preskorn, S.H. (1989) Tricyclic antidepressants: The whys and hows of therapeutic drug monitoring. J. Clin. Psych. 50 (Suppl) 34–42.
74. Orsulak, P.J. (1989) Therapeutic monitoring of antidepressant drugs: guidelines updated. Ther. Drug. Monit. *11*, 497–507.
75. Biggs, J.T., Spiker, D.G., Petit, J.M., Ziegler, V.E. (1977) Tricyclic antidepressant overdose. J. Am Med. Assoc. *238*, 135–138.
76. Petit, J.M., Spiker, D.G., Ruwitch, J.F., Ziegler, V.E., Weiss, A.N., Biggs, J.T. (1977) Tricyclic antidepressant plasma levels and adverse effects after overdose. Clin. Pharm. Ther. *21*, 47–51.
77. Crome, P. and Braithwaite, R.A. (1978) Relationship between clinical features of tricylic antidepressant poisoning and plasma concentrations in children. Arch. Dis. Child. *53*, 902–905.
78. Hulten, B.-A., Adams, R., Askenasi, R., Dallos, V., Dawling, S., Volans, G., Heath, A. (1992) Predicting severity of tricyclic antidepressant overdose. Clin. Toxicol. *30*, 161–170.
79. Boehnert, M.T. and Lovejoy, F.H. (1985) Value of the QRS duration versus the serum drug level in predicting seizures and ventricular arrythmias after an acute overdose of tricyclic antidepressant. N. Eng. J. of Med. *313*, 474–479.
80. Bailey, D.N., Van Dyke, C., Langou, R.A., Jatlow, P.I. (1978) Tricyclic antidepressants: plasma levels and clinical findingsin overdose. Am. J. of Psych. *135*, 1325–1328.
81. Amitai, Y., Erickson, T., Kennedy, E.J., Leikin, J.B., Hryorczuk, D.O., Noble, J., Hanashiro, P.K. and Frischer, H. (1993) Tricyclic antidepressants in red cells and plasma: correlation with impaired intraventricular conduction in acute overdose. Clin. Pharm. Ther. *54*, 219–227.
82. Bailey, D. (1979) Tricyclic antidepressants: interpretation of blood and tissue levels in fatal overdose. J. Anal. Tox. *3*, 43–46.
83. Apple, F.S. (1989) Post-mortem tricyclic antidepressant concentrations assessing cause of death using parent drug to metabolite ratio. J. Anal. Tox. *13*, 197–198.
84. Pounder, D.J. and Jones, G.R. (1990) Postmortem drug redistributin – a toxicology nightmare. Forensic Sci. Int. *45*, 253–36.

3.7 Beta-Blocking Drugs

M.S. Leloux

3.7.1 Introduction

Beta-blocking drugs (fig. 1) are versatile cardiovascular drugs which are used for the pharmacotherapeutical treatment of e.g. angina pectoris, hypertension, hyperthyroidism, migraine and anxiety states (1, 2). They are weak bases, due to the secundary amino-function in the isopropylaminopropanol-side-chain, which is ty-

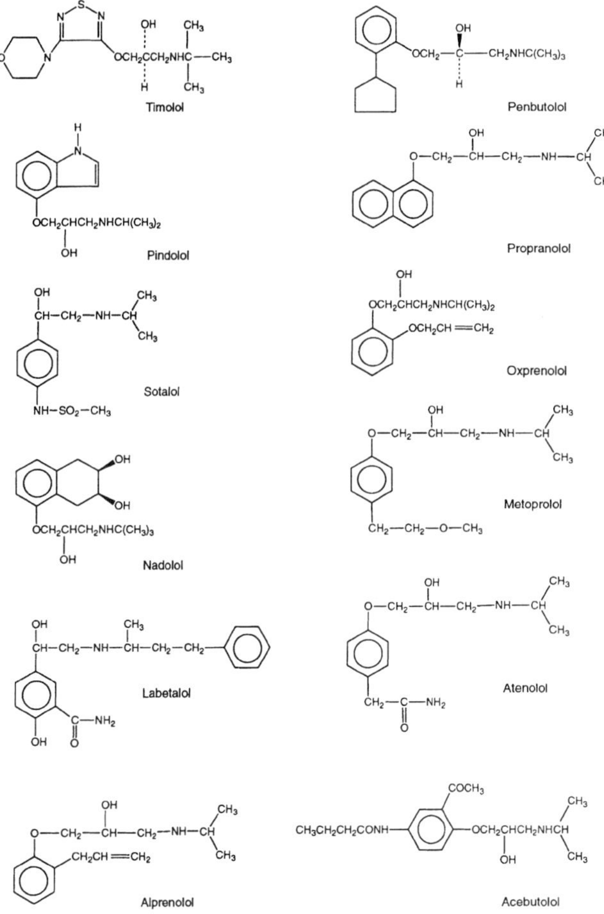

Figure 1. Structure of Respective β-Blocking Drugs.

pically linked via an ether-bridge to an aryl-function, such as a naphtyl-moiety in the case of propranolol, being the most well-known and prescribed beta-blocking drug since 1964. The pharmacokinetic fate of beta-blocking drugs depends on their polarity: polar beta-blocking drugs, such as atenolol and sotalol, are excreted almost unchanged by the kidneys, whereas non-polar beta-blocking drugs, such as alprenolol and propranolol are extensively metabolized by the liver before they enter the systemic circulation. Metabolic conversion of beta-blocking drugs includes phase-I reactions, such as oxidative deamination of the aminopropanol side chain and hydroxylation reactions of the aromatic nuclei, as well as phase-II reactions, such as conjugations with glucuronic acid (3–6).

Bioanalytical assays for the quantitative analysis of individual beta-adrenoceptor blocking drugs in biological specimens, e.g. plasma and urine, are of considerable value in pharmacokinetic and pharmacodynamic studies. Bioanalytical methods for the screening of several beta-adrenoceptor blocking drugs are important for doping research and other aspects of analytical toxicology.

Generally, a bioanalytical method for beta-blocking drugs consists of a typical sample pre-treatment, i.e. an isolation of the compound(s) of interest from the biological specimens and a concomitant clean-up from endogenous constituents, followed by instrumental analysis. As a rule, a chromatographic separation method prior to detection is inserted into the analytical scheme. In both cases, the instrumental assay should be rapid and should permit the routine analysis of a large number of samples per day. For the same reason, the sample pre-treatment method should be simple and fast. Current techniques for the bioanalysis of beta-adrenoceptor blocking drugs and metabolites include extraction procedures, gas-chromatography, high-performance liquid chromatography and mass spectrometry. Furthermore, scientific attention is increasingly focused on the determination of individual enantiomers of these drugs in plasma, rather than on the bioanalysis of the unresolved racemates. In this section, the bioanalysis of beta-blocking drugs will be outlined.

3.7.2 Sample Pre-Treatment Methods for Beta-Blocking Drugs

Conjugates of beta-blocking drugs in urine or serum are rapidly hydrolyzed by either strong acid in the presence of heat (7, 8) or by enzymes such as arylsulfatase and ß-glucuronidase (9). Using the first approach, artefacts may be formed and an anti-oxidant, e.g. cysteine, is usually being added to prevent oxidation. The second pre-isolation-step generally includes liquid-liquid or solid-phase extraction methods. Liquid-liquid extraction methods for beta-blocking drugs, including their main metabolites, involve an adjustment of the pH of the biological specimen to a pH of 9.0 or higher, using sodium hydroxide, a boronate or phosphate buffer, or sodium carbonate, and subsequent extraction with (mixtures from) diethylether, dichloromethane, dibutylacetate, dichloroethane, n-hexane etc (10).

Lately, solid phase extraction (SPE) using bonded phases has gained more interest, due to its easiness, highly reproducible results, clean extracts, high recoveries, good precision and the possibility of automatization (11–13). The choice of several commercially available bonded phases showing different polarities greatly enhances its potential as a versatile extraction technique. An important drawback of this methodology is its relatively high cost with respect to a conventional liquid-liquid extraction. However these expenses may be reduced by re-conditioning the solid-phase cartridge several times (14). For beta-blocking drugs, recoveries ranging from 60–100% have been found using octadecyl-, cyanopropyl and ethyl-bonded silicas for their extraction from biological specimens (15–20).

Other solid-phase extraction approaches, such as Extrelut® solid phase extraction cartridges containing diatomeous earth, are also described for the extraction of beta-blockers from biological fluids (9).

3.7.3 Gas Chromatography of Beta-Blocking Drugs

So far, several gas-chromatographic analytical methods have been published for beta-blocking drugs. In general, the bifunctional polar groups of their aminopropanol side-chain are converted to less polar derivatives by means of specific derivatization reactions, such as acylation, silylation, transboronation, oxazolidine formation, and combinations. The combination of an acylation-derivatization and an electron capture detection (ECD) is the principal analytical procedure for the gas-chromatographic determination of beta-blocking drugs. Several reagents are used, e.g. acetic acid anhydride (8, 21) or trifluoroacetic anhydride (22–30) for both screening and quantiative analysis of individual beta-blocking drugs, such as practolol, atenolol, oxprenolol and metoprolol in biological specimens, pentafluoropropionic anhydride (31–36), heptafluorobutyric anhydride, (37–46), trifluoroacetylimidazole (47, 48), N-heptafluorobutyrylimidazol (49, 50) and N-methyl-bistrifluoroacetamide (MSTFA) (51). A disadvantage of the use of the perfluoroacyl derivatives for the analysis of biological fluids is that their low selectivity and high reactivity with endogenous substances may result in a high background (52). Additionally, the use of a catalyst, e.g. trimethylamine or pyridine, is needed to shorten the reaction time of the derivatization reaction. As a result, an additional washing-step is needed to remove this catalyst, which might decrease the reproducibility of the method and is time-consuming. Silylation reagents, such as N-trimethylsilylimidazol (53–55) for the analysis of nadolol in plasma and serum, N- bis (trimethylsilyl)trifluoroacetamide (BSTFA) (56–59) for the analysis of timolol, tertalol, 4-hydroxy-tertalol and bufuralol including its main metabolites in biological fluids, have been described. An important advantage of silylation reagents is their relative ease of derivative formation. However, a disadvantage of the procedure is that absolutely dry reaction conditions are needed. Moreover, the frequent use of silyl derivatives and the direct injection of the reaction mixture onto the column usually causes cumulation of by-products such as silicium dioxide in the gaschromatographic detector, which may seriously decrease sensitivity. Transboronation has been applied by some authors (19, 60–62) who used alkylboronic acids for the determination of several beta-blokking drugs. The use of the noxious compound phosgene as a derivatizing reagent prior to gas-chromatographic analysis has been reviewed by Gyllenhaal and Vessmann (63). Another approach for the derivatization of the bifunctional groups of beta-blocking drugs involves the formation of N-TFA, O-TMS-derivatives of beta-blocking drugs using N-methyl-N-trimethylsilyltrifuoroacetamide (MSTFA) and N-methyl-bis-trifluororacetamide (MPTFA) which is recommended for the doping analysis of beta-blocking compounds (7, 19, 64, 65).

3.7.4 High Performance Liquid Chromatography of Beta-Blocking Drugs

In recent years, HPLC-methods have become increasingly popular for the routine determination of beta-blocking drugs in biological matrices. The detection method of choice generally involves a spectrophotometric or spectrofluorimetric procedure, since beta-blocking drugs contain the structural requirements which are needed for UV- and/or fluorimetric detection. The early literature on HPLC-assays for beta-blockers is covered by several excellent reviews (10, 66). For the bioanalysis of beta-blocking drugs using reversed-phase high-performance liquid chromatography, mobile phases consisting of methanol- or acetonitrile-buffer (pH 2.4–5) are generally being used. In a few cases, small amounts of an alkylamine, e. g. trimethylamine or nonylamine, have been added to improve the chromatography of the eluting beta-blocking drugs. The addition of an alkylamine not only minimizes the interaction between the basic drugs and the weakly acidic residual silanol groups of the HPLC-stationary phase, but it also influences the ionization of the basic drugs which in turn influences the peak shape (67).

In reversed-phase ion-pair chromatography, a counter-ion has been added to the mobile phase. The mobile phase further consists of an acidic methanol or acetonitrile-water mixture. The acid makes the beta-blocking drug cationic, thereby allowing the compound to be ion-paired with the counter-ion. Several counter-ions have been evaluated: heptane sulfonic acid (68–75), N-octane sulphonic acid (74–81), dodecane sulphonic acid (82–84) and camphor sulphonic acid (85, 86). Finally, a cation-exchange stationary phase for HPLC (Partisil 10-SCX) and a mixture of acetonitrile-water-diethylamine-85% orthophosphoric acid (20:80:0.2:0.15, v/v) were used for the determination of propranolol and nadolol in serum, urine and saliva (87).

3.7.5 Mass Spectrometry of Beta-Blocking Drugs

Nowadays, mass spectrometry is an important and powerful analytical tool for compound confirmation and metabolite identification. Gas chromatography-electron impact (EI) mass spectrometry has been used in a study of trifluoroacetylated and acetylated beta-blocking drugs and their O-trimethylsilyl, transboronation, TMS/TFA, heptafluorobutyric anhydride and pentafluoropropionic anhydride derivatives respectively (7, 8, 19, 22, 34, 46, 52, 62, 65, 88, 89). While positive chemical ionization $(CI)^+$ mass spectrometry of TMS/TFA-derivatives of beta-blockers has been studied recently (90), the negative chemical ionization $(CI)^-$ mode has received little attention for the investigation of structural and analytical problems of these drugs (91). Its use for the detection of beta-blocking compounds after derivatization with pentafluoropropionic anhydride (31) and heptafluorobutyric anhydride (40) has been demonstrated. Thermospray liquid chromatography-mass spectrometry deals with a soft form of ionization and generates molecular ions with little fragmentation directly post-HPLC. Its use has been limited to the plasma determination of labetalol (92) and some polar beta-blocking drugs in urine (93).

3.7.6 Chiral Separation of Beta-Blocking Drugs

During the last few years, a considerable progress in the field of instrumental methods for enantiomer separation on the analytical scale has been made. Several literature reviews dealing with this subject have been published by some leading authorities in this field (94–103). Typical conventional methods for measuring optical activity, e.g. polarimetry, optical rotatory dispersion and circular dichroism have mostly been replaced by chromatographic separation methods, such as gas-chromatography (GC) and high-performance liquid chromatography (HPLC). All chromatographic chiral separation methods utilize the principle of two separate, virtually inmiscible phases and a chiral discriminator or selector in the stationary phase (GC, HPLC, TLC) and/or mobile phase (HPLC, TLC).

Basically, indirect chromatographic separation methods involve the formation of either enantiomeric derivatives with a non-chiral reagent and their subsequent separation with a chiral stationary phase, or the creation of diastereomic derivatives using a chiral reagent and their subsequent separation with a non-chiral (or chiral) stationary phase. Unfortunately, such derivatization reactions prior to chromatography are not only time-consuming, but the availability of the reagents might be low, the costs of optically pure reagents might be high, and the occurrence of side-effects, e.g. racemizations and other changes of original enantiomeric ratios, might be evident as well. Direct chiral chromatographic separation methods involve either the use of a chiral stationary phase (CSP) with different affinities for the enantiomers to be separated, or the use of a non-chiral stationary phase in combination with the addition of a chiral reagent to the mobile phase. The chiral separation of beta-blocking drugs using gas-chromatography was applied by several investigators. A chiral derivatization reagent combined with a normal stationary phase has been described (104) using N-heptafluorobutyryl-1-prolylchloride (or N-trifluoroacetyl-1-prolylchloride) and BSTFA as the derivatization reagent and a capillary or packed 3% OV-225 column, also after a derivatization with R-(+)-1-phenylethyl-isocyanate (105). The use of a non-chiral derivatization-reagent and a chiral stationary phase was also evaluated (106, 107). In both cases, a capillary XE-phenylethylamide-valine-(R)-α-stationary phase was used. Derivatization was performed by phosgene to form cyclic oxazolidines. High-performance liquid chromatography may be considered to be a more popular analytical tool for the chiral separation of beta-blocking drugs in biological fluids (108). Petterson and co-workers (109–111) introduced a very elegant enantiospecific ion-pair HPLC approach for beta-blocking drugs (e.g.

alprenolol, metoprolol, propranolol and oxprenolol) using a polar stationary phase and an organic mobile phase containing (+)-10-camphorsulphonate or N-benzoxycarbonyl-glycyl-L-proline (ZGP) as the respective counterions. Gupta et al. (112) used a mobile phase containing d-10-camphorsulphonic acid as the ion-pairing reagent and tert.-butylamine as the competing base, and a Zorbax cyano stationary phase for the chiral separation of propranolol and derivatives. The application of d-10-camphorsulphonate-ion-pair HPLC for pharmacokinetic studies has also been demonstrated (113). Hermansson (114) described the use of enantiospecific ion-pair HPLC in combination with a chiral derivatization, to improve the resolution of propranolol enantiomers. The derivatization reagent was tert.-butoxycarbonyl-L-alanine or tert.-butoxycarbonyl-L-leucine. The mobile phase consisted of a 0.02 M N,N-dimethyloctylamine and 25% (v/v) acetonitrile in phosphate-buffer pH 3.0. The stationary phase was a Lichrosorb RP-18 reversed-phase column. No racemization occurred. However, the entire derivatization procedure was quite time-consuming. Several procedures, involving both chiral derivatization reagents and reversed phase HPLC have been investigated with respect to their suitability for the chiral separation of beta-blocking drugs. The use of a few R,R-tartaric acids (113) or symmetrical anhydrides of tert.-butoxy-carbonyl-L-leucine (115), N-trifluoroacetyl-(−)-prolylchloride (TPC) (117, 118), some isothiocyanates, e.g. 2,3,4,6-tetra-O-acetyl-β-D-glucopyranosylisothiocyanate (GITC) or 2,3,4-tri-O-acetyl-α-D-rabinopyranosylisothiocyanate (AITC) (119, 120), to form the respective thiourea of beta-blocking drugs. Isocyanates, e.g. R-(+)-1-phenylethylisocyanate (121, 122), S-(−)-1-phenylethylisocyanate (123), S-(−)-α-methylbenzyl-isocyanate (124, 125), R-(−)-(1-naphthyl)-ethylisocyanate (126) and S-(+)-(1-naphthyl)ethyl-isocyanate (127) have also been used for the derivatization of the enantiomers of beta-blockers in plasma and urine. In these studies, reversed-phase HPLC using aqueous methanol-diethylamine mobile phases was combined with a fluorescence-detector. In several cases, the methods have been applied to pharmacokinetic studies.

The formation of non-chiral derivatives in combination with a chiral stationary phase was successfully tried by several investigators (128–130), combining either phosgene to form the oxazolidine diasteriomers of beta-blocking drugs prior to a chromatographic separation using an α_1-glyco-protein chiral stationary phase or a Pirkle Type I-A-stationary phase, consisting of an aminopropyl packing modified with (R)-N-(3,5-dinitrobenzoyl)-phenylglycine and a hexane-isopropanol-acetonitrile (97:3:1, v/v)-mobile phase. Several chiral stationary phases for the direct enantiospecific assay of beta-blocking drugs, such as silica-bonded α_1-acid glycoprotein columns (131) and a tris (3,5-dimethylphenylcarbamate) cellulose chiral stationary phase have been evaluated and applied for pharmacokinetic studies (132–139). A typical chiral chromatographic separation of a few beta-blocking drugs using this enantiospecific stationary phase is shown in figure 2.

3.7.7 Concluding Remarks

In current pharmacology, trends towards multi-disciplinary research can be seen. Accordingly, bioanalytical science may be faced as an important tool. As a conse-

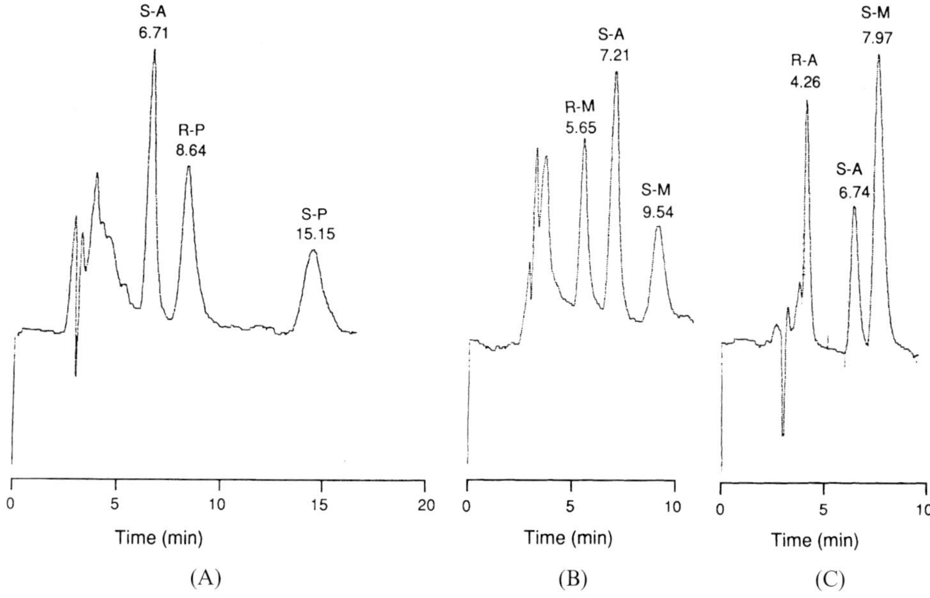

Figure 2. Chiral separation of propranolol, metoprolol and alprenolol, using a tris (3,5-dimethylphenylcarbamate)-cellulose chiral stationary phase and a n-hexane/2-propanol/diethylamine = 75/25/0.05 v/v mobile phase with a flow of 1.0 ml/min and fluorescence detection at 225 nm/295 nm.
R-P and S-P = R- and S-propranolol (A)
R-M and S-M = R- and S-metoprolol (B)
R-A and S-A = R- and S-alprenolol. (C)
S-metoprolol serves as an internal standard for R- and S-alprenolol; S-alprenolol serves as an internal standard for R- and S-metoprolol and R- and S-propranolol. For more detailed information refer to reference 139.

quence, the need for adequate qualitative and/or quantitative routine assays for beta-blocking drugs that combine specificity, sensitivity, reliability and speed is growing. For the qualitative screening of several beta-blockers, e.g. in forensic chemistry, the combination of gas-chromatography with mass-spectrometry will be most promising. Recent experience in combined electron impact and chemical ionization mass spectrometry greatly improves the reliability of mass spectrometric detection. The discriminative power of current high-performance liquid chromatography detection systems (e.g. ultraviolet or fluorescence detection) is not high enough for identification purposes. However, these detection systems may sufficiently meet the requirements made by verification problems, such as in doping cases. It is expected that coupling HPLC with either diode-array ultraviolet detection or mass spectrometry will seriously enhance the application of HPLC for identification of (unknown) beta-blockers. For the quantitative determination of beta-blockers and/or their metabolites in biological fluids, the use of high-performance liquid chromatography rather than gas-chromatography is recommended, because HPLC does not require derivatization reactions prior to analysis. Sample preparation may preferably occur using solid-phase extraction technologies, rather than conventional liquid-liquid extractions, due to high recoveries and reproducibilities. Both the speed of such sample pre-treatment procedures and the ease of automation in routine analysis counterbalance the relatively high demands on time and skill of the analyst developing the method.

For the quantitative determination of enantiomers of beta-blockers, the application of direct chiral high-performance liquid chromatographic assays is most promising. A further increase into the development of cheap commercially available chiral stationary phases for HPLC is expected.

References

1. Thien, Th., Man in Veld, A. J. and Lenders, J. W. M. (1989) Antihypertensia: β-adrenoceptor antagonisten Ned. Tijdschr. Geneesk. *133*, 484–489.
2. Frishman, W. H. (1981) β-adrenoceptor antagonists: new drugs and new indications New Engl. *305*, 500–506.
3. Pfeiffer, S. and Zimmer, C. (1975) Biotransformation und Pharmakokinetik von β-rezeptorenblockern. Die Pharmazie *30*, 625–633.
4. Johnsson, G. and Regardh, C. G. (1976) Clinical Pharmacokinetics of β-adrenoceptor blocking drugs. Clin. Pharmacokin. *1*, 233–263.
5. Routledge, P. A. and Shandm, D. G. (1979) Clinical pharmacokinetics of propranolol. Clin. Pharmacokin. *4*, 73–90.
6. Riddell, J. G., Harron, D. W. G. and Shanks, R. G. (1987) Clinical pharmacokinetics of β-adrenoceptor antagonists – an update. Clin. Pharmacokin. *12*, 305–320.
7. Nolteersting, E. et al. (1987) Paper presented at the 5th Cologne Workshop in Dope Analysis, Cologne, September 6–11.
8. Maurer, H., and Pfleger, K. (1986) Identification and differentiation of beta-blockers and their metabolites in urine by computerized gas chromatography-mass spectrometry. J. Chromatogr. *382*, 147–165.
9. Delbeke, F. T., Debackere, M., Desmet, N. and Maertens, F. (1988) Qualitative gas chromatographic and gas chromatographic-mass spectrometric screening for β-blockers in urine after solid-phase extraction using Extrelut-1 columns. J. Chromatogr. *426*, 194–201.
10. Ahnoff, M., Ervik, M., Lagerström, P. O., Persson, B. A. and Vessman, J. (1985) Drug level monitoring : cardiovascular drugs. J. Chromatogr. *340*, 73–128.
11. McDowall, R. D., Pearce, J. C. and Murkitt, G. S. (1986) Liquid-solid sample preparation in drug analysis. J. Pharm. Biomed. Anal. *4*, 3–21.
12. Doyle, E., McDowall, R. P., Murkitt, G. S., Picot, V. S. and Rogers, S. J. (1990) Two systems for the automated analysis of drugs in biological fluids using HPLC. J. Chromatogr. *527*, 67–77.
13. Van Horne, K. C. and Good, T. (1983) Automated on-stream syringeless injection in HPLC. Am. Lab. 116–124.
14. Stubbs, C., Skinner, M. F. and Kaufer, I. (1990) Reusability of solid-phase extraction columns during the HPLC analysis of some macrolide antibiotics in serum and urine. Chromatographia *29*, 31–34.
15. Verghese, C., McLeod, A. and Shand, D. (1983) Rapid high-performance liquid chromatographic method for the measurement of atenolol in plasma using UV detection. J. Chromatogr. *275*, 367–375.
16. Hoyer, G. L. (1988) Improved high-performance liquid chromatography method for the analysis of serum sotalol. J. Chromatogr. *427*, 181–187.
17. Gupta, R. N., Haynes, R. B., Logan, A. G., MacDonald, L. A., Pickersgill, R. and Achber, C. (1983) Liquid-chromatographic determination of nadolol in plasma. Clin. Chem. *29*, 1085–1087.
18. Harrison, P. M., Tonkin, A. M. and McLean, A. J. (1985) Simple and rapid analysis of atenolol and metoprolol in plasma using solid-phase extraction and high-performance liquid chromatography. J. Chromatogr. *389*, 429–433.
19. Leloux, M. S. Jong, E. G. de and Maes, R. A. A. (1989) Improved screening method for beta-blockers in urine using solid-phase extraction and capillary gas chromatography-mass spectrometry. J. Chromatogr. *488*, 357–363.
20. Bates, J., Carey, P. F. and Godward, R. E. (1987) Combined use of an automated sample processor and a polymer-based high-performance liquid chromatography column to determine the pharmacokinetics of labetalol in man. J. Chromatogr. *395*, 455–461.

21. De Boer, A.G., Breimer, D.D. and Gubbens-Stibbe, J.M. (1980) Assay of propranolol in human and rat plasma by capillary gas chromatography with nitrogen selective detection. Pharm. Weekbl. Sci. *Ed. 2*, 101–105.
22. Garteiz, D.A. and Walle, T. (1972) Electron impact fragmentation studies of β-blocking drugs and their metabolites by GC-mass spectrometry. J. Pharm. Sci. *61*, 1728–1731.
23. Walle, T. (1974) GLC determination of propranolol, other β-blocking drugs, and metabolites in biological fluids en tissues. J. Pharm. Sci. *62*, 1885–1891.
24. Walle, T., Morrison, J., Walle, K. and Conradi, E. (1975) Simultaneous determination of propranolol and 4-hydroxypropranolol in plasma by mass fragmentography. J. Chromatogr. *114*, 351–359.
25. Degen, P.H. and Riess, W. (1976) Simplified method for the determination of oxprenolol and other β-receptor blocking agents in biological fluids by gas-liquid chromatography. J. Chromatogr. *121*, 72–75.
26. Saelens, D.A., Walle, T. and Privitera, P.J. (1976) Quantitative determination of propranolol, propranolol glycol and N-desisopropylpropranolol in brain tissue by electron-capture gas chromatography. J. Chromatogr. *123*, 185–192.
27. Ervik, M., Kylberg-Hanssen, K. and Lagerström, P.O. (1980) Electron-capture gas chromatographic determination of atenolol in plasma and urine, using a simplified procedure with improved selectivity. J. Chromatogr. *182*, 341–347.
28. Desager, J.P. and Harvengt, C. (1975) An improved gas-liquid chromatographic method for determining practolol in plasma and urine. J. Pharmacol. *27*, 52–55.
29. Kinney, C.D. (1981) Determination of metoprolol in plasma and urine by gas-chromatography with electron-capture detection. J. Chromatogr. *225*, 213–218.
30. Jack, B.D. and Riess, W.J. (1974) Determination of low levels of oxprenolol in blood or plasma by gas-liquid chromatography. J. Chromatogr. *88*, 173–176.
31. Cartoni, G.P., Ciardi, M., Giarrusso, A. and Rosati, F. (1988) Detection of β-blokking drugs in urine by capillary column gas chromatography-negative ion chemical ionization mass spectrometry. J. High Res. Chrom. Commun. *11*, 528–532.
32. Kates, R.E. and Jones, C.L. (1977) Rapid GLC determination of propranolol in human plasma samples. J. Pharm. Sci. *66*, 1490–1492.
33. Delbeke, F.T., Debackere, M., Desmet, N. and Maertens, F. (1988) Comparative study of extraction methods for the GC and GC-MS screening of urine for beta-blocker abuse. J. Pharm. Biomed. Anal. *6*, 827–835.
34. Bernard, N., Cuisinaud, G., Jozefczak, C., Seccia, M., Ferry, N. and Sassard, J. (1980) Sensitive gas chromatographic method for the determination in plasma of FM 24 1-(2-exo-bicyclo[2,2,1]hept-2-ylphenoxy)-3-[1-methylethyl)amino]-2-propanol, a new β-adrenoceptor blocking agent. J. Chromatogr. *183*, 99–103.
35. Zak, S., Honc, F. and Gilleran, T.G. (1980) A sensitive gas chromatographic method for the determination of metoprolol in human plasma. Anal. Lett. *13*, 1359–1371.
36. Higuichi, S., Urano, C. and Kawamura, S. (1985) Determination of plasma protein binding of propafenone in rats, dogs and humans by highly sensitive gas chromatography-mass spectrometry. J. Chromatogr. *341*, 305–311.
37. Bianchetti, G., Ganansia, J. and Morselli, P.L. (1979) Sensitive gas chromatographic method for the determination in blood and urine of SL 75212 [4-(2 cylopropylmethoxyethyl)-1-phenoxy-3-isopropylaminopropa-2-ol], a new β-adrenoceptor blocking agent. J. Chromatogr. *176*, 134–139.
38. Sioufi, A., Colussi, D. and Mangoni, P. (1983) Gas chromatographic determination of oxprenolol in human plasma. J. Chromatogr. *278*, 185–188.
39. Gaudry, D., Wantiez, D., Metayer, J.P. and Richard, J. (1985) Simultaneous determination of oxprenolol and $^2H_6$-labelled oxprenolol in human plasma by gas chromatography/mass spectrometry. Biomed. Mass. Spectrom. *12*, 269–273.
40. Gaudry, D., Wantiez, D., Richard, J. and Metayer, J.P. (1985) Simultaneous determination of metoprolol and deuterium – labelled metoprolol in human plasma by gas chromatography-mass spectrometry. J. Chromatogr. *339*, 404–409.
41. Ganansia, J., Gillet, G., Padovani, P. and Bianchetti, G. (1983) Improved determination of betaxolol in biological samples by capillary column gas chromatography. J. Chromatogr. *275*, 183–188.
42. Randinitis, E.J., Nelson, C. and Kinkel. A.W. (1984) Gas chromatographic deter-

mination of bevantolol in plasma. J. Chromatogr. *308*, 345–349.
43. Malbica, J.O. and Monson, K.R. (1975) New and expedient determination of atenolol in biological samples. J. Pharm. Sci. *64*, 1992–1994.
44. Scales, B. and Copsey, P.B. (1975) The gas chromatographic determination of atenolol in biological samples. J. Pharm. Pharmacol. *27*, 430–433.
45. DeBruyne, D., Kinsun, H., Moulin, M.A. and Bigot, M.C. (1979) Improved electron-capture GLC determination of alprenolol and oxprenolol in serum using a wall-coated open tubular column. J. Pharm. Sci. *68*, 511–512.
46. DiSalle, E., Baker, K.M.; Bareggi, S.R., Watkins, W.D., Chidsey, C.A., Frigerio, A. and Morselli, P.L. (1973) A sensitive gas chromatographic method for the determination of propranolol in human plasma. J. Chromatogr. *84*, 347–353.
47. Guerret, M. (1980) Determination of pindolol in biological fluids by an electron-capture gas-liquid chromatographic method on a wall-coated open tubular column. J. Chromatogr. *221*, 387–392.
48. Guerret, M., Lavene, D. and Kiechel, J.R. (1980) Determination of pindolol in biological fluids by electron-capture GLC. J. Pharm. Sci. *69*, 1191–1193.
49. Quaterman, C.P., Kendall, M.J. and Jack, D.B. (1980) Determination of metoprolol metabolites in plasma and urine by electron-capture gas-liquid chromatography. J. chromatogr. *183*, 92–98.
50. Tocco, D.J., de Luna, F.A. and Duncan, A.E.W. (1975) Electron-capture GLC determination of timolol in human plasma and urine. J. pharm. Sci. *64*, 1879–1881.
51. Ervik, M., Hoffman, K.J. and Kylberg-Hanssen, K. (1981) Selected ion monitoring of metoprolol and two metabolites in plasma and urine using deuterated internal standards. Biomed. Mass Spectrom. *8*, 322–326.
52. Poole, C.F., Johansson, L. and Vessman, J. (1980) Formation of electron-capturing derivatives of alprenolol by transboronation. Application to the determination of alprenolol in plasma. J. Chromatogr. *194*, 365–377.
53. Ribick, M., Ivashkiv E., Jemal, M. and Cohen, A.I. (1986) Use of an inexpensive mass-selective detector for the high-sensitivity gas chromatographic determination of nadolol in plasma. J. Chromatogr. *381*, 419–423.
54. Funke, P.T., Malley, M.F., Ivashkiv, E. and Cohen, A.I. (1978) Determination of serum nadolol levels by GLC-selected ion monitoring mass spectrometry : comparison with a spectrofluometric method. J. Pharmac. Sci. *67*, 653–657.
55. Cohen, A.I., Declin, R.G., Ivashkiv, E., Funke, P.T. and McCormick, T. (1984) Determination of orally co-administered nadolol and its deuterated analogue in human serum and urine by gas chromatography with selected-ion monitoring mass spectrometry. J. Pharm. Sci. *73*, 1571–1575.
56. Carlin, J.R., Walker, R.W., Davies, R.O., Ferguson, R.T. and Vandenheuvel, W.J.A. (1980) Capillary column GLC-mass spectrometric assay with selected ion monitoring for timolol and $[^{13}C_3]$ timolol in human plasma. J. Pharmac. Sci. *69*, 1111–1115.
57. Efthymiopoulos, C., Staveris, S., Weber, F., Koffel, J.C. and Jung, L. (1987) Simultaneous quantitative determination of tertatolol and its hydroxilated metabolites in human plasma and urine by gas chromatography mass spectrometry. J. Chromatogr. *421*, 360–366.
58. Francis, R.J., East, P.B., McLaren, S.J. and Larman, J. (1976) Determination of bufuralol and its metabolites in plasma by mass fragmentography and by gas chromatography with electron-capture detection. Biomed. Mass Spectr. *3*, 281–285.
59. Staveris, S., Blaise, P., Efthymiopoulos, C., Schneider, M., Jamet, G., Jung, L. and Koffel, J.C. (1985) Quantitative determination of tertalolol in biological fluids by gas chromatography-mass spectrometry. J. Chromatogr. *339*, 97–103.
60. Munro, G., Hunt, J.H., Rowe, L.R. and Evans, M.B. (1978) The quantitative analysis of the diastereoisomers of salmefamol and labetalol by gas-liquid chromatography. Chromatographia *11*, 440–446.
61. Yamaguchi, T., Morimoto, Y., Sekine, Y. and Haschimoto, M. (1982) Determination of β-adrenergic blocking drugs as cyclic boronates by gas chromatography with nitrogen-selective detection. J. Chromatogr. *239*, 609–615.
62. Zamecnik, J. (1990) Use of cyclic boronates for GC/MS screening and quantitation of beta-adrenergic blockers and some bronchodilators. J. Anal. Toxicol. *14*, 132–136.
63. Gyllenhaal, O. and Vessmann, J. (1988) Phosgene as a derivatizing reagent prior to gas and liquid chromatography. J. Chromatogr. *435*, 259–269.

64. Francis, R.J., East, P.B., McLaren, S.J. and Larman, J. (1976) Determination of bufuralol and its metabolites in plasma by mass fragmentography and by gas chromatography with electron-capture detection. J. Biomed. Mass. Spectr. *3*, 281–285.
65. Lho, D.S., Hong, J.K., Peak, H.K., Lee, J.A. and Park, J. (1990) Determination of phenolalkylamines, narcotic analgesics, and beta-blockers by gas chromatography/mass spectrometry. J. Anal. Toxicol. *14*, 77–83.
66. Mehta, A.C. (1988) Direct separation of drug enantiomers by high-performance liquid chromatography with chiral stationary phases. J. Chromatogr. *426*, 1–13.
67. Smet, M. de and Massart, D.L. (1989) HPLC analysis of basic drugs series. III. High-performance liquid chromatography of basic drugs on cyanopropyl-bonded phases: practical aspects. Trends Anal. Chem. *8*, 96–101.
68. Yee, Y.G., Rubin, P. and Blaschke, T.F. (1979) Atenolol determination by high-performance liquid chromatography and fluorescence detection. J. Chromatogr. *171*, 357–362.
69. Weddle, O.H., Amick, E.N. and Mason, W.D. (1978) Rapid determination of atenolol in human plasma and urine by high-performance liquid chromatography. J. Pharm. Sci. *67*, 1033–1035.
70. Gillilan, R.B. and Mason, W.D. (1933) Revised HPLC determination of atenolol in plasma and urine. Anal. Lett. *16*, 941–949.
71. Mazzo, D.J. (1984) Simultaneous determination of maleic acid and timolol by high-performance liquid chromatography. J. Chromatogr. *229*, 503–507.
72. Rosseel, M.T., Belpaire, F.M., Bekaert, I. and Bogaert, M.G. (1982) High-performance liquid chromatographic determination of metoprolol in plasma. J. Pharm. Sci. *71*, 114–115.
73. Lennard, M.S. and Silas, J.H. (1983) Rapid determination of metoprolol and α-hydroxymetoprolol in human plasma and urine by high-performance liquid chromatography. J. Chromatogr. *272*, 205–209.
74. Gluth, W.P., Sörgel, F. and Gelmacher-von-Mallinckrodt, M. (1984) HPLC-assay of the beta-blocking agents oxprenolol, alprenolol, sotalol and atenolol. in : Maes, R.A.A. (ed) Topics in Forensic and Analytical Toxicology, Elsevier Science Publishers B.V., Amsterdam, *Vol. 20*, 113–118.
75. Lefebvre, M.A., Girault, J. and Fourtillan, J.B. (1981) β-blocking agents : determination of biological levels using high-performance liquid chromatography. J. Liq. Chromatogr. *4*, 483–500.
76. Meredith, P.A., McSharry, D., Elliot, H.L. and Reid, J.I. (1981) The determination of labetalol in plasma by high-performance liquid chromatography using fluorescence detection. J. Pharmacol. Meth. *6*, 309–314.
77. Skärkäinen, S. (1984) High-performance liquid chromatographic determination of sotalol in biological fluids. J. Chromatogr. *336*, 313–319.
78. Sa'sa, S. (1988) Determination of atenolol and its related compounds by ion-pair high-performance liquid chromatography. J. Liq. Chromatogr. *11*, 929–942.
79. Tsuei, S.E., Thomas, J. and Moore, R.G. (1980) Quantification of oxprenolol in biological fluids using high-performance liquid chromatography. J. Chromatogr. *181*, 135–140.
80. Yamashita, K., Motohashi, M. and Yashiki, T. (1990) Sensitive high-performance liquid chromatographic determination of propranolol in human plasma with ultraviolet detection using column switching combined with ion-pair chromatography. J. Chromatogr. *527*, 196–200.
81. Kintz, P., Lohner, S., Tracqui, A., Mangin, P., Lugnier, A. and Chaumont, A.J. (1990) Rapid HPLC determination of bisoprolol in human plasma. Fresenius J. Anal. Chem. *336*, 517.
82. Meffin, P.J., Harapat, S.R., Yee, Y.G. and Harrison, D.C. (1977) High-pressure liquid chromatographic analysis of drugs in biological fluids. V. Analysis of acebutolol and its major metabolites. J. Chromatogr. *138*, 183–191.
83. Lemmer, B., Ohm, T. and Winkler, H. (1984) Determination of the beta-adrenoceptor blocking drug sotalol in plasma and tissues of the rat by high-performance liquid chromatography with ultraviolet detection. J. Chromatogr. *309*, 187–192.
84. Winkler, H. and Lemmer, B. (1984) Determination of the beta-adrenoceptor blocking drug bupranolol in plasma and tissues of the rat by high-performance liquid chromatography with ultra-violet detection. J. Chromatogr. *309*, 193–197.
85. Holt, D.W., Bamra, R., Thorley, K.J., Fowler, M.B. and Jackson, G. (1982) Proceedings of the British Pharmacological Society, March–April.
86. Nagakura, A. and Kohei, H. (1982) Simultaneous determination of bunitrolol and its

metabolite in biological fluids, plasma and urine. J. Chromatogr. *232*, 137–143.

87. Piotrovskii, V.K., Belolipetskaya, V.G., El'Man, A.R. and Metelitsa, V.I. (1983) Ion-exchange high-performance liquid chromatography in drug assay in biological fluids. J. Chromatogr. *278*, 469–474.
88. Lee, C.R. and Esnaud, H. (1988) Determination of melatonin by GC-MS : problems with solid phase extraction (SPE) columns. Biomed. Environm. Mass. Spectrom. *15*, 677–679.
89. Gyllenhaal, O. and Hoffmann, K.J. (1984) Simultaneous determination of metoprolol and metabolites in urine by capillary column gas chromatography as oxazolidineone and trimethylsilyl derivatives. J. Chromatogr. *309*, 317–328.
90. Leloux, M.S. and Maes, R.R.A. (1990) The use of electron impact and positive chemical ionization mass spectrometry in the screening of beta-blockers and their metabolites in human urine. Biomed. Environm. Mass Spectrom. *19*, 137–142.
91. Budzikiewicz, H. (1986) Negative chemical ionization (NCI) of organic compounds. Mass. Spectrom. Rev. *5*, 345–380.
92. Lant, M.S., Oxford, J. and Martin, L.E. (1987) Automated sample preparation on-line with thermospray high-performance liquid chromatography-mass spectrometry. J. Chromatogr. *394*, 223–230.
93. Leloux, M.S., Niessen, W.M.A. and Van der Hoeven, R.A.M. (1991) Thermospray liquid chromatography/mass spectrometry of polar β-blocking drugs. Biol. Mass Spectrom. *20*, 647–649.
94. Lochmüller, C.H. and Souter, R.W. (1975) Chromatographic resolution of enantiomers. Selective review. J. Chromatogr. *113*, 283–302.
95. Tamegai, T., Ohmae, M., Kawabe, K. and Tomoeda, M. (1979) Separation of optical isomers as diastereomeric derivatives by high-performance liquid chromatography. J. liq. Chromatogr. *2*, 1229–1250.
96. Blaschke, G. (1980) New analytical methods. 17. Chromatographic racemate separation. Angew. Chem. *92*, 14–25.
97. Däppen, H., Arm, H. and Meyer, V.R. (1986) Applications and limitations of commercially available chiral stationary phases for high-performance liquid chromatography. J. chromatogr. *373*, 1–20.
98. Lindner, W. (1987) Recent development in HPLC enantioseparation – A selected review. Chromatographic *24*, 97–107.
99. Armstrong, D.W. (1987) Optical isomer separation by liquid chromatography. Anal. Chem. *59*, 89A–91A.
100. Armstrong, D.W. and Han, S.M. (1988) Enantiomeric separations in chromatography. CRC Critical Rev. Anal. Chem. *19*, 175–224.
101. Petterson, C. (1988) Liquid chromatographic separation of enantiomers using chiral additives in the mobile phase. Trends Anal. Chem. *7*, 209–217.
102. Hermansson, J. (1989) Enantiomeric separation of drugs and related compounds based on their interaction with α_1-acid glycoprotein. Trends Anal. Chem. 251–259.
103. Lindner, W. and Hirschböck, I. (1984) Tartaric acid derivatives as chiral sources for enantioseparation in liquid chromatography. J. Pharm. Biomed. Anal. *2*, 183–189.
104. Caccia, S., Chiabrando, C., De Ponte, P. and Fanelli, R. (1978) Separation of beta-adrenoceptor antagonist enantiomers by high resolution capillary gas chromatography. J. Chromatogr. Sci. *16*, 543–546.
105. Thompson, J.A., Tsuru, M., Lerman, C.L. and Holtzman, J.L. (1982) Procedure for the chiral derivatization and chromatographic resolution of R-(+)and S(−)-propranolol. J. Chromatogr. *238*, 470–475.
106. Gyllenhaal, O., König, W.A. and Vessman, J. (1985) Eniantiomer separation of metoprolol and its analogues and metabolites by capillary column gas chromatography after derivatization with phosgene. J. Chromatogr. *350*, 328–331.
107. König, W.A., Ernst, K. and Vessman, J. (1984) Improved gas chromatographic enantiomer separation of pharmaceutically relevant amino alcohols as oxazolidinones. J. Chromatogr. *294*, 423–426.
108. Feitsma, K.G. and Drenth, B.F.H. (1988) Chromatographic separation of enantiomers. Pharm. Weekbl. Sci. Ed. *10*, 1–11.
109. Petterson, C. and Josefsson, M. (1986) Chiral separation of aminoalcohols by ion-pair chromatography. Chromatographia *21*, 321–326.
110. Petterson, C. and Schill, G. (1982) Chiral resolution of aminoalcohols by ion-pair chromatography. Chromatographia *16*, 192–197.
111. Petterson, C. and Schill, G. (1986) Separation of enantiomers in ion-pair chromatographic systems. J. Liq. Chromatograph. *9*, 269–290.
112. Gupta, M.B., Hubbard, J.W. and Midha, K.K. (1988) Separation of enantiomers of

derivatised or underivatised propranolol by means of high-performance liquid chromatography. J. Chromatogr. *424*, 189–194.

113. Leemann, T., Daver, P., Balant, L. and Buri, P. (1984) Direct determination of R- and S-metoprolol in biological fluids by ion-pair-HPLC. Experienta *40*, 647.

114. Hermansson, J. (1982) Separation and quantitation of (R)- and (S)-propranolol as their diasteromeric derivatives in human plasma by reversed-phase ion-pair chromatography. Acta Pharm. Suec. *19*, 11–24.

115. Lindner, W., Leitner, Ch. and Uray, G. (1984) Liquid chromatographic separation of enantiomeric alkanolamines via diastereomeric tartaric acid monoesters. J. Chromatogr. *316*, 605–616.

116. Hermansson, J. and von Bahr, C. (1982) Determination of (R)- and (S)-alprenolol and (R)- and (S)-metoprolol as their diastereomeric derivatives in human plasma by reversed-phase liquid chromatography. J. Chromatogr. *227*, 113–127.

117. Silber, B. and Riegelman, S. (1980) Stereospecific assay for (−) – and (+) – propranolol in human dog plasma. J. Pharmacol. Exp. Ther. *215*, 643–648.

118. Hermansson, J. and von Bahr, C. (1980) Simultaneous determination of d- and l-propranolol in human plasma by high-performance chromatography. J. Chromatogr. *221*, 109–117.

119. Sedman, A. J. and Gal, J. (1983) Resolution of the enantiomers of propranolol and other beta-adrenergic antagonists by high-performance liquid chromatography. J. Chromatogr. *278*, 199–203.

120. Gal, J. (1984) R-α-methylbenzyl isothiocyanate, a new and convenient chiral derivatizing agent for the separation of enantiomeric amino compounds by high-performance liquid chromatography. J. Chromatogr. *314*, 275–281.

121. Gulaid, A. A., Houghton, G. W. and Boobis, A. R. (1985) Separation of acebutolol and diacetolol diastereomers by reversed-phase high-performance liquid chromatography. J. Chromatogr. *318*, 393–397.

122. Wilson, M. J. and Walle, T. (1984) Silica gel high-performance liquid chromatography for the simultaneous determination of propranolol and 4-hydroxypropranolol enantiomers after chiral derivatization. J. Chromatogr. *310*, 424–430.

123. Pflugmann, G., Spahn, H. and Mutschler, E. (1987) Rapid determination of the enantiomers of metoprolol, oxprenolol and propranolol in urine. J. Chromatogr. *416*, 331–339.

124. Hsyu, L. H. and Giacomini, K. M. (1986) High-performance liquid chromatographic determination of the enantiomers of β-adrenoceptor blocking agents in biological fluids I : studies with pindolol. J. Pharm. Sci. *75*, 601–605.

125. Chin, S. K., Hui, A. C. and Giacomini, K. (1989) High-performance liquid chromatographic determination of the enantiomers of β-adrenoceptor blocking agents in biological fluids. II. studies with atenolol. J. Chromatogr. *489*, 438–445.

126. Gübitz, G. and Mihellyes, S. (1984) Optical resolution of β-blocking agents by thin-layer chromatography and high-performance liquid chromatography as diastereomeric R-(−)-1-(1-naphthyl)ethylureas. J. Chromatogr. *314*, 426–466.

127. Piquette-Miller, M., Foster, R. T., Pasutto, F. M. and Jamali, F. (1990) Stereospecific high-performance liquid chromatography assay of acebutolol in human plasma and urine. J. Chromatogr. *526*, 129–137.

128. Hermansson, J. (1985) Resolution of racemic aminoalcohols (β-blockers), amines and acids as enantiomeric derivatives using a chiral α,-acid glycoprotein column. J. Chromatogr. *325*, 379–384.

129. Weller, O., Shulze, J. and König, W. A. (1987) Separation of racemic pharmaceuticals by high-performance chromatography on silica gel modified with carbohydrate residues. J. Chromatogr. *403*, 263–270.

130. Wainer, I. W., Doyle, K. H., Donn, K. H. and Powell, J. R. (1984) The direct enantiomeric determination of (−)- and (+)-propranolol in human serum by high-performance liquid chromatography on a chiral stationary phase. J. Chromatogr. *306*, 405–411.

131. Persson, B. A., Balmer, K. and Lagerström, P. O. (1990) Enantioselective determination of metoprolol in plasma by liquid chromatography on a silica-bonded α_1-acid glycoprotein column. J. Chromatogr. *500*, 629–636.

132. Okamoto, Y., Aburatani, R., Hatano, K. and Hatada, K. (1988) Optical resolution of racemic drugs by chiral HPLC on cellulose and amylose tris (phenylcarbamate) derivatives. J. Liq. Chromatogr. *11*, 2147–2163.

133. Okamoto, Y., Kawashima, M., Aburatani, R., Hatada, K., Nishivama, T. and Masuda, M. (1986) Optical resolution of β-

blockers by HPLC on cellulose triphenylcarbamate derivatives. Chem. Lett. 1237–1244.

134. Krstulovic, A. M., Fouchet, M. H., Burke, J. T., Gillet, G. and Durand, A. (1988) Direct enantiomeric separation of betaxolol with applications to analysis of bulk drug and biological samples. J. Chromatogr. *452*, 477–485.

135. Straka, R. J., Lalonde, R. L. and Wainer, I. W. (1988) Measurement of underivatized propranolol enantiomers in serum using a cellulose-tris (3,5-dimethylphenylcarbamate) high-performance liquid chromatographic (HPLC) chiral stationary. Pharm. Res. *5*, 187–189.

136. Rutledge, D. R. and Garrick, C. (1989) Rapid high-performance liquid chromatographic method for the measurement of the enantiomers of metoprolol in serum using a chiral stationary phase. J. Chromatogr. *497*, 181–190.

137. Ching, M. S., Lennard, M. S., Gregory, A. and Trucker, G. T. (1989) Measurement of underivatized metoprolol enantiomers in human plasma by high-performance liquid chromatography with a chiral stationary phase. J. Chromatogr. *497*, 313–319.

138. Hartmann, C., Kraus, D., Spahn, H. and Mutschler, E. (1989) Simultaneous determination of (R)- and (S)-celiprolol in human plasma and urine : high-performance liquid chromatography assay on a chiral stationary phase with fluorimetric detection. J. Chromatogr. *496*, 387–396.

139. Leloux, M. S. (1992) Rapid chiral separation of metoprolol in plasma – application to the pharmacokinetics/pharmacodynamics of metoprolol enantiomers in the conscious goat. Biomed. Chromatogr. *6*, 99–105.

3.8 Analgesics, Antipyretics and Non-Steroidal Anti-Inflammatory Agents

B. Widdop

3.8.1 Opioid Analgesics

The term opioid describes a group of drugs with properties similar to those of opium or morphine. Their main use is in pain relief, although opioids have other pharmacological properties. Opium refers to the juice of the poppy, **Papaver Somniferum**, which has been in use medicinally for centuries. More than 20 alkaloids have been isolated from opium; morphine and codeine are the most important. By the end of the 19th Century the use of pure alkaloids had largely superseded crude opium in medical practice. This exacerbated the existing problem of opium abuse which had spread from the far-east to the USA following an influx of opium – smoking Chinese labourers and was compounded by the introduction of the hypodermic syringe for parenteral administration of drugs. Opioid addiction, particularly to diamorphine (heroin), remains a significant social and medical problem in many countries to the present day.

The phenomenon of opioid addiction fueled a search for potent analgesics devoid of this property. This met with limited success and opioid drugs such as methadone, meperidine, pentazocine and buprenorphine, which are chemically dissimilar to morphine, have been found to induce physical dependence and have a liability for abuse.

3.8.1.1 Pharmacological properties of opioid analgesics

Opioids exert their major effects on the central nervous system and the gastro-intestinal tract. The range of properties is diverse and includes analgesia, drowsiness, mood changes, respiratory depression, decreased gastro-intestinal mobility, nausea, vomiting and changes to the endocrine and autonomic nervous systems (1).

3.8.1.2 Toxic effects of opioid analgesics

Therapeutic use of opioids is associated with a wide range of adverse effects such as nausea and vomiting, dysphoria, pruritus, constipation and increased pressure on the biliary tract. Skin rashes, notably urticaria can occur and anaphylactoid reactions following intravenous morphine have been reported (2).

Morphine and related opioids must be used with great caution in patients with hepatic disease where decreased metabolic capacity leads to increased bioavailability and cumulative effects (3). Administration of morphine and codeine to patients in renal failure can produce severe respiratory depression and sedation (4). Similarly, their use can exacerbate conditions where the respiratory reserve is already compromised such as emphysema, kyphoscoliosis and chronic ear pulmonale. Patients with hypotension or head injury are also more susceptible to adverse effects from opioid therapy.

Intravenous abuse of opioids carries grave risks of infective disorders such as septicaemia, tetanus, hepatitis and acquired immune deficiency syndrome (AIDS) in addition to anaphylaxis, endocarditis, nephropathy and leukoencephalopathy (5, 6).

A common feature of all opioid drugs is the development of tolerance and physical dependence with repeated use. Tolerance induces the opioid addict to increase the dosage to maintain a "high" and doses many times higher than the normal therapeutic regime may be taken. Abrupt removal of the drug precipitates the well-known "withdrawal syndrome" which, although extremely distressing and prolonged, is rarely life-threatening. Nevertheless, fear of withdrawal is a key factor in reinforcing opioid abuse.

Acute opioid poisoning may follow deliberate overdose, as in a suicide attempt, or may be accidental. Intravenous drug abusers may inject an illicit supply of, for example, heroin which is far more concentrated than that previously tolerated. Prolonged abstinence from opioid abuse causes a fall in tolerance such that individuals who inject their previously normal dose suffer fatal consequences. The practise of smuggling packages of illicit drugs across international boundaries by swallowing them before transit is increasing. Serious poisoning or death can occur when the packages break open in the intestines (7).

The classical signs of opioid poisoning are depressed consciousness or coma, pinpoint pupils and respiratory depression. Since overdosage is often associated with addiction, evidence of needle marks supports the diagnosis. Administration of a small intravenous dose of the opioid – antagonist, naloxone, produces a dramatic reversal of these symptoms and is an invaluable diagnostic tool as well as a first-line treatment measure.

3.8.2 Applications of Toxicological Analyses for Opioid Analgesics

3.8.2.1 Clinical applications

Opioid abuse is widespread and many countries have developed non-punitive policies to treat and rehabilitate addicted individuals. These usually involve referral to a specialized drug dependence treatment clinic providing medical, psychiatric and social support. When first interviewed, the patient's account of drug use should be verified by a urine analysis for opioids and other commonly abused drugs. Having established opioid addiction, the progress of any treatment measures to reduce dependence is monitored by periodic urine analyses. For example, many physicians still favour prescribing methadone as a substitute for heroin; regular urine checks to evaluate the patients compliance have long been a feature of these methadone maintenance programmes.

An alternative is to detoxify patients in specialist in-patient units by gradual reduction of the opioid dose whilst supporting psychological and physical withdrawal symptoms. Again, urine tests are useful to verify that the patient is not taking alternative opioids.

Rehabilitation units aim to nurse previously addicted patients back into society and operate strict rules of complete abstinence from abused substances. Periodic urine screening with dismissal of non-compliant patients helps to maintain a drug-free environment.

Plasma opioid measurements have played little role in drug-dependence therapy, although recent work has shown a good relationship between dose and plasma methadone concentration (8). This may prove helpful in adjusting methadone dosages in patients experiencing withdrawal symptoms (9).

In treating chronic pain, it is usual to titrate the dose against the degree of pain relief achieved and in normal circumstances plasma drug concentrations need not be monitored. However, in patients particularly susceptible to adverse reactions (see section 3.8.1.2), the value of measuring parent drug and active metabolites in plasma is now gaining acceptance (10, 11).

3.8.2.2 Forensic applications

Uncontrolled opioid abuse is generally paralled by a chaotic home environment in which young children suffer neglect. This may result in their being removed into more responsible care, until a parent has proved either that they are compliant with and stabilized on methadone maintenance therapy or that they have been opioid-free for several months. Routine urine testing for opioids and other abused drugs helps to sustain or refute these claims.

Urine screening for opioids and other abuse drugs is an integral feature of programmes designed to limit the impact of drug abuse in the armed forces and industry (12). Positive findings can lead either to refusal of a job application or dismissal from employment.

Although blood measurements of opioid drugs are not essential in managing acute poisoning, if death ensues, post-mortem analyses of blood and tissues are used to establish the cause of death. These also form a vital part of the forensic evidence when foul play is suspected.

3.8.3 Individual Opioid Drugs

3.8.3.1 Heroin (diamorphine)

3.8.3.1.1 Usage

Heroin is a more potent analgesic than morphine, but has a shorter duration of action. Doses of up to 5 mg as the hydrochloride salt may be given intravenously to relieve very severe pain. The drug can also be administered by intramuscular or

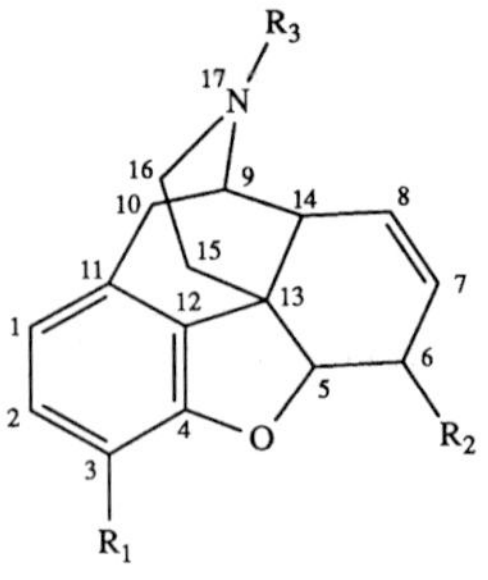

	R_1	R_2	R_3
Morphine	OH	OH	CH_3
Heroin	$OCOCH_3$	$OCOCH_3$	CH_3
Codeine	OCH_3	OH	CH_3
Buprenorphine*	OH	OCH_3	$CH_2 - \triangleleft$

* Additional changes:
(1) Single bond between C_7 and C_8
(2) *Endo*-etheno bridge betweenC_6 and C_{14}
(3) 1-Hydroxy-1,2,2-trimethylpropyl substituent on C_7

Figure 1. Chemical Structures of Morphine and Related Opioids.

subcutaneous injection of 5 to 10 mg; similar doses can be given orally. Tolerant heroin addicts take up to 200 mg daily by intravenous injection or nasal insufflation. Due to its high abuse potential, the clinical use of heroin is strictly controlled in many countries.

3.8.3.1.2 Metabolism and excretion

Blood esterases rapidly hydrolyse heroin to 6-monoacetyl morphine (half-life 9min) and this metabolite is further degraded to morphine in the liver (half-life 38 min). Monoacetyl morphine and morphine itself are believed to account for most of the opioid activity of heroin (13). Very little heroin is excreted unchanged; the bulk of the dose is eliminated in urine as morphine and its conjugates with a small proportion as 6-monoacetylmorphine.

3.8.3.1.3 Analysis

Plasma levels of heroin have little value due to its extremely short half-life and it is more practical to measure morphine and its metabolites. Gas-chromatography/mass-spectrometry (GC/MS) has been used to detect 6-monoacetylmorphine in urine as a means of discriminating between heroin use and ingestion of codeine, morphine or poppy-seeds, but the absence of this metabolite does not rule out heroin abuse (14). Application of this principle to hair samples provides much more reliable evidence (15).

3.8.3.2 Morphine

3.8.3.2.1 Usage

Morphine is usually given by subcutaneous or intramuscular injection of the sulphate salt in doses of 5 to 10 mg/70 kg. Slow release oral preparations are also available which are administered typically in 60 mg doses 12 hourly to provide prolonged relief of severe pain.

3.8.3.2.2 Metabolism and excretion

The oral bioavailability of morphine ranges from 15 to 64% and reflects extensive first pass metabolism by the liver. A small fraction (ca. 5%) is N-demethylated to the less active normorphine metabolite whereas the remainder is converted to morphine-3-glucuronide, morphine-6-glucuronide, morphine-3-ethereal sulphate and morphine-3, 6-diglucuronide (16). About 10% of the morphine dose is excreted as free morphine in the urine over 72 hours, 75% is present as morphine-3-glucuronide together with much smaller amounts of the other conjugates.

3.8.3.2.3 Blood concentrations

Single intravenous doses of morphine given to adults (0.125 mg/kg) gave an average serum concentration of 0.44 mg/L at 30 seconds with a rapid fall to 0.02 mg/L by 2 hours post-dosing (17). Adults given the same dose intramuscularly achieved average peak serum levels of 0.07 mg/L between 10 to 20 minutes later which declined to 0.02 mg/L at 4 hours (18).

Cancer patients receiving oral controlled release morphine sulphate in daily doses of 180 to 460 mg had peak levels of 0.042–0.308 mg/L (free morphine) 1.512–7.960 mg/L (morphine-3-glucuronide) and 0.441–2.737 mg/L (morphine-6-glucuronide) (19). A similar group of patients receiving either oral or subcutaneous morphine sulphate (oral dose range: 0.37–6.82 mg/kg; subcutaneous dose range: 0.22 to 3.60 mg/kg) achieved median plasma concentrations of morphine, morphine-3-glucuronide and morphine-6-glucuronide of 36 mg/L, 1035 mg/L and 142 mg/L respectively (20).

Post-mortem blood morphine concentrations are notoriously difficult to interpret since personal susceptibility to the drug's toxic effects is so variable. For example, in a study of 52 heroin fatalities, free morphine blood concentrations ranged from 0.05 to 2.05 mg/L (21). Measurement of the ratio of free to total morphine in blood is a useful guide to the time lapse between heroin injection and death, being always greater than 50% in rapid death (21, 22).

3.8.3.2.4 Analysis

Due to the prevalence of heroin and, to a much lesser extent, morphine as an abused drug, tests for the presence of morphine and its metabolites in urine have found widespread application. Methods involving solvent extraction followed by thin-layer chromatography are still popular and have the advantage of being capable of detecting other abused opioids such as codeine and methadone at the same time (23). Since morphine glucuronide conjugates extract poorly into organic solvents, liberation of free morphine by acid or enzyme hydrolysis prior to extraction achieves much greater detection sensitivity (24).

The need for massive screening programs for military and work-place personnel, particularly in the United States, stimulated the development of immunoassays for detecting urine opioids. Commercial kits based on radioimmunoassay (RIA), enzyme immunoassays (EIA) and fluorescence polarized immunoassay (FPIA), dominate this market (25). These have varying degrees of specificity and for medico-legal purposes a confirmatory analysis by, for example, gas-chromatography – mass spectrometry, is always advisable. For serum morphine measurements, the DPC Coat-a-Count system is highly selective for free morphine and can detect as little as 1 ng/ml with ease (26).

Gas-liquid – chromatographic methods to measure morphine in biological samples invariably require derivatisation with silylating or acetylating reagents to improve its elution profile (27–31). Flame ionization detection (FID) of silylated morphine is too insensitive for therapeutic concentrations of morphine in plasma; nitrogen-phosphorus detection (NPD) of the trimethylsilyl derivative or electron capture detection (ECD) of a halogenated derivative using, for example, heptofluorobutyric anhydride are more than adequate. Mass spectrometric analysis offers very high sensitivity and unequivocal identification of morphine and other opiates in biological fluids (32–36), but for other than medico-legal purposes more conventional techniques will suffice.

High Performance Liquid Chromatography (HPLC) avoids the complexities of sample preparation and derivatisation of GLC methods for morphine and is most widely used. Methods using reversed phase (37) or ion pair separations (38) or chromatography on unmodified silica (39) columns have been developed. Detection of morphine and its congeners by UV or visible spectrophotometry is prone to interference and, for low sensitivity levels, fluorescence or electrochemical detection is preferable (40). More recently, the emphasis has shifted towards methods which simultaneously determine morphine and its principal metabolites in plasma (morphine-3-glucuronide, morphine-6-glucuronide and normorphine) using fluorescence detection (41) or a combination of fluorescence and electrochemical detectors in series (Bansal-personal communication).

3.8.3.3 Codeine

3.8.3.3.1 Usage

Codeine is used pharmaceutically as the sulphate or phosphate salt and is present in a multitude of over-the-counter preparations. Its analgesic potency is considerably less than morphine, but it is an effective antitussive agent. Total daily oral doses range from 60–240 mg.

3.8.3.3.2 Metabolism and excretion

Codeine undergoes O-demethylation to morphine and N-demethylation to norcodeine. Approximately 10% is excreted as free codeine in a 24 hour urine with 32–46% as a glucuronide conjugate. Conjugated morphine accounts for 5–13% of the 24 hour excreted dose with 10–21% as conjugated norcodeine together with traces of free morphine and free norcodeine (42).

3.8.3.3.3 Blood concentrations

Oral administration of 60 mg codeine phosphate produces peak plasma concentrations of around 0.135 mg/L at 1 hour after ingestion, which decline with a half-life of 2.4 hours. Plasma morphine concentrations following this dosage average a peak of 0.007 mg/L at 1.5 hours (43). Blood codeine levels in fatalities range from 1 to 9 mg/L (44), but there are reports of levels as high as 48 mg/L (45).

3.8.3.3.4 Analysis

Screening methods for morphine are generally applicable to codeine also (23, 25). Codeine, and most probably its glucuronide conjugate, have higher cross-reactivity than morphine in the Roche Diagnostics radioimmunoassay for morphine and the Syva opiate enzyme immunoassay. Acid hydrolysis of urine samples can greatly increase the extraction recovery into organic solvents (46).

GLC or GC/MS methods for morphine in biological fluids often cater for the simultaneous determination of codeine (28, 29, 31–34). Derivatisation is required to achieve acceptable chromatography. By the same token, HPLC methods for morphine quantification often measure codeine at the same time (47–49).

3.8.3.4 Methadone

3.8.3.4.1 Usage

The clinical use of methadone is limited to narcotic maintenance treatment of heroin addicts. The 1-isomer accounts for almost all its pharmacological activity. Maintenance patients receive up to 180 mg daily as a linctus; tablet, parenteral and suppository formulations are also available.

3.8.3.4.2 Metabolism and excretion

Methadone undergoes mono and di-N-demethylation to unstable metabolites which cyclize spontaneously to form 2-ethylidene-1,5-dimethyl-3,3-diphenylpyrrolidine (EDDP) and 2-ethyl-5-methyl-3,3-diphenylpyrroline (EMDP) (50). EDDP and EMDP are the major urinary excretion products of methadone. The proportions of these products excreted in relation to the methadone dose is governed to a large extent by urine pH and urine volume and individual variation in metabolism rate.

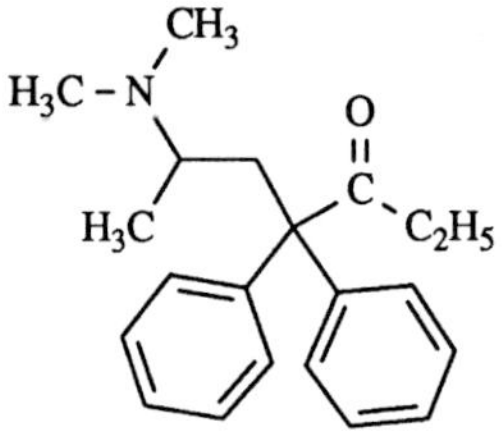

Figure 2. Chemical Structure of Methadone.

3.8.3.4.3 Blood concentrations

Following a single oral dose of 15 mg of methadone, peak plasma concentrations averaged 0.075 mg/L at 4 hours (51). Chronic oral administration of methadone linctus in doses ranging from 5 to 100 mg daily gave steady state plasma concentrations of 0.02 to 0.9 mg/L and, with few exceptions, there was a close correlation between dosage and plasma level (8). Withdrawal symptoms became apparent when the plasma methadone concentration fell below 0.05 mg/L. In fatal cases blood methadone concentrations vary widely from 0.4 to 1.8 mg/L (52).

3.8.3.4.4 Analysis

Methadone and its main metabolic products are readily detected in urine by thin-layer chromatographic methods (23). Gas-chromatographic detection either on packed columns or capillary columns is often used to confirm the TLC results since separation of methadone, EDDP and EMDP is easily accomplished (24, 53–56). The metabolities are excreted into urine for longer periods than the parent drugs. So far, immunoassays for methadone in urine have no cross-reactivity towards EDDP or EMDP (25) and therefore the chromatographic methods can discern methadone use in immunoassay-negative samples.

For plasma or blood methadone assays, gas-chromatography with flame-ionisation detection has proved most popular. Better sensitivity can be gained by using electron-capture detection following oxidation of methadone to benzophenone (57), but the ability to measure metabolites is lost. An alternative is to use nitrogen-phosphorus detection (58). GC/MS methods for methadone and its metabolites are also available (59), but required only in legal cases.

A sensitive HPLC method for methadone in plasma using a silica column and a non-aqueous mobile phase has been reported (8).

3.8.3.5 Propoxyphene

3.8.3.5.1 Usage

Propoxyphene is closely related in chemical structure to methadone, but has only mild narcotic analgesic properties, being less potent than codeine. The l-isomer is inactive as an analgesic, but finds use as an antitussive agent. Combinations of d-propoxyphene with aspirin, codeine or paracetamol are prescribed on a large scale for the relief of chronic pain. Numerous fatalities have occurred following intentional overdose (60).

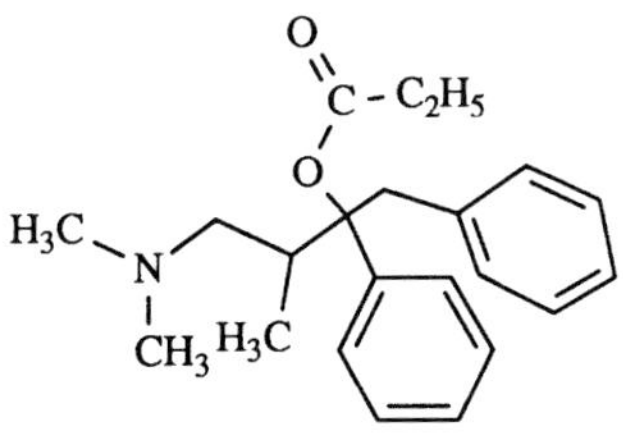

Figure 3. Chemical Structure of Propoxyphene.

3.8.3.5.2 Metabolism and excretion

N-demethylation of propoxyphene produces norpropoxyphene which, though less active, has a longer half-life and accumulates in plasma. Further N-demethylation gives dinorpropoxyphene which is converted by dehydration to cyclic dinorpropoxyphene. After a single oral dose of propoxyphene, the 20 hour urine contained the parent drug (1.1%), norpropoxyphene (13.2%) and dinorpropoxyphene (0.7%) (61). The remainder of the dose was unaccounted for, but ring hydroxylation and glucuronide formation are likely pathways.

3.8.3.5.3 Blood concentrations

A single oral dose of 130 mg of propoxyphene hydrochloride gave peak plasma levels at 2 hours of 0.23 mg/L with a peak level of 0.27 mg/L for norpropoxyphene at 4 hours (62). Chronic daily doses of 195 mg of the hydrochloride gave average plasma levels of 0.42 mg/L propoxyphene and 1.45 mg/L norpropoxyphene 2 hours after the last dose (63). Plasma propoxyphene concentrations greater than 1.0 mg/L are generally associated with serious toxicity (64).

3.8.3.5.4 Analysis

Dextropropoxyphene abuse is known to occur and a homogeneous enzyme immunoassay (EMIT) is available for urine samples. Nordextropropoxyphene is unstable, especially in strong basic conditions, and rearranges to give nordextropropoxyphene amide which is stable. This product gives a characteristic streak or base-ball pattern on TLC analysis of urine samples (23).

Gas-chromatographic methods are plagued by the problem of thermal decomposition of dextropropoxyphene (65) and various attempts to avoid this include injection with a silylating agent (66), hydrolysis of the ester linkage (67), derivative formation (68) and use of more polar stationary phases (69). With capillary columns the injection technique is crucial to analytical performance and on-column injection with nitrogen-phosphorus detection affords accurate and simultaneous measurement of dextropropoxyphene and nordextropropoxyphene in plasma at therapeutic concentrations (70).

HPLC avoids the problem of thermal decomposition, but with UV detection a low wavelength (210 nm) is needed to gain sensitivity at therapeutic levels (71). This introduces the danger of non-selective detection and other workers have recommended electrochemical detection (69, 71).

3.8.3.6 Meperidine (Pethidine)

3.8.3.6.1 Usage

Meperidine is approximately ten times less potent than morphine and has a shorter duration of action. Tablet, linctus and parenteral formulations are used in single doses of 50–150 mg and daily doses of 150–1200 mg.

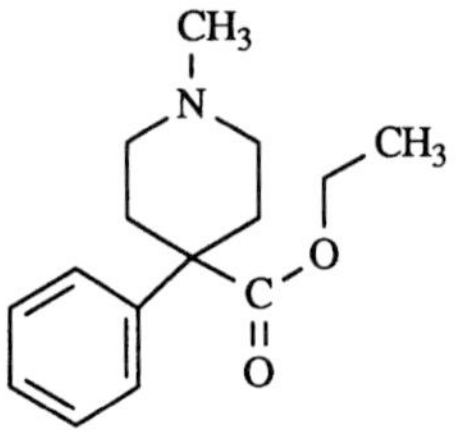

Figure 4. Chemical Structure of Meperidine (Pethidine).

3.8.3.6.2 Metabolism and excretion

Meperidine is N-demethylated to normeperidine. Both compounds are de-esterified to give meperidinic and normeperidinic acids which account for about 42% and 23% of the dose as conjugates in urine. Approximately 7% of a meperidine dose appears in urine unchanged with 17% as normeperidine. Excretion of these compounds is pH-dependent.

3.8.3.6.3 Blood concentrations

An oral dose of 100 mg of meperidine hydrochloride gives average peak plasma levels of 1.7 mg/L 1.3 hours after dosing (72). The same dose given intramuscularly gives average peak plasma concentrations of 0.31 mg/L at 1 hour which decline to 0.21 mg/L after 3 hours (73). An intravenous injection of 50 mg of meperidine hydrochloride gives average serum concentrations of 0.52 mg/L at 12 minutes and 0.18 mg/L by 1 hour (74). Overdose concentrations in blood range from 8–20 mg/L (meperidine) and 8–30 mg/L (normeperidine) after oral ingestion. After intravenous injection the equivalent ranges are 1–8 and 0–7 mg/L (75).

3.8.3.6.4 Analysis

Meperidine is readily analysed in biological samples by gas-liquid chromatography. Various methods involving flame-ionisation (76), nitrogen-phosphorus (77), electron capture (78) or mass-spectrometric detection (79) have been reported. Often, derivatization of normeperidine prior to chromatography has been employed in these procedures. HPLC has also proved a suitable means of measuring meperidine and normeperidine in serum and urine (80).

3.8.3.7 Diphenoxylate

3.8.3.7.1 Usage

Diphenoxylate is a derivative of meperidine used for symptomatic treatment of diarrhoea. It is used in combination with sub-therapeutic amounts of atropine to discourage drug abuse. Oral doses of 5 to 10 mg are given 6 hourly. Typical opioid effects occur at high dosage (40 to 60 mg) and in acute overdose, symptoms and treatment parallel those of morphine.

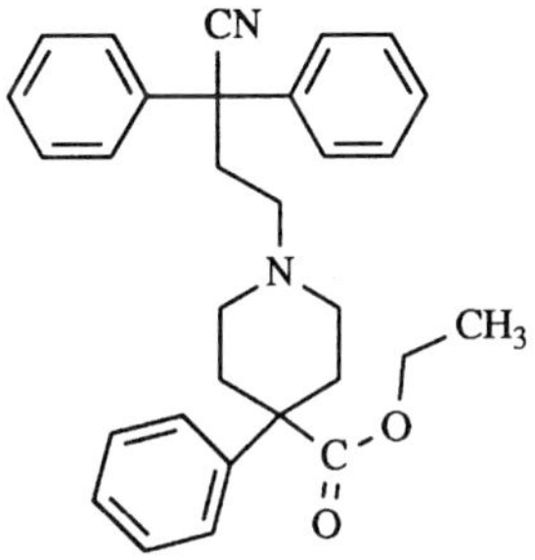

Figure 5. Chemical Structure of Diphenoxylate.

3.8.3.7.2 Metabolism and excretion

Diphenoxylate is extensively metabolized and less than 0.1 % of a dose is excreted in the urine with 0.8 % as the active metabolite, diphenoxylic acid. The bulk of a dose is excreted in the faeces, mainly in the form of hydroxylated and conjugated metabolites.

3.8.3.7.3 Blood concentrations

After a 5 mg oral dose of diphenoxylate hydrochloride, peak plasma levels of 0.01 mg/L occur at 2 hours. The major metabolite, diphenoxylic acid, which has 5 times the antidiarrheal activity of the parent drug, has peak plasma levels of 0.04 mg/L also at 2 hours (81). In a fatal poisoning case involving a child who died 43 hours after ingestion, the post-mortem blood diphenoxylate concentration was 0.34 mg/L (82).

3.8.3.7.4 Analysis

Assay methods for biological samples have concentrated on determining diphenoxylic acid, and mass-spectrometric (83, 84) and radioimmunoassay (85) techniques have been described.

3.8.3.8 Fentanyl

3.8.3.8.1 Usage

Fentanyl and its congeners, alfentanil and sufentanil, are potent narcotic analgesics used in general anaesthesia. Single doses of 25–100 μg are administered either by intravenous or intramuscular injection as the citrate salt. Methylfentanyl (α-methylfentanyl; "China White") has equipotency with fentanyl and has been abused in parts of the USA. The much more potent 3-methylfentanyl is also a street drug. Both derivatives are classified as "designer drugs".

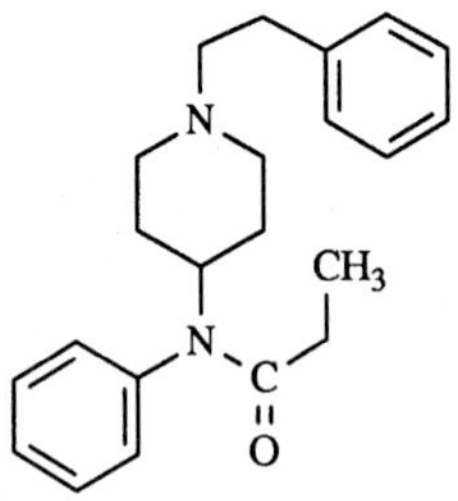

Figure 6. Chemical Structure of Fentanyl.

3.8.3.8.2 Metabolism and excretion

Oxidative N-dealkylation of fentanyl yields norfentanyl and despropionylfentanyl both of which have been detected in plasma (86). Less than 7% of the dose is eliminated over 72 hours in urine as unchanged fentanyl with 26 to 55% as norfentanyl together with unknown quantities of hydroxylfentanyl and hydroxynorfentanyl (87). The metabolites are considered inactive.

3.8.3.8.3 Blood concentrations

Volunteers given a 6.4 μg/kg intravenous dose of fentanyl developed initial average plasma levels of 18 μg/L declining to around 1 μg/L after 1.5 hours (88). Levels greater than 20 μg/L induce loss of conciousness (89, 90).

3.8.3.8.4 Analysis

A highly specific and sensitive radioimmunoassay for fentanyl in serum has been developed (91). Gas-chromatographic methods for the determination of fentanyl and its derivatives in biological samples have used electron-capture, nitrogen-phosphorus and mass-spectrometric detection (92–94). These correlate well with radioimmunoassay methods (95). High performance liquid chromatography has also been applied to measure fentanyl and alfentanil in plasma (96).

3.8.3.9 Buprenorphine

3.8.3.9.1 Usage

Buprenorphine is a powerful partial agonist analgesic related in structure to morphine (fig. 1). It is used in the treatment of acute and chronic pain and administered parenterally or sublingually in doses of 0.3 to 0.6 mg. Misuse of buprenorphine by heroin addicts has been reported (97, 98). Oral overdose has no serious toxic effects since buprenorphine is almost completely metabolized during first pass through the liver (99).

3.8.3.9.2 Metabolism and excretion

Buprenorphine undergoes N-dealkylation and 3-O-glucuronidation in the liver. N-dealkylbuprenorphine is also conjugated to form the glucuronide (100–102). About 70% of a buprenorphine dose is excreted faecally and only small amounts appear in urine mainly as buprenorphine glucuronide, N-dealkylbuprenorphine glucuronide and dealkylbuprenorphine.

3.8.3.9.3 Blood concentrations

Patients receiving sublingual buprenorphine doses of 1.2 mg daily achieved plasma levels of 0.73 to 0.97 nmol/L (buprenorphine), 1.41 to 1.86 nmol/L (buprenorphine glucuronide) and 0.62 to 0.90 nmol/L (N-dealkylbuprenorphine) (104).

3.8.3.9.4 Analysis

Due to the low dosage of buprenorphine, its analysis by conventional techniques is difficult. A commercial I^{125}-labelled radioimmunoassay is widely used to detect abuse of the drug in urine specimens. The antibodies cross-react strongly with buprenorphine glucuronide, but a selective extraction procedure enables measurement of free buprenorphine in plasma after therapeutic doses (104).

HPLC methods using electrochemical detection have been applied with some success to the measurement of buprenorphine and dealkylbuprenorphine in urine samples after therapeutic dosage (105, 106). Detection limits of approximately 150 pg/ml (0.15 μg/L) for buprenorphine and 250 pg/ml (0.25 μg/L) for N-dealkylbuprenorphine were achieved. A gas-chromatographic method involving formation of the heptafluorobutyryl derivative of buprenorphine and electron-capture detection claimed a detection limit in plasma and urine of 0.5 μg/L, but no data were presented on samples from either patients or volunteers (107). A GC/MS method has proved demonstrably more sensitive and it was possible to analyse plasma concentrations of buprenorphine and desalkylbuprenorphine for up to 24 hours and urine concentrations for over 7 days after parenteral injection of a 0.6 mg dose (108).

3.8.4 Antipyretic and Non-Steroidal Anti-Inflammatory Agents

This is a diverse group of drugs, often unrelated in chemical structure, which nevertheless share certain therapeutic properties and side-effects. The prototype is aspirin and therefore compounds in this group are often referred to as "aspirin-like drugs".

Their therapeutic activity is thought to depend largely on an ability to inhibit the biochemical synthesis of prostaglandins and related autacoids. Prostaglandins are known to participate in the pathogenesis of inflammation and fever (109, 110).

3.8.4.1 Pharmacological properties of antipyretic and non-steroidal anti-inflammatory drugs

These compounds have varying degrees of analgesic, antipyretic and anti-inflammatory activities. For example, paracetamol is analgesic and antipyretic, but has very weak anti-inflammatory effects. Their analgesic value is limited to pain of low to moderate intensity, particularly that associated with inflammation or post-operative surgery. Unlike the opioid analgesics, they do not induce physical dependence and exert no change on sensory perception other than pain. The major clinical use of these drugs is in the treatment of rheumatoid arthritis, osteoarthritis and ankylosing spondylitis.

3.8.4.2 Toxic effects of antipyretic and non-steroidal anti-inflammatory drugs

In therapeutic use, these drugs share a number of adverse effects, most commonly a propensity to induce gastric or intestinal ulceration. This can lead to a secondary anaemia from the resultant blood loss. Disturbances of platelet function and prolongation of gestation or spontaneous labour also occur.

While kidney damage is rarely caused by the chronic use of an individual aspirin-like drug, chronic abuse of analgesic mixtures has been linked to renal injury including papillary necrosis and chronic interstitial nephritis (111).

Following overdosage, the features of poisoning range from the serious metabolic disturbances of salicylates and the hepatotoxicity of paracetamol through to the inconsequential effects of the propionic acid derivatives.

3.8.5 Applications of Toxicological Analyses for Antipyretic and Non-Steroidal Anti-Inflammatory Drugs

With the exception of aspirin (112), there is no evidence that blood concentrations of these agents parallel their therapeutic effects and therefore blood assays have no role in therapeutic drug monitoring other than to establish compliance. The value of serum/plasma concentration measurements of salicylates (113) and paracetamol (114) in the diagnosis and subsequent management of poisoning with these drugs has been established for many years. For other members of the group, blood analyses serve mainly to confirm a diagnosis of poisoning.

3.8.6 Individual Antipyretic and Non-Steroidal Anti-Inflammatory Drugs

3.8.6.1 Aspirin (acetylsalicylic acid)

3.8.6.1.1 Usage

Aspirin is the most commonly used therapeutic agent either alone or in combination with numerous other drugs.

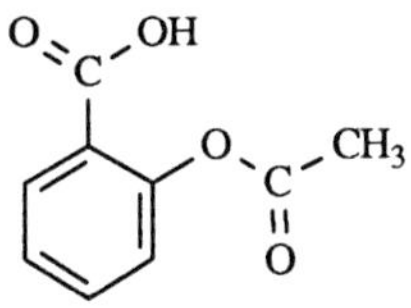

Figure 7. Chemical Structure of Acetylsalicylic Acid (Aspirin).

The usual oral dose for adults to relieve mild pain and fever is 325–975 mg daily. Its use in children is now largely contra-indicated due to evidence of a link with Reye's syndrome (115). In the treatment of rheumatoid arthritis, oral doses of 3 g to 5 g daily are administered. Other forms of salicylate are employed topically as oils and liniments (e.g. methyl salicylate or oil of wintergreen) and as keratolytic agents (salicylic acid).

3.8.6.1.2 Metabolism and excretion

Aspirin is almost totally hydrolysed by blood and liver esterases to salicylic acid which is the active principle. A single dose is eliminated in the urine as salicylic acid (5 %), salicyluric acid (80 %), salicyl phenolic glucuronide (10 %) and salicyl acyl glucuronide (5 %) (116).

3.8.6.1.3 Blood concentrations

Aspirin itself (acetylsalicylic acid) has a short half-life in blood (15 minutes) and plasma concentrations after doses of 600 to 900 mg are less than 10 mg/L. Salicylic acid serum concentrations after oral doses of 1 g peak at 2 hours (range 31–114 mg/L) (117). Arthritic patients maintained on doses of around 3 g daily develop steady-state serum salicylate concentrations of 44 to 330 mg/L (118). Serum salicylate concentrations greater than 500 mg/L are associated with toxicity (113).

3.8.6.1.4 Analysis

Rapid measurement of plasma salicylate concentrations is vital to confirm a diagnosis of salicylate poisoning and assess its severity.

Hospital-based biochemists favour colorimetric methods which involve the formation of a purple complex with ferric salts in weakly acid solution and absorbance measurement at 540 nm (119). The most widely used method includes mercuric chloride in the colour reagent to precipitate proteins which are a source of interference (120). Spuriously high results can occur with plasma samples from patients with diabetic ketoacidosis and other conditions such as Reye's syndrome where an excess of keto-acids react with ferric ions to give a coloured complex (121). Interferences by the presence of overdose concentrations of other drugs (e. g. phenothiazines (122), diflunisal and dichloralphenazone (123)) have been reported, but the speed and simplicity of the assay far outweighs these disadvantages. An enzymatic kit is available commercially. This is based on the conversion of salicylic acid to catechol in the presence of salicylate mono-oxgenase and NADH, which then reacts with 4-aminophenol at high pH to give a blue product (124). This is a highly specific and simple two stage procedure which requires only 50 μl of sample. This and the Trinder method (120) compare very favourably with HPLC in terms of accuracy and precision (123). Methods involving differential ultra-violet spectroscopy (125) and fluorimetry (126) require solvent extraction steps to reduce interference and offer no advantage in the overdose situation. More elaborate methods using either GLC (127) or HPLC (128, 129) aim to differentiate acetyl salicylic acid, salicylic acid and its conjugated metabolites for pharmacokinetic studies or therapeutic investigations. While representing an "overkill" in overdose work, HPLC methods which simultaneously measure serum paracetamol and salicylate concentrations either after solvent extraction (130) or by direct serum injection (131) can play a useful role in this context.

3.8.6.2 Paracetamol (acetaminophen)

3.8.6.2.1 Usage

Paracetamol has eclipsed aspirin as the most popular over-the-counter analgesic due to its lower potential for side-effects. It is also dispensed in combination with other drugs such as codeine, dihydrocodeine and dextropropoxyphene. Paracetamol has analgesic and antipyretic properties, but has no anti-inflammatory effects. In pure form, oral doses of up to 4 grams daily are administered. Although generally safe in therapeutic amounts, overdosage can lead to severe hepatotoxicity; ingestion of approximately 20 grams can prove fatal in adults.

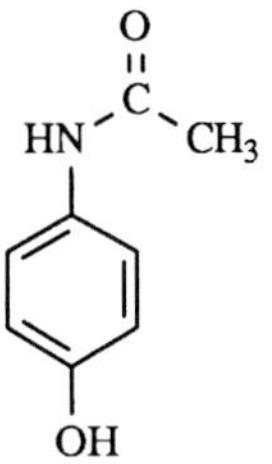

Figure 8. Chemical Structure of Paracetamol (Acetaminophen).

3.8.6.2.2 Metabolism and excretion

After therapeutic dosage, 2% of paracetamol is excreted unchanged in the urine together with a glucuronide conjugate (45–55%), a sulphate (20–30%) and 15–55% as cysteine and mercapturic acid conjugates. In overdosage, saturation of conjugation pathways occurs, and stores of glutathione, which is needed to form cysteine and mercapturic acid conjugates, become depleted. A reactive intermediate formed by N-hydroxylation of paracetamol (N-acetyl-benzoquinoneimine) is then free to react with the sulphydryl groups of hepatic proteins resulting in necrosis. Antidotal treatment with N-acetyl cysteine to replenish the glutathione stores has proved effective if given during the early stages of poisoning (132).

3.8.6.2.3 Blood concentrations

Plasma paracetamol concentrations 6 hours after a 324 mg oral dose averaged 4.2 mg/L (133). Serum concentrations ranging from 4.8 to 12.7 mg/L were found at 30 minutes in 4 volunteers who took 1300 mg (134). In overdose patients, plasma paracetamol concentrations greater than 300 mg/L 4 hours after ingestion indicate the probability of hepatic damage developing (135). A nomogram linking plasma drug concentrations, time since overdose and hepatotoxicity helps to decide the need for antidotal therapy (136).

3.8.6.2.4 Analysis

The need for measurement of serum paracetamol in emergency toxicology stimulated the development of rapid and simple colorimetric methods which entailed no solvent extraction step. Some of these involved hydrolysis of paracetamol conjugates prior to colour development (137, 138) and gave grossly misleading results (139). A more reliable method (140) converted paracetamol to a nitroso derivative which formed an intense yellow phenate ion in alkaline solution and correlated well with HPLC methods (139). Some interference from salicylate could be reduced by modifying the measurement wavelength (141). A novel technique whereby paracetamol is cleaved by aryl acylamide amidohydrolase to 4-aminophenol which then reacts with orthocresol and ammoniacal copper sulphate to give a blue dye (142) is now available in kit form and has been modified to operate on automated clinical analysers (143). This method is highly selective, precise and accurate and, together with commercial immunoassays using EMIT or FPIA technology (144), has largely superseded the older optical procedures in emergency work.

Chromatography methods offer increased selectivity and specificity. A number of GLC procedures, most of which use prior derivatisation, have been published (145–147) for use in clinical situations. HPLC methods using reversed phase ion-exchange chromatography (148, 149) are capable of separating and measuring paracetamol and its major metabolites in biological samples after therapeutic dosage with a high degree of precision and accuracy.

3.8.6.3 Pyrazolan derivatives

This group includes phenylbutazone, oxyphenbutazone, antipyrine, amidopyrine and the most recent addition, azapropazone. Antipyrine and amidopyrine are rarely used clinically. Phenylbutazone and azapropazone use has declined considerably and in many countries is confined to cases of ankylosing spondylitis or acute gout under hospital supervision.

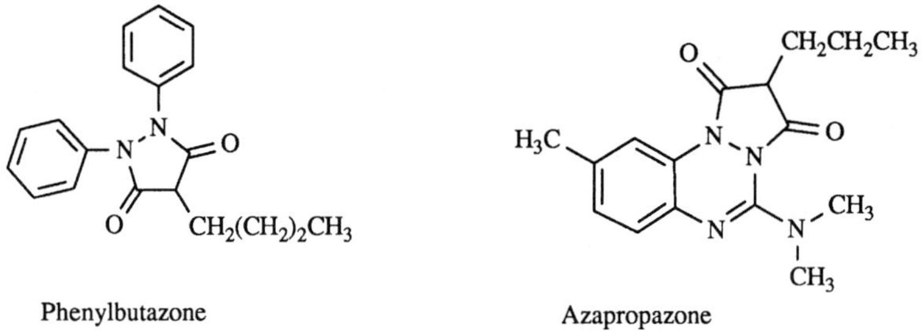

N N O O $CH_2(CH_2)_2CH_3$

Phenylbutazone

O $CH_2CH_2CH_3$ H_3C N N O N N CH_3 CH_3

Azapropazone

Figure 9. Chemical Structures of Pyrazolan Derivatives.

3.8.6.3.1 Phenylbutazone

3.8.6.3.1.1 Usage

Phenylbutazone is available as the free acid in 100 mg tablets and the typical oral dosage is 100–400 mg.

3.8.6.3.1.2 Metabolism and excretion

Phenylbutazone is hydroxylated to form oxyphenbutazone which has equivalent anti-inflammatory activity. Hydroxylation of the n-butyl side chain to an inactive product also takes place. A dose of phenylbutazone is slowly excreted in the urine over several days, mainly as conjugated metabolites (150).

3.8.6.3.1.3 Blood concentrations

A single oral 400 mg dose of phenylbutazone gave peak plasma concentrations of around 60 mg/L shortly after ingestion, with levels of 1–3 mg/L still detectable after

16 days (151). In overdose cases, plasma phenylbutazone concentrations in excess of 600 mg/L have been recorded (152).

3.8.6.3.1.4 Analysis

Gas-liquid chromatographic methods with prior methylation (151) and without derivatisation (153, 154) have been described.

HPLC methods, some of which estimate in addition the major metabolites of phenylbutazone, have also been developed (155–157).

3.8.6.3.2 Azapropazone

3.8.6.3.2.1 Usage

This pyrazolan derivative is reputedly less toxic than phenylbutazone and has a similar spectrum of pharmacological activity. It is used in the treatment of rheumatoid arthritis, osteoarthritis and gout. The usual daily oral dose is 1200 mg in divided doses.

3.8.6.3.2.2 Metabolism and excretion

Azapropazone is partially metabolized to the 6-hydroxy derivative and about 20% of a dose appears in the urine in this form with 65% as the unchanged drug.

3.8.6.3.2.3 Blood concentrations

Patients given 1200 mg of azapropazone daily for 5 days achieved steady-state plasma concentrations of 40.4 to 104.3 mg/L.

3.8.6.3.2.4 Analysis

HPLC is the preferred method (158).

3.8.6.4 Indole derivatives

Indomethacin and its less toxic congener, sulindac, have potent anti-inflammatory and analgesic-antipyretic properties.

3.8.6.4.1 Indomethacin

3.8.6.4.1.1 Usage

Due to the high incidence of severe side effects, indomethacin is not used routinely as an analgesic or antipyretic. It has proved effective in treating ankylosing spondylitis, osteoarthritis and acute gout. Oral doses of 100 to 200 mg daily are administered.

3.8.6.4.1.2 Metabolism and excretion

Approximately 50% of an oral dose of indomethacin is O-demethylated, 10% forms a glucuronide conjugate and a small portion is N-deacylated. The metabolites are believed to be inactive. Up to 20% of the dose is excreted unchanged in the urine. A considerable amount undergoes enterohepatic circulation.

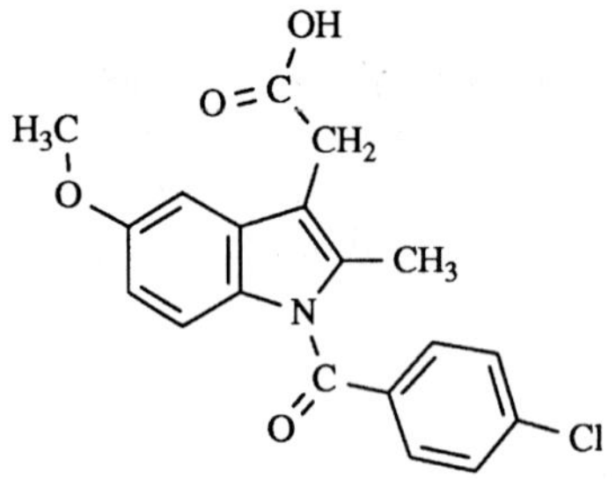

Figure 10. Chemical Structure of Indomethacin.

3.8.6.4.1.3 Blood concentrations

Single oral doses of 50 mg gave an average peak plasma level of 1.9 mg/L between 1 to 4 hours after administration (159). Steady-state plasma concentrations of between 0.31 to 0.63 mg/L were achieved 6 days after giving oral doses of 25 mg three times a day to 5 subjects (160). There is no correlation between indomethacin plasma concentrations and clinical effects. Toxicity is associated with plasma levels greater than 5 mg/L.

3.8.6.4.1.4 Analysis

Indomethacin chromatographs poorly on GLC systems, although this can be improved by prior methylation (161). HPLC is a better approach and methods using UV detection (162) or fluorescence detection (163) have been reported, the latter being sensitive to therapeutic plasma concentrations.

3.8.6.4.2 Sulindac

3.8.6.4.2.1 Usage

Sulindac is used for the same medical conditions as indomethacin and oral doses of 300 to 400 mg are taken.

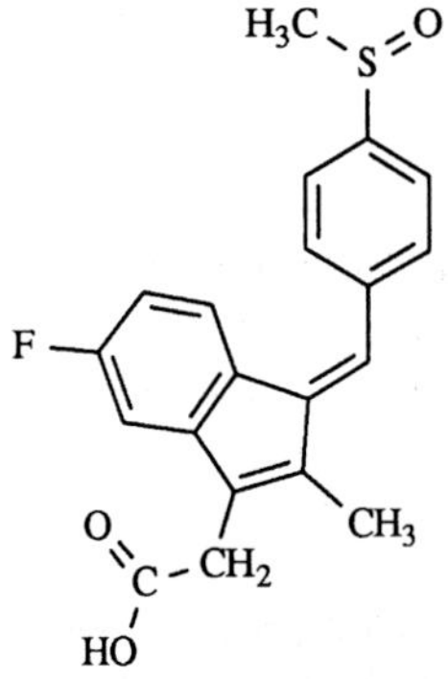

Figure 11. Chemical Structure of Sulindac.

3.8.6.4.2.2 Metabolism and excretion
Sulindac itself is inactive and is metabolised to a pharmacologically active sulphide metabolite and by oxidation to an inactive sulphone. These metabolites and the parent drug undergo conjugation with about 20% of a dose excreted as sulindac conjugate in the urine. Neither free nor conjugated sulphide metabolite is found to any significant extent in the urine (164, 165).

3.8.6.4.2.3 Blood concentrations
Average peak plasma concentrations of sulindac of 4 mg/L occurred 1 hour after a single 200 mg oral dose; at 2 hours average peak plasma concentrations for the sulphide and the sulphone metabolites were 3 mg/L and 2 mg/L respectively (165).

Chronic dosing with 400 mg of sulindac gave steady-state plasma concentrations of 5 mg/L for sulindac, 7 mg/L for the sulphide metabolite and 2.6 mg/L for the sulphone metabolite (166). In a case of sudden death after sulindac overdose, a blood level of 130 mg/L was reported (167).

3.8.6.4.2.4 Analysis
Sulindac is best assayed in biological fluids by chromatographic methods. Both GLC (161, 167) and HPLC methods (162, 168) have been described. Gas-chromatography/mass spectrometry has been used as a sensitive and selective means to measure sulindac and its metabolites (167).

3.8.6.5 Propionic acid derivatives

This is a relatively new group of drugs which are, in the main, better tolerated than other aspirin-like drugs. Fenoprofen, ketoprofen, naproxen and ibuprofen are the most widely used, the latter now being available without prescription in several countries including the United States.

3.8.6.5.1 Fenoprofen

3.8.6.5.1.1 Usage
Fenoprofen is marketed as the calcium salt in capsules and tablets containing 200 to 600 mg of the active drug. In treating mild to moderate pain, doses of 200 mg are given every 4 to 6 hours. Up to 3200 mg daily in divided doses is given to treat rheumatoid arthritis or osteoarthritis.

3.8.6.5.1.2 Metabolism and excretion
Fenoprofen is mainly hydroxylated to 4-hydroxyfenopropfen and the parent drug and metabolite form glucuronide conjugates. These account for approximately 90% of an oral dose excreted in the 24 hour urine. Less than 2% of a dose appears unchanged in the urine and no metabolites have been found in plasma (168).

3.8.6.5.1.3 Blood concentrations
Subjects given a single oral dose of 600 mg of fenoprofen achieved peak plasma concentrations after 1 to 2 hours of about 50 mg/L (169). In overdose cases, levels

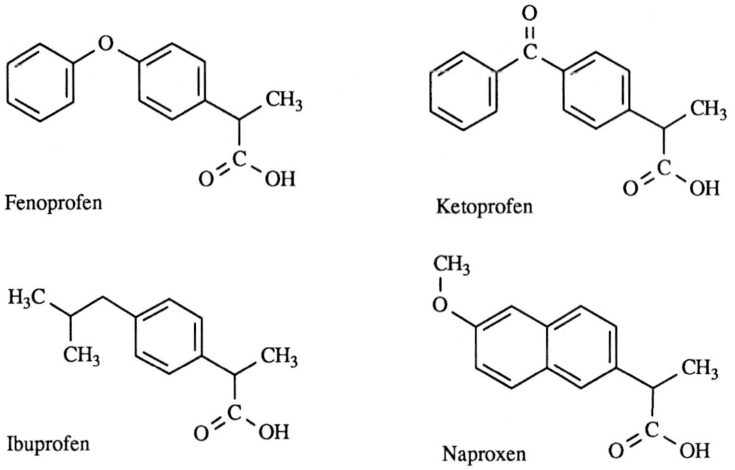

Fenoprofen

Ketoprofen

Ibuprofen

Naproxen

Figure 12. Chemical Structures of Propionic Acid Derivatives.

of between 31 to 828 mg/L have been determined, usually accompanied by only mild toxic effects (170).

3.8.6.5.1.4 Analysis

Fenoprofen is readily analysed in serum by HPLC in both therapeutic and overdosage (170–172). A method which uses an identical extraction, chromatographic and detection system can, by varying the internal standard, detection wavelength and mobile phase characteristics, accommodate measurement of diflunisal, indomethacin, mefanamic acid and piroxicam in addition to fenoprofen and related drugs (170).

3.8.6.5.2 Ketoprofen

3.8.6.5.2.1 Usage

Ketoprofen is used for similar purposes as fenoprofen, but is more potent. Daily oral doses range from 75 to 900 mg.

3.8.6.5.2.2 Metabolism and excretion

Ketoprofen is eliminated primarily in the urine as the glucuronide conjugate or its conjugated metabolites. Less than 1% of a dose appears unchanged in the urine (173, 174).

3.8.6.5.2.3 Blood concentrations

Peak plasma ketoprofen concentrations ranging from 1.9 to 8.4 mg/L are reached between 1 to 1.5 hours after dosing (175). Following overdose, plasma levels of up to 114 mg/L have occurred (170).

3.8.6.5.2.4 Analysis

A GLC method using electron capture detection after prior methylation has been published, but is labour intensive with respect to sample extraction (176). Most other methods have been based on HPLC, using no internal standard (177) or naproxen as the internal standard (174, 178). Methods which avoid using another commonly prescribed drug as the internal standard are preferable (170).

3.8.6.5.3 Ibuprofen

3.8.6.5.3.1 Usage

For mild to moderate pain ibuprofen is given in doses of 400 mg every 4 hours. Daily doses of up to 2400 mg may be used to treat rheumatoid arthritis or osteoarthritis.

3.8.6.5.3.2 Metabolism and excretion

The major portion of a dose is excreted as 2-carboxyibuprofen and 2-hydroxyibuprofen. Up to 9% of a dose is eliminated as unchanged ibuprofen in the 24 hour urine (179).

3.8.6.5.3.3 Blood concentrations

An oral dose of 400 mg of ibuprofen gave peak plasma concentrations of 17 to 30 mg/L occurring 1 to 1.5 hours after dosing (180). In overdose, plasma ibuprofen concentrations ranged from 220 to 840 mg/L (181).

3.8.6.5.3.4 Analysis

GLC methods without (182) or with derivatisation (183, 184) and flame ionization detection are available. Electron capture detection of a halogenated derivative has also been described (185). HPLC methods are more convenient and several have been published (170, 186, 187).

3.8.6.5.4 Naproxen

3.8.6.5.4.1 Usage

Naproxen is a naphthyl acetic acid derivative widely used in the treatment of rheumatoid arthritis, osteoarthritis and ankylosing spondylitis. The usual oral dose is 250 to 375 mg daily.

3.8.6.5.4.2 Metabolism and excretion

Naproxen undergoes demethylation to inactive O-desmethylnaproxen, and conjugates of the parent drug and of this metabolite are the major urinary excretion products (188). Previous reports that up to 10% of a naproxen dose is excreted unchanged in the urine are thought to be erroneous and attributed to the ease by which the naproxen conjugates are hydrolysed during specimen storage (189).

3.8.6.5.4.3 Blood concentrations

Subjects given a single 250 mg oral dose of naproxen achieved an average peak plasma concentration of 31 mg/L at 3 hours (190). Peak levels greater than 120 mg/L occurred when 1500 mg of the drug was administered daily for 5 days. A patient who took 25 g of naproxen experienced only mild adverse effects and developed a serum concentration of 414 mg/L. In other overdose cases, plasma naproxen levels of up to 700 mg/L have been measured (170).

3.8.6.5.4.4 Analysis

Naproxen can be assayed in plasma as a methyl derivative by gas-chromatography with flame-ionization detection (190). HPLC offers a simpler alternative and suitable methods have used reverse phase systems with either UV (170, 178) or fluorescence detection (191).

3.8.6.6 Fenamates

The fenamates are derivatives of N-phenylanthranilic acid of which mefenamic acid has been most widely employed. Other congeners include meclofenamic, flufenamic, tolfenamic and etofenamic acids.

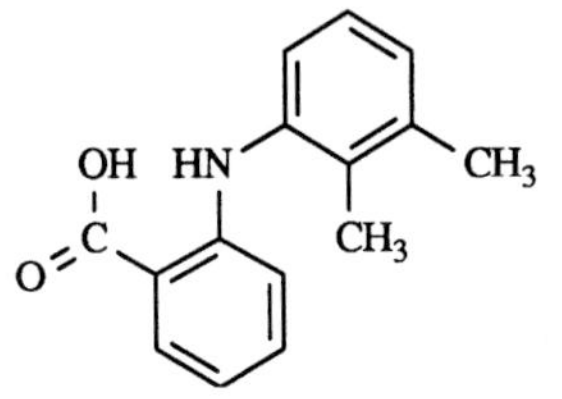

Figure 13. Chemical Structure of Mefenamic Acid.

3.8.6.6.1 Mefenamic acid

3.8.6.6.1.1 Usage

For acute pain, mefenamic acid is given in oral doses of 500 mg initially, followed by 250 mg every 6 hours. Similar doses are given to relieve arthritic conditions.

3.8.6.6.1.2 Metabolism and excretion

Mefenamic acid forms 3-hydroxymethylmefenamic acid by oxidation of the 3-methyl group. Conjugation to glucuronic acid also takes place and the mefenamic acid conjugate concentration in plasma exceeds that of the parent drug. Similar levels of free and conjugated 3-hydroxymethylmefenamic acid occur in the plasma. About 50% of an oral dose of mefenamic acid is excreted in urine as the conjugate with 25% as conjugated 3-hydroxymethylmefenamic acid and 20% as conjugated and free 3-carboxymefenamic acid (192).

3.8.6.6.1.3 Blood concentrations

Subjects receiving 1000 mg of mefenamic acid achieved average peak plasma concentrations of 10 mg/L at 2 hours. Levels of around 20 mg/L occurred on a daily dose of 4000 mg (192). Central nervous system toxicity develops in overdose patients with plasma mefenamic acid levels greater than 70 mg/L (193).

3.8.6.6.1.4 Analysis

Older methods involving fluorimetry (192) or gas-chromatography with flame-ionisation (194) or electron-capture detection (195) have been superseded by HPLC procedures. The most sensitive and reproducible of these uses reverse phase columns with UV detection (170, 196).

3.8.6.7 Piroxicam

3.8.6.7.1 Usage

Piroxicam is a relatively new anti-inflammatory agent with a long half-life and equivalent potency to aspirin, indomethacin and naproxen. For the relief of arthritic conditions a single daily dose of 20 mg is administered.

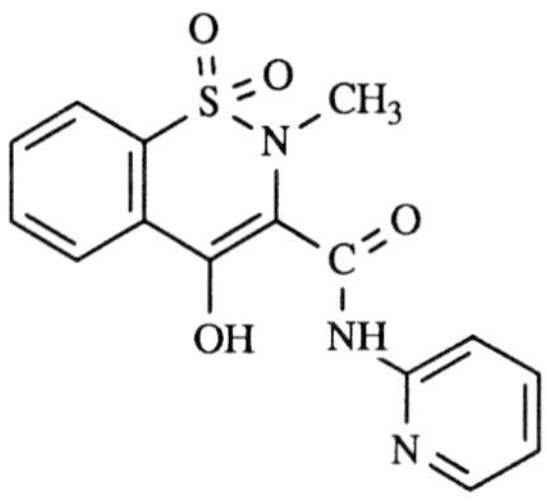

O
S=O
N
CH_3
C=O
OH
NH
N

Figure 14. Chemical Structure of Piroxicam.

3.8.6.7.2 Metabolism and excretion

Piroxicam is hydroxylated to 5-hydroxypiroxicam which is converted to other products by conjugation, hydrolysis, decarboxylation and N-demethylation (197). Approximately 60 % of a dose is excreted in the urine as metabolites with less than 5 % as unchanged piroxicam. Faecal excretion accounts for about 30 % of the dose.

3.8.6.7.3 Blood concentrations

Male subjects given 40 mg oral doses of piroxicam developed plasma levels in the range 3.4 to 6.4 mg/L between 1 to 3 hours after dosing (198). In a later study, male subjects given 20 mg of piroxicam achieved peak concentrations of 1.5 to 2.0 mg/L and female subjects 2.0 to 3.0 mg/L (199). In chronic therapy with 20 mg piroxicam daily, steady-state plasma levels range from 3 to 7 mg/L (200). Following overdose, plasma levels of up to 37 mg/L have been measured (170).

3.8.6.7.4 Analysis

Methods which rely on solvent extraction and ultra-violet spectrophotometric or fluorescence measurements are insensitive and non-specific (201). HPLC methods are much preferred and include liquid-liquid extraction procedures (201) or on-line solid-phase extraction (203) coupled with reverse phase column chromatography with UV detection.

References

1. Martin, W.R., (1983) Pharmacology of opioids. Pharmacol. Rev. *35*, 283–383.
2. Gilman, A.G., Goodman, L.S., Rall, T.W., Murad, F. (Eds) (1985) The Pharmacological Basis of Therapeutics, 7th ed. Macmillan Publishing Co, New York.
3. Novik, D.M., Kreek, M.J., Fanizza, A.M. et. al (1981) Methadone disposition in patients with chronic liver disease. Clin. Pharmacol. Ther. *30*, 353–363.
4. McQuay, H., Moore, A. (1984) Be aware of renal function when prescribing morphine. Lancet *2*, 284.
5. Lewis, J.W., Groux, N., Elliot, P. et al.. (1980) Complications of attempted central venous injections performed by drug abusers. Chest, *78*, 4.
6. Walters, E.C., Stam, F.C., Lousberg, R.J. et al.. (1982) Leucoencephalopathy after inhaling heroin pyrolysate. Lancet *2*, 1233–1237.
7. Widdop, B. & Caldwell, R. (1991) The operation of a hospital laboratory service for the detection of drugs of abuse. In: The Analysis of Drugs of Abuse (Gough, T.A., ed.) John Wiley & Sons. New York.
8. Wolff, K., Sanderson, M., Hay, A.W.M. et al. (1991) Methadone concentrations in plasma and their relationship to drug dosage. Clin. Chem. *37*/2, 205–209.
9. Wolff, K., Hay, A., Raistrick, D. (1991) High dose methadone and the need for drug measurements in plasma. Clin. Chem. *37*/*9*, 1651–1664.
10. Mazoit, J., Sandouk, P., Zetlanoi et al. (1987) Pharmacokinetics of unchanged morphine in normal and cirrhotic patients. Anesth. Anal. *66*, 293–298.
11. Osbourne, R.J., Joel, S.P., Selvin, M.L. (1986) Morphine intoxication in renal failure: the role of morphine-6-glucuronide Brit. Med. J. *292*, 1548–1549.
12. Segura, J. and de la Torre, R. (eds.) (1992) Current Issues of Drug Abuse Testing: First International Symposium CRC Press, Boco Raton, Ann Arbor, London.
13. Inturrisi, C.E., Schultz, M., Shin, S. et al. (1983) Evidence from opiate binding studies that heroin acts through its metabolites. Life Sci 33 (suppl. 1), 773–778.
14. Cone, E.J., Welch, P., Mitchell, J.M. et al. (1991) Forensic drug testing for opiates: I. Detection of 6-acetylmorphine in urine as an indicator of recent heroin exposure; drug and assay considerations and detection times. J. Anal. Toxicol. *15* (1), 1–7.
15. Goldenberg, B.A., Caplan, Y.H., Maguire, T. et al. (1991) Testing human hair for drugs of abuse: III Identification of heroin and 6-acetylmorphine as indicator of heroin use. J. Anal. Toxicol. *15* (5), 226–231.
16. Yeh, S.Y., Gorodetzky, S.W., Krebs, H.A. (1977) Isolation and identification of morphine 3- and 6-glucuronide, morphine 3,6-diglucuronide, morphine 3-ethereal sulfate, normorphine and normorphine 6-glucuronide as morphine metabolites in humans. J. Pharm. Sci. *66*, 1288–1293.
17. Aitkenhead, A.R., Vater, M., Acholo, K. et al. (1984) Pharmacokinetics of single dose i.v. morphine in normal volunteers and patients with end-stage renal failure. Brit. J. Anaesth. *56*, 813–818.
18. Berkowitz, B.A., Ngai, S.H., Yang, J.C. et al. (1975) The disposition of morphine in surgical patients. Clin. Pharm. Ther. *17*, 629–635.
19. Douglas, S., Coates, P.E., Scoble, J. et al. (1989) Linearity of morphine pharmacokinetics using controlled release morphine sulphate tablets of 60 and 100 mg strengths in patients. In: The Edinburgh symposium on pain control and medical education (ed. RG Twycross) Royal Society of Medicine Services Internation Congress and Symposium Series No: 149. Royal Society of Medicine Services Limited, London.
20. Peterson, G.M., Randall, C.T.C., Paterson, J. (1990) Plasma levels of morphine and morphine glucuronides in the treatment of cancer pain; relationship to renal function and route of administration Eur. J. Clin. Pharmacol. *38*, 121–124.
21. Staub, C., Jeanmonod, R., Fryc, O. (1990) Morphine in post-mortem blood: its importance for the diagnosis of deaths associated with opiate addiction. Int. J. Leg. Med. *104*, 39–42.
22. Spiehler, V. and Brown, R. (1987) Unconjugated morphine in blood by radioimmunoassay and gas chromatography-mass spectrometry. J. Forensic Sci *32*, 906–916.
23. Berry, D.J. and Grove, J. (1971) Improved chromatographic techniques and their interpretation for the screening of urine from drug dependent subjects. J. Chromatogr. *61*, 111–123.
24. Widdop, B. (1986) Hospital Toxicology. In: Clarke's Isolation and Identification of Drugs. 2nd Edition. Ed. AC Moffat The Pharmaceutical Press, London pp. 3–34.

25. Widdop, B. Drugs of Abuse In: The Immunoassay Handbook Ed. DG Wild Macmillan, London (in press).
26. Hand, C.W., Moore, R.A., McQuay, H.J. et al. (1987) Analysis of morphine and its major metabolites by differential radioimmunoassay. Ann. Clin. Biochem. *24*, 153–160.
27. Nicolau, G., Lear, G.V., Kaul, B. et al. (1977) Determination of morphine by electron capture gas-liquid chromatography. Clin. Chem. *23*, 1640–1643.
28. Dahlstrom, L., Paalzow, L. and Edlund, P.O. (1977) Simultaneous determination of codeine and morphine in biological samples by gas chromatography with electron capture detection. Acta Pharmacol. et Toxicol. *41*, 273–279.
29. Edlund, P.O. (1981) Determination of opiates in biological samples by glass capillary gas chromatography with electron capture detection. J. Chromatogr. *206*, 109–116.
30. Wasels, R., Belleville, F., Paysant, P. et al. (1989) Determination of morphine in plasma by gas chromatography using a macrobore column and thermionic detection after Extrelute column extraction: Application to follow-up morphine treatment in cancer patients. J. Chromatogr. *489*, 411–418.
31. Lee, H.M. and Lee, C.W. (1991) Determination of morphine and codeine in blood and bile by gas-chromatography with a derivatization procedure. J. Anal. Toxicol. *15* (4), 182–187.
32. Lhoest, G. and Mercier, M. (1980) Gas-chromatography – selected ion monitoring applied to the determination of codeine in plasma. Pharm. Acta Helv. *55*, 316–319.
33. Saady, J.J., Narasimhachari, N. and Blanke, R.V. (1982) Rapid simultaneous quantitation of morphine, codeine and hydromorphone by GC-MS. J. Anal. Toxicol. *6*, 235–237.
34. Paul, B.D., Mell, L.D., Mitchell, J.M. et al. (1985) Simultaneous identification and quantitation of codeine and morphine in urine by capillary gas-chromatography and mass spectrometry. J. Anal. Toxicol. *9*, 222–226.
35. Berjamo, A.M., Ramos, I., Fernandez, P. et al. (1992) Morphine determination by gas chromatography/mass spectrometry in human vitreous humor and comparison with radioimmunoassay. J. Anal. Toxicol. *16*, 372–374.
36. Schuberth, J. and Schuberth, J. (1988) Gas-chromatographic – mass. spectrometric determination of morphine, codeine and 6-monoacetylmorphine in blood extracted by solid phase. J. Chromatogr. *490*, 444–449.
37. Schneider, J.J. and Ravenscroft, P.J. (1989) Determination of morphine in plasma by high-performance liquid chromatography with fluorescence detection. J. Chromatogr. *497*, 326–329.
38. Svensson, J., Rane, A., Sawe, J. et al. (1982) Determination of morphine, morphine-3-glucuronide and (tentatively) morphine-6-glucuronide in plasma and urine using ion-pair high-performance liquid chromatography. J. Chromatogr. *230*, 427–432.
39. Zoer, J., Virgili, P. and Henry, J.A. (1986) High-performance liquid chromatographic assay for morphine with electrochemical detection using an unmodified silica column with a non-aqueous ionic eluent. J. Chromatogr. *382*, 189–197.
40. Tagliaro, F., Franchi, D., Dorizzi, R. et al. (1989) High performance liquid chromatographic determination of morphine in biological samples: an overview of separation methods and detection techniques J. Chromatogr. *488*, 215–228.
41. Glare, P.A., Walsh, T.D. and Pippenger, C.E. (1991) A single, rapid method for the simultaneous determination of morphine and its principal metabolites in plasma using high performance liquid chromatography and fluorometric detection. Ther. Drug. Monit. *13*, 226–232.
42. Adler, T.K., Fujimoto, J.M., Way, E.L. et al. (1955) The metabolic fate of codeine in man. J. Pharm. Exp. Ther. *114*, 251–262.
43. Findlay, J.W.A., Fowle, A.S.E., Butz, R.F. et al. (1978) Plasma codeine and morphine concentrations after therapeutic oral doses of codeine – containing analgesics Clin. Pharm. Ther. *24*, 60–68.
44. Nakamura, G.R., Griesemer, E.C. and Noguchi, T.T. (1976) Antemortem conversion of codeine to morphine in man. J. For. Sci. *21*, 518–524.
45. Peat, M.A. and Sengupta, S. (1977) Toxicological investigations of cases of death involving codeine and dihydrocodeine. For. Sci. *9*, 21–32.
46. Goenechea, S., Kobbe, K. and Goebel, K.J. (1978) Verhalten von codeine und codeine-6-glucuronid bei der hydrolyse mit salzsaeure. Arz. Forsch. *28*, 1070–1071.
47. Posey, B.L. and Kimble, S.N. (1983) Simultaneous determination of codeine and morphine in urine and blood by HPLC. J. Anal. Toxicol. *7*, 241–245.

49. Chen, Z.R., Boechner, F. and Somogyi, A. (1989) Simultaneous determination of codeine, norcodeine and morphine in biological fluids by high performance liquid chromatography with fluorescence detection. J. Chromatogr. *491*, 367–378.

50. Pohland, A., Boaz, H.E. and Sullivan, H.R. (1971) Synthesis and identification of metabolites resulting from the biotransformation of dl-methadone in man and the rat. J. Med. Chem. *14*, 194–197.

51. Inturrisi, C.E. and Verebely, K. (1972) Disposition of methadone in man after a single oral dose. Clin. Pharm. Ther. *13*, 923–930.

52. Manning, T., Bidanset, J.H., Cohen, S. et al. (1976) Evaluation of the Abuscreen for methadone. J. For. Sci. *21*, 112–120.

53. Caldwell, R. and Challenger, H. (1989) A capillary gas-chromatographic method for the identification of drugs of abuse in urine samples. Ann. Clin. Biochem. *5*, 430–443.

54. Inturrisi, C. and Verebely, K. (1972) A gas-liquid chromatographic method for the quantitative determination of methadone in human plasma and urine. J. Chromatogr. *65*, 361–369.

55. Thompson, B.C. and Caplan, Y.H. (1977) A gas-chromatographic method for the determination of methadone and its metabolites in biological fluids and tissues. J. Anal. Toxicol. *1*, 66–69.

56. Greizerstein, H.B. and McLaughin, I.G. (1983) Sensitive method for the determination of methadone in small blood samples. J. Chromatogr. *264*, 312–315.

57. Hartvig, P. and Naslund, B. (1975) Electron-capture gas chromatography of methadone after oxidation to benzophenone. J. Chromatogr. *111*, 347–354.

58. Jacob, P., Rigod, J.F., Pond, S.M. et al. (1981) Determination of methadone and its primary metabolite in biologic fluids using gas chromatography with nitrogen phosphorus detection. J. Anal. Toxicol. *5*, 292–295.

59. Kang, G.I. and Abbott, F.S. (1982) Analysis of methadone and metabolites in biological fluids with gas-chromatography – mass spectrometry. J. Chromatogr. *231*, 311–319.

60. Finkle, B.S. (1984) Self-poisoning with dextropropoxyphene and dextropropoxyphene compounds: the USA experience. Human Toxicol. *3*, 115S–134S.

61. McMahon, R.E., Sullivan, H.R., Due, S.L. et al. (1973) The metabolite pattern of d-propoxyphene in man. The use of heavy isotopes in drug disposition studies. Life Sci. *12*, 463–473.

62. Verebely, K. and Inturrisi, C.E. (1974) Disposition of propoxyphene and norpropoxyphene in man after a single oral dose. Clin. Pharm. Ther. *15*, 302–309.

63. Verebely, K. and Inturrisi, C.E. (1973) The simultaneous determination of propoxyphene and norpropoxyphene in human biofluids using gas-liquid chromatography. J. Chromatogr. *75*, 195–205.

64. Schon, J., Angelo, H., Dam, W. et al. (1978) Pharmacokinetics of dextropropoxyphene in acute poisoning. Arch. Toxicol. (suppl 1): 343–346.

65. Millard, B.J., Sheinin, E.B. and Benson, W.R. (1980) Thermal decomposition of propoxyphene during GLC analysis. J. Pharm. Sci. *69*, 1177–1179.

66. Christensen, H. (1977) Determination of dextropoxyphene and nordextropoxyphene in autopsy material. Acta Pharmacol. Toxicol. *40*, 289–297.

67. Norheim, G. (1976) Determination of dextropoxyphene and nordextropoxyphene in autopsy material by gas-chromatography. Arch. Toxicol. *36*, 89–95.

68. Cleeman, M. (1977) Gas chromatographic determination of propoxyphene and norpropoxyphene in plasma. J. Chromatogr. *132*, 287–294.

69. Flanagan, R.J., Ramsey, J.D. and Jane I (1984) Measurement of dextropropoxyphene and nordextropropoxyphene in biological fluids. Human Toxicol. *3*, 103S–114S.

70. Shin, H., Dongseok, L. and Jongesi, P. (1989) Simultaneous determination of propoxyphene and norpropoxyphene in biological samples by gas-chromatography using an on-column injection technique with a fused-silica capillary column. J. Chromatogr. *491*, 448–454.

71. Flanagan, R.J., Johnston, A., White, A.S.T. et al. (1989) Pharmacokinetics of dextropropoxyphene and nordextropropoxyphene in young and elderly volunteers after single and multiple dextropropoxyphene dosage. Br. J. Clin. Pharmac. *28*, 463–469.

72. Mather, L.E. and Tucker, G.T. (1976) Systemic availability of orally administered meperidine. Clin. Pharm. Ther. *20*, 535–540.

73. Mather, L.E., Lindop, M.J., Tucker, G.T. et al. (1975) Pethidine revisited: plasma concentrations and effects after intramus-

cular injection. Brit. J. Anaesth. *47*, 1269–1275.

74. Stambaugh, J.E., Wainer, I.W., Sanstead, J.K. et al. (1976) The clinical pharmacology of meperidine – comparison of routes of administration. J. Clin. Pharm. *16*, 245–256.
75. Siek, T.J. (1978) The analysis of meperidine and normeperidine in biological specimens. J. For. Sci. *23*, 6–13.
76. Evans, M.A. and Harbison, R.D. (1977) Micromethod for determination of meperidine in plasma. J. Pharm. Sci. *66*, 599–600.
77. Jacob, P., Rigod, J.F., Pond, S.M. et al. (1982) Gas chromatographic analysis of meperidine and normeperidine: determination in blood after a single dose of meperidine. J. Pharm. Sci. *71*, 166–168.
78. Hartvig, P. and Fagerlund, C. (1983) A simplified method for the determination of pethidine and norpethidine after derivatization with trichloroethylchloroformate. J. Chromatogr. *274*, 355–360.
79. Verbeeck, R.K., James, R.C., Taber, D.F. et al. (1980) The determination of meperidine, normeperidine and deuterated analoges in blood and plasma by gas chromatography – mass spectrometry selected ion monitoring. Biomed Mass Spec. *7*, 58–60.
80. Meatherall, R.C., Guay, D.R.P. and Chalmers, J.L. (1985) Analysis of meperidine and normeperidine in serum and urine by high performance liquid chromatography. J. Chromatogr. *338*, 141–149.
81. Karim, A., Ranney, R.E., Evensen, K.L. et al. (1972) Pharmacokinetics and metabolism of diphenoxylate in man. Clin. Pharm. Ther. *13*, 407–419.
82. Al Ragheb, S.A., Dajani, B.M., Salhab, A.A. et al. (1982) A case of fatal Lomotil overdosage. Med. Sci. Law. *22*, 210–214.
83. Haskins, N.J., Ford, G.C., Grigson, S.J.W. et al. (1978) An assay for diphenoxylic acid in urine using an automated gas-chromatograph – mass spectrometer system. In: Quantitative Mass Spectrometry in Life Sciences II (AP de Leenheer et al., eds) Elsevier, New York. pp. 287–293.
84. Ford, G.C., Haskins, N.J., Palmer, R.F. et al. (1976) The measurement of diphenoxylic acid in plasma following administration of diphenoxylate. Biochemical Mass Spectrometry *3*, 45–47.
85. Jackson, L.S. and Stafford, J.E. (1987) The evaluation and application of a radioimmunoassay for the measurement of diphenoxylic acid, the major metabolite of diphenoxylate hydrochloride (Lomotil), in human plasma. J. Pharmacol. Methods. *18(3)*, 189–197.
86. Van Rooy, H.H., Vermeulen, N.P.E. and Bovill, J.B. (1981) The assay of fentanyl and its metabolites in plasma of patients using gas chromatography with alkali flame ionisation detection and gas chromatography – mass spectrometry. J. Chromatogr. *223*, 85–93.
87. Goromaru, T., Matsuura, H., Yoshimura, N. et al. (1984) Identification and quantitative determination of fentanyl metabolites by gas-chromatography – mass spectrometry. Anesthesiology *61*, 73–77.
88. McClain, D.A. and Hug, C.C. (1980) Intravenous fentanyl kinetics. Clin. Pharm. Ther. *28*, 106–114.
89. Lunn, J.K., Stanley, T.H., Eisele, J. et al. (1979) High dose fentanyl anesthesia for coronary artery surgery: plasma fentanyl concentrations and influence of nitrous oxide on cardiovascular response. Anesth. Anal. *58*, 390–395.
90. Hug, C.C. (1984) Pharmacokinetics and pharmacodynamics of narcotic analgesics. In: Pharmacokinetics of Anaesthesia (Prys-Roberts C. and Hug, C.C. – eds) Blackwell, Oxford pp. 187–234.
91. Schuttler, J. and White, P.F. (1984) Optimization of the radioimmunoassays for measuring fentanyl and alfentanil in human serum. Anaesthesiology. *61*, 315–320.
92. Hammargren, W.R. and Henderson, G.L. (1988) Analysing normetabolites of the fentanyls by gas chromatography/electron detection. J. Anal. Toxicol. *12(4)*, 183–191.
93. Kowalski, S.R., Gourley, G.K., Cherry, D.A. et al. (1987) Sensitive gas liquid chromatography method for the determination of fentanyl concentrations in blood. J. Pharmacol Methods. *18*, 347–355.
94. Watts, V. and Caplan, Y. (1988) Determination of fentanyl in whole blood at subnanogram concentrations by dual capillary column gas-chromatography with nitrogen sensitive detectors and gas chromatography/mass spectrometry. J. Anal. Toxicol. *12(5)*, 246–254.
95. Woestenborghs, R.J., Stanski, D.R., Scott, J.C. et al. (1987) Assay methods for fentanyl in serum: gas-liquid chromatography versus radioimmunoassay. Anaesthesiology. *67 (1)*, 85–90.
96. Kumar, K., Morgan, D.J. and Crankshaw, D.P. (1987) Determination of fentanyl and

alfentanil in plasma by high performance liquid chromatography with ultra-violet detection. J. Chromatogr. *419*, 464–468.

97. Strang, J. (1985) Abuse of buprenorphine. Lancet *2*, 725.
98. Talwar, K. and Watson, I. D. (1991) Incidence of drug abuse in the east end of Glasgow. Proc. Int. Assoc. of Forensic Toxicologists. Glasgow August 1989. Aberdeen, Scotland: Aberdeen Univ. Press. pp. 357–360.
99. Banks, C. D. (1979) Overdosage of buprenorphine: case report. New Zealand Medical Journal *89*, 255–256.
100. Bartlett, A. J., Lloyd-Jones, J. G., Rance, M. J. et al. (1980) The radioimmunoassay of buprenorphine. Eur. Jnl. Clin. Pharmacol. *18*, 339–345.
101. Brewster, D., Humphrey, M. J. and McLeavy, M. A. (1981) Biliary excretion, metabolism and enterohepatic circulation of buprenorphine. Xenobiotica. *11*, 189–196.
102. Cone, E. J. et al. (1984) The metabolism and excretion of buprenorphine in humans. Drug Metabolism and Disposition. *12*, 577–581.
103. Blom, Y., Bondesson, U. and Anggard, E. (1985) Analysis of buprenorphine and its N-dealkylated metabolite in plasma and urine by selected – ion monitoring. J. Chromatogr. *338*, 89–98.
104. Hand, C. W., Baldwin, D., Moore, R. A. et al. (1986) Radioimmunoassay of buprenorphine with iodine label: analysis of buprenorphine and metabolites in human plasma. Ann. Clin. Biochem. *23*, 47–53.
105. Debrabandere, L., Van Boven, M. and Daenens, P. (1991) High performance liquid chromatography with electrochemical detection of buprenorphine and its major metabolite in urine. J. Chromatogr. *564*, 557–566.
106. Debrabandere, L., Van Boven, M. and Daenens, P. (1991) Analysis of buprenorphine in urine specimens (1991) J. Forensic. Sci. *37(1)*, 82–89.
107. Martinez, D., Jurado, M. C. and Repetto, M. (1990) Analysis of buprenorphine in plasma and urine by gas chromatography. J. Chromatogr. *528*, 459–463.
108. Blom, Y. and Bondesson, U. (1985) Analysis of buprenorphine and its dealkylated metabolite in plasma and urine by selected-ion monitoring. J. Chromatogr. *338*, 89–98.
109. Vane, J. R. and Botting, R. (1987) Inflammation and the mechanism of action of anti-inflammatory drugs. FASEB J. *1*, 89–96.
110. Dascombe, M. J. (1985) The Pharmacology of Fever. *25*, 327–373.
111. Maher, J. F. (1984) Analgesic nephropathy. Am. J. Med. *76*, 345–348.
112. Seymour, R. A., Rawlins, M. D. et al. (1982) Efficacy and pharmacokinetics of aspirin in post-operative dental pain. Brit. J. Clin. Pharmacol. *13*, 807–710.
113. Done, A. K. (1960) Salicylate intoxication. Pediatrics *26*, 800–807.
114. Prescott, L. F. and Critchley, J. A. J. H. (1983) The treatment of acetaminophen poisoning. Annu. Rev. Pharmacol. Toxicol. *23*, 87–101.
115. Starko, K. M. and Mullick, F. G. (1983) Hepatic and cerebal pathology findings in children with fatal salicylate intoxication: further evidence for a causal relation between salicylate and Reye's syndrome. Lancet *1*, 326–329.
116. Davison, C. (1971) Salicylate metabolism in man. Ann. N.Y. Acad. Sci. *179*, 249–268.
117. Hollister, G. E. and Kanter, S. L. (1965) Studies of delayed medication. IV Salicylates. Clin. Pharm. Ther. *6*, 5–13.
118. Gupta, N., Sarkissian, E. and Paulus, H. E. (1975) Correlation of plateau serum salicylate level with rate of salicylate metabolism. Clin. Pharm. Ther. *18*, 350–355.
119. Keller, W. J. (1947) A rapid method for the determination of salicylates in serum or plasma. Am. J. Clin. Pathol. *17*, 415–417.
120. Trinder, P. (1954) Rapid determination of salicylate in biological fluids. Biochem. J. *57*, 301–303.
121. Kang, E. S., Todd, T. A., Capaci, M. T. et al. (1983) Measurement of true salicylate concentration in serum from patients with Reye's syndrome. Clin. Chem. *29*, 1012–1014.
122. Frings, C. S. (1973) Drug screening. CRC Crit. Rev. Clin. Lab. Sci. *4*, 357–382.
123. Jarvie, D. R., Heyworth, R. and Simpson, D. (1987) Plasma salicylate analysis: a comparison of colorimetric, HPLC and enzymatic techniques. Ann. Clin. Biochem. *24*, 364–373.
124. Chubb, S. A. P., Campbell, R. S., Ramsey, J. R. et al. (1986) An enzyme mediated colorimetric method for the measurement of salicylate. Clin. Chim. Acta. *155*, 209–220.
125. Williams, L. A., Linn, R. A. and Zak, B. (1959) Differential ultra-violet spectropho-

tometric determination of serum salicylates. J. Lab. Clin. Med. *53*, 156–162.

126. Chirigos, M.A. and Udenfriend, S. (1959) A simple fluorimetric procedure for determining salicylic acid in biological tissues. J. Lab. Clin. Med. *54*, 769–772.
127. Rance, M.J., Jordan, B.J. and Nichols, J.D. (1975) A simultaneous determination of acetylsalicylic acid, salicylic acid and salicylamide in plasma by gas liquid chromatography. J. Pharm. Pharmac. *27*, 425–429.
128. Cham, B.E., Johns, D., Bochner, F. et al. (1979) Simultaneous liquid chromatographic quantitation of salicylic acid, salicyluric acid and gentisic acid in plasma. Clin. Chem. *25*, 1420–1425.
129. Mays, D.C., Sharp, D.E., Beach, C.A. et al. (1984) Improved method for the determination of aspirin and its metabolites in biological fluids by HPLC. J. Chromatogr. *311*, 301–309.
130. Yu, S.S., Osterloh, J., Nishiki-Finley, R. et al. (1986) Simultaneous HPLC analysis for serum salicylate, acetaminophen and theophylline. Clin. Chem. *32*, 1063.
131. Dawson, C.M., Wang, T.W.M., Rainbow, S.J. et al. (1988) A non-extraction HPLC method for the simultaneous determination of serum paracetamol and salicylate. Ann. Clin. Biochem. *25*, 661–667.
132. Prescott, L.F. (1981) Treatment of severe acetaminophen poisoning with intravenous acetyl cysteine. Arch. Int. Med. *141*, 386–389.
133. Thomas, B.H., Coldwell, B.B., Zeitz, W. et al. (1972) Effects of aspirin, caffeine and codeine on the metabolism of phenacetin and acetaminophen. Clin. Pharm. Ther. *13*, 906–910.
134. Fletterick, C.G., Grove, T.H. and Hohnadel, D.C. (1979) Liquid-chromatographic determination of acetaminophen in serum. Clin. Chem. *25*, 409–412.
135. Prescott, L.F., Wright, N., Roscoe, P. et al. (1971) Plasma paracetamol half-life and hepatic necrosis in patients with paracetamol overdosage. Lancet *1*, 519–522.
136. Rumack, B.H. and Peterson, R.G. (1978) Acetaminophen overdosage: incidence, diagnosis and management in 416 patients. Pediatrics *62*, Part 2 (Suppl.) 898–903.
137. Welch, R.M. and Conney, A.H. (1965) Simple method for the quantitative determination of N-acetyl-p-aminophenol in urine. Clin. Chem. *11*, 1064.
138. Wilkinson, G.S. (1976) Rapid determination of plasma paracetamol. Ann. Clin. Biochem. *13*, 435–437.
139. Stewart, M.J., Adriaenssens, P.I., Jarvie, D.R. et al. (1979) Inappropriate methods for the determination of plasma paracetamol. Ann. Clin. Biochem. *16*, 89–95.
140. Glynn, J.P. and Kendal, S.E. (1975) Paracetamol measurements. Lancet *1*, 1147–1148.
141. Mace, P.F.K. and Walker, G. (1976) Salicylate interference with plasma paracetamol measurement. Lancet *2*, 1362.
142. Hammond, P.M., Scawen, M.D., Atkinson A. et al. (1984) Development of an enzyme based assay for acetaminophen. Anal. Biochem. *143*, 152–157.
143. Morris, H.C., Overton, P.D., Ramsey, J.R. et al. (1990) Development and validation of an automated enzyme assay for paracetamol (acetaminophen) Clin. Chim. Acta. *187*(2): 95–104.
144. Edinbora, L.E., Jackson, G.F., Jortani, S.A. et al. (1991) Determination of serum acetaminophen in emergency toxicology: evaluation of newer methods: Abbott TDX and second derivative spectrophotometry. Clin. Toxicol. *29* (2): 241–255.
145. Prescott, L.F. (1971) The gas-liquid chromatographic estimation of phenacetin and paracetamol in plasma and urine. J. Pharm. Pharmac. *23*, 111–115.
146. Wahl, K.C. and Rejent, T.A. (1978) Quantitative gas chromatographic determination of acetaminophen using trimethylalininium hydroxide as a derivatizing agent. J. For. Sci. *23*, 14–20.
147. Huggett, A., Andrews, P. and Flanagan, R.J. (1981) Rapid micro-method for the measurement of paracetamol in blood plasma or serum using gas liquid chromatography with flame-ionisation detection. J. Chromatogr. *209*, 67–76.
148. Kamali, F. and Herd, B. (1990) Liquid-liquid extraction and analysis of paracetamol (acetaminophen) and its major metabolites in biological fluids by reversed-phase ion-pair chromatography. J. Chromatogr. *530* (1): 222–225.
149. Aquilar, M.I., Hart, S.J. and Calder, I.C. (1988) Complete separation of urinary metabolites of paracetamol and substituted paracetamols by reversed-phase ion-pair high-performance liquid chromatography. J. Chromatogr. *426* (2): 315–333.
150. Aarbakke, J., Bakke, O.M., Milde, E.J. et al. (1977) Disposition and oxidative meta-

bolism of phenylbutazone in man. Eur. J. Clin. Pharm. *11*, 359–366.

151. Midha, K.K., McGilveray, I.J. and Charette, C. (1974) GLC determination of plasma concentration of phenylbutazone and its metabolite, oxyphenbutazone. J. Pharm. Sci. *63*, 1234–1239.
152. Prescott, L. F., Critchley, J. A. J. H. and Balali-Mood M. (1980) Phenylbutazone overdose: abnormal metabolism associated with hepatic and renal damage. Brit. Med. J. *2*, 1106–1107.
153. Sionfi, A., Caudal, F. and Marfil, F. (1978) GLC determination of phenylbutazone in human plasma. J. Pharm. Sci. *67*, 243–245.
154. Budd, R.D. (1982) Gas chromatographic determination of Butazolidin (phenylbutazone) in biological fluids. J. Chromatogr. *243*, 368–371.
155. Aarons, L. and Higham, C. (1980) An improved HPLC assay for monitoring phenylbutazone and its two major oxidised metabolites in plasma. Clin. Chim. Acta. *105*, 377–382.
156. Alvineric, M. (1980) Reversed phase high-performance liquid chromatography of phenylbutazone in body fluids. J. Chromatogr. *181*, 132–134.
157. Marunaka, T., Shibata, T., Minami, Y. et al. (1980) Simultaneous determination of phenylbutazone and its metabolites in plasma and urine by high-performance liquid chromatography. J. Chromatogr. *183*, 331–338.
158. Spahn, H. and Mutschler, E. (1983) HPLC determination of azapropazone in human plasma. Arch. Pharm. *316*, 42.
159. Kwan, K.C., Breault, G.O., Umbenhauer, E.R. et al. (1976) Kinetics of indomethacin absorption, elimination and enterohepatic circulation. J. Pharmacokinet. Biopharm. *4*, 255–280.
160. Alvan, G., Orme, M., Bertilsson, L. et al. (1975) Pharmacokinetics of indomethacin. *18*, 364–373.
161. Sharp, M.E. (1987) A rapid screening method for acidic and neutral drugs in blood by high-resolution gas chromatography. J. Anal. Tox. *11*, 8–11.
162. Shimek, J.L., Rao, N.G.S. and Khalil, S.K.W. (1981) High performance liquid chromatographic analysis of tolmetin, indomethacin and sulindac in plasma. J. Liq. Chrom. *4*, 1987–2013.
163. Bernstein, M.S. and Evans, M.A. (1982) High performance liquid chromatography – fluorescence analysis for indomethacin and metabolites in biological fluids. J. Chromatog. *229*, Biomed. Appl. *18*, 179–187.
164. Hucker, H.B., Stauffer, S.C., White, S.D. et al. (1973) Physiologic disposition and metabolic fate of a new anti-inflammatory agent in the rat, rhesus monkey and man. Drug Met. Disp. *1*, 721–736.
165. Duggan, D.E., Hare, L.E., Ditzler, C. et al. (1977) The disposition of sulindac. Clin. Pharm. Therap. *21*, 326–355.
166. Swanson, B.N., Boppana, V.K., Vlasses, P.H. et al. (1982) Sulindac disposition when given once or twice daily. Clin. Pharm. Ther. *32*, 397–403.
167. Archibald, J.T. (1984) A case report of suspected sulindac poisoning. Can. Soc. For. Sci. J. *17*, 71–76.
168. Rubin, A., Rodda, B.E., Warrick, P. et al. (1972) Physiological disposition of fenoprofen in man II: plasma and urine pharmacokinetics after oral and intravenous administration. J. Pharm. Sci. *61*, 739–745.
169. Waife, S.O., Cruber, C.M., Rodda, B.E. et al. (1975) Problems and solutions to single dose testing analgesics: comparison of propoxyphene, codeine and fenoprofen. Int. J. Clin. Pharm. *12*, 301–304.
170. Streete, P.J. (1989) Rapid high-performance liquid chromatographic methods for the determination of overdose concentrations of some non-steroidal anti-inflammatory drugs in plasma or serum. J. Chromatogr. *495*, 179–193.
171. Bopp, R.J., Farid, K.Z. and Nash, J.F. (1981) High performance liquid chromatographic assay for fenoprofen in human plasma. J. Pharm. Sci. *70*, 507–509.
172. Katogi, Y., Ohmura, T. and Adachi, M. (1983) Simple and rapid determination of fenoprofen in plasma by high-performance liquid chromatography. J. Chromatogr. *278*, 475–477.
173. Delbarre, F., Roucayrol, J.C., Amor, B. et al. (1976) Pharmacokinetic study of ketoprofen in man using the tritiated compound. Scand. J. Rheum (Suppl. 14): 45–52.
174. Kaye, C.M., Sankey, M.G. and Holt, J.E. (1981) A high pressure liquid chromatographic method for the assay of ketoprofen in plasma and urine and its application to determining the urinary excretion of free and conjugated ketoprofen following oral administration of Orudis to man. Brit. J. Clin. Pharm. *11*, 395–398.

175. Upton, R.A., Williams, R.L., Guentert, T.W. et al. (1981) Ketoprofen pharmacokinetics and bioavailability based on an improved sensitive and specific assay. Eur. J. Clin. Pharm. *20*, 127–133.
176. Stenberg, F., Johnsson, T.E., Nilsson, B. et al. (1979) Determination of ketoprofen in plasma by extractive methylation and electron capture gas chromatography. J. Chromatogr. *177*, 145–148.
177. Ballerini, R., Cambi, A., Del Soldado, P. et al. (1979) Determination of ketoprofen by direct injection of deproteinized body fluids into a high-pressure liquid chromatographic system. J. Pharm. Sci. *68*, 366–368.
178. Upton, R.A., Bushkin, T.W., Guentert et al. (1980) Convenient and sensitive high-performance liquid chromatography assay for ketoprofen, naproxen and other allied drugs in plasma or urine. J. Chromatogr. *190*, 119–128.
179. Mills, R.F.N., Adams, S.S., Cliffe, E.E. et al. (1973) The metabolism of ibuprofen. Xenobiotica *3*, 589–598.
180. Collier, P.S., D'Arcy, P.F., Harron, D.W.G. et al. (1978) Pharmacokinetic modelling of ibuprofen. Brit. J. Clin. Pharm. *5*, 528–530.
181. Court, H., Streete, P.J. and Volans, G.N. (1981) Overdose with ibuprofen causing unconsciousness and hypotension. Br. Med. J. *282*, 1073.
182. Heikkinen, L. (1984) Silica capillary gas chromatographic determination of ibuprofen in serum. J. Chromatogr. *307*, 206–209.
183. Kaiser, D.G. and Vangiessen, G.J. (1974) GLC determination of ibuprofen in plasma. J. Pharm. Sci. *63*, 219–221.
184. Midha, K.K., Cooper, J.K., Hubbard, J.W. et al. (1977) A rapid and simple GLC procedure for determinations of plasma concentrations of ibuprofen. Can J. Pharm. Sci. *12*, 29–31.
185. Kaiser, D.G. and Martin, R.S. (1978) Electron capture GLC determination of ibuprofen in serum. J. Pharm. Sci. *67*, 627–630.
186. Aravind, M.K., Micelli, J.N. and Kauffmann, R.E. (1984) Determination of ibuprofen by high performance liquid chromatography. J. Chromatogr. *308*, 350–353.
187. Shah, A. and Jung, D. (1986) Rapid and simple determination of the major metabolites of ibuprofen in biological fluids by high-performance liquid chromatography. J. Chromatogr. *378*, 232–236.
188. Segre, E.J. (1975) Naproxen metabolism in man. J. Clin. Pharm. *15*, 316–323.
189. Upton, R.A., Buskin, J.N., Williams, R.L. et al. (1980) Negligible excretion of unchanged ketoprofen, naproxen and probenecid in urine. J. Pharm. Sci. *69*, 1254–1257.
190. Weber, S.S., Bankhurst, A.D., Mroszezak, E. et al. (1981) Effect of mylanta on naproxen biovailability. Ther. Drug. Mon. *3*, 75–83.
191. Shimek, J.L., Rao, N.G.S. and Wahba Khalil, S.E. (1982) An isocratic high-performance liquid chromatographic determination of naproxen and desmethylnaproxen in human plasma. J. Pharm. Sci. *71*, 436–439.
192. Glazko, A.J. (1968) Pharmacology of fenamates. III Metabolic disposition. Ann. Phys. Med. 9 (Suppl.): 23–36.
193. Balali-Mood, M., Critchley, J.A.J.H., Proudfoot, A.T. et al. (1981) Mefenamic acid overdose. Lancet *1*, 1354–1356.
194. Dursci, L.J. and Hackett, L.P. (1978) Gas-liquid chromatographic determinations of mefenamic acid in human serum. J. Chromatogr. *161*, 340–342.
195. Bland, S.A., Blake, J.W. and Ray, R.S. (1976) Mefenamic acid blood and urine levels in the horse determined by derivative gas-liquid chromatography electron capture. J. Chrom. Sci. *14*, 201–203.
196. Poirier, J.-M., Lebot, M. and Cheymol, G. (1992) Rapid and sensitive liquid chromatographic assay of mefenamic acid in plasma. Ther. Drug Monit. *14*, 322–326.
197. Twomey, T.M. and Hobbs, D.C. (1978) Biotransformation of piroxicam by man. Fed. Proc. *37*, 271.
198. Hobbs, D.C. and Twomey, T.M. (1979) Piroxicam pharmacokinetics in man: aspirin and antacid interaction studies. J. Clin. Pharm. *19*, 270–281.
199. Richardson, C.J., Block, K.L.N., Ross, S.G. (1985) Effects of age and sex on piroxicam disposition. Clin. Pharm. Ther. *37*, 13–18.
200. U.S. Pharmacopeial Convention (1987) Drug Information for the Health Care Provider, 7th Ed. Rockville, Maryland, p. 303.
201. Ishizaki, T., Nomura, T. and Abe, T. (1979) Pharmacokinetics of piroxicam, a new nonsteroidal anti-inflammatory agent, under fasting and post-prandial states in man. J. Pharm. Biopharm. *7*, 369–381.

202. Milligan, P. A. (1992) Determination of piroxicam and its major metabolites in the plasma, urine and bile of humans by high-performance liquid chromatography. J. Chromatogr. *576*, 121–128.

203. Saeed, K. and Becher, M. (1991) On-line solid-phase extraction of piroxicam prior to its determination by high-performance liquid chromatography. J. Chromatogr. *567*, 185–193.

3.9 Antihistamines

H. Brandenberger

3.9.1 General Remarks

Histamine or 2-(4-imidazolyl)ethylamine is a normal body constituent, formed by decarboxylation of histidine. In physiological concentration, it exhibits various activities. It stimulates, i.e., gastric secretion and regulates the permeability of blood vessels. Many illnesses or body affections, including allergies and inflamations, produce locally elevated, undesired concentrations of histamine and similar acting substances.

Antihistamines are pharmaceuticals which displace histamine competitively from their receptors, therefore suppressing or canceling the undesired effects. They are sold both by prescription and over-the-counter for the treatment of colds and allergies, as remedies against motion sickness and as sedatives. High doses can produce excitatory and/or depressant effects of the central nervous system, which are, however, not very characteristic. In addition, damaged brains and kidneys could be observed in autopsies of people with histories of chronic abuse of antihistamines.

Already therapeutic doses of antihistamines may cause a dry mouth, drowsiness, headaches, nausea, tachycardia, urinary retention and nervousness. Toxic doses can lead to disorientation, hallucinations, tremor, convulsions and coma. These symptoms often vary from patient to patient.

The Merck Index (1) lists around 70 antihistamines. It classifies them into aminoalkyl ethers (–O–C–C–N=), ethylenediamines (=N–C–C–N=), piperazines, other alkylamine derivatives, phenothiazines, and about a dozen differently structured members, mostly also tricyclics. In this place, it will of course not be possible to deal with such a large number of drugs individually. This is also hardly necessary, since only a limited number have gained toxicological importance. In tables 1 and 4 to 8, we have listed a substantial part of the drugs on the market with their formulas. It is evident that the structural similarities of the members in the different subgroups permit use of the same chemical techniques for detection and quantification. We will therefore illustrate recommendable analytical procedures on selected examples. Readers familar with analytical chemistry will then easily be able to extrapolate the methods to the structurally related drugs.

Since the antihistamines are, in most countries, freely available (over-the-counter drugs), and since many of them have a strongly sedating and also remarkable hypnotic action, they are often used as substitutes for hypnotics placed under medical or narcotic control. This has led to severe intoxications, some with lethal outcome. A computer search of the pertinent literature (2) has shown that diphenhydramine (compound 1 in Table 1) leads the list of intoxications caused by antihistamines. This agrees with the experiences from our own analytical control work. It seems therefore justified to pay special attention to this specific drug.

Table 1. Antihistamines with Aminoalkylether Structure

A) Diphenyl-methoxy derivatives

1.	Diphenhydramine, Benzhydramine, Benadryl, Benocten, Dabylene, Dolestan, Felben, Sedopretten, Valdrene, Wehydryl	$(Ph)(Ph)CH-O-CH_2-CH_2-N(CH_3)(CH_3)$
2.	Orphenadrin, Mephenamin, Banflex, Biorphen, Disipal, Norflex	$(o\text{-}CH_3-Ph)(Ph)CH-O-CH_2-CH_2-N(CH_3)(CH_3)$
3.	Neo-Benodine, Toladryl	$(p\text{-}CH_3-Ph)(Ph)CH-O-CH_2-CH_2-N(CH_3)(CH_3)$
4.	Medrylamine, Histaphen, Postafen	$(p\text{-}CH_3O-Ph)(Ph)CH-O-CH_2-CH_2-N(CH_3)(CH_3)$
5.	Ambodryl, Bromazine, Bromodiphenhydramine, Histabromamine	$(p\text{-}Br-Ph)(Ph)CH-O-CH_2-CH_2-N(CH_3)(CH_3)$
6.	Mephenhydramine, Alphadryl, Alfadril, Spofa 325	$(Ph)(Ph)C(CH_3)-O-CH_2-CH_2-N(CH_3)(CH_3)$
7.	Chlorphenoxamine, Clorevan, Systral, Contristaminene, Phenoxene, Phenoxine	$(p\text{-}Cl-Ph)(Ph)C(CH_3)-O-CH_2-CH_2-N(CH_3)(CH_3)$
8.	Embramine, Bromadryl, Mebryl, Mebrophenhydramine	$(p\text{-}Br-Ph)(Ph)C(CH_3)-O-CH_2-CH_2-N(CH_3)(CH_3)$
9.	Diphenazoline, Antadril	$(Ph)(Ph)CH-O-CH_2-C\langle{}^{NH-CH_2}_{=N-CH_2}\rangle$ (ring: $NH-CH_2$, CH_2-N)
10.	Diphenpyraline, Diphenylpyrilene, Anginosan, Allerzine, Anti-Hist, Diafen, Hispril, Histalert, Histryl, Kolton, Lergoban, Lyssipoll	$(Ph)(Ph)CH-O-CH\langle{}^{CH_2-CH_2}_{CH_2-CH_2}\rangle N-CH_3$

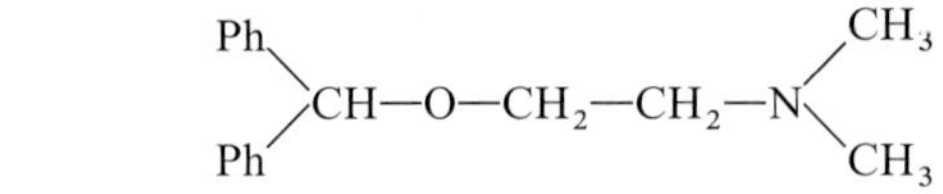

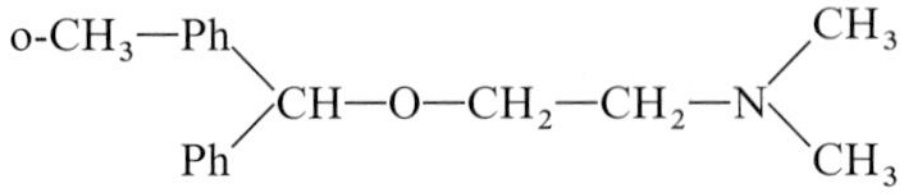

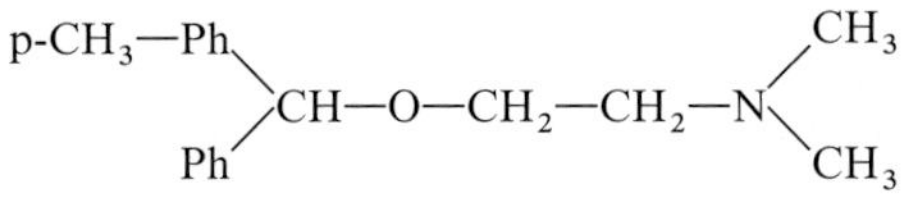

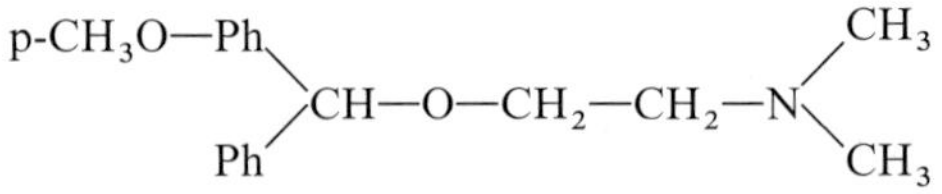

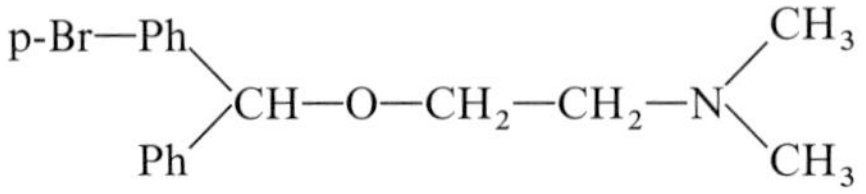

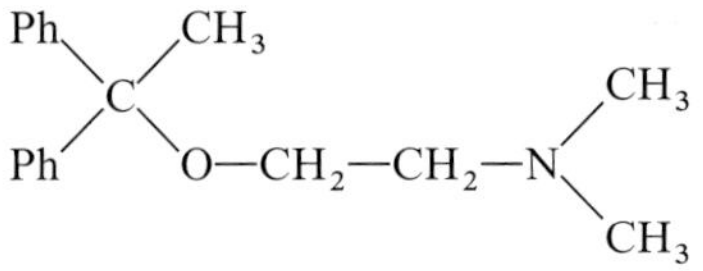

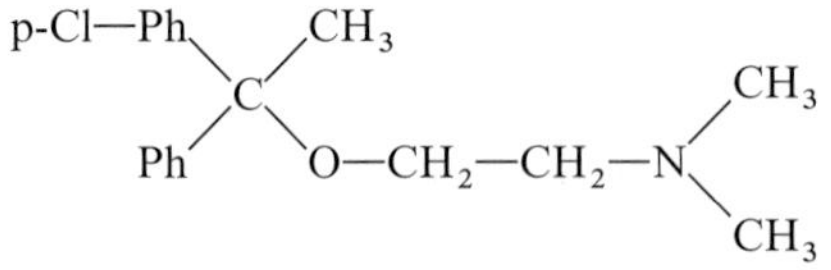

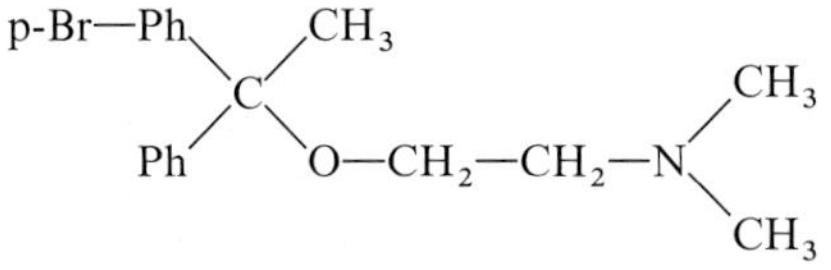

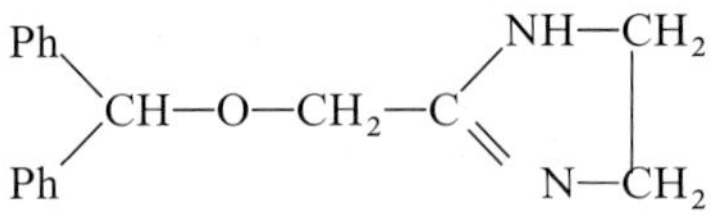

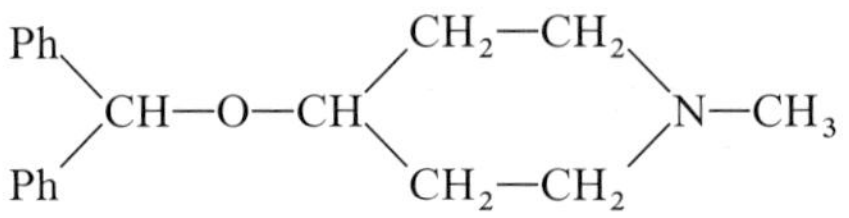

Table 1. (continued)

11. Clemastine, Meclastine, Mecloprodin, Aloginal, Anhistan, Lecasol, Piloral, Reconin, Tavegil, Tavist, Trabest	$(p\text{-Cl-Ph})(Ph)(CH_3)C-O-CH_2-CH_2-CH$ (ring: $CH-N(CH_3)-CH_2-CH_2-CH_2$)
B) Phenyl-pyridyl-methoxy derivatives	
12. Carbinoxamine, Allergefon, Clistin, Ciberon, Hislosine, Histex, Lergefin, Polistin, T-Caps, Ziriton	$(p\text{-Cl-Ph})(Py)CH-O-CH_2-CH_2-N(CH_3)_2$
13. Doxylamine, Histadoxylamine, Bendectin, Debendox, Decapryn, Hoggar N, Mereprine, Nethaprin, Sedaplus, Syndol, Unisom	$(Ph)(Py)(CH_3)C-O-CH_2-CH_2-N(CH_3)_2$
C) Different derivatives	
14. Phenyltoxolamine, Phenoxadrine, Dutin, Bristamine	$o\text{-}(Ph-CH_2)-Ph-O-CH_2-CH_2-N(CH_3)_2$

3.9.2 Diphenhydramine

Diphenhydramine is one of the first substances found to have antihistamine properties. It can give relief to many allergic reactions and possesses anti-emetic properties. But the drug has also a strong sedative and even a hypnotic action. It has been sold under a large number of brand names. Only a few are listed in table 1, which also shows the chemical formula of the compound. The aminoethyl ether is substituted with a diphenylmethyl residue at the oxygen and with 2 methyl groups at the nitrogen. The resulting molecule is a strong base with relatively low polarity. It can be extracted from aqueous solution already with quite nonpolar organic solvents such as hexane containing some alcohol. Pharmaceutical preparations contain diphenhydramine usually as its hydrochloride. Mixtures of the drug with the potent hypnotic methaqualone have been market runners for years and are responsible for many lethal intoxications.

Diphenhydramine is dosed relatively high. Between 75 and 200 mg are given daily. This leads to therapeutic blood concentrations between 0.1 to 0.5 μg/ml. Blood levels over 1 μg/ml are usually associated with toxic effects (3, 4). The estimated minimal lethal dose is 2 to 3 g (3, 5). In fatal intoxications, blood levels until 50 μg/g and considerably higher tissue concentrations have been measured (3). Since the drug is well metabolized, the urine content of intact diphenhydramine in cases of therapeutic dosage is only low. At higher dosages, the urine ratio of unchanged drug to metabolites increases drastically.

The most efficient procedure for the isolation of diphenhydramine from body fluids consists of ether extraction (the use of methyl-tert.butyl ether is recommended) after alkalization to pH 10 or higher. By re-extracting the drug into small volumes of acidified water, it can be purified and concentrated. Only in intoxications, but not with therapeutic dosages, this often permits detection of the benzylic UV-absorption bands around 255 nm (max. at 257 nm), despite the low absorption coefficient ($\varepsilon_{257} = 433$). The drug can of course be detected by TLC with the usual spray reagents for tertiary amines (3). Trace analysis and detection of metabolites are best carried out by GC. The column choice is not critical, as long as phases with relatively low polarity are selected. Whenever possible, mass spectrometric detection should be used, especially in urine analysis, since a large number of metabolites may be present. It would not be convenient and also not always reliable to try to identify them exclusively on the basis of their retention data.

Table 2 lists the metabolites of diphenhydramine which can originate from a stepwise degradation of the amine side chain by oxidative deamination, red-ox processes, and decarboxylation. Among the metabolites are the secondary amine (numbered with 1), the primary amine (comp. 2), two aldehydes (comp. 3 and 7), two acids (comp. 4 and 8), two alcohols (comp. 5 and 9), an ether (comp. 6) and a ketone (comp. 10). In the urine, the acids and alcohols are to a large part present in coupled form, usually bound to glycine, glycuronic acid or sulfate. Any analysis for diphenhydramine and all its metabolites must therefore be preceded by an enzymatic or an acid hydrolysis. If that is omitted, only the glycine conjugates may be coextracted, not the sulfates and glycuronates. The free acids obtained after hydrolysis will of course accumulate in the acid extract and the metabolites without alkaline or acid function in the neutral extract.

A second important metabolic pathway of diphenhydramine is the hydroxylation of benzyl rings. Both mono- and di-hydroxy-derivatives have been found from the parent drug as well as from the metabolites listed in table 2. They also will partly undergo conjugation with glucuronic acid, which necessitates a hydrolysis step before extraction. The hydroxyl derivatives of the parent compound and its 2 demethylation products have amphoteric properties.

A third and less important metabolic transformation to be considered is N-oxide formation. It does not seem to be a major pathway in the case of diphenhydramine, but it is with other drugs listed in table 1.

The mass spectrometric degradation of diphenhydramine is easy to understand. It is illustrated in table 3 and explained in footnotes to the table. The parent drug as well as its metabolites listed in table 2 can be detected by searching for ions resulting from fragmentations in B and to some extent also in D and E. The functional groups can be identified from the presence of low mass ions resulting from fragmentation in A which usually furnishes the base ion. The molecular ions are seldom visible, except in metabolites 6 and 10. A search for traces of diphenhydramine and metabolites may be started by GC with mass specific recording of the ions with m/z 167, 165 and 152 (for group analysis) and extended by running

Table 2. Metabolic Degradation of Diphenhydramine Side Chain

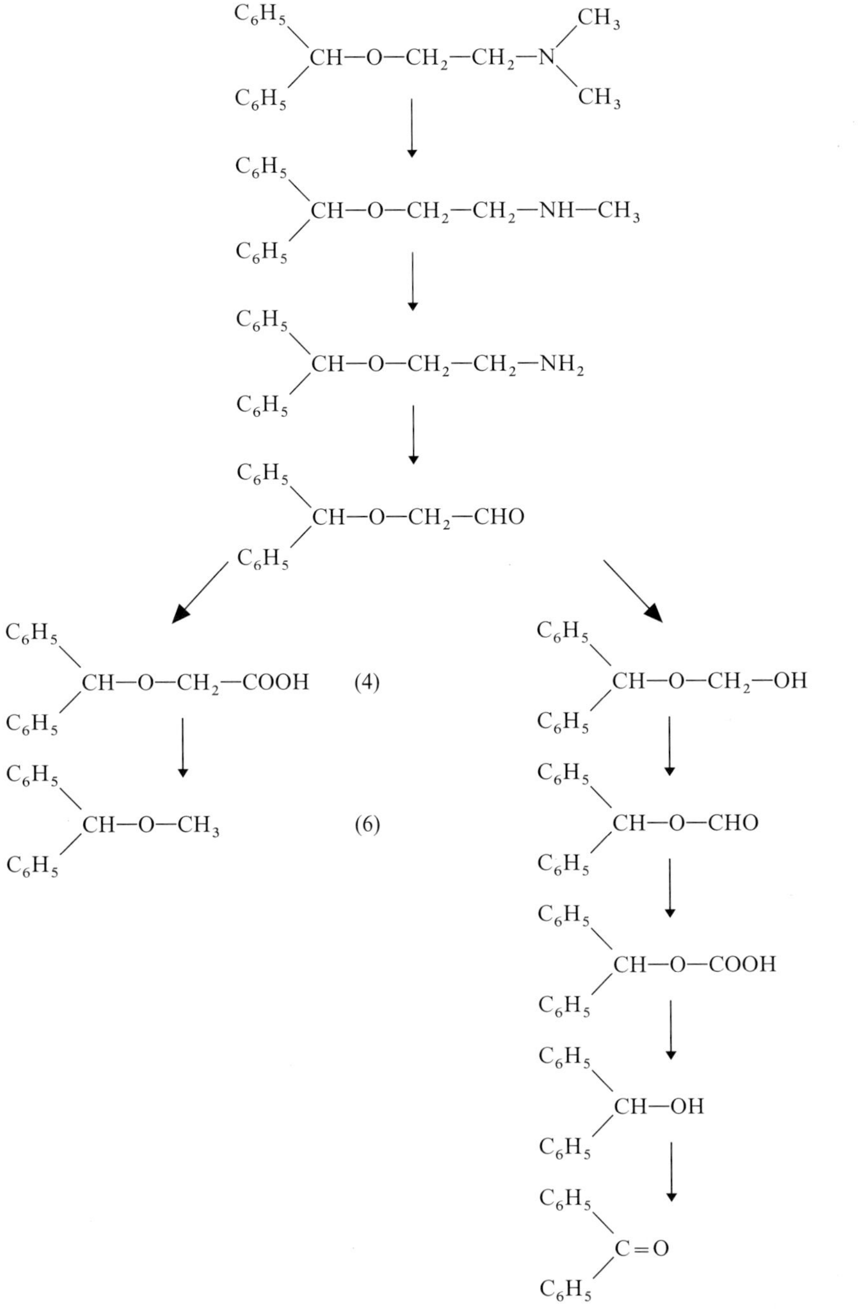

$$(C_6H_5)_2CH-O-CH_2-CH_2-N(CH_3)_2 \quad (0)$$

↓

$$(C_6H_5)_2CH-O-CH_2-CH_2-NH-CH_3 \quad (1)$$

↓

$$(C_6H_5)_2CH-O-CH_2-CH_2-NH_2 \quad (2)$$

↓

$$(C_6H_5)_2CH-O-CH_2-CHO \quad (3)$$

↙ ↘

$$(C_6H_5)_2CH-O-CH_2-COOH \quad (4)$$

↓

$$(C_6H_5)_2CH-O-CH_3 \quad (6)$$

$$(C_6H_5)_2CH-O-CH_2-OH \quad (5)$$

↓

$$(C_6H_5)_2CH-O-CHO \quad (7)$$

↓

$$(C_6H_5)_2CH-O-COOH \quad (8)$$

↓

$$(C_6H_5)_2CH-OH \quad (9)$$

↓

$$(C_6H_5)_2C{=}O \quad (10)$$

The aldehydes 3 and 7 are relatively unstable.

Table 3. The Mass Spectrometric Fragmentation of Diphenhydramine

```
          B                                          B
$C_6H_5$  |       A           CH3         77         |       A          15
   \      |       |          /              \        |       |         /
    CH—O+CH2+CH2—N                            13—16+14+14—14
   X  |   |        X                          X   |   |       X
C6H5  \   |       / CH3                   77   \   |      /     15
       D  E      C                              D  E     C
```

Fragmentation in A furnishes the base ion m/z 58, which indicates the presence of a tertiary amine.

Fragmentation in B splits off the diphenylmethyl residue. The resulting ion with m/z 167 undergoes cyclization to m/z 165 (five-membered ring) and splits off the CH-group. This yields ion m/z 152.

Fragmentation in C releases a methyl group. In combination with the fragmentation in B, it yields also an ion with m/z 73. The low intensity ion m/z 227 results from the loss of both methyl groups (formation of the primary amine).

Fragmentation in D releases a phenyl residue (m/z 77). Together with the fragmentation in B, a tropylium ion is formed (C_7H_7, m/z 91), together with a fragmentation in E the ion C_6H_5—C=O (m/z 105).

Of special value in the detection of diphenhydramine and its metabolites in GC with selected ion recording are the 2 ions from the fragmentation in B, eventually in combination with the ion resulting from the fragmentation in E. Fragmentation in A helps to identify the nature of the amine group.

full scan spectra and checking especially the low mass region for the presence of functional groups.

The retention data of diphenhydramine and a substantial number of its metabolites (6), as well as their mass spectra (7), are recorded in the literature. Unfortunately, the only compilation of mass spectral data which includes many metabolites (7) does not cover the low mass range region. This hinders a recognition of functional groups and often falsifies a computerized library search.

Analysis for diphenhydramine and its metabolites is also possible by on-line combination of GC with FTIR. The ether group exhibits a sharp and strong absorption band around 1100 cm^{-1}. The intensities of gas phase amine bands, however, are extremely low. They can seldom be seen. This renders compound identification difficult.

Furthermore, analysis of diphenhydramine and metabolites has been carried out by HPLC. But since the UV-absorption of the compounds is low and not specific, both sensitivity and specificity of such an approach are insufficient.

3.9.3 Other Aminoalkyl Ethers

As already stated, analysis for the other aminoalkyl ethers (most of them are aminoethyl ethers) can be carried out by the same methods as used for diphenhydramine, provided that some differences are taken into consideration, especially in respect to UV-detection and interpretation of mass spectra.

The UV-spectra of drugs with unsubstituted phenyl groups are practically identical: benzylic absorption bands with maxima near 257 nm and low molar absorption coefficients of a few hundred only. The phenyl-pyridyl-methoxy derivatives show the pyridyl UV-absorption around 262 nm with ε between 5'000 and 10'000. They are therefore more likely to be detected by UV. Drugs with p-substituted phenyl groups show additional bands, usually with stronger absorption, as for example compound 5 (table 1), which possesses a UV-maximum at 230 nm with an ε near 15'000. But this does not necessarily imply that these spectra can be better detected in extracts of body fluid. At the low wavelengths, they may be obscured by UV-absorbing impurities. The drug doses are also often considerably lower.

The mass spectrometric fragmentation of diphenylmethoxy- and phenyl-pyridyl-methoxy derivatives is similar. Since a pyridyl ring counts 1 mass unit (and if protonated 2 mass units) more than a phenyl ring, the search for not ring-substituted members and their metabolites can be based on ions in the same mass range as the search for diphenhydramine and its metabolites, that is on the masses m/z 165 to 169. In the presence of substituents, these tracer ions can be found at correspondingly higher masses, i.e. for the methylated compound 2 in table 1 at m/z 181 and 165 (ring formation by loss of CH_3 and H). The amine side chain usually furnishes the base ion: m/z 58 for all dimethylaminoethyl derivatives, m/z 44 for the mono-desmethyl metabolites, m/z 30 for the di-desmethyl metabolites, m/z 98 and 99 for the methylpiperidine derivative 10, m/z 84 for the methylpyrrolidine derivative 11 (table 1).

Compound 14 in table 1 is an exception, both with respect to UV absorption and mass spectrum. Its UV-maximum lies at 270 nm with inflexion at 276 nm. The 2 strongest ions in the mass spectrum are m/z 58 (dimethylaminoethyl residue) and 255 (molecular ion). This permits an easy detection.

All antihistamines listed in table 1 undergo metabolic transformations. A good number of metabolites have been detected, i.e. (just as examples):

- in case of compound 2 the N-oxide, the two desmethyl-metabolites, as well as 2-methyl-benzhydrol and 2-methyl-benzhydroxy-oxyacetic acid, both in conjugated form;
- in case of compound 5 the N-oxide, the two desmethyl-metabolites, p-bromo-benzhydrol and p-bromo-benzophenone.

3.9.4 Antihistamines with Alkylamine Structure

Table 4 lists 5 drugs belonging to this sub-class. The first is a diphenylmethyl-derivative, the other 4 are phenyl-pyridyl-derivatives. The unsubstituted pheniramine, its p-Cl- and its p-Br-derivative are the most important drugs of the group. While the dosage for pheniramine may be as high as 150 mg daily, only 8 to 32 mg from the two other drugs are given. After therapeutic dosages, blood concentrations may reach 0.2 µg/ml for pheniramine, but 0.02 µg/ml only for the halogenated compounds. Their action is considerably stronger.

All listed drugs are relatively non-polar substances and typical bases. They can be extracted from alkaline solution with hydrocarbons or ether-type solvents. In TLC, they can be detected by the typical reagents for tertiary amines. In intoxications, the UV spectra may signalize a possible presence of one of the drugs in the extract. Compound 1 exhibits maxima at 260, 265 and 273 nm, the first 2 with an

Table 4. Antihistamines with Alkylamine Structure

1. Tolpropamine, Pragman, Tylagel	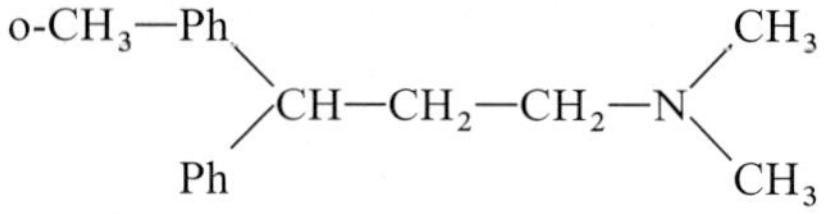 $(o\text{-}CH_3\text{—}Ph)(Ph)CH\text{—}CH_2\text{—}CH_2\text{—}N(CH_3)_2$
2. Pheniramine, Avil, Daneral, Inhiston, Trimeton, Tripoton	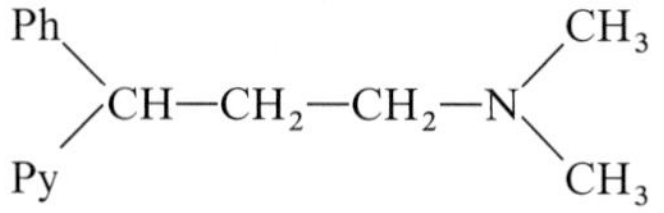 $(Ph)(Py)CH\text{—}CH_2\text{—}CH_2\text{—}N(CH_3)_2$
3. Chlorpheniramine, Allerclor, Allergisan, Antagonate, Chlor-Trimeton, Chlor-Tripolon, Histalen, Lorphen, Polaronil, Teldrin	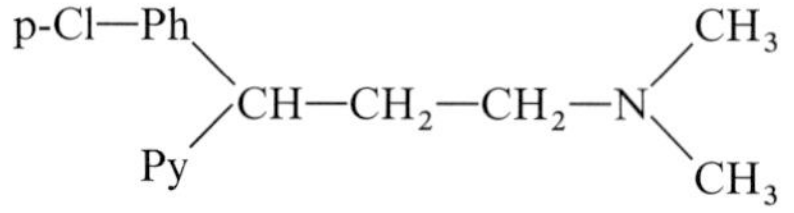 $(p\text{-}Cl\text{—}Ph)(Py)CH\text{—}CH_2\text{—}CH_2\text{—}N(CH_3)_2$
4. Brompheniramine, Dimegan, Dimetane, Ebalin, Ilvin, Nagemid, Veltane	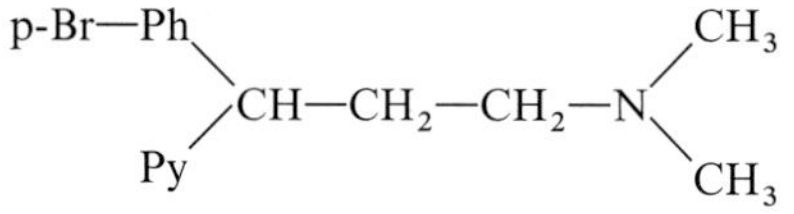 $(p\text{-}Br\text{—}Ph)(Py)CH\text{—}CH_2\text{—}CH_2\text{—}N(CH_3)_2$
5. Triprolidine, Actidil, Actidilon, Alleract, Venen, Pro-Entra	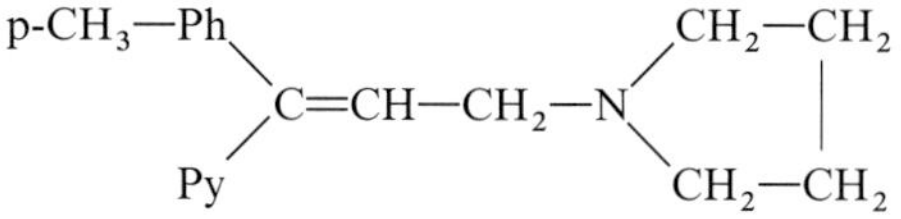 $(p\text{-}CH_3\text{—}Ph)(Py)C{=}CH\text{—}CH_2\text{—}N\langle CH_2\text{—}CH_2\text{—}CH_2\text{—}CH_2\rangle$ (pyrrolidine ring)

ε near 600. Compounds 2 to 4 show typical pyridyl UV absorption bands from 260 to 270 nm with an ε near 8000 (acid solution). Compound 5 with its exocyclic double bond absorbs at the longest wavelength and with highest intensity. Its UV-maximum is 290 nm with an ε near 10'000.

Qualitative analysis (detection and identification) as well as quantifications are best carried out by GC, if possible in combination with MS. Our recommendations given for diphenhydramine can be followed. Again, the amino group will be characterized by low mass ions (m/z 58 for the dimethylamino residue, m/z 44 for the monomethylamino metabolite). They may, however, not be base peaks. Some of the ions obtained by fragmentation of the CH_2-chain are usually more intense. They are (listed in the order of descending intensity):

- m/z 169, 168, 170 and 167 for pheniramine,
- m/z 203, 205, 204 and 202 for chlorpheniramine, and
- m/z 247, 249, 248 and 250 for brompheniramine.

They represent the phenyl-pyridyl-methyl-fragment, which also appears in protonated form.

The mass spectrum of compound 5 in table 4 can easily be understood. The molecule fragments between CH and CH_2, yielding the N-ring with m/z 84. Ions with higher intensity are formed by cleavage of the CH_2-N bond, yielding m/z 209,

208 and 207. In addition, the molecular ion with mass 278 is also visible. It may be useful as tracer ion in GC with mass specific detection. The ions m/z 193 and 194 result from a loss of the N-terminal ring with m/z 84.

The approach to the analysis of antihistamines with alkylamine structure is, as we can see, very similar to a search for aminoalkyl ethers. This holds also for the metabolites. For pheniramine and its halogenated analogs, the mono- and di-desmethyl-metabolites have been observed in urine extracts, in the case of brompheniramine also 3-bromophenyl-3-pyridyl-propionic acid in form of its glycine conjugate.

3.9.5 Antihistamines with Ethylenediamine Structure

The commercially important members of this sub-group of antihistamines are listed in table 5. They are strongly alkaline compounds. Their polarity is slightly higher than that of aminoethyl ethers and alkylamino derivatives. Nevertheless, all drugs extract well from alkaline solution with ether-type solvents. Here again, we recommend a purification and concentration step by back-extraction into diluted acid (i.e. H_2SO_4 or HCl). The UV-spectra of these concentrates should be recorded at 2 (or even 3) pH-values, as described in chapter 3.1.8.

Table 5. Antihistamines with Ethylenediamine Structure

Compound	Structure
1. Phenbenzamine, Antergan, Bridal Lergitin	$(Ph{-}CH_2)(Ph)N{-}CH_2{-}CH_2{-}N(CH_3)_2$
2. Histapyrrodine, Calcistin, Dormistan, Luvistin	$(Ph{-}CH_2)(Ph)N{-}CH_2{-}CH_2{-}N$ (ring: $CH_2{-}CH_2{-}CH_2{-}CH_2$)
3. Methaphenilene, Diatrin, Enstamine, Nilhistine	(thienyl: CH–CH, CH, C, S)$-CH_2$, Ph$>N{-}CH_2{-}CH_2{-}N(CH_3)_2$
4. Cetoxim, Febramine	$(Ph{-}CH_2)(Ph)N{-}CH_2{-}C(=NOH){-}NH_2$
5. Tripelenamine, Pyrinamine, Pyribenzamine, Resistamine	$(Ph{-}CH_2)(Py)N{-}CH_2{-}CH_2{-}N(CH_3)_2$

Table 5. (continued)

	Names	Structure
6.	Mepyramine, Pyrilamine, Anthisan, Histan, Histatex, Histosol, Neo-Antergan, Stamine	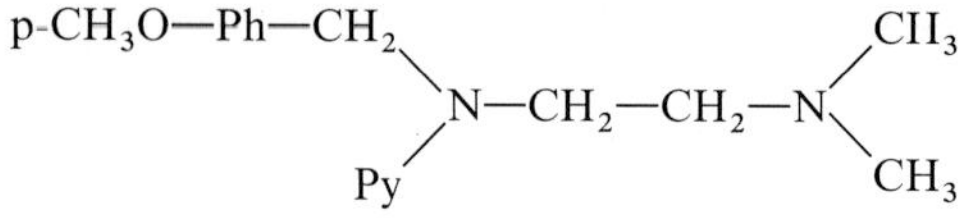 $p\text{-}CH_3O\text{—}Ph\text{—}CH_2(Py)N\text{—}CH_2\text{—}CH_2\text{—}N(CH_3)_2$
7.	Chloropyramine, Halopyramine, Synpen, Chlorotripelennamine	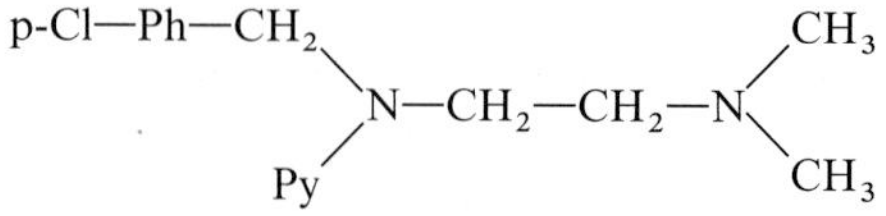 $p\text{-}Cl\text{—}Ph\text{—}CH_2(Py)N\text{—}CH_2\text{—}CH_2\text{—}N(CH_3)_2$
8.	Bromtripelennamine, Hibernon	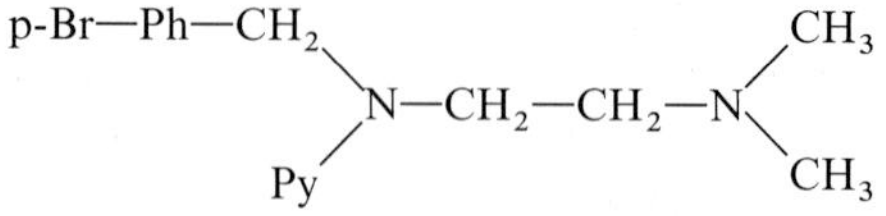 $p\text{-}Br\text{—}Ph\text{—}CH_2(Py)N\text{—}CH_2\text{—}CH_2\text{—}N(CH_3)_2$
9.	Methapyrilene, Tenalin, Thenylene, Thenyldiamine, Dormin, Lullamin, Restryl, Sleepwell	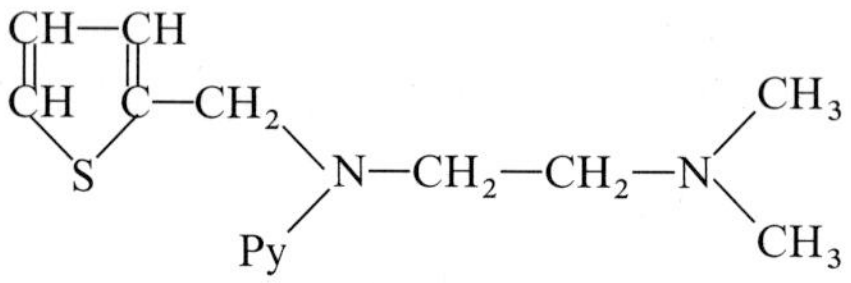 CH—CH / CH C—CH_2 / S (thiophene ring); $(Py)N\text{—}CH_2\text{—}CH_2\text{—}N(CH_3)_2$
10.	Chlorpyrilene, Chloromethapyrilene, Chlorothen, Tagathen, Thenclor	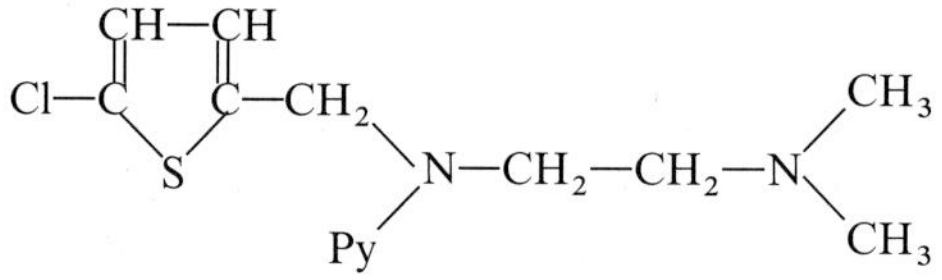 CH—CH / Cl—C C—CH_2 / S (thiophene ring); $(Py)N\text{—}CH_2\text{—}CH_2\text{—}N(CH_3)_2$
11.	Methafurylene, Foralamin	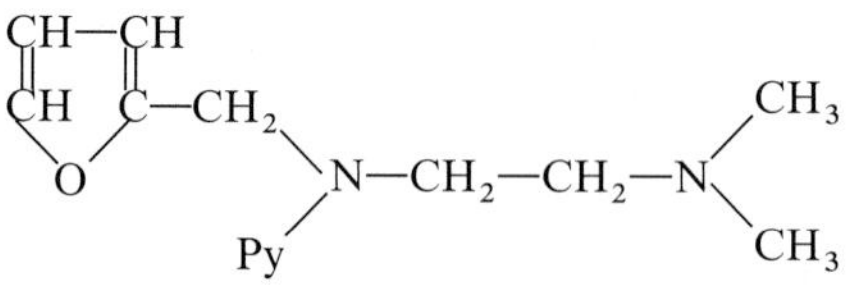 CH—CH / CH C—CH_2 / O (furan ring); $(Py)N\text{—}CH_2\text{—}CH_2\text{—}N(CH_3)_2$

The first 3 drugs show a UV absorption band near 250 with an ε near 4000 nm, and a less intense band near 295 nm. All pyridyl-substituted drugs exhibit 2 much stronger bands, one between 237 and 239 nm (acid solution, ε-values between 14'000 and 20'000), the second one between 312 and 314 nm (intensity only about half). On alkalization, the band at the lower wavelength (but not that at the higher wavelength) undergoes a marked bathochromic shift of 5 to 10 nm. This may be used as a preliminary indication for the presence of a member of this group of antihistamines. It contrasts with the insignificant bathochromic shift of the aminoethyl ethers and the small hypsochromic shift given by the aminoalkyl drugs.

Identification of the individual compounds is again possible by DC, HPLC, GC and even capillary electrophoresis. We recommend the on-line combination of GC with MS. The fragmentation patterns in EI-MS are illustrated in the following scheme:

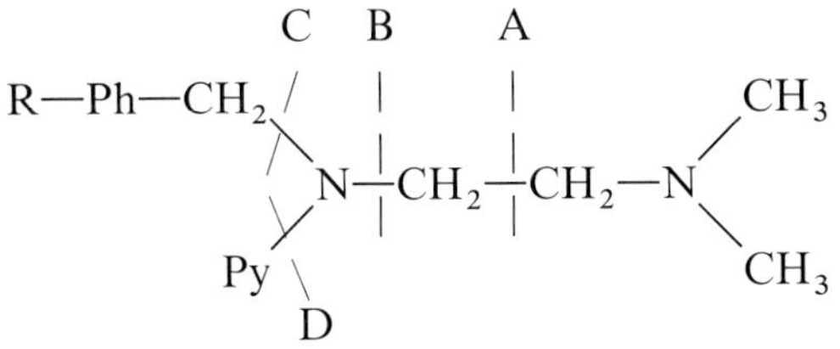

```
        C   B     A
R—Ph—CH2 /  |     |         CH3        Ph = Phenyl
       \    |     |        /           Py = Pyridyl
        N—|—CH2—|—CH2—N               R  = p-Substituent
       /    |     |        \
    Py \                    CH3
        D
```

Fragmentation in A furnishes m/z 58. Usually it is base ion. In the mono-desmethyl metabolite, it is of course replaced by m/z 44. The other part of the molecule (i.e. m/z 197 in example 5) is often also visible.

Fragmentation in B yields m/z 72 and 71 (from the dimethylamino side of the molecule), as well as $[M\text{-}71]^+$ and $[M\text{-}70]^+$ (singly or doubly protonated ions from the aromatic side of the molecule).

Fragmentation in C leads to a very intense ion from the benzyl-substituent. In compound 4 it is the tropylium ion m/z 91, in compound 6 the base ion m/z 121, in compounds 9 to 11 the heterocyclic substituent with the adjoining methylene.

Fragmentation in D finally furnishes the pyridyl group in nonprotonated (m/z 78) and in protonated (m/z 79) form, usually with only low intensity.

GC with mass specific recording of m/z 58, also combined with a recording of m/z 71 and 72, is a good approach for revealing the presence of traces of these antihistamines (as well as of other dimethylaminoethyl derivatives), but not their metabolites. For single compound identification, the presence of ions from the fragmentations in B and C must be checked by mass scan.

3.9.6 Antihistamines with Piperazine Moiety

3.9.6.1 Hydroxyzine

Compound 6 in table 6 has occupied our laboratory repeatedly. It is not only prescribed as antihistamine, but also as tranquilizer. Furthermore, it is an ingredient, together with allobarbital and brallobarbital, of the combination drug Vesparax. This preparation has been widely abused in the drug scene and is responsible for a considerable number of fatalities.

Recommended therapeutic doses of hydroxyzine are 75 to 400 mg per day. A single oral dose will lead to plasma concentrations of 0.07 to 0.09 μg/ml. In fatalities, blood concentrations of near 40 μg/ml have been measured with even higher concentrations in brain and liver (8). In cases of massive intoxication, a presence of the drug may be already detectable by UV in the acid back-wash of organic extracts from body fluids. The compound shows the 3 benzylic bands (258, 263, 270 nm) with low extinction and a stronger absorption at 232 nm with ε near 15'000. This unspecific spectrum may permit quantifications after the drug has been identified by another technique, be it by DC with alkaloid spray reagents, by HPLC, GC or

Table 6. Antihistamines with Piperazine Structure

Name	Structure
1. Cyclizine, Marzine, Marezine, Nautazine, Valoid, Neo-Devomit	Ph, Ph $>\mathrm{CH{-}N}<(\mathrm{CH_2{-}CH_2})_2>\mathrm{N{-}CH_3}$
2. Chlorcyclizine, Alergicide, Di-Paralene, Histantin, Histachlorazine, Perazil, Trihistan	$\mathrm{p\text{-}Cl{-}Ph}$, Ph $>\mathrm{CH{-}N}<(\mathrm{CH_2{-}CH_2})_2>\mathrm{N{-}CH_3}$
3. Cinnarizine, Aplactan, Aplexal, Apotomin, Artate, Carecin, Cerebolan, Cerepar, Cinnacet, Cinaperazine, Cinnageron, Denapol, Dimitron, Giganten, Midronal, Sedatromin, Stugeron, Stutgeron	Ph, Ph $>\mathrm{CH{-}N}<(\mathrm{CH_2{-}CH_2})_2>\mathrm{N{-}CH_2{-}CH{=}CH{-}Ph}$
4. Meclizine, Meclozine, Bonamine, Bonine, Calmonal, Navicalm, Peremesin, Veritab	$\mathrm{p\text{-}Cl{-}Ph}$, Ph $>\mathrm{CH{-}N}<(\mathrm{CH_2{-}CH_2})_2>\mathrm{N{-}CH_2{-}Ph{-}m{-}CH_3}$
5. Clocinizine, Cliocinizine, Denoral	$\mathrm{p\text{-}Cl{-}Ph}$, Ph $>\mathrm{CH{-}N}<(\mathrm{CH_2{-}CH_2})_2>\mathrm{N{-}CH_2{-}CH{=}CH{-}Ph}$
6. Hydroxyzine, Atarax, Aterax, Alamon, Durrax, Orgatrax, Quiess, Vistaril, Paxistil, Tranquizine	$\mathrm{p\text{-}Cl{-}Ph}$, Ph $>\mathrm{CH{-}N}<(\mathrm{CH_2{-}CH_2})_2>\mathrm{N{-}CH_2{-}CH_2{-}O{-}CH_2{-}CH_2{-}OH}$
7. Cetrizine, Virlix, Zirtek, Zyret	$\mathrm{p\text{-}Cl{-}Ph}$, Ph $>\mathrm{CH{-}N}<(\mathrm{CH_2{-}CH_2})_2>\mathrm{N{-}CH_2{-}CH_2{-}O{-}CH_2{-}COOH}$

capillary electrophoresis. The highest specificity and sensitivity are again obtained by GC-MS.

The spectrum obtained by EI-MS is easy to understand (9). The molecular ion and its chlorine-37 isotope are weak but clearly visible (m/z 374 and 376). Loss of the hydroxymethyl group by fragmentation in β-position to both oxygens furnishes an even less intensive ion pair m/z 343 and 345), fragmentation of the side chain in β to the oxygen and nitrogen the more intensive ion pair m/z 299 and 301. The base ion m/z 201 with isotope 203 stems from the chlorinated diphenylmethyl residue. By ring formation with loss of HCl, it converts to m/z 165.

Not much is known about the metabolism of hydroxyzine. It can be assumed that the terminal hydroxymethyl group is oxidized to a carboxy group, that is to compound 7 in table 6. Except for its amphoteric nature, which must be accounted for in the extraction, this drug (and metabolite) possesses similar analytical properties to hydroxyzine.

3.9.6.1 Further members with piperazine moiety

All members of this sub-group are diphenylmethyl derivatives. This reduces the usefulness of UV for their analytical determination, since aromatic bands have a weaker absorption than pyridyl bands. In case of the p-chloro-phenyl derivatives 2 and 4 (table 6) an additional band at lower wavelength (232 nm with higher intensity), in the presence of conjugated exocyclic double bonds (compounds 3 and 5) an additional band at longer wavelength, appear.

All substances of this sub-group are strong bases with low polarity. Identification is possible by DC (with alkaloid spray reagents), HPLC, GC and capillary electrophoresis. Once more we recommend the use of GC-MS. This combination will identify not only the drugs but also their metabolites. The molecular ions are not intensive but usually clearly visible. The non chloro-substituted diphenylmethane derivatives yield m/z 167 and 165, the chlorinated derivatives m/z 165, resulting from a loss of HCl by cyclization of the fragment m/z 201. The 2 first drugs on table 6 show m/z 99 as base ion (the methylated piperazine ring) and m/z 56 as ion with next highest intensity. They metabolize extensively to their nor-compounds which yield base peaks with m/z 85 (piperazine residue).

Cyclizine is used as an anti-emetic and antihistamine, chlorocyclizine and cinnarizine are typical antihistamines and also prescribed in peripherial arterial disease. Therapeutic doses are usually between 50 and 200 mg. Their toxicological importance is rather limited.

3.9.7 Antihistamines with Phenothiazine Structure

3.9.7.1 General remarks

Initially, phenothiazines have been introduced into the pharmaceutical market as antihistamines. But soon it became evident that most members of this class have interesting neuroleptic properties, often with a marked sedating action. As a rough rule of thumb, it can be generalized that the members with an aminopropyl side chain in position 10 are mainly useful as neuroleptics or sedatives, those with an aminoethyl side chain as antihistamines. For a review of neuroleptics, we refer to chapter 3.6, which treats pharmacokinetics, metabolism, as well as analytical problems connected with this compound class. The methods for determining phenothiazines containing only 2 carbons in the side chain bridge between the 2 nitrogens are of course very similar to those for the identification of the members with 3 carbons in the bridge. This holds especially for the problems connected with extraction, analysis by TLC, detection and determination by GC with flame ionization,

Table 7. Antihistamines with Phenothiazine Structure

```
   CH—CH                    Pheno-
 //     \\                  thiazine
CH        CH                ring
  \      /
   C====C
  /      \
S         N—R10
  \      /
   C====C
  /      \
CH        CH
  \\     //
   CH—CR2
```

```
   CH—CH                    1-Aza-
 //     \\                  pheno-
CH        CH                thiazine-
  \      /                  ring
   C====C
  /      \
S         N—R10
  \      /
   C====C
  /      \
CH        N
  \\     //
   CH—CH
```

A) Normal phenothiazines	R^{10}	R^2
1. Fenethazine, Anergan, Anergen, Ethysine, Etisine, Lysergan, Rutergan,	$—CH_2—CH_2—N(CH_3)_2$	H
2. Promethazine, Phenergan, Atosil, Diprazinium, Histantil, Prozamine, Progan, Promine, Prothazine, Remsed, Sominex	$—CH_2—CH(CH_3)—N(CH_3)_2$	H
3. Isopromethazine, Isophenergan, Isomethazine, Diprazin	$—CH(CH_3)—CH_2—N(CH_3)_2$	H
4. Ahistan, Histantin	$—CO—CH_2—N(CH_3)_2$	H
5. Pyrathiazine, Parathiazine, Pyrrolazote, Rolazote, Abergic	$—CH_2—CH_2—N$ (ring: $CH_2—CH_2$ / $CH_2—CH_2$)	H
6. Etymemazine, Ethotrimeprazine, Ethylisobutrazine, Diquel, Nuital, Sergetyl	$—CH_2—CH(CH_3)—CH_2—N(CH_3)_2$	$—CH_2—CH_3$
B) 1-Aza-phenothiazines		
7. Isothipendyl, Andantol, Andanton Nilergex, Theruhistin	$—CH_2—CH(CH_3)—N(CH_3)_2$	

electron capture or thermionic detectors, as well as analysis by HPLC. These methods are discussed by R. Whelpton (chapter 3.6). In this place, we will just list the most important antihistamines with phenothiazine structure (table 7), compare these structures with those of the other antihistamine classes, and discuss the UV-, fluorescence- and mass spectrometric properties of the compounds, which are of importance for their analytical detection and quantification. The drug promethazine, probably the most often used antihistamine with phenothiazine structure, will be given special attention.

Table 7 shows the formulas of the best known antihistamines with phenothiazine or 1-aza-phenothiazine structure. Our unconventional drawing of the formulas points out the structural relationship with other antihistamines. Phenothiazines with aminoethyl chain are actually a special group of the antihistamines with ethylenediamine structure. The similarity extends to metabolism and mass spectrometric degradation pattern, but not to UV-absorption and fluorescence, since the phenothiazine ring and not the aliphatic part of the molecule is responsible for these properties.

Compounds 1, 2, 3 and 7 in table 7 are typical dimethylaminoethyl derivatives with very similar analytical properties. In compound 5, the dimethylamino end group is replaced by a pyrrolidyl residue. Compound 4 is an exception; it contains a carbonyl between the 2 nitrogens. This affects the spectrometric data and may also influence the metabolic transformations. Another type of exception is compound 6. It contains 3 carbons in the side chain bridge, as well as a substituent in position 2 of the ring system. We have included it to illustrate that the border line between antihistamines, neuroleptics and sedatives is often fluid. Even though ethymemazine is listed as antihistamine, its commercial name Nuital reflects its sedating action. It is also used as a veterinary tranquilizer.

3.9.7.2 Promethazine

This compound has typical antihistaminic and some antiemetic action. It is or has been sold under more than twenty brand names such as Atosil, Diphergan, Fenazil, Phenergan, Protazine and Thiergan, just to name a few. Combined with 8-chlorotheophylline, it is marketed under the name Avomine.

The daily dose of promethazine is 20 to 75 mg. Around 2% is excreted unchanged in the urine (a much larger percentage of course in intoxications), the rest is metabolized. The main metabolite seems to be the sulfoxide, but desmethyl-promethazine has also been found, as well as glucuronides of hydroxylated compounds. After therapeutic doses of 25 mg, peak blood levels near 0.005 μg/l have been measured after 2 hours (10). In fatal intoxications, blood concentrations between 2 and 6 μg/ml were determined, together with 1 to 3 μg/ml of sulfoxide and around 1 μg/ml of N-monodesmethyl-promethazine (11). In urine, the sulfoxide concentration is considerably higher than the contents of unchanged and desmethylated drug.

The UV-spectrum of promethazine is typical for phenothiazines without exocyclic double bonds. In acid solution, it shows a sharp maximum at 249 nm with ε 29'000, and a low, flat absorption band at 298 nm. Alkalization causes a bathochromic shift of 5 respectively 7 nm.

Promethazine possesses a weak native fluorescence with activation maximum at 320 and emission maximum at 450 nm. It extends over the entire pH range from 1 to 14. By treatment with permanganate (13) or hydrogen peroxide (14), the parent compound can be oxidized to the sulfoxide possessing much stronger fluorescence, with activation maximum at 340 and emission maximum at 375 nm (12, 15). It is strongly pH-dependent and should be measured in acid solution. Since the sulfoxide is a main metabolite of the drug, a fluorescence test of a urine extract can serve as a useful check for an ingestion of promethazine or another phenothiazine.

The EI mass spectrum of promethazine shows the molecular ion, usually as a peak of medium intensity. The amino end of the molecule furnishes m/z 72 and 73 as base peak and peak of second highest intensity, resulting from a degradation in β-position to both nitrogens. As less intense ions, the condensed ring system is also represented by m/z 198 and 199, as well as m/z 213 from the β-degradation. Interesting results can be obtained with negative ion CI mass spectrometry. They are discussed below.

3.9.7.3 Recommendations for a search for antihistamines with phenothiazine or aza-phenothiazine structure

3.9.7.3.1 Preliminary tests

As a preliminary test for detecting phenothiazines, we recommend the FPN color reaction (16). By slowly adding this heavy reagent to a small test tube with a few ml of urine, 2 layers are formed and a color, i.e. orange for promethazine, appears at the interface.

Urine extracts should also be checked for fluorescence, i.e. by irradiating a few drops in a watch glass with short and/or long wavelength mercury-lamp UV-light. Fluorescence from phenothiazines does not disappear on acidification, as does that from salicylic acid. This test can be extended to other biological extracts after oxidation by addition of two drops of H_2O_2.

3.9.7.3.2 Extraction

As pathway for the identification of phenothiazines, we recommend extraction at alkaline pH (10-11) with methyl-tert.butyl ether (rather than diethyl ether, since peroxides oxidize the phenothiazine ring). Less polar solvents such as hydrocarbons containing some alcohol can also be used, but may not be able to co-extract all metabolites. An acid back-wash concentrates the extracted bases and permits recording of the UV-spectrum in acid solution. It is advisable to re-run the spectrum after alkalization, before re-extracting the bases into the organic phase. Due to their high absorption, the presence of phenothiazines is usually signalized by the UV-spectra, at least in cases of overdosage.

3.9.7.3.3 UV absorption of phenothiazines

All members without exocyclic double bonds in conjugation to the ring system show, in aqueous acid solution, an asymmetric sharp UV-maximum in the neighborhood of 250 nm with an ε near 30'000 (higher in presence of auxochromes) and, in addition, an approximately eight times lower, flat absorption band between 298 and 312 nm. Alkalization usually causes bathochromic shifts, small only for the maxima at 250 nm (up to 5 nm), larger usually (up to 20 nm) for the 300 nm absorption bands.

1-Aza-phenothiazines like isothipendyl (comp. 7 in table 7) show the same UV-absorption characteristics. But basically different spectra can be found in phenothiazines with exocyclic double bond in conjugation to the heterocyclic ring system. Such compounds possess, as a rule, 3 UV-maxima: near 235 nm, near 270 nm (ε-values for both between 20'000 and 35'000), and between 320 and 360 nm (a flat band or shoulder, hardly visible). Since these compounds are not real antihistamines, it would go too far to discuss their spectra in detail in this chapter.

3.9.7.3.4. Fluorescence of phenothiazines

Already in 1957 (17), Udenfriend has shown that the phenothiazine chlorpromazine possesses a weak-, its metabolic oxidation product chlorpromazine sulfoxide a much stronger and different fluorescence. The same holds for most of the other phenothiazines. By oxidizing the basic drugs (in acid solution) with permanganate (13) or hydrogen peroxide (14), they can be converted to the strongly fluorescent sulfoxides (15). While the phenothiazines possess only 2 excitation maxima, the sulfoxides show 4. Their emission maxima lie at lower wavelengths (between 380 and 440 nm) than those of the corresponding basic drugs (between 350 and 500 nm) (12). Procedures for detecting and measuring phenothiazines on the basis of these fluorescence properties have been described (12, 15, 18). But often, the presence of co-extracted fluorescent endogenous substances limits the sensitivity of the methods.

3.9.7.3.5 Chromatographic methods

The identification possibilities by DC, GC and HPLC are well covered in chapter 3.6. We just want to emphasize that DC is favored by the availability of many excellent spray reagents (19). The strong UV-absorption and fluorescence of phenothiazines are good bases for HPLC with UV- or fluorescence recording. The 2 nitrogen atoms and the high electron affinity of the molecules are favorable prerequisites for obtaining high sensitivity in GC with thermionic detection or GC with electron capture. An analytical chemist can therefore choose from many good possibilities in his search for phenothiazines. It is interesting to combine them or to run them in parallel.

3.9.7.3.6 Mass spectrometry

If it comes to compound identification, the combination of GC with MS is capable of furnishing the most and best information in a minimum of time. We have already illustrated the EI-fragmentation for promethazine. It can easily be extrapolated to the other members of the compound class. Molecular ions of phenothiazines are more stable than those of most of the other antihistamines; they usually appear in the EI-spectra and point out the molecular mass.

Interesting information can be obtained by using negative ion mass spectrometry in the CI-mode. At higher source pressures (one torr or above), there is little fragmentation. However, with negative chemical ionization at low source pressure (between 10^{-2} and 10^{-3} torr), an interesting fragmentation, completely different from that in EI, can be observed (see references 34–38 of chapter 1.7). It can be steered to yield the phenothiazine or aza-phenothiazine nucleus with m/z 198 to 201 as base ions. GC with mass recording of these intense negative ions is a sensitive and practical method for trace detection of all members with phenothiazine nucleus (20, 21). For the identification of single compounds, especially for metabolites, we can recommend the use of the "dual-MS" technique, a combination of positive EI-MS with negative CI-MS, already described in chapters 1.7 and 3.1 (see references 40–42 in chapter 1.7, or 27 and 33 in chapter 3.1).

3.9.8 Concluding Remarks

Antihistamines are a rather complex class of pharmaceuticals. We have tried to stress the similarities and the differences in structure and analytical properties between the members of the different sub-classes. Antihistamines are frequently used and abused. They can often be found unexpectedly in specimens submitted for analysis, usually not alone, but together with different types of drugs. A "general search" for unspecified pharmaceuticals is therefore a good approach for detecting, among others, also the compounds of this drug class. A central role in such a search may be alloted to GC-MS (or other combinations of chromatographic techniques with MS). For detecting the antihistamines with phenothiazine structure, the incorporation of UV- and fluorescence spectrometry into the analytical scheme seems essential.

References

1. The Merck Index: An Encyclopedia of Chemicals, Drugs and Biologicals, twelfth edition (1996). Merck & Co., Inc., Rahway, New Jersey.
2. We are indebted to Prof. Dr. Dr. Max von Clarmann, Josef-Wiesberger-Str. 6a, 85540 Haar, Germany, for this search.
3. Clarke's Isolation and Identification of Drugs in pharmaceuticals, body fluids and post-mortem materials, second edition (1986), pp. 557–558 (Moffat, A. C., Jackson, J. V., Moos, M. S. & Widdop, B., Eds.). The Pharmaceutical Press, London.
4. Uges, D. R. A. (1987) Therapeutic drug monitoring and clinical toxiocology by Dutch hospital pharmacists. Ziekenhuis facmacie *3*, suppl. 1, pp. 73–77. See this book part 4.
5. Brandenberger, H. (1988) Wichtige exogene Vergiftungen und ihre differentialdiagnostische Erkennung. In: Differentialdiagnose Innerer Krankheiten (Siegenthaler, W., ed.), 16th., Chap. *39*. George Thieme, Stuttgart & New York.
6. De Zeeuw, R., Franke, J., Maurer, H. & Pfleger, K. (1992) Gas Chromatographic Retention Indicies of Toxicologically Relevant Substances on Packed or Capillary Columns with Dimethylsilicone Stationary Phases, 3rd ed. (Report XVIII of the DFG Commission for Clinical-Toxicological Analysis and Spec. Issue of the TIAFT Bulletin). VCH, Weinheim & New York.
7. Pfleger, K., Maurer, H. & Weber, A. (1992) Mass Spectral and GC Data of Drugs, Poisons, Pesticids, Pollutants and their Metabolites, Parts I, II & III, 2nd revised and enlarged ed., VCH, Weinheim & New York.
8. Reference 3, pp. 673–674.
9. Brandenberger, H. (1970) Die Gaschromatographie in der toxikologischen Analytik.

Verbesserung von Selektivität und Empfindlichkeit durch Einsatz multipler Detektoren. Pharmaceutica Acta Helvetiae *45*, 394–413.
10. DiGregorio, G.J. & Ruch, E. (1980) Human whole blood and parotid saliva concentrations of oral and intramuscular promethazine. Pharm. Sci. *69*, 1457–1459.
11. Allender, W.J. & Archer, A.W. (1984) Liquid chromatographic analysis of promethazine and its major metabolites in human postmortem material. J. Forensic Sci. *15*, 515–526.
12. Udenfriend, S. (1969) Fluorescence Assay in Biology and Medicine, Volume II. Part V: Sedatives and Tranquilizers, pp. 518–525. Academic Press, New York & London.
13. Mellinger, T.J. & Keeler, C.E. (1963) Spectrofluorometric identification of phenothiazine drugs. Anal. Chem. *35*, 554–558.
14. Ragland, J.B. & Kinross-Wright, V.J. (1964) Spectrofluorometric measurement of phenothiazines. Anal. Chem. *36*, 1356–1359.
15. Mellinger, T.J. & Keeler, C.E. (1964). Factors influencing spectrofluorometry of phenothiazine drugs. Anal. Chem. *36*, 1840–1847.
16. Widdop, B. (1986) in reference 3: Hospital toxicology and drug abuse screening, chapter Phenothiazines, p. 25.
17. Udenfriend, S. (1962) Fluorescence Assay in Biology and Medicine, Volume I, pp. 435–437. Academic Press, New York & London.
18. White, Ch. E. & Argauer, R.J. (1970). Fluorescence Analysis, a practical approach, pp. 219–220. Marcel Dekker, Inc., London.
19. Stevens, H.M. (1986) in reference 3: Colour tests, pp. 128–147.
20. Brandenberger, H. & Ryhage, R. (1977) Possibilities of negative ion mass spectrometry in forensic and toxicological analysis. Mass Fragments *1*, an LKB-Publication, & Intern. Bioanalytical Forum at Guildford University, in: Assays of Drugs and other Trace Compounds in Biological Fluids, Vol. II (Reid, E., ed.), pp. 173–184. North-Holland Publ. Co, Amsterdam.
21. Brandenberger, H. (1979) Present state and future trends of negative ion mass spectrometry in toxicology, forensic and environmental chemistry. In: Recent Developments in Mass Spectrometry in Biochemistry and Medicine, Vol. *2* (Frigerio, A., ed.), pp. 227–255. Plenum Publ. Corp., New York.

3.10 Volatiles Used and Abused as Anesthetics

H. Brandenberger

3.10.1 General Remarks

Volatile anesthetics act after reaching a lipophilic environment (1). They must therefore possess a good fat solubility. This explains why most of them are low molecular weight hydrocarbons or ethers, non-halogenated or halogenated.

Only a limited number of the volatiles with anesthetic and/or narcotic properties are used as medical anesthetics. But many widely distributed organic solvents have similar effects and are often responsible for prolonged unconsciousness or fatalities. Since only very few of them are listed in part 2 of our book (chapters 2.4 and 2.5), these solvents will also be discussed in this chapter. A joint treatment of medical anesthetics and other volatiles with anesthetic and/or narcotic properties is also justified in consideration of their identical analytical detection methods.

Since chapter 2.5 contains considerable information on the toxicity of volatile solvents and anesthetics, we can refrain from discussing those aspects and limit ourselves to a systematic chemical review and outline of the analytical possibilities.

Table 1 lists the better known volatiles with anesthetic properties. The noble gas xenon is mentioned more as a curiosity. If used in sufficiently high concentration, it has definitely an anesthetic effect. But medicine has not shown too much interest in xenon. One reason may be its high price. The next compound on the list, nitrous oxide or laughing gas, can look back to the longest active life of all anesthetics. This is partly due to its low toxicity with few side effects and to its analgesic property (1). All following compounds are hydrocarbons or ethers, to a large part halogenated. We have included the important chlorinated solvents (entries 5 to 18). They are so widely distributed in industry, trade and agriculture, that they are in almost everbody's reach. Most intoxications with volatile anesthetics/narcotics that we have encountered during over 25 years of clinical and forensic toxicological analysis, have been caused by such chlorinated solvents, inhaled or taken orally. They certainly deserve our special attention. Entries 19–21 are examples of chlorofluorohydrocarbons (Freons) with negligible anesthetic and narcotic properties. They lead over to the medically-used halogenated hydrocarbons (entries 22–24) with minor significance as solvents. The ethers at the end of the table are, except for the last few, widely used solvents. All of them have been and some still are administered as medical anesthetics.

Table 1. Volatile Anesthetics and Related Compounds

a) Inorganic compounds		
Xenon (a noble gas)	Xe	bp − 108 °C
Dinitrogen monoxide Nitrous oxide, Laughing gas	N_2O	bp − 88 °C inh. analgesic & anesthetic
b) Hydrocarbons		
Ethene, Ethylene	$CH_2{=}CH_2$	bp − 102 °C flammable
Cyclopropane, Trimethylene	$\underline{CH_2-CH_2-CH_2}$ (ring)	bp − 33 °C flammable
c) Halogenated Hydrocarbons		
Monochloromethane, Methyl chloride	CH_3Cl	bp − 24 °C flammable refrigerant
Dichloromethane, Methylene dichloride Methylene chloride	CH_2Cl_2	bp + 40 °C nonflammable solvent paint stripper
Trichloromethane, Chloroform	$CHCl_3$	bp + 62 °C nonflammable solvent
Tetrachloromethane, Carbon tetrachloride, Perchloromethane	CCl_4	bp + 77 °C nonflammable solvent
Monochlorethane, Ethylchloride, Anesthetic, Chloryl, Narcotile, Kelene	CH_2Cl-CH_3	bp + 12 °C flammable top. anesthetic refrigerant
1,1-Dichloroethane, assym. Dichloroethane, Ethylidene chloride	$CHCl_2-CH_3$	bp + 57 °C flammable solvent narcotic
1,2-Dichloroethane, sym. Dichloroethane, Ethylene dichloride, Brocide	CH_2Cl-CH_2Cl	bp + 84 °C flammable solvent
1,1-Dichloroethene, assym. Dichloroethylene, Vinylidene chloride	$CCl_2{=}CH_2$	bp + 32 °C flammable intermediate
1,2-Dichloroethene, sym. Dichloroethene, Acetylene dichloride, Dioform	$CHCl{=}CHCl$	bp cis + 60 °C bp trans + 47 °C solvent narcotic
1,1,1-Trichloroethane, Methylchloroform, Chloroethene	CCl_3-CH_3	bp + 74 °C nonflammable solvent
1,1,2-Trichloroethane, Vinyl trichloride	$CHCl_2-CH_2Cl$	bp + 114 °C nonflammable solvent

Table 1. (continued)

Trichloroethylene, Trichloroethene, Tri, Trichloran, Trilene, Trichloren, Chlorylen	$CCl_2=CHCl$	bp +87 °C nonflammable solvent analgesic
1,1,2,2-Tetrachloroethane, sym. Tetrachloroethane, Acetylene tetrachloride, Cellon, Bonoform	$CHCl_2-CHCl_2$	bp +146 °C nonflammable solvent strong narcotic
Tetrachloroethene, Tetrachloroethylene, Perchloroethylene, Nema, Perclene, Tetracap, Tetropil	$CCl_2=CCl_2$	bp +121 °C nonflammable solvent, dry cleaning & de-greasing agent
Trichlorofluoromethane, Freon 11, Frigen 11, Arcton 11	CCl_3F	bp +24 °C nonflammable refrigerant
Dichlorodifluoromethane, Freon 12, Frigen 12, Arcton 12, Halon	CCl_2F_2	bp −30 °C nonflammable refrigerant
1,2-Dichloro-1,1,2,2-tetrafluoroethane, Freon 114, Frigen 114, Arcton 114, Cryoflurane	$CClF_2-CClF_2$	bp +4 °C nonflammable refrigerant
2-Bromo-1,1,1,2-tetra-fluoro ethane, Tefluorane, Terflurane	$CF_3-CHFBr$	bp +6 °C nonflammable inh. anesthetic
2-Bromo-2-chloro-1,1,1-trifluoro ethane, Halothane, Fluothane, Rhodialothan	$CF_3-CHClBr$	bp +50 °C nonflammable inh. anesthetic
3-Bromo-1,1,2,2-tetra-fluoro propane, Halopropane, Tebron	$CHF_2-CF_2-CH_2Br$	bp +74 °C nonflammable inh. anesthetic
d) Ethers		
Ethoxyethane, Diethyl ether, Ethyl ether, Anesthetic ether, Sulfuric ether	$CH_3-CH_2-O-CH_2-CH_3$	bp +35 °C solvent flammable inh. anesthetic
1-Methoxypropane, Methyl propyl ether	$CH_3-O-CH_2-CH_2-CH_3$	bp +39 °C flammable inh. anesthetic
Divinyl ether, Vinyl ether, Vinethene, Vinesthene	$CH_2=CH-O-CH=CH_2$	bp +28 °C flammable inh. anesthetic
e) Halogenated Ethers		
2,2,2-Trifluoroethyl-vinyl ether, Fluro-xene, Fluoromar	$CF_3-CH_2-O-CH=CH_2$	bp +43 °C explosive inh.anesthetic

Table 1. (continued)

2,2-Dichloro-1,1-difluoroethyl methyl ether, Methoxyflurane, Penthrane, Pentrane	$CH_3-O-CF_2-CHCl_2$	bp +105°C nonflammable inh. anesthetic
2-Chloro-1,1,2-trifluoroethyl difluoromethyl ether, Methylflurether, Enflurane, Ethrane, Efrane, Alyrane	$CHF_2-O-CF_2-CHClF$	bp +57°C nonflammable inh. anesthetic
1-Chloro-2,2,2-trifluoroethyl difluoromethyl ether, 2-Chloro-2-(difluoromethoxy)-1,1,1-trifluoroethane, Isoflurane, Forane, Forene, Aerrane	$CHF_2-O-CHCl-CF_3$	bp +48°C nonflammable solvent for fluorinated materials inh. anesthetic
1,2,2,2-Tetrafluoroethyl difluoromethyl ether, Desflurane	$CF_3-CHF-O-CH_2F$	bp +24°C nonflammable inh. anesthetic (experimental)
1,1,1,3,3,3-Hexafluoropropyl fluoromethyl ether, Sevoflurane, Sevofrane	$(CF_3)_2CH-O-CH_2F$	bp +59°C nonflammable inh. anesthetic (experimental)
f) Others		
Tribromoethyl alcohol, Tribromoethanol, Bromethol, Ethobrom, Narcolan, Avertin	CBr_3-CH_2-OH	mp 79–82°C marketed in amylene hydrate solution as inh. anesthetic

3.10.2 Analytical Considerations

3.10.2.1 Sampling procedures

Of the 30 odd compounds listed in table 1, a good half dozen are gases (at room temperature and atmospheric pressure), the others are volatile liquids with relatively low boiling points. This makes them ideally suited for gas chromatographic analysis. Almost all can be sampled using the head space technique. If properly planned, this eliminates the work for sample preparation in the analysis of biological fluids or tissues. We used to provide the pathologists with special glass jars, whose covers could be replaced in our laboratory, after freezing the content, by covers with in-built silicon membranes. After warming the jars to room temperature (or an elevated temperature), head space gas can be removed with 1 or 2 ml syringes and directly injected into a chromatographic column. With such a simple, clean and safe procedure, the danger of contamination can be eliminated, and no evaporation losses occur. After the search for volatiles, the sample material is still available – practically untouched – for further investigations (analysis for drugs and other non-volatiles).

In a search for the few higher boiling compounds on the list, or in case of very low concentrations of volatiles, it may be necessary to submit the biological materials to a distillation (super-heated steam). Relatively large volumes of the distillate (several μl) can be injected directly into the column, or the distillate can be extracted with a small volume of a low boiling solvent such as hexane. But this is seldom necessary, since the concentrations of the solvents in the body are usually high enough to permit sampling by the head space technique.

3.10.2.2 Gas chromatographic separations

The choice of chromatographic column is not much of a problem. For separating the gases or low boiling solvents, cross-linked resins (Porapak Q or Chromosorb 102, see chapter 2.3) can be recommended. Such solid phases have also been made available as capillaries. For most of the volatiles, there is a large choice of columns or capillaries, i.e. with the polar phases Carbowax 20 M or phenylsilicones, but also with the nonpolar methylsilicones. A large number of retention data is reported in the literature (2, 3). Some can be found in our chapters 2.3, 2.4 and 2.5, especially in Figure 1 of chapter 2.3. With knowledge of the structural formula, polarity and vapor pressure (respectively boiling point) of a compound, its relative range of retention on a given column can usually be anticipated.

3.10.2.3 Identification of separated compounds

For the analysis of aqueous solutions (distillates) or humid head space gases, flame ionization is the logical simple detection device. However, its lack of specificity implies that identification is based on retention data and that reference compounds must be available. Before the days of GC-MS, we have tried to increase the information content of detection by incorporating an electron capture detector into the system, to be used in parallel to flame ionization. This not only improves the detection limits for all halogenated compounds, but the ratio of electron capture response to flame ionization response furnishes a good indication about number and maybe even type of halogen atoms in an eluting compound. We call these ratios the REA-values (from relative electron affinity) of specific compounds (4, 5). The measured values depend of course strongly on the type of detectors used and must always be related to a reference compound, just like retention times. Table 2 shows that each chlorine atom raises the REA-value by a factor of about 10. Bromine atoms have a larger effect.

A disadvantage of electron capture detectors is their sensitivity to water. They can not be used with aqueous solutions or with large volumes of moist head space gases.

The detector which yields the best information and is not affected by humidity is of course a mass spectrometer. Figures 1 to 4 give examples for the identification of eluting compounds by conventional positive EI-MS. But we also employ MS in the negative ion CI-mode (6–9) as briefly described in chapter 1.7. With ionization

Table 2. GC of Volatile Chlorinated Hydrocarbons: Relative Retention Times on Porapak Q and Relative Electron Affinities (Quotients of ECD- to FID-response). CH_2Cl_2 = 1 and 100.

Compound	RRT (Porapak Q)	REA
CH_3Cl	0.26	2.4×10^1
CH_2Cl_2	1.00	1.0×10^2
$CHCl_3$	2.40	1.0×10^3
CCl_4	3.70	1.0×10^4
CH_3Cl	0.26	2.4×10^1
CH_3Br	0.49	0.5×10^2
CH_3J	1.10	2.4×10^3
CH_2Cl-CH_2Cl	3.10	1.2×10^2
$CHCl_2-CH_3$	2.20	3.0×10^2
$CHCl=CHCl$	1.80	0.4×10^2
$CCl_2=CH_2$	1.50	2.0×10^2
$CHCl_2-CH_2Cl$	5.60	0.2×10^3
CCl_3-CH_3	3.30	0.7×10^3
$CCl_2=CHCl$	3.00	0.4×10^3

gases favoring electron attachment, even better sensitivities for halogenated compounds are obtained than with conventional electron capture detectors. But the best basis for identifying eluting compounds is our so-called "Dual-MS" method, a combination of positive EI-MS with negative CI-MS (10–14), also described in

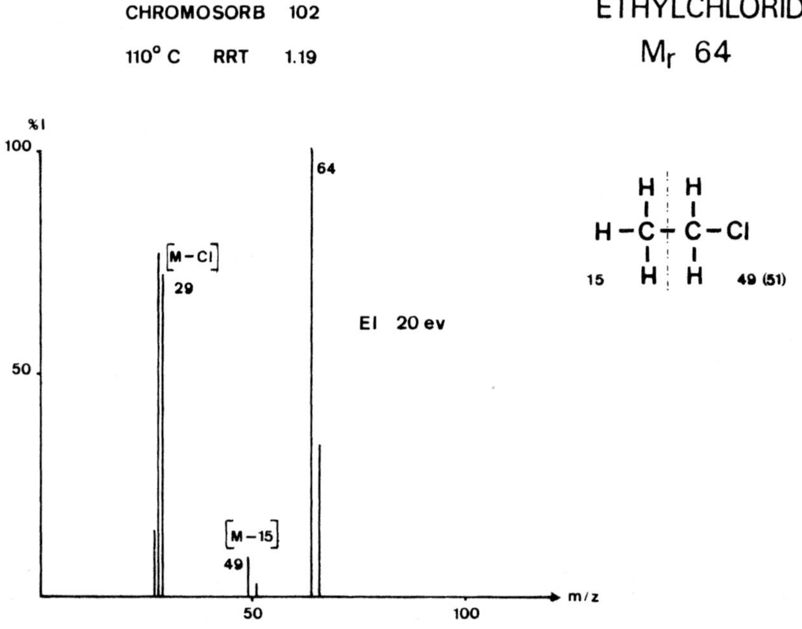

Figure 1. EI Mass Spectrum of Ethylchloride (20 eV).

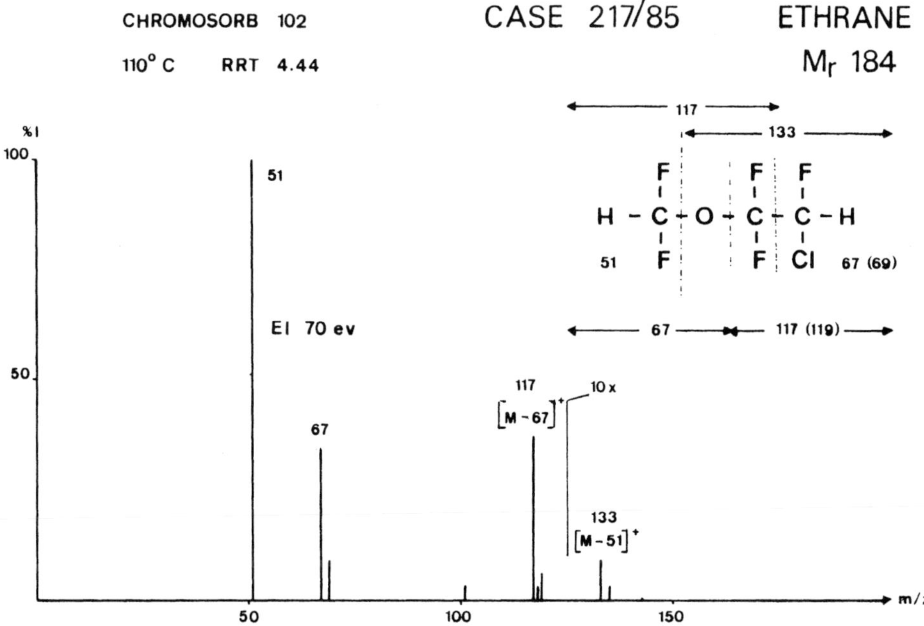

Figure 2. Identification of Ethrane in a Blood Sample by GC with EI-MS (70 eV).

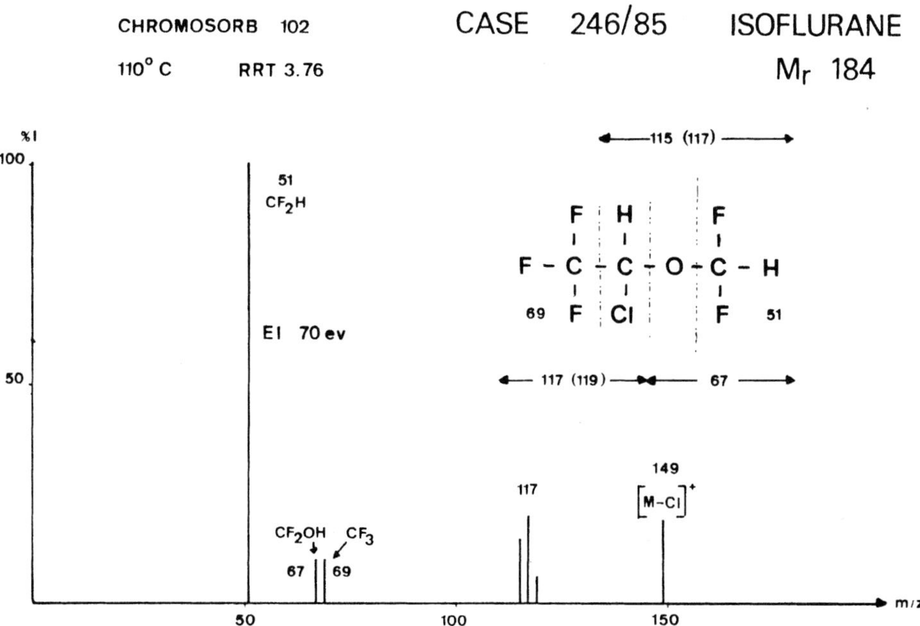

Figure 3. Identification of Isoflurane in a Blood Sample by GC with EI-MS (70 eV).

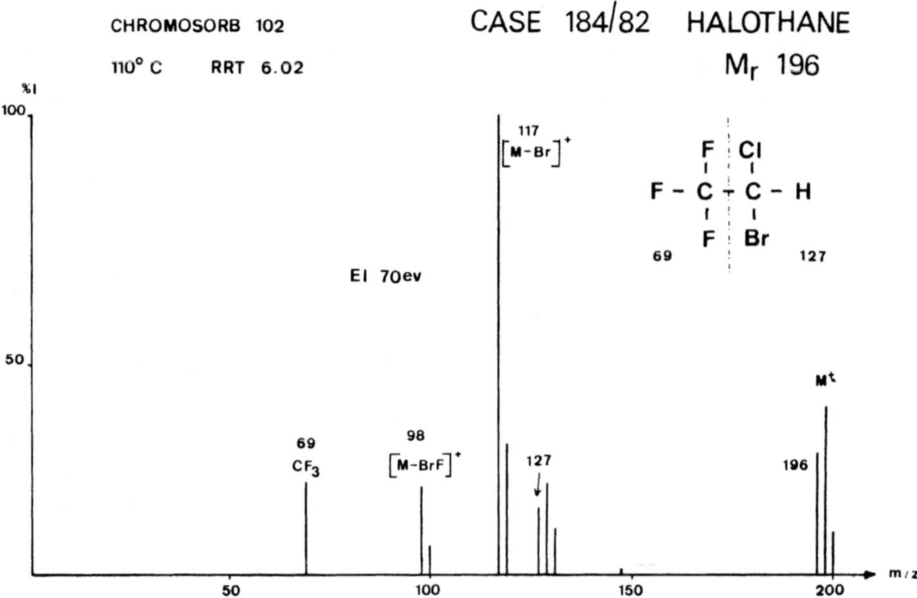

Figure 4. Identification of Halothane in a Blood Sample by GC with EI-MS (70 eV).

chapter 1.7. Figures 5 and 6 illustrate the power of such a dual identification system (from ref. 12). The negative ion spectra reveal at first sight a possible presence of Cl or Br in the molecule and may also show an analytically useful negative ion fragmentation (fig. 5). The information is complementary to that from EI-MS, not

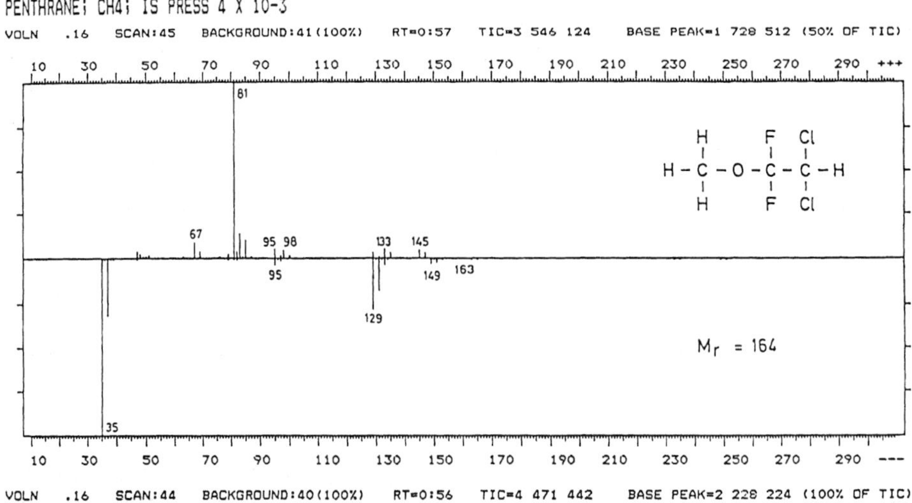

Figure 5. Dual Mass Spectrum of Penthrane: Simultaneously recorded Positive EI Mass Spectrum and low-pressure Negative CI Spectrum of Penthrane.

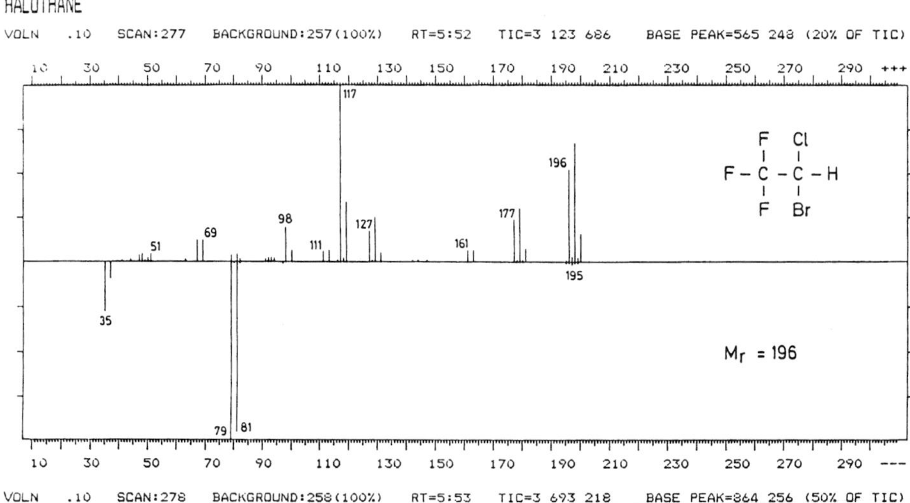

Figure 6. Dual Mass Spectrum of Halothane: Simultaneously recorded Positive EI Mass Spectrum and low-pressure Negative CI Spectrum of Halothane.

just a duplication of results. Interpretation of the negative ion spectra is usually easy. The positive EI spectra permit a library search. However, our "Dual-MS" furnishes so much information, that it is seldom necessary to consult mass spectral data banks, at least not for such simple compounds as volatile solvents and anesthetics.

2.10.2.4 Trace detection

Electron capture detectors are often used for trace analysis of halogenated compounds. But a negative ion mass spectrometer can do a better job. It has a superior detection sensitivity and yields much more specific information. It permits injection of humid gases or aqueous solutions. In searching for traces of chlorine- or bromine-containing volatiles, for example, the spectrometer can be used to monitor the appearance of the negative ions with masses 35 and 37 (chlorine isotopes) or 79 and 81 (bromine isotopes). We do not know a more sensitive technique for the trace detection of halogenated solvents and anesthetics than negative ion CI-MS. It can reveal sub-pg quantities (fmoles).

3.10.3 Some Important Volatile Solvents and Anesthetics

3.10.3.1 Nitrous oxide

Nitrous oxide (N_2O) is rapidly absorbed and quickly eliminated by the body. The onset and end of its physiological effects can therefore be obtained within a few minutes. Since the solubility of N_2O in the blood is relatively low, high concentrations have to be used. Oxygen containing below 50 % N_2O has an analgesic effect, but does not lead to narcosis. This is ideal in the fields of dentistry and childbirth. But for a deep narcosis, N_2O-concentrations of 50 to 75% (only 50 to 25% O_2) are required. This can cause problems during longer anesthesias. In such cases, N_2O is commonly employed as starting anesthetic, to be later replaced by a more potent medicament. For such combinations, much less of the second anesthetic is needed than if it were used alone, thus reducing the danger of toxic side effects, since the toxicity of N_2O is lower than that of other anesthetics.

One reason for the low toxicity of N_2O may be the fact that it is practically not metabolized, except for an insignificant reduction to N_2 in the intestines. The main part is excreted unchanged, mostly exhaled.

In our toxicological control work, we have never encountered lethal intoxications with N_2O. However, it should be kept in mind that the gas can easily escape detection. In chromatographic analysis, it behaves like the inert gases O_2, N_2 and CO. The same columns as for the analysis of these gases (i.e. in controlling CO-intoxications) can therefore be employed. We recommend molecular sieve 5A at 70 °C (5) or cross-linked synthetic resins such as Chromosorb 101 or 102 at or below 40 °C. If the concentrations are sufficiently high, thermoconductivity detection can be used as for the inert gases, otherwise MS detection. In EI-MS, N_2O yields positive ions with the masses 44 (100%), 30 (30%), 28 (10%), 16 and 14. In low pressure negative CI, the fragment anions with masses 16 and 30 dominate (below 10^{-2} torr, O^- is base anion, $[NO]^-$ at or over 10^{-2} torr). The masses 32 (O_2^-) and 40 ($[N_2O]^-$) have only little intensity (14).

3.10.3.2 Gaseous hydrocarbons

Cyclopropane is a ten times stronger anesthetic than N_2O and can therefore be used in lower concentrations. The onset of physiological action is also very fast. The inconvenience of this agent is its explosiveness. It has therefore not become popular for prolonged applications. The same holds for ethylene.

To detect these 2 compounds, we recommend the same procedure as for N_2O, that is head space sampling with gas chromatographic analysis on columns with molecular sieve or cross-linked resins. With a flame ionization detector, the elution of hydrocarbons is not obscured by the presence of inert blood gases, which do not yield flame ionization signals.

3.10.3.3 Volatile chlorinated hydrocarbons (solvents)

A large number of chlorinated hydrocarbon solvents has been synthesized and used in industry and trade. Some of them are or have been produced in huge quantities and belong to the synthetic chemicals with highest production quotas. There is hardly an industrial firm, a household or an agricultural business, which does not possess one or several chlorinated hydrocarbons, as solvents, paint strippers, degreasing agents, spot cleaners or typewriter cleaning fluids, just to name a few applications. Table 1 (entries 5 to 18) covers a substantial part, but not all, members of this class of compounds. Only the more important ones are discussed in the following paragraphs.

Under atmospheric conditions and ambient temperature, **methyl chloride** is a gas with sweetish odor. It is used for extractions at low temperature (aromas), in the production of silicones and methyl cellulose, as a refrigerant and in plant protection. A considerable number of chronic intoxications have been reported. Refrigerator repairmen have suffered from such exposures, since the gas has a toxic action even at concentrations too low to be detected by odor. The toxic effects of methyl chloride, described in chapters 2.4 and 2.5, are mainly resposible for its replacement by fluorochloro hydrocarbons (Freons). In medicine, methyl chloride is used as a topical, but not as an inhalation anesthetic.

Methylene dichloride has a strong narcotic effect but is much less toxic than methyl chloride. It is an excellent solvent for oils, resins and synthetic polymers. It is used as a paint stripper and for a great number of extractions, even in the food industry (caffeine, soy oils). Between 1920 and 1930, it has also been tested as a volatile anesthetic, but later abandoned due to some unexpected incidents. The toxicity and metabolism of the compound are discussed in chapter 2.5.3. We just want to add that this solvent is becoming increasingly important. It has replaced more toxic chlorinated hydrocarbons in industrial applications. In our laboratory, it has completely supplanted chloroform as an extraction solvent for amphoteric compounds, since it does not show the disturbing decomposition to HCl.

The toxicological problems involved with the metabolic decomposition of CH_2Cl_2 to CO are discussed in chapter 2.5.3.2.4.

Chloroform has been administered as an anesthetic in surgery, but later abandoned because of its toxic effects. Its industrial importance has also diminished, but it is still employed as a solvent for oils, fat, waxes, alkaloids, rubber and resins. It has been banned from use in drugs, cosmetics and in the food field, since it is assumed to be a carcinogen. Prolonged or repeated exposures to low concentrations of chloroform vapor have produced toxic symptoms. In our work, the solvent could repeatedly be indentified as toxic agent in homocide or suicide. The reason for the choice of this solvent may have been, in part, its pharmaceutical name "chloroformium pro narcosi" or "chloroformium anestheticum", and the reason for the success of the undertaking, the fact that a fast onset of narcosis already occurs at vapor concentrations as low as 1%. The minimal lethal dose for oral intake has been estimated at 7 g (15). In lethal intoxications, blood levels between 10 and 50 μg $CHCl_3$ per g blood could be found (16).

Carbon tetrachloride is the most toxic of the 4 chlorinated methanes. But nevertheless, it has been extensively used as a fire extinguisher and dry cleaning agent. It is a good solvent for resins, rubber, waxes, varnishes, lacquers, fat and oil. It has been administered in medicine and veterinary medicine as an antihelmintic. It can lead to chronic and acute intoxications as a result of oral intake, inhalation and absorption through the skin. Continued exposures to vapor concentrations as low as 100 ppm can already injure blood cells, liver and kidney (fat degeneration), as well as the nervous system. Concentrations over 1000 ppm can produce acute intoxications. Lethal intoxications have resulted in connection with fire extingushers or use as a cleaner in a confined space. The symptoms are usually delayed, so that many CCl_4-intoxications may not be recognized.

Blood levels in the range of 20 μg/g and considerably higher levels in tissues, especially liver, have been reported for lethal intoxications. An ingestion of only

4 ml of CCl_4 must be considered as minimal lethal dose. In spite of its high toxicity and easy accessibility, relatively few acute intoxications with CCl_4 have been reported. This may be due to the relatively low vapor pressure, if compared with other chlorinated methanes. It may be helpful to add that skin contact can lead to dermatitis (defatting action), that the compound is considered to be carcinogenic, and that it is not compatible with alcohol.

Ethyl chloride is less toxic than methyl chloride. It is used as an alkylating reagent (tetraethyl lead), solvent, refrigerant, and in medicine and veterinary medicine as a topical anesthetic. High vapor concentrations can cause narcosis. The presence of ethyl chloride in blood is signalized in the analysis for blood alcohol described in our chapter 2.3. It can easily be verified by GC-MS (fig. 1). Chapter 2.3.7 lists also a case of criminal abuse of ethyl chloride with lethal consequences.

All **dichloroethanes** and **-ethenes** are highly toxic. They are used as industrial solvents and intermediates. They can cause respiratory irritations and, in high concentrations, narcosis.

Trichloroethylene is the chlorinated hydrocarbon that we have most often met in our toxicological work. It has been responsible for many intoxications with long narcotic stages, some with lethal outcome. The solvent has been used extensively as extracting agent for fat (from bones, fish residues), caffeine (decaffeination), nicotine, and also for degreasing textiles and metals. In the household, it may be present in typewriter cleaning fluids and spot removers. Thus it is readily accessible. "Tri" has had medical use as an anesthetic, especially in dentistry and childbirth (Chlorylene). An euphoric prenarcotic state must be accountable for its widespread abuse in "solvent sniffing", especially among young people. Some intoxications that we had to investigate were caused by accidental oral intake of the solvent. The cleaning fluid had been stored in unlabeled household bottles (in one case a beer bottle) and was mistaken for a refreshing drink. In such intoxications, already 5 to 7 ml of trichloroethylene can be lethal (15).

Only a small part of the solvent is excreted unchanged. The bulk is metabolized (by way of chloral) to trichloroethanol and trichloroacetic acid; both are excreted in urine partly in their free and partly in their conjugated form. For more information, especially on the analytical determination of metabolites, we refer to the discussion on chloral in chapter 3.3.3. In lethal intoxications caused by oral intake of trichloroethylene, blood concentrations of 25 μg/ml and up have been found (16). In lethal intoxications due to solvent sniffing, the concentrations were usually considerably lower (16).

Today, trichloroethylene extraction of foodstuffs and feed has largely been abandoned for the following reasons:

Way back in 1916, it could be shown in England that the aplastic anemia of cattle could be provoked by feeding trichloroethylene-extracted soybean meal, but not by incorporating some trichloroethylene into the feed (17). During and after World War II, the fat in soybean was again extracted with trichloroethylene and the extraction residues fed to farm animals. They developed the same illness symptoms (18–20, which list further references). Feeding trichloroethylene-extracted meat residues to calves also led to aplastic anemia. It took a good bit of research to establish that the solvent combines with the sulfhydro groups of free or bound cysteine to form S-dichlorovinylcysteine or the corresponding peptide derivatives

(Formulas I or II). The highly reactive chlorine atom(s) on the double bond is (are) held responsible for the illness symptoms (21).

$$\text{I} \quad \begin{array}{l} CH_2-S-CCl=CHCl \\ | \\ CH-NH_2 \\ | \\ COOH \end{array} \qquad \text{II} \quad \begin{array}{l} CH_2-S-CH=CCl_2 \\ | \\ CH-NH_2 \\ | \\ COOH \end{array}$$

In 1967, we called attention to the fact that trichloroethylene is also used to extract human food. We showed that after extracting coffee with the C-14-labeled solvent and removing all extract and solvent, some radioactivity remained in the decaffeinated product (22). Most coffee manufacturers have subsequently abandoned the trichloroethylene extraction processes, first in favor of methylene chloride extraction and, at a later date, by extraction with liquid carbon dioxide.

1,1,1-Trichloroethane and **1,1,2-Trichloroethane** are also popular solvents and cleaning fluids, i. e. for metals. They are nonflammable and insoluble in water like trichloroethylene, and since the chlorine atoms do not sit on double-bonded carbons, they could replace "Tri" in extraction processes.

Both compounds have narcotic properties, but did not find applications as medical anesthetics. With respect to toxic doses and blood level data, they resemble trichloroethylene. Especially 1,1,1-trichloroethane metabolizes mainly to trichloroethanol and trichloroacetic acid.

Tetrachloroethene and **tetrachloroethane** are, at higher concentrations, potent narcotics. But they have not been used medically due to their toxicity. Tetrachloroethane is probably the most toxic among the chlorinated aliphatic hydrocarbons. Its minimal lethal dose is estimated at 3 ml.

Both compounds are excellent solvents. Tetrachloroethane is used for dissolving fat, oil, resins, rubber, phosphorus and sulfur. It is an important synthetic intermediate, especially for the manufacture of other chlorinated hydrocarbons. Tetrachloroethene (Perchloroethylene, Perklone) is a degreasing agent for metals and a dry cleaning agent. It was also used for coffee decaffeination. But since it seems likely that it can react, like trichloroethylene, with the sulfhydryl groups of free and bound cysteine, this process also had to be abandoned. Tetrachloroethene has found medical use as an antihelmintic.

Both compounds are absorbed from the gastro-intestinal tract, by the lungs and through the skin. Both are excreted mainly unchanged and at a very slow rate, to a large part with the expired air. Only a low percent of tetrachloroethene is metabolized to trichloroacetic acid (and probably also trichloroethanol).

3.10.3.4 Freons

The 3 Freons listed in table 1 as examples of simple fluorochloro hydrocarbons have only minor toxic effects and are not used as anesthetics. They only exhibit narcotic properties at very high concentrations. They have been extensively used as refrigerants and aerosol propellants, but are going to be banned, since they are considered to be in part responsible for the ozone hole in our atmosphere. We have listed them in table 1, since they may be detected in a search for volatiles with the GC-columns mentioned before, and since they lead over to the halogenated hydrocarbon anesthetics.

3.10.3.5 Halogenated hydrocarbon anesthetics

It is interesting to learn that all 3 more modern hydrocarbon anesthetics (Terfluorane, Halothane and Halopropane) possess, in contrast to the aforementioned Freons, 1 bromine atom in the molecule. In many countries, Halothane has been the volatile anesthetic of choice for many years. The toxicity problems involved are extensively discussed in chapter 2.5.3.2 of this book. Due to these toxic effects, many hospitals do not use Halothane any longer.

For years, we have often encountered Halothane, occasionally also Terfluorane and Halopropane, in blood samples secured during or after an anesthesia. They can be detected in the analysis for blood alcohol by GC, as described in chapter 2.3. For an incontestable identification, we have usually confirmed such preliminary results by GC-MS (fig. 4). Negative ion MS is even more sensitive and shows the presence of the bromine atom at first sight (fig. 6). For trace detection of all 3 compounds, we reommend GC with mass specific recording of the isotopic bromide anions with the masses 79 and 81.

3.10.3.6 Ethers

Three different non-halogenated ethers have been used in medical anesthesia (table 1). All 3 are easily flammable and explosive. This is one reason why they have been to a large part replaced, even the most often used and best known diethyl ether or anesthesia ether, which is briefly discussed at the end of chapter 2.4.

Diethyl ether is a good general solvent and readily accessible. It still occupies the analytical toxicologist, since it is fairly often abused for suicides and homocides. In the course of an analysis for alcohol, which is carried out as a routine part of the investigation of unexpected deaths, the presence of ether is automatically signalized by our computerized GC method described in chapter 2.3.7.

Diethyl ether is not metabolized in the body. It is in part exhaled and in part excreted with the urine, which can therefore also serve as analytical specimen.

3.10.3.7 Halogenated anesthetics with ether structure

All the halogenated ethers used as medical anesthetics and listed in table 1 possess 2 or more fluorine atoms in the molecule, but, in contrast to the halogenated hydrocarbon anesthetics, no bromine. The first on the list (table 1), Fluoromar, is not totally inflammable. It can explode and is therefore not preferable to diethyl ether or divinyl ether which have similar physical properties. The 2 last mentioned entries, Desflurane and Sevoflurane, are still in the experimental stage. Penthrane, Enflurane (Ethrane) and Isoflurane (Forane) have been and still are widely used in many countries. As mentioned in chapter 2.5, these compounds have some of the same toxic effects as the halogenated hydrocarbons such as Halothane, since they can metabolize to yield fluoride ions. The formation of fluoride seems to be more pronounced with Penthrane than with Ethrane or Isofluorane (1).

All 6 halogenated ethers listed in table 1 can be detected easily by GC. If they are present in sufficiently high concentrations (during or immediately after an anesthesia), they will be revealed already in a search for alcohol and other volatiles as described in chapter 2.3. We usually confirmed such preliminary information by GC-MS, either in the positive EI-mode (see figs. 2, 3) or in our Dual-MS-mode (fig. 5).

3.10.4 Concluding Remarks

Detection and dosage of volatile anesthetics and related solvents in the body is certainly one of the simplest and least time consuming jobs in a toxicological laboratory. If the institution is also carrying out blood alcohol determinations, this routine procedure can be set up not only for a quantitative determination of ethyl alcohol, but also for calling attention to the presence of other low boiling substances including most of the volatile anesthetics. A confirmation of such indications is easily possible by combined GC-MS. We especially recommend the incorporation of negative ion MS into the detection scheme.

The direct search for most anesthetics and related volatiles is possible with head space sampling and GC or GC-MS. If the analytical specimens (body fluids or tissues) are packed in special glass jars, as described in 3.10.2.1, work for sample preparation can be eliminated and the analytical specimens remain practically untouched for further investigations.

References

1. Schaer, H. (1982) Pharmakologie für Anästhesisten und Intensivmediziner. Verlag Hans Huber, Bern, Stuttgart & Wien.
2. De Zeeuw, R., Franke, J.P., Maurer, H.H. & Pfleger, K. (1992) Gas Chromatographic Retention Indices of Toxicologically Relevant Substances on Packed or Capillary Columns with Dimethylsilicone Stationary Phases, 3rd Ed. (Report XVIII of the DFG Commission for Clinical-Toxicological Analysis, Spec. Issue of the TIAFT Bulletin). VCH Weinheim & New York.
3. De Zeeuw, R., Franke, J.P., Machata, G., Möller, M.R., Müller, R.K., Graefe, A., Tiess, D., Pfleger, K. & Geldmacher-von Mallinckrodt, M. (1992) Gas Chromatographic Retention Indices of Solvents and other Volatile Substances for Use in Toxicological Analysis (Report XIX of the DFG Commission for Clinical-Toxicological Analysis, Spec. Issue of the TIAFT Bulletin). VCH Weinheim & New York.
4. Brandenberger, H. (1970) Die Gas-Chromatographie in der toxikologischen Analytik. Verbesserung von Selektivität und Empfindlichkeit durch Einsatz multipler Detektoren. Pharmazeutica Acta Helvetiae *45*, 394–413.
5. Brandenberger, H. (1974) Toxicology. In: Clinical Biochemistry, Principles and Methods (Curtius, H.Ch. & Roth, M., eds.), pp. 1425–1467. Walter de Gruyter, Berlin & New York.
6. Brandenberger, H. (1978) Negative ion mass spectrometry, a new tool for the forensic toxicologist. In: Instrumental Applications in Forensic Drug Chemistry, Proceedings of the International Symposium of the US Drug Enforcement Administration in Arlington, Virginia (Klein, M., Kruegel, A.V. & Sobol, S.P., eds.), pp. 48–59. US Government Printing Office, Washington, D.C.
7. Brandenberger, H. & Ryhage, R. (1978) Possibilities of negative ion mass spectrometry in forensic and toxicological analysis. In:

Blood Drugs and other Analytical Challenges. Vol. 7 of Methodological Surveys in Biochemistry (Reid, E., ed.), pp. 173–184. Ellis Horwood Ltd, Chichester, a John Wiley division.

8. Brandenberger, H. (1979) Present state and future trends of negative ion mass spectrometry in toxicology, forensic and environmental chemistry. In: Recent Developments in Mass Spectrometry in Biochemistry and Medicine, Vol. 2 (Frigerio, A., ed.), pp. 227–255. Plenum Publishing Corp., New York.
9. Brandenberger, H. (1983) Negative ion mass spectrometry by low pressure chemical ionization; inherent possibilities for structural analysis. Journal of Mass Spectrometry and Ion Physics *47*, 213–216, and in: Mass Spectrometric Advances 1982, Proceedings of the 9th International Mass Spectrometry Conference in Vienna (Schmid, E. R., Varmuza. K. & Fogy, I., eds.), part C pp. 213–216. Elsevier, Amsterdam, Oxford & New York.
10. Brandenberger, H. (1982) Negative ion mass spectrometry by low pressure chemical ionization and its combination with conventional positive ion mass spectrometry by electron impact. In: Proceecings of the 30th Annual Conference on Mass Spectrometry and Allied Topics, Honolulu, pp. 468–469. Ed. by American Society for Mass Spectrometry.
11. Brandenberger, H. (1984) Negative Ion Mass Spectrometry. Fresenius Z. Anal. Chem. *317*, 634–635.
12. Brandenberger, H., Brunner-Miersch, M., Fluri, T., Frangi-Schnyder, D., Leuenberger-West, F. & Stoll, C. (1987) Identification of additional volatiles in blood samples submitted for alcohol analysis. In: "TIAFT 22", Reports on Forensic Toxicology, Proceedings of the 1985 International Meeting of the Internatinal Association of Forensic Toxicologists in Rigi Kaltbad, Switzerland; Brandenberger, H. & R., eds.) pp. 75–88. Publ. by Branson Research, CH-6356 Rigi Kaltbad.
13. Brandenberger, H. (1987) Low Pressure negative chemical ionization mass spectrometry and dual mass spectrometry: A review of the work since 1977. In: "TIAFT 22", Reports on Forensic Toxicology, Proceedings of the 1985 International Meeting of the International Association of Forensic Toxicologists in Rigi Kaltbad, Switzerland (Brandenberger, H. & R., eds.) pp. 437–453. Publ.by. Branson Research, CH-6356 Rigi Kaltbad.
14. Brandenberger, H. (1980) Negative ion mass spectrometry by low-pressure chemical ionization. In: Recent Developments in Mass Spectrometry in Biochemistry and Medicine, Vol. 6 (Frigerio, A. & McCamish, M., eds.), Elsevier, Amsterdam, Oxford & New York.
15. Brandenberger, H. (1988) Wichtige exogenen Vergiftungen und ihre differentialdiagnostische Erkennung. In: Differentialdiagnose innerer Krankheiten (Siegenthaler, W., ed.), 16.ed., chap. 39. Georg Thieme Verlag, Stuttgart & New York.
16. Bonnichsen, R. & Maehly, A. C. (1966) Poisoning by volatile compounds II. Chlorinated hydrocarbons. Forensic Sciences *11*, 414–427.
17. Stockman, S. (1916) Cases of poisoning in cattle by feeding on meal from soy bean after extraction of the oil. J. Comp. Path. & Therap. *29*, 95–107.
18. Barth, K. (1956) Ist mit Trichloraethylen extrahiertes Sojaschrot für die Ernährung landwirtschaftlicher Nutztiere verwendbar? Dissertation, Landw. Hochschule Hohenheim.
19. Seto, T.A., Schultze, M.O., Perman, V., Bates, F.W. & Sautter, J.H. (1958) Properties of the toxic factor in trichlorethylene-extracted soybean-oil meal. J. agr. Food Chemistry *6*, 49–54.
20. Seto, A. T. (1958) Chemical nature of a toxic factor which causes aplastic anemia in calves. Dissertation University of Minnesota, Univ. Microfilms Mic. *58–1140*.
21. McKinney, L. L., Weakley, F. B., Campell, R. E., Eldridge, A. C. & Cowan, J. C. (1957) Toxic protein from trichloroethylene-extracted soybean meal. J. Amer. Oil Chemist's Soc. *34*, 461–466.
22. Brandenberger, H. & Bader, H. (1967) Über den Einbau von Trichloräthylen in Inhaltsstoffe des Kaffees bei dessen Decoffeinierung. Helv. Chimica Acta *50*, 463–466.

3.11 Digitalis (= Cardiac) Glycosides

J. Hallbach and H. Vogel

3.11.1 General Remarks

3.11.1.1 Definition of digitalis glycosides

"Digitalis" and „digitalis compounds" are terms that encompass the entire group of cardiac glycoside inotropic drugs (those influencing the contractility of myocardial muscular tissue) and chronotropic drugs (those affecting the rate of rhythmic movements such as the heart beat). Digitalis glycosides form one of the most beneficial group of drugs available to aid the failing heart. These glycosides are the drugs of choice for the treatment of congestive heart failure and certain disturbances in cardiac rhythm. Digoxin is clinically the most commonly used digitalis glycoside. Cardiac glycosides are obtained from 11 plant families, those of therapeutical importance are obtained from Digitalis purpurea Linne (Fam. Scrophulariaceae: digitoxin, digitalis, gitalin), from Digitalis lanata Ehrhart (Fam. Scrophulariaceae: digoxin, digitoxin, lanatoside C, deslanoside, acetyldigitoxin), Strophantus gratus (ouabain), and Acokanthera schimperi (ouabain) (1).

3.11.1.2 Chemical structure and physico-chemical properties

Digitalis glycosides have a characteristic condensated ring structure (aglycone or genin) to which one or more sugars are coupled. The aglycone portion of the glycoside consists of a steroid nucleus and an α, β-unsaturated five- or six-membered lactone ring at the C17 position of the steroid nucleus. The hydroxyl groups at C3 and 14 are in the β-configuration. The sugars are attached usually through the C3

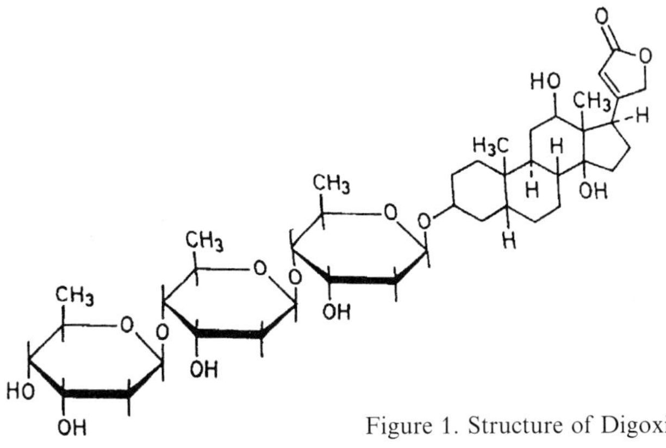

Figure 1. Structure of Digoxin.

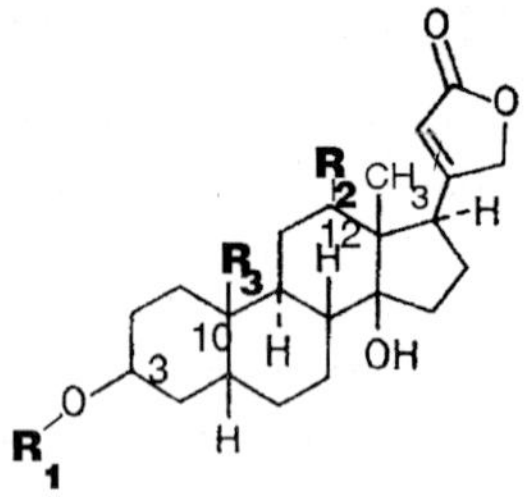

Compound	R1 (C3)	R2 (C12)	R3 (C10)
Digoxin	tridigitoxose	OH	CH_3
β-Methyl-digoxin	4-Methyl-tridigitoxose	OH	CH_3
Digitoxin	tridigitoxose	H	CH_3
Deslanoside	tridigitoxose-glucose	OH	CH_3
Lanatoside C	didigitoxose-acetyldigitoxose-glucose	OH	CH_3
Ouabain	6-deoxy-α-mannose	OH	CH_2OH
		OH additional at C1, 5, 11	

Figure 2. Structure of Key Cardiac Glycosides.

hydroxyl. The structure of digoxin is given in figure 1. Figure 2 illustrates the structure differences of key cardiac glycosides.

Physico-chemical properties of digitalis glycosides are given in table 1 (2).

Table 1. Physico-Chemical Properties of Digitalis Glycosides (modified)

Compound	MW	M.p.	Slb	UV_{max}
Digoxin	780.9	240	− water + $MeOH/CHCl_3$	220
Digitoxin	764.9	> 232	− water + $CHCl_3/MeOH$	279
Lanatoside A	969.1	> 245		
Lanatoside B	985.1	> 245		
Lanatoside C	985.1	> 210	(+) water + dioxan	230
g-Strophanthine	728.8	178.8	+ water + methanol	219
k-Strophanthine-α	548.7			
k-Strophanthine-β	702.8		+ water + EtOH	

MW = molecular weight, M.p. = melting point. Slb = solubility, UV_{max} = max. UV absorption.

3.11.2 Uses and Route of Exposure

Digoxin is one of the drugs most widely used in current health care delivery, accounting for greater than 21 million prescriptions in North America in 1990. Moreover, cardiac glycosides are the most prescribed drugs in Germany.

3.11.2.1 Therapeutic use of cardiac glycosides

The main pharmacological property of the cardiac glycosides is their ability to increase the force and velocity of myocardial systolic contraction (positive inotropic action). The currently accepted mechanism of action of digoxin and related compounds is its inhibitory action on the Na,K-transporting ATPase (EC 3.6.1.37) transmembrane pump of sarcolemma. The sodium-potassium-ATPase pump is stimulated by the addition of potassium to the extracellular fluid and of sodium to the intracellular fluid. It is inhibited by lowering the extracellular potassium concentration or the intracellular sodium concentration (3). It is also inhibited by digitalis, which binds to a receptor site on the amino-terminal portion of the Na,K-ATPase alpha 1 subunit. High and low affinity active forms of the Na,K-ATPase are coexisting and may contribute to different toxicity profiles of digitalis derivatives. The therapeutically relevant range is between 25% and 50% inhibition of Na,K-ATPase. Since the pump moves three sodium ions outward for each two potassium ions moved inward, the inhibition results in the accumulation of intracellular sodium ion and a reduction of negative charge inside the cell and, hence, depolarization. In heart muscle, the result is increased contractility. Perhaps digitalis binding to the sodium-potassium-ATPase pump is thought to cause displacement of bound calcium ions. The resulting free calcium ions exhibit a positive inotropic effect, resulting in a more forceful contraction of the myocardium.

In patients with congestive heart failure, increased myocardial contractility and cardiac output reduce sympathetic tone, thus slowing increased heart rate and causing diuresis in edematous patients. In contrast, glycoside-induced slowing of heart rate in patients without congestive heart failure is negligible and is primarily attributable to vagal (cholinergic) and sympatholytic effects on the sinoatrial node.

The major indication for digoxin use is the combination of congestive heart failure and atrial fibrillation. In tachysystolic atrial fibrillation, in atrial flutter, in paroxysmal atrial fibrillation, and in manifest myocardial failure (III and IV NYHA), cardiac glycosides are usually effective and necessary. Another indication for use of these drugs is chronic congestive failure with sinus rhythm that has failed to respond to diuretics alone. In patients with severe myocardial failure with low blood pressure, often a combination of cardiac glycosides with unloading drugs is indicated.

3.11.2.2 Absorption, distribution, metabolism, influences on the pharmacodynamic behaviour, and elimination

Absorption: The bioavailability or completeness of absorption of digoxin varies markedly (55% to more than 90%) from one preparation to the next, and many reports of toxic digoxin concentrations can be attributed to a patient's changing from one formulation to another. For adults, typical loading doses are 12 to 20 μg/kg digoxin and 10–20 μg/kg digitoxin. The maintenance dose should be adjusted on the basis of the effective loading dose and estimated renal function in case of digoxin and its derivatives. Typical doses are 125 to 500 μg/d for digoxin and 10% of loading dose for digitoxin.

Intestinal motility affects the absorption of digoxin tablets. Therefore, drugs that slow intestinal peristalsis, such as the anti-cholinergics, tend to increase bioavailability. Peak concentration following an oral digoxin dose is usually reached in 60 to 90 minutes after administration, while intravenous administration gives peak concentrations immediately. However, peak concentration does not reflect tissue digoxin concentration and thus pharmacological effects. Digitoxin is only administered orally and shows peak serum concentrations after approximately 3 to 6 hours. Cymarin (k-strophanthin-alpha) is absorbed orally to 47%, k-strophanthoside (k-strophanthin-gamma) to 16%, and ouabain (g-strophanthin) to only 1.4%. Intravenous digoxin has high utility for instances in which digitalization must be rapidly accomplished or when oral digitalization is not feasible. Myocardial uptake of i.v. digoxin is rapid and extensive, yet, serum half-life is as long as in oral administration.

Distribution: The distribution of digoxin is aptly described by a two-compartment open pharmacokinetic model. The first very rapid decrease in concentration, mainly a result of dilution in blood, takes a few minutes. During the distribution or α-phase, the drug equilibrates between the central and peripheral compartment, particularly skeletal muscle and myocardium. During digitalization approximately 13% of skeletal muscle digitalis glycoside receptors become occupied. Considering the large skeletal muscle mass, this indicates that the skeletal Na,K-ATPase pool constitutes a major volume of distribution for digoxin during digitalization. The ratio of the concentrations in plasma and heart reportedly is between 1:30 and 1:200 for digoxin. Digoxin has an extremely wide apparent volume of distribution (500–700 l). This large V_d is an indicator of digoxin's significant binding to tissues. Lipophilic drugs like quinidine or theophylline may displace digoxin from storage sites and increase plasma concentration. In plasma, however, only 15 to 25% of digoxin is bound to proteins, whereas digitoxin is protein bound to a much greater extent (> 90%). This has to be considered when interpreting digitoxin serum levels.

Metabolism: Some ten years ago it was believed that relatively polar glycosides such as digoxin, deslanoside and ouabain were not metabolized appreciably. Today it is known that metabolism can be extensive and can involve reduction of the lactone ring to form the R-dihydrodigoxin epimer and the stepwise removal of sugar molecules, followed by epimerization of 3β-hydroxyl to the 3α-(epi) position and conjugation to give the polar metabolites 3-epi-glucuronide and 3-epi-sulfate (1). Metabolism of digoxin occurs mainly in the liver. Normally, however, most digoxin passes from the body unaltered, with less than 5% being metabolized. Quantitatively the most abundant metabolites are polar and can average 26% of a single digoxin dose in plasma 6 hrs after drug admission. Stepwise removal of the sugar residues leads to a stepwise loss of cardioactivity, while reduction of the lactone ring results in almost total loss of pharmacological activity. Table 2 gives the percentage cardioac-

Table 2. Cardioactivity of Digoxin Metabolites and Immunological Cross Reactivity

Metabolite	Activity relative to digoxin, %	Cross reactivity, %
Dihydrodigoxin	2–6	< 1 (ACS)
Dihydrodigoxigenin	2	2.6 (TDx)
Digoxigenin	4–21	97 (TDx) < 1 (ACS)
Digoxigenin mono-digitoxiside	66	181 (TDx) 68 (ACS)
Digoxigenin bis-digitoxiside	77	146 (TDx) 90 (ACS)

tivity of digoxin metabolites and shows the cross-reactivities in two commercially available immunoassays. When interpreting digoxin serum levels, it should be kept in mind that metabolites may make up an important component of the serum digoxin concentration determined by immunological methods.

Biotransformation of digitoxin involves the oxidative cleavage of dual deoxy sugars, that depends from the cytochrome P_{450} IIIA activity (4), to form digitoxigenin bisdigitoxoside and digitoxigenin monodigitoxoside, whereby the cleavage of the terminal sugar will be the rate-limiting step. The cleavage products are conjugated to glucuronic acid. Only 8% of the administered dose was found in the urine unchanged (5). Digitoxin and its cleavage products can be hydroxylated to the corresponding digoxin compounds. However, metabolism to digoxin and its metabolites is thought to contribute little to the pharmacological effects of digitoxin.

Influences on the pharmacodynamic behaviour: Factors influencing myocardial sensitivity to cardiac glycosides include hypo- and hyperkalemia, hypercalcemia, hypo- and hypermagnesemia, hypothyroidism, acid-base disorders, myocardial ischemia, hypoxemia, underlying heart disease, and pulmonary disease.

Drugs influencing digitalis serum level: Administration of itraconazole for the treatment of osteomyelitis resulted in a statistically significant increase in the half-life of digoxin that may necessitate reduction of the digoxin dose by almost 60%. Quinidine, amiodarone, and spironolactone reduce the renal clearance of digoxin, without affecting its biliary clearance. Drugs inducing hepatic cytochrome P_{450} enzymes, e.g. phenobarbital and rifampicin, have been shown to reduce digitoxin plasma concentration.

Elimination: The elimination half-life of digoxin in healthy test subjects reportedly varies between 20 and 45 hrs, although a significantly longer $t_{1/2}$ has been reported for patients with a reduction in renal clearance. Accordingly, not only the digoxin maintenance dose, but the digoxin loading dose as well must be adjusted in this condition. Renal clearance of digoxin occurs by glomerular filtration, tubular reabsorption, and tubular secretion, with the latter accounting for half the amount of digoxin in urine. Digoxin secretion at the apical membrane of renal tubular cells seems to be an active P-gylcoprotein mediated transport that can be inhibited by verapamil. Hypothyroidism may decrease the renal clearance of digoxin. In the majority of patients at least 75% of digoxin are excreted unchanged in urine by the kidneys, while in some patients up to 60% may be excreted as metabolites, largely dihydrodigoxin. Digoxin and its metabolites are also present in bile. The hepatic elimination may be increased substantially in patients with renal impairment. Additionally, the small intestine is a major site of absorption of digoxin, making enterohepatic recycling a significant factor in digoxin pharmacokinetics. Elimination half-life of digitoxin covers the range from 3 to more than 15 days. In spite of extensive metabolism in the liver, the majority of the metabolites of a given digitoxin dose is excreted in urine. Hepatic disease has little effect on the elimination of digitoxin since the liver has a reserve capacity for its metabolism. 21% of a given dose of cymarin is excreted by the kidneys, mainly as conjugated metabolites, with a half-life of elimination at 23 hrs. Oral k-strophanthoside is excreted only to 6% as the unchanged drug, whereas after i.v.-injection 70% is excreted as the unchanged drug with a half-life of elimination of 99 hrs. Intravenously given ouabain is eliminated to 80% unchanged with a total renal excretion of 33% of the given dose and with a half-life of elimination of 23 hrs.

3.11.3 Toxicity for Man

3.11.3.1 Mechanism of toxic action

Therapeutic doses of glycosides normalize heart rate in patients with congestive heart failure and do not influence heart rate significantly in normal persons. Digitalis intoxication causes ventricular dysrhythmias and tachycardia, pacemaker shifts, and conduction disturbances in the sinoatrial node, and decrease in contractile force, but the mechanisms underlying these changes have not been fully explained. It can be observed that ouabain and digoxin quickly intoxicate the sinoatrial node, whereas strophanthidin does not, possibly because of the lack of a sugar moiety (6).

3.11.3.2 Overdosing

Digitals intoxication is a frequent iatrogenic effect in patients on treatment with digoxin. In ten percent of patients who died suddenly of cardiac disease, an overdosage of digoxin was found by digoxin measurement in postmortem serum samples (7). This reflects the narrow therapeutic range of digoxin and the multiple mechanisms that can lead to intoxication: overdose itself, drug interactions, hypokalemia, hypomagnesemia, volume depletion, renal insufficiency, and chronic disease states.

3.11.3.3 Acute toxic effects

The manifestations of digoxin intoxication can include cardiac, gastrointestinal, and neurological symptoms. Cardiotoxicity of digitalis produces commonly an abnormal electrocardiogram showing ventricular or atrial arrythmias, with or without some degree of concurrent atrioventricular (AV) block (8). In severe intoxication, repeated episodes of asystole alternating with ventricular fibrillation have been observed. Other arrythmias, considered more diagnostic of digitalis toxicity, include paroxysmal atrial tachycardia with AV block, bidirectional ventricular tachycardias, and atrial fibrillation with slow ventricular rate. Gastrointestinal symptoms include nausea, vomitting, diarrhea, and anorexia. Headaches, dizziness, trigeminal neuralgia, disturbances of color vision and auditory hallucinations represent neurological manifestations. Advanced age ($\geq$ 80 years) seems to be an independent risk factor for digoxin toxicity.

3.11.3.4 Delayed clinical effects

It has been reported that van Gogh suffered from epilepsy, for which he was treated with digitalis, as was usually done in the late 19th century (9). Therefore van Gogh's fascination with the color yellow might be due to a chronic toxic effect of digitalis besides schizophrenia.

3.11.3.5 Biochemical effects

Marked thrombocytopenia was observed in digitoxin overdose, which was normalised within 12 days after discontinuation of digitoxin treatment (10).

3.11.3.6 Treatment of overdosing and intoxication

The most important step in the management of toxicity due to any of the cardiac glycosides is its recognition (8). Since 20 years digoxin-specific Fab antibody fragments are available to treat patients with advanced, life-threatening digitalis toxicity of all commonly used digitalis preparations. In contrast to cardiac pacing, Fab immunotherapy was not associated with any serious adverse effects and tended to be more effective (11). In infants and young children, FAB-therapy is recommended if a serum digoxin concentration of greater than 6.4 nmol/l (5 μg/l) has been measured, accompanied by a life-threatening arrhythmia, hemodynamic instability, hyperkalemia, or rapidly progressive toxicity (12). Adolescents, who are more sensitive to the toxic effects of digoxin, may require treatment with Fab already at lower serum concentrations. Digoxin and digoxin immune Fab, its antidote, are eliminated by the kidney. The concentration-time profile of Fab appears to be similar to the concentration-time profile of total digoxin. Therefore renal dysfunction may necessitate prolonged clinical monitoring. During Fab-treatment, total digitalis serum concentrations and renal digitalis clearance should rise, indicating redistribution of drug from tissue to serum and urinary elimination of Fab-bound digitalis. Digitoxin intoxication needs prolonged and repeated i. v. infusions of high quantities of anti-digitalis Fab antibodies. In contrast to commercial glycoside product poisoning with homemade foxglove extracts, the use of Fab only showed temporary improvements, but not a shortened clinical course (13). In severe intoxication, cardiopulmonary resuscitation and adrenaline administration may be necessary. Supportive conventional therapy requires the maintenance of plasma potassium levels greater than or equal to 4 mmol/l, reversal of decompensated heart failure or overt myocardial ischemia, attention to serum magnesium levels and the patient's acid-base status.

3.11.4 Analyses

3.11.4.1 Determination of digitalis compounds

Various HPLC methods for the routine quality assurance of pharmaceutical formulations and analysis of leaf powders are described, also suitable for the determination of derivatives like β-methyldigoxin (14) and lanatoside C (15), and secondary cardiac glycosides in Digitalis purpurea such as gitoxin or gitaloxin. These methods may be adapted to determine digitalis compounds from stomach content or other sources.

3.11.4.2 Determination of cardiac glycosides in blood and urine by chromatography

While as good as no GC or GC/MS methods for the determination of cardiac glycosides in biological fluids have been established, various high pressure liquid chromatographic procedures have been described. However, they have not been widely applied in the routine clinical chemistry laboratory, owing to their marked lack of sensitivity, which would necessitate lengthy sample preparation and large sample size.

From the digitalis glycosides, only digitoxin is referred in the DFG/TIAFT report (16) on gas chromatographic retention indices. In GC/MS, the digitoxose sugar residues are unstable under the conditions of assay. Therefore only the aglycone part of the molecule may be detectable.

3.11.4.2.1 Thin-layer chromatography of urine in standardized systems

Various digitalis glycosides (digoxin, digitoxin, lanatoside C, proscillaridine, strophanthin) are detectable in the DC systems evaluated by DFG/TIAFT (17). In contrast, the commercially available toxi-lab system (DRG Instruments, Marburg, FRG) does not find any digitalis glycosides. Thin layer chromatography can also be useful to detect cardiotoxic glycosides from other plants, such as oleander Nerium, in blood and tissue homogenates (18).

3.11.4.2.2 HPLC with immunological detection

Separation of digoxin, digitoxin, their cardioactive metabolites and their semi-synthetic derivatives (β-methyl- and β-acetyldigoxin) was achieved on a 10 μm Nucleosil C_{18} column. A complex gradient of acetonitrile/water was needed to reach full peak resolution (19). This method allows easy preparative work with collection of fractions for final determination by RIA after evaporation to dryness and dissolution in 1 ml of blank serum. This method could be adapted for measuring cardiac glycosides in human tissues in forensic medicine. By preliminary HPLC, many substances inherent in biological media that potentially cross-react with the antibody are exluded. The specificity and reliability of the measurement of individual compounds can thus be improved (20, 21). Results for samples from patients taking digoxin, obtained with FPIA (Abbott) and HPLC/FPIA, showed substantial discrepancies for 20% of the samples.

Interferences in digitoxin immunoassays from its metabolites and endogenous compounds can also be overcome by using HPLC fractionation (5). Urine excretion of a given dose was found to consist of 8.1% digitoxin itself, 3.4% digitoxigenin bisdigitoxoside, 0.2% digitoxigenin monodigitoxoside, 1.2% digitoxigenin, and 1.7% digoxin. The cross reactivities with the digitoxin antibody were 23% for digoxin, 78% for digitoxigenin, 20% for the mono- and 91% for the bis-digitoxoside-derivative, respectively. In a pooled serum of a patient, a digitalis-like immunoreactivity (DLIF) concentration of as much as 189 μg/l – measured with the TDx digitoxin assay (Abbott) – was found. DLIF was extracted with Bond-elut C–18 and gave, on a reversed phase gradient elution, an immunologically detectable peak distinct from digitoxin itself (22).

3.11.4.2.3 HPLC with fluorescence detection

Combination of HPLC and radioimmunoassay is time consuming. Therefore, direct detection of glycosides was attempted. Direct UV-detection was unsuccessful owing to the lack of sensitivity. Consequently, pre- and post-column derivatization techniques have been introduced and postcolumn derivatization was carried out after hydrolysis to the aglycone (23). But also these techniques were found not to be sensitive enough for use in routine investigations of human urine samples. Another attempt for quantification of digitalis in urine was derivatization as the 3,5-dinitrobenzoyl ester. A methodology that seems sensitive enough for measurement of digoxin in plasma at therapeutic concentrations consists of involving fluorogenic post-column derivatization (24, 25). Using an improved extraction procedure (26) and fluorescence enhancement by post-column addition of concentrated hydrochloric acid, 1.5 ng of digoxin could be detected in a 3 ml sample. This chromatographic method separates digoxin from its metabolites (digoxigenin, digoxigenin monodigitoxoside, digoxigenin bisdigitoxoside, dihydrodigoxigenin and dihydrodigoxin) prior to derivatization, allowing for quantification of digoxin in the presence of metabolites. Also cross-reactivity from steroids (dehydroepiandrosterone-3-sulfate, cortisone, cortisol, deoxycortisone, Δ^4-androstene-3,17-dione, progesterone and glycochenodeoxycholic acid) was not observed. This method was compared with FPIA and RIA testing and provided more accurate values. However, HPLC with fluorescence detection seems not to be useful under routine conditions, due to use of concentrated hydrochloric acid, relatively low recovery of digoxin (less than 80 %), time-consuming sample preparation and lack of sensitivity compared with immunoassays and HPLC-immunoassay procedures (26).

3.11.4.2.4 HPLC with electrochemical detection

Human serum is an extremely complex matrix and contains a lot of electroactive constituents. Reported is the derivatization of digoxin and its metabolites in pyridine using 3,5-dinitrobenzoyl chloride (27) with complete esterification of all primary and secondary alcohol moieties in the digitalis derivatives. Electrochemical reduction on a single glassy carbon electrode gave a limit of detection of 2.2 ng of derivatized digoxin (representing 0.98 ng of digoxin). A dual electrode procedure, where 3,5-dinitrobenzoyldigoxin was reduced at the first electrode and the reduced product re-oxidized at the second electrode, provided a maximum sensitivity of 0.39 ng of digoxin.

3.11.4.3 Quantification of cardiac glycosides in blood by immunoassay

3.11.4.3.1 Principles

Up to now, digoxin and digitoxin in body fluids are routinely measured almost exclusively through the use of various immunoassay procedures, i.e. radioimmunoassay (RIA), enzyme multiplied immunoassay (EMIT), fluorescence polarization immunoassay (FPIA), chemiluminescence immunoassay (LIA), recombinant enzyme immunoassay (CEDIA), and others.

3.11.4.3.2 Specificity of different assays

Digoxin antibodies frequently are generated against the steroid moiety of digoxin. As a consequence, digoxigenin bis- and monodigitoxoside and digoxigenin itself all react with the antibodies, whereas dihydrodigoxigenin and dihydrodigoxin – in which C22 is reduced – show little or no cross-reactivity (1). The cross-reactivity of the metabolites of digoxin with the antidigoxin antibodies is well known and is usually stated for most commercial antisera (table 2). What has not been appreciated previously is that in more than 50% of patients, digoxin undergoes considerable to extensive metabolism, and measurement of „digoxin" concentrations in serum with nonspecific antibodies can lead to problems in interpreting results. This is probably responsible for the lack of a consensus as to the correlation of serum digoxin concentration with either its therapeutic or toxic effects. In addition, some of the conclusions of drug/drug interactions for digoxin may need to be reevaluated once a specific immunoassay is available, because these drugs may influence the extent to which digoxin is metabolized via a particular pathway, thereby altering the metabolite concentration and affecting quantification when nonspecific assays are used.

In current immunoassays for digoxin there is cross-reactivity not only with many of the digoxin metabolites but also with many apparently less-related or even structurally unrelated compounds, among them a group of steroids that includes progesterone, cortisone, and others, as for example spironolactone and its metabolite canrenone. Some of these compounds, for example some drugs, do not interfere in the immunoassay itself. However, they might be converted to compounds or else elicit release of substances that do interfere in the immunoassay system studied.

Combination of immunoassay, e.g. EMIT, with HPLC cleanup of NADH prior to amperometric detection (28) instead of the conventional photometric quantitation may increase specificity of immunoassay procedures. HPLC and column switching were used to separate electroactive NADH from the complex serum matrix, because in serum, interference to NADH amperometric measurement occurs both electrochemically and by passive adsorption at the electrode surface. The relative standard deviation of this method was 2.3% at a concentration of 2.2 ng/ml digoxin. Comparison with radioimmunoassay (ARIA HT, Becton Dickinson) of 46 patient's samples showed good correlation, with the least-squares correlation coefficient being 0.94 ($y = 0.99 \times + 0.24$).

3.11.4.3.3 Determination of digoxin

RIA methods have more and more been overcome by immunoassays with non-radioactive labeling of digoxin tracer molecules or antibodies with enzymes, fluorophors, luminophors, enzyme fragments by recombinant DNA techniques (CEDIA), and europium chelators. The Abbott TDx fluorescence polarization immunoassay (Abbott, Wiesbaden, FRG) exhibits a sufficient CV-profile (14.5 to 4.9% over the therapeutic concentration range) (29), as well as good correlation with RIA, but shows significant cross-reactivity with digitoxin (table 3) and protein dependence of digoxin quantitation. Whereas the majority of the commercially available immunoassays for digoxin need a manual sample preparation for protein precipitation, the ACS 180 digoxin chemiluminescence assay (Ciba Corning Diagnostics, Fernwald, FRG) is fully mechanised. Using this procedure for a comparison study with

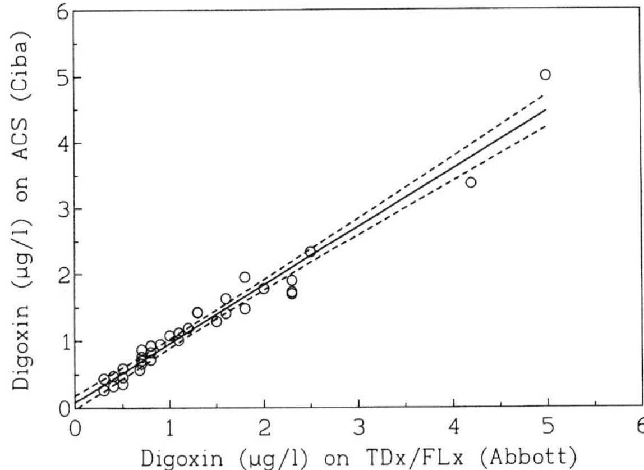

Figure 3. Comparison of TDx/FLx and ACS Digoxin Assay.

35 patient samples, we found an acceptable correlation (fig. 3) with the TDx assay (30) with the least squares correlation coefficient being 0.96 ($y = 0.88 \times - 0.09$). Lower values for the ACS may result from the lack of cross-reactivity with digitoxin (table 3), and with lower sensitivity to DLIF. Between-run precision was found in

Table 3. Cross Reactivity of Digitoxin with Digoxin Assays

Digitoxin (nmol/l)	Apparent Digoxin			
	TDx/FLx		ACS	
	(nmol/l)	(%)	(nmol/l)	(%)
13.4	1.3	10	0.0	0
17.9	1.2	7	0.0	0
21.9	1.4	6	0.0	0
23.2	1.8	8	0.2	< 1
30.6	1.8	6	0.1	< 1
51.8	3.1	6	0.1	< 1

the same range as with other assays (table 4). Determination of spiked QC samples usually showed coefficients of variation in the 11–21 % range, irrespective of the technique used. Re-analysis of patient samples, randomly selected from routine at reference laboratories, gave best agreement between routine and reference results

Table 4. Between-Run Precision of ACS 180 Digoxin Assay

N	Mean Digoxin (nmol/l)	Between-run % CV
20	0.71	15.6
20	1.52	8.2

for samples originally measured by FPIA (r = 0.96) (31). Digoxin-specific Fab antibody fragments, used in the treatment of severe digitalis intoxication, cause a marked interference with some immunoassays for digoxin, while for the ACS assay, no interference has been detected up to 4.6 mg/l digoxinspecific Fab.

The free digoxin concentration in serum is measurable after preparation of a protein-free ultrafiltrate (32). This technique also may be useful in evaluating serum digoxin concentrations in patients with abnormal plasma protein concentrations. A natural protein-poor sample material is saliva. Salivary digoxin is lower in stimulated saliva, suggesting that salivary digoxin concentration decreases with increased saliva production rate. The best correlation was found between stimulated saliva and serum digoxin concentrations (33).

3.11.4.3.4 Determination of digitoxin

Digitoxin is measurable by FPIA, CEDIA and a solid phase enzyme immunoassay based on the competition principle (Enzymun-Test Digitoxin, Boehringer Mannheim, FRG). This assay showed a measuring range from 4 to 60 μg/l and interassay precision from 2 to 7 % in different laboratories. Under routine conditions we found for the TDx digitoxin assay (Abbott, Wiesbaden, FRG) between-run CV's in the range from 6.1 % (mean concentration 17.8 nmol/l, N = 51) to 13.8 % (mean concentration 8.3 nmol/l, N = 51). The interference of endogenous compounds and metabolites can be overcome by solvent extraction, reversed phase HPLC and radioimmunoassay (5).

3.11.4.3.5 Other digitalis glycosides

Beta-methyldigoxin (metildigoxin) becomes demethylated to digoxin in the liver and can be detected with digoxin immunoassays.

Thevetin B from Thevetia neriifolia can be determined with digitoxin immunoassays because thevetin B genin is structurally identical to digitoxigenin. Also intoxication with a homemade foxglove extract showed serum immunoassay results demonstrating a digitoxin-like glycoside (13).

3.11.5 Other Biological Analyses

3.11.5.1 Potassium and sodium

Hypokalemia can potentiate the effects of cardiac glycosides on the heart, whereas high potassium concentrations give the ability to overcome the inhibition of sodium pump activity by the cardiac glycosides. In order to discriminate between toxic and non-toxic reactions of patients, intraerythrocytic sodium and potassium concentrations can be determined beside plasma digoxin levels (34). Digoxin intoxication is associated with elevated intraerythrocytic sodium and reduced potassium. Sensitivity and positive predictive values for diagnosing toxicity in chronically digitalized patients were found to be higher for intraerythrocytic sodium determinations compared with plasma digoxin values.

3.11.5.2 Magnesium

Clinicians are more attuned to avoiding hypokalemia than hypomagnesemia, although both may contribute to the toxic effects of digitalis (35). In a study, plasma magnesium deficiency was the most frequent electrolyte disturbance in relation to digoxin toxicity. Furthermore, in the presence of magnesium deficiency, digoxin toxicity developed at relatively low serum digoxin concentrations (36).

3.11.6 Influence of Endogenous Digitalis Like Factors

Digoxin immunoreactivity can be measured in plasma and urine (37) from individuals known never to have received the drug (concentrations as high as 10 µg/l have been observed). Like the cardiac glycosides, this substance (cardiodigin) increases cardiac contractility, constricts blood vessels, and reduces the renal tubular reabsorption of sodium (38). Commercially available digoxin immunoassays detect DLIF to a very differing extent giving inaccurate digoxin estimates (39). For example, cross-reactivity of DLIF in sera from pregnant women, newborns, patients undergoing hemodialysis and patients with renal insufficiency was generally lower in the CEDIA assay than with FPIA (40). Furthermore, addition of digoxin to sera from untreated patients suggested that DLIF interference may be less pronounced in sera of patients undergoing digoxin therapy compared to untreated persons (40). DLIF may be a natriuretic hormone that should not be confused with atrial natriuretic peptide. Atrial natriuretic peptides (ANP) cause a natriuresis, diuresis, and drop in blood pressure, but they neither cross react with digoxin antibodies nor inhibit Na,K-ATPase. In contrast, DLIF has been isolated from plasma, placenta, adrenal cortex (strongest immunohistochemical reactivity) after inhibition of adrenal steroid biosynthesis, and other tissues; it cross reacts with digoxin antibodies, inhibits Na,K-ATPase, competes with ouabain binding to its receptor, and also causes a natriuresis and diuresis (1). Some digoxin-like factors in human plasma may have a dietary source, since elevated plasma digoxin and Na,K-ATPase inhibiting activity increased after ingestion of some herb teas (41). DLIF was also detected in human milk, where it seems not to be bound to milk protein. A DLIF compound was also found in a chinese medicine „kyushin", with toad venom as ingredient (42). DLIF seems to be a slow-acting, heat-resistant small molecule attached to protein. DLIF's from plasma and urine show chromatographic and biological heterogeneity, but produce a dominant fragment at m/z 532 with FAB mass spectral analysis. Dehydroepiandrosterone sulfate, progesterone, cortisol, and lignans (enterolactone) (43) for example are discussed as a DLIF compound, but account for less than 25% of total DLIF. Furthermore, addition of albumin completely abolished any detectable interaction of these compounds with both anti-digoxin antibodies or Na,K-ATPase (44). Manifestation of digitalis like immunoreactivity can be observed in renal insufficiences, neonates, pregnant women, and patients with liver disease (cirrhosis). Elimination of endogenous digoxin-like immunoreactive factors requires use of a highly specific digoxin or digitoxin antibody, or separation of DLIF before immunoassay. Extraction of digoxin on a column of derivatized silicagel (EMIT procedure) seems appropiate to eliminate detectable DLIF (39), while protein precipitation with 5-sulfosalicylic acid (TDx procedure), in contrast, left significant amounts of DLIF in the samples, most probably because this procedure disrupted protein-DLIF binding. Also the use of disposable C18 low-pressure columns may improve the specificity of immunoassays. With C18 reverse-phase HPLC two DLIF's could be purified from human urine. The less polar DLIF was undistinguishable from digoxin in HPLC, NMR and fast atom bombardment mass spectrometry (37).

3.11.7 General Interpretation of Results

The usual therapeutic range for digoxin is 1.3 to 2.6 nmol/l (1.0 to 2,0 μg/l), and for digitoxin 13 to 33 nmol/l (10 to 25 μg/l). To get best correlation of serum and tissue concentrations, serum samples should be taken after equilibrium time, e.g. at least 8 hours after the latest dose of any therapeutic cardiac glycoside. Steady state conditions are normally achieved for digoxin within 5 to 7 days, but will increase depending upon the degree of reduction of renal function. In contrast, steady state concentrations of digitoxin will be reached after approximately one month. Usually used immunoassays for digoxin and digitoxin measure total glycoside concentrations. Since over 90% of digitoxin is protein bound, one must be aware of diseases which lower serum albumin concentration or drugs which may cause a decrease in protein binding and therefore an increased free effective concentration.

Increased digoxin concentrations above the therapeutic range are much more common (6.7% of a total of 280000 studied digoxin levels) (45) than manifestations of digoxin intoxication (46). An apparent excess of serum digoxin may result from improper sampling. In fact about 30% of routinely sampled specimens in the toxic range were obtained before steady state had occured. Peak concentrations of digoxin are seen 2–3 hrs after oral dose, with a maximal therapeutic effect apparent in 4–6 hrs. After intravenous administration, the maximal effect is seen in 1–3 hrs.

Detection of digitalis intoxication remains difficult, because of the frequency of improper sampling. An elevated serum digoxin or digitoxin concentration often does not serve as a reliable indicator of toxicity. Furthermore, patients show a marked degree of variability in the sensitivity to the toxic effects of digitalis. The signs and symptoms of digitalis intoxication (see 3.11.3.3), in addition, may also occur in patients with congestive heart failure and underlying coronary atherosclerosis who are not receiving a cardiac glycoside.

Strategically, in any case of a suspected intoxication with cardiac glycosides, immunoassays for both, digoxin and digitoxin, should be carried out in serum or plasma. Since not all cardiac glycosides are detectable with these assays because of lacking cross-reactivity, only the examination of urine with TLC or HPLC and fluorescence detection can exclude the absorption of cardiac glycosides with a high degree of security.

For situations with unexpected high digoxin concentration, a strategy (47) was developed to distinguish true digoxin from DLIF and other artifacts of digoxin measurement. Nonspecific interferences will frequently show variable reproducibility of test results and nonlinear dilution characteristics. To test this possibility, different test kits for digoxin and variable dilutions should be examined. Enhancement of incubation time and temperature should reduce interferences from substances with weaker binding to the digitalis specific antibodies. Digoxin is only weakly protein bound, whereas DLIF is strongly protein bound. Therefore ultrafiltration or extraction should remove DLIF. At least chromatographic methods can separate digoxin from interfering substances and DLIF. For testing the biological activity of apparent digoxin, the inhibition of Na,K-ATPase can be assayed.

References

1. Soldin, S.J. (1986) Digoxin – issues and controversies. Clin. Chem. *32*, 5–12.
2. Müller, R.K., Demme, U., Gastmeier, G., Geldmacher-von Mallinckrodt, M., Giebelmann, R. & Twele, R. (1991) In: Toxicological Analysis (Müller, R.K., ed.) pp. 283–284; 327–328; 409; 419–420, Ver. Gesundheit, Berlin.
3. Katz, A.M. (1985) Effects of digitalis on cell biochemistry: sodium pump inhibition. J. Am. Coll. Cardiol. *5*, 16A–21A.
4. Eberhart, D.C., Gemzik, B., Halvorson, M.R. & Parkinson, A. (1991) Species difference in the toxicity and cytochrome P450 IIIA-dependent metabolism of digitoxin.
5. Santos, S.R.C.J., Kirch, W. & Ohnhaus, E.E. (1987) Simultaneous analysis of digitoxin and its clinically relevant metabolites using highperformance liquid chromatography and radioimmunoassay. J. Chromatography *419*, 155–164.
6. Gonzales, M.D. & Vassale, M. (1993) Role of oscillatory potential and pacemaker shifts in digitalis intoxication of the sinoatrial node. Circulation *87*, 1705–1714.
7. Eriksson, M., Lindquist, O. & Edlund, B. (1984) Serum levels of digoxin in sudden cardiac deaths. Z. Rechtsmed. *93*, 29–32.
8. Kelly, R.A. & Smith. T.W. (1992) Recognition and management of digitalis toxicity. Am. J. Cardiol. *69*, 108–118.
9. Wolf, P.L. (1994) If clinical chemistry had existed then ... Clin. Chem. *40*, 328–335.
10. Gschwantler, M., Gulz, W., Brownstone, E., Feichtenschlager, T., Pulgram, T., Schrutka-Kolbl, C. & Weiss, W. (1993) Digitoxin-induzierte Thrombozytopenie. Wien. Klin. Wochenschr. *105*, 500–502.
11. Taboulet, P., Baud, F.J., Bismuth, C. & Vicaut, E. (1993) Acute digitalis intoxication – is pacing still appropriate? J. Toxicol. Clin. Toxicol. *31*, 261–273.
12. Woolf, A.D., Wenger, T., Smith, T.W. & Lovejoy, F.H. (1992) The use of digoxin-specific Fab fragments for severe digitalis intoxication in children. N. Engl. J. Med. *326*, 1739–1744.
13. Rich, S.A., Libera, J.M. & Locke, R.J. (1993) Treatment of foxglove extract poisoning with digoxin-specific Fab fragments. Ann. Emmerg. Med. *22*, 1904–1907.
14. Fujii, Y.I. & Yamazaki, M. (1988) Micro highperformance liquid chromatographic determination of cardiac glycosides in β-methyldigoxin and digoxin tablets. J. Chromatogr. *448*, 157–164.
15. Orosz, F., Nuridsany, M. & Ovadi, J. (1986) Isolation and quantitative determination of some cardioactive gylcosides from Digitalis lanata by high-performance liquid chromatography. Anal. Biochem. *156*, 171–175.
16. DFG, Dt. Forschungsgemeinschaft, TIAFT, The International Association of Forensic Toxicologists. (1992) Gas chromatographic retention indices of toxicologically relevant substances on packed or capillary columns with dimethylsilicone stationary phases (prepared by R.A. De Zeeuw) VCH Verlagsgesellschaft, Weinheim, Basel, Cambridge, New York.
17. DFG, Dt. Forschungsgemeinschaft, TIAFT, The International Association of Forensic Toxicologists. (1992) Thin layer chromatographic R_f values of toxicologically relevant substances on standardized systems (prepared by R.A. De Zeeuw) VCH Verlagsgesellschaft, Weinheim, Basel, Cambridge, New York.
18. Blum, L.M. & Rieders, F. (1987) Oleandrin distribution in a fatality from rectal and oral Nerium oleander extract administration. J. Anal. Toxicol. *11*, 219–221.
19. Plum, J. & Daldrup, T. (1986) Detection of digoxin, digitoxin, their cardioactive metabolites and derivatives by high-performance liquid chromatography – radioimmunoassay. J. Chromatography *377*, 221–231.
20. Vetticaden, S.J. & Chandrasekaran, A. (1990) Chromatography of cardiac glycosides. J. Chromatogr. *531*, 215–234.

21. Stone, J.A. & Soldin, S.J. (1988) Improved liquid chromatographic/immunoassay of digoxin in serum. Clin. Chem. *34*, 2547–2551.
22. Pickert, A. & Stein, W. (1991) Rapid detection of „digitoxin-like immunoreactive substances" in serum. Lab. med. *15*, 19–21.
23. Jakobsen, B. (1986) Determination of digoxin, digoxigenin and dihydrodigoxigenin in urine by extraction, derivatization and high-performance liquid chromatography. J. Chromatography *382*, 349–354.
24. Kwong, E. & Mc Erlane, K.M. (1986) Analysis of digoxin at therapeutic concentrations using high-performance liquid chromatography with post-column derivatization. J. Chromatography *381*, 357–363.
25. Reh, E. (1988) Determination of digoxin in serum by on-line immunoadsorptive clean-up high-performance liquid chromatographic separation and fluorescence-reaction detection. J. Chromatography *433*, 119–130.
26. Embree, L. & Mc Erlane, K.M. (1989) Development of a high-performance liquid chromatographic-post-column fluorogenic assay for digoxin in serum. J. Chromatography *496*, 321–334.
27. Embree, L. & Mc Erlane, K.M. (1990) Electrochemical detection of the 3,5-dinitrobenzoyl derivative of digoxin by high-performance liquid chromatography. J. Chromatogr. *526*, 439–446.
28. Wright, D.S., Halsall, H.B. & Heineman, W.R. (1986) Digoxin homogenous enzyme immunoassay using high-performance liquid chromatographic column switching with amperometric detection. Anal. Chem. *58*, 2995–2998.
29. Ferreri, L.F., Raisys, V.A. & Opheim, K.E. (1984) Analysis of digoxin concentrations in serum by fluorescence polarization immunoassay: an evaluation. J. Anal. Toxicol. *8*, 138–140.
30. Schmolke, M. & Hallbach, J. Unpublished data.
31. Clinical Pharmacology and Toxicology Study Group, Italian Society for Clinical Biochemistry (1991) Interlaboratory variability in drug assay: a comparison of quality control data with reanalysis of routine patient samples. II: Digoxin. Ther. Drug. Monit. *13*, 140–145.
32. Hursting, M.J., Raisys, V.A., Opheim, K.E., Bell, J.E., Trobaugh, G.B. & Smith, T.W. (1987) Determination of free digoxin concentrations in serum for monitoring Fab treatment of digoxin overdose. Clin. Chem. *33*, 1652–1655.
33. Miles, M.V., Miranda-Massari, J.R., Dupuis, R.E., Mill, M.R., Zaritsky, A.L., Nocera, M. & Lawless, S.T. (1992) Determination of salivary digoxin with a dry strip immunometric assay. Ther. Drug. Monit. *14*, 249–254.
34. Pedersen, K.E., Koldkjaer, O., Berndtzs, N.H., Hvidt, S., Kjaer, K., Midtskov, C. & Sanchez, G. (1983) The diagnostic value of determination of intraerythrocytic sodium and potassium concentrations versus plasma digoxin concentration in digoxin intoxication. Acta Med. Scand. *213*, 357–362.
35. Whang, R., Oei, T.O. & Watanabe, A. (1985) Frequency of hypomagnesemia in hospitalized patients receiving digitalis. Arch. Intern. Med. *145*, 650–656.
36. Young, I.S., Goh, E.M., McKillop, U.H., Stanford, C.F., Nicholls, D.P. & Trimble, E.R. (1991) Magnesium status and digoxin toxicity. Br. J. Clin. Pharmacol. *32*, 717–721.
37. Goto, A., Ishiguro, T., Yamada, K., Ishii, M., Yoshioka, M., Eguchi, C., Shimora, M. & Sugimoto, T. (1990) Isolation of a urinary digitalis-like factor indistinguishable from digoxin. BBRC *173*, 1093–1101.
38. Haddy, F.J. (1987) Endogenous digitalis-like factor or factors. New England J. Med. *316*, 621–623.
39. Skogen, W.F., Rea, M.R. & Valdes, R. (1987) Endogenous digoxin-like immunoreactive factors eliminated from serum samples by hydrophobic silica-gel extraction and enzyme immunoassay. Clin. Chem. *33*, 401–404.
40. Schlebusch, H., Jarausch, J. & Domke, I. (1992) Endogenous digoxin-like immunoreactive factors (DLIF) as measured by the CEDIA digoxin assay and a fluorescence polarization immunoassay. Wien. Klin. Wochenschr. (Suppl.) *191*, 59–66.
41. Longerich, L., Johnson, E. & Gault, M.H. (1993) Digoxin-like factors in herbal teas. Clin. Invest. Med. *16*, 210–218.
42. Lin, C.S., Lin, M.C., Chen, K.S., Ho, C.C., Tsai, S.R., Ho, C.S. & Shieh, W.H. (1989) A digoxin-like immunoreactive substance and atrioventricular block induced by a Chinese medicine „kyushin". Jpn. Circ. J. *53*, 1077–1080.
43. Fagoo, M., Braquet, P., Robin, J.P., Esanu, A. & Godfraind, T. (1986) Evidence that mammalian lignans show endogenous digitalis-like activities. Biochem. Biophys. Res. Commun. *134*, 1064–1070.

44. Lau, B.W. & Valdes, R. (1988) Criteria for identifying endogeneous compounds as digoxin-like immunoreactive factors in humans. Clin. Chim. Acta *175*, 67–77.
45. Howanitz, P.J. & Steindel, S.J. (1993) Digoxin therapeutic drug monitoring practices. A College of American Pathologists Q-Probes study of 666 institutions and 18, 679 toxic levels. Arch. Pathol. Lab. Med. *117*, 684–690.
46. Wieland, H. (1992) Practical aspects of monitoring cardiac drugs in the blood. Wien. Klin. Wochenschr. (Suppl.) *191*, 48–51.
47. Doolittle, M.H., Lincoln, K. & Graves, S.W. (1994) Unexplained increase in serum digoxin: A case report. Clin. Chem. *40*, 487–492.

3.12 Cyclosporins

D. W. Holt

3.12.1 Introduction

Cyclosporin A (ciclosporin, cyclosporin, cyclosporine, Sandimmun®) is an immunosuppressant drug used to prevent graft rejection following transplant surgery. It is used as a first-line treatment following kidney, liver, bone-marrow, heart, heart/lung and pancreas transplantation. Recently, it has been used in the treatment of a wide variety of autoimmune diseases including psoriasis, rheumatoid arthritis, primary cirrhosis, Crohn's disease and uveitis.

It is produced by the fungus *Tolypocladium inflatum* Gams, which was originally isolated from soil samples collected from the Hardanger Vidda in Norway (1). At least 25 naturally occurring cyclosporins are produced by this fungus; all are composed of 11 amino acids. Approximately 750 synthetic or semi-synthetic analogues have been produced and tested for immunosuppressive activity, in a search for a drug with a more acceptable profile of side-effects than cyclosporin A (2).

Cyclosporin A is a neutral, lipophilic, cyclic peptide with a molecular weight of 1203. Overall, it possesses the greatest pharmacological efficacy, compared with the natural and synthetic cyclosporins, when tested in *in vitro* or *in vivo* models of immunosuppression. The only other naturally occurring cyclosporin with immunosuppressive properties which are similar in potency to those of cyclosporin A is cyclosporin G. In this, latter, compound, norvaline replaces α-aminobutyric acid at amino acid position 2. Other naturally occurring cyclosporins which have been shown to exert a strong immunosuppressive effect *in vivo* are cyclosporins C, D and M and dihydrocyclosporin D (1).

Over the last 12 years cyclosporin A has had a profound effect on the development and results of organ transplantation. However, its use is associated with some serious adverse effects which limit its application in some clinical settings. In addition, it is not always easy to judge the pharmacological efficacy of the drug in patients. As a result of these problems many clinicians use measurements of the drug in blood as a guide to optimising therapy (3).

3.12.2 Measurement

Almost all the data on the measurement of cyclosporins relate to cyclosporin A and its metabolites. There are a number of methodological problems associated with the measurement of cyclosporin A, concerning the choice of both the sample matrix and the analytical technique. However, for routine clinical monitoring, there

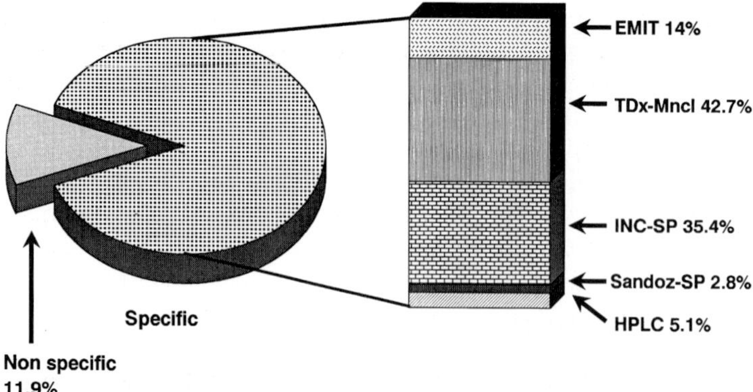

Figure 1. Choice of Assay Methodology Amongst Participants in the UK Cyclosporin Quality Assessment Scheme (August 1993).
HPLC = high-performance liquid chromatography, Sandoz-SP = Sandoz Sandimmun®-Kit radioimmunoassay with specific monoclonal antibody, INC-SP = INCSTAR CYCLO-Trac® SP radioimmunoassay with specific monoclonal antibody, TDx-Mncl = Abbott fluorescence polarization immunoassay with monoclonal antibody and EMIT = Syva enzyme multiplied immunoassay with specific monoclonal antibody.

is now good agreement amongst laboratory workers and clinicians on the use of blood as the sample matrix and the use of a methodology selective for the parent compound (4, 5). However, there is a disparity between the results given by a variety of immunoassay techniques and those produced by high-performance liquid chromatography (HPLC) (6–9).

Figure 1 illustrates recent data for the choice of assay specificity and the choice of assay method, amongst those laboratories using a methodology with a high specificity for cyclosporin A. In figure 2 the relative performance of these assays for the measurement of blood pools from kidney and liver transplant patients receiving cyclosporin A are displayed. The results given by HPLC are consistently lower than those given by the immunoassays. The reason for the discrepancy between these

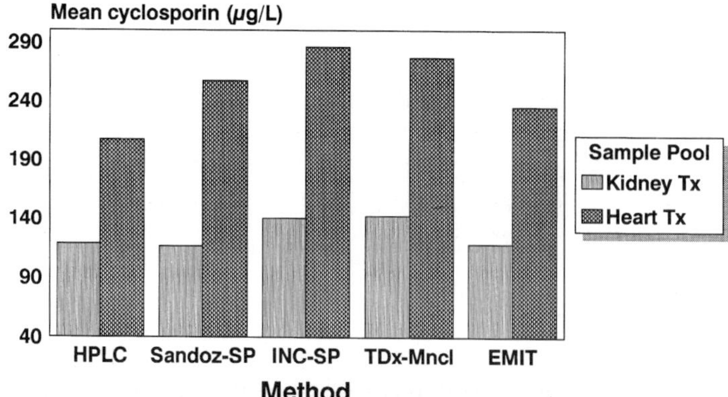

Figure 2. Mean Results for the Measurement of Cyclosporin A in Pooled Blood Samples from Renal Transplant (Tx) and Heart Transplant Patients.
Methods Key as for figure 1. Data are from the UK Cyclosporin QA Scheme.

assay techniques is not clear. The least selective antibody used in these immunoassay techniques is that used in the fluorescence polarization assay (TDx-Mncl), but the other monoclonal antibodies all have a very high specificity for the parent compound, with almost no cross-reactivity with the metabolites of the drug. Recent data suggest that the results for samples of known value are lower by HPLC compared with those produced by immunoassay kits. This difference may be due to such factors as drug recovery in the extraction systems used by the HPLC techniques, and this may contribute to the between-method differences for patient samples (8). In most clinical settings, these differences are not important for the management of patients receiving the drug, and the application of these measurements will be discussed below.

Cyclosporin A is extensively metabolised by the liver, predominantly by the hepatic cytochrome P-450 III A isoenzyme (10). At least 14 cyclosporin A metabolites have been characterised and other known metabolites await structural analysis (11). The original nomenclature for these metabolites is based on their elution order on a chromatographic system and, therefore, the numbering did not follow a systematic approach. A revised nomenclature, which includes identification of the amino acid(s) modified and the type of chemical modification, has been proposed, and is now in general use (12).

In man, there are three primary metabolites, formed by hydroxylation of amino acids 1 and 9, and N-demethylation of amino acid 4; in the conventional nomenclature they are known as M-17, M-1 and M-21, respectively (1). A diagrammatic representation of the formation of the principal metabolites is shown in figure 3a.

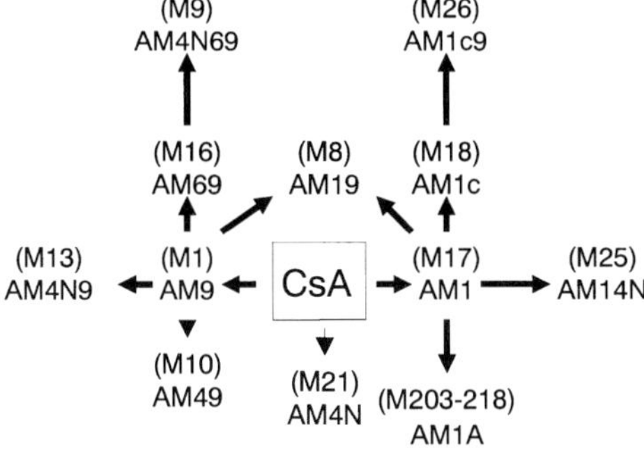

Figure 3a. Formation of Principal Metabolites of Cyclosporin A.
Old nomenclature is shown in parenthesis. New nomenclature follows the convention that hydroxylation is indicated by the number of the amino acid, an N following a number indicates the site of demethylation, c following a number indicates cyclization and A following a number the formation of a carboxylic acid.

Of the other cyclosporins only cyclosporin G and dihydrocyclosporin D have been used in formal clinical studies, but on a relatively small scale. As a result, there has been limited interest in the measurement of these drugs or their metabolites. For the measurement of cyclosporin G, both nonspecific and specific radioimmunoassays for cyclosporin A have been modified, by the use of cyclosporin G standards, and HPLC assays have been developed (2, 13, 14). Recently, the isolation and cha-

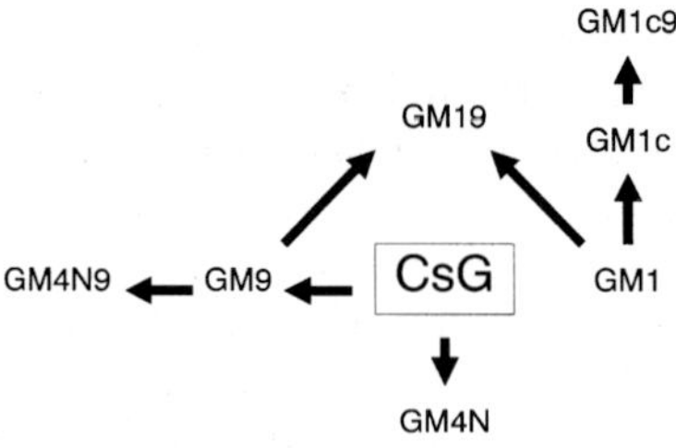

Figure 3b. Formation of the Metabolites of Cyclosporin G Identified to Date. Nomenclature as in figure 1a.

racterisation of cyclosporin G metabolites formed in man has been described (15). The formation of the metabolites identified to date is illustrated in figure 3b.

For studies aimed at resolving the metabolites of the cyclosporins and elucidating their pharmacological significance, the only methodology suitable is HPLC. Several published methods have sufficient power to effect a separation of the metabolites of cyclosporin A found in blood, urine or bile, but the chromatography run-times are so long as to preclude their application to routine monitoring of the drug in a large number of samples (16, 17). An example of the chromatographic separation of cyclosporin A which can be achieved by HPLC following sample extraction is shown in figure 4 (18).

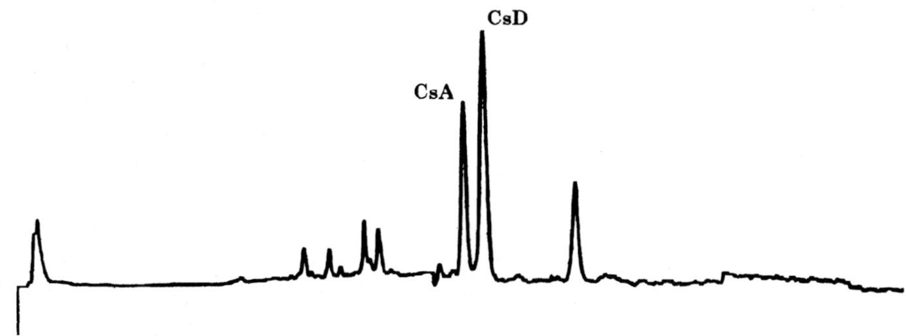

Figure 4. Separation of Cyclosporin A and Internal Standard Cyclosporin D, Using Sample Extraction and HPLC with Column Switching (18).
Cyclosporin A elutes at approximately 14 minutes and the cycle time between injections is approximately 22 minutes. The chromatogram was produced following the chromatography of an extract of a blood sample from a patient receiving cyclosporin; the concentration of cyclosporin in the sample was 150 µg/L. (Reproduced by kind permission of Dr. G. Schumann, Medizinische Hochschule, Hannover, Germany).

3.12.3 Clinical Problems

A limiting factor in the use of cyclosporin A is its profile of side-effects. The most commonly cited problem is the nephrotoxic effect of the drug. This manifests as multiple effects on the kidney (19). There is a proximal tubulopathy which results in reduced magnesium reabsorption and uric acid excretion. Low serum magnesium concentrations are common in cyclosporin A treated patients and this hypomagnesaemia has been shown to result in low endomyocardial tissue concentrations of the cation. The tubular effects are reversible on cessation of therapy. There are also vascular effects which produce vasoconstriction of renal afferent vessels resulting in increased serum creatinine and urea concentrations. These effects are also reversible. More serious is an afferent arteriolar vasculopathy. Damage may occur to the endothelial cells and smooth muscle cells leading to an irreversible occlusion and obliteration of the arteriole.

Some clues as to the mechanism of the acute renal vascular toxicity are given by the response to the drug pentoxifylline. This drug affects vascular endothelium by increasing production of vasodilator I and E prostaglandins and decreasing thromboxane A_2 production. The response to the drug is an increase in blood flow and thrombolysis. Bone marrow transplant patients experience a particularly high incidence of acute renal failure, associated with the concomitant use of both cyclosporin A and nephrotoxic antibiotic agents. One pilot study in five bone marrow transplant patients has shown that co-administration of pentoxyfylline with cyclosporin and amphotericin B resulted in a rapid fall in serum creatinine, not seen in a control group who did not receive pentoxifylline (20). Thus, the vascular effects of this, latter, drug may reverse those which limit the use of cyclosporin A.

Another endogenous compound implicated in the aetiology of cyclosporin A-induced nephrotoxicity is the powerful pressor molecule endothelin. Studies in rats have shown a marked rise in circulating endothelin concentrations associated with a fall in glomerular filtration rate, in animals receiving cyclosporin A (21). Similarly, the drug has been shown to induce an increase in renal endothelin receptor density, in association with a reduction in renal function (22). These data point to a possible involvement of endothelin in the increased renal vascular resistance observed in cyclosporin A-induced nephrotoxicity.

An additional clinically important adverse effect of cyclosporin A is the development of hypertension, which can be seen in the absence of renal abnormalites (23). The underlying mechanisms are largely undefined, although disruption of both endothelium dependent and independent dilatation has been suggested. A recent study has shown that, in man, the release of the endothelium derived relaxing factor – nitric oxide – is not impaired by cyclosporin A (24). Again, endothelin may be involved in the development of this adverse effect. A recent study has shown an association between cyclosporin A therapy, the onset of hypertension and circulating endothelin concentrations following liver transplantation (25).

Whilst the nephrotoxic effects can be shown to be dose related in animal models, knowledge of the daily dose in patients is a poor predictor of toxicity, since a number of factors may exacerbate the adverse effects of the drug. These include pre-operative and intra-operative ischaemia, other drug therapy and intercurrent illness. Added to this, the symptoms and signs of under dosage, leading to graft rejection, are not always easy to distinguish. As a result, measurements of the drug, principally in blood, have become a routine part of optimising therapy.

In renal transplant patients, a number of studies have demonstrated an association between low concentrations of the drug and graft rejection, whilst high concentrations have been associated with nephrotoxicity, especially during the early months after transplantation (26–29). There are similar findings in the other major transplant indications such as heart, liver and bone marrow. However, there is an overlap between the concentrations measured in each group, partly because the clinical end-points of rejection and toxicity are not always easy to define.

Between-subject differences in the pharmacokinetics of the drug may also play a part in the development of nephrotoxicity. There is some evidence that total exposure to the drug, as reflected by the area under the concentration-time curve (AUC), is related to the decline in renal function (19). Thus, some groups have advocated the use of profile concentration monitoring as a guide to therapy, or the use of a limited sampling strategy capable of reflecting the AUC (30–32). High peak blood concentrations of the drug may also be associated with subsequent poor renal function (33). In a group of paediatric patients who had received a liver transplant, it has been shown that dividing the daily dose into three doses, rather than two, lowered the peak concentration and improved the glomerular filtration rate (34). These examples provide a rationale for continued monitoring of the drug in an effort to avoid long-term renal damage.

Endogenous markers of renal dysfunction have been investigated, in an attempt to distinguish cyclosporin toxicity from kidney graft rejection, but they lack specificity. Hence, urinary neopterin concentrations rise during infection, as well as during rejection episodes (35), and the electrophoretic analysis of urinary proteins shows increased concentrations of proteins indicative of glomerular or tubular damage in almost all cyclosporin treated patients (36).

3.12.4 Activity of the Cyclosporins and Their Metabolites

A major point of contention regarding the cyclosporins, in particular cyclosporin A, has been whether the metabolites contribute significantly to the pharmacological or toxicological properties of the drug. Complicating factors in this debate have been the limited supply of pure metabolites, discrepancies between *in vitro* tests of immunological activity, and the lack of an exact animal model to reproduce the nephrotoxicity seen in man.

Signs of renal dysfunction similar to those seen in man can be induced in the spontaneously hypertensive rat (37). One study investigated both the activity and toxicity of cyclosporin A metabolites administered to this strain of rat (38). Synthesised metabolite 17, an extract of bile from patients receiving cyclosporin A or the pure drug, were administered intraperitoneally into groups of rats. A control group received the vehicle used for intravenous cyclosporin A. Only the pure drug caused a fall in ^{51}Cr-EDTA clearance (50%) following intraperitoneal administration and there were similar findings when the test substances were administered acutely by intravenous infusion. A limiting factor in this study was the failure to establish the kinetics of the test substances following intraperitoneal dosage, in particular the extent of absorption. Using four *in vitro* tests, the immunosuppressive activity of metabolite 17 was 4–10 fold less than the parent compound, whilst that of the bile extract was unmeasurable. These discrepancies between tests for *in vitro* activity of the metabolites are typical of other studies. However, it is generally considered that even those metabolites found in high concentrations in humans contribute little to the therapeutic effect of cyclosporin A (11).

In a more rigorous test of the potential toxicity of cyclosporin A metabolites, spontaneously hypertensive rats received pure synthesised metabolites M17, its oxidation product 203–218, M18 and M21; the route of administration was subcutaneous (39). Only a group of rats receiving cyclosporin A showed significant signs of renal or hepatic dysfunction, or histopathological alterations consistent with the drug's toxicity. Again, the study could be criticised in terms of the relative exposures of the animals to the drug or its metabolites, but it was concluded that the absorption of metabolites had been sufficient to demonstrate their lack of toxicity *in vivo*.

The pharmacological activity of the cyclosporins has been linked with the known binding of cyclosporin A to the cytosolic protein cyclophilin. This protein catalyses the *cis-trans* isomerization of peptidyl-proline bonds (40). It has been postulated that inhibition of this enzymatic activity by cyclosporin A could interfere with the correct folding of transcription factors and, thus, lymphocyte activation.

Initial experimental evidence seemed to confirm the central role of cyclophilin in the activity of cyclosporin A. When a limited range of cyclosporin analogues were studied, binding to cyclophilin was related to their immunological activity (41). It has also been shown that the major cytosolic binding protein for the macrolide immunosuppressant FK 506 has similar enzymatic activity to cyclophilin (42). Thus, cyclophilin is a major cytosolic receptor for cyclosporin A, can bind other cyclosporins and is responsible for the intracellular accumulation of the compound.

Whether cyclophilin is the central mediator of the immunosuppressive and toxicological properties of cyclosporin A has been questioned by the findings of a recent study (43). A series of 61 cyclosporin analogues were tested for their *in vitro* im-

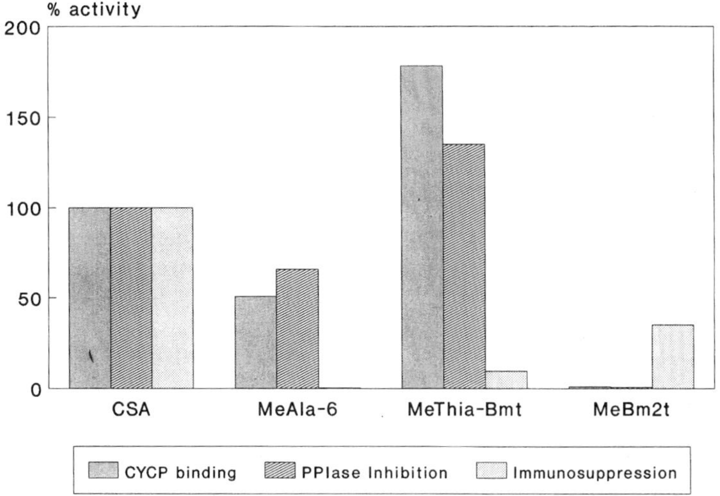

Figure 5. Cyclophilin (CYCP) Binding, Peptidyl-Prolyl Isomerase (PPIase) Inhibition and *in vitro* Immunosuppressive Effects of Three Cyclosporin Analogues in Comparison with Cyclosporin A. Data from reference 43.

munosuppressive properties, their ability to inhibit peptidyl-prolyl isomerase activity and their ability to induce renal pathology in an animal model. Although, in general, there was a good correlation between cyclophilin binding and immunosuppressive activity, there were some notable exceptions. Some analogues could interact with cyclophilin and inhibit enzymatic activity, but were less immunosuppressive than expected. Others bound little to cyclophilin, did not inhibit enzymatic activity, but did show immunosuppressive activity. These properties are illustrated in figure 5 for three analogues – MeAl-6, MeThia-Bmt and $MeBm_2t$, in comparison with cyclosporin A. It was also noted that, at concentrations of the analogue MeAl-6 which should have saturated cyclophilin, the compound did not demonstrate antagonism to the immunosuppressive properties of cyclosporin A.

In the animal model used, the analogues giving rise to nephrotoxic effects were also those with significant immunosuppressive activity. Toxicity was not related to the binding of an analogue to cyclophilin or its peptidyl-prolase inhibitory effects. It is also worth noting that the immunosuppressive agent rapamycin, structurally related to FK506, is able to bind to the FK-binding protein and inhibit its enzymatic activity. However, it acts as an antagonist of the immunosuppressive properties of FK506 (44). In turn, FK506 has been shown to antagonise the immunosuppression induced by cyclosporin (45). Thus, the relationship between the binding of the cyclosporins to cyclophilin, and the effects of these compounds on immunosuppression and renal pathology is less clear than the initial data on cyclophilin binding had suggested.

3.12.5 Other Biological Effects of the Cyclosporins

Initially, cyclosporin A was tested for its antiparasitic effects, properties shared with some other derivatives, including dihydrocyclosporin. Cyclosporin A also has some antifungal activity (1).

A pharmacological effect exciting considerable interest is the ability of cyclosporin A and some of its analogues to overcome multidrug-resistance. After prolonged treatment with anti-cancer drugs, tumour cells may become resistant to their effects. Multidrug-resistance is thought to be associated with the overexpression of a class of transmembrane glycoproteins – the P-glycoproteins – which lower the intracellular concentration of anti-cancer drugs below their minimum effective concentration, by actively pumping them from the cell.

Several drugs, including verapamil, erythromycin, chloroquine, quinidine and amiodarone, have been shown to decrease multidrug-resistance. Both cyclosporin A and cyclosporin G have been shown to reverse multidrug-resistance in an *in vitro* model, and were much more effective than either verapamil or amiodarone (46). Of particular interest is the observation that a nonimmunosuppressive cyclosporin analogue (SDZ PSC 833) is more than ten times as active as cyclosporin A in its ability to sensitise multidrug-resistent cells *in vitro* (47). These observations hold-out the possibility of increasing the efficacy of the anti-cancer therapy currently available.

3.12.6 Conclusions

Cyclosporin A has had an enormous impact on the development of solid organ transplantation and now seems set to make a major contribution to the treatment of a broad spectrum of autoimmune diseases (48). Its use is limited by some serious adverse effects which have stimulated an interest into an investigation of the immunosuppressive and toxiological properties of many compounds with a similar structure. To date, only one other cyclosporin has been studied in man, namely cyclosporin G. The results are equivocal in terms of its likely benefits over cyclosporin A.

Other cyclosporins are known to be in development but there are no data available on which to make any predictions of their place in immunosuppressive therapy. Several other compounds are also in development and are showing early promise as alternatives to, or adjuncts to, cyclosporin A (49–53). There seems little doubt that the result of this intense activity will be the discovery of a compound with an immunosuppressive potency equal to, or greater than, that of cyclosporin A, but with less toxicity.

In the meantime, clinicians strive to optimise therapy with a drug which has prolonged graft survival in many thousands of patients, using both routine clinical chemistry measurements and a diverse selection of assays designed to measure the drug in blood.

Addendum

Methodological issues relating to the measurement of cyclosporin A have been summarised in a recent review (54). There has been renewed interest in the clinical application of cyclosporin G in kidney transplantation; a multicentre study is underway in North America (55). Modifications of methods for the measurement of cyclosporin A to the measurement of cyclosporin G have been described (56). The major development in cyclosporin therapy is the modification of the oral formulation. A new formulation, *Sandimmun Neoral*®, has been developed which has markedly improved absorption characteristics compared with the original formulation (57, 58). Initial studies in renal transplant patients are impressive, and it is anticipated that the improved consistency of oral absorption will enhance the value of blood concentration monitoring (59).

References

1. Borel, J. (1989) The cyclosporins. Transplant. Proc. *21*, 810–815.
2. Jeffery, J. R. (1991) Cyclosporine analogues. Clin. Biochem. *24*, 15–21.
3. Holt, D. W., Fashola, T. O. A. & Johnston, A. (1991) Monitoring cyclosporin: is it still important? Immunology Letters *29*, 99–104.
4. Kahan, B. D., Shaw, L. M., Johnston, A., Grevel, J., Pitty, M.-H. & Holt, D. W. (1990) Consenus Document: Hawk's Cay Meeting on Therapeutic Drug Monitoring of Cyclosporine. Transplant. Proc. *22*, 1357–1361.
5. Shaw, L. M., Yatscoff, R. W., Bowers, L. D., Freeman, D. J., Jeffrey, J. R., Keown, P. A., McGilveray, I. J., Rosano, T. G. & Wong, P.-Y. (1990) Canadian Consensus Meeting on Cyclosporine: Report of the Consenus Panel. Clin. Chem. *36*, 1841–1846.
6. Johnston, A., Marsden, J. T. & Holt, D. W. (1989) The continuing need for quality assessment of cyclosporine measurement. Clin. Chem. *35*, 1309–1312.
7. Holt, D. W., Marsden, J. T. & Johnston, A. (1990) Quality assessment of cyclosporine

measurements: comparison of current methods. Transplant. Proc. *22*, 1234–1239.

8. Johnston, A., Cullen, G. & Holt, D.W. (1991) Quality assurance for cyclosporin assays in body fluids. Annals Acad. Med. Singapore *20*, 3–8.
9. Johnston, A., Holt, D.W. (1993) External assessment scheme for cyclosporin. Scand. J. Clin. Invest. *53*, (Suppl. 212) 48–53.
10. Combalert, J., Fabre, I., Dalet, I., Derancourt, J., Cano, J.P. & Maurel, P. (1989) Metabolism of cyclosporin A. IV. Purifiaction and identification of the rifampicin-inducible human liver cytochrome P450 (cyclosporin A oxidase) as a product of the P450IIIA gene family. Drug Metab. Dispos. *17*, 197–207.
11. Yatscoff, R.R., Rosano, T.G. & Bowers, L.D. (1991) The clinical significance of cyclosporine metabolites. Clin. Biochem. *24*, 23–35.
12. Kahan, B.D., Shaw, L.M., Holt, D.W., Grevel, J. & Johnston, A. (1990) A consensus document: Hawk's Cay meeting on therapeutic drug monitoring of cyclosporine. Clin. Chem. *36*, 1510–1516.
13. Wenk, M., Bindschedler, M., Costa, E., Zuber, M., Vozeh, S., Thiel, G., Abisch, E., Keller, H.P., Beveridge, T. & Folath, F. (1988) Pharmacokinetics of cyclosporine G in patients with renal failure. Transplantation *45*, 558–561.
14. McBride, J.H., Kim, S.S., Danovitch, G.M., Rodgerson, D.O., Reyes, A.F., Ota, M.K. (1993) Whole blood cyclosporin G in renal transplant recipients determined by two immunoassays and liquid chromatography. Clin. Chem. *39*, 1415–1419.
15. Copeland, K.R. & Yatscoff, R.W. (1991) The isolation, structural characterisation and immunosuppressive activity of cyclosporin G metabolites. Therapeutic Drug Monitoring *13*, 281–288.
16. Bowers, L.D. & Singh, J. (1987) A gradient HPLC method for the quantitation of cyclosporine and its metabolites in blood and bile. J. Liquid. Chromatogr. *10*, 411–420.
17. Christians, U., Schlitt, H.J., Bleck, J.S., Schiebel, H.M., Hownatzki, R., Maurer, G., Strohmeyer, S.S., Schottmann, R., Wonigeit, K., Pichlmayr, R. & Sewing, K.-F. (1988) Measurement of cyclosporine and 18 metabolites in blood, bile and urine by high-performance liquid chromatography. Transplant. Proc. *20*, (Suppl. 2) 609–613.
18. Schumann, G., Oellerich, M., Wonigeit, K. & Wrenger, M. (1986) Ciclosporin: HPLC routine method with column switching. Fresemius. Z. Anal. Chem. *324*, 328–329.
19. Mason, J. Renal side-effects of cyclosporine. (1990) Transplant Proc. *22*, 1280–1283.
20. Bianco, J.A., Almgren, J., Kern, D.L., Ballard, B., Roark, K., Andrews, F., Nemunaitis, J., Shields, T. & Singer, J.W. (1991) Evidence that oral pentoxifylline reverses acute renal dysfunction in bone marrow transplant recipients receiving amphotericin B and cyclosporine: results of a pilot study. Transplantation *51*, 925–927.
21. Kon, V., Sugiura, M., Inagami, T., Harvie, B.R., Ichikawa, I. & Hoover, R.L. (1900) Role of endothelin in cyclosporin-induced glomerular dysfunction. Kidney International *37*, 1487–1491.
22. Nambi, P., Pullen, M., Contino, L.C. & Brooks, D.P. (1990) Upregulation of renal endothelial receptors in rats with cyclosporine A-induced nephrotoxicity. Eur. J. Pharmacol. *187*, 113–116.
23. Bennett, W.M. & Porter, G.A. (1988) Cyclosporin-associated hypertension. Am. J. Med. *85*, 131–133.
24. O'Neil, G.S., Chester, A.H., Kushwaha, S., Rose, M., Tadjkarimi, S. & Yacoub, M.H. (1991) Cyclosporin treatment does not impair the release of nitric oxide in human coronary arteries. Br. Heart J. *66*, 212–216.
25. Lerman, A., Click, R.L., Narr, B.J., Weiner, R.H., Krom, A.F., Textor, S.C. & Burnett, J.C. (1991) Elevation of plasma endothelin associated with systemic hypertension in humans following orthotopic liver transplantation. Transplantation *51*, 646–650.
26. Moyer, T.P., Post, G.R., Sterioff, S. & Anderson, C.F. (1988) Cyclosporine nephrotoxicity is minimised by adjusting dosage on the basis of drug concentration in blood. Mayo Clin. Proc. *63*, 214–247.
27. Phillips, T.M., Karmi, S.A., Frantz, S.C. & Henriques, H.F. (1988) Absorption profiles of renal allograft recipients receiving oral doses of cyclosporine: a pharmacokinetic study. Transplant. Proc. *20* (2, Suppl. 2), 457–461.
28. Holt, D.W., Marsden, J.T., Johnston, A., Bewick, M. & Taube, D.H. (1986) The relationship between cyclosporin blood concentrations and renal allograft dysfunction. Br. Med. J. *293*, 1057–1059.
29. Lindholm, A., Dahlqvist, R., Groth, G.G. & Sjoqvist, F. (1990) A prospective study of cyclosporine concentration in relation to its therapeutic effect and toxicity after renal

transplantation. Br. J. Clin. Pharmacol. *30*, 443–452.

30. Grevel, J., Napoli, K.L., Gibbons, S., Kahan, B.D. (1990) Area-under the curve monitoring of cyclosporine therapy: performance of different assay methods and their target concentrations. Ther. Drug. Monit. *12*, 8–15.
31. Johnston, A., Sketris, I., Marsden, J.T., Galustian, C.G., Fashola, T., Taube, D., Pepper, J., Holt, D.W. (1990) A limited sampling strategy for the measurement of cyclosporine AUC. Transplant. Proc. *22*, 1345–1346.
32. Lindholm, A., Welsh, M., Rutzky, L., Kahan, B.D. (1993) The adverse impact of high cyclosporine clearance rates on the incidence of acute rejection and graft loss. Transplantation *55*, 985–993.
33. Fashola, T.O.A., Johnston, A., Counihan, P., Pepper, J.R. & Holt, D.W. (1991) Blood cyclosporin absorption profiles and renal function in heart transplant recipients. Br. J. Clin. Pharmacol. *31*, 236–237P (Abstract).
34. Paradis, K., O'Regan, S., Seidman, E., Laberge, J.-M., Dupuis, C., Gaudreault, P. & Rasquin-Weber, A. (1991) Improvement in true glomerular filtration rate after cyclosporine fractionation in pediatric liver transplant recipients. Transplantation *51*, 922–925.
35. Reibnegger, G., Aichberger, C., Fuchs, D., Hausen, A., Spielberger, M., Werner, E.R., Margreiter, R. & Wachtehr, H. (1991) Posttransplant neopterin excretion in renal allograft recipients – a reliable diagnostic aid for acute rejection and a predictive marker of lon-term graft survival. Transplantation *52*, 58–63.
36. Hossein-Nia, M., Devlin, J.J., Fagbohun, M., Tredger, J.M., Williams, R., Johston, A. & Holt, D.W. (1991) Urinary proteins as a marker of drug-induced renal damage following treatment with cyclosporine or FK-506. Transplant. Proc. *23*, 3150–3152.
37. Ryffel, B., Siegl, H., Petric, R., Muller, A.M., Hauser, R. & Mihatsch, M.J. (1986) Nephrotoxicity of cyclosporine in spontaneously hypertensive rats effects on blood pressure and vascular lesions. Clin. Nephrol. *25*, (Suppl. 1) 193–198.
38. Ryffel, B. (1988) Biologic significance of cyclosporine metabolites. Transplant. Proc. *20*, (2, Suppl. 1) 575–584.
39. Donatsch, P., Rickenbacher, U., Ryffel, B. & Brouillard, J.-F. (1990) Sandimmun metabolites: their potential to cause adverse reactions in the rat. Transplant. Proc. *22*, 1137–1140.
40. Takahashi, N., Hayano, T. & Suzuki, M. (1989) Peptidyl-prolyl cis-trans isomerase is the cyclosporin A binding protein cyclophilin. Nature *337*, 473–475.
41. Quesniaux, V.F., Schieier, J.M., Wenger, R.M., Hiestand, P.C., Harding, M.W. & Van Regenmortal, M.H.V. (1987) Cyclophilin binds to the region of cyclosporine involved in its immunosuppressive activity. Eur. J. Immunol. *17*, 1359–1365.
42. Harding, M.W., Galat, A., Uehling, D.E. & Schreiber, S.L. (1989) A receptor for the immunosuppressant FK506 is a cis-trans peptidyl-proplyl isomerase. Nature *341*, 758–760.
43. Sigal, N.H., Dumont, F., Durette, P., Siekierka, J.J., Peterson, L., Rich, D.H., Dunlap, B.E., Staruch, M.J., Melino, M.R., Koprak, S.L., Williams, D., Witzel, B. & Pisano, J.M. (1991) Is cyclophilin involved in the immunosuppressive and nephrotoxic mechanism of action of cyclosporin A? J. Exp. Med. *173*, 619–628.
44. Dumont, F.J., Melino, M.R., Staruch, M.J., Koprak, S.L., Fischer, P.A. & Sigal, N.H. The immunosuppressive macrolide FK506 and rapamycin act as reciprocal antagonists in murine T-cells. (1990) J. Immunol. *144*, 1418.
45. Vathsala, A., Goto, S., Yoshimura, N., Stepkowski, S., Chou, T.-C. & Kahan, B.D. (1991) The immunosuppressive antagonism of low doses of FK506 and cyclosporine. Transplantation *52*, 121–128.
46. Boesch, D., Gaveriaux, C. & Loor, F. (1991) Reversal of multidrug-resistance in CHO cells by cyclosporin A and other resistance modifying agents. J. Cell Pharmacol. *2*, 92–98.
47. Gaveriaux, C., Boesch, D., Jachez, B., Bollinger, P., Payne, T. & Loor, F. (1991) SDZ PSC 833, a non-immunosuppressive cyclosporin analog, is a very potent multidrug-resistance modifier. J. Cell. Pharmacol. *2*, 37–46.
48. Bach, J.F. (1992) Immunointervention in autoimmune diseases from cellular selectivity to autoantigen specificity. J. Autoimmunity *5*, (Suppl. A) 3–10.
49. Macleod, A.M. & Thomson, A.W. (1991) FK 506: an immunosuppressant for the 1990s? Lancet *337*, 25–27.
50. Stepkowski, S.M., Chen, H., Daloze, P. & Kahan, B.D. (1991) Rapamycin, a potent

immunosuppressive drug for vascularised heart, kidney, and small bowel transplantation in the rat. Transplantation *51*, 22–26.

51. Platz, K.P., Sollinger, H.W., Hullett, D.A., Eckhoff, D.E., Eugui, E.M. & Allison, A.C. (1991) RS-61443 – a new, potent, immunosuppressive agent. Transplantation *51*, 27–31.
52. Reichenspurner, H., Hilderbrandt, A., Human, P.A., Boehm, D.H., Rose, A.G., Odell, J.A., Reichart, B. & Schorlemmer, H.U. (1990) 15-deoxyspergualin for induction of graft nonreactivity after cardiac and renal allotransplantation in primates. Transplantation *50*, 181–185.
53. First, M.R. (1992) Transplantation in the nineties. Transplantation *53*, 1–11.
54. Holt, D.W., Johnston, A., Roberts, N.B., Tredger, J.M., Trull, A.K. Methodological and clinical aspects of cyclosporin monitoring. Report of the Association of Clinical Biochemists Task Force. *Annal. Clin. Biochem.* 1994; *31*: 420–446.
55. Henry, M.L., Tesi, R.J., Elkhammas, E.A., Ferguson, R.M. A randomized, prospective, double-blinded trial of cyclosporine vs OG 37–325 in cadaveric renal transplantation – a preliminary report. *Transportation* 1993; *55*: 748–52.
56. Yatscoff, R.W., Langman, L.J., LeGatt, D.F. Cross-reactivites of cyclosporin G (NVa^2 cyclosporin) and metabolites in cyclosporin A immunoassays. *Clinical Chemistry* 1993; *39*: 1089–92.
57. Holt, D.W., Mueller, E.A., Kovarik, J.M., van Bree, J.B., Kutz, K. The pharmacokinetics of Sandimmun Neoral: A new oral formulation of cyclosporine. *Transplant Proc* 1994; *26*: 2935–2939.
58. Holt, D.W., Johnston, A. Pharmacokinetics and monitoring of cyclosporin A. In: Immunosuppressive Drugs. Developments in antirejection therapy. Eds. Thomson A.W. and Starzl, T.E. Edward Arnold, London, 1994; 37–45.
59. Kovarik, J.M., Mueller, E.A., van Bree, J.B., Arns, W., Renner, E., Kutz, K. Withinday consistency in cyclosporine pharmacokinetics from a microemulsion formulation in renal transplant patients. *Ther Drug Monit* 1994; *16*: 232–237.

3.13 Anticoagulants

M. Geldmacher – v. Mallinckrodt

3.13.1 General Remarks

The coagulation of blood entails the formation of insoluble fibrin by the interaction of more than a dozen proteins in a cascading series of proteolytic reactions (fig. 1).

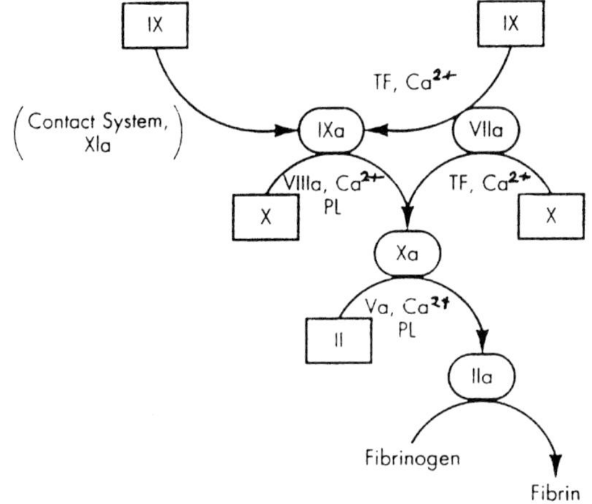

Figure 1. Major Reactions of Blood Coagulation (2).
Boxes enclose the coagulation factors and the ovals the active proteases. PL = platelets of phospholipids; TF = tissue factor; Va = activated factor V; VIIIa = activated factor VIII.

At each step a clotting factor (table 1) undergoes limited proteolysis and itself becomes an active protease. This clotting factor enzyme activates the next clotting factor, until ultimately an insoluble fibrin clot is formed (1, 2).

Two separate pathways lead to the formation of activated factor X and the activation of prothrombin (factor II). In the intrinsic system, all the protein factors necessary for coagulation are present in the circulating blood. In the extrinsic system, lipoproteins called tissue thromboplastin (factor III), which are not present in the circulating blood, activate blood coagulation at the level of factor X. Both pathways must be intact for adequate hemostasis (1, 2).

Blood coagulation can be inhibited in many ways, e.g. by removing Ca^{++} (factor IV), which plays an important role in the whole system, by complex formation with citric acid. In practice, there are only two groups of substances which are used

Table 1. Blood Clotting Factors

Factor	Common Synonyms
I	Fibrinogen
I′	Fibrin monomer
I″	Fibrin polymer
II	Prothrombin
III	Tissue thromboplastin
IV	Calcium^{2+}
V	Labile factor
VII	Proconvertin
VIII	Antihemophilic globulin, AHG
IX	Plasma thromboplastin antecedent, PTA
X	Stuart factor
XI	Plasma thromboplastin antecedent
XII	Hagemann factor
XIII	Fibrin-stabilizing factor
HMW-K	Hight-molecular-weight kininogen, Fitzgerald factor
Pre-K	Prekallikrein, Fletcher factor
Ka	Kallikrein
PL	Platelet phospholipid

therapeutically for inhibiting coagulation in vivo:
- Heparin and related compounds
- Coumarin derivatives.

3.13.2 Heparin

3.13.2.1 Chemical structure

Heparin is a complex linear polysaccharide of 60,000 to 100,000 daltons. It is strongly acidic because of its content of covalently linked sulfate and carboxylic acid groups. Since heparin from mammalian tissue sources is in limited supply, semisynthetic heparin has been prepared (2).

3.13.2.2 Uses and route of exposure

Heparin occurs intracellularly in tissues that contain mast cells. It cannot be detected in plasma under normal circumstances.

Low doses of heparin have been employed for primary prophylaxis in patients who are undergoing surgery to prevent postoperative venous thrombosis. Heparin in conventional dosis has been used as secondary prophylaxis to prevent the extension or recurrence of venous thrombi, thrombophlebitis, or pulmonary emboli. Heparinisation of the patient or the extracorporal device is applied in cardiovascular surgery and hemodialysis to prevent blood coagulation in heart-lung machines and dialyzers.

In the absence of a suitable chemical assay, standardization of a sample of heparin is based on comparison in vitro with a known standard in a non-specific assay of anticoagulant activity. The USP unit of heparin is the quantity that will prevent 1.0 ml of citrated sheep plasma from clotting for 1 hour after the addition of 0.2 ml of a diluted $CaCl_2$ solution.

Heparin must be administered parenterally. The therapeutical plasma range is 0.2–0.6 units/ml (3). A 10,000-unit bolus of heparin administered intravenously to a 70 kg patient results in an initial plasma concentration of heparin of about 3 units/ml, and anticoagulant activity disappears with a half-life of 1.5 hours.

The individual dosis of heparin depends on the aims of the therapy and the way of application. As an example, for continuous intravenous infusion the average-sized patient receives in 24 hours 24,000 units of heparin. This amount keeps a measure of blood clotting, the activated partial thromboplastin time (aPTT) about 1.5–2.0 times that of the patients pretreatment value. Therapy is routinely monitored by the aPTT (1, 2).

3.13.2.3 Metabolism and elimination

Heparin crosses membranes poorly because of its polarity and molecular size. It is thus not absorbed from the gastrointestinal tract, but has to be given parenterally. When given intravenously, the anticoagulant activity of heparin disappears from the blood by a dose dependent kinetic. The half-life in the plasma for usual doses lies between 1–3 hours. Heparin has a volume of distribution of 0.06 l/kg (3). Heparin seems to be cleared and degraded primarily by the reticuloendothelial system. The metabolites, mostly desulfurated uroheparin, are excreted in the urine; only a small amount of undegraded heparin appears in the urine after administration of large doses intravenously. In patients with renal failure or with hepatic cirrhosis, the half-life of the anticoagulant activity of heparin is significantly longer than in normal subjects (2).

Heparin does not cross the placenta and is not secreted in the breast milk (2).

3.13.2.4 Toxicology

3.13.2.4.1 Mechanism of action

The anticoagulant effect of heparin given intravenously is essentially immediate. Heparin acts indirectly by means of a plasma cofactor, antithrombin III. This cofactor inhibits several activated coagulation factors, including thrombin, Xa, IXa, XIa, XIIa and kallicrein. Inhibition of thrombin and Xa probably accounts for most of the anticoagulant effects of heparin. High doses of heparin can interfere with platelet aggregation (1, 2).

3.13.2.4.2 Toxicity

If commercial preparations of heparin are given, side effects are infrequent. Hypersensitivity reactions are known, including chills, fever, urticaria, or anaphylactic shock.

Increased loss of hair and reversible alopecia have been reported during heparin application. Osteoporosis and spontaneous fractures occurred in patients who have recieved 15,000 units or more of heparin for over three months (4). Heparin causes transient mild thrombocytopenia in 25 % of the patients. In a few patients severe thrombocytopenia has been observed.

The chief complication of therapy with heparin is hemorrhage.

Hemorrhage, severe thrombocytopenia and death have occurred even in patients recieving "low-dose" heparin therapy.

Reports on suicide or homicide with heparin are very rare. There is one report (3) on a 45 year old man who had injected himself subcutaneously 10 ampoules, each containing 25,000 units of heparin, and ingested four hundred 5 mg tablets of warfarin. The next day he was brought to the hospital with severe bleeding. The heparin overdose was reversed with protamine sulfate. He further got vitamin K and factors II, IX, and X because of the intake of warfarin.

Contraindications:
There is a rather long list of contraindications for heparin therapy, like consuming of large doses of ethanol and many diseases like hemophilia, purpura etc. (1, 2).

Interactions:
Drugs that affect platelet function (e.g. aspirin, nonsteroidal anti-inflammatory drugs, dipyridamole, dextran) as well as thrombolytic agents (streptokinase, urokinase) may increase the risk of hemorrhage (3).

Heparin Resistance:
Occasionally, patients' activated partial prothrombin time (aPTT) will not be prolonged to 1.5 times their normal mean unless very high doses of heparin (> 50.000 units per day) are administered. Some of these patients have very short aPTT values already prior to treatment due to an increased concentration of factor VIII.

Other reasons may be accelerated clearance of heparin, acquired antithrombin deficiency (less than 25 % of normal of this inhibitor in plasma) accompanying hepatic cirrhosis, nephrotic syndrome, or disseminated intravascular coagulation (2).

3.13.2.5 Management of intoxication

Mildly excessive anticoagulant effects of heparin are treated by discontinuation of the drug. If the effects are severe and bleeding occurs, intravenous administration of a specific antagonist may be indicated. Protamine is available for this purpose.

Protamines are proteins of low molecular weight which are strongly basic because of their high content of arginine. In vivo, protamine inhibits the anticoagulant effect of heparin. The requirement for protamine can be determined directly in vitro by "titration" of the patients blood with protamine (2, 3).

3.13.2.6 Relevant laboratory analyses

Anticoagulant therapy with heparin must be monitored. Methods for the determination of the heparin blood level or of heparin and its metabolites in urine are not available in most hospital laboratories.

The best parameter is the determination of the partial thromboplastin time (PTT) or the activated partial thromboplastin time (aPTT) which has replaced the whole blood clotting time used previously. Suitable methods can be found in the textbooks of Clinical Biochemistry. The therapeutic range is 1.5–2.0 times the reference value and depends on the method used.

If high doses of heparin are given, also the thrombocytes should be controlled.

3.13.3 Coumarin Derivatives

3.13.3.1 Chemical structure

The parent molecules from which the oral coagulant coumarins are derived are 4-hydroxycoumarin and indan-1,3-dione. The chemical structures for some anticoagulant coumarin derivatives are given in figure 2. Some like warfarin have an asymmetrical carbon atom in the substituent at the 3 position, and the available preparations of the drug are mixtures of the two optical isomers.

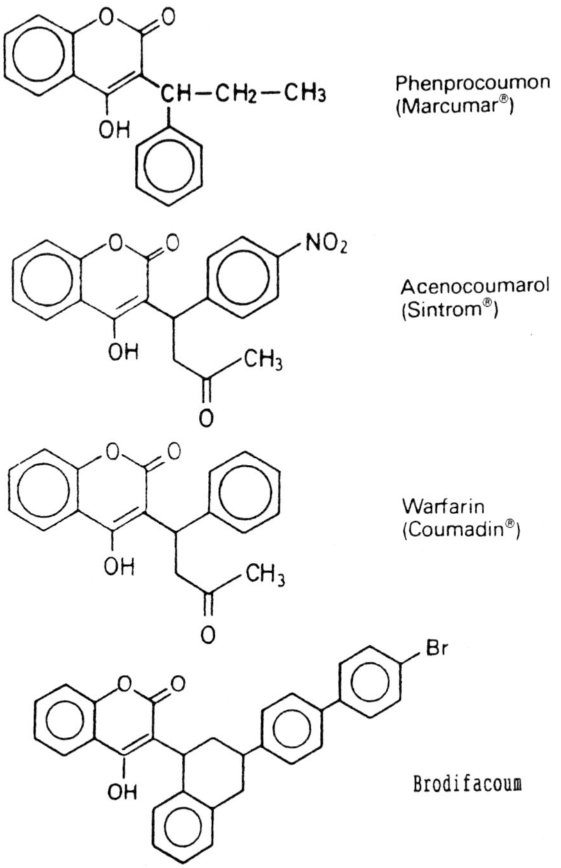

Phenprocoumon (Marcumar®)

Acenocoumarol (Sintrom®)

Warfarin (Coumadin®)

Brodifacoum

Figure 2. Chemical Structures of Some Coumarin Derivatives.

3.13.3.2 Uses and route of exposure

Coumarin derivatives are used as anticoagulant drugs in daily oral doses as given in table 2. A reduction of the prothrombin time to about 15–25 % of the normal range is desirable in most cases.

Table 2. Examples for Dosage and Duration of Effects for Some Coumarin Anticoagulants (8)

Substance	Dosage	Duration of effect (days)
Acecoumarol	1. day: 3–4 × 4 mg the following days: 1–2 × 4 mg per day	1–3
Phenprocoumon	1. day: 3–4 × 3 mg the following days: 1–2 × 3 mg per day	7–10
Warfarin	1. day: 3–4 × 5 mg the following days: 1–2 × 5 mg per day	3–5

The compounds are also used as rodent poisons (22) and are found in commercial animal baits. Their effectiveness depends on the fact that the action is cumulative. When the deficiency of the clotting factors becomes pronounced, the death of the animal occurs as a result of internal hemorrhage.

Industrial exposure is generally by inhalation or absorption through the skin. Oral anticoagulants have been used to malinger or create factitious disease, accomplish murder, commit suicide, induce abortion; they have also been dispensed in error or ingested by mistake.

Fatal intoxication (accidental, suicidal, homicidal) with coumarin derivatives in man is an infrequent event since a single dose is usually insufficient to cause significant depression of prothrombin time.

3.13.3.3 Metabolism and elimination

Warfarin is taken as an example. Warfarin is absorbed orally, by inhalation and via the skin. Racemic warfarin sodium is rapidly and completely absorbed in the gastrointestinal tract, and peak concentrations in plasma are reached within 1 hour after ingestion. The bioavailability of warfarin potassium is significantly less than that of warfarin sodium.

Racemic warfarin in the circulation is almost totally bound (99 %) to albumin during long term therapy, which largely prevents its diffusion into red blood cells, cerebrospinal fluid, urine and breast milk. The half-life of racemic warfarin administered by intravenous bolus injection is about 40 h.

In man, the dextrowarfarin enantiomorph is metabolized by side chain reduction to a secondary alcohol, whereas levowarfarin is metabolized by oxidation of the ring, primarily to 7-hydroxywarfarin. These metabolic products are to some extent conjugated with glucuronic acid, undergo an enterohepatic circulation, and are ultimately excreted in the urine and the feces. Urinary excretion accounted for 16–43 % of a single dose, apparently as the 7-hydroxy compound (1, 2, 5, 6, 7, 8).

3.13.3.4 Toxicology

3.13.3.4.1 Mechanism of action

The major pharmacological effect of anticoagulant coumarin derivatives is inhibition of blood clotting by interference with the hepatic posttranslational modification of the vitamin K-dependent clotting factors (II, VII, IX and X). These compounds are antagonists of vitamin K and are often called indirect anticoagulants because they act only in vivo.

The therapeutic effect is delayed for 8–12 h after oral or intravenous administration of racemic warfarin, because it results from an altered balance between the partially inhibited rates of modification and unaltered rates of degradation of the four proteins.

The only sigificant difference that exists in the inherent ability of various coumarin anticoagulants to produce and maintain hypoprothrombinemia is their half-life (c.f. table 2). Larger initial doses of these compounds hasten the onset of hypoprothromibinemia only to a certain limited extent.

Oral anticoagulants pass the placenta. Birth defects and abortion may occur (1, 2, 3).

3.13.3.4.2 Toxicity

Therapeutic doses of warfarin decrease the total amount of each vitamin K dependent coagulation factor made by the liver by 30–40 %. As a consequence, blood clotting activity is reduced, which leads to internal hemorrhage in overdose. Therapeutic doses see table 2.

The MAK value (Maximum Concentration Value in the Workplace) 1995 for warfarin is 0.5 mg/m^3 (9).

Inhalation or dermal absorption of warfarin was believed responsible for multiple acute hemorrhagic episodes in an woman who frequently prepared rat poison for domestic purposes (10).

An amount of 567 mg warfarin taken as a suicidal gesture over a six-day period resulted in a state of intoxication that was sucessfully treated by vitamin K administration. A death due to warfarin was reported that resulted from the ingestion of 1 g of warfarin over a 13-day period (10). Further case reports see (3).

Interactions:
A wide range of drugs interact with vitamin K antagonists through a variety of mechanisms to enhance or to reduce their anticoagulant effect.

Potentiation interactions e. g. by aspirin or salicylates lead to episodes of bleeding which may be fatal. Other drugs that increase the chance of bleeding are disulfiram, phenylbutazone and other related anti-inflammatory drugs, uricosuric agents and the acute intake of ethanol (1, 2, 11).

Drugs which stimulate the activity of liver microsomal enzyme systems increase the metabolism of vitamin K antagonists and decrease the response to oral anticoagulants. Such drugs include barbiturates and some other sedative-hypnotics like gluthetimide, ethchlorvynol, meprobamate and griseofulvin (1,2).

Anticoagulant resistance:
An inherited resistance to anticoagulants of the coumarin and indanedione groups has been detected in rodents. In colonies of wild rats, the resistance appears to have arisen by mutation.

The first human case of warfarin-resistance was a man who required twenty times the usual dose to produce an adequate anticoagulant effect. Analysis of the family history indicated that an autosomal dominant gene is responsible for the trait. The metabolism of the drug is normal, but the requirement for vitamin K is markedly increased in both species (2, 12).

Superwarfarins:
Because of the emergence of warfarin resistance, new potent long-acting anticoagulants (e.g. brodifacoum) have been synthezised and are now commercially available. This has led to accidental and purposeful human ingestions, some of them fatal (13).

3.13.3.4.3 Toxic effects

Hemorrhage is the main unwanted effect caused by therapy with oral anticoagulants. Anticoagulant therapy must always be monitored by determination of the prothrombin time (Quick time), and the patient must be observed carefully for development of bleeding. In order of decreasing frequency, complications include ecchymoses, hematuria, uterine bleeding, melena or hematochezia, epistaxis, gingival bleeding, hemoptysis, and hematemesis. Particularly serious bleeding episodes include compression neuropathy following brachial artery puncture for arteriographic or blood-gas studies, intraperitoneal hemorrhage from rupture of corpus luteum, retroperitoneal hemorrhage with compression femoral neuropathy, hemopericardium even in the absence of myocardial infarction or pericarditis, intracranial hemorrhage, adrenal hemorrhage, and necrosis of skin and breast. After uptake of higher than therapeutic doses, the same symptoms, which may be fatal, will occur (1, 2).

3.13.3.5. Management of intoxication

In the case of an acute oral intake of an overdose, gastric lavage should be performed.

If bleeding occurs, administration of the specific antagonist, vitamin K, is indicated. Vitamin K must be applied until the prothrombin time is approximately in the normal range. In cases of intoxication with "superwarfarins" this may be necessary for weeks (13). The effect of vitamin K administration, i.e. a sufficient improvement of prothrombin time, takes 6–12 hours, a complete regeneration of the vitamin K dependent clotting factors 36–48 hours, since it is achieved by de-novo synthesis of prothrombin.

In cases with life threatening bleedings, a prothrombin-complex concentrate containing the vitamin K depending clotting factors (II, VII, IX, X) must be applied as soon as possible. It shows an immediate effect. The application of blood transfusions, plasma or "fresh frozen plasma" is also recommended (2, 3).

3.13.3.6 Relevant laboratory analyses

3.13.3.6.1 Biochemical analyses

The therapy with oral anticoagulant coumarin derivatives must be monitored by the prothrombin time, which is also an essential laboratory evaluation of anticoagulant activity in suspected oral anticoagulant poisoning.

This is a parameter for which in most hospital laboratories routine methods are available (see textbooks of Clinical Biochemistry). Since results can differ, depending on the different thromboplastin preparations used, it is mandatory to carry out quality control with commercially available control sera.

3.13.3.6.2 Toxicological analyses

For the qualitative identification and quantitative determination of coumarin derivatives in biological materials only advanced methods can be recommended.

Most coumarin compounds can be extracted from an acidified medium with ether. Problems can arise because of the high protein binding in serum or plasma.

Some therapeutic and toxic concentrations in serum are given in table 3.

Table 3. Therapeutic and Toxic Concentrations of Some Coumarin Derivatives in Serum

Substance	Therapeutic concentration (mg/l)		Toxic symptoms (mg/l)	
Acecoumarol				
min*	0.03–0.09	(20)	0.1–0.15*	(20)
max**	0.1 –0.5	(20)		
Brocifacoum	–		0.03	(21)
			0.167	(19)
			0.786	(13)
Dicoumarol	8–30	(20)	40–50	(20)
Phenprocoumon	1–3	(20)	> 5	(20)
	0.6–5.0	(17)	16–20	(14)
Warfarin	1–7	(20)	> 10	(20)

* concentration before another therapeutic dose has been given
** peak concentration

a) Gas chromatography

It has to be considered that many of the coumarin compounds contain no nitrogen. Therefore detection should be done by a FID instead of a NPD (14).

Retention indices for many coumarin derivatives and their metabolites on OV–1 or SE 30 are given in (14).

Details on a method for the quantitative determination of phenprocoumon by gas chromatography have been published by (15):

The qualitative identification had been carried out with GC/MS after derivatisation with propionic acid anhydride (16).

The quantitative determination was performed by gaschromatography after derivatisation with propionic acid anhydride, using benzbromaron as an internal standard. Phenprocoumon had been extracted with dichloroethane from an acidified plasma sample.

b) HPLC

High performance liquid chromatography, HPLC, is the method of choice for the quantitative determination of coumarin derivatives and their metabolites in serum or plasma, e.g. for phenprocoumon (17) or warfarin (18).

For the simultaneous determination of eight anticoagulant rodenticides (brodifacoum, bromadiolone, chlorofacicone, coumafuryl, coumatetralyl, diphacinone, difenacoum,warfarin) in serum, a liquid chromatographic method is available (19).

The anticoagulants were extracted from serum with acetonitrile. Extracts were applied to solid phase extraction columns, which contained mixed packings, and were subjected to reversed-phase liquid chromatography. Hydroxycoumarins were detected by fluorescence at an excitation wavelength of 318 nm and emission wavelength of 390 nm. Indandiones were detected by UV absorption at 285 nm.

Recovery was > 69 %. The within-run precision (CV) ranged from 2.4 to 8.6 %, the between-run precision (CV) from 1.5 to 12.2 %. Hydroxycoumarin rodenticides were detected at 1 ng/ml, indandiones at 10 ng/ml.

References

1. O'Reilly, R.A. (1985) Anticoagulant, antithrombotic, and thrombolytic drugs. In: Goodman and Gilman The pharmacological basis of therapeutics (Goodman Gilman, A., Goodman, L.S., Rall, T.W., Murad, F., eds.) pp 1338–1359, Macmillan, New York, Toronto, London.
2. Majerus, P.W., Broze, G.J., Miletich, J.P. & Tollefsen, D.M. (1992) Anticoagulant, thrombolytic, and antiplatelet drugs. In: Goodman & Gilman The pharmacological basis of therapeutics (Goodman Gilman, A., Rall, T.W., Nies, A.S., Taylor, P., eds) pp 1311–1331, McGraw-Hill, New York, St. Louis, San Francisco.
3. Ellenhorn, M.J. & Barceloux, D.G. (1988) Medical toxicology: Diagnosis and treatment of human poisoning. Elsevier, New York, pp 209–228.
4. Megard, M., Cuche, M., Grapeloux, A., Bojoly, C. & Meunier, P.J. (1982) Ostéoporose de héparinothérapie: Analyse histomorphométrique de la biopsie osseuse. Une observation. Nouv. Presse Méd. *11*, 261–264.
5. Hackett, L.P., Ilett, K.F. & Chester, A. (1985) Plasma warfarin concentrations after a massive overdose. Med. J. Aust. *142*, 642–643.
6. O'Reilly, R.A. (1987) Warfarin metabolism and drug-drug interactions. Adv. Exp. Med. Biol. *214*, 205–212.
7. Egan, D., O'Kennedy, R., Moran, E., Cox, D., Prosser, E. & Thornes, R.D. (1990) The pharmacology, metabolism, analysis, and applications of coumarin and coumarin-related compounds. Drug Metab. Rev. *22*, 503–529.
8. Forth, W., Henschler, D. & Rummel, W. (1987) Allgemeine und spezielle Pharmakologie und Toxikologie. 5. ed. BI Wissenschaftsverlag Mannheim, Wien, Zürich.
9. Deutsche Forschungsgemeinschaft (1995) List of MAK and BAT values. VCH Verlagsgesellschaft Weinheim.
10. Baselt, R.C. & Cravey, B.S. (1989) Disposition of toxic drugs and chemicals in man. 3.ed. Year Book Medical Publishers, Chicago, London, Boca Raton, Littleton, Mass.

11. Hirsh, J. (1991) Oral anticoagulant drugs. N. Engl. J. Med. 1865–1875.
12. O'Reilly, R. A. (1971) Vitamin K in hereditary resistance to oral anticoagulant drugs. Am. J. Physiol. *221*, 1327–1330.
13. Rough, C. R., Triplett, D. A., Murphy, M. J., Felice, L. J., Sadowski, J. A. & Bovill, E. G. T. (1991) Superwarfarin ingestion and detection. Am. J. Hematol. *36*, 50–54.
14. DFG/TIAFT (Deutsche Forschungsgemeinschaft/The International Association of Forensic Toxicologists) (1992) Gas chromatographic retention indices of toxicologically relevant substances on packed or capillary columns with dimethylsilicone stationary phases. VCH Verlagsgesellschaft Weinheim.
15. Meyer, von, L., Kauert, G. & Drasch, G. (1982) Phenprocoumon-Missbrauch. Deutsche Apotheker Z. *122*, 179–181.
16. Meyer, von, L., Kauert, G. & Drasch, G. (1980) The detection of coumarin anticoagulants in biological fluids by Gas Liquid-Mass Spectrometry. in: Forensic Toxicology (Oliver, J. S., ed.) Croom Helm, London, pp 245–251.
17. Petersen, D., Barthels, M., Schumann, C. & Büttner, J. (1993) Concentrations of Phenprocoumon in serum and serum water determined by high performance liquid chromatography in patients on oral anticoagulant therapy. Haemostasis *23*, 83–90.
18. Ueland, P. M., Kvalheim, G. & Lonning, P. E. (1985) Determination of warfarin in human plasma by high performance liquid chromatography and photoiodide array detector. Ther. Drug Monitor. *7*, 329–335.
19. Chalermchaikit, T., Felice, L. J. & Murphy, M. J. (1993) Simultaneous determination of eight anticoagulant rodenticides in blood, serum, and liver. J. Anal Toxicol. *17*, 56–61.
20. Uges, D. R. A. (1990) Orientierende Angaben zu therapeutischen und toxischen Konzentrationen von Arzneimitteln und Giften in Blut, Serum oder Urin. VCH Verlagsgesellschaft, Weinheim.
21. Helmuth, R. A., McCloskey, D. W., Doedens, D. J. & Hawley, D. A. (1989) Fatal ingestion of a brodifacoum-containing rodenticide. Lab. Med. *20*, 25–27.
22. Pelfrem, A. F. (1991) Anti-Vitamin K compounds in: Handbook on pesticide toxicology (Hayes, W. J. jr & Laws, E. R., eds.), Academic Press, San Diego, New York, Boston.

3.14 Alkaloids

Y. Kuroiwa and T. Yoshida

3.14.1 Introduction

Plants are the most economically viable sources of many commercially important chemicals produced as secondary metabolites, including many pharmaceuticals, especially alkaloidal compounds. The term alkaloid "alkali-like" is thus applied to the large group of basic nitrogen-containing naturally occurring compounds of botanical origin. However, some alkaloids, such as ricinine, obtainable from the seeds of *Ricinus communis*, and colchicine, obtainable from *Colchicum autamnale*, are rather neutral compounds. In addition, with respect to the origins of alkaloid, bufotenine can be isolated from the plant *Piptadenia*, the mushroom *Amanita* and also from the toad *Bufo vulgaris* and related species. Likewise, many natural products corresponding to the alkaloids have also been isolated from animals and other sources, such as extremely toxic steroid alkaloids, batrachotoxins from *Phyllobates* and samandarine from *Samandra*, and homobatrachotoxin from the birds *Pitohui* (1). Even in mammals, there are some published papers indicating that morphine and codeine are endogenously found and synthesized *in vivo* (2–4).

The plants containing alkaloids are distributed throughout the world. Of the plants, *Solanaceae, Rubiaceae, Legminosae, Papaveraceae, Menispermaceae, Berberidaceae, Ranunculaceae* and *Lauraceae* are generally rich in alkaloid contents. The characteristics of distribution of alkaloids is such that nicotine is distributed in a wide variety of plants, and that alkaloidal plants generally contain various alkaloids along with their respective biosynthetic pathways. Prototype alkaloids, such as morphine in opium, quinine in *Cinchona* and nicotine in tabacco characterize the respective plant.

The nitrogen atom of alkaloids is generally present in heterocyclic rings, such as in pyrrole, pyridine, pyrrolidine, quinoline and isoquinoline, but also is found in alipathic side chains, i.e. in ephedrine.

The alkaloids have been classified in groups according to their chemical skeleton (such as indole, isoquinoline, piperidine, etc.). In some instances, however, alkaloids of the same botanical origin (such as aconite alkaloids, opium alkaloids and ipecacuanha alkaloids) and alkaloids of importance from a special point of view (drugs of abuse) have been treated separately.

The toxic properties of some plants are well known, but those of many are relatively obscure and difficult to ascertain. Plants containing toxic alkaloids are of concern not only to humans but also to animals, although exposure to such plants are rare. Serious plant intoxications occur through experimenting, food-gathering purposes, in tasting and ingesting unknown plants or plants morphologically similar to known edible wild plants. This sometimes fatal mistake seems to occur repeatedly with plants which are represented not only by the common carrot but also the water hemlock. A great many types of plants contain alkaloids that cause anticholinergic symptoms in human beings. Most of these are found in the family *Solanaceae*, which includes the genera *Atropa, Datura, Hyoscyanamus, Lycium* and *Solanum*. The principal alkaloids in these plants include solanine, atropine and scopolamine. The alkaloid content of each species and each plant varies greatly and depends on such parameters as the time of year, the available moisture and the temperature. For this reason it is very difficult to predict toxicity in relation to the amount and the origin of plant material. Recreational abusers of mushrooms and *Datura stramonium* (jimson weed), for example, are unable to titrate the dose of ingested substance because of this tremendous biological variability and are therefore prone to severe anticholinergic poisonings.

To date, over five thousand alkaloids have been isolated and structurally determined. Many modern drugs are either constituents of plants or modifications of known plant chemical products. The alkaloids in general produce strong physiological activities with relatively small doses on animal organisms, and many of them are poisonous. Thus alkaloids and their synthetic derivatives are of considerable importance in pharmacology, medicine and toxicology. With respect to the high content and to the specific pharmacological activity, a plant's primary alkaloid is referred to as the "main alkaloid" and other alkaloids in the same plant as "by-alkaloids". Since many alkaloids have specific pharmacological activities, they offer a clue to develop new drugs. Even for alkaloids which are not utilizable due to high toxicity, the chemical structure is often important for drug-design as a model compound. Such chemical modification of the original alkaloid leads to the development of a new drug. Therefore, to date many alkaloids and their derivatives have been utilized as pharmaceuticals such as narcotic analgesics, antitussives, CNS stimulants, local anesthetics, anti-cancer drugs, anti-malarials, and so forth. Based on such important roles of alkaloids in the treatments of human diseases, great efforts have continuously been paid to finding novel effective compounds from plants and to develop *in vitro* culture techniques and systems of plants to provide for good medicinal supplies.

In turn, alkaloids are the most important class of compound in toxicology because of their wide distribution in plants, pharmaceuticals, agrochemicals, and so on, and of many cases of accidental, homicidal and suicidal poisoning.

Only a limited number of these alkaloids have been subjected to toxicological analysis. Specifically, the alkaloids used as medicines, agrochemicals, narcotics and related compounds, and highly toxic alkaloids, such as aconitine, and hallucinogenic alkaloids, are of great importance in toxicology.

Drugs with sedative, mood-altering and analgesic properties similar to morphine are termed narcotics. This term has also been applied for legislative purposes to any drug that produces psychological and physical craving. Pharmacologically diverse drugs such as LSD, cocaine and cannabis have been regulated under narcotic control laws even though they are not derived from opium. There are many narcotics and stimulants distributed in the world, which are either naturally occurring, semi-synthesized or totally synthesized. Narcotics and other dependence-forming compounds have been nearly uniformly regulated throughout the world according to the recommendations of the WHO Expert Committee on Drug Dependence. Yet, clandestine laboratories continue to produce and distribute narcotics and stimulants worldwide.

Knowledge of alkaloid chemistry and of the biosynthetic pathways of alkaloids has been accumulated extensively to date. Additionally, the analytical methods for alkaloids up to 1990 have been compiled and edited as books (5–8). Analytical methods and other subjects for individual alkaloids can also be found in a book edited by the Pharmaceutical Society of Great Britain (9).

This chapter will therefore describe mainly recent progress in toxicology and toxicological analysis of alkaloids except cocain and lysergic acid and their related compounds which will be presented in other chapters in this book.

The alkaloids generally subjected to toxicology and toxicological analysis are classified in table 1.

Table 1. A Classification of Alkaloids and Related Compounds

A. aromatic amines	E. opiates (papaverine group)	H. indole ring systems
amphetamine	cotarnine	aspidopermine
capsaicine	cupaverine	brucine
cathine	narceine	physostigmine
ephedrine	narcotine	strychnine
hordenine	papaverine	yohimbine
		yohimbine
B. pyrrol-pyridine	F. opiates (morphine group)	
alkaloids	codeine	I. ergot alkaloids
apoatropine	dihydrocodeine	dihydroergocristine
atropine	ethylmorphine	
cocaine		dihydroergotamine
homatropine	hydrocodone	ergocristine
nicotine	hydromorphone	ergocritinine
pseudococaine	morphine	ergometrinine
scopolamine	oxycodone	ergonovine
scopoline	thebaine	
sparteine		ergotamine
tropacocaine	G. rauwolfia alkaloids	ergotaminine
	ajmaline	
C. quinoline ring	neoajmaline	J. miscellaneous
systems	raubasine	alkaloids
cinchonine	rauwolscine	aconitine
cupreine		colchicine
quinidine	reserpine	gelsemine
quinine	reserpinine	pilocarpine
	sarpagine	solanine
D. isoquinoline ring systems	serpentine	veratridine
apomorphine	serpentinine	veratrine
berberine		psilocybin
boldine		psilocin
bulbocapnine		
cephaeline		
emetine		
hydrastinine		

3.14.2 Isolation and Identification of Alkaloids

In general, alkaloids are soluble in acids, ether, chloroform, alcohol and other organic solvents. Most of the alkaloids are crystallizable with the exception of nicotine. They form salts with acids and dissolve in the sap of the plant. Of the acids, acetic acid, citric acid, malic acid and oxalic acid are common, although some specific acids are found to form salts with alkaloids in plants, specifically quinic acid in *Cinchona succirubra*, aconitic acid in *Aconitum* and meconic acid in *Papaverum*. The presence of these acids is thus important for identifying the crude materials in toxicology, especially in the case of opium.

Because alkaloid research has been an important field for centuries, numerous papers, reviews and books have been published to date (4–9). In this chapter, therefore, we simply describe the isolation, identification, and qualitative and quantitative methods for alkaloid analysis.

3.14.2.1 Isolation of alkaloids

An important goal of alkaloid analysis is the isolation of the alkaloids present in the material to be investigated in their genuine form while avoiding artefacts.

The removal of non-alkaloidal compounds during the isolation and purification of the alkaloids is a great problem. Specifically, such problems are encountered in pharmacological and toxicological investigations, where the alkaloids are usually found as salts in complex mixtures of water-soluble compounds of all kinds of lipids.

The extraction and isolation of alkaloids can be carried out in various ways depending on the nature of the alkaloids in question and the material in which they are found. The prolonged contact of alkaloids with strong mineral bases may lead to alterations in many alkaloids, ammonia is therefore commonly preferred for their isolation. Ammonia is sufficiently basic to liberate most of the common alkaloids without much risk of undesirable reactions, and is volatile enough for subsequent removal.

During the extraction, isolation and analysis of alkaloids, it should be borne in mind that the alkaloid stability is sensitive to light, pH and even to solvents. The commonly used organic solvents in alkaloid research may produce artefacts, such as photochemical decomposition in chloroform solutions, peroxidation by with peroxide contaminated ethers, reaction with ketones, etc. Thus, great care should be paid to the choice of solvent.

Heat will destroy morphine, cocaine and aconitine, and thus caution is necessary during isolation. Urine is the specimen of choice for detection. Add 1 drop of diluted hydrochloric acid prior to evaporation of solvent to prevent loss of the alkaloids.

The primary purpose of sample preparation is to provide a sample free of contaminants that may interfere with analysis, damage the analytical instruments, or affect resolution and reproducibility of chromatographic columns. Therefore, the choice of methods and extents of sample preparation may be dependent on subsequent analytical methods. Specifically, toxicological sample types to be analyzed for alkaloids are widely divergent and may include plants and other biological samples, especially body fluids (urine, plasma, serum, blood), tissues, organs, feces, gastric gavages, hairs, etc. Thus, different approaches to sample preparation are required. Additionally, recently developed sophisticated analytical instrumentation also calls for utmost care in sample preparation to obtain adequate and good recovery in toxicological analysis.

For the toxicological analysis of alkaloids, the so-called Stas and Otto method of solvent extraction is the usual method of choice.

This original method has been widely used, adapted and modified by various investigators. The solvent extraction method for alkaloids is still routinely used today, but it requires experience and training in order to obtain good results.

Briefly, the alkaloids must be extracted at an optimum alkaline pH value. Especially alkaloids having amphoteric character, such as morphine, should be extracted at an optimum pH of $8.5 \sim 9.0$ into chlorinated solvent. Other alkaloids are normally extracted at an alkaline pH with organic solvents.

However, optimum extraction conditions for alkaloids must be determined by trial and error to give maximum recoveries from each specimen. Additionally, for urine specimens, acid or enzymatic hydrolysis prior to extraction is necessary to cleave the conjugated metabolite(s).

Recently, various solvent-solid extraction procedures have been developed. Many disposable extraction cartridge columns with various structured resins have been made commercially available and applied to alkaloid research and toxicological analysis. These include XAD-2, XAD-4, Exrelut, Clin Elut, Sep-Pak C-18, Bond Elut, Bio-Beads SX-2 and SX-3, and so forth. These methods have the advantages of simplicity, rapidness, high recovery and reproducibility. However, the limited retention volume and number of samples are relatively disadvantageous.

3.14.2.2 General qualitative analysis of alkaloids

3.14.2.2.1 Sedimentation reaction

There are several specific sedimentation reagents for qualitative analysis of alkaloids as listed in Table 2. Almost all of the alkaloids react with these sedimentation reagents and form sediments. Detection limits of the alkaloids with these sedimentation reagents generally range from 1 to 10 μg. Of these sedimentation reagents, Wagner and Meyer reagents are relatively sensitive to the alkaloids.

Table 2. Some Typical Sedimentation Reagents for Alkaloids

Wagner reagent	Tannic acid reagent
Warme reagent	Dragendorff reagent
Meyer reagent	Sonnenschein reagent
Sheibler reagent	Picric acid

In addition to these reagents, various sedimentation reactions of the alkaloids have been utilized. These reactions include: sediment formation of free bases reacting with basic reagents; binding with oxidative acids; halogenations; complex formation with metallic compounds; binding with acidic organic compounds.

With careful technique and controlled comparison with known alkaloids, an unknown alkaloid can be identified by microscopic examination of typical crystals as to colour, structure, and general appearance.

3.14.2.2.2 Colour reaction

Colour reactions with various reagents are also very useful for detecting alkaloids; however, in contrast to the sedimentation reactions, colour reactions are not specific and general tests for alkaloids. Some alkaloids can react with almost all of the colour reagents, but cocaine and caffeine do not react with any of them. Detection limits of the alkaloids with the colour reagents generally range from 1 to 20 μg when heating treatment is employed in the reactions. Some colour reactions are fairly sensitive to specific alkaloids, such as the Vitali reaction for atropine and the murexide reaction for caffeine and other purine bases. Table 3 represents some typical colour reagents utilized for the detection of alkaloids.

Combinations of various colour reagents and comparison of the observed colours with known alkaloids are very useful for identification.

The colour reagents have been also used as spray reagents after thin layer chromatography.

Table 3. Some Typical Colour Reagents for Alkaloids

Marquis reagent	Erdman reagent
Frohde reagent	Mecke reagent
Mandelin reagent	Kalo reagent, Potassium iodoplatinate reagent

3.14.2.2.3 Thin layer chromatography (TLC)

Thin layer chromatography (TLC) is one of the more useful and effective techniques for the isolation and identification of many toxic agents and has been applied to the toxicological analysis of alkaloids. TLC techniques are simple, rapid, reliable, inexpensive to perform, and allow a wide range of compounds to be separated. They can also be used to identify metabolites in relation to parent compounds, a further assistance in identification.

The alkaloids can be separated, detected and identified by TLC with simultaneous development of the reference standards.

Various developing solvents have been tried and applied for separating the alkaloids. After developing a TLC plate in various solvents, each alkaloid can be detected by spray reagents, such as the potassium iodoplatinate or Dragendorff test solution. It is also helpful to observe the blue fluorescence under ultraviolet irradiation at 3650 A before spraying the detection reagents.

Many alkaloids produce blue-(series) or orange-coloured spots with the potassium iodide platinate or Dragendorff test solution, respectively. The limits of detection by these reagents are a few μg or less. Iodoplatinate reagent has the advantage that it usually produces different colours with different alkaloids, whereas Dragendorff reagent gives only slight differences in colour. In addition, the iodoplatinate reagent is non-destructive and the alkaloids can be recovered after application of the spray reagent. The sensitivity of the reagent for alkaloids is in the range of 0.01 – 1 µg.

Sample preparation is less important in TLC than in other chromatographic techniques, as a TLC plate is used only once. In TLC, it is not as important as in other chromatographic techniques to remove non-alkaloidal compounds from the sample to be analysed prior to the analysis. However, for the analysis of extremely small amounts of alkaloids in mixtures with other compounds from biological material, extensive purification and removal of contaminants from the sample may be necessary.

Quantitative analysis of alkaloids in TLC can be performed either directly on the plate or indirectly, after elution of the alkaloid from the spot on the plate.

The recent development of chromatographic materials has led to a broader application of TLC in toxicology. Methods using high quality adsorbents and layers have been developed as a new field termed high-performance TLC (HPTLC). Layers with optimized thickness and other properties have been prepared using smaller particles (5 – 10 µm) with uniform size distribution. HPTLC permits high-resolution and high sensitivity. HPTLC plates are commercially available. For the quantitative application of HPTLC, *in situ* densitometry and spot elution techniques and related instrumentation have been greatly improved.

TLC can also be used for chiral separations. Recent development of stationary phases in TLC, such as chemically bonded ones, are being used in TLC and HPTLC, primarily in reversed-phase partition chromatography.

3.14.3 General Qualitative and Quantitative Methods for Alkaloid Analysis

3.14.3.1 Gas chromatography (GC)

The alkaloids can be detected and identified with gas chromatography (GC). GC is sensitive, accurate, rapid, and simple for the separation, identification, and quantification of many compounds. In order to increase sensitivity and resolution, various stationary phases, detectors and derivatizing methods have been developed. Detectors such as FID, NPD and ECD are generally employed for the determination of alkaloids.

Owing to the low volatility and thermal instability of many alkaloids, derivatization is performed to endow them with better gas chromatographic properties. Derivatization is also done to give alkaloids better detection properties for quantitative determinations.

To date, many derivatizing methods and reagents have been developed and successfully applied for the determinations of alkaloids. Derivatizations such as trimethylsilylation, alkylation, heptafluorobutylation, acetylation, propionylation and pentafluoropropylation have been employed by many investigators.

Derivatization can improve specificity and sensitivity. Derivatization can be carried out before the injection of the sample, especially by converting polar compounds into non-polar compounds, or by means of flash heater derivatization. Derivatization may in many cases increase the resolution and the differences in retention times of compounds, and the derivatives may give extra information during the identification of unknown compounds. On-column derivatization in GC has also been carried out by many investigators.

After the introduction of glass capillary columns instead of packed columns for GC analysis, capillary GC has been extended greatly and applied for the determination of alkaloids. Capillary GC generally allows the determination of alkaloids down to the low nanogram range with and without derivatization, and also effectively resolves various alkaloids. Therefore, sensitive and simultaneous determination of alkaloids can be performed by capillary GC in combination with various detectors. The fused silica and leached glass columns coated with polysiloxanes (OV-1, OV-101, OV-73 and SE-54) and polyethylene glycols (Superox 20M) provide a good choice for capillary GC of alkaloids.

3.14.3.2 High performance liquid chromatography (HPLC)

3.14.3.2.1 Separation techniques

Today, good separations and quantitative determinations of alkaloids have been extensively and successfully carried out using high-performance liquid chromatography (HPLC). As some alkaloids, such as ergot alkaloids, have relatively high molecular weights, low vapor pressures and thermal instability, their GC analysis is often unsatisfactory or impossible. Such disadvantages are not found in HPLC. HPLC analysis of alkaloids has so far been performed by means of ion-exchange, reversed-phase, ion-pair and straight-phase chromatography.

Although the ionic properties of alkaloids would make them suitable objects for ion-exchange chromatography, this technique has only found limited application in the analysis of alkaloids.

Bonded-phase columns in the reversed-phase mode have been mainly used in HPLC. The organic mobile phase is usually methanol, acetonitrile, or tetrahydrofuran. Various bonded-phases (stationary phase), principally made from bonding an organochlorosilane group to a silica substrate, have been developed and successfully applied to alkaloid analysis.

Retention times of the alkaloids can be altered by changes in the mobile phase, i.e. the ionic strength, the nature of the counter ion, the pH and the addition of an organic solvent to the mobile phase.

However, in several cases the reversed-phase materials have been found to be unsuitable for the analysis of basic compounds. The best results will generally be obtained with a stationary phase with the highest possible coverage of the silanol groups.

To overcome the problem of tailing in reversed-phase HPLC, salts can also be added to the mobile phase (ion suppression). Ammonium carbonate, sodium acetate and sodium phosphate have been used for this purpose. Ion-pairing has also proved to be successful in alkaloid analysis.

In order to reduce tailing on reversed-phase materials, basic mobile phases can be used. However, the stability of the chemically bonded groups is limited above pH 8.5. The stability of some reversed-phase materials for various amines has been studied. The addition of long alkyl chain amines in low concentrations to the mobile phase improves peak shape in the analysis of basic compounds. It has been shown that an increase in chain length of the amine additive results in a significant improvement of peak shapes of basic compounds. The beneficial effect of the addition of amines to the mobile phase is likely to be due to the masking of free silanol groups by the amines.

An advantage of reversed-phase chromatography in the analysis of alkaloids in biological fluids is that analysis can be carried out directly without laborious sample clean-up procedures. However, the use of a precolumn to avoid a too rapid deterioration of the HPLC column is advisable. When using aqueous salt solutions in reversed-phase chromatography, one has to be aware of the risk of corrosion of stainless steel columns.

Microparticulate macroporous polymer resins also have been used in the analysis of alkaloids. A disadvantage of macroporous polymer resins is that they are not as rigid as the reversed-phase materials based on silica gel. In addition, they may shrink or swell slightly, depending on the composition of the mobile phase. On the other hand, they are more stable than chemically bonded phases on silica gel. They can be used in the entire pH range and have a column life of more than two years without loss in efficiency.

Cyclodextrins have been used as mobile phase additives for the chiral separation of enantiomers and diastereomers in HPLC. Preparative chromatographic separation of enantiomers has been recently extensively reviewed (10).

3.14.3.2.2 Detection methods

UV is the primary means of detection in the HPLC analysis of alkaloids. Fixed wavelength (254 and/or 280 nm) and multiwavelength detectors have been widely employed with a sensitive in the ng-range or lower.

Fluorescence detection is highly specific and often extremely sensitive. However, its use is limited to fluorescent alkaloids. Thus, many investigators have made alkaloids fluorescent by derivatization. Various chemical reactions can be used for this purpose, such as oxidation to give fluorescent oxidation products, coupling with fluorescent groups, i.e. dansyl groups, and ion-pairing with fluorescent ions, such as picrate, β-naphthalenesulfonate and 9,10-dimethoxyanthracene sulfonate. Pre- and post-column derivatization techniques have been developed by various investigators.

Electrochemical detection has also been successfully applied to the analysis of some alkaloids with very low detection limits.

Photodiode array detection, which allows detection over the whole UV-VIS spectrum, has been successfully applied to the determination of alkaloids, as well as rapid scan UV/Visible detectors. They can be very useful for the qualitative and quantitative determination and for establishing a data-base.

3.14.3.3 GC-mass spectrometry (GC-MS) and HPLC-MS

A recent great advance in toxicological analysis is the coupling of mass spectrometry (MS) to GC and HPLC. GC-MS and HPLC-MS have also revolutionized the study of alkaloids. MS gives important information on the possible structure of the original compound, possible molecular weight and so forth. These two techniques are therefore the most effective methods for the identification of alkaloids and their metabolites. These techniques have been extensively and successfully applied to metabolic studies of alkaloids as well.

3.14.3.4 Immunoassay

Radioimmunoassay and enzyme immunoassay for the determination of alkaloids in biological samples and plant extracts have been developed and successfully utilized in routine analysis and pharmacokinetic and toxicokinetic studies. These methods are highly sensitive, specific and simple. It is often possible to determine alkaloids without prior extraction from the sample, and to carry out the assay on very small sample volumes.

Radioimmunoassay, however, has disadvantages associated with the use of radiolabelled compounds, such as instability, potential health hazards, rather higher cost of labels and counting instrumentation in addition to the disposal of radioactive materials.

To date, enzyme immunoassay has been the main alternative to radioimmunoassay, because it has overcome the disadvantages encountered. Enzyme immunoassays utilizing fluorescent, chemiluminescent and bioluminescent labels have been developed, and can be as or more sensitive than radioimmunoassay. Cross reactivity and interference from sample constituents must always be considered for such a highly sensitive assay. On the other hand, when simultaneous determination of alkaloids in biological samples is necessary, immunoassay must be replaced by other alternatives, such as chromatographic methods. Despite the few disadvantages of immunoassay, this method is attractive for the determination of important alkaloids which serve as modern therapeutic drugs.

3.14.4 Alphabetical Listing of Toxic Alkaloids – Occurrence, Metabolism and Analysis

3.14.4.1 Aconitum alkaloids

Occurence and usage: Aconitum alkaloids, aconitine, mesaconitine, hypaconitine, jesaconitine etc., occur naturally in the roots of *Aconitum species*. The dried roots of *Aconitum japonica* and related plants have been used for centuries as oriental medicine "bushi". Of the aconitum alkaloids, aconitine and mesaconitine are highly toxic and thus of toxicological concern, because they can cause accidental, homicidal and suicidal deaths. All parts of plants contain toxic alkaloids, but the alkaloid content and composition vary throughout the year.

Toxicity: Lethal dose of aconitine is around 5 mg. Aconitum poisoning produces numbness and tingling of the tongue, throat and mouth, nausea, salivation, and general weakness. Cardiotoxicity of serious aconitum poisoning is very important, but is commonly complicated by hypotension, shock, conduction delays, dysrhythmias and arrhythmias. Even in cases of severe poisoning, patients do not lose consciousness.

Metabolism and excretion: The metabolic fate of aconitum alkaloids in the body has not been clarified in detail. It is thought that aconitine and related compounds are deacetylated to give benzoylaconines, and further hydrolyzed to aconines by debenzoylation. The toxicity of these benzoylaconines in mice was decreased to $^{1}/_{50}-^{1}/_{1000}$ that of the mother compounds (11). In the surviving aconitum-poisoned patient, aconitine and mesaconitine were excreted into urine up to 7 days after intoxication (12). However, total amounts of aconitum alkaloids excreted were far less than expected, based on the ingested dose.

Analysis: Qualitative determination of aconitum alkaloids by HPLC has been reported (13). However, the sensitivity and specificity of the determination of these aconitum alkaloids in biological specimens are not satisfactory. Recently, GC-SIM has been used for the detection of aconitine in biological samples from aconite-intoxicated cadavers (12). By using this method, it is possible to detect levels of aconitine and related aconitum alkaloids over 10 pg. Trimethylsililates of aconitine, mesaconitine and hypaconitine are heat-stable; therefore, this method can be applied to the toxicokinetic study of aconitum intoxication.

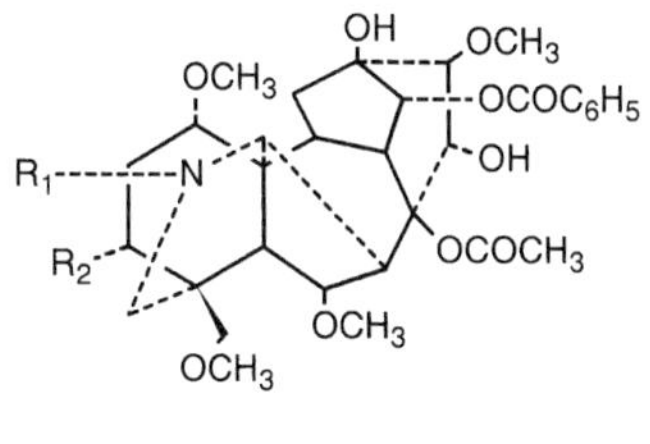

Compounds	R_1	R_2	***MW***
Aconitine	C_2H_5	OH	**645**
Mesaconitine	CH_3	OH	**631**
Hypaconitine	CH_3	H	**615**

Figure 1. Chemical Structures of Aconitum Alkaloids.

3.14.4.2 Belladonna alkaloids, atropine and scopolamine

Occurrence and usage: Atropine (*dl*-hyoscyamine) is derived from certain plants, especially *Atropa belladonna* and *Datura stramonium*, which also contain scopolamine. It is a potent anti-cholinergic agent, attributable to the *l*-isomer (hyoscyamine), which is present in plants and changes to the *dl*-form (atropine), and has been used for centuries as a drug and poison.

3.14.4.2.1 Atropine

Usage: Atropine is used as a preanesthetic medication to reduce salivary and bronchial secretions, and for the relaxation of the gastrointestinal tract. It is also used to produce mydriasis in ophthalmic prodedures. In toxicology, atropine is used as an antidote to poisoning by cholinesterase inhibitors such as organophosphate and carbamate insecticides via intravenous doses of 1–6 mg.

Toxicity: Atropine decreases gastric and other body secretions, and leads to dilation of the eyes. Atropine doses exceeding 10 mg cause moderate to severe symptoms of toxicity, and doses greater than 50 mg can be fatal. Physostigmine has been used as a specific antidote for atropine poisoning.

Atropine administered intravenously to adults disappears from blood within few hours. Its half life has been shown to increase in young children and in the elderly. Metabolism and excretion: Atropine is metabolized to noratropine, atropine-N-oxide, tropine and tropic acid and excreted into urine in the order described (14). Ethnic differences in response to atropine with respect to heart rate have been shown between Chinese and white subjects (15).

Analysis: Vitali reaction is fairly specific for detecting atropine and scopolamine. Atropine has been analyzed in biological and other specimens by radioimmunoassay (16), GC (17), GC-MS after trimethylsilylation or after hydrolysis and formation of heptafluorobutyryl derivatives (18), and HPLC (19–21).

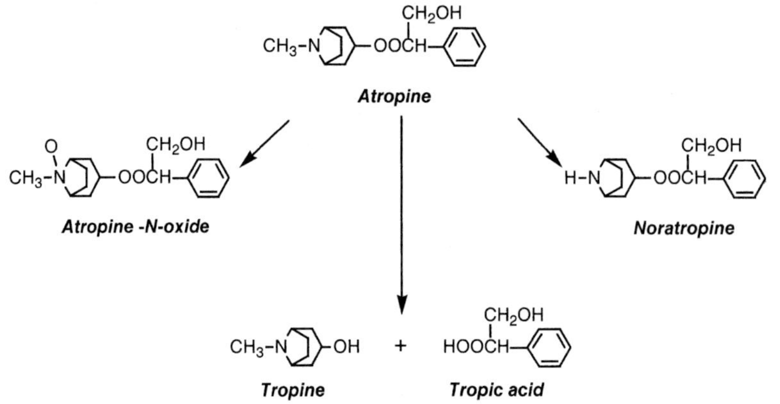

Figure 2. Chemical Structure and Metabolic Fate of Atropine.

3.14.4.2.2 Scopolamine

Usage: Scopolamine has an antimuscarinic property and central nervous system depressant activity, and is used as a preanesthetic medication, for treatment of Parkinsonism and gastrointestinal disorders, and for prevention of motion sickness.

Toxicity: Scopolamine overdoses may cause double vision, dryness of the throat, dilated pupils, skin flushing, hallucinations, coma and death. Scopolamine has been shown to readily cross the placenta after maternal administration, and thus care should be taken for newborn toxicity.

Metabolism and excretion: Previously, the major metabolite of scopolamine was shown to be a glucuronide in animals. However, recent studies have shown that there are various metabolic pathways of scopolamine, such as hydroxylation, dehydrogenation, N-demethylation, N-oxidation in addition to glucuronide formation in animals with marked species and individual differences (22). In humans, about 4–5 % of a single oral dose of scopolamine is excreted unchanged in the urine within 2 days (23).

Analysis: Scopolamine radioreceptor assay (23) and enzyme immunoassay (24) have been reported. GC, GC-MS and HPLC analysis of scopolamine have also been reported as described for atropine.

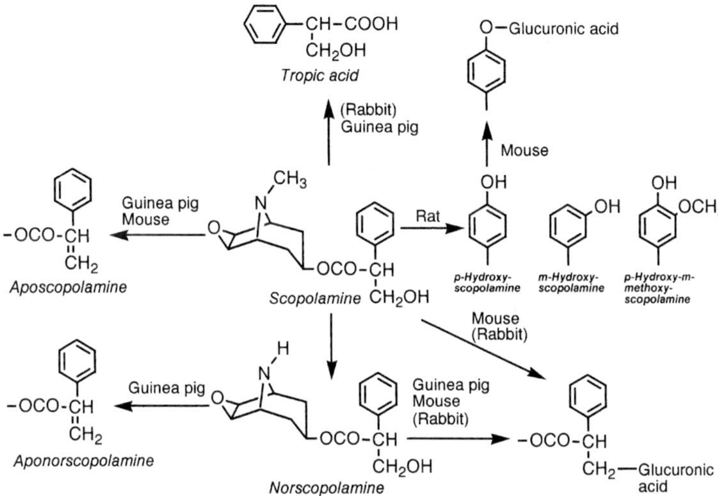

Figure 3. Chemical Structure and Metabolic Fate of Scopolamine.

3.14.4.3 Cinchona alkaloids, quinidine and quinine

Occurrence and usage: About 30 or more alkaloids have been isolated from *Cinchona*. Of the famous Cinchona alkaloids, quinine and quinidine occur together in *Cinchona* bark in concentrations of 0.25 % to 3.0 %. Both are stereoisomers of each other and are important alkaloids in medicine. Quinine and quinidine have played historically impotant roles in the treatment of malaria and arrhythmia, respectively. Quinine inhibits the growth of the malaria protozoa "plasmodium". The structural side chain double-bond and its reductive product lead to the inhibition of plasmodium growth. Its epi-structure does not show such an inhibitory effect on growth. Quinidine is a widely prescribed anti-arrhythmic drug for long-term management of supraventricular and ventricular arrhythmias. The recent importance of Cinchona alkaloids in toxicological concerns is a result of their frequent presence as adulterants in illicit heroin.

Toxicity: Quinine and quinidine intoxication has been termed "cinchonism". The symptoms and signs include (1) disturbed hearing, tinnitus, decreased auditory acuity and vertigo, (2) disturbed vision, blurred vision, disturbed colour perception (yellow vision), photophobia, diplopia scotomata and contracted visual fields, and (3) other central nervous system manifestations, headache, fever, apprehension, excitement, confusion and delirium. Quinidine is extremely irritating to the gastrointestinal tract. Even therapeutic doses of quinidine can cause nausea, vomiting, abdominal pain and diarrhea. Quinidine also causes cardiovascular toxicity. Hypokalcemia and metabolic acidosis are common with overdoses of quinidine.

Metabolism and excretion: Quinine and quinidine are metabolized primarily by hydroxylation in the liver to several more polar metabolites (25–30). Some of the metabolites are believed to be pharmacologically active and could contribute to the toxicity. About 17% of a dose is eliminated unchanged in the 24-hr urine during chronic therapy; about the same amount is present as 3-hydroxyquinine and about one-third of that amount as 2'-hydroxyquinine.

Plasma concentrations of quinine of 10 mg/l or greater generally result in toxicity in persons receiving chronic therapy with the drug. In non-fatal acute poisoning, plasma quinine concentrations ranged from 6.8–26 mg/l. Blood quinine concentrations of 0.4–11 mg/l were noted in 186 cases of death due to overdosage of heroin that was adulterated with quinine.

Quinidine is oxidized to 2'-quinidinone, 3-hydroxyquinidine, quinidine-N-oxide and quinidine-10,11-dihydrodiol (25, 27, 28). The quinidine metabolites quinidinone and 3-hydroxyquinidine have been shown to have antiarrhythmic activity similar to that of quinidine. These metabolites are present in urine in lower concentrations than quinidine whose unchanged excretion form amounts to about 20%. The urinary excretion of quinidine is highly pH dependent. Quinidine-N-oxide is a major urinary metabolite of the drug. The range of quinidine concentrations reported in cases of significant intoxications is 5.1 to 28.4 mg/l.

Recently, quinidine has been applied in the characterization of cytochrome P-450 species involved in the metabolism of drugs or chemicals both *in vivo* and *in vitro*, because this compound is a specific inhibitor of P-450 2D6 (31, 32).

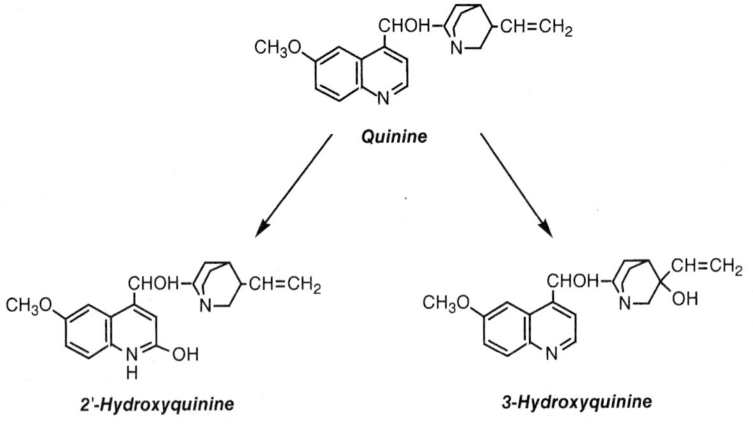

Quinine

2'-Hydroxyquinine

3-Hydroxyquinine

Figure 4. Chemical Structure and Metabolic Fate of Quinine.

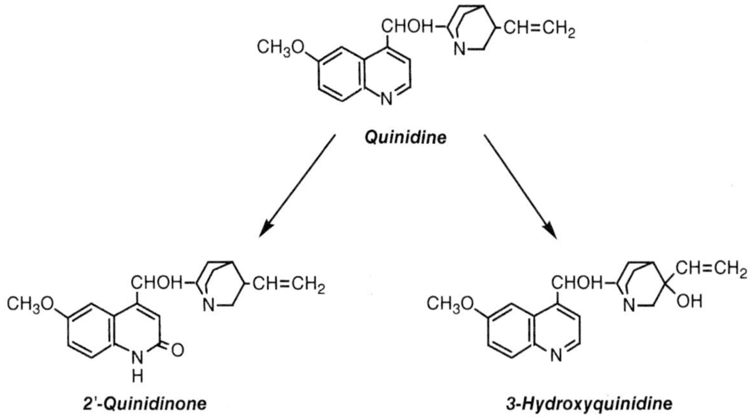

Quinidine

2'-Quinidinone ***3-Hydroxyquinidine***

Figure 5. Chemical Structure and Metabolic Fate of Quinidine.

Analysis: The determination of quinidine and quinine has been accomplished by fluorometry, GC with FID or NPD. Recently, HPLC is the choice for quinine and quinidine determination together with their metabolites (33, 34). An enzyme immunoassay is commercially available. HPLC-MS has also been reported to determine cinchona alkaloids in *in vitro* culture (35, 36).

3.14.4.4 Colchicum alkaloids, colchicine

Occurrence and usage: Colchicine occurs naturally in the bulb and seeds of *Colchicum autumnale*. The nitrogen atom of colchicine is in the form of an amide and does not show any characteristics of basicity. Thus, in a strict sense, colchicine is not an alkaloid. However, its biosynthetic pathway clearly indicates that colchicine is one of the isoquinoline alkaloids. Colchicine is a potent inhibitor of cellular mitosis and is used in the treatment of gout. Specifically, colchicine is the drug of choice for acute gouty arthritis. Its use has recently been extended to the treatment of immune or inflammatory diseases, such as Behcet's syndrome and primary biliary cirrhosis. The inhibitory effect of colchicine on cell division has been also utilized as a useful tool for plant mutagenesis.

Toxicity: Colchicine overdosage is often manifested by nausea, vomiting, diarrhea, confusion, fever, shock, respiratory distress, hematuria, renal failure, metabolic acidosis and cardiovascular collapse (37). Thrombocytopenia, granulocytopenea, myopathy, neuropathy, and alopecia may develop in the later stages of acute poisoning (38).

Death cases have been reported due to intentional intake of colchicine at the dose of 7.5 mg or 20 mg. Even in lethal intoxications, colchicine blood concentrations decreased rapidly to extremely low levels.

Metabolism and excretion: Cancer and gout patients eliminate about 4 % of the dose as unchanged drug, while normal asthmatic subjects eliminate about 28 % in this manner; another 1 ~ 13 % is present in the 48 hr urine as unidentified metabolites. The detailed metabolic fate of colchicine is not clarified, although it is believed to be metabolized by deacetylation (39). Leukocytes have been shown to be pharmacologically active disposition sites (40).

Analysis: Colchicine is sensitive to light, thus in many analytical steps the sample solution should be protected from light exposure.

Colchicine has been quantitatively determined in biological specimens by fluorometry (41), radioimmunoassay (42), HPLC (43, 44) and GC-MS (45). A HPLC method for colchicine determination in human plasma and urine has been shown to be unsuitable for pharmacokinetic studies but suitable for the relatively high levels found in cases of overdose (44).

3.14.4.5 Conium alkaloids

Occurrence and usage: Conium is termed "koneion" in the book written by Dioscorides. Coniine and related alkaloids including N-methylconiine, coniceine, conhydrin etc. which have relatively simple chemical structures occur naturally in *Conium maculatum* and other plants. Conium has been known as a poisonous plant from very ancient times. It is known as a toxin used by the ancient Greek philosopher Socrates. The common name is poison hemlock. The term "hemlock" has been loosely applied to several plants and trees that vary in toxcity, such as the water hemlock plant (Cicuta) and the ground hemlock (Taxus). Poisonings usually result from mistaken identification by people foraging for natural food. Conium poisoning has been described in historical accounts in Greek. Conium poisonings to domestic animals are also important from a toxicological viewpoint. Coniin is volatile and is lost slowly from *Conium* while drying.

Toxicity: Conium alkaloids are structurally related to nicotine and function somewhat similarly. Thus, coniin stimulates the central nervous system, then depresses the system, and leads to death by paralyzing the respiratory center. It is also known to paralyze peripheral motor nerves. Conium alkaloids intially produce tremors, ataxia, mydriasis, nausea, vomiting, and sore throat, followed by cardiorespiratory depression (bradycardia, paralysis, coma) and ascending paralysis. Death results from respiratory failure. Conium toxicity is usually not accompanied by convulsions.

Metabolism and excretion: Coniin is well absorbed by the gastrointestinal tract. However, little information is available on the metabolic fate of conium alkaloids.

Analysis: Coniin is separated and quatitated by overpressed TLC (46). It can be also determined by GC, HPLC and GC-MS (9).

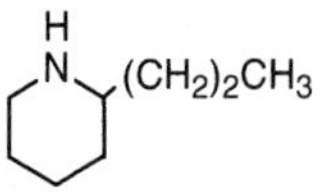

H
N $(CH_2)_2CH_3$

Figure 6. Chemical Structure of Coniine.

3.14.4.6 Ergot alkaloids

Occurrence and usage: Ergot alkaloids, the derivatives of the tetracyclic compound 6-methylergonine, ergotoxine, ergotamine etc., occur in the rye parasitized by *Claviceps purprea*. Other *Claviceps* species, *C. paspali* and *C. cinera*, are also known to parasitize the grains of variously cultivated and wild grasses with resultant animal disease. Historically, *C. purprea* infection of rye grain has been closely associated with human illness world-wide, and the scloretum of this fungal species has become the source of many pharmacologically active compounds and important principles.

The naturally occurring ergot alkaloids can be classified as clavine alkaloids and lysergic acid derivatives. The lysergic acid alkaloids can be divided into water-soluble alkaloids (including ergometrine) and water-insoluble or peptide alkaloids [the eragotamine group (ergotamine, ergosine) and the ergotoxine group (ergocristine, ergocryptine, ergocornine)].

Lysergic acid derivatives are found not only in fungi but also in higher plants, known under the name "Morning Glory".

Usage: The pharmaceutically and toxicologically important group of ergot alkaloids is derived from *d*-lysergic acid. The naturally occurring alkaloids, the semi-synthetic derivatives, as well as the dihydro compounds are used in therapy. The semi-synthetic alkaloid *d*-lysergic acid diethylamide (LSD) has strong hallucinogenic effects and its abuse is widespread. The details on LSD will be described in another chapter. Ergot alkaloids have been used therapeutically in obstetrics, endocrinology, migraine and other vascular headaches, venostasis, senile cerebral insufficiency, and in Parkinson's disease. However, the pharmacological activities of ergot alkaloids are extremely varied and complex. Ergot has been used to control uterine hemorrhage and for treatment of migraine headaches. Ergot extract may vary in intensity and properties due to variance of consituents.

Toxicity: The term "ergotism" is used to describe a patient who clinically presents ergot toxicity (47, 48). Although no human epidemics due to ingestion of contaminated grain have been reported recently, future human exposure remains a great concern and livestock poisoning is an ongoing reality because of a possible world food shortage. Additionally, ergotism is still a continued medical concern in modern therapeutic applications of various natural and semisynthetic ergot alkaloids, especially associated with the treatment of migraine (49, 50).

The allusions to ergot toxicity have been known from pre-Christian history. Convulsive and mixed convulsive/gangrenous presentations were recorded in the literature of 11th century. Cures for ergotism by pilgrimages to the shrine of St. Anthony are documented, where an incidental change of diet literally "cured the patient". Thus, ergotism is often named "holy fire" or "St. Anthony's". Ergotism has continued into the twentieth century despite the recognition of a causal relation between the ingestion of "ergot of rye" and clinical disorders. Several outbreaks of ergotism have been reported in Europe and India. Ergotism has been divided into "gangrenous" and "convulsive" forms. In epidemics, one form or the other seems to predominate, although mixed presentations in a single epidemic may occur. It is still not clear whether variable toxic contents of sclerota, dose effects or population susceptibilities are considerations.

Toxic symptoms of ergot poisoning have been clarified somewhat after utilization of specific alkaloids as follows. Acute: burning pain in abdomen, vomiting, great thirst, diarrhea. Delayed: edema of face and extremities, abnormal sensation of the skin, tonic contraction of muscles, slow weak pulse, tinnitus, miosis, blurred vision, intestinal and uterine contraction, loss of speech and inability to walk, convulsions and mania, skin gangrene after some time, death due to respiratory and cardiac failure. Chronic: spasmodic or convulsive type from ingesting food containing ergot, indefinite pains, disturbed digestion, itching and numbness; anesthesia of arms, legs, toes and fingers; painful tetanic spasms, contraction of limbs, mental weakness, dementia. Limbs get cold and become dark in colour; fingers and toes dry and gangrenous. Cataracts are common; internal organs become gangrenous. The administration of ergot alkaloids to individuals with sepsis, hepatic damage, vascular disease, renal disease may cause acute effects as a result of underlying disease states or impaired ability to metabolize the drug.

Metabolism and excretion: Ergotamine and dihydroergotamine are both extensively metabolized in the liver, and excreted mainly into the bile (47, 48). Ergotamine may sequester in various tissues, which probably accounts for its long-lasting effects despite a short plasma half-life of about 2 hrs. Thus, the decreased hepatic function due to the reduced metabolism of ergotamine may accumulate and increase its toxicity.

Analysis: Ergotamine and other ergot alkaloids have been analyzed by HPLC (51, 52), MS (53) and GC-MS (54). Immunoassays for ergotamine and other ergot alkaloids have also been reported (55–57).

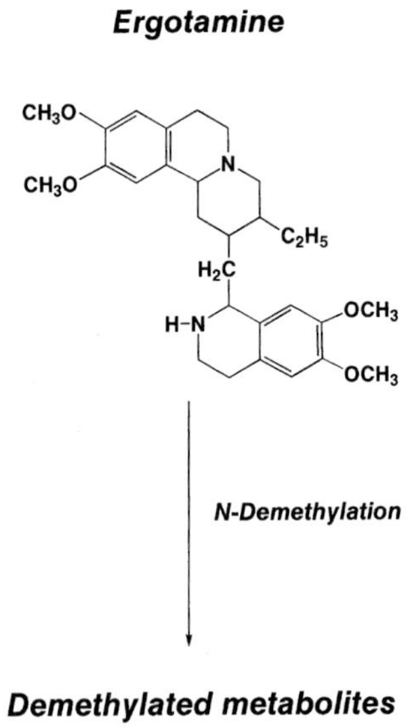

Ergotamine

CH_3O CH_3O N C_2H_5 H_2C H-N OCH_3 OCH_3

N-Demethylation

Demethylated metabolites

Figure 7. Chemical Structure and Metabolic Fate of Ergotamine.

3.14.4.7 Glycoalkaloids

Occurrence and usage: Glycoalkaloids occur naturally in the species of *Solanum*. The glycoalkaloids all contain the same alkamine aglycone, solanidine, with different composition of the sugar chain. Of these glycoalkaloids, solanine is most common with respect to its toxicity to humans.

Solanine is removed by boiling, but not baking potatoes, because it is water soluble.

Toxicity: Human poisoning by solanine occurs after consumption of green or stressed potatoes (58–60). However, because of its bitter taste, solanine poisoning is rare except in times of food shortage. Generally, 100 ~ 200 mg glycoalkaloids per kg potatoes has been accepted as the upper safe limit. However, there are indications that solanine and related compounds can accumulate in tissues. Since little information is available concerning the sub or chronic toxicity of glycoalkaloids in this respect, further toxicological study will be required.

Solanine poisoning causes gastrointestinal and neurologic symptoms depending on the amounts, together with anticholinergic syndroms. Vomiting, headache, flushing, diarrhea, anorexia, and malaise are common symptoms.

In animals, toxic effects of solanine are different from species to species and depend on the route of administration. Man is more sensitive to solanine than are other animals. The hamster might serve as an animal model for solanine toxicity, because it seems to be suspectible to the glycoalkaloid (61).

Metabolism and excretion: Solanine is poorly absorbed from the gastrointestinal tract. Elimination occurs rapidly in the feces and to a lesser extent in urine as studied in fed rats. However, excretion ratios between feces and urine are also different among animal species.

Analysis: Solanine has been simply measured spectrometrically at 600 nm. Solanine and related compounds have also been quantitated by GC (62, 63) and HPLC (64, 65). Radioimmunoassay of potato steroidal alkaloids in human serum and saliva has also been reported (66).

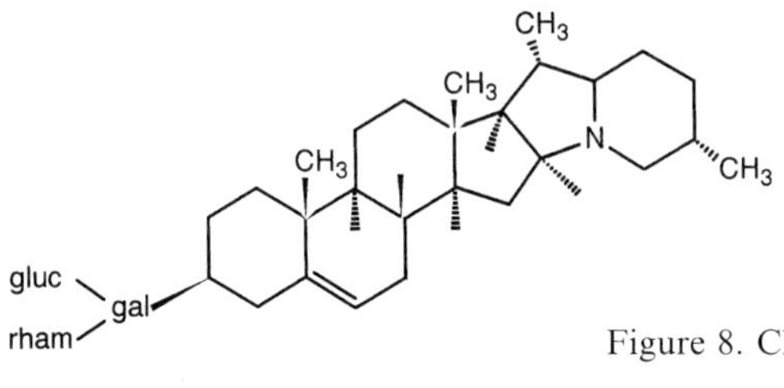

gluc
gal
rham
CH_3
CH_3
CH_3
N
CH_3

Figure 8. Chemical Structure of Solanine.

3.14.4.8 Nicotinum alkaloids, nicotine

Occurrence and usage: Nicotine is present in amounts of 0.5% to 8.0% by weight in tobacco leaves. The compound was first isolated in 1828. The free base is a liquid that slowly darkens on exposure to air. Nicotine is one of the highly toxic alkaloids that causes stimulation of autonomic ganglia and the central nervous system.

Nicotine can be steam-distilled under strong alkaline conditions.

It can be absorbed through skin and through all portals. Death usually occurs within a matter of minutes. It causes tachyphylaxis. Mild poisoning may appear as a result of overindulgence of tobacco. Nicotine is not a prescribed drug, although it is widely consumed throughout the world as tobacco.

Toxicity: Nicotine causes stimulation followed by depression of the central nervous system. Toxic doses of nicotine produce paralysis of the central nervous system including the respiratory center, leading to tachycardia, hypertension, curare-like effect on muscle, abdominal pain, disturbed vision, constriction of blood vessels. Other symptoms include labored breathing, irregular pulse, mental confusion, partial or complete unconsciousness and clonic convulsions followed by collapse with muscular relaxation. Smaller quantities produce nausea, salivation, vomiting, purging, strong typical cigarette odor on breath and in blood, powerful abdominal corrosion, burning sensation in mouth and throat. Above 0.5 mg/dl blood may be lethal.

Metabolism and excretion (67–71): Nicotine is rapidly absorbed from the gastrointestinal tract, the respiratory tract, and intact skin. Nicotine is readily distributed throughout tissues and it is also found in the milk of lactating women who smoke. Nicotine is metabolized to cotinine, which is further converted to hydroxycotinine and a ring cleavage product. Nicotine-1-N-oxide and norcotinine are also found as metabolites. There are sex and species differences in nicotine metabolism and excretion (70), in addition to the similar differences between tabacco smokers and nonsmokers. Metabolic fate of nicotine is of continued concern to various investigators with respect to its pharmacological and toxicological effects, because it is assumed to play a key role in tobacco habit and addiction. Recent progress in analytical methods has led to a further identification of various nicotine metabolites. Some subjects were found to be deficient in nicotine-N-oxide (67), which is catalyzed by flavin-containing monooxygenase. The present state of knowledge concerning the metabolism and disposition of nicotine in human and experimental animals has been collectively reviewed by Kyerematan and Vessel (71).

Urinary excretion of unchanged nicotine accounts for only a small percent of the administered dose. However, urinary excretion of nicotine is enhanced by acidification of the urine, whereas cotinine excretion is less affected by pH changes.

Analysis: Nicotine is isolated by alkaline-steam distillation or alkaline-chloroform or ether extraction. Nicotine has a strong characteristic odor, burning taste, and causes blood to become dark and to have a sharp odor like stale tobacco. UV spectrometry (very sensitive): maximum absorption at 260 nm in 0.1 N sulfuric acid solution ($E1_{cm}$, 490). Nicotine, cotinine and their metabolites have been determined simultaneously by GC (72–74) and GC-MS (75) in biological samples. HPLC and LC-MS have also been utilized for the simultaneous determination of nicotine and its metabolites (76). Recent developments of the coupling of MS to GC and HPLC make it possible to identify and quantitate nicotine and its metabolites. The toxicological analysis of very low concentrations of nicotine in biological specimens requires careful precautions to avoid interference from exogenous nicotine.

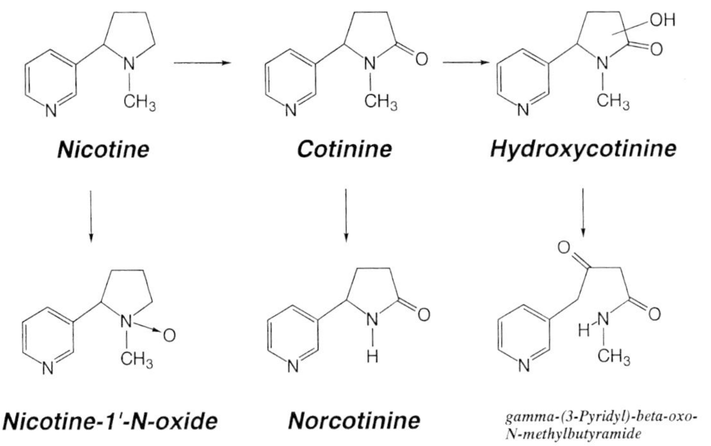

Nicotine → Cotinine → Hydroxycotinine

Nicotine-1'-N-oxide | Norcotinine | gamma-(3-Pyridyl)-beta-oxo-N-methylbutyramide

Figure 9. Chemical Structure and Metabolic Fate of Nicotine.

3.14.4.9 Opium alkaloids

Occurrence and usage: Opium alkaloids occur naturally in *Papaver somniferum*. Opium is obtained from the milky exudate of incised unripe seed capsules of the poppy plant. Almost 20 opium alkaloids including phenanthrenes (morphine, codeine, thebaine etc.) and benzylisoquinolines (papaverine, noscapine etc.) are found. These opium alkaloids play important roles in the treatment of human diseases, as drugs such as analgesics, antitussives, vasodilators, and so forth. Opium is historically a very old drug, and has been available and utilized for thousands of years. Specifically, the narcotic effect of opium plays a therapeutically important role in curing pains from diseases; inversely, this effect has led to addiction to the drug and evoked social disasters worldwide. Many natural, semisynthesized and totally synthesized opium alkaloids are available for therapeutical purposes, but also for illicit use. Because of such worldwide therapeutical or illicit use, opium alkaloids and their related compounds are subject to continued medical, social and toxicological concerns.

Analysis: Because of the important role of opium alkaloids in the treatment of human diseases, numerous papers concerning the quantitative analysis of alkaloids including morphine, heroin and codeine have been published to date. Specifically, simultaneous determinations of these alkaloids and their metabolites in biological specimens are of great therapeutical and toxicological concern. The quantitative assay of individual opium alkaloids performed by GC, HPLC, GC-MS and HPLC-MS has been reported in many papers (77–90). Recent developments of HPLC techniques allow the simultaneous determination of morphine, codein and heroin and their metabolites. A relatively simple method for simultaneous quantitations of codeine, morphine and 6-monoacetylmorphine in blood has been described (90). Commercial kits are available and are excellent for the detection of morphine (opiates) and related drugs.

Specimens of choice for toxicological analysis of morphine and its derivatives are plasma, urine and bile. Bile contains much more morphine per unit volume but is very difficult to extract.

In addition to opium alkaloids, it is also important to identify meconic acid which is present specifically in opium. Meconic acid can be identified easily with ferric chloride reagent. There is a report showing the simultaneous determination of opium alkaloids together with meconic acid using HPLC (89).

Poppy seeds may contain morphine, codeine and other narcotic alkaloids. Thus, care should be taken for urine analysis of morphine and related compounds in toxicology (77–81).

3.14.4.9.1 Codeine

Usage: Codeine has been used as a narcotic analgesic and an antitussive drug.
Metabolism and excretion: Codeine is partly metabolized to morphine in man, as is ethylmorphine (92). Therefore, in cases where morphine is found in blood or urine, the presence of codeine and ethylmorphine should be determined in order

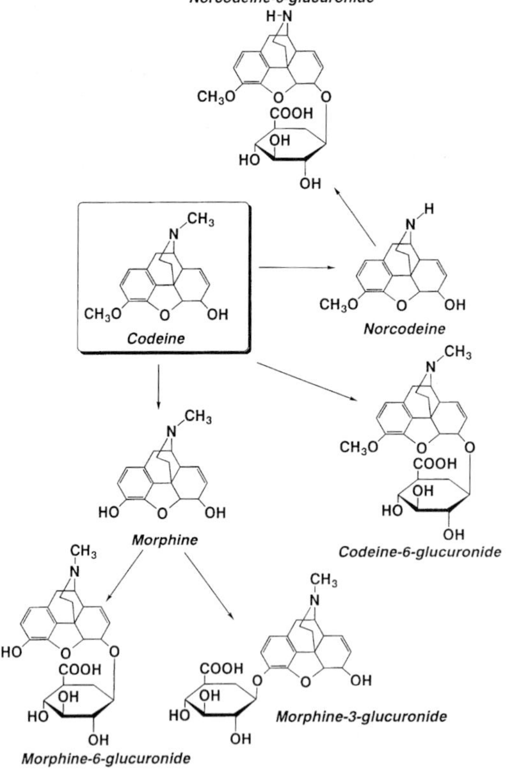

Figure 10. Chemical Structure and Metabolic Fate of Codeine.

to confirm or exclude the intake of these drugs. A method for simultaneously measuring these three drugs is therefore of great practical value. Recent evidence indicates that the metabolic conversion of codeine to morphine is catalyzed by cytochrome P-450 2D (6, 82, 83) which is deficient in several percent of the European population. The analgesic effect of codeine thus seems to depend, in large part, on its conversion to morphine, and codeine should be much less effective in people with this cytochrome P 450 deficiency. There are species differences in codeine metabolism among experimental animals (84).

3.14.4.9.2 Heroin (diacetylmorphine)

Usage: Heroin is not a naturally occurring alkaloid, but a very important alkaloid-derived compound in toxicology. Heroin was first synthesized by acetylation of morphine in 1874. Soon after its medical use as an analgesic, heroin was recognized to produce heavy addiction and to cause overdose fatalities. Thereafter, heroin addiction spread worldwide. Heroin is one of the most abused compounds worldwide, and its illicit use remains a subject of social concern. It is the most important compound in toxicology to date.
Metabolism and excretion: Heroin is absorbed from all sites administered. It is very lipophilic, thus leading to rapid penetration through the blood-brain barrier. Pharmacokinetic study of heroin has revealed that its half-life in blood is less than 20 min. Heroin is rapidly metabolized in man almost completely to 6-acetylmorphine and morphine (91). The morphine metabolite then conjugates almost entirely (90 %) to its glucuronide. Evidence for the illicit use of heroin includes its presence in the urine and hair. However, because of its rapid and extensive metabolism into morphine and its glucuronides, it is very hard to identify heroin in human urine. There is evidence that 6-acetylmorphine is detectable in the urine of abusers. Since 6-acetyl-

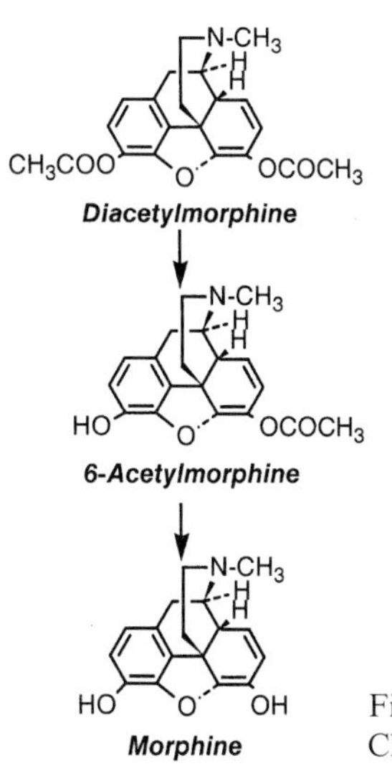

Figure 11.
Chemical Structure and Metabolic Fate of Diacetylmorphine (Heroin).

morphine is not a metabolite of morphine, the detection of this metabolite in the urine could corroborate the illicit use of heroin. It has been shown that heroin has little affinity for the opiate receptor in brain tissue, and thus 6-acetylmorphine and morphine account for most of the narcotic activity of heroin.

3.14.4.9.3 Morphine

Usage: Morphine was first isolated in 1803 as the first opium alkaloid isolated. Medical use of morphine (opium) has been described in the ancient Mesopotamian age. The name morphine was taken from Morpheus, a goddess in Greek mythology. Morphine still has a therapeutically important role in curing pains from diseases, usually by subcutaneous or intramuscular injection. Morphine is now available for oral administration as well-controlled delivery tablets. Morphine is a prototype of the narcotic analgesics, and has been used as a leading compound to develop more effective and less toxic analgesics by semi-synthesis from opium (heroin, oxycodon, hydromorphone), and full synthesis (propoxyphene, methadone, meperidine).

Some of these compounds are commonly prescribed as major analgesics for relief of intense pain. Opiates are also utilized therapeutically for sedation, preanesthetic medication and anesthesia in the hospital setting, and as antitussives and antidiarrheals in ambulatory medicine.

Morphine is recommended by the World Health Organization (WHO) as the treatment of choice for moderate-to-severe cancer pain.

Toxicity: Adverse or toxic effects of morphine include pupillary constriction, constipation, urinary retention, nausea, vomiting, hypothermia, drowsiness, dizziness, apathy, confusion, respiratory depression, hypotension, cold and clammy skin, coma, and pulmonary edema. Chronic usage of morphine leads to tolerance and addiction.

Naloxone is the specific antagonist of choice and is very effective. It can reverse a fall in blood pressure and decrease pulse rate and cardiac arrhythmia when produced by morphine or its derivatives. Loss of superficial and deep reflexes, corneal and gag reflexes, and pupillary constriction return to normal within 5 min. Since this antagonist prevents the action of morphine, it can also provoke withdrawal symptoms in those persons addicted to morphine or its derivatives.

Metabolism and excretion (91–93): Morphine is partly metabolized to normorphine via N-demethylation. The majority of morphine is converted to morphine 3-glucuronide. Very small amounts of morphine-6-glucuronide, morphine-3-ethereal sulfate, and morphine-3,6-diglucuronide are also excreted into urine. Elimination of morphine is thus largely controlled by the formation of its 3- and 6-glucuronides. Uraemic patients have significantly elevated plasma concentrations of morphine-6-glucuronide and severe respiratory depression in these patients may be due to this metabolite. There is increasing evidence that morphine-6-glucuronide is an active metabolite (86, 87, 95–99), a significant contributor to analgesia during chronic morphine treatment.

Normorphine is a pharmacologically active metabolite, and has neurotoxic effects (93). The formation of morphinone is also toxic to the body. Thus, formation of these metabolites might be involved in tolerance and toxicity of morphine (100).

Renal failure causes the increase in blood morphine-6-glucuronide concentration, thus leading to drug intoxication (94).

Since WHO recommended morphine as the choice of drug for the treatment of cancer pain, extensive pharmacokinetic studies have been carried out using radioimmunoassay or HPLC (93). However, interindividual differences and the formation of the active metabolites morphine-6-glucuronide and normorphine still require further development of sensitive and simple simultaneous determinations of morphine and its metabolites. Recently, ethnic differences in response to morphine have been indicated.

Evidence is accumulating that the consumption of poppy seed foods results in morphine and codeine concentrations as high as 0.100 and 0.007 mg/l in serum, and 4.5 and 0.2 mg/l in urine (77–81). Therefore, in the case of toxicological detection of morphine and related compounds, analysts should consider whether these compounds originate from poppy seed foods or from illicited use.

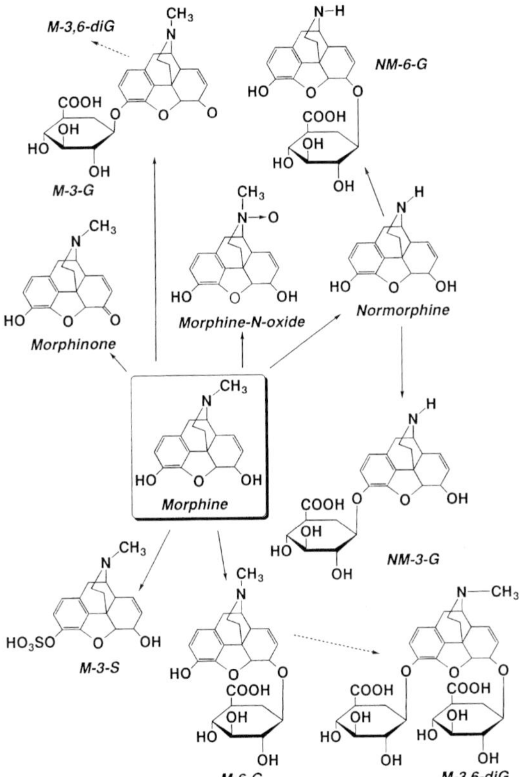

Figure 12. Chemical Structure and Metabolic Fate of Morphine.

3.14.4.9.4 Papaverine

Usage: Papaverine is an isoquinoline alkaloid that is present in opium to the extent of about 1 % by weight. Papaverine is a smooth muscle relaxant and is used clinically as a vasodilator and antispasmodic.

Toxicity: Papaverine causes headache, dizziness, nausea, sweating, diarrhea, and jaundice as general adverse effects.

Metabolism and excretion (101–104): Papaverine is mainly converted to phenolic metabolites via O-demethylation, which are excreted as glucuronide conjugates in urine. About 60% of an oral dose of papaverine is eliminated in 24hr, with less than 1% as unchanged drug. Conjugated metabolites include 3′-hydroxy-, 4′-hydroxy-, 6-hydroxy-, 7-hydroxy, and a trace amount of 4′,6-dihydroxy-papaverine.

Analysis: The determination of papaverine has been accomplished by UV spectrometry, GC equipped with FID or NPD (105), and HPLC (106), depending on the amounts and sample specimens.

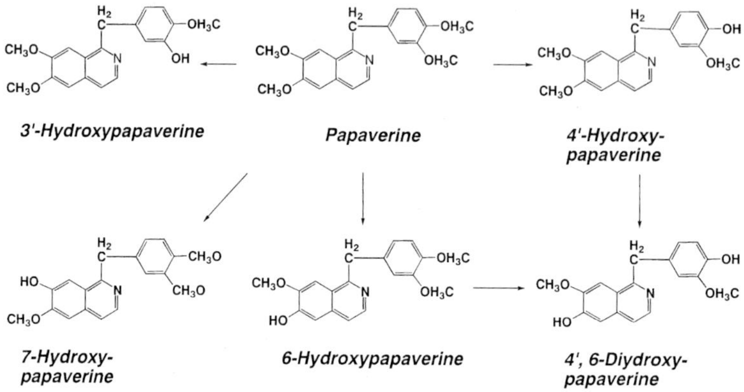

3′-Hydroxypapaverine — Papaverine — 4′-Hydroxy-papaverine

7-Hydroxy-papaverine — 6-Hydroxypapaverine — 4′, 6-Diydroxy-papaverine

Figure 13. Chemical Structure and Metabolic Fate of Papaverine.

3.14.4.10 Psilocybe alkaloids

Occurrence and usage: Psilocin and psilocybin are naturally occurring indole alkaloids in the mushroom *Psilocybe mexicana*, and some others. These so-called "magic-mushrooms" were used in ancient Aztec relegious rites. Psilocin and psilocybin have hallucinogenic properties, which have features similar to those of LSD and bufotenine, and thus their separation and identification have been dealt with in connection with the analysis of drugs of abuse.

Toxicity (107, 108): Intoxication of psilocin and psilocybin cause perceptual alterations and illusions, including changes in touch, taste and odor. At high doses, they cause hallucinations and loss of contact with reality.

Metabolism and excretion: Psilocybin is rapidly metabolized by dephosphorylation, catalyzed by alkaline phosphatase in the body, to form psilocin. Mammalian tissues and plasma are also capable of rapidly dephosphorylating psilocybin, the renal enzyme having higher activity than that of the liver and other tissues. Psilocin is further metabolized to 4-hydroxyindoleacetic acid. The effect of psilocybin on the central nervous system is thought to be exerted mainly by psilocin. The behavioural effects closely follow the increase in brain levels of psilocin. Tissue distribution of psilocybin and psilocin has been studied extensively in animals.

Analysis: For the detection of psilocin and pscilocybin, potassium indoplatinate and p-dimethylaminobezaldehyde reagents have been used with dark purple colour in the former, and blue and purple-blue colours for psilocin and psilocybin, respectively, with the latter reagent. The quantitative analysis of psilocin and psilocybin has been carried out by GC (9), HPLC and TLC (109–110), and GC-MS (9).

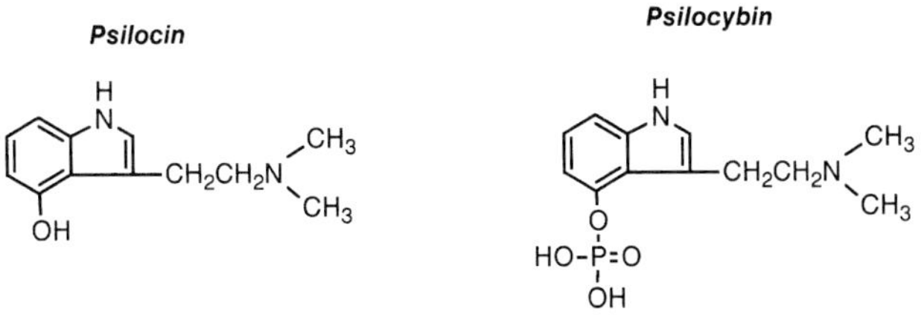

Psilocin

Psilocybin

Figure 14. Chemical Structures of Psilocin and Psilocybin.

3.14.4.11 Rauwolfia alkaloids

Occurrence and usage: The Rauwolfia alkaloids reserpine, reserpidine, serpentine, deserpine, rescinamide, ajmaline and etc., occur in the roots of *Rauwolfia serpentina*. Powdered dried whole root preparations, partially purified alkaloid fractions of the plants and pure alkaloids are commercially available.

3.14.4.11.1 Reserpine

Usage: Reserpine has been widely used for the treatment of hypertension in recent years and as a neuroleptic (major tranquilizer). It is also a very useful tool in experimental studies to produce so-called reserpinized animals.

Toxicity: The unwanted adverse effects of reserpine are likely due to the depletion of norepinephrine and dopamine. Chronic usage of rauwolfia alkaloids, mainly reserpine, causes nasal congestion, central nervous system related symptoms (depression, drowsiness, fatigue etc.), cardiovascular system disturbance (bradycardias, hypotensions etc.), parkinsonism and other severe adverse effects. However, fatalities have not been reported. There are no antidotes.

Metabolism and excretion: Reserpine is readily absorbed through oral and intramuscular routes. Reserpine is rapidly and extensively metabolized by hydrolysis and O-demethylation into 3,4,5-trimethoxybenzoic acid and methyl reserpate (111, 112). Reserpic acid, syryngomethyl reserpate and syringic acid are also formed. The elimination half-life of reserpine in blood is biphasic with 4.5 hr in the first phase and 271 hrs in the second phase. Rauwolfia alkaloids are widely distributed in the brain, liver, spleen, kidney, and other tissues.

Analysis: Reserpine has been analyzed by GC (113), but its low volatility leads to poor sensitivity. Rather, HPLC is the analytical method of choice for reserpine and its metabolites (114, 115).

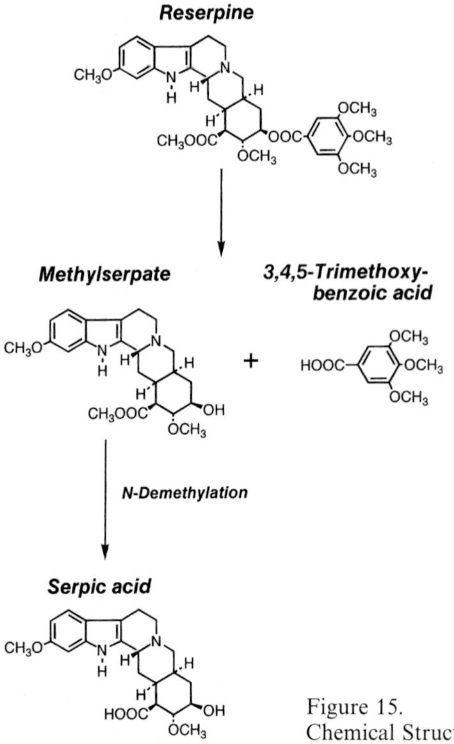

Reserpine

Methylserpate

3,4,5-Trimethoxy-benzoic acid

N-Demethylation

Serpic acid

Figure 15.
Chemical Structure and Metabolic Fate of Reserpine.

3.14.4.11.2 Ajmaline

Usage: Ajmaline is a potent antiarrhythmic drug. It is used worldwide for the treatment of tachycardia in preexcitation syndromes and ventricular arrhythmia.

Toxicity: Ajmaline depresses the conductivity of the heart, and at high doses can cause heart block. It may produce a negative inotropic effect at very high doses. High doses of ajmaline may also cause cardiac arrhythmias, coma and death. Adverse neurological effects include eye twitching, convulsions and respiratory depression. It may also cause hepatotoxicity and agranulocytosis.

Metabolism and excretion: Only a few percent of ajmaline are excreted unchanged into urine. Ajmaline is metabolized by mono- and dihydroxylation, N-demethylation, N-oxidation, oxidation of hydroxyl groups and a combination of these metabolic pathways (116). Ajmaline N-oxide is probably the only active metabolite. Ajmaline metabolism cosegregates with oxidative polymorphic sparteine/debrisoquine/dextromethorphan metabolism in humans (117–119).

Analysis: Radioimmunoassay has been reported for the determination of ajmaline in plant extracts (120). HPLC assay has been used for the determination of ajmaline in biological samples (118, 119).

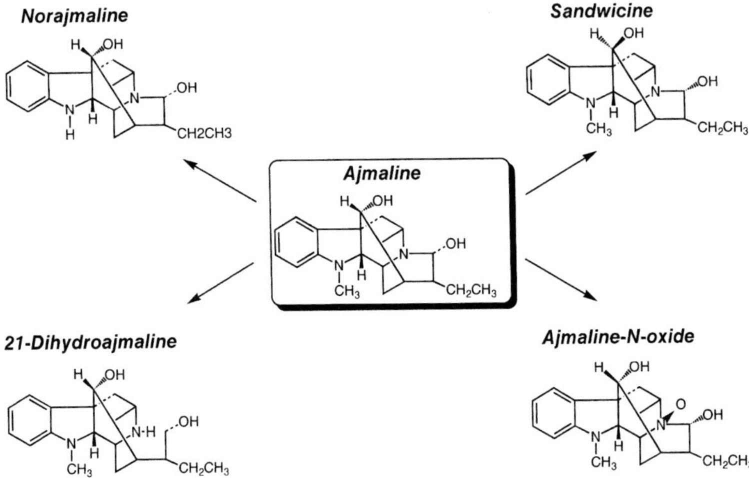

Figure 16. Chemical Structure and Metabolic Fate of Ajmaline.

3.14.4.12 Strychnos alkaloids

About 300 strychnos alkaloids are known. Of these alkaloids, strychnine and brucine are very important. They have been analyzed most extensively by TLC, as present in plant material or plant extracts, in pharmaceutical preparations, in toxicological analysis or as adulterants in drugs of abuse.

Occurrence and usage: Strychnine, brucine and other alkaloids occur in the seeds of *Strychnos nux vomica* L. Strychnine has been used as a stimulant for centuries. It is also used as a rodenticide. Strychnine and brucine produce a very bitter taste. The sensitivity to taste can be as low as 1 : 2000. As strychnine is very toxic, accidental and suicidal intoxication or death cases do often occur. They are potent central nervous system stimulants and convulsants, acting by the selective blockade of post-synaptic neuronal inhibition. Strychnine causes hypersensivity of all senses, and muscles of the neck and face become rigid. The body of a patient may be "arched" backwards. Convulsions may last a minute and more. Prolonged convulsions can lead to coma and death due to apnea and anoxia.

Toxicity: Strychnine-caused death may occur within 15 min, or delayed up to 10 hours or more, during or following a massive convulsion with exhaustion of the phrenic and interacostal muscles, and resulting apnea (121, 122).

Metabolism and excretion: Strychnine is readily absorbed from the gastrointestinal tract and rapidly metabolized by the liver. However, little information is available regarding the metabolic fate of strychnine in humans. Strychnine is not cumulative and is rapidly eliminated from the blood and tissues. It can be detected in the urine within 30 min and up to 12 hours. Metabolic pathways of strychnine in experimental animals have been extensively studied and various metabolites, such as 16-, 2- or 22-hydroxylated-, 21, 22-epoxydated- and N-oxidated-metabolites, have been found (123–126). There are species differences in the metabolism of strychnine (125).

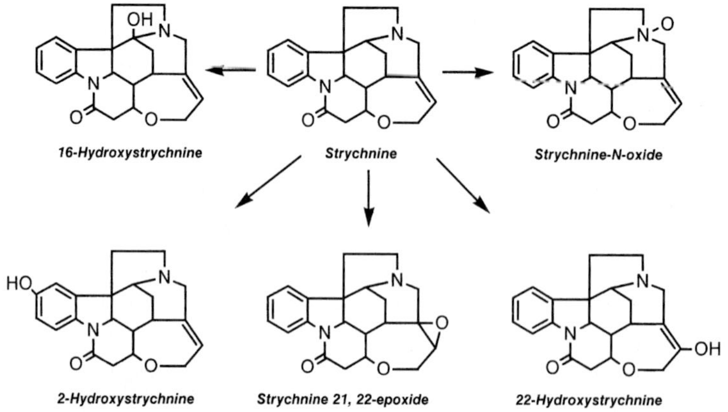

16-Hydroxystrychnine Strychnine Strychnine-N-oxide

2-Hydroxystrychnine Strychnine 21, 22-epoxide 22-Hydroxystrychnine

Figure 17. Chemical Structure and Metabolic Fate of Strychnine.

Analysis: Urine, liver, stomach, intestine and brain are specimens of choice. Blood may sometimes be negative even in fatal poisonings (127).

Strychnine has been analyzed by GC(123) and HPLC (124–126).

3.14.4.13 Xanthines

3.14.4.13.1 Caffeine

Occurrence and usage: Caffeine is a weekly basic alkaloid that occurs naturally in coffee, cocoa beans and tea leaves. Two caffeine-like alkaloids with purine skeletons, theobromine and theophylline, are also present in these plants. Chronic usage of caffeine is believed to result in tolerance and habituation to this agent. Toxicologically, caffeine is of importance as an adulterant to drugs of abuse.

Toxicity: As the lethal dose of caffeine is very large (about 10 g), very few death cases have been reported by its use, in contrast to other alkaloids. Caffeine alone rarely causes cardiac toxicity, although cases of supraventricular tachycardias have been reported. There are reports of severe caffeine intoxication with tachyarrhythmias followed by cardiovascular collapse and death in children. Caffeine potentiates the effects of sympathomimetic drugs and might contribute to adverse cardiac events.

Although the toxicity of caffeine is very low, the most significant modern toxicological concern about this compound has probably been the consumption by pregnant women and the effects on fetal development. A cohort study indicated a higher incidence of spontaneous abortion in subjects who consumed caffeine in moderate-to-heavy (> 151 mg/day) compared with light amounts (< 150 mg/day) or none at all (128).

Metabolism and excretion: Plasma half-life of caffeine has been shown to be around several hours by collecting data after oral administration.

Caffeine is extensively metabolized by N-demethylation and oxidation of the 8-carbon to uric acid derivatives. Several metabolites are found in urine, such as paraxanthine, 7-methylxanthine, 1-methylxanthine, 1,3-dimethyluric acid, and 1-

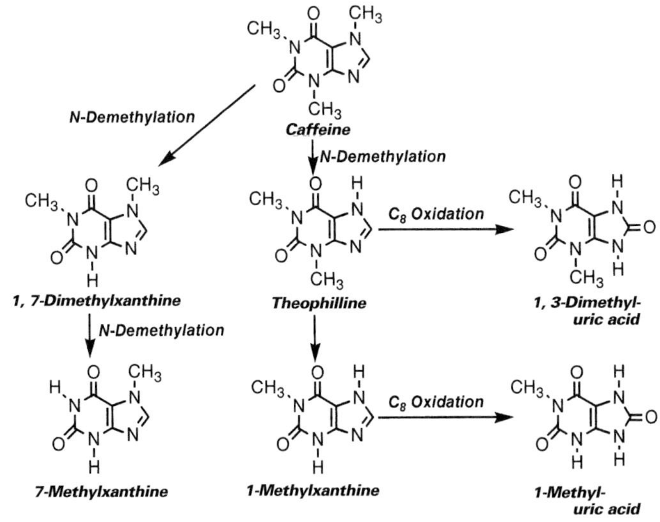

Figure 18. Chemical Structure and Metabolic Fate of Caffeine.

methyluric acid (129–131). Newborns generally have low ability to metabolize caffeine and excrete up to 85% unchanged, with a plasma half-life of about 4 days. The ability to metabolize the drug reaches adult levels by the age of one year (130).

There are species differences in caffeine metabolism (131–133): in humans 3-demethylation predominates, in monkeys 7-demethylation predominates, and rats demonstrate a third major pathway of 1-demethylation to theobromine.

The metabolic patterns of caffeine in humans have been shown to be useful indicators for predicting the presence of cytochrome P-450 1A species, important drug-metabolizing enzymes, in the liver (134, 135). Additionally, caffeine has been shown to be a good probe drug for phenotyping N-acetyltransferase polymorphism (135–138). Caffeine metabolism has also been shown to be a good indicator for the assessment of liver function (139, 140).

Analysis: Murexide reaction is fairly sensitive for qualitative determination of methylxanthines. Caffeine, its analogues and their metabolites are preferentially separated and quantitated by HPLC (141–144).

3.14.4.13.2 Theophylline

Usage: Theophylline is one of the most important naturally occurring therapeutic drugs and has been used as a bronchodilator in the control of asthma. It also has cardiotonic, diuretic and respiratory properties. The problem of therapeutic usage of theophylline is that its effective plasma concentration is in the narrow range of 8–20 μg/ml, and at levels higher than 20 μg/ml serious side effects may occur. Therefore, therapeutic monitoring of theophylline is required during treatment with this drug.

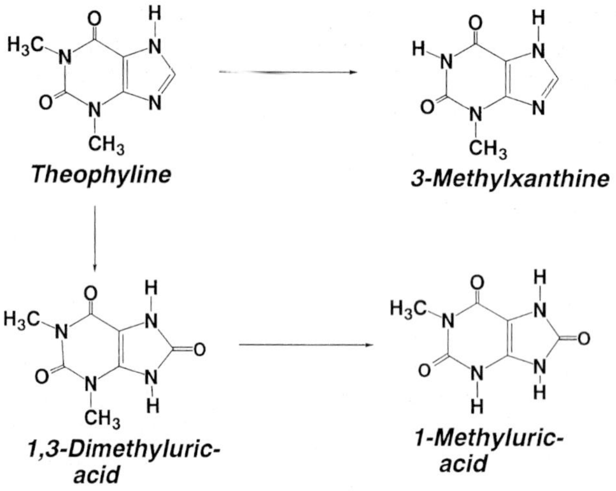

Theophyline

3-Methylxanthine

1,3-Dimethyluric-acid

1-Methyluric-acid

Figure 19. Chemical Structure and Metabolic Fate of Theophylline.

Toxicity: Theophylline causes nausea, vomiting, epigastric pain, headache and grand mal depending on the plasma concentration over 20 μg/ml. Overdoses and intentional acute ingestion of theophylline produce emesis, tachycardia, seizure and respiratory arrest.

Metabolism and excretion: There is a large age-dependent difference in theophylline metabolism (145). Theophylline is oxidized to 1,3-dimethyluric acid, and the parent compound and this metabolite are further metabolized by N-demethylation to 3-methylxanthine and 1-methyluric acid, respectively, and excreted into urine. A systematic review of interspecies and interindividual differences in metabolism and pharmacokinetics as well as their relation to pathophysiological conditions has recently been written by Gaspari and Bonati (146). Theophylline metabolism differs greatly in infants and adults (145). Even in infants, it has been shown that there is a dramatic change in theophylline disposition during the first year of life (147). Theophylline is frequently used for a variety of indications throughout infancy; thus, a better understanding of the time profile for metabolic and pharmacokinetic changes of the drug is essential for its safe and effective use in infancy.

Analysis: The determination of theophylline has usually been accomplished together with caffeine and related compounds. Liquid chromatographic techniques are now in routine use for the determination of theophylline and its metabolites as described for caffeine. An immunoassay method is also commercially available.

3.14.5 Future Perspectives of Alkaloids Toxicology and Toxicological Analysis

Alkaloids are of great importance in toxicology with respect to their worldwide distribution in plants, general high toxicity, and therapeutic and other usages. Searches for highly toxic alkaloids present in small amounts in plants and other sources continue. A newly discovered highly toxic alkaloid may provide new insights into pharmacological and toxicological regulatory systems in the body. Based on the successful usages of various groups of alkaloids as medicines, finding novel pharmacological effects of alkaloids may produce new lead-compounds. In turn, such findings would necessitate the development of analytical methods for pharmacokinetic and toxicokinetic studies. An example: the use of taxol as an anti-cancer drug in phase II clinical studies may lead to Taxus poisoning, and may require further development of methods for toxicological analysis of the drug and alkaloids in the plant Taxus (yew) (148).

Metabolic fates of many highly toxic alkaloids, such as aconitine and related alkaloids, are still not known in detail even in animals. Therefore, pharmacokinetic and toxicokinetic studies on these alkaloids should be performed in the future to improve clinical treatment of intoxicated patients. For this purpose, however, further developments of highly sensitive analytical methods are required. Recent advances in the development of state-of-the-art techniques such as GC-MS, HPLC-MS and tandem MS are very useful for metabolic studies. These methods make it possible to simultaneously determine and identify the drug and its metabolites.

With few exceptions, drugs and chemicals ingested into the body are mainly metabolized by cytochrome P-450-containing drug metabolizing enzymes, and also other hydrolyzing enzymes, and excreted into urine and/or bile after conjugation reactions. Recently, interindividual and ethnic differences in drug metabolism have received a lot of attention in clinical pharmacology (pharmacogenetic studies). Such genetic differences in drug metabolizing enzymes, especially cytochrome P–450 species, have also been revealed with the alkaloids codeine, ajmaline and nicotine. Because many alkaloids and their derivatives are utilized extensively as modern therapeutic drugs, toxicologists should consider in future studies, whether interindividual differences in the toxic effects of alkaloids are related to such genetic factors. Additionally, drug metabolizing enzymes are subject to change under various pathophysiological conditions. A correlation between an interindividual change in a drug metabolizing enzyme under such pathophysiological conditions and the appearance of unwanted adverse effects produced by various natural and semisynthesized alkaloid drugs remains to be determined in future studies of toxicokinetics.

After the discovery of the excretion of opium alkaloids in human hair (149), evidence has been accumulating that many basic substances including alkaloids and other compounds are present in the hair (149–152). Therefore, hair analysis for toxic compounds is now an important field of study in forensic toxicology. Hair analysis will also be a new diagnostic tool for detecting drugs of abuse as well as therapeutic drugs. Detailed study with hair could help determine the approximate time of ingestion of toxic alkaloids. Although hair analysis has made great progress, many problems remain to be solved regarding (1) the kind of drugs which can be detected in hair, (2) whether there is any correlation between the doses administered and hair drug concentrations, (3) whether the drugs in hair are retained and to what extent, (4) what kind of factor(s) determines the incorporation rates of drugs from the blood stream into hair, (5) whether the drugs change chemically or biochemically in the hair, (6) whether it is possible to elucidate past drug usages, (7) whether there is any correlation between pathophysiological conditions and the excretion of drugs into hair, (8) whether there are interindividual or ethnic differences in the excretion of drugs into hair.

On the other hand, forensic toxicologists are involved in making chemical fingerprints or impurity profiles of illicit drugs such as heroin, cocaine and so forth, in order to identify the origin or sources of the drugs. Many alkaloids have been shown to be adulterants in those illicit drugs (153–158). Some alkaloids, such as strychnine, which is occasionally found to be one of the adulterants, are highly toxic. Thus, it is suspected that some deaths or intoxications can be attributed to such toxic alkaloid adulterants. Together with the establishment of chemical finger prints or impurity profiles and corresponding data-bases, toxicologists should be concerned with toxic interaction between illicit drugs and adulterants. The definitions of such toxic interactions are future subjects of study in toxicology. A better understanding of toxic interactions will readily enable forensic toxicologists to deduce the cause of death or intoxication from illicit drugs.

Recent progress in analytical methodology and instrumentation has also greatly promoted micro-toxicological analysis, resulting in serious problems in identifying an objective compound because of the simultaneously increased sensitivity toward contaminating substances.

The developments of analytical procedures and analytical instruments may provide social problems in the toxicological evaluation of obtained data. Commercially available poppy seeds are used in cooking. Several reports have shown that poppy seeds contain substantial amounts of opium alkaloids such as morphine and codeine (77–81). In addition, urinary excretion of morphine and codeine has been detected in normal volunteers after ingesting foods containing poppy seeds. These findings suggest that one should be aware of possible urinary excretion of such opium alkaloids when evaluating results from a forensic or toxicological viewpoint. Thus, toxicologists should always consider whether the obtained analytical data are within or over the cut-off levels. The establishment of cut-off levels for illicit drugs, specifically morphine and related alkaloids, will be required.

Finally, the discovery and development of specific antidotes for toxic alkaloids and other drugs is also a subject for further studies. At present, there are some promising antidotes, such as atropine for organophosphate or carbamate poisoning, and naloxone for morphine and related compounds. An inhibition of enteric hydrolytic enzymes, such as β-glucuronidase and β-glucosidase, may reduce or protect enteric toxicity or other tissue toxicity from the excreted conjugated active metabolites by inhibiting hydrolysis and stimulating their excretion from the body. Still great efforts should be made to discover mechanism-based or function-based antidotes for toxic alkaloids and drugs.

The better understanding of toxicology and toxicological analysis of alkaloids could lead to the protection from and treatment of these health endangering compounds.

References

1. Dumbacher, J.P., Beeler, B.M., Spande, T.F., Martin Garaffo. H. and Daly, J.W. (1992) Homobatrachotoxin in the genus Pitohui: Chemical defense in Birds. Science, *258*, 799–801.
2. Donnerer, J., Oka, K., Brossi, A., Rice, K.C. and Spector, S. (1986) Presence and formation of codeine and morphine in the rat. Proc. Natl. Acd. Sci. USA, *83*, 4566–4567.
3. Kosterlitz, H.W. (1987) Biosynthesis of morphine in the animal kingdom. Nature, *330*, 606.
4. Matsubara, K., Fukushima, S., Akane, A., Kobayashi, S. and Shino, H. (1992) Increased urinary morphine, codeine and tetrahydropapaveroline in a parkinsonian patient undergoing 1–3, 4-dihydroxyphenylalanine therapy: a possible biosynthetic pathway of morphine from 1–3, 4-dihydroxyphenylalanine in humans. J. Pharmacol. Exp. Ther., *260*, 974–978.
5. Chromatography of alkaloids – part A: thin-layer chromatography; Journal of chromatography library, Vol. 23 A (1983) Baerheim Svendsen, A. and Verpoorte, R., Elsevier Scientific Publishing company INC.
6. Chromatography of alkaloids – Part B: gas-liquid chromatography and high-performance liquid chromatography. Journal of chromatography library Vol. 23 B (1984) Verpoort, R. and Baerheim Svendsen, A., Elsevier Scientific Publishing company INC.
7. Chromatographic analysis of alkaloids; Chromatographic sciencs series, Vol. 53, ed. by Popl. M., Fahnrich, J. and Tatar. V., Marcel Dekker Inc., New York and Basel, 1990.
8. Holder, A.T. (1986) The analysis of narcotics; in Analytical methods in human toxicology, Part 2, ed. by Curry, A.S., Verlag Chemie, Weinheim, pp. 241–288.
9. Clark's isolation and identification of drugs. ed. by The Pharmaceutical Society of Great Britain, second edition, The Pharmaceutical Press, London, 1986.
10. Francotte, E. and Junker-Buchheit, A. (1992) Preparative chromatographic separation of enantiomers. J. Chromatogr., *576*, 1–45.
11. Sato, H., Yamada, C., Konno, C., Ohizumi, Y., Endo, K. and Hikino, H. (1977) Pharmacological actions of aconitine alkaloids. Tohoku J. Exp. Med., *128*, 175–187.
12. Mizugaki, M., Ohyama, Y., Kimura, K., Ishibashi, M., Ohno, Y., Uchima, E., Nagamori, H. and Suzuki, Y. (1988) Analysis of aconitum alkaloids by means of gas chromatography/selected ion monitoring. Jap. J. Toxicol. Environ. Health, *34*, 359–365.
13. Hikino, H., Murakami, M., Konno, C. and Watanabe, H., (1983) Determination of aconotine alkaloids in *Aconitum* roots. Planta Med. *48*, 67–71.
14. Vander Meer, M.J., Hudndt, H.K.L. and Muller, F.O. (1986) The metabolism of atropine in man. J. Pharm. Pharmacol., *38*, 781–784.
15. Zhou, H.H., Adedoyin, A. and Wood, A.J.J. (1992) Differing effect of atropine on heart rate in Chinese and white subjects. Clin. Pharmacol. Therap., *52*, 120–124.
16. Berghem, L., Berghem, U., Schildt, B. and Sorbo, B. (1980) Plasma atropine concentrations determined by radioimmunoassay after single-dose i.v. and i.m. administration. Br. J. Anesthesiol., *52*, 597–601.
17. Martinsen, A., Hiltunen, K. and Huhtikangas, A. (1992) Nitrogen-phosphorus detector potential in tropane alkaloid gas chromatography. Phytochem. Anal., *3*, 69–72.
18. Eckert, M. and Hinderling, P.H. (1981) Atropine: a sensitive gas chromatography-mass spectrometry assay and pharmacokinetic studies. Agents and Actions, *11*, 520–545.
19. Mandal, S., Naqvi, A.A. and Thakur, R.S. (1991) Analysis of some tropane alkaloids in plants by mixed column high-performance liquid chromatography. J. Chromatogr., *547*, 468–471.
20. Oshima, T., Sagara, K., Hirayama, F., Mizutani, T., He, L., Tong, Y., Chen, Y. and Itokawa, H. (1991) Combination of ion-pair column switching in high-performance liquid chromatography of tropane alkaloids. J. Chromatogr., *547*, 175–183.
21. Papadoyannis, I.N., Samanidou, V.F., Theodoridis, G.A., Vasilikiotis, G.S., van Kempen, G.J.M. and Beelen, G.M. (1993) A simple and quick solid phase extraction and reversed phase HPLC analysis of some tropane alkaloids in feedstuffs and biological samples. J. Liquid Chromatogr., *16*, 975–998.
22. Wada, S., Yoshimitsu, T., Kaga, N., Yamada, H., Oguri, K. and Yoshimura, H. (1991) Metabolism *in vivo* of the tropane

alkaloid, scopolamine, in several mammalian species. Xenobiotica, *21*, 1289–1300.
23. Cintron, N.M. and Chen, Y. (1987) A sensitive radioreceptor assay for determining scopolamine concentrations in plasma and urine., J. Pharm. Sci., *76*, 328–332.
24. Hagemann, K., Piek, K., Stoeckigt, J. and Weiler, E.W. (1992) Monoclonal antibody-bases enzyme immunoassay for the quantitative determination of the tropane alkaloid, scopolamine. Planta Med., *58*, 68–72.
25. Brodie, B.B., Baer, J.E. and Craig, L.C. (1951) Metabolic products of the cinchona alkaloids in human urine. J. Biol. Chem., *188*, 567–581.
26. Watabe, T. and Kiyanaga, K. (1972) The metabolic fate of quinine in rabbits. J. Pharm. Pharmacol., *24*, 625–630.
27. Carrol, F.I., Smith, D., Wall, M.E. and Moreland, C.G. (1974) Carbon-13 magnetic resonance study. Structure of the metabolites of orally administered quinidine in humans. J. Med. Chem., *17*, 985–987.
28. Barrow, S.E. and Taylor, A.A. (1980) HPLC separation and isolation of quinidine and quinine metabolites in rat urine. J. Chromatogr., *181*, 219–226.
29. Liddle, C., Graham, G.G., Christopher, R.K., Bhuwapathanapun, S. and Dufield, A.M. (1981) Identification of new urinary metabolites in man of quinine using methane chemical ionization gas chromatography-mass spectrometry. Xenobiotica, *11*, 81–87.
30. Bolaji, O.O., Babalola, C.P. and Dixon, P.A.F. (1991) Characterization of the principal metabolite of quinine in human urine by ^{1}H-n.m.r. spectroscopy. Xenobiotica, *21*, 447–450.
31. Kobayashi, S., Murray, S., Watson, D., Sesardic, D., Davies, D.S., and Boobis, A.R. (1989) The specificity of inhibition of debrisoquine 4-hydroxylase activity by quinidine and quinine in the rat is the inverse of that in man. Biochem. Pharmacol., *38*, 2795–2799.
32. Brosen, K., and Gram, L.F. (1989) Clinical significance of the sparteine/debrisoquine oxidation polymorphism. Eur. J. Clin. Pharmacol., *36*, 537–547.
33. Mihaly, G.M., Hyman, K.M., Smallwood, R.A. and Hardy, K.J. (1987) High-performance liquid chromatographic analysis of quinine and its diastereoisomer quinidine. J. Chromatogr., *415*, 177–182.
34. MacKichan, J.J. and Shields, B.J. (1987) Specific high-performance liquid chromatographic determination of quinidine in serum, blood and urine. Therap. Drug Monitor., *9*, 104–112.
35. Giroud, C., Van der Leer, T., Van der Heijden, R., Verpoorte, R., Heeremans, C.E.M., Niessen, W.M.A. and Van der Greef, J. (1991) Thermospray liquid chromatography/mass spectrometry (TSP LC/MS) analysis of the alkaloids from Cinchona in *in vitro* cultures. Planta Med., *57*, 142–148.
36. Abdulrahman, S., Harrison, M.E., Welham, K.J., Baldwin, M.A., Phillipson, J.D. and Roberts, M.F. (1991) High-performance liquid chromatographic-mass spectrometric assay of high-value compounds for pharmaceutical use from plant cell tissue culture: Cinchona alkaloids. J. Chromatogr., *562*, 713–721.
37. Stapzynski, J.S., Rothstein, R.J., Gaye, W.A. and Niemann, J.T. (1981) Colchicine overdose: report of two cases and review of the literature. Ann. Emerg. Med., *10*, 354–369.
38. Kuncl, R.W., Duncan, G., Watson, D., et al. (1987) Colchicine myopathy and neuropathy. New Eng. J. Med., *316*, 1562–1568.
39. Walaszek, E.J., Kocsis, J.J., Leory, G.V. and Geiling, E.M.K. (1960) Studies on the excretion of radioactive colchicine. Arch. Int. Pharm., *125*, 371–382.
40. Chappey, O.N., Niel, E., Wautier, J.-L., Hung, P.P., Dervichian, M., Cattan, D. and Schermann, J.M. (1993) Colchicine disposition in human leukocytes after single and multiple oral administration. Clin. Pharmacol. Therap., *54*, 360–367.
41. Arai, T. and Okuyana, T. (1975) Fluorometric assay of tubulin-colchicine complex. Anal. Biochem., *69*, 443.
42. Schermann, J.M., Boudet, L., Pontikis, R., (1980) A sensitive radioimmunoassay for colchicine. J. Pharm. Pharmacol., *32*, 800–803.
43. Lhermitte, M., Bernier, J.L., Mathieu, D. Mathieu-Nolf, M., Erb, F. and Roussel, P. (1985) Colchicine quantitation by HPLC in human plasma and urine. J. Chromatogr. Biomed. Appl., *342*, 416–423.
44. Fernandez, P., Bermejo, A.M., Tabernero, M.J., Lopez-Rivadulla, M. and Cruz, A. (1993) Determination of colchicine in biological fluids by reverse-phase HPLC. Variation of colchicine levels in rats. Forensic Sci. Int., *59*, 15–18.
45. Clevenger, C.V., August, T.F. and Shaw, L.M. (1991) Colchicine poisoning: report

of fatal case with body fluid analysis by GC-MS and histopathologic examination of postmortem tissues. J. Anal. Toxicol., *15*, 151–154.

46. Jacques, P., Nicole, G. and Claude, V. (1991) Separation of alkaloids in plants extracts by overpressed layer chromatography with ethyl acetate as mobile phase. J. Planar Chromatogr., *4*, 392–396.
47. Orton, D.A. and Richardson, R.J. (1982) Ergotamine absorption and toxicity. Postgrad. Med. J., *58*, 5–11.
48. McGuigan, M.A. (1984) Ergot alkaloids. Clin. Toxicol. Rev., *6*, 1–2.
49. Weaver, R., Philips, M. and Vacek, J.L. (1989) St Anthony's Fire: A medieval disease in modern times: case history. Angiology, *40*, 929–932.
50. Magee, R. (1991) Saint Anthony's fire revisited: Vascular problems associated with migraine medication. Med. J. Aust., *154*, 145–149.
51. Edlund, P.O. (1981) Determination of ergot alkaloids in plasma by high-performance liquid chromatography and fluorescence detection. J. Chromatogr., *226*, 107–115.
52. Lin, L.A. (1993) Detection of alkaloids in foods with a multi-detector high-performance liquid chromatographic system. J. Chromatogr., *632*, 46. 69–78.
53. Haering, N., Schubert, R., Settlage, J.A. and Sanders, S.W. (1985) The measurement of ergotamine in human plasma by triple sector quadrupole mass spectrometry with negative ion chemical ionization. Biom. Mass Spectrom., *12*, 197–199.
54. Feng, N., Minder, E.I., Grampp, T. and Vonderschmit, D.J. (1992) Identification and quantitation of ergotamine in human plasma by gas chromatography-mass spectrometry. J. Chromatogr., *575*, 289–294.
55. Shelby, R.A. and Kelley, V.C. (1990) An immunoassay for ergotamine and related alkaloids. J. Agric.Food Chem., *38*, 1130–1134.
56. Shelby, R.A. and Kelley, V.C. (1991) Detection of ergot alkaloids in tall fescue by competitive immunoassay with a monoclonal antibody. Food Agric. Immunol., *3*, 169–177.
57. Shelby, R.A. and Kelley, V.C. (1992) Detection of ergot alkaloids from Claviceps species in agricultural products by competitive ELISA using a monoclonal antibody. J. Agric. Food Chem., *40*, 1090–1092.
58. Jadhave, S.J., Sharma, R.P. and Salunkhe, D.K. (1981) Naturally occurring toxic alkaloids in foods. CRC Crit. Rev. Toxicol., *9*, 21–104.
59. Dalvi, R.R. and Bowie, W.C. (1983) Toxicology of solanine: An overview. Vet. Hum. Toxicol., *25*, 13–15.
60. Slanina, P. (1990) Solanine (glycoalkaloids) in potatoes: toxicological evaluation. Food Chem. Toxicol., *28*, 759–761.
61. Baker, D., Keeler, R. and Gaffield, W. (1987) Lesions of potato sprout and extracted potato sprout alkaloid toxicity in Syrian hamsters. Clin. Toxicol., *25*, 199–208.
62. Bushway, R.J., McGann, D.F. and Bushway, A.A. (1984) Gaschromatographic method for the determination of solanidine and its application to a study of feed-milk transfer in the cow. J. Agric. Food Chem., *32*, 548–551.
63. Lawson, D.R., Erb, W.A. and Miller, A.R. (1992) Analysis of Solanum alkaloids using internal standardization and capillary gas chromatography. J. Agric. Food Chem., *40*, 2186–2191.
64. Carman Jr., A.S., Kuan, S.S., Ware, G.M., Francis Jr., O.J. and Kirschenheuter, G.P. (1986) Rapid high-performance liquid chromatographic determination of the potato glycoalkaloids – solanine and chaconine –. J. Agric, Food Chem., *34*, 279–282.
65. Friedman, M. and Levin, C.E. (1992) Reversed-phase high-performance liquid chromatographic separation of potato glycoalkaloids and hydrosis products on acidic columns. J. Agric. Food Chem., *40*, 2157–2163.
66. Harvey, M.H., Morris, B.A., McMillan, M. and Marks, V. (1985) Measurement of potato steroidal alkaloids in human serum and saliva by radioimmunoassay. Human Toxicol., *4*, 503–512.
67. Ayesh, R. Al-Waiz, M., Crothers, M.J., Cholerton, S., Idle, J.R. and Smith, R.L. (1988) Deficient nicotine N-oxidation in two sisters with trimethylaminuria. Br. J. Clin. Pharmacol., *25*, 664p–665p.
68. Cashman, J.R., Park, S.B., Yang, Z.-C., Wrighton, S.A., Jacob III, P. and Benowitz, N.L. (1992) Metabolism of nicotine by human liver microsomes; stereoselective formation of trans-nicotine N'-oxide. Chem. Res. Toxicol., *5*, 639–646.
69. Kyerematen, G.A., Morgan, M., Warner, G., Martin, L.F. and Vessel, E.S. (1990) Metabolism of nicotine by hepatocytes. Biochem. Pharmacol., *40*, 1747–1756.
70. Nwosu, C.G. and Crooks, P.A. (1988) Species variation and stereoselectivity in the

metabolism of nicotine enantiomers. Xenobiotica, *18*, 1361–1372.

71. Kyerematen, G.A. and Vessel, E.S. (1991) Metabolism of nicotine. Drug Metab. Rev., *23*, 3–41.
72. Voncken, P., Shepers, G. and Schafer, K.H. (1989) Capillary gas chromatographic determination of trans–3'-hydroxycotinine simultaneously with nicotine and cotinine in urine and blood samples. J. Chromatogr., *479*, 410–418.
73. Zhang, Y., Jacob III, P. and Benowitz, N.L. (1990) Determination of nornicotine in smokers' urine by gas chromatography following reductive alkylation to N'-propylnicotine. J. Chromatogr., *525*, 349–357.
74. Paviainen, M.T., Puhakainen, E.V.J., Laitikainen, R., Savolainen, K., Harranen, J. and Barlow, D. (1990) Nicotine metabolites in the urine of smokers. J. Chromatogr., *525*, 193–202.
75. Norbury, C.G. (1987) Simplified method for the determination of plama cotinine using gas chromatography-mass spectrometry. J. Chromatogr., *414*, 449–453.
76. McManus, K.T., deBethizy, J.D., Garteiz, D.A., Kyerematen, G.A. and Vessel, E.S. (1990) A new quantitative thermospray LC-MS method for nicotine and metabolites in biological fluids. J. Chromatogr. Sci., *28*, 510–516.
77. Bjerver, K., Jonsson, J., Nilsson, A., Schuberth, J. and Schuberth, J. (1982) Morphine intake from poppy seed food. J. Pharm. Pharmacol., *34*, 798–801.
78. Fritschi, G. and Prescott Jr., W.R., (1985) Morphine levels in urine subsequent to poppy seed consumption. Foren. Sci. Int., *27*, 111–117.
79. Hayes, L.W., Krasselt, W.G. and Mueggler, P.A. (1987) Concentrations of morphine and codeine in serum and urine after ingestion of poppy seeds. Clin. Chem., *33*, 806–808.
80. ElSohly, H.N., Stanford, D.F., Jones, A.B., ElSohly, M.A., Snyder, H. and Pedersen, C. (1988) Gas chromatographic/mass spectrometric analysis of morphine and codeine in human urine of poppy seed eaters. J. Foren. Sci., *33*, 347–356.
81. ElSohly, H.N., ElSohly, M.A. and Stanford, D.F. (1990) Poppy seed ingestion and opiates urinalysis: a closer look. J. Anal. Toxicol., *14*, 308–310.
82. Mortimer, O., Persson, K., Ladona, M.G., Spalding, D., Zanger, U.M. and Meyer, U.A. (1990) Polymorphic formation of morphine from codeine in poor and extensive metabolizers of dextromethorphan: Relationship to the presence of immunoidentified cytochrome P-450 2D6. Clin. Pharmacol. Ther., *47*, 27–35.
83. Yue, Q.-Y., Sawe, J. (1992) Interindividual and interethnic differences in codeine metabolism. In Kalow, W. ed. Pharmacogenetics of Drug Metabolism. New York: Pergamon Press, 721–727.
84. Oguri, K., Hanioka, N. and Yoshimura, H. (1990) Species differences in metabolism of codeine: urinary excretion of codeine glucuronide, morphine-3-glucuronide and morphine-6-glucuronide in mice, rats, guinea pigs and rabbits. Xenobiotica, *20*, 683–688.
85. Gjerde, H., Fongen, U., Gundersen, H. and Christophersen, A.S. (1991) Evaluation of a method for simultaneous quantification of codeine, ethylmorphine and morphine in blood. Forensic Sci. Int. *51*, 105–110.
86. Osborne, R., Joel, S., Trew, D. and Slevin, M. (1990) Morphine and metabolite behavior after different routes of morphine administration: Demonstration of the importance of the active metabolite morphine-6-glucuronide. Clin. Pharmacol. Ther., *47*, 12–19.
87. Portenoy, R.K., Thaler, H.T., Inturrisi, C.E., Friedlander-Klar, H. and Foley, K.M. (1992) The metabolite morphine-6-glucuronide contributes to the analgesia produced by morphine infusion in patients with pain and normal renal function. Clin. Pharmacol. *51*, 422–431.
88. Svensson, J.-O., Rane, A., Sawe, J. and Sjoqvist, F. (1982) Determination of morphine, M3G and (tentatively)morphine-6-glucuronide in plasma and urine using ion-pair high-performance liquid chromatography, J. Chromatogr., *230*, 427–432.
89. Ayyangar, N.R. and Bhide, S.R. (1988) Separation of eight alkaloids and meconic acid and quantitation of five principal alkaloids in gum opium by gradient reversed-phase high-performance liquid chromatography. J. Chromatogr., *436*, 455–465.
90. Schuberth, J. and Schuberth, J. (1989) Gas chromatographic-mass spectrometric determination of morphine, codeine and 6-acetylmorphine in blood extracted by solid phase. J. Chromatogr., *490*, 444–449.
91. Bpoerner, U., Abbott, S. and Roe, R.L. (1975) The metabolism of morphine and heroin in man. Drug Metab. Rev., *4*, 39–73.

92. Brunk, F. and Delle, M. (1975) Morphine metabolism in man. Clin. Pharmacol. Ther., *16*, 51–57.
93. Glare, P.A. and Walsh, T.D. (1991) Clinical pharmacokinetics of morphine. Therap. Drug Monitor., *13*, 1–23.
94. Osborne, R., Joel, S., Grebenik, K., Trew, D. and Slevin,M. (1993) The pharmacokinetics of morphine and morphine glucuronides in kidney failure. Clin. Pharmacol. Therap., *54*, 158–167.
95. Oguri, K., Yamada-Mori, I., Shegezane, J., Hirano, T. and Yoshimura, H. (1987) Enhanced binding of morphine and nalorphin to opioid delta receptor by glucuronate and sulfate conjugations at the 6-position, Life Sci., *41*, 1548–1464.
96. Pasternak, G.W., Bodnar, R.J., Clark, J.A. and Inturrisi, C.E. (1987) Morphine-6-glucuronide, a potent mu agonist. Life Sci., *41*, 2845–2849.
97. Abbott, F.V. and Palmour, R.M. (1988) Morphine-6-glucuronide: analgesic effects and receptor binding profile in rats. Life Sci., *43*, 1685–1695.
98. Paul, D., Standifer, K.M., Inturrisi, C.E. and Pasternak, G.W. (1989) Pharmacological characterization of morphine-6-glucuronide, a very potent morphine metabolite. J. Pharmacol. Exp. Ther., *251*, 477–483.
99. Osborne, R., Thompson, P., Joel, S., Trew, D., Patel, N. and Slevin, M. (1992) The analgesic activity of morphine-6-glucuronide. Br. J. Clin. Pharmacol., *34*, 130–138.
100. Kumagai, Y., Todaka, T. and Toki, S. (1990) A new metabolic pathway of morphine: in vivo and in vitro formation of morphinone and morphinone-glutathione adduct in guinea pig. J. Pharmacol. Exp. Therap., *255*, 504–510.
101. Belpaire, F.M., Bogaert, M.G., Rosseel, M.T. and Anteunis, M. (1975) Metabolism of papaverine I. Identification of metabolites in rat bile. Xenobiotica, *5*, 413–420.
102. Belpaire, F.M. and Bogaert, M.G. (1975) Metabolism of papaverine II. Species differences. Xenobiotica, *5*, 421–429.
103. Belpaire, F.M., Rosseel, M.T. and Bogaert, M.G. (1978) Metabolism of papaverine. IV. Urinary elimination of papaverine metabolites in man. Xenobiotica, *8*, 297–300.
104. Ritschel, W.A. and Hammer, G.V. (1977) Pharmacokinetics of papaverine in man. Intern. J. Clin. Pharmacol., *15*, 227–229.
105. Bellia, V., Jacob, J. and Smith, H.T. (1978) Determination of papaverine in blood samples by gas chromatography using a flame-ionization and a nitrogen-phosphorus detector. J. Chromatogr., *161*, 231–235.
106. Hoogewijs, G., Michotte, Y., Lambrecht, J. and Massart, D.L. (1981) High-performance liquid chromatographic determination of papaverine in whole blood. J. Chromatogr., *226*, 423–430.
107. Benjamin, C. (1979) Persistent psychiatric symptoms after eating psilocybin mushrooms. Br. Med. J., *1*, 1319–1221.
108. Peden, N.R., Pringle, S.D. and Crooks, J. (1982) The problem of psylocibin abuse. Human Toxicol., *1*, 417–424.
109. Thomson, B.M. (1980) Analysis of psilocybin and psilocin in mushroom extracts by reversed-phase high performance liquid chromatography. J. Forensic Sci., *25*, 779–785.
110. Beug, M.W. and Bigwood, J. (1981) Quantitative analysis of psilocybin and psilocin in *Psilocybe baeocystis* by high-performance liquid chromatography and by thin layer chromatography. J. Chromatogr., *207*, 379–385.
111. Maass, A.R., Jenkins, B., Shen, Y. (1969) Studies on absorption, excretion and metabolism of ^{3}H-reserpine in man. Clin. Pharmacol. Therap., *10*, 366–377.
112. Stitzel, R.E. (1977) The biological fate of reserpine. Pharmacol. Rev., *28*, 179–205.
113. Ardrey, R.E. (1981) Gas-liquid chromatographic retention indices of 1318 substances of toxicological interest on SE-30 or OV-1 stationary phase. J. Chromatogr., *220*, 195–252.
114. Lang, J.R., Stewart, J.T. and Honigberg, I.L. (1983) Post-column air-segmentation photochemical reactor for the fluorometric detection of reserpine by liquid chromatography. J. Chromatogr., *264*, 144–148.
115. Wang, J. and Bonakdar, M. (1986) Liquid chromatography of reserpine and rescinamine using electrochemical detection. J. Chromatogr., *382*, 343–348.
116. Köppel, C., Tenczer, J. and Arndt, I. (1989) Metabolic disposition of ajmaline. Eur. J. Drug Metab. Pharmacokin., *14*, 309–316.
117. Hori, R., Okumura, K., Inui, K.I., Yasuhara, M., Yamada, K., Sakurai, T. and Kawai, C. (1984) Quinidine-induced rise in ajmaline plasma concentration. J. Pharm. Pharmacol., *36*, 202–204.

118. Köppel, C., Wagemann, A., Hansen, G.R. and Müller, C. (1992) Monitoring of ajmaline in plasma with high-performance liquid chromatography. J. Chromatogr., *575*, 87–91.
119. Padrini, R., Piovan, D., Gaglione, E., Zordan, R. and Ferrari, M. (1992) Determination of lorajmine and its metabolite ajmaline in plasma and urine by a new high-performance liquid chromatographic method. Therap. Drug Monitor., *14*, 391–396.
120. Arens, H., Deus-Neumann, B. and Zenk, M.H. (1987) Radioimmunoassay for the quantitative determination of ajmaline. Planta Med., *22*, 179.
121. Edmunds, M., Sheehan, T.M.T. and Vant Hoff, W. (1986) Strychnine poisoning: clinical and toxicological observations on a non-fatal case. Clin. Toxicol. *24*, 245–255.
122. Winek, C.L., Wahba, W.W., Esposito, F.M., et al. (1986) Fatal strychnine level. J. Anal. Toxicol. *10*, 120–121.
123. Mishima, M., Tanimoto, Y., Oguri, K. and Yoshimura, H. (1985) Metabolism of strychnine *in vitro*. Drug Metab. Dispos., *13*, 716–721.
124. Oguri, K., Tanimoto, Y., Mishima, M. and Yoshimura, H. (1989) Metabolic fate of strychnine in rats. Xenobiotica, *19*, 171–178.
125. Tanimoto, Y., Ohkuma, T., Oguri, K. and Yoshimura, H. (1990) Species differences in metabolism of strychnine with liver microsomes of mice, rats, guinea pigs, rabbits and dogs. J. Pharmacobio-Dyn., *13*, 136–141.
126. Tanimoto, Y., Ohkuma, T., Oguri, K. and Yoshimura, H. (1991) A novel metabolite of strychnine, 22-hydroxystrichnine. Xenobiotica, *21*, 395–402.
127. Oliver, J.S., Smith, H. and Watson, A.A. (1979) Poisoning of strychnine Med. Sci. Law, *19*, 134–137.
128. Srisuphan, W. and Bracken, M. (1986) Caffeine consumption during pregnancy and association with late spontaneous abortion. Am. J. Obstet. Gynecol., *154*, 14–20.
129. Cornish, H.H. and Christman, A.A. (1957) A study of metabolism of theobromine, theophylline, and caffeine in man. J. Biol. Chem., *228*, 315–323.
130. Aldridge, A., Aranda, J.V. and Neims, A.H. (1979) Caffeine metabolism in the newborn. Clin. Pharmacol. Ther., *25*, 447–453.
131. Grant, D.M., Tang, B.K. and Kalow, M. (1983) Variability in caffeine metabolism. Clin. Pharmacol. Ther., *33*, 591–602.
132. Gilbert, S.G., Stavric, B., Klassen, R.D. and Rice, D.C. (1985) The fate of chronically consumed caffeine in the monkey (*Macaca fascicularis*). Fundam. Appl. Toxicol., *5*, 578–587.
133. Latini, R., Bonati, M., Marzi, E. and Garattini, S. (1981) Urinary excretion of an uracilic metabolite from caffeine by rat, monkey, and man. Toxicol. Lett., *7*, 267–272.
134. Campbell, M.E., Grant, D.M., Inaba, M. and Kalow, W. (1987) Biotransformation of caffein, paraxanhine, theophilline and theobromine by polycyclic aromatic hydrocarbon-inducible cytochrome(s) P–450 in human liver microsomes. Drug Metab. Dispos., *15*, 237–249.
135. Butler, M.A., Lang, N.P., Young, J.F. (1992) Determination of CYP1A2 and NAT2 phenotypes in human populations by analysis of caffeine urinary metabolites. Pharmacogenetics, *2*, 167–127.
136. Tang, B.K., Kadar, D., Qian, L., Iriah, J. Yip, J. and Kalow, W. (1991) Caffeine as a metabolic probe: validation of its use for acßetylator phenotyping. Clin. Pharmacol. Ther., *49*, 648–657.
137. Bechtel, Y.C., Bonati-Pellie, C., Poisson, N., Magnette, J. and Bechtel, P.R. (1993) A population and family study of N-acetyltransferase using caffeine urinary metabolites. Clin. Pharmacol. Ther., *54*, 134–141.
138. Joeres, R., Klinker, H., Heusler, H., Epping, J., Zilly, W. and Richter, E. (1988) Influence of smoking on caffeine elimination in healthy volunteers and in patients with alcoholic liver cirrhosis. Hepatology, *8*, 575–579.
139. Cheng, W.S., Murphy, T.L., Smith, M.T., Cooksley, W.G.E., Halliday, J.W. and Powell, L.W. (1990) Dose-dependent pharmacokinetics of caffeine in humans: Relevance as a test of quantitative liver function. Clin. Pharmacol. Therap., *47*, 516–524.
140. Wahlander, A., Mohr, S. and Paumgartner, G. (1990) Assessment of hepatic function-comparison of caffeine clearance in serum and saliva during the day and at night. J. Hepatol., *10*, 129–137.
141. Casoli, P. and Verne, H. (1990) High-performance liquid chromatographic determination of methylxanthines in canine serum, gastric and pacreatic juice. Biomed. Chromatogr., *4*, 209–213.

142. Miceli, J. N. and Chapman, W. (1990) Measurement of caffeine metabolites in urine using diode-array detection HPLC. J. Liq. Chromatogr., *13*, 2239–2251.
143. Parra, P., Limon, A., Ferre, S., Guix, T. and Jane, F. High-performance liquid chromatographic separation of caffeine, theophylline, theobromine and paraxanthine in rat brain and serum. (1991) J. Chromatogr., *570*, 185–190.
144. Tanaka, E. (1992) Simultaneous determination of caffeine and its primary demethylated metabolites in human plasma by high-performance liquid chromatography. J. Chromatogr., *575*, 311–314.
145. Grygiel, J. J. and Birkett, D. J. (1980) Effect of age on patterns of theophylline metabolism. Clin. Pharmacol. Ther., *24*, 456–462.
146. Gaspari, F. and Bonati, M. (1990) Interspecies metabolism and pharmacokinetic scaling of theophylline disposition. Drug Metab. Rev., *22*, 179–207.
147. Kraus, D. M., Fisher, J. H., Reitz, S. J., Kecskes, S. A., Yeh, T. F., McCulloch, K. M., Tung, E. C., and Cwik, M. J. (1993) Alterations in theophylline metabolism during the first year of life. Clin. Pharmacol. Ther., *54*, 351–359.
148. Van Ingen, G., Visser, R., Peltenburg, H., Van der Ark, A. M. and Voorman, M. (1992) Sudden unexpected death due to Taxus poisoning. A report of five cases, with review of the literature. Forensic Sci., Intern., *56*, 81–87.
149. Baumgartner, A. M., Jones, P. F., Baumgartner, W. A. and Black, C. T. (1979) Radioimmunoassay of hair for determining opiate-abuse histories. J. Nucl. Med., *20*, 748–752.
150. Nakahara, Y., Takahashi, K., Shimamine, M. and Saitoh, A. (1992) Hair analysis for drugs of abuse. IV. Determination of total morphine and confirmation of 6-acetylmorphine in monkey and human hair by GC/MS. Arch. Toxicol., *66*, 669–674.
151. Poet, T. S., Martinez, F. and Watson, R. R. (1992) Effect of murine retrovial infection on hair and serum levels of cocaine and morphine. Foren. Sci. Intern., *54*, 29–38.
152. Kintz, P., Kiffer, I., Messer, J. and Mangin, P. (1993) Nicotine analysis in neonates' hair for measuring gestational exposure to tobacco. J. Forens. Sci., *38*, 119–123.
153. Moore, J. M., Allen, A. C. and Cooper, D. A. (1984) Determination of manufacturing impurities in heroin by capillary gas chromatography with electron capture detection after derivatization with heptafluorobutyric anhydride. Anal. Chem. *56*, 642–646.
154. Moore, J. M., Allen, A. C. and Cooper, D. A. (1986) Determination of neutral manufacturing impurities in heroin by capillary gas chromatography with electron capture detection after reduction with lithium alminum hydride and derivatization with heptafluorobutyric anhydride. Anal. Chem., *58*, 1003–1007.
155. Barnfield, C., Burns, S., Byrom, D. L. and Kemmenoe, A. V. (1988) The routine profiling of forensic heroin samples. Forensic Sci. Intern., *39*, 107–117.
156. Neumann, H. (1990) Comment on the routine profiling of illicit heroin samples. Forensic Sci. Intern., *44*, 85–87.
157. Chiarotti, M., Fucci, N. and Furnari, C. (1991) Comparative analysis of illicit heroin samples. Forensic Sci. Int., *50*, 47–55.
158. O'Neil, P. J. and Pitts, J. E. (1992) Illicitly imported heroin products (1984 to 1989): Some physical and chemical features indicative of their origin. J. Pharm. Pharmacol., *44*, 1–6.

3.15 Stimulants

D. de Boer, J. A. Seppenwoolde-Waasdorp and R. A. A. Maes

3.15.1 Testing in General

3.15.1.1 Introduction

An established approach to primary testing for drugs of abuse is to screen samples using immunological procedures (1). Although other screening methodologies have been developed, the possibility to screen large volumes of samples in a sensitive and rapid way and at a reasonable cost makes immunological procedures a most attractive methodology. In order to confirm the suspected presence of a drug and/or respective metabolite(s) in biological specimens, mass spectrometry is the preferred identification technique. This review includes a brief summary of aspects of the stimulants cocaine and amphetamines abuse testing and reports some recent developments and insights.

3.15.1.2 Screening Assays in General

Besides a few chromatographic assays, several types of commercial immunoassays are available for drugs of abuse testing in general and for cocaine and amphetamines testing in particular. The selection of the right type depends on the specific needs and objectives of a drug testing program. Important aspects which have to be considered are the operational characteristics and the ability to detect confirmable positive samples (1). Drug testing programs, which require e.g. the analysis of large numbers of samples, are typically performed in a laboratory by automated assays using defined types of instrumentation. Programs however, which involve analysis outside a laboratory (on-site testing), need inexpensive, rapid and portable assays, which also must be easy to perform and should require little training. Some on-site testing demands a monoanalyte procedure, while others a multianalyte assay. Especially immediate emergency testing for clinical toxicological purposes calls for a rapid multianalyte assay.

3.15.1.2.1 Instrumental immunoassays

Specific instruments needed for the performance of a typical immunoassay are e.g. a gamma-counter for RIA (radioimmunoassay), a fluorescence polarization analyzer for FPIA (fluorescence polarization immunoassay) and a typical spectrophotometer for EMIT® (enzyme multiplied immunoassay technique). The first judgement of instrumental test results is according to programmed criteria, so that objectivity is ensured. Normally the obtained data are recorded and therefore available afterwards.

Abbott FPIA methodology can be performed on different Abbott analyzers, principally all using the same reagents. The TDx® analyzer is dedicated to perform clinical therapeutic and toxicological assays and the ADx® analyzer to drugs of abuse assays. A six-point calibration-curve is applied and the results are quantitatively expressed in ng compound equivalents per ml. An important advantage is the stability of the calibration curve (> 1 month). One disadvantage is the limited instrumental workload, and therefore Abbott Laboratories have developed recently the HTDx® system, which consists of several TDx® analyzers controlled by a central computer. The TDxFLx® analyzer is an updated and modernized TDx® analyzer.

The EMIT® technique can be adapted to a wide variety of analyzers. Depending on the analyzer, a certain type of reagents must be used. The EMIT® st™ assay is for on-site drug testing using a Syva EMIT® st™ spectrophotometer, the EMIT® d.a.u.™ and EMIT® I assays are for the use on the Syva ETS® analyzer with limited instrumental workload, and the EMIT® II assay for analyzers with a large workload. The EMIT® II is the improved successor of EMIT® 700 (2).

RIA methodology suffers from an increasing number of practical drawbacks, e.g. short reagent shelf-life, special reagent handling and waste disposal requirements. One advantage is the possibility to analyze a large number of samples within a relatively short period, but nowadays some non-radioactive isotopic methodologies also can fulfill this requirement.

3.15.1.2.2 Non-instrumental immunoassays

The non-instrumental immunoassays allow visual detection of the label and do not require instrumentation. The drawback of these assays based on visual recognition is the subjective nature of identifying the results (3). If a record of the test result is desired, a photocopy of the whole slide can be made. The success of non-instrumental assays in general will especially depend on the possibility to perform critical evaluations.

Of the assays mentioned, the Triage™ assay is an unique immunoassay. It utilizes ASCEND™ MultImmunoassay technology, which permits the use of multiple drug-specific monoclonal antibodies within each drug class (4). This way the specificity for a broad range of metabolites can be guaranteed.

The available multianalyte assays are all non-instrumental, allowing to screen for more than one drug simultaneously. The number of drugs, which can be detected by Advisor® (5), AbuSign™ DOA 4 and Triage™ (4), are 5, 4 and 7 respectively.

3.15.1.2.3 Chromatographic assays

The Quik Test™ assay was the first non-instrumental assay and consists of a column extraction and elution of the extract onto potassium iodoplatinate impregnated test paper (6). The Toxi-Lab® procedure concerns a standardized method of TLC using fiberglass chromatography combined with several color reactions (7, 8). The REMEDi® drug profiling system is an automated multicolumn HPLC system combined with fast scanning UV-detection (9).

3.15.1.2.4 General considerations

Correct interpretation and reporting in drug abuse testing is of course very important. Especially using immunoassays, misinterpretation of results is a particular concern. Such assays therefore have to be well evaluated. Metabolism of the drugs of abuse has to be studied extensively, the specificity and cross-reactivity of the antibody has to be known and cut-off levels have to be established.

Cut-off or threshold levels are applied for screening tests in order to define a minimum concentration level at which a sample is considered to be positive. These values are used for several reasons, such as compensating for the broad specificity of immunoassays and minimizing the possibility of detecting cases of unknown ingestion or passive exposure (2). The demand for lower cut-off levels for drug abuse testing has resulted in the development of antibodies which have a better defined specificity, especially near cut-off levels (10, 11).

Several factors which affect the methodology and which may influence the outcome of an immunoassay are already known. For instance, in order to obtain a false-negative result, addicts may add adulterants to their sample. These are substances that either interfere with the label or the binding characteristics of the assay. Examples designed to disturb the EMIT® methodology are adding bleach, vinegar or salts to urine (12). On the other hand, subjects taking high-dose vitamin supplements may cause false-positive results by FPIA due the presence of fluorescent riboflavin (11). At the laboratory also, problems may be introduced. Reagent dilution is e.g. a popular means of cost reduction (13) but also may be the introduction of unwanted variations (11). However, in order to avoid such misinterpretations, the problem has be identified. Collection site personnel and laboratory staff have to be very alert.

3.15.2 Cocaine

3.15.2.1 Introduction

The use of cocaine has been very common in South America for ages. Since decades its recreational use is also widespread in the rest of the world. Meanwhile, testing of persons for cocaine abuse is well established and is not uncommon in today's society.

Cocaine may be administered in different forms (coca leaves, cocaine paste, cocaine hydrochloride or cocaine free base) through various routes, e.g. oral, intranasal, intravenous and smoking (14).

3.15.2.2 Metabolism and distribution

3.15.2.2.1 Metabolism

Cocaine is extensively metabolized in the human body by several routes. The principal pathway is spontaneous hydrolysis or hydrolysis by plasma and liver esterases (15), which mainly result in benzoylecgonine (16) and ecgonine methyl ester (17) and to a lesser extent also in ecgonine (18, 19). A quantitatively less important

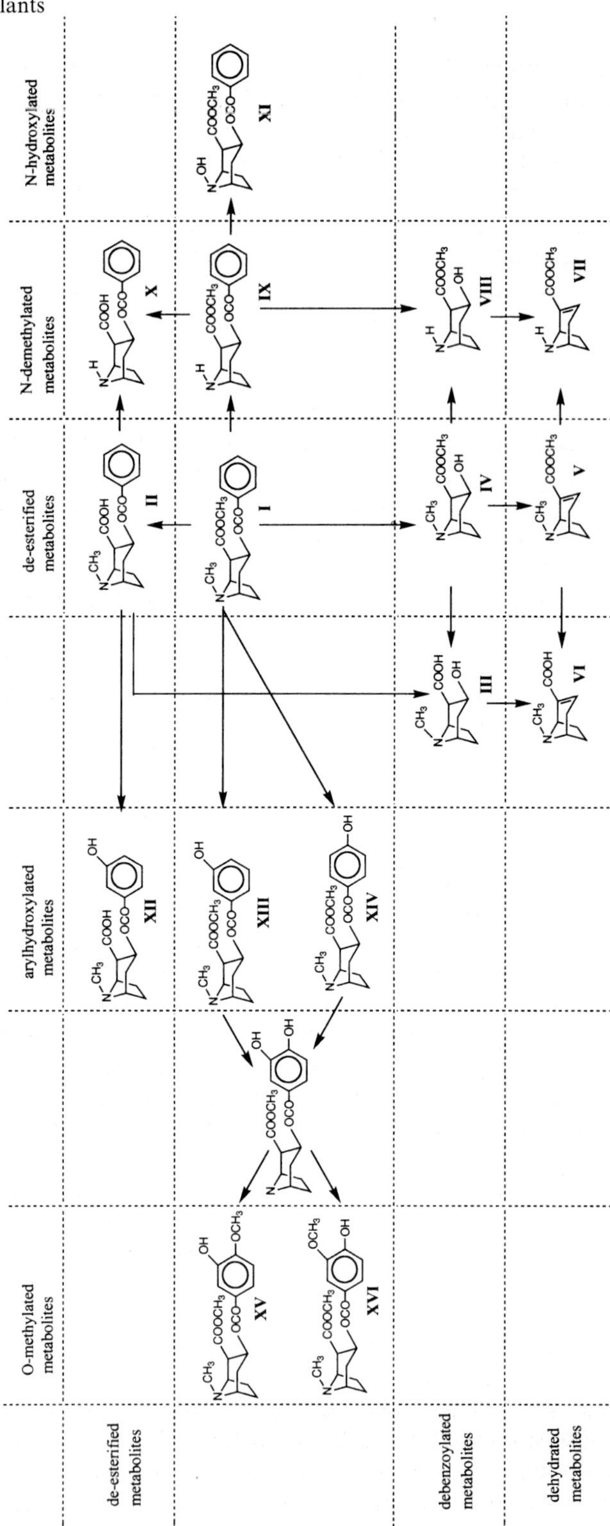

Figure 1. Metabolism of cocaine in humans (see table 1 for the names of the metabolites).

pathway, which resembles phenytoin metabolism (20), leads to catechol- and methylated catechol-like metabolites (19, 21, 22). These aryl-hydroxylated metabolites are excreted as conjugates of sulphate or glucuronic acid. A second minor pathway includes *N*-demethylation (23), which may be followed by *N*-oxidative biotransformation, resulting in potent hepatotoxic metabolites (24). The *N*-demethylated metabolite norcocaine is one of the few pharmacologically active cocaine metabolites (25). The third minor pathway, dehydrobenzoylation, leads to ecgonidine-like metabolites (19, 26). The exact biochemical pathways and the inter-pathway relations of the metabolism of cocaine in humans are not known yet. A possible scheme is presented in figure 1.

In the presence of the alcohols ethanol or isopropanol, transesterification of cocaine and not esterification of benzoylecgonine (27) leads to the respective cocaethylene (28) or cocaisopropylene (29) metabolites. Further metabolism leads to the corresponding transesterified cocaine metabolites (28–30). Transesterification may have some toxicological consequences. Cocaethylene e. g. has been found to be more lethal in mice than cocaine or ethanol alone (31). Moreover, because ethanol also inhibits the carboxyesterase-catalyzed hydrolysis, the cocaine concentration increases and as a result also the degree of *N*-oxidative metabolism and thus of the hepatotoxicity (24).

In most illicit cocaine samples, cinnamoylcocaine is present (32). If such a sample is taken, cinnamoylcocaine and its hydrolyzed metabolite cinnamoylecgonine also can be found in urine (19).

3.15.2.2.2 Biological specimens

Cocaine and its metabolites can be found in all kinds of biological specimens, such as blood (33–39), saliva (40–42), urine (7, 16, 17, 21–23, 28–30, 33, 38, 43, 45, 46), sweat (47), bile (26), meconium (36, 48), amniotic fluid (49) and hair (50–55).

Less than 5 % of a dose is excreted unchanged in urine (16). Approximately 26 % to 49 % of the metabolism results in urinary benzoylecgonine and ecgonine methyl ester (17, 56, 57). Other metabolites, which may be excreted mainly in urine and for a small fraction in e. g. saliva, hair and bile, are shown in table 1 and account for the rest of the percentage of metabolism. The relevance of the presence of cocaine or certain metabolites is presented in table 2 and can be illustrated by describing some examples.

1) In postmortem blood, the ecgonine methyl ester also arises from *in vitro* hydrolysis of cocaine. That can be used to estimate perimortem cocaine concentrations, which in their turn can be helpful to interpret a case history (34) within some limitations (35).

2) In meconimum, *m*-hydroxybenzoylecgonine is a quantitatively important metabolite. Because this metabolite proved to be immunoreactive, it is responsible for the discrepancy between benzoylecgonine concentrations in meconium as measured by immunoassay and gas chromatography/mass spectrometry (GC/MS) (48).

3) Because of a concern in the ability to distinguish between cocaine abuse and environmental contamination of hair (52), the metabolites norcocaine and cocaethylene have been suggested as characteristic markers (50). However, some caution

Table 1. Structures of Cocaine, Cinnamoylcocaine and Possible Metabolites as Reported in Human Biological Specimens and Respective References

metabolite no. name	substituent R_1	R_2	R_3	R_4	R_5	references
cocaine	(see structures 1A or 1B)					
I benzoylecgonine methyl ester	CH_3	CH_3	COC_6H_5	H	H	(16)
hydrolyzedcocaine metabolites						
II benzoylecgonine	CH_3	H	COC_6H_5	H	H	(16)
III ecgonine	CH_3	H	H	–	–	(16, 19)
IV ecgonine methyl ester	CH_3	CH_3	H	–	–	(17, 56)
dehydrobenzoylated cocaine metabolites	(see structure 1C)					
V ecgonidine methyl ester	CH_3	CH_3	H	–	–	(26, 19)
VI ecgonidine	CH_3	H	H	–	–	(19)
VII norecgonidine methyl ester	H	CH_3	H	–	–	(19)
N-demethylated cocaine metabolites						
VIII norecgonine methyl ester	H	CH_3	H	H	H	(19)
IX benzoylnorecgonine methyl ester	H	CH_3	COC_6H_5	H	H	(17, 23)
X benzoylnorecgonine	H	H	COC_6H_5	H	H	(18)
N-hydroxylated cocaine metabolite						
XI *N*-hydroxybenzoylnorecgonine methyl ester	OH	CH_3	COC_6H_5	H	H	(24)
aryl hydroxylated cocaine metabolites						
XII *m*-hydroxybenzoylecgonine	CH_3	H	COC_6H_5	OH	H	(19)
XIII *m*-hydroxybenzoylecgonine methyl ester	CH_3	CH_3	COC_6H_5	OH	H	(22, 26)
XIV *p*-hydroxybenzoylecgonine methyl ester	CH_3	CH_3	COC_6H_5	H	OH	(22)
XV *m*-hydroxy-*p*-methoxybenzoyl-ecgonine methyl ester	CH_3	CH_3	COC_6H_5	OCH_3	H	(21)
XVI *p*-hydroxy-*m*-methoxybenzoyl-ecgonine methyl ester	CH_3	CH_3	COC_6H_5	H	OCH_3	(21)
transesterificated cocaine metabolites						
XVII benzoylecgonine ethyl ester	CH_3	CH_2CH_3	COC_6H_5	H	H	(28)
XVIII benzoylecgonine isopropyl ester	CH_3	$CH_2(CH_3)_2$	COC_6H_5	H	H	(29)
XIX ecgonine ethyl ester	CH_3	CH_2CH_3	H	–	–	(28)
XX ecgonine isopropyl ester	CH_3	CH_2CH_3	H	–	–	(29)
XXI *m*-hydroxy-*p*-methoxybenzoyl-ecgonine ethyl ester	CH_3	CH_2CH_3	COC_6H_5	OCH_3	H	(30)
XXII *p*-hydroxy-*m*-methoxybenzoyl-ecgonine ethyl ester	CH_3	CH_2CH_3	COC_6H_5	H	OCH_3	(30)
XXIII benzoylnorecgonine ethyl ester	H	CH_2CH_3	COC_6H_5	H	H	(58)
cinnamoylcocaine and characteristic metabolite (see structure 1A)						
XXIV cinnamoylecgonine methyl ester	CH_3	CH_3	$COCH=CHC_6H_5$	H	H	(19)
XXV cinnamoylecgonine	CH_3	H	$COCH=CHC_6H_5$	H	H	(19)

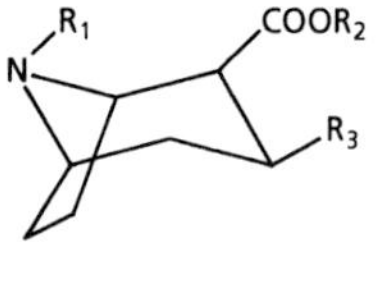

R1, COOR2, N, R3 — R1, COOR2, N, OCO, OR5, OR4 — R1, COOR2, N, R3

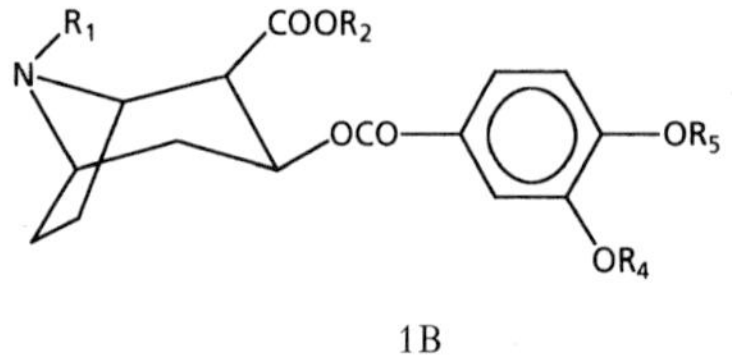

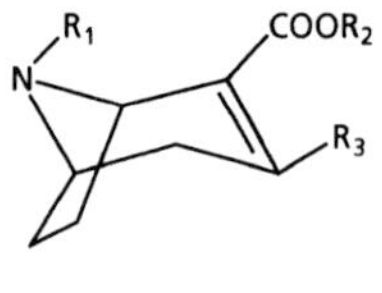

structure 1A 1B 1C

has to be taken into account, as e.g. cocaethylene has been found in illicit cocaine samples (59).

Table 2. Comparison of Cocaine Abuse Testing Issues for Different Biological Specimens

issues	biological specimen serum	plasma	post-mortem blood	saliva	urine
relative invasiveness	high	high	n. a.	low	high/low
main compounds excreted in general	BZE > COC > EME	EZE > COC > EME	EME > BZE > COC	COC > BZE > EME	BZE = EME > COC
type of measured time period	moment	moment	n. a.	moment	cumulative
cumulative time period	–	–	–	–	hours
detection period	½–1 days	½–1 days	n. a.	½–1 days	2–3 days
	sweat	meconium	ammiotic	hair	
relative invasiveness	low	high/low	high	low	
main compounds excreted in general	COC > BZE = EME	HBZE = BZE > EME	BZE > COC = EME	COC > BZE > EME	
type of measured time period	cumulative	cumulative	cumulative	cumulative	
cumulative time period	weeks	months	months	months-years	
detection period	weeks	months	months	months-years	

BZE = benzoylecgonine
COC = cocaine
EME = ecgonine methyl ester
HBZE = *m*-hydroxybenzoylecgonine
n. a. = not applicable

3.15.2.3 Detection methodologies

3.15.2.3.1 Commercially available screening assays

3.15.2.3.1.1 Instrumental immunoassays

For cocaine abuse testing, the available instrumental assays (Table 3) are regularly adapted to the actual requirements, and up until now, are performing quite satisfactorily. The trend to replace RIA methodology for non-radioactive isotopic techniques is illustrated by the fact that fully automated EMIT® II and Online® assays are preferred over the Abuscreen® testing kits.

3.15.2.3.1.2 Non-instrumental immunoassays

As far as evaluated for cocaine abuse testing (table 3), the non-instrumental immunoassay Triage™ has been well received. The EZ-SCREEN® and ONTRAK® assay have been subject to critical remarks. The other non-instrumental assays as mentioned in Table 3 have not been marketed or extensively evaluated yet.

Table 3. Available Commercial Assays for the Testing of Cocain Abuse

name of assay system	type of methodology	manufacturer	references
monoanalyte immunoassay systems			
instrumental assays			
TDx®, TDxFLx®, ADx® or HTDx®	FPIA	Abbott Laboratories	(1, 3, 7, 8, 39, 43–45, 48, 49, 55, 60, 64, 65)
CEDIA®	EIA	Boehringer Mannheim Corporation	(10)
Coat-A-Count®	RIA	Diagnostic Products Corporation	(7, 8, 38, 40, 48, 60)
Double Antibody	RIA	Diagnostic Products Corporation	(7, 8, 38)
Abuscreen®	RIA	Roche Diagnostic Systems	(1–3, 7, 8, 45, 48, 64)
Online®	KIMS	Roche Diagnostic Systems	(1, 10, 48)
EMIT® st™	EIA	Syva Company	(7, 8, 65)
EMIT® d.a.u.™ or EMIT® I	EIA	Syva Company	(7, 8, 44, 46, 48, 60, 64, 65, 67)
EMIT® II	EIA	Syva Company	(1, 2, 10, 48, 60, 67)
non-instrumental assays			
EZ-SCREEN®	EIA	Environmental Diagnostics	(60, 61)
accu-PINCH™	EIA	Hycor Biomedical	(62)
AbuSign™	agglutination assay	Princeton BioMedical	none
ONTRAK®	agglutination assay	Roche Diagnostics	(3, 60, 65, 66)
multianalyte immunoassay systems			
non-instrumental assays			
Advisor®	agglugination assay	Abbott Laboratories	(5)
Triage™	ASCEND™ MultImmunoassay	Biosite Diagnostics	(4, 60, 63)
AbuSign™ DOA 4	agglutination assay	Princeton BioMedical	none
chromatographic systems			
Quick Test™	–	Keystone Diagnostics	(6–8)
REMEDi®	HPLC/UV	Bio-Rad Laboratories	(9, 60)
Toxi-Lab®	TLC	Marion Laboratories	(7, 8, 60)

3.15.2.3.1.3 Antibody specificity

The antibodies applied in different immunoassays for detecting abuse of cocaine differ in specificity (8). The antibodies are directed against benzoylecgonine, with some exceptions which are directed to cocaine itself. The unique immunoassay Triage™ assay, which employs multiple antibodies for different compounds, will give results for cocaine abuse testing comparable with common immunoassays techniques, because until now a single antibody is used (63).

In general, cross-reactivity studies showed a good specificity (64). However, ingestion of therapeutic doses of aspirin caused false-negative results for some cocaine EMIT® assays (67). This effect is not an antibody-related phenomenon, but is somehow connected to the EMIT® methodology.

In some special cases, discrepancies between immunoassays and confirmation methods can be explained by the presence of typical metabolites, which in those cases are quantitatively important and cross-react with benzoylecgonine; e.g. *m*-hydroxy-benzoylecgonine in meconium (48).

In general it can be concluded that the specificity of cocaine abuse tests based on immunoassays is satisfactory.

3.15.2.3.1.4 Cut-off levels of immunoassays

Although various immunoassays specify a cut-off level of 300 ng benzoylecgonine equivalents per ml of urine (table 3), the future trend is that the required cut-off value will be 150 ng/ml. Some assays already fulfill this requirement (8, 45).

3.15.2.3.1.5 Chromatographic assays

After critical evaluations the Quik Test™ assay did not show the required accuracy and specificity (6–8, 68), especially for the detection of benzoylecgonine. The performance of the Toxi-Lab® assay and the REMEDi® drug profiling system are reliable, because of the low frequency of interferences (i.e. coelutions) (60), but compared to immunoassays, the sensitivity is still not satisfactory (7, 8). The advantage of these chromatographic assays, however, is the possibility to detect the parent compound cocaine and various other drugs simultaneously. In this respect the REMEDi® drug profiling system is considered as a useful complementary technique in emergency systems (9).

3.15.2.3.2 Confirmative testing

For cocaine abuse testing, the preferred identification technique is hyphenated GC/MS. Because of the type of GC columns applied and because of their volatility, GC/MS analysis of cocaine metabolites requires derivatization. Cocaine itself is not derivatized by the procedures used.

Several types of derivatives using different kinds of reagents have been investigated. Commonly used are the *N,O*-propyl (69), *N,O*-trifluoroacetyl (23, 55), *N*-pentafluoropropionyl-*O*-pentafluoropropyl (29, 36, 46, 48, 54), *N*-pentafluoropropionyl-*O*-1,1,1,3,3,3-hexafluoroisopropyl (19, 33), *N,O*-trimethylsilyl (43, 47, 70) and *N,O-tert.*-butyldimethylsilyl (51) derivatives. Regarding the mass spectrometric properties of the derivatives, the main difference is an increase of the final molecular weight of the respective derivative, which may result in fragments at a higher *m/z* range where the frequency of interferences of other fragments is lower, resulting in an increase in sensitivity. The fragmentation process in the electron impact (EI) ionization mode is not influenced by the kind of derivative.

3.15.2.3.3 Storage conditions

Urine specimens in drug testing programs have been and are the first choice of biological specimen. Because cocaine may hydrolyze to benzoylecgonine spontaneously, storage of urine can be critical for the concentrations of cocaine and this metabolite (71). The other major metabolite ecgonine methyl ester results from an enzymatic hydrolysis of which the enzymes are not present in urine.

Depending on the storage conditions, degradation of cocaine can be prevented for at least a month (72). The optimal storage conditions for urine specimens with cocaine and benzoylecgonine were determined to be at $-15\,^{\circ}C$ and a pH of 5.0. Ascorbic acid was found to be a favorable agent to achieve this acidic pH value. Unsilanized glass as a container material and storage in the dark were also favorable. This way degradation can be prevented for at least 110 days (73).

3.15.3 Amphetamines

3.15.3.1 Introduction

Amphetamine and the amphetamines-like compounds are sympathomimetic amines whose biological effects include central nervous system, anorectic, hyperthermic and cardiovascular effects (74). They have been abused for their stimulant and anorectic effects and are in general administered orally.

The structure of amphetamines in general is based primarily on the phenylisopropylamine structure (table 4). Through subclassification, the amphetamine-like compounds can be divided in "classical" amphetamines ($R_2 = H$; $R_3 = H$), phentermines ($R_2 = H$; $R_3 = CH_3$), ephedrines ($R_2 = H$; $R_3 = OH$) and miscellaneous. Examples of the last category are fencamfamin, the phenmetrazines and propylhexedrine (Table 4). The amphetamines which belong to the designer drugs are described in chapter 3.16.

Table 4. Structures of Subclasses of Amphetamine-Like Compounds and Several Examples

A. amphetamines ($R_2 = H$; $R_3 = H$)			
amphetamine	$R_1 = H$	$R_4 = H$	$R_5 = H$ 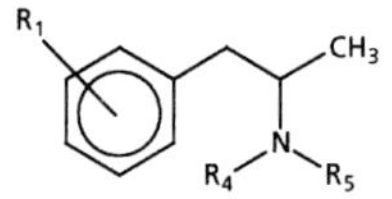
dimethylamphetamine	$R_1 = H$	$R_4 = CH_3$	$R_5 = CH_3$
fenfluramine	$R_1 =$ *meta*-CF_3	$R_4 = H$	$R_5 = C_2H_5$
methamphetamine	$R_1 = H$	$R_4 = H$	$R_5 = CH_3$
B. phentermines ($R_2 = H$; $R_3 = CH_3$)			
chlorphentermine	$R_1 =$ *para*-Cl	$R_4 = H$	$R_5 = H$ 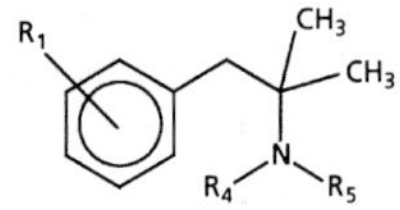
mephentermine	$R_1 = H$	$R_4 = H$	$R_5 = CH_3$
phentermine	$R_1 = H$	$R_4 = H$	$R_5 = H$

R₁ CH₃ N R₄ R₅

R₁ CH₃ CH₃ N R₄ R₅

Table 4. (continued)

C. ephedrines ($R_2 = OH$; $R_3 = H$)				
ephedrine	$R_1 = H$	$R_4 = H$	$R_5 = CH_3$	
ethylephedrine†	$R_1 = H$	$R_4 = C_2H_5$	$R_5 = CH_3$	
methylephedrine	$R_1 = H$	$R_4 = CH_3$	$R_5 = CH_3$	
norephedrine‡	$R_1 = H$	$R_4 = H$	$R_5 = H$	
D. miscellaneous				
fencamfamin	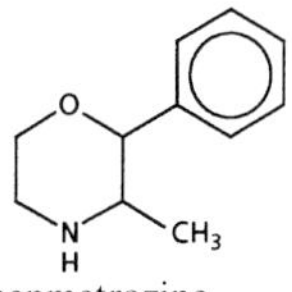	phenmetrazine	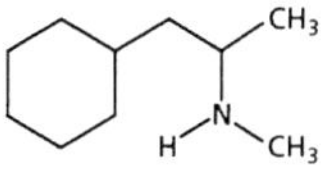	propylhexedrine

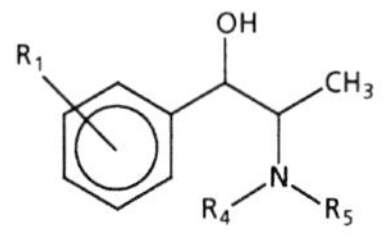

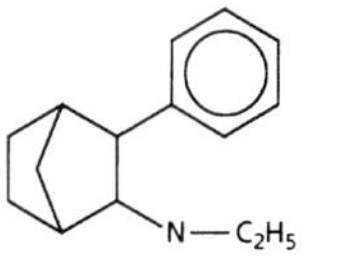

† etafedrine
‡ phenylpropanolamine

The "classical" amphetamines have one chiral center resulting in two enantiomeric forms. The ephedrines have two chiral centra and exist in two diastereomeric and four enantiomeric forms. The phentermines are non-chiral compounds. Typical pharmacological activity can be attributed to one enantiomeric form (74).

3.15.3.2 Metabolism and distribution

3.15.3.2.1 Metabolism

Amphetamines are metabolized in humans through different routes (75). The common pathways include aryl- and α-C-hydroxylation, *N*-dealkylation, deamination and carboxylation, some of them combined with conjugation (figure 2). Conjugation occurs with glucuronic acid, sulfate or glycine. Depending on the substituents present in the phenylisopropylamine base structure, certain pathways may be blocked. The chlorine atom at the *para*-position in chlorphentermine (76) or the trifluoromethyl-group in fenfluramine (77) e. g. may prevent arylhydroxylation. On the other hand, the kind of alkyl substituent may influence the degree of *N*-dealkylation (78, 79).

The *N*-hydroxy metabolites are intermediates in the route leading to hippuric acid conjugates (figure 2). During analytical procedures this metabolite is rapidly oxidized to a nitroso compound and, if the solutions are alkaline, to a nitro compound (80).

In combination with ethanol use, a specific pathway leads to 1,3-dimethyl-1,2,3,4-tetrahydroisoquinoline (81). This type of metabolite is the result of condensation with the ethanol metabolite acetaldehyde and is potentially neurotoxic.

In man, deamination is the main metabolic route. Metabolism of amphetamine-like compounds, however, display genetic polymorphism, which may result in an increased ratio of parent compound to metabolite concentration (82). Metabolism of amphetamines in humans is enantioselective (78).

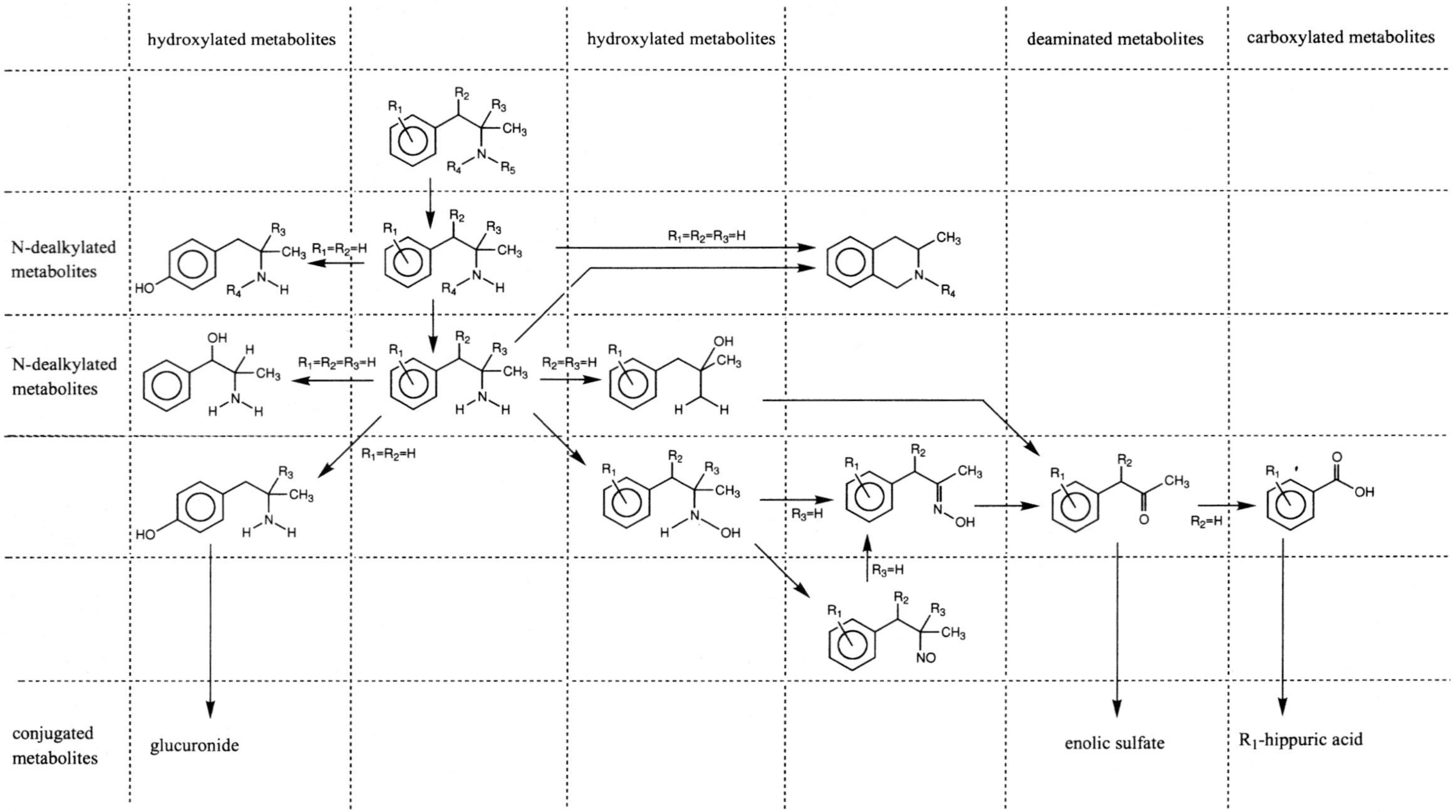

Figure 2. Metabolism of amphetamines in humans (R is a general symbol for a substituent if not specified; see table 4 for examples).

3.15.3.2.2 Biological specimens

Amphetamines can be found in all kinds of biological specimens, such as blood (38, 83, 84), saliva (85), urine (38, 86–90), sweat (85), breast milk (91), meconium (92), hair (55, 85, 93) and nails (85).

The excretion of amphetamines in urine is pH and volume dependent (78). Under acidic conditions the excretion is greater than under basic conditions. As the urinary pH profiles of individuals vary from day to day and are influenced by e. g. the diet, the intra- and intersubject differences may be great. Approximately 90 % of a total dose of amphetamine is excreted in urine over 3 to 4 days, of which 40 % to 70 % as unchanged amphetamine (14, 78).

3.15.3.3 Detection methodologies

3.15.3.3.1 Commercially available screening assays

3.15.3.3.1.1 Instrumental immunoassays

For amphetamines abuse testing the available instrumental assays (table 5) vary regarding their performance and have to be used with the known limitations. The main problems are the small structural differences within the group of amphetamines on one hand and the impossibility to produce a highly specific antibody for the most important amphetamines on the other hand. It has led to the search for more specific antibodies and the introduction of different categories of antibodies. Within their limitations these assays perform well.

Based on specificity, the extensively evaluated instrumental immunoassays for amphetamines can be classified in 4 categories:

- A) Assays specific for amphetamine and methamphetamine in varying degrees and also with some cross-reactivity for other amphetamine-like compounds.
- B1) Assays with well-defined specificity for amphetamine.
- B2) Assays with well-defined specificity for methamphetamine.
- C) Dual assays with well-defined specificity for amphetamine and methamphetamine.

The available RIA immunoassays have a well-defined specificity for either amphetamine or methamphetamine (category B) (table 5). The FPIA and EMIT® assays have antibodies which are specific for amphetamines with varying cross-reactivity for ephedrines (category A), or which have a well-defined specificity for both amphetamine and methamphetamine (category C) (table 5). The Online® and CEDIA® assay are considered to belong to category C.

3.15.3.3.1.2 Non-instrumental immunoassays

The non-instrumental assays for amphetamines abuse testing are difficult to classify, because a limited number of evaluation studies are available (table 5). The Triage™ assay can be considered to perform very well. Although apparently specific, some precautions have to be taken in special cases. Postmortem decomposition processes resulting in an increase of urinary tyramine concentrations for instance may cause false positive amphetamine results when using Triage™ assay (95). As for cocaine abuse testing, the EZ-SCREEN® has been subject to critical remarks and the other

Table 5. Available Commercial Assays for the Testing of Amphetamine Abuses

name of assay system	type of methodology	manufacturer	references
monoanalyte immunoassay systems			
instrumental assays			
TDx®, TDxFLx®, ADx® or HTDx®	FPIA category A	Abbott Laboratories	(3, 55, 60, 82, 90)
	FPIA category C	Abbott Laboratories	(82, 84, 86, 88, 89, 94)
CEDIA®	EIA category C	Boehringer Mannheim Corporation	(10)
Coat-A-Count®	RIA category B1	Diagnostic Products Corporation	(89)
	RIA category B2	Diagnostic Products Corporation	(38, 84, 89)
Double Antibody	RIA category B1	Diagnostic Products Corporation	(82)
	RIA category B2	Diagnostic Products Corporation	(38)
Abuscreen®	RIA category B1	Roche Diagnostic Systems	(3, 82, 89)
	RIA category B2	Roche Diagnostic Systems	(82, 89)
Online®	KIMS category C	Roche Diagnostic Systems	(10)
EMIT® st™	EIA category A	Syva Company	none
EMIT® d.a.u.™ or EMIT® I	EIA category A	Syva Company	(82)
	EIA category C	Syva Company	(60, 82, 86, 87, 89)
EMIT® II	EIA category A	Syva Company	(60)
	EIA category C	Syva Company	(10, 87, 90)
non-instrumental assays			
EZ-SCREEN®	EIA	Environmental Diagnostics	(60, 61)
AbuSign™	agglutination assay category B1	Princeton BioMedical	none
	agglutination assay category B2	Princeton BioMedical	none
ONTRAK®	agglutination assay	Roche Diagnostics	(3, 60)
multianalyte immunoassay systems			
non-instrumental assays			
Advisor®	agglugination assay	Abbott Laboratories	(5)
Triage™	ASCEND™ MultImmunoassay	Biosite Diagnostics	(4, 60, 63)
AbuSign™ DOA 4	agglutination assay	Princeton BioMedical	none
chromatographic systems			
Quick Test™	–	Keystone Diagnostics	(6)
REMEDi®	HPLC/UV	Bio-Rad Laboratories	(9, 60)
Toxi-Lab®	TLC	Marion Laboratories	(60)

non-instrumental assays mentioned in table 5 have not been marketed or extensively evaluated yet.

3.15.3.3.1.3 Antibody specificity

The different immunoassays (table 5) which have been developed for the detection of amphetamines vary in specificity and have varying degrees of cross-reactivity to other compounds. The antibodies, which have been produced, are primarily directed against (*S*)-(+)-amphetamine or (*S*)-(+)-methamphetamine. Information on the immunogen structures used and the specificities of the antibodies obtained have allowed insight in structure-specificity (11, 96). The assays intended e.g. to detect either (*S*)-(+)-amphetamine or (*S*)-(+)-methamphetamine with minimal cross-reactivity, employ immunogens with amphetamine or methamphetamine derivatized via the *para*-position of the phenyl ring. Such assays theoretically show minimal cross-reactivity with other secondary or tertiary amines and respective (*R*)-(−)-enantiomers, but may strongly cross-react with phenyl ring substituted analogs, including the 3,4-methylenedioxyamphetamines. On the other hand, assays intended for detection of both (*S*)-(+)-amphetamine and (*S*)-(+)-methamphetamine employ amphetamine, rather than methamphetamine, derivatized via its amino group as in immunogen. Such assays theoretically show minimal cross-reaction with other tertiary amines, phenyl-substituted amphetamine or methamphetamine and respective (*R*)-(−)-enantiomers. Specificity can be further directed by employing *N*-(aminoalkyl)-amphetamines as immunogens. Using *N*-(3-aminopropyl)-amphetamine (97) and *N*-(4-aminobutyl)-amphetamine (98), antibodies were obtained with a 7 × and 20 × higher affinity for methamphetamine than for amphetamine, respectively.

The development of assays with monoclonal antibodies instead of polyclonal antibodies means that they have better defined specificity, rather than an increased specificity (11). The monoclonal antibody in the EMIT® I assay (category C) for instance, unexplained cross-reacted with ranitinide (99), with metabolite(s) of chlorpromazine (100, 101) and of brompheniramine (100) and with thioridazine, pipothiazide, fluspirilene and/or respective metabolite(s) (101), while the polyclonal version did not. The most recent monoclonal EMIT® II assay (category C) did, up until now, not show this surprising cross-reactivity, but still gave positive results with several amphetamines including (*R*)-(−)-enantiomers (87).

Unexpected cross-reactivity of the Abuscreen® (category B1), EMIT® and FPIA (category A) assays with trimethobenzamide (102) can be explained by the fact that it replaces the analyte with the label attached, rather than the analyte itself (11). A similar explanation has been proposed (11) regarding cross-reactivity in the EMIT® I assay (category C) for labetalol metabolite(s) (103).

Some apparent cross-reactivity is in fact due the formation of amphetamines as metabolites of the drugs involved. Besides clobenzorex (104) and famprofazone (105), several examples are given in chapter 1.4 on Doping Analysis. This phenomenon is rather an interpretation problem and requires additional confirmation techniques.

Some of the described cross-reactivity of FPIA or EMIT® is inherent to the application of their homogeneous set-up (assays not requiring a separation step). The heterogeneous assays with an additional separation step seem to be more specific, especially radioactive isotopic assays (11).

3.15.3.3.1.4 Cut-off levels of immunoassays

The cut-off levels vary with the category to which the assay belongs. At the moment the various immunoassays belonging to category C specify at least a cut-off level of 1000 ng (*S*)-(+)-amphetamine or (*S*)-(+)-methamphetamine equivalents per ml of urine (Table 5), but the future trend is that the required cut-off level will be 500 ng/ml (10). The category A FPIA and EMIT® assays have a cut-off level of 300 ng (*S*)-(+)-amphetamine equivalents per ml of urine.

3.15.3.3.1.5 Chromatographic assays

As for cocaine abuse testing, the performances of the Toxi-Lab® assay and of the REMEDi® drug profiling system are specific for amphetamines abuse testing, but lack the required sensitivity (9, 60). The ratinidine interference problem with the EMIT® I (category C) assay was successfully studied with the REMEDi® drug profiling system (99), illustrating its complementary value.

3.15.3.3.2 Confirmative testing

As for cocaine abuse testing, the preferred identification technique is hyphenated GC/MS. The use of methanol or ethanol as the injection solvent for the GC/MS analysis of primary and secondary amines may yield imines and the use of methanol for secondary amines additional methylation (106). However, if derivatization is applied, this problem is avoided. Common derivatives which have been used for GC/MS analysis are the *N,O*-acetyl (88), *N,O*-trichloroacetyl (107), *N,O*-trifluoroacetyl (82, 85, 93), *N,O*-heptafluorobutyryl (86, 108–110), *N,O*-4-carbethoxyhexafluorobutyryl (108, 110, 111), *N*-perfluorooctanoyl (83), *N*-trifluoroacetyl-*O*-trimethylsilyl (112) and *N,O-tert.*-butyldimethylsilyl derivatives (113).

In order to avoid incorrect interpretation of the presence of methamphetamine in a urine specimen, when is a product of thermal transformation of ephedrines (114), periodate degradation of the interfering ephedrines is recommended (115). The pH at which the periodate step is performed has to be carefully controlled, because at pH > 9.1 methamphetamine may transform to amphetamine (116). Some manufacturers recommend to perform the periodate step in addition to their assay in order to eliminate false-positive results due to ephedrines (117). The periodate step even has been incorporated into a non-commercial RIA for the dual detection of amphetamine and methamphetamine (118).

Since over-the-counter medicines may contain (*R*)-(−)-methamphetamine (90), which is not considered to be a drug of abuse, a enantioselective confirmation method is necessary in order to avoid false-positive results for (*S*)-(+)-methamphetamine (119). Derivatization of amphetamines with *N*-trifluoroacetyl-(*S*)-(−)-prolyl chloride offers the possibility to perform enantioselective GC/MS analysis (120).

For amphetamines, Fourier Transform Infrared Spectroscopy (FTIR) has also been evaluated as a confirmation technique. The main drawback of infrared procedures in the past has been their rather low sensitivity in a biological matrix, but new developments and the combination with chromatography has made FTIR a valuable alternative or complement technique to GC/MS (121). The serial GC/MS/FTIR combination is even a more powerful identification procedure (122).

3.15.3.3.3 Storage conditions

In urine no significant storage problems are known for amphetamines. It has been reported for instance that amphetamine and methamphetamine are stable at −17 °C for 45 days (72) and up to 12 months at −20 °C (71).

3.15.4 Conclusions

Immunological procedures are the most attractive methodologies for cocaine and amphetamines abuse testing. Important requirements for correct interpretation of results of an immunological methodology, such as metabolism and distribution of the drugs involved, already have been studied. Also the knowledge regarding specificity of antibodies, cross-reactivities and other pitfalls of the commercially available assays are evaluated and updated regularly. Especially for amphetamines abuse testing by immunoassays, many limitations are known, a phenomenon which is inherent in the heterogeneity of the group of amphetamines. Within the known limitations, most of the instrumental immunoassays for cocaine and amphetamine abuse testing perform very well. In this respect the non-instrumental immunoassay devices for on-site-testing, which have been developed and marketed recently, still have to be studied extensively. The success of immunological procedures in general and the non-instrumental assays in particularly will depend on further critical evaluations. The demand for lower cut-off levels for drug abuse testing will also need additional studies of confirmation procedures, of which hyphenated GC/MS techniques are preferred.

References

1. Armbruster, D.A., Schwarzhoff, R.H., Hubster, E.C. and Liserio, M.K. (1993) Enzyme immunoassay, kinetic microparticle immunoassay, radioimmunoassay, and fluorescence polarization immunoassay compared for drugs-of-abuse screening. Clin. Chem. *39*, 2137–2146.
2. Armbruster, D.A., Schwarzhoff, R.H., Pierce, B.L. and Hubster, E.C. (1994) Method comparison of EMIT 700 and EMIT II with RIA for drug screening. J. Anal. Toxicol *18*, 110–117.
3. Armbruster, D.A. and Krolak, J.M. (1992) Screening for drugs of abuse with the Roche ONTRAK assays. J. Anal. Toxicol *16*, 172–175.
4. Buechler, K.F., Moi, S., Noar, B., MacGrath, D., Villela, J., Clancy, M., Shenhav, A., Colleymore, A., Valkirs, G., Lee, T., Bruni, J.F., Walsh, M., Hoffmann, R., Ahmuty, F., Nowakowski, M., Buechler, J., Mitchell, M., Boyd, D., Stiso, N. and Anderson, R. (1992) Simultaneous detection of seven drugs of abuse by the Triage panel for drugs of abuse. Clin. Chem. *38*, 1678–1684.
5. Parsons, R.G., Kowal, R., Le Blond, D., Yue, V.T., Neargarder, L., Bond, L., Garcia, D., Slater, D. and Rogers, P. (1993) Multianalyte assay system developed for drugs of abuse. Clin. Chem. *39*, 1899–1903.
6. Cone, E.J. and Menschen, S.L. (1987) Lack of validity of the KDI Quik Test™ drug screen for detection of benzoylecgonine in urine. J. Anal. Toxicol *11*, 276–277.
7. Cone, E.J., Menschen, S.L., Paul, B.D., Mell, L.D. and Mitcell, J. (1989). Validity testing of commercial urine cocaine metabolite assays: I. Assay detection times, individual excretion patterns, and kinetics after cocaine administration to humans. J. Forensic Sci. *34* 15–31.
8. Cone, E.J. and Mitchell, J. (1989) Validity testing of commercial urine cocaine metabolite assays: II. Sensitivity, specificity, accurancy and confirmation by gas chromatography/mass spectrometry. J. Forensic Sci. *34*, 32–45.
9. Demedts, P., Wauters, A., Franck, F. and Neels, H. (1994) Evaluation of the REMEDi® drug profiling system. Eur. J. Clin. Chem. Clin. Biochem. *32*, 409–417.
10. Armbruster, D.A., Hubster, E.C., Kaufman, M.S. and Ramon, M.K. (1995) Cloned enzyme donor immunoassay (CEDIA) for drugs-of-abuse screening. Clin. Chem. *41*, 92–98.
11. Colbert, D.L. (1994) Drug abuse screening with immunoassays: unexpected cross-reactivities and other pitfalls. Br. J. Biomed. Sci. *51*, 136–146.
12. Mikkelsen, S.L. and Ash, K.O. (1988). Adulterants causing false negatives in illicit drug testing. Clin. Chem. *34*, 2333–2336.
13. Asselin, W.M. and Leslie, J.M. (1992). Modification of EMIT assay reagents for improved sensitivity and cost effectiveness in the analysis of heomlyzed whole blood. J. Anal. Toxicol. *16*, 381–388.
14. Busto, U., Bendayan, R. and Sellers, E.M. (1989). Clinical pharmacokinetics of non-opiate abused drugs. Clin. Pharmacokin. *16*, 1–26.
15. Stewart, D.J., Inaba, T., Lucassen, M. and Kalow, W. (1979). Cocaine metabolism: cocaine and norcocaine hydrolysis by liver

and serum esterases. Clin. Pharmacol. Ther. *25*, 464–468.
16. Fish, F. and Wilson, W.D.C. (1969). Excretion of cocaine and its metabolites in man. J. Pharm. Pharmacol. *21*, 135S–138S.
17. Inaba, T., Stewart, D.J. and Kalow, W. (1978). Metabolism of cocaine in man. Clin. Pharmacol. Ther. *23*, 547–552.
18. Misra, A.L., Nayak, P.K., Block, R. and Mulé, S.J. (1975). Estimation and disposition of ^{3}H-benzoylecgonine and pharmacological activity of some cocaine metabolites. J. Pharm. Pharmacol. *27*, 784–786.
19. Zhang, Y. and Foltz, R.L. (1990). Cocaine metabolism in man: identification of four previously unreported cocaine metabolites in human urine. J. Anal. Toxicol. *14*, 201–205.
20. Midha, K.K., Hindmarsh, K.W., MacGilveray, I.J. and Cooper, J.K. (1977). Identification of urinary catechol and methylated catechol metabolites of phenytoin in humans, monkeys and dogs by GLC and GLC-mass spectrometry. J. Pharm. Sci. *66*, 1596–1602.
21. Smith, R.M., Poquette, M.A. and Smith, P.J. (1984). Hydroxymethoxybenzoylmethylecgonine: new metabolites of cocaine from human urine. J. Anal. Toxicol. *8*, 29–34.
22. Smith, R.M. (1984). Arylhydroxy metabolites of cocaine in the urine of cocaine users. J. Anal. Toxicol. *8*, 35–37.
23. Jindal, S.P., Lutz, T. and Vestergaard, P. (1978). Mass spectrometric determination of cocaine and its biologically active metabolite, norcocaine, in human urine. Biomed. Mass Spectrom. *5*, 658–663.
24. Roberts, S.M., Harbison, R.D. and James, R.C. (1991). Human microsomal *N*-oxidative metabolism of cocaine. Drug Metab. Dispos. *19*, 1046–1051.
25. Borne, R.F., Bedford, J.A., Buelke, J.L., Craig, C.B., Hardin, T.C., Kibbe, A.H. and Wilson, M.C. (1977). Biological effects of cocaine derivatives I: improved synthesis and pharmacological evaluation of norcocaine. J. Pharm. Sci. *66*, 119–120.
26. Lowry, W.T., Lomonte, J.N., Hachett, D. and Garriott, J.C. (1979). Identification of two novel cocaine metabolites in bile by gas chromatography and gas chromatography/mass spectrometry in a case of acute intravenous cocaine overdose. J. Anal. Toxicol. *3*, 91–95.
27. Hearn, W.L., Flynn, D.D., Hime, G.W., Rose, S., Cofina, J.C., Mantero-Atienza, E., Wetli, C.V. and Mash, D.C. (1991). Cocaethylene: an unique cocaine metabolite displays high affinity for the dopamine transporter. J. Neurochem. *56*, 698–701.
28. Rafla, F.K. and Epstein, R.L. (1979). Identification of cocaine and its metabolites in human urine in the presence of ethyl alcohol. J. Anal. Toxicol. *3*, 59–63.
29. Wu, A.H.B., Onigbinde, T.A., Johnson, K.G. and Wimbish, G.H. (1992). Alcohol-specific cocaine metabolites in serum and urine of hospitalized patients. J. Anal. Toxicol. *16*, 132–136.
30. Smith, R.M. (1984). Ethyl esters of arylhydroxy- and arylhydroxymethoxycocaines in the urine of simultaneous cocaine and ethanol users. J. Anal. Toxicol. *8*, 38–42.
31. Hearn, W.L., Rose, S., Wagner, J., Ciarleglios, A. and Mash, D.C. (1991). Cocaethylene is more potent than cocaine in mediating lethality. Pharmacol. Biochem. & Behaviour *39*, 531–533.
32. Moore, J.M. (1973). Identification of cis- and transcinnamoylcocaine in illicit cocaine seizures. J. Assoc. Off Anal. Chem. *55*, 1199–1205.
33. Aderjan, R.E., Schmitt, G., Wu, M. and Meyer, C. (1993). Determination of cocaine and benzoylecgonine by derivatization with iodomethane-D_3 or PFPA/HFIP in human blood and urine using GC/MS (EI or PCI mode). J. Anal. Toxicol. *17*, 51–55.
34. Isenschmid, D.S., Levine, B.S. and Caplan, Y.H. (1992). The role of ecgonine methyl ester in the interpretation of cocaine concentrations in postmortem blood. J. Anal. Toxicol. *16*, 319–324.
35. Logan, K. and Peterson, K.L. (1994). The origin and significance of ecgonine methyl ester in blood samples. J. Anal. Toxicol. *18*, 124–125.
36. Abusada, G.M., Abukhalaf, I.K., Alford, D.D., Vinzon-Bautista, I., Pramanik, A.K., Ansari, N.A., Manno, J.E. and Manno, B.R. (1993). Solid-phase extraction and GC/MS quantification of cocaine, ecgonine methyl ester, benzoylecgonine, and cocaethylene from meconium, whole blood and plasma. J. Anal. Toxicol. *17*, 353–358.
37. Farré, M., De la Torre, R., Lorente, M., Lamas, X., Ugena, B., Segura, J. and Camí, J. (1993). Alcohol and cocaine interactions in humans. J. Pharmacol. Exp. Ther. *266*, 1364–1373.
38. Appel, T.A. and Wade, N.A. (1989). Screening of blood and urine for drugs of abuse

utilizing diagnostic products corporation's Coat-A-Count® radioimmunoassay. J. Anal. Toxicol. *13*, 274–276.

39. Yee, H.Y., Nelson and Papa, V.M. (1993). Measurement of benzoylecgonine in whole blood using the Abbott ADx® analyzer. J. Anal. Toxicol. *17*, 84–86.
40. Cone, E.J. and Weddington, W.W. (1989). Prolonged occurence of cocaine in human saliva and urine after chronic use. J. Anal. Toxicol. *13*, 65–68.
41. Thompson, L.K., Yousefnejad, D., Kumor, K., Sherer, M. and Cone, E.J. (1987). Confirmation of cocaine in human saliva after intravenous use. J. Anal. Toxicol. *11*, 36–38.
42. Kato, K., Hillsgrove, M., Weinhold, L., Gorelick, D.A., Darwin, W.D. and Cone, E.J. (1993). Cocaine and metabolite excretion in saliva under stimulated and nonstimulated conditions. J. Anal. Toxicol. *17*, 338–341.
43. Ortuño, J., De la Torre, R., Segura, J. and Camí, J. (1990). Simultaneous detection in urine of cocaine and its main metabolites. J. Pharm. Biomed. Anal. *8*, 8–12.
44. Poklis, A. (1987). Evaluation of TDx cocaine metabolite assay. J. Anal. Toxicol. *11*, 228–230.
45. Baugh, L.D., Allen, E.E., Liu, R.H., Langner, J.G., Fentress, J.C., Chadha, S.C., Forster, Cook, L. and Walia, A. (1991). Evaluation of immunoassay methods for the screening of cocaine metabolites in urine. J. Forensic Sci. *36*, 79–85.
46. Verebey, K. and DePace, A. (1989). Rapid confirmation of enzyme multiplied immunoassay technique (EMIT®) cocaine positive urine samples by capillary gasliquid chromatography/nitrogen phosphorus detection (GLC/NPD). J. Forensic Sci. *34*, 46–52.
47. Cone, E.J., Hillsgrove, M.J., Jenkins, A.J., Keenan, R.M. and Darwin, W.D. (1994). Sweat testing for heroin, cocaine and metabolites. J. Anal. Toxicol. *18*, 298–305.
48. Steele, B.W., Bandstra, E.S., Wu, N.C., Hime, G.H. and Hearn, W.L. (1993). *m*-Hydroxybenzoylecgonine: an important contributor to the immunoreactivity in assays for benzoylecgonine in meconium. J. Anal. Toxicol. *17*, 348–352.
49. Ripple, M.G., Goldberger, B.A., Caplan, Y.H., Blitzer, M.G. and Schwartz, S. (1992). Detection of cocaine and its metabolites in human amniotic fluid. J. Anal. Toxicol. *16*, 328–331.
50. Cone, E.J., Yousefnejad, D., Darwin, W.D. and Maguire, T. (1991). Testing human hair for drugs of abuse. II. Identification of unique cocaine metabolites in hair of drug abusers and evaluation of decontamination procedures. J. Anal. Toxicol. *15*, 250–255.
51. Harkey, M.R., Henderson, G.L. and Zhou, C. (1991). Simultaneous quantification of cocaine and its major metabolites in human hair by gas chromatography/chemical ionization mass spectrometry. J. Anal. Toxicol. *15*, 260–265.
52. Welch, M.J., Sniegoski, L.T., Allgood, C.C. and Habram, M. (1993). Hair analysis for drugs of abuse: evaluation of analytical methods, environmental issues, and development of reference materials. J. Anal. Toxicol. *17*, 389–398.
53. DiGregorio, G.J., Barberi, E.J., Ferko, A.P. and Ruch, E.K. (1993). Prevalence of cocaethylene in the hair of pregnant women. J. Anal. Toxicol. *17*, 445–446.
54. Möller, M.R., Fey, P. and Rimbach, S. (1992). Identification and quantification of cocaine and its metabolites, benzoylecgonine and ecgonine methyl ester, in hair of Bolivian coca chewers by gas chromatography/mass spectrometry. J. Anal. Toxicol. *16*, 291–296.
55. Kintz, P., Ludes, B. and Mangin, P. (1992). Detection of drugs in human hair using Abbott ADx, with confirmation by gas chromatography/mass spectrometry (GC/MS). J. Forensic Sci. *37*, 328–331.
56. Ambre, J.J., Ruo, T.I., Smith, G.L., Bakkes, D. and Smith, C.M. (1982). Ecgonine methyl ester, a major metabolite of cocaine. J. Anal. Toxicol. *6*, 26–29.
57. Ambre, J.J., Fishman, M. and Ruo, T.I. (1984). Urinary excretion of ecgonine methyl ester, a major metabolite in humans. J. Anal. Toxicol. *8*, 23–25.
58. Dean, R.A., Harper, E.T., Dumaual, N., Stoeckel, D.A. and Bosron, W.F. (1992). Effect of ethanol on cocaine metabolism: Formation of cocaethylene and norcocaethylene. Toxicol. Appl. Pharmacol. *117*, 1–8.
59. Janzen, K. (1992). Concerning norcocaine, ethylbenzoylecgonine and the identification of cocaine use in human hair. J. Anal. Toxicol. *16*, 402.
60. Ferrara, S.D., Tedeschi, L., Frison, G., Brunsini, G., Castagna, F., Bernardelli, B. and Soregaroli, D. (1994). Drugs-of-abuse testing in urine: statistical approach and ex-

perimental comparison of immunochemical and chromatographic techniques. J. Anal. Toxicol. *18*, 278–291.

61. Schwartz, R.H., Bogema, S. and Thorne, M.M. (1990). Evaluation of the EZ-SCREEN enzyme immunoassay test for detection of cocaine and marijuana metabolites in urine specimens. Pediatr. Emerg. Care *6*, 147–149.
62. Anderson, A.J., Vekich, A.J. and Scott, M. (1990). Evaluation of a rapid qualitative EIA test for drugs of abuse (Abstract). Clin. Chem. *36*, 1021.
63. Wu, A.H.B., Wong, S.S., Johnson, K.G., Callies, J., Shu, D.X., Dunn, W.E. and Wong, S.H.Y. (1993). Evaluation of the Triage system for emergency drugs-of-abuse testing in urine. J. Anal. Toxicol. *17*, 241–245.
64. Baselt, R.C. and Baselt, D.R. (1987). Little cross reactivity of local anesthetics with Abuscreen, EMIT d.a.u., and TDx immunoassays for cocaine metabolite. Clin. Chem. *33*, 747.
65. Schwartz, J.G., Zollars, P.R., Okorodudu, A.O., Carnahan, J.J., Wallace, J.E. and Briggs, J.E. (1991). Accuracy of common drug screen tests. Am. J. Emerg. Med. *9*, 166–170.
66. Welch, E., Fleming, L.E., Peyser, I., Greenfield, W., Steele, B.W. and Bandstra, E.S. (1993). Rapid screening of urine in a newborn nursery. J. Pediatr. *123*, 468–470.
67. Wagener, R.E., Linder, M.W. and Valdes, R.V. (1994). Decreased signal in EMIT assays of drug of abuse in urine after ingestion of aspirin: potential for false-negative results. Clin. Chem. *40*, 608–612.
68. Wells, H. (1988). A Discussion of the KDI Quik Test™ drug screen. J. Anal. Toxicol. *12*, 111.
69. Joern, W.A. (1994). Unexpected volatility of barbiturate derivatives: an extractive alkylation procedure for barbiturates and benzoylecgonine. J. Anal. Toxicol. *18*, 423.
70. Jindal, S.P. and Lutz, T. (1986). Ion cluster techniques in drug metabolism: use of a mixture of labeled and unlabeled cocaine to facilitate metabolite identification. J. Anal. Toxicol *10*, 150–155.
71. Dugan, S., Bogema, S., Schwarz, R.W. and Lappas, N.T. (1994). Stability of abuse in urine samples stored at $-20\,^{\circ}$C. J. Anal. Toxicol *18*, 391–396.
72. Paul, B.D., McKinley, R.M., Walsh, J.K., Jamir, T.S. and Past, M.R. (1993). Effect of freezing on the concentration of drugs of abuse in urine. J. Anal. Toxicol. *17*, 378–380.
73. Hippenstiel, M.J. and Gerson, B. (1994). Optimization of storage conditions for cocaine and benzoylecgonine in urine: a review. J. Anal. Toxicol. *18*, 104–109.
74. Hoffman, B.B. and Lefkowitz, R.J. (1990). Catecholamines and sympathomimetic drugs. pp 210–213 & Jaffe, J.H. Drug addiction and abuse. pp 539–545. In: Goodman & Gilman's The Pharmacological basis of therapeutics. Goodman Gilman A et al. (eds). Pergamon Press, New York.
75. Smith, R.L. and Dring, L.G. (1970). Patterns of metabolism of β-phenylisopropylamines in man and other species. In: International symposium on amphetamines and related compounds. Costa, E. and Garattini, S. (eds) pp. 121–140, Raven Press, New York.
76. Møller, Nielsen, I. and Dubnick, B. (1970). Pharmacology of chlorphentermine. In: International symposium on amphetamines and related compounds. Costa, E. and Garattini, S. (eds) pp. 63–73, Raven Press, New York.
77. Le Douarec, J.C. and Neveu, C. (1970). Pharmacology and biochemistry of fenfluramine. In: International symposium on amphetamines and related compounds. Costa, E. and Garattini, S. (eds) pp. 75–105, Raven Press, New York.
78. Beckett, A.H. and Brookes, L.G. (1970). The effect of chain and ring substitution on the metabolism, distribution and biological action of amphetamines. In: International symposium on amphetamines and related compounds. Costa, E. and Garattini, S. (eds) pp. 109–120, Raven Press, New York.
79. Vree, T.B. and Van Rossum, J.M. (1970). Kinetics of metabolism and excretion of amphetamines in man. In: International symposium on amphetamines and related compounds. Costa, E. and Garattini, S. (eds) pp. 165–190, Raven Press, New York.
80. Beckett, A.H. and Bélanger, P.M. (1974). The mechanism of metabolic *N*-oxidation of phentermine and chlorphentermine to their hydroxylamino- and nitrosocompounds. J. Pharm. Pharmac. *6*, 558–560.
81. Makino, Y., Ohta, S., Tasaki, Y., Tachikawa, O., Kashiwasake, M. and Hirobe, M. (1990). A novel and neurotoxic tetrahydroisoquinoline derivative in vivo: formation of 1,3-dimethyl-1,2,3,4-tetrehydroisoqui-

noline, a condensation product of amphetamines, in brains of rats under chronic ethanol treatment. J. Neurochem. *55*, 963–969.

82. Moody, D.E., Ruangyuttikarn, W. and Law, M.Y. (1990). Quinidine inhibits in vivo metabolism of amphetamine in rats: impact upon correlation between GC/MS and immunoassay findings in rat urine. J. Anal. Toxicol *14*, 311–317.
83. Gjerde, H., Hasvold, I., Pettersen, G. and Christophersen, A.S. (1993). Determination of amphetamine and methamphetamine in blood by derivatization with perfluorooctanoyl chloride and gas chromatography/mass spectrometry. J. Anal. Toxicol. *17*, 65–68.
84. Simonick, T.F. and Watts, V.W. (1992). Preliminary evaluation of the Abbott TDx for screening of d-methamphetamine in whole blood specimens. J. Anal. Toxicol. *16*, 115–118.
85. Suzuki, S., Inoue, T., Hori, H. and Inayama, S. (1989). Analysis of methamphetamine in hair, nail, sweat and saliva by mass fragmentography. J. Anal. Toxicol. *13*, 176–178.
86. Przekop, M.A., Manno, J.E., Kunsman, G.W., Cockerman, K.R. and Manno, B.R. (1991). Evaluation of the Abbott ADx amphetamine/methamphetamine II abused drug assay: comparaison to TDx, EMIT, and GC/MS methods. J. Anal. Toxicol. *15*, 323–326.
87. Dasgupta, A., Saldana, S., Kinnaman, G., Smith, M. and Johansen, K. (1993). Analytical performance evaluation of EMIT® II monoclonal amphetamine/methamphetamine assay: more specificity than EMIT® d.a.u.™ monoclonal amphetamine/methamphetamine assay. Clin. Chem. *39*, 104–108.
88. Maurer, H.H. and Kraemer, T. (1992). Toxicological detection of selegiline and its metabolites in urine using fluorescence polarization immunoassay (FPIA) and gas chromatography-mass spectrometry (GC-MS) and differentiation by enantioselective GC-MS of the intake of selegiline from abuse of methamphetamine or amphetamine. Arch Toxicol *66*, 675–678.
89. D'Nicuola, J., Jones, R., Levine, B. and Smith, M.L. (1992). Evaluation of six commercial amphetamine and methamphetamine immunoassays for cross-reactivity to phenylpropanolamine and ephedrine in urine. J. Anal. Toxicol. *16*, 211–213.
90. Poklis, A., Jortani, S.A., Brown, C.S. and Crooks, C.R. (1993). Response of the EMIT II amphetamine/methamphetamine assay to specimens collected following use of Vicks inhalers. J. Anal. Toxicol. *17*, 284–286.
91. Steiner, E., Villén, T., Halberg, M. and Rane, A. (1984). Amphetamine secretion in breast milk. Eur. J. Clin. Pharmacol. *27*, 123–124.
92. Dahlem, P., Bucher, H.U., Cuendet, D., Mieth, D. and Gautschi, K. (1993). Pravalenz von drogen in meconium. Monatsschr. Kinderheilkd. *141*, 237–240.
93. Nakahara, Y., Takahaski, K., Shimamine, M. and Takeda, Y. (1990). Hair analysis for monitoring drug of abuse. II. Determination of methamphetamine in hair by isotope dilution gas chromatography/mass spectrometry. J. Forensic Sci. *36*, 70–78.
94. Cody, J.T. and Schwarzhoff, R. (1993). Fluoresence polarization immunoassay detection of amphetamine, methamphetamine, and illicit amphetamine analogues. J. Anal. Toxicol. *17*, 26–30.
95. Röhrich, J., Schmidt, K. and H. Bratzke (1994). Application of the novel immunoassay TRIAGE™ to a rapid detection of antemortem drug abuse. J. Anal. Toxicol. *18*, 407–414.
96. Suttijitpaisal, P. and Ratanabanangkoon, K. (1992). Immunoassays of amphetamines: immunogen structure vs antibody specificity. Asian Pac. J. Allergy Immunol. *10*, 159–164.
97. Mongkolsirichaikul, D., Tarnchompoo, B. and Ratanabanangkoon, K. (1993). Development of a latex agglunation inhibition reaction test for amphetamines in urine. J. Immunol. Meth. *157*, 189–195.
98. Cheng, L.T., Kim, S.Y., Chung, A. and Castro, A. (1973). Amphetamines: new radioimmunoassay. FEBS Lett. *36*, 339–342.
99. Poklis, A., Hall, K.V., Still, J. and Binder, S.R. (1991). Ranitidine interference with the monoclonal EMIT d.a.u. amphetamine/methamphetamine immunoassay. J. Anal. Toxicol. *15*, 101–103.
100. Olsen, K.M., Gulliksen, M. and Christophersen, A.S. (1992). Metabolites of chlorpromazine and brompheniramine may cause false-positive urine amphetamine results with monoclonal EMIT® d.a.u.™ assay. Clin. Chem. *38*, 611–612.
101. Crane, T., Dawson, C.M. and Ticker, T.R. (1993). Falsepositive results form the Syva EMIT™ d.a.u. monoclonal amphetamine

assay as a result of antipsychotic drug therapy. Clin. Chem. *39*, 549.
102. Jones, R., Klette, K., Kuhlman, J.J., Levine, B., Smith, M.L., Watson, C.V. and Selavka, C.M. (1993). Trimethobenzamide cross-reacts in immunoassays of amphetamine/methamphetamine. Clin. Chem. *39*, 699–700.
103. Poklis A (1992). Unvailability of drug metabolite reference material to evaluate false-positive results for monoclonal EMIT d.a.u. of amphetamine. J. Clin. Chem. *38*, 2560.
104. Tarver, J.A. (1994). Amphetamine-positive drug screens from use of clobenzorex hydrochlorate. J. Anal. Toxicol. *18*, 183.
105. Yoo, Y., Chung, H. and Choi, H. (1994). Urinary methamphetamine concentration following famprofazone administration. J. Anal. Toxicol. *18*, 265–268.
106. Clark, C.R., DeRuiter, J. and Noggle, F.T. (1992). GC-MS identification of amine-solvent condensation products formed during analysis of drugs of abuse. J. Chromatogr. Sci. *30*, 399–404.
107. Hornbeck, C.L. and Czarny, R.J. (1989). Quantitation of methamphetamine and amphetamine in urine by capillary GC/MS. Part I. Advantage of trichloroacetyl derivatization. J. Anal. Toxicol. *13*, 144–149.
108. Thurman, E.M., Pedersen, M.J., Stout, R.L. and Martin, T. (1992). Distinguishing sympathomimetic amines from amphetamines and methamphetamine in urine by gas chromatography/mass spectrometry. J. Anal. Toxicol. *16*, 19–27.
109. Jones, J.B. and Mell, L.D. (1993). A simple wash procedure for improving chromatography of HFAA derivatizated amphetamine extracts for GC/MS analysis. J. Anal. Toxicol. *17*, 447.
110. Wu, A.H.B., Onigbinde, T.A. and Wong, S.S. (1992). Identification of methamphetamines and over-the-counter sympathomimetic amines by full-scan GC-ion trap MS with electron impact and chemical ionization. J. Anal. Toxicol. *16*, 137–141.
111. Czarny, R.J. and Hornbeck, C.L. (1989). Quantitation of methamphetamine and amphetamine in urine by capillary GC/MS. Part. II. Derivatization with 4-carbethoxyhexafluorobutyryl chloride. J. Anal. Toxicol. *13*, 257–262.
112. Donike, M. (1973). Acelyring mit bis(acylamiden). *N*-methyl-bis(trifluoroactamid) und bis(trifluoroactamid), zwei neue reagenzien zur trifluoroacetylering. J. Chromatogr. *78*, 273–279.
113. Melgar, R. and Kelly, R.C. (1993). A novel GC/MS derivatization method for amphetamines. J. Anal. Toxicol. *17*, 399–402.
114. Hornbeck, C.L., Carrig, J.E. and Czarny, R.J. (1993). Detection of a GC/MS artifact peak as methamphetamine. J. Anal. Toxicol. *17*, 257–263.
115. Elsohly, M.A., Stanford, D.F., Sherman, D., Shah, H., Bernot, D. and Turner, C.E. (1992). A procedure for eliminating interferences from ephedrine and related compounds in the GC/MS analysis of amphetamine and methamphetamine. J. Anal. Toxicol. *16*, 109–111.
116. Paul, D.P., Past, M.R., McKinley, R.M., Foreman, J.D., McWhorter, L.K. and Snyder, J.J. (1994). Amphetamine as an artifact of methamphetamine during periodate degradation of interfering ephedrine, pseudoephedrine, and phenylpropanolamine: an improved procedure for accurate quantification of amphetamines in urine. J. Anal. Toxicol. *18*, 331–336.
117. Spiehler, V.R., Maugh, K. and El Shami, S. (1993). Elimination of ephedrine and pseudoephedrine cross-reactivity in the Coat-A-Count methamphetamine radioimmunoassay. J. Anal. Toxicol. *17*, 125–126.
118. Ward, C., McNally, A.J., Rusyniak, D. and Salamone, S.J. (1994). ^{125}I Radioimmunoassay for the dual detection of amphetamine and methamphetamine. J. Forensic Sci. *39*, 1486–1496.
119. Hornbeck, C.L. and Czarny, R.J. (1993). Retrospective analysis of some L-methamphetamine/L-amphetamine urine data. J. Anal. Toxicol. *17*, 23–25.
120. Cody, J.T. and Schwarzhoff, R. (1993). Interpretation of methaphetamine and amphetamine enantiomer data. J. Anal. Toxicol. *17*, 321–326.
121. Kalasinsky, K., Levine, B., Smith, M.L., Magluilo, J. and Schaefer, T. (1993). Detection of amphetamine and methamphetamine in urine by gas chromatography/fourier transform infrared spectroscopy. J. Anal. Toxicol. *17*, 359–364.
122. Platoff, G.E., Hill, D.W., Koch, T.R. and Caplan, Y.H. (1992). Serial capillary gas chromatography/fourier transform infrared spectrometry/mass spectrometry (GC/IR/MS): qualitative and quantitative analysis of amphetamine, methamphetamine, and related analogues in human urine. J. Anal. Toxicol. *16*, 389–397.

3.16 Designer Drugs

G. L. Henderson

3.16.1 Introduction

The term "Designer Drugs" was used originally to describe analogs of the potent narcotic analgesic fentanyl. These drugs were synthesized in clandestine laboratories in the early 1980's, sold on the street as heroin substitutes under names like "China White" and "Synthetic Heroin", and eventually were responsible for numerous overdose deaths (1, 2). Because these analogs are very potent and have chemical structures unlike opiates, they were not easily detected by conventional toxicological screening tests. In addition, until the U.S. drug laws were changed, their unique structures prevented them from being classified as illegal or restricted drugs.

The term "Designer Drugs" has since been popularized and is now used to describe nearly any new drug that emerges from clandestine laboratories for illicit use. Thus, a more useful and functional definition of the term "Designer Drugs" today might be – any new drug with abuse potential whose unique chemical structure makes it unlikely to be detected by routine toxicological screens. With this later definition in mind, this chapter considers three groups of drugs – the fentanyls, the ring-substituted amphetamines, and the aminorex derivatives – all of which have been manufactured in illicit laboratories and associated with drug overdose deaths. In addition, the drug gamma hydroxy butyrate (GHB) is included. Although this drug has been available as a non-prescription drug for decades, it has been associated with a recent series of poisonings. It is included in this chapter because it is rather difficult to analyze and some have suggested that the poisonings are the result of its interaction with other designer drugs. The overall objective of this chapter, therefore, is to discuss the analysis of newer drugs of abuse which may be important toxicologically, but for which methods for their detection may not be readily available.

3.16.2 The Fentanyls

3.16.2.1 Toxicological importance

The fentanyls were the first family of drugs to which the term "Designer Drugs" was applied. In the last decade, over 12 different derivatives of the potent narcotic analgesic fentanyl have been synthesized by clandestine laboratories and over 100 deaths have been attributed to the abuse of these drugs (2). The fentanyls are chemically quite different from all other opiates, yet they produce all the effects of heroin (e.g., euphoria and analgesia), as well as all the side effects (e.g., pin-point pupils and respiratory depression). Experienced heroin users report that the effects of the fentanyls are similar to, and are an acceptable substitute for, heroin (3). When used regularly, they produce tolerance and physiological dependence. The typical fentanyl overdose death may look much like a "sudden heroin death"; that is, the victim will appear to have died very shortly after using the drug. However, no drug is likely to be detected by routine toxicological examination. Because the fentanyls are so potent and are present in such small amounts in body fluids and tissues (even in powder and paraphernalia samples), highly sensitive and specific screening tests must be used to detect either the parent drugs, their precursors, or metabolites.

3.16.2.2 Chemistry

The fentanyls were developed in the 1960's by the Janssen pharmaceutical company for clinical use as analgesics. This large family of substituted piperidines contains over 200 analogs which are known to be pharmacologically active and many more are theoretically possible (4, 5, 6). Figure 1 shows the chemical formulae of the two major subdivisions of the fentanyl series, the 4-substituted fentanyls and the 4,4-disubstituted fentanyls, as well as the structures of the two primary metabolites of the parent drug fentanyl.

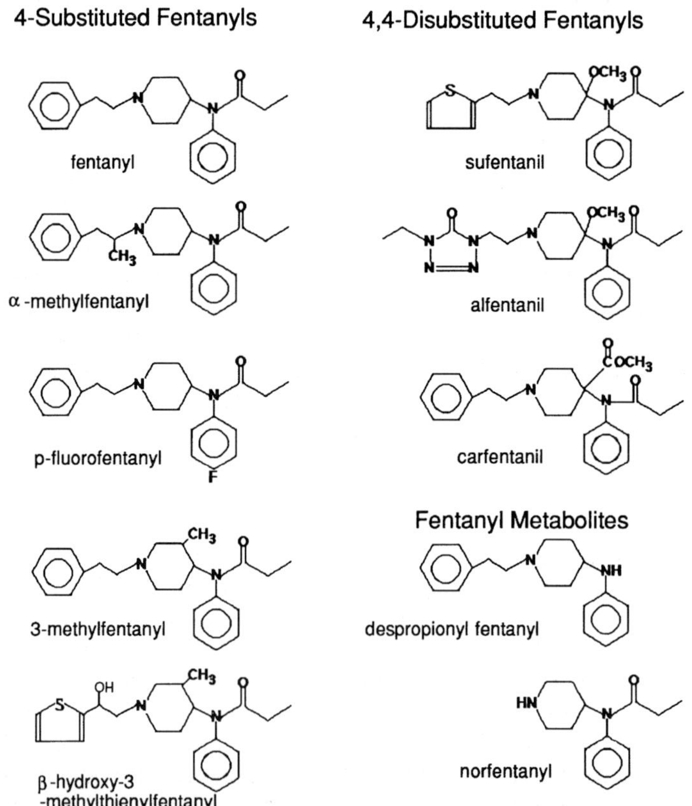

Figure 1. Chemical Structures of 4-Substituted Fentanyls, 4,4-Disubstituted Fentanyls, and the Two Primary Metabolites Norfentanyl and Despropionyl Fentanyl.

All five of the 4-substituted fentanyls shown in figure 1 have been identified in street samples sold as heroin. Sufentanil and alfentanil, which are clinically important analgesics, are examples of 4,4-di-substituted fentanyls. The other disubstituted analog, carfentanil, is used as an immobilizing agent for animals. To date, there have been no di-substituted fentanyls found in illicit samples.

The structure-activity relationships of the fentanyls are relatively simple – most structures are active and changes in the basic fentanyl structure generally modify only potency or duration of action, not the intrinsic opiate-like character. Potency can be increased by simple alkyl substitutions at the α- or β-position of the phenethyl sidechain and at the 3- and 4-positions of the piperidine ring. Hydroxylation of the phenethyl sidechain also increases potency. In addition, the benzene ring of the phenyl sidechain can be replaced by heterocycles such as thiophene or furan.

The 4-substituted fentanyls are surprisingly easy to synthesize. Also, clandestine chemists have discovered novel synthetic pathways not reported in the literature and have been able to produce them from common and readily available chemicals. For example, 3-methylfentanyl, an analog many hundreds of times as potent as heroin, can be easily synthesized from phenethylamine, methyl acrylate and methyl methacrylate via a Dieckman condensation. Thus, it is likely that the fentanyls will persist as illicit substitutes for heroin because of their ease of synthesis and their difficulty in detection.

3.16.2.3 Metabolism

The fentanyls are very lipid soluble, distribute rapidly throughout the body, and reach the brain quickly regardless of the route of administration. They are rapidly and extensively metabolized, generally to more polar, pharmacologically inactive metabolites that are excreted in the urine and feces. Oxidative N-dealkylation to normetabolites is the major metabolic route for these drugs (7, 8), while amide hydrolysis to despropionyl metabolites occurs to a lesser extent, as does ring and sidechain-hydroxylation (9, 10). Less than 10% of an administered dose of fentanyl will be eliminated in the form of unchanged parent drug (11).

3.16.2.4 Blood and tissue levels

The fentanyls distribute rapidly to peripheral tissues and the resultant blood concentrations are quite low, generally in the low nanogram range. In overdose victims, parent drug concentrations, measured by radioimmunoassay, may range from 0.2 to greater than 50 ng/ml. The threshold drug concentration for respiratory depression is thought to be approximately 2 ng/ml. Respective values for the more potent analogs are correspondingly much lower. For example, lethal doses of potent analogs like 3-methyl fentanyl are likely to be in the low microgram range. Urine drug concentration of the fentanyls, measured by radioimmunoassay, may range from 0.2 to greater than 800 ng/ml (1, 11, 12, 13, 14). Even powder and paraphernalia samples associated with overdose deaths have concentrations of active drug generally less than 1% (15).

3.16.2.5 Analytical considerations

The fentanyls represent a considerable challenge for the analyst. They are chemically unrelated to other opiates and thus will not cross-react with typical opiate immunoassay screening tests. Also, because they are present in subnanogram concentrations in biological fluids and tissues, these drugs are unlikely to be detected by thin layer chromatography screening tests. Even GC/MS analysis for the fentanyls is impeded by the fact that there are no class-specific ions that can be considered diagnostic. Finally, because of the large number of analogs possible, absolute confirmation can be difficult and time consuming.

3.16.2.6 Extraction methods

The most generally useful extraction solvent system for the fentanyls consists of heptane-isoamylol (98.5 : 1.5). The solvent system can be used to extract a variety of samples including blood, urine or powder samples. Extracting 1-ml-alkalized samples with this solvent system will yield a very clean extract with a recovery of approximately 80%. The amount of amylol can be increased slightly to improve the extraction efficiency for the more polar metabolites. Extracts may be further purified by differential pH extraction (16) and/or solid phase columns (17). Recovery can be improved by silanizing all glassware and using Teflon-lined caps on all extraction tubes.

3.16.2.7 Screening methods

Radioimmunoassay

A sensitive and highly specific radioimmunoassay for the parent drug fentanyl has been available since the drug was introduced into clinical medicine (18). The antisera do not cross react with other drugs of abuse and can be used to detect picogram amounts of fentanyl in urine and blood (19, 20, 21, 22, 23). This radioimmunoassay is available in solid-phase form from the Diagnostics Products Corporation (Los Angeles, CA). Calibrators in this assay kit range from 0.25 to 7.0 ng/ml and the recommended cut-off is 0.5 ng/ml. Modified drug-free horse urine is used as a diluent to minimize matrix effects and to facilitate this assay use in equine drug testing (fentanyl produces stimulation in horses and is used as a doping agent). Even powder samples can be conveniently screened by this assay (15).

The analyst faces a dilemma in using this assay's recommended cut-off of 0.5 ng/ml. Pharmacologically or toxicologically significant effects of the fentanyls may be observed with blood concentrations as low as 0.1 ng/ml and even in overdose cases, blood levels may be below 0.5 ng/ml (1, 2, 24). Therefore, in certain cases the analyst may find it necessary to extend the range of the fentanyl assay below the manufacturer's recommended cutoff. However, as with other immunoassays, matrix effects or large concentrations of weakly cross-reacting drugs can produce a false positive, thus the requirement for confirmation by an alternate method, preferably mass spectroscopy, becomes even more important.

The immunoassay described above will detect 4-substituted analogs of fentanyl (which includes all the illicit fentanyls found to date) but will not detect the 4,4-disubstituted piperidines. Radioimmunoassays specific for these derivatives (e.g., sufentanil and alfentanil) may be obtained from Janssen Pharmaceutica (Beerse, Belgium).

3.16.2.8 Confirmation methods

Gas chromatography-nitrogen specific detection
A number of gas chromatographic techniques with nitrogen specific detection have been developed for the fentanyls using both packed and capillary columns. At least three internal standards have been used: papaverine (25), the p-methoxyphenyl derivative of fentanyl (16), and R 38527, a Janssen compound which is an analog of alfentanil, but with a propylene bridge, rather than a ethylene bridge, separating the tetrazol and piperidine rings (17). The fentanyl metabolites (norfentanyl and despropionyl) can be analyzed by preparing acetyl derivatives (25). The column most often reported in the literature is 3% OV-17 although the more recent use of capillary columns with splitless injection may improve sensitivity and reproducibility at low concentrations by resolving fentanyl from potentially interfering peaks (26, 27). Most gas chromatographic methods report detection limits down to 0.1 ng/ml for urine, serum, or blood samples.

Gas chromatography – electron capture detection
The normetabolites of the fentanyls are found in urine at up to tenfold higher concentrations than the parent drug. Thus, one useful approach to confirming urine samples which have very low concentrations of the parent drug is to derivatize the sample (with heptafluorobutyric anhydride for example), then analyze by gas chromatography with electron capture detection (28). The normetabolites may also be derivatized with pentafluoropropionic anhydride (29).

High-performance liquid chromatography
High-performance liquid chromatography (HPLC) can be used both as a separation technique prior to mass spectrometry and as a method for the direct analysis of the fentanyls in powder or biological samples. A microflow HPLC method with mass spectrometry – chemical ionization detection has been reported for the analysis of fentanyl (30). This procedure uses a 5-cm × 0.05-mm i.d. piece of polytetrafluoroethylene tubing packed with 10-μm silica ODS SC-01 as the analytical column, with a mobile phase consisting of 40% acetonitrile in water.

A fast and efficient technique for analyzing up to 26 different fentanyl analogs in powder samples has been reported using reversed-phase isocratic HPLC (31). The system employed a 5-cm × 25-cm stainless steel column packed with 10 μm C18 and a mobile phase containing phosphate buffer : acetonitrile : tetrahydrofuran : methanol (81 : 10 : 5 : 4). Recently, a HPLC method for alfentanil as well as fentanyl in plasma has been reported with a detection limit of 1 ng/ml (32). Alkalinized 1-ml plasma samples are extracted with either heptane, for fentanyl, or heptane : isoamylol (98.5 : 1.5), for alfentanil prior to HPLC analysis with UV detection.

Mass spectrometry
The mass spectral analysis of the fentanyls should be approached carefully. Despite the fact that the fentanyl analogs are often simple variations of the parent molecule, there are no class-specific ions for the fentanyls. Also, some analogs have identical molecular weights and common ions, but quite different chemical structures. For example, the molecular ion for both α-methylfentanyl and 3-methylfentanyl is m/z 350 and another anion common to both drugs is m/z 259. Similarly, fentanyl and thienylfentanyl both share a common molecular ion at m/z 336 as well as ions at m/z 245, 189, and 146.

Absolute confirmation of powder samples can present a special problem if they contain isomeric mixtures. The resultant mass spectra may vary depending on relative abundance of the enantiomers and the chromatographic conditions used. For example, the relative abundance of the *cis* and *trans* isomers of 3-methylfentanyl will vary depending on the synthetic route and reaction conditions. Resolution of these two isomers will vary depending on the chromatographic conditions used and, as a result, the mass spectra of 3-methylfentanyl in powder samples may vary considerably from laboratory to laboratory, even when the powder samples contain relatively large amounts of the drug.

Finally, some fentanyl compounds may decompose or rearrange during mass spectral analysis. For example, the major metabolite of fentanyl in the horse, a malonanalinic acid derivative, is thermally unstable and decarboxylates readily under GC conditions to form despropionyl fentanyl (33). Thus, when horse urine is analyzed by RIA, the results will suggest that the parent drug is present. However, GC/MS analysis will show only norfentanyl present. This paradox can be resolved, however, by using LC/MS analysis which will show the presence of carboxylic acid metabolite.

Despite these limitations, there are an increasing number of GC/MS techniques reported for the analysis of the fentanyls and their metabolites (9, 27, 30, 34, 35), and deuterated fentanyl and norfentanyl are now available for use as internal standards. An excellent collection of mass spectral data, as well as infrared and nuclear magnetic resonance spectra, for the fentanyls has been published by the U.S. Drug Enforcement Administration (DEA) Special Testing and Research Laboratory (36).

3.16.3 Ring-Substituted Amphetamines

3.16.3.1 Toxicological importance

Only a decade ago, the ring-substituted amphetamines (sometimes known by their acronyms MDA, MDMA or MDEA) were of interest to only a few mental health professionals. The unique properties of these drugs, that of producing expanded emotional insight and empathy, prompted early investigators to study these drugs as potential psychotherapeutic adjuncts. Today, these same properties, coupled with the more commonly known effects of all amphetamine compounds, mood-elevation and CNS stimulation, has made this class of compounds increasingly popular as recreational drugs especially among today's youth. The most commonly used drug

MDMA is known by a variety of popular names which include: Ecstasy, XTC, E, X, or Adam. Compared with their parent drugs amphetamine and methamphetamine, the ring-substituted analogs have been associated with remarkably few acute overdose deaths. However, experiments in laboratory animals have shown that these drugs can be toxic to brain dopamine and serotonin nerve cells. As a result, their manufacture and sale is illegal in most of the world.

3.16.3.2 Chemistry

The ring-substituted amphetamines constitute a very large family of drugs with unique psychoactive properties. Shulgin has recently published a comprehensive review of the synthesis, chemical-physical properties, and pharmacological properties of these unique drugs (37). Chemical structures for the derivatives most commonly reported and some major derivatives are shown in figure 2.
The structure activity of the phenethylamines is complex; however, the addition of a methylenedioxy moiety on the 3,4 positions of the phenyl ring seems to confer some rather unusual pharmacological properties. These derivatives retain some of the stimulant properties of amphetamine but, in addition, have some of the psychedelic properties of mescaline. Some have postulated a unique site or mechanism of action for MDMA-like compounds and suggest they are a new class of drugs called "entactogens" or "empathogens".

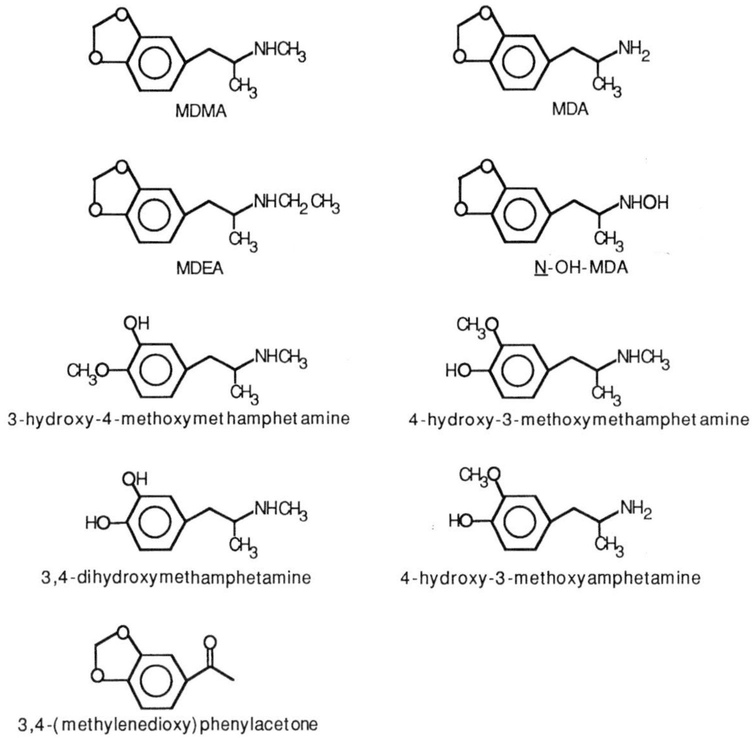

MDMA

MDA

MDEA

N-OH-MDA

3-hydroxy-4-methoxymethamphetamine

4-hydroxy-3-methoxymethamphetamine

3,4-dihydroxymethamphetamine

4-hydroxy-3-methoxyamphetamine

3,4-(methylenedioxy)phenylacetone

Figure 2. Chemical Structures of Ring-Substituted Amphetamines and Metabolites.

3.16.3.3 Metabolism

Relatively little is known about the metabolism of the ring-substituted amphetamines in man; however, it is thought that they are degraded like the amphetamines by N-dealkylation, deamination, and subsequent oxidation of the side chain. The methylenedioxy ring is oxidatively cleaved, then O-dealkylated to form a series of hydroxylated compounds which include 3-hydroxy-4-methoxymethamphetamine, 4-hydroxy-3-methoxymethamphetamine, 4-hydroxy-3-methoxyamphetamine, 3,4-dihydroxymethamphetamine and 3,4-(methylenedioxy)phenylacetone (38, 39, 40, 41). Both parent drug and metabolites are eliminated primilary in the urine. The hydroxylated metabolites are excreted primarily as glucuronide or sulfate conjugates.

3.16.3.4 Blood and tissue levels

Despite the widespread non-medical use of the ring-substituted amphetamines there have been relatively few deaths and no consistent pathological findings; thus, there is relatively little post-mortem toxicological data. The most common cause of death appears to be cardiac arrhythmia; however, hyperthermia and dehydration are also thought to be causally related. In the few MDMA-associated deaths reported in the literature, blood levels range from 0.9 μg/ml to 2 μg/ml.

3.16.3.5 Analytical considerations

Because significant amounts of the ring-substituted amphetamines are thought to be excreted unchanged in urine, this is the specimen of choice for toxicological examination. Like the other amphetamines, however, the ring-substituted analogs may not extract from biofluids efficiently and they may be difficult to chromatograph without derivatization. Direct aqueous derivatization of the polar hydroxylated metabolites might be one way to circumvent these problems (41).

3.16.3.6 Screening tests

Immunoassays
To minimize cross-reactivity with the numerous over-the-counter medications containing a phenethylamine moiety, most manufacturers maximize the specificity of their amphetamine assays. In fact, most "amphetamine" immunoassays are targeted at methamphetamine and cross-react rather weakly even with amphetamine. Thus, most commercially available amphetamine immunoassays exhibit only slight cross-reactivity with the ring-substituted amphetamines and therefore are not likely to be useful as screening tools for MDMA (42, 43, 44, 45).

3.16.3.7 Confirmation tests

HPLC
An HPLC-UV method for the simultaneous identification of amphetamine, methamphetamine, 3,4-methylenedioxyamphetamine (MDA) and 3,4-methylenedioxy-

methamphetamine (MDMA) in urine has been reported by Tedeschi et al. (46). It involves extraction of the analogs from urine using Extrelut® 3 columns, derivatization with sodium 1,2-naphthoquinone-4-sulphonate to form chromophoric UV-VIS derivatives, and ion-pair reversed-phase HPLC analysis with eluent monitoring at 480 nm. Practical detection limits were reported to be 40–60 ng/ml for all derivatives.

Noggle et al. have described a reversed-phase liquid chromatographic method consisting of a C_{18} stationary phase and an aqueous acidic mobile phase which can be used to differentiate MDMA from butanamine, a new psychotherapeutic drug (47). These two compounds differ only in the position of a single methyl group and thus have quite similar infrared spectra and mass spectral fragmentation patterns with a common base peak at m/z 58.

Gas chromatography-mass spectroscopy

A number of GC/MS procedures have been reported recently for the detection of the ring-substituted amphetamines, their metabolites and other phenethylamines. A method for quantitating MDA and MDMA in very small samples (200 µl) of whole blood has been described by Fitzgerald et al. (48). The analytes are isolated from blood by liquid-liquid extraction, then derivatized with the chiral reagent N-trifluoroacetyl-L-prolyl chloride. Separation, identification and quantitation of diastereomeric derivatives is achieved by gas chromatography-mass spectrometry with an analytical range of 0.12 ng to 48 ng injected on-column.

Gan et al. have reported a method for quantitating amphetamine, methamphetamine, and MDMA in urine samples (49). The drugs are extracted using Bond Elut® Certify bonded silica cartridges, then derivatized with trichloroacetic anhydride. Quantitation is accomplished using GC/MS analysis with selected ion monitoring and d5-amphetamine as the internal standard for amphetamine and d9-methamphetamine as the internal standard for methamphetamine and MDMA. With a sample size of 2 ml, the lowest detectable concentration is about 50 ng/ml.

A GC/MS chiral assay for simultaneous quantitation of MDMA and three of its hydroxylated metabolites in biological specimens has been described by Lim et al. (41). It employs direct aqueous derivatization with N-heptafluorobutyryl-S-prolyl chloride, capillary chromatographic separation of the diastereomeric derivatives and detection by electron capture negative ion chemical ionization mass spectrometry. The assay is reported to be linear from 5 to 1000 ng/ml for each enantiomer. Alternatively, MDMA and its hydroxylated metabolites may be analyzed by gas chromatography-mass spectrometry with perfluorotributylamine-enhanced ammonia positive-ion chemical ionization (40).The assay is linear from 2 to 1000 ng/ml.

3.16.4 Aminorex Derivatives

3.16.4.1 Toxicological importance

Aminorex and its derivatives first appeared as drugs of abuse in 1987, when 4-methylaminorex was detected in paraphernalia and body fluids associated with a suspected drug overdose death (50). Although aminorex and 4-methylaminorex have appeared on the streets only sporadically since then and have been suspected as the causative agent in only a few overdose deaths, these drugs may have significant abuse potential because they can be synthesized easily from very common (and legal) starting materials. Aminorex and 4-methylaminorex are touted as producing pharmacological effects somewhere between cocaine and amphetamine and may be sold under attractive names like "EU4EA". "EU4uh", or "U4EA". Occasionally, they may be sold as "Ice" or "Blue Ice", terms which may cause confusion because they are terms also used for smokeable (i. e., free-base) methamphetamine.

Chronic toxicity may be an additional concern with these compounds because aminorex was withdrawn from clinical use as an appetite suppressant in the 1960's after some patients developed fatal pulmonary hypertension (51, 52, 53, 54). Whether 4-methylaminorex has similar toxicity is not known. The aminorex derivatives produce all the effects and side effects of central nervous stimulants (restlessness, increased alertness, dry mouth, loss of appetite, insomnia and analgesia) (55, 56, 57). Acute toxicity results from over-stimulation of the central nervous system which produces seizures, loss of consciousness, respiratory depression, and death.

3.16.4.2 Chemistry

Aminorex and 4-methylaminorex are oxazolines that were first synthesized by McNeil laboratories in the early 1960's and found to be potent appetite suppressants and CNS stimulants (58). Their chemical structures and the structures of the most probable major metabolites are shown in figure 3.

These may be viewed as cyclic phenylethylamines with biological activity similar to CNS stimulants like amphetamines and pemoline. In fact, the structure-activity relationships for aminorex compounds are similar to those for the amphetamines (59). Potency is increased when halogens are substituted on the phenyl ring and the para-fluoro derivative is the most potent of the analogs tested. However, unlike

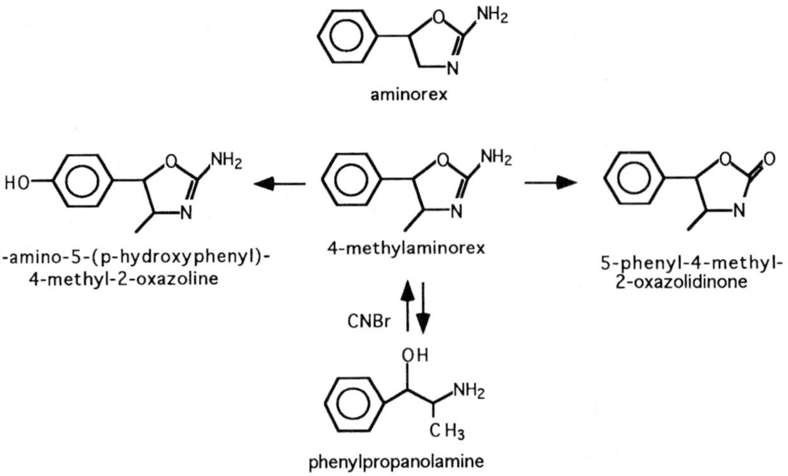

aminorex

2-amino-5-(p-hydroxyphenyl)-4-methyl-2-oxazoline

4-methylaminorex

5-phenyl-4-methyl-2-oxazolidinone

CNBr

phenylpropanolamine

Figure 3. Chemical Structures of Aminorex, 4-Methylaminorex, and Three Possible Metabolites. Phenylpropanolamine is Both the Synthetic Precursor and a Possible Metabolite.

the amphetamines in which only the δ isomer is active as a stimulant, all stereoisomers of the aminorex compounds appear to be active (60, 61, 62).

The aminorex compounds can be synthesized easily by reacting the appropriate amino alcohol with cyanogen bromide. 4-Methylaminorex is prepared in one step with a greater than 50% yield from phenylpropanolamine (norephedrine). The final product crystallizes easily as the free base. The *cis* and *trans* isomers are formed in a ratio of about 7:1 (63). Aminorex is prepared from 2-amino-1-phenylethanol as a starting material. Synthesis of the oxazolines can be accomplished via a number of synthetic routes; however, they are generally more complicated and result in lower yields.

3.16.4.3 Metabolism

There is limited information on the metabolism of aminorex and 4-methylaminorex; however, their metabolic fate appears to be similar to that of the amphetamines in that they are eliminated primarily in urine as unchanged drug. There may be some minor oxidative deamination to the corresponding oxazolidinone followed by hydroxylation at the para position. Also, some slight conversion of 4-methylaminorex back to phenylpropanolamine may occur. There appears to be no significant glucuronidation (63, 64). Studies in experimental animals suggest that the methyl group of 4-methylaminorex may provide some protection against metabolic attack and yield a metabolic profile somewhat different than that of aminorex (63, 64).

3.16.4.4 Blood and tissue levels

The aminorex derivatives are approximately equipotent with the amphetamines, thus, the general range of concentrations of these drugs in biological fluids are likely to be similar. However, because tolerance develops to both classes of drugs, the range of concentrations encountered is likely to be very large. In the one 4-methylaminorex overdose death reported in the literature, the drug concentration in blood was 21.3 μg/ml and 12.3 μg/ml in urine (50).

3.16.4.5 Analytical considerations

For the analyst, the aminorex derivatives present the same set of problems as do the amphetamines. Both classes of drugs have chemical-physical properties similar to the multitude of over-the-counter preparations and both classes may require derivatization prior to gas chromatography. However, unlike the amphetamines, there are no sensitive and specific immunoassay screening tests.

3.16.4.6 Extraction methods

The aminorex compounds can be isolated from biological fluids and tissues by a variety of extraction schemes. Davis used differential solvent extraction to remove 4-methylaminorex from the urine of an overdose victim (50). Urine was made alkaline with $NaHCO_3$:Na_2CO_3 (3:1), then extracted with hexane:ethylacetate (70:30). The organic layer was removed and extracted with 1 N HCl. The aqueous layer was made alkaline again with Na_2CO_3, then extracted with $CHCl_3$. Smith and Kidwell proposed alkalizing urine with 1.0 N NaOH, then extracting it with chloroform:isopropanol (65). Solid phase extraction has also been used as a extraction and clean-up system (63). C8 cartridges are conditioned with methanol followed by distilled water. The biological sample is applied to the column which is then washed with 10 ml distilled water followed by 10 ml of 7% acetonitrile. The aminorex compounds are then eluted with 2.5 ml of methanol.

3.16.4.7 Screening tests

Thin layer chromatography (TLC)
4-Methylaminorex and its metabolites can be analyzed using the TLC system of O'Brien et al. for identifying a variety of phenylethylamines (66). Extracts from biological fluids are spotted on silica gel thin-layer-plates, then developed with a solvent system containing acetone plus 5 µg/ml concentrated NaOH. Rf values for norephedrine and 4-methylaminorex are 0.59 and 0.16, respectively. Norephedrine reacts with ninhydrin to give a pink color, while 4-methylaminorex reacts with acidified iodoplatinate to produce a purple-brownish color (63).

3.16.4.8 Confirmation tests

Mass spectrometry
Davis and Brewster have described a GC/MS method for analyzing 4-methylaminorex using a gas chromatograph equipped with a 10-µ methyl silicone narrow bore capillary column interfaced with a mass selective detector (50). Smith and Kidwell proposed derivatization with pentafluoropropanyl anhydride prior to GC/MS analysis (65) and reported a limit of detection of 50 ng/ml for their method.

3.16.5 Gamma-Hydroxybutyrate

3.16.5.1 Toxicological importance

Gamma hydroxybutyrate (also known as GHB, sodium oxybate, 4-hydroxy butyrate) is both a natural brain metabolite and a substance with a remarkably diverse spectrum of pharmacological effects. These include: sleep induction (67, 68), increasing brain dopamine and glucose levels (69), stimulating the release of growth hormone and prolactin (70), modulation of cellular metabolism (71) and producing a "high" somewhat similar to ethanol. Proposed medical uses of GHB include treatment of Parkinsonism, ethanol withdrawal, ischemic conditions, and narcolepsy. GHB has been available as an over-the-counter medication for sleep induction and sold in health food stores as an aid to bodybuilding for decades, apparently without untoward effects. However, in 1990, the United States Food and Drug Administration was notified of approximately 70 cases of GHB-related poisonings. These poisonings were the result of illegal manufacture and distribution of GHB products and were characterized by gastrointestinal symptoms, central nervous system and respiratory depression, and seizures (72, 73, 74). Although no deaths were reported, many patients required emergency room treatment including ventilator support. These events prompted the FDA to ban the marketing of GHB in the US. No direct causal relationship has been established definitively for GHB, and interaction with other psychoactive drugs has been suggested as a contributing factor. More epidemiological data and toxicological findings (including blood drug levels) may clarify the picture.

3.16.5.2 Chemistry

GHB is distributed as the sodium salt in powder or tablet form and is usually dissolved in water before ingestion. The lactone form (gamma-hydroxybutyrolactone, GBL) is a clear liquid, more active pharmacologically as a sleep inducer, but has not been identified in illicit samples as yet. GHB has been sold under a variety of common names including Gamma Hydroxybutyric Acid, Sodium Oxybate, Gamma Hydroxybutyrate Sodium, Gamma-OH, 4-Hydroxybutyrate, Gamma Hydrate, and Somatomax PM. Typically, GHB is taken in doses of 1–3 g. The threshold dose for any significant pharmacological effect is about 10 mg/kg (producing amnesia and hypotonia) while doses in excess of 50 mg/kg are considered toxic and will produce increasingly severe respiratory depression, seizure-like activity, and coma (75).

3.16.5.3 Metabolism

Chemical formulae for GHB, GBL, metabolic precursors and probable metabolites are shown in figure 4.

GHB is well absorbed orally and crosses the blood-brain-barrier readily. The primary source for endogenous GHB in brain is considered to be gamma-aminobutyric acid (GABA) (67, 71, 76); however, in peripheral tissues and organs, which have low GABA concentrations, alternate sources such as 1,4-butanediol may exist (77). A possible catabolite of GHB resulting from β oxidation is trans-γ-hydroxycrotonic acid (HCA) (78). Finally, there may be some slight possible interconversion of GHB to the γ-lactone form which may complicate the quantitative analysis of GHB. GHB is metabolized to semi succinate aldehyde, then succinate and eventually to carbon dioxide and water. None of the known GHB metabolites are pharmacologically active.

3.16.5.4 Blood and tissue levels

No deaths, thus no lethal blood levels, have been reported for GHB. Endogenous concentrations of GHB in various mammalian brains range from 1–4 nmoles/g (76). In human brain, endogenous GHB levels are reported to be from 1.8–20 nmoles/g and 0.1 nmoles/ml in human cerebrospinal fluid (79). Van der Pol found

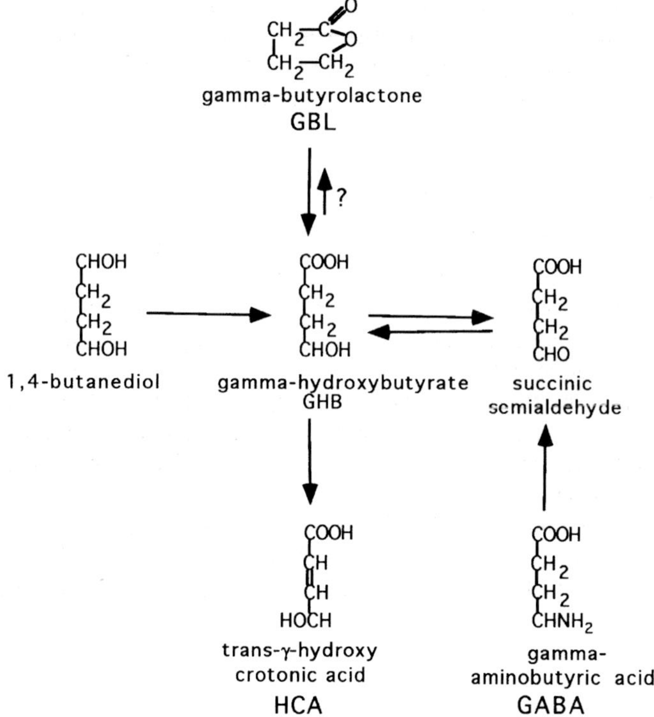

Figure 4. Anabolic and Catabolic Pathways for Gamma-Hydroxybutyrate.

that GHB has a half-life of approximately 1 hour (78 min in the dog) and a volume of distribution of approximately 0.35 l/kg (80). Using these pharmacokinetic values and assuming a typical dose to be approximately 2.5 grams, the toxicologist is likely to encounter GHB concentrations in blood specimens in the order of 100 μg/ml.

3.16.5.5 Analytical considerations

Although GHB is present in relatively high concentrations in biological fluids, the analysis of this compound is still a challenge for the toxicologist. GHB is very polar and water soluble, and therefore, may be difficult to isolate from biological fluids and tissues. Thus, most reported analytical methods involve conversion of GHB to GBL prior to analysis. However, this conversion is not quantitative and there may be some hydrolysis back to the butyrate. The presence of two functional groups on the GHB molecule necessitates a two-step derivatization procedure which is non-quantitative. The low molecular weight of GHB makes it susceptible to interfering artifacts from other small molecular weight compounds present in biological samples. Finally, there is some disagreement as to the accuracy of differential assays for GHB and GBL. GBL is rapidly hydrolyzed to GHB in plasma with a half-life of about 1 min at 37 °C (81) and most studies have found that endogenous levels of GBL are approximately one-tenth those of GHB (82). However, when blood is spiked with GHB alone, significant concentrations of GBL are often found (82). Whether this GBL is formed from metabolic conversion from GHB or is an artifact generated during sample preparation has not been clarified.

3.16.5.6 Extraction methods

GHB is most efficiently isolated from a biological matrix by first converting it to GBL. Typically, a sample is heated in the presence of a strong mineral acid and the lactone extracted with an organic solvent like $CHCl_3$. In one of the first methods proposed for GHB, Giarman and Roth (83) suggested precipitating plasma proteins using an equal volume of 20% trichloroacetic acid, centrifuging, heating the supernatant for 15 min at 80 °C, then extracting with $CHCl_3$. However, even with successive $CHCl_3$ extractions, recovery is only about 60%. To extract GBL to the exclusion of GHB, ice-cold trichloroacetic acid is used for protein precipitation and the heating step is omitted.

A rapid GC/MS method with minimal sample preparation has been proposed in which brain tissue is homogenized with 4 volumes of 0.1 perchloric acid (0 °C), neutralized with $KHCO_3$, centrifuged, then derivatized directly without extraction (78).

3.16.5.7 Screening tests

There are no sensitive or specific screening tests available for either GHB or GBL. Even spectrophotometric methods may be of limited use because they are subject to interference from other endogenous compounds and are not very sensitive.

3.16.5.8 Confirmation tests

Gas Chromatography – flame ionization detection
The first gas chromatography methods for GHB analyzed the drug as the lactone directly without derivatization. In the method of Giarman and Roth (83), GHB was converted to GBL with trichloroacetic acid followed by heating for 15 min. Van der Pol suggested adding concentrated H_2SO_4 and eliminating the heating step (80). Both methods used flame ionization detection, γ-valerolactone as the internal standard, and reported detection limits of approximately 20 µg/ml. A critical review of both of these methods and a discussion of the accuracy of differential assays for GHB and GBL can be found in the communication by Lettieri and Fung (82).

Gas chromatography – electron capture detection
Doherty et al. developed a rather elaborate electron capture gas chromatography method for GHB and GBL in brain tissue (84). This method also involved converting GHB to GBL followed by extraction with $CHCl_3$. Tissue samples were homogenized in 1 N $HClO_4$ and the precipitate removed by centrifugation. The supernatant was heated at 85 °C for 15 min and then extracted with 2 volumes $CHCl_3$. GBL and

δ-valerolactone (the internal standard) were methylated with 14% BF_3 in methanol, then further derivatized with heptafluorobutyric anhydride (HFBA) in pyridine. The reaction mixture was semi-purified by thin-layer chromatography prior to GC analysis on a 24 ft. glass column packed with 3% OV-1 on Cromasorb-W. At a column temperature of 150 °C and a gas (5% methane-95% argon) flow rate of 35 ml/min, the retention time for the GBL derivative was 6.5 min and 12.5 min for the γ-valerolactone internal standard. Overall efficiency of this method was low (only 15% of the starting GBL was derivatized) and large interfering peaks were sometimes encountered; however, this method was used to establish a concentration of GHB in rat brain of approximately 2 nmoles/g.

Mass spectrometry-electron impact

GC methods for GHB using electron impact mass spectrometry have been described by Eli et al. (85) and Doherty et al. (84). The samples were prepared as described above (84) and quantitation was achieved by GC with electron capture detection. Absolute confirmation of the analyte molecule was attempted using mass spectrometry in the electron impact mode, but no molecular ion for derivatized GBL (m/z 314) could be detected. There was a significant ion at m/z 283, however, which corresponds to the loss of an -OCH_3 group. The base peak for GBL and the internal standard was at m/z 169, which corresponds to C_3F_7 and is not specific for these compounds. Thus, one of the limitations of GC/MS analysis using the electron impact mode is that GHB fragments badly and quantitation must be accomplished using small peaks of the spectrum.

Mass spectrometry-negative ion

A negative ion mass spectrometry method which may produce less mass fragmentation (thus more accurate quantitation) has been described by Ehrhardt et al. (78). This method is useful also for the analysis of trans-γ-hydroxycrotonic acid (HCA), a possible GHB catabolite. Rat brain extracts are derivatized first with pentafluorobenzyl bromide (PFB-Br), then silylated with N-methyl N-*tert*-butyldimethylsilyltrifluoroacetamide (MTBSTFA) prior to chromatography on a 25 m OV-1 fused silica capillary column (i.d. 0.32 nm) and analysed using a mass spectrometer modified for negative ion detection with ammonia as the reagent gas. The authors reported that under these mass spectrometric conditions, only one peak occurs in the spectra, which is the anion resulting from the loss of the pentafluorobenzyl group. When deuterated GHB was used as an internal standard, the detection limit was reported to be 5 pg per injection.

References

1. Henderson, G.L. (1983) Blood concentrations of fentanyl and its analogs in overdose victims. Proc. West. Pharm. Soc. *26*, 287–290.
2. Henderson, G.L. (1988) Designer drugs: Past history and future prospects. J. Forensic. Sci. *33*, 569–575.
3. LaBarbera, M. and Wolfe, T. (1983) Characteristics, attitudes and implications of fentanyl use based on reports from self-identified fentanyl users. J. Psychoactive Drugs. *15*, 293–301.
4. Janssen, P.A.J. (1985) The development of the new synthetic narcotics. In: Opioids in

Anesthesia. Estafanous, ed. pp. 37–44. Butterworth Publishers. Boston.

5. Van Bever, W.F.M., Niemegeers, C.J.E., Janssen, P.A.J. (1974) Synthetic Analgesics. Synthesis and pharmacology of the diastereomers of N-[3-methyl-1-(2-phenyethyl)-4-piperidyl]-N-phenylpropanamide and N-[3-methyl-1-(1-methyl-2-phenethyl)-4-piperidyl]-N-phenyl-propanamide. J. Med. Chem. *17*, 1047–1051.
6. Van Bever, W.F.M., Niemeggers, C.J.E., Schellekens, K.H.L. and Janssen, P.A.J. (1976) N-4-substituted 1-(2-arylethyl-4-peperidinyl-N-phenylpropanamieds, a novel series of extremely potent analgesics with unusually high safety margin. Arzneim. Forsch. *26*, 1548–1551.
7. Van Wijngaarden, I. and Soudijn, W. (1968) The metabolism and excretion of the analgesic fentanyl (#4263). Life Science. *7*, 1239–1244.
8. Maruyama, Y. and Hosoya, E. (1969) Studies on the fate of fentanyl. Keio J. Med. *18*, 50–70.
9. Goromaru, T., Matsuura, H., Furuta, S., Baba, N., Yoshimura, T., Miyawaki, T. and Sameshina, T. (1982) Identification of fentanyl metabolites in rat urine by gas chromatography-mass spectroscopy with stable-isotope tracers. Drug Metab. Dispos. *10*, 542–546.
10. Goromaru, T., Matsuura, H., Yoshimura, N., Miyawaki, T., Sameshima, T., Miyao, J., Furuta, T. and Baba, S. (1984) Identification and quantitative determination of fentanyl metabolites in patients by gas chromatography-mass spectrometry. Anesthesiology. *61*, 73–77.
11. Hess, P., Steibler, G. and Herz, A. (1972) Pharmacokinetics of fentanyl in man and rabbit. Eur. J. Clin. Pharmacol. *4*, 137–141.
12. Garriott, J.C., Rodriquez, R. and DiMaio, V.J. (1984) A death from fentanyl overdose. J. Anal. Tox. *8*, 288–289.
13. Gillespie, T.J., Gandolfi, A.J., Davis, T.P. and M.R.A. (1982) Identification and quantification of alpha-methylfentanyl in post mortem specimens. J. Anal. Tox. *6*, 139–142.
14. Matejezyk, R. (1988) Fentanyl related overdose. J. Anal. Tox. *12*, 236–238.
15. Henderson, G.L., Harkey, M.R. and Jones, A.D. (1989) Rapid screening of fentanyl (China White) powder samples by solid-phase radioimmunoassay. J. Anal. Toxicol. *14*, 172–175.
16. Gillespie, T.J., Gandolfi, A.J., Maiorino, R.M. and Vaughn, R.W. (1981) Gas chromatographic determination of fentanyl and its analogs in human plasma. J. Anal. Toxicol. *5*, 133–137.
17. Woestenborghs, R., Michielsen, L. and Heykants, J. (1981) Rapid and sensitive gas chromatographic method for the determination of alfentanil and sufentanil in biological samples. J. Chrom. *224*, 122–127.
18. Henderson, G.L., Frincke, J., Leung, C.Y., Torten, M. and Benjamini, E. (1975) Antibodies to fentanyl. J. Pharmacol. Exp. Ther. *192*, 489–496.
19. Michiels, M., Hendricks, R. and Heykants, J. (1977) A sensitive radioimmunoassay for fentanyl. Eur. J. Clin. Pharmacol. *12*, 153–158.
20. Phipps, J.A., Sabourin, M.A., Buckingham, W. and Strumin, L. (1983) Measurement of plasma fentanyl concentrations: Comparison of three methods. Can. Anaesth. Soc. J. *30*, 162–165.
21. Schleimer, R., Benjamini, E., Eisele, J. and Henderson, G.L. (1978) Pharmacokinetics of fentanyl as determined by radioimmunoassay. Clin. Pharmacol. *23*, 188–194.
22. Schuttler, J. and White, P.F. (1984) Optimization of the radioimmunoassays for measuring fentanyl and alfentanil in human serum. Anesthesiology. *61*, 315–320.
23. Woods, W.E., Tai, H.H., Tai, T., Wood, H., Barios, H., Blake, J.W. and Tobin, T. (1986) High-sensitivity radioimmunoassay screening method for fentanyl. Am. J. Vet. Res. *10*, 2180–2183.
24. Henderson, G.L. (1989) Designer Drugs: The California Experience. In: Clandestinely Produced Drugs, Analogs and Precusors. (Klein, Sapienza, McClain Jr. and Kahn, eds.) pp. 7–20. U.S. Department of Justice, Drug Enforcement Administration. Washington, D.C.
25. Van Rooy, H.H., Vermeulen, N.P.E. and Bovill, J.C. (1981) The assay of fentanyl and its metabolites in plasma of patients using gas chromatography with alkalai flame ionization detection and gas chromatography-mass spectroscopy. J. Chromatog. *223*, 85–93.
26. Weldon, S.T., Perry, D.F., Cork, R.C. and Gandolfi, A.J. (1985) Detection of picogram levels of sufentanil by capillary gas-chromatography. Anesthesiology. *63*, 684–687.
27. Watts, V. and Caplan, Y. (1988) Determination of fentanyl in whole blood at subnanogram concentrations by dual capillary column chromatography with nitrogen-sensitive detectors, and gas chromatography/

mass spectrometry. J. Anal. Tox. *12*, 246–254.

28. Moore, J.M., Allen, A.C., Cooper, D.A. and Carr, S.M. (1986) Determination of fentanyl and related compounds by capillary gas-chromatography with electron capture detection. Anal. Chem. *58*, 1656–1660.
29. Hammargren, W.R. and Henderson, G.L. (1988) Analyzing normetabolites of the fentanyls by gas chromatography/electron capture detection. J. Anal. Tox. *12*, 183–191.
30. Henion, J.D. (1981) Continuous monitoring of total micro LC eluant by direct liquid introduction LC/MS.J. Chromatography. *19*, 57–64.
31. Lurie, I.S., Allen, A.C. and Issag, H.J. (1984) Reversed-phase high-performance liquid chromatographic separation of fentanyl homologues and analogues. I. An optimized isocratic chromatographic system utilizing absorbance ratioing. J. Liq. Chrom. *7*, 463–473.
32. Kumar, K. and Morgan, D.J. (1987) Determination of fentanyl and alfentanil in plasma by high-performance liquid chromatography with ultra violet detection. J. Chrom. *419*, 468–474.
33. Frincke, J.M. and Henderson, G.L. (1981) The major metabolite of fentanyl in the horse. Drug Metab. Dispos. *8*, 425–427.
34. Shen-Nan, L., Tseng-Pu, F.W., Caprioli, R.M. and Mo, B.P.N. (1981) Determination of plasma fentanyl by GC-MS spectrometry and pharmacokinetic analysis. J. Pharm. Sci. *70*, 1276–1278.
35. Cheng, M.T., Kruppa, G.H., McLafferty, F.W. and Cooper, D.A. (1982) Structural information from tandem mass spectrometry for China White and related fentanyl derivatives. Anal. Chem. *54*, 2204–2207.
36. Cooper, D., Jacob, M. and Allen, A. (1986) Identification of fentanyl derivatives. J. Foren. Sci. *31*, 511–528.
37. Shulgin, A.J. and Shulgin, A. (1991) PHIKAL, A Chemical Love Story, Transform Press, Berkeley, CA.
38. Lim, H.K. and Foltz, R.L. (1989) Identification of metabolites of 3,4-(methylenedioxy) methamphetamine in human urine. Chem. Res. Toxicol. *2*, 142–143.
39. Lim, H.K. and Foltz, R.L. (1991) In vivo formation of aromatic hydroxylated metabolites of 3,4-(methylenedioxy) methamphetamine in the rat: identification by ion trap tandem mass spectrometric (MS/MS and MS/MS/MS) techniques. Biol. Mass. Spectrom. *20*, 677–686.
40. Lim, H.K., Zeng, S., Chei, D.M. and Foltz, R.L. (1992) Comparative investigation of disposition of 3,4-(methylenedioxy)methamphetamine (MDMA) in the rat and the mouse by a capillary gas chromatography-mass spectrometry assay based on perfluorotributylamine-enhanced ammonia positive ion chemical ionization. J. Pharm. Biomed. Anal. *10*, 657–665.
41. Lim, H.K., Su, Z. and Foltz, R.L. (1993) Stereoselective disposition: enantioselective quantitation of 3,4-(methylenedioxy) methamphetamine and three of its metabolites by gas chromatography/electron capture negative ion chemical ionization mass spectrometry. Biol. Mass. Spectrom. *22*, 403–411.
42. Cody, J.T. (1990) Detection of D,L-amphetamine, D,L-methamphetamine, and illicit amphetamine analogs using Diagnostig Products Corporation's amphetamine and methamphetamine radioimmunoassay. J. Anal. Toxicol. *14*, 321–324.
43. Cody, J.T. (1990) Cross-reactivity of amphetamine analogues with Roche Abuscreen radioimmunoassay reagents. J. Anal. Toxicol. *14*, 50–53.
44. Cody, J.T. and Schwarzhoff, R. (1993) Fluorescence polarization immunoassay detection of amphetamine, methamphetamine, and illicit amphetamine analogues. J. Anal. Toxicol. *17*, 23–33.
45. Kunsman, G.W., Manno, J.E., Cockerham, K.R. and Manno, B.R. (1990) Application of the Syva EMIT and Abbott TDx amphetamine immunoassays to the detection of 3,4-methylene-dioxymethamphetamine (MDMA) and 3,4-methylene-dioxyethamphetamine (MDEA) in urine. J. Anal. Toxicol. *14*, 149–153.
46. Tedeschi, L., Frison, G., Castagna, F., Giorgetti, R. and Ferrara, S.D. (1993) Simultaneous identification of amphetamine and its derivatives in urine using HPLC-UV. Int. J. Legal Med. *105*, 265–269.
47. Noggle, F., Jr., Clark, C.R., Andurkar, S. and DeRuiter, J. (1991) Methods for the analysis of 1-(3,4-methylenedioxyphenyl)-2-butanamine and N-methyl-1-(3,4-methylenedioxyphenyl)-2-propanamine (MDMA). J. Chromatogr. Sci. *29*, 103–106.
48. Fitzgerald, R.L., Blanke, R.V., Glennon, R.A., Yousif, M.Y., Rosecrans, J.A. and Poklis, A. (1989) Determination of 3,4-methylenedioxyamphetamine and 3,4-methylenedioxymethamphetamine enantiomers in whole blood. J. Chromatogr. *490*, 59–69.

49. Gan, B.K., Baugh, D., Liou, R.H. and Walia, A.S. (1991) Simultaneous analysis of amphetamine, methamphetamine, and 3,4-methylenedioxymethamphetamine (MDMA) in urine samples by solid-phase extraction, derivatization, and gas chromatography/mass spectrometry. J. Forensic Sci. *36*, 1331–1341.
50. Davis, F.T. and Brewster, M.E. (1988) A fatality involving U4Euh, A cyclic derivative of phenylpropanolamine. J. Forensic Sci. *33*, 549–553.
51. Follath, F. Burkart and Schweizer, W. (1971) Drug-induced pulmonary hypertension? Brit. Med. J. *1*, 265–266.
52. Gurtner, H.P. (1985) Aminorex and pulmonary hypertension, A review. Cor Vasa. *27*, 160–171.
53. Kay, J.M., Smith, P. and Heath, D. (1971) Aminorex and the pulmonary circulation. Thorax. *26*, 262–270.
54. Widgren, S. (1977) Pulmonary hypertension related to aminorex intake. Current Topics in Pathology. *64*, 1–64.
55. Yelnosky, J. and Katz, R. (1963) Sympathomimetic actions of *cis*-2-amino-4-methyl-5-phenyl-2-oxazoline. J. Pharmacol. Exp. Therap. *141*, 180–184.
56. Yelnosky, J., Henson, R.J., Mundy, J. and Mitchell, J. (1966) Cardiovascular effects of aminorex, a new anorexigenic agent. Arch. Int. Pharmacodyn. Ther. *164*, 412–418.
57. Roszkowski, A.P. and Kelley, N.M. (1963) A rapid method for assessing drug inhibition of feeding behavior. J. Pharmacol. Exp. Therap. *140*, 367–374.
58. Poos, G.I., Carson, J.R., Rosenau, J.D., Roszkowski, A.P., Kelley, N.M. and McGowin, J. (1963) 2-Amino-5-aryl-2-oxazolines. Potent new anorexic agents. J. Med. Chem. *6*, 266–272.
59. Glennon, R.A. and Misenheimer, B. (1990) Stimulus properties of a new designer drug: 4-Methylaminorex ("U4Euh"). Pharmacol. Biochem. Behav. *35*, 517–521.
60. By, A.W., Dawson, B.A., Lodge, B.A. and Sy, W.W. (1989) Spectral distinction between *cis*- and *trans*-4-methylaminorex. Forensic. Sci. Internatl. *43*, 83–91.
61. Carson, J.R., Poos, G.I. and Almond, H.R.J. (1965) 2-Amino-5-aryl-oxazolines tautomerism, stereochemistry and an unusual reaction. J. Org. Chem. *30*, 2225–2228.
62. Hiltman, R. and Kronesberg, H.G. (1980) Stereochemische Untersuchungen über Arzneimittel, 8. Mitt. phenyl-3-imino-perhydro-3H-oxazolo (3.4-a) pyridine, blutdrucksteigernder Wirksamkeit. Eur. J. Med. Chem. *15*, 111–117.
63. Cheu, Y.T. (1988) The metabolism of 4-methyaminorex in the rat. Master's Thesis. University of California, Davis.
64. Peters, L. and Gourzis, J.T. (1974) Pharmacological, toxicological and clinical observations with aminorex in the United States. In: Adipostas-Kreislauf Anorektika Obesity-Circulation. Blankart, ed. pp. 61–77. Hans Huber Publishers. Vienna.
65. Smith, F.P. and Kidwell, D.A. (1989) Designer Amphetamines-Their Synthesis and Detection. U.S. Naval Research Laboratory Report.
66. O'Brien, B.A., Bonicamp, J.M. and Jones, D.W. (1982) Differentation of amphetamine and its major hallucinogenic derivatives using thin-layer chromatography. J. Anal. Toxicol. *6*, 143–147.
67. Roth, R.H. and Giarman, N.J. (1968) Evidence that central nervous system depression by 1,4-butanediol is mediated through a metabolite of gammahydroxybutyrate. Biochem. Pharmacol. *17*, 735–739.
68. Besseman, S.P. and Skolnik, S. (1964) Gamma-hydroxybutyrate and gamma-butyrolactone concentration in rat tissues during anesthesia. Science. 143–147, 1045–1047.
69. Gessa, G.L., Vargin, L., Crabai, F., Boero, G.C., Caboni, F. and Camba, R. (1966) Selective increase of brain dopamine induced by gammahydroxybutyrate. Life Sci. *5*, 1921–1930.
70. Takahara, J., Yunoki, S., Yakushiji, W., Yamauchi, J., Yamane, Y. and Ofuji, T. (1977) Stimulatory effects of gamma-hydroxybutyric acid on growth hormone and prolactin release in humans. J. Clin. Endocrinol. Metab. *44*, 1014–1017.
71. Mamelak, M. (1989) Gammahydroxybutyrate: andogenous regulator of energy metabolism. Neurosci. Biobehav. Rev. *13*, 187–198.
72. Center for Disease Control (1990) Multistate outbreak of poisonings associated with illicit use of gamma hydroxy butyrate. Morbidity and Mortality Weekly Report. *39*, 861–863.
73. (1991) Gamma hydroxy butyrate poisoning. The Medical Letter. *33*, 8.
74. (1991) From the Centers for Disease Control. Multistate outbreak of poisonings associated with illicit use of gamma hydroxy butyrate. JAMA. *265*, 447–448.
75. (1990) Drug information for the health care professional. USP Dispensing Information. Vol. 1 B, United States Pharmaceutical Convention, Inc.

76. Roth, R.H. and Giarman, N.J. (1970) Natural occurrence of gammahydroxybutyrate in mammalian brain. Biochem. Pharmacol. *19*, 1087–1093.
77. Barker, S.A., Snead, O.C., Poldrugo, F., Liu, C., Fish, F.P. and Settine, R.L. (1985) Identification and quantitation of 1,4-butanediol in mammalian tissues: An alternative biosynthetic pathway for gamma-hydroxybutyric acid. Biochem. Pharmacol. *34*, 1849–1852.
78. Ehrhardt, J.D., Vayer, P. and Maitre, M. (1988) A rapid and sensitive method for the determination of γ-hydroxybutyric acid and trans-γ-hydroxycrotonic acid in rat brain tissue by gas chromatography/mass spectrometry with negative ion detection. Biomed. Environ. Mass. Spectrom. *15*, 521–524.
79. Doherty, D., Hattox, S.E., Snead, O.C. and Roth, R.H. (1978) Identification of endogenous γ-hydroxybutyrate in human and bovine brain and its regional distribution in human, guinea pig and rhesus monkey brain. J. Pharmacol. Exp. Ter. *207*, 130–139.
80. van der Pol, W., van der Klejin, E. and Lauw, M. (1975) Gas chromatographic determination and pharmacokinetics of 4-hydroxybutyrate in dog and mouse. J. Pharmacokinet. Biopharm. *3*, 99–113.
81. Roth, R.H. and Giarman, N.J. (1966) γ-Butyrolactone and γ-hydroxybutyric acid-I. Distribution and metabolism. Biochem. Pharmacol. *15*, 1333–1348.
82. Lettieri, J.T. and Fung, H. (1978) Evaluation and development of gas chromatographic procedures for the determination of γ-hydroxybutyric acid and γ-butyrolactone in plasma. Biochemical Medicine. *20*, 70–80.
83. Giarman, N.J. and Roth, R.H. (1964) Differential estimation of gamma-butyrolactone and gamma-hydroxybutyric acid in rat blood and brain. Science. *145*, 583–584.
84. Doherty, J.D., Snead, O.C. and Roth, R.H. (1975) A sensitive method for quantitation of γ-butyrolactone in brain by electron capture gas chromatography. Anal. Biochem. *69*, 268–277.
85. Eli, M. and Cattabeni, F. (1983) Endogenous γ-hydroxybutyrate in rat brain areas: Post mortem changes and effects of drugs interfering with γ-hydroxybutyric acid metabolism. J. Neurochem. *41*, 524–530.

Part 4
Blood Level Data

4.1 Blood Level Data

D. R. A. Uges

4.1.1 Introductory Remarks

In therapeutic drug monitoring (TDM) and clinical and forensic toxicology, serum concentrations of drugs and toxic compounds in patients are compared with so called reference values.

Ideally, such reference drug data should be found in patients under the same condition and with the same age, illness and medication as the patient involved. Moreover, the clinical effects of our reference group has to be established. Unfortunately there are large intra- and interindividual differences between patients. These differences are even greater between patients and healthy volunteers. Nevertheless, for those who know the relevant pharmacokinetic and pharmacodynamic factors that can influence the serum concentrations of drugs, a list of reliable therapeutic and toxic reference values can be very valuable.

The values which are included in the table were obtained from the literature or found in clinical practice by means of modern methods. The procedures applied are subject to careful quality control and there is consensus among the numerous testers who were involved about the range of concentrations indicated. TDM and clinical toxicology for all compounds mentioned in the table were determined in Dutch hospital laboratories.

Therapeutic levels are steady state concentrations which need to be reached for the drug to exert a significant clinical benefit without causing unacceptable side-effects.

Toxic levels are serum concentrations above which unacceptable, concentration dependent, side- or toxic effects might appear.

It has to be taken into account that these values will never be static and might change with advancing knowledge or with other (therapeutic) use of the compounds.

Use of drug data without sufficient knowledge about the patient, pharmacokinetics, pharmacodynamics and the right interpretation could be dangerous for the patient.

4.1.2 List of Therapeutic and Toxic Reference Values

Explanatory notes to table 1
B = whole blood, heparinized or EDTA, S = serum, U = urine, P = plasma

Reference concentration in mg/l (µg/ml) during steady state.
T = trough level, just before drug administration.
P = peak level, 1–2 hours after drug administration.

Minimum level or range at which concentration dependent side effects or toxic effects were noticed.

– = If no values are given, toxic concentrations are not available.

T = trough level, just before drug administration.

P = peak level, 1–2 hours after drug administration.

Table 1. List of Therapeutic and Toxic Reference Values

Parent Compound metabolite	Re-marks	Mat.	Ref. Concentration Therapeutic mg/l T = trough P = peak	Ref. Concentration Toxic mg/l T = trough P = peak
Acebutolol			0.5–1.25	–
diacetol		S	0.65–4.5	–
Acenocoumarol		S	T 0.03–0.09 P 0.1–0.5	T 0.1–0.15
Acetaldehyde		B	0–30	100–125
Acetazolamide	1	S	4–20	25
Acetone	2	B	(5) 10–20	200–400
Acetylsalicylic acid	3			
salicylic acid	59	S	30–300	400–500 child 300
Acyclovir	4	S	T 0.5–1.5 P 5–15	–
Alimemazine		S	0.05–0.4	0.5
Allobarbital		S	5–10	20–30
Allopurinol			P 1–5	–
oxypurinol		S	5–15	20
Alprazolam		S	0.02–0.06	0.1–0.4
Alprenolol			0.05–0.1	T 0.1 P 1
hydroxyalprenolol		S	0.04–0.065 sum 0.1–0.2	sum 0.25–0.3
Aluminium	5	S	0–0.02	0.1–0.15
Amantadine		S	0.3–0.6	1
Amikacin		S	T 1–4(8) P 15–25(30)	T 10 P 35
4-Aminopyridine		S	0.025–0.075	0.15–0.2
Amiodarone			1–2.5 T 0.5–2	3
desethylamiodarone		S	sum (1–5)	sum 5–8
Amitriptyline			0.05–0.2	–
nortriptyline		S	sum 0.1–0.25	sum 0.5
Amobarbital		S	1–5	10
Amoxicillin	6	S	T 0.5–1 P 5–15	–
Amphetamine		S	0.05–0.15	0.2–1
Amphotericin B		S	T 0.025–1 P 1.5–3.5	T 5–10
Ampicillin	6	S	T 0.02–1 P 2–20	–
Amsacrine	7	S	0.1–0.5 T 0.03 P 0.15–5.5	–
Aprindine		S	0.75–2.5	2
Arsenic	8	B	0.002–0.07	0.1–0.25
Arsenic		U	0–0.1	0.2–1
Atenolol		S	0.2–0.6	2
Azathioprine	9		P 0.05–0.3	–
mercaptopurine		S	0.04–0.3	1–2
Aztreonam		S	T 1–10 P 50–250	–
Barbital		S	10–40	60–80
Biperiden		S	0.05–0.1	–
Bismuth		B	0–0.05	0.1
Bisoprolol		S	0.01–0.06	–
Brallobarbital		S	4–8	10
Bromazepam		S	0.08–0.17	0.25–0.5

Table 1. (continued)

Parent Compound metabolite	Re-marks	Mat.	Ref. Concentration Therapeutic mg/l T = trough P = peak	Ref. Concentration Toxic mg/l T = trough P = peak
Bromide	10	S	0–30 therapeutic 75–100	500–1000
Bromisoval		S	10–20	30–40
Bromperidol	11	S	0.002–0.02	–
Bupivacaine	12	S	0.25–3 P 1–4	4–8
Butobarbital		S	5–15	20
Cadmium		B	0–0.0065	0.015–0.05
Cadmium		U	0–0.005/g creatinine	–
Caffeine	13	S	8–15	20–30 (50)
Camazepam		S	0.1–0.6	2
Carbamazepine	14	S	4.5–9	12–15
Carbamazepine + epoxide			0.5–3	15
Carbon monoxide	15	B	up to 5%	25–30%
Carboplatin free fraction		P		
platinum		S	P 10–25	T 0.1–0.2
Carbromal		S	5–10	15–20
Cefaclor	6	S	oral 13–35; iv to 900	–
Cefotaxime	6	S	T 0.5–2 P 10–50 iv to 225	–
Cefsulodin	6	S	20–100	–
Ceftazidime	6	S	T 20–40 P 50–200	–
Ceftriaxone	6	S	15–75	–
Cefuroxime	6	S	T 0.5–1 P 10–60 iv to 180	–
Cephaloridine	6	S	T 0.5–1 P 10–50	–
Cephamandole	6	S	T 0.5–5 P 10–40	–
Cephradine	6	S	T 0.5–1 P 20–50	–
Chloral hydrate				
trichloroethanol		S	5–15	40–70
Chloramphenicol		S	5–15 T 5–10 P 10–20 (25)	25 T 10
Chlordiazepoxide			0.7–2	3.5–10
demoxepam		S	0.3–2.8	–
Chlorophenoxyacetic acid		S	–	200
Chloroquine		S	0.02–0.4	0.5–1
Chlorpromazine		S	0.05–0.5 child 0.04–0.08	0.5–1
Chlorprothixene		S	0.03–0.3	0.7
Cimetidine		S	0.5–1	1.25
Ciprofloxacin		S	0.4–4 T 0.05–0.5 P 1–5	–
Cisplatin free fraction		P	P 1–5	T 0.1
platinum		S	P 10–25	T 0.1–0.2
Clobazam	16		0.1–0.4	–
N-desmethylclobazam		S	2–4	–
Clomipramine			0.05–0.15	0.4
desmethylclomipramine		S	sum 0.15–0.3	sum 0.4–0.5
Clonazepam		S	0.03–0.06	0.12
Clopenthixol (zu)		S	0.005–0.05 (–0.1)	0.15–0.3
Clorazepic acid	17	S	–	–
nordazepam		S	sum 0.25–0.8	sum 2
Cloxacillin	6	S	5–30 P 85	–
Clozapine	18	S	0.2–0.6 (0.8)	0.8–1.3
Cobalt		B	0.0001–0.0022	–
Cocaïne	19	S	0.05–0.3	0.9
Codeine		S	T 0.01–0.05 P 0.05–0.25	0.3–0.5

Table 1. (continued)

Parent Compound metabolite	Re-marks	Mat.	Ref. Concentration Therapeutic mg/l T = trough P = peak	Ref. Concentration Toxic mg/l T = trough P = peak
Colchicine		S	0.0003–0.0024 P 0.003	0.005
Co-trimoxazole	20	S/U	–	–
Cyanide		B	0.001–0.012(–0.015)	0.5
Cyclizine			0.1–0.25	0.75
norcyclizine		S	0.005–0.025	–
Cyclobarbital		S	5–10	10–15
Cyclosporine A	21	B	T 0.1–0.4	T 0.4–0.5
Cytarabine (Ara C)	22	S	0.05–0.5	–
Dantrolene		S	0.4–1.5 T 0.3–1.4 P 1–3	–
Dapsone		S	0.5–5	10–20
Desipramine		S	0.075–0.25	0.5
Dexfenfluramine		S	0.03–0.06	0.15–0.25
(dextro)Propoxyphene			0.1–0.75	1
norpropoxyphene		S	0.1–0.15	2
Diazepam			0.125–0.75	1.5–5
norazepam	23	S	0.2–0.6(1.8)	2
Diazinon		S	–	0.05–0.1
Diazoxide		S	10–50	50–100
Dibenzepin			T 0.025–0.15 P 0.1–0.25	–
desmethyldibenzepin		S	sum 0.2–0.4	sum 3
Dichlorophenoxyacetic acid		S	–	200
Diclofenac		S	T 0.05–0.5 P 0.1–2.5	–
Dicoumarol		S	8–30	50–70
Digoxin	24	S	T 0.0007–0.0022	T 0.002–0.004
Diltiazem		S	0.05–0.3	0.8
Dimethadione		S	500–1000	1000
Dinitrocresol	25	S	1–5	30–60
Diphenhydramine		S	0.025–0.11	0.2–2
Dipyridamole	26	S	1–2 T 0.1–1	4
Diquat		S/U	–	0.1
Disopyramide	27		2.5–7	
nordisopyramide		S		sum 8–10
Disulfiram		S	0.1–0.3	0.5–5
diethyldithiocarbamate		S	0.3–1.4	
Dosulepin			0.05–0.15(0.4)	
desmethyldosulepine			0.1–0.2	0.75
dosulepine-S-oxide		S	0.04–0.4	0.65–2
Doxepin		S	0.05–0.25	
nordoxepin		S	sum 0.2–0.35	0.5–2
Doxorubicin		S	0.006–0.02	–
Doxycycline		S	5–10	30
Ephedrine		S	0.02–0.1	1
Epirubicin		S	0.01–0.05	–
Erythromycin		S	0.5–6 T 0.5–1 P 4–12	12–15
Ethambutol		S	0.5–5	6–10
Ethanol		B	0–25	1000–2000
Ethosuximide		S	40–100	150
Ethylene glycol		S	–	200
Etomidate		S	0.5–1	
Etoposide		S	T 2–6 P 8–14	–

Table 1. (continued)

Parent Compound metabolite	Re-marks	Mat.	Ref. Concentration Therapeutic mg/l T = trough P = peak	Ref. Concentration Toxic mg/l T = trough P = peak
Fenfluramine		S	0.05–0.15 (0.1–0.12)	0.3–0.7
Flecainide		S	T 0.45–0.9 P 0.75–1.25	T 1–1.5
Flucloxacillin		S	5–15	
Fluconazol		S	T 1.5–15	20
Flucytosine		S	T 25–50 P 50–100	100
Flunitrazepam		S	0.005–0.015	0.05
Fluorine	28	S/U	0.5 T 0.08–0.15	T 0.5–2
5-Fluorouracil	29	S	0.05–0.3	0.4–0.6
Fluoxetine		S	0.15–0.5	–
norfluoxetine	30	S	0.1–0.5	–
Flupentixol	31	S	0.001–0.015	–
Fluphenazine	32	S	0.005–0.015	0.05–0.1
Flurazepam	33		0.0005–0.028	0.15
desalkylflurazepam		S	0.04–0.15	sum 0.2–0.5
Fluvoxamine	30	S	0.05–0.25	–
Ganciclovir		S	0.5–5 T 0.2–1 P 5–12.5	T 3–5 P 20
Gentamicin	34	S	T 0.1–1.5 P 5–10 (15)	T 2 P 12
Glutethimide		S	2–12	20
Gold		S	3–8	10–15
Haloperidol	35	S	0.005–0.04	0.05–0.1
Heptabarbital		S	2–5	10
Heptobarbital		S	50–100	125–150
Hexapropymate		S	2–5	10–20
Hexobarbital		S	4–10	15
Hydrochlorothiazide		S	0.05–0.45	–
Hydroxychloroquine		S	T 0.1–0.4 P 0.5–2.0	–
Hydroxyzine		S	P 0.05–0.09	0.1
Ibuprofen		S	15–30	100
Imipenem		S	T 0.5–2 P 20–75	–
Imipramine			0.045–0.15	0.4–0.5
desipramine		S	0.075–0.25 sum 0.15–0.3	0.5; sum 0.4–0.6
Indomethacin		S	0.8–2.5	4–6
Isoniazid		S	T 0.2–1 P 3–10	20
Isopropyl alcohol		B	–	200–400
acetone		B	–	400
Ketanserin		S	0.015–0.2 P 0.08–1	–
Ketazolam			0.001–0.02	–
nordazepam	23	S	0.2–0.6	1–2
Ketoconazole		S	T 0.3–0.5 P 3–10	–
Labetalol		S	0.025–0.2	0.5–1
Lead	36	B	up to 0.3	0.4–0.45
Levomepromazine		S	0.03–0.15	0.5
Lidocaine (Lignocaine)		S	1.5–5(6)	7–14
monoethylglycinexyliclide (MEGX)		S	0.07–0.175	–
Lithium		S	4–10	14
Lorazepam		S	0.02–0.25	0.3–0.5
Lormetazepam	37	S	0.001–0.01	–

Table 1. (continued)

Parent Compound metabolite	Re-marks	Mat.	Ref. Concentration Therapeutic mg/l T = trough P = peak	Ref. Concentration Toxic mg/l T = trough P = peak
Maprotiline		S	0.075–0.3	0.5
desmethylmaprotiline		S	sum 0.1–0.4	sum 0.75–1
Medazepam		S	0.01–0.15 P 0.1–0.5	0.6
nordazepam		S	0.2–0.6	1–2
MEGX (liver test)		S	T 0.070–0.175	0.05
Mepivacaïne		S	2–4 (5.5)	6–10
Meprobamate		S	10–25	50
Meptazinol		S	0.01–0.1	–
6-Mercaptopurine	38	S	0.03–0.08	1–2
Mercury	39	B/U	0–0.01	0.1–0.3
Mesuximide		S	0.04–0.08	–
N-desmethylmesuximide		S	10–30	40
Methadone	40	S	0.1–0.75 (1.1)	1–2
Methamphetamine		S	0.01–0.05	0.2–1
Methanol		B	–	200
Methaqualone		S	0.5–3	5–8
Methotrexate	41	S	active 0.005	T 0.2 (48 hrs)
7-hydroxymethotrexate	41	S	–	–
Methylenedioxymethyl-amphetamine		S	0.1–0.35	0.5
Metoclopramide		S	0.05–0.13	0.1–0.2
Metoprolol		S	0.05–0.5	0.75–1
Metronidazole		S	10–30	200
Mexiletine		S	1–2 T 0.5–2	1.5–3
Mianserin			0.02–0.09	–
desmethylmianserin		S	sum 0.04–0.125	sum 0.3–0.5
Miconazol		S	0.5–5	–
Midazolam	42	S	0.08–0.25	1–1.5
Moclobemide		S	1 P 2–2.5	–
Morphine	43	S	0.08–0.12	0.15–0.5
Nalidixic acid		S	10–30	50
		U	50–200	
Naproxen	44	S	25–75	–
Netilmicin	45	S	T 1–3 P 6–10(12)	T 4 P 12(15)
Nicotine	46		sum T 0.001–0.275	
cotinine		S	sum P 0.025–0.35	sum 0.3–1
Nitrazepam	47	S	0.03–0.12	0.2–0.5
Nitrofurantoin		S	0.5–2	3–4 (child 2)
		U	10–400	
Nomifensine	48	S	T 0.02–0.06 P 0.2–0.6	0.8–0.1
Nordazepam		S	0.2–0.8 (1.8)	1.5–2
Norfloxacin		S	0.5–5	–
Nortriptyline		S	0.075–0.25	0.5
Ofloxacin		S	1–6 T 0.05–5 P 1–7	–
Opipramol		S	0.05–0.2	0.5–2
Orphenadrine	49	S		
tofenacin		S	sum 0.05–0.2	sum 0.5–1
Oxazepam		S	1–2	3–5
Oxcarbazepine				
hydroxycarbamazepine		S	12–30 (40)	50

Table 1. (continued)

Parent Compound metabolite	Re-marks	Mat.	Ref. Concentration Therapeutic mg/l T = trough P = peak	Ref. Concentration Toxic mg/l T = trough P = peak
Oxprenolol		S	(0.05) 0.1–1	2–3
Oxycodone		S	0.01–0.1	0.2
Papaverine		S	0.2–0.6	–
Paracetamol	50	S	10–20	T 75–100 P 100–150
Paraldehyde	51	S	30–300	400
acetaldehyde		S/B	–	100–125
Paraquat		S/U	–	0.05
Parathion	52	S	–	0.01–0.05
Paroxetine		S	0.01–0.075 P 0.03–0.15	–
Pefloxacin		S	T 0.1–6 P 5–10	25
Penbutolol		S	0.3–0.7	–
Penicillin (benzyl)	6	S	1–12	–
Pentazocine		S	0.05–0.2	1
Pentobarbital	53	S	1–10 (25–40)	10
Perazine		S	0.025–0.1	0.5
Periciazine		S	0.005–0.03	0.1
Perphenazine	54	S	0.0004–0.03	0.05
Pethidine		S	0.2–0.8	1–5
Phenacetin		S	5–20	50
Phenazone	55	S	5–25	50–100
Phenobarbital		S	20–40	60–80
Phenprocoumon		S	1–3	5
Phensuximide		S	4–10 P 10–20	80
Phenylbutazone		S	50–150	250–500
Phenytoine		S	8–18 baby 6–14	20 baby 15
free fraction		S	0.2–2	2
Pimozide		S	0.001–0.02	–
Pindolol		S	0.01–0.07 (0.15)	0.7
Pipamperone		S	0.1–0.4	0.5–0.6
Piperacillin		S	T 1–5 P 20–70	–
Pipothiazine		S	0.001–0.06	0.1
Piroxicam		S	7–20	–
Platinum		S	0.5–5 P 10–30	30 T 10
Prazepam		S	0.01–0.04	–
nordazepam		S	0.2–0.8	1–2
Prazosin		S	0.001–0.075	–
Primidone			T 5–12	15
phenobarbital		S	20–40	60–80
Probenecid	56	S	40–60; 100–150	–
Procaïnamide			4–8	10–15
N-acetylprocaïnamide		S	2–12 sum T 10–30	sum 40
Procaïne		S	5–15	20–40
Prochlorperazine		S	0.01–0.04	0.2–0.3
Promazine		S	0.1–0.4	2–3
Promethazine		S	0.1–0.4	1–2
Propafenone	57		(0.2) 0.4–1.1 (1.6)	1.1–2
norpropafenone		S	0.07–0.7	–
Propericiazine		S	T 0.005 P 0.05	0.1
Propofol		B	2–5	–
Propranolol	58	S	T 0.05–0.15 P 0.1–0.3	1–2

Table 1. (continued)

Parent Compound metabolite	Re-marks	Mat.	Ref. Concentration Therapeutic mg/l T = trough P = peak	Ref. Concentration Toxic mg/l T = trough P = peak
Propylene glycol		S	0.05–0.5	1000–2000
Protionamide		S	T 0.5 P 3–8	–
Protriptyline		S	0.07–0.17 (0.38)	0.5–1
Pyrazinamide		S	30–75	–
Quinidine		S	2.5–5	6–10
Quinine		S	2.5–9.5	10
Ranitidine		S	0.15–0.5	–
Rifabutin		S	T 0.05 P 0.15	–
Rifampicin				
desacetylrifampicine		S	sum T 0.1–1 P 4–10	sum T 2–4 P 12–20
Salicylic acid	59	S	30–300	400–500
Secobarbital		S	1–5	6–10
Silver	60	B	0–0.005	0.06–0.6
Sotalol		S	0.8–5	–
Strychnine		S	–	0.075–0.1
Sulfamethoxazole		S	40–60 (100–200)	200
Sulfonamides		S/U	T 35–75	200
Sulindac		S	0.5–5	–
– sulindac-sulfide		S	1–4	–
Sulpiride		S	0.04–0.6 (P 0.15–0.75)	–
Sultiame		S	1–12	12–15
Temazepam		S	0.3–0.8 T 0.02–0.1	1
Tetracycline		S	5–10 T 1–5	30
Thallium	61	B	–	0.1–0.5
Thalidomide		S	–	–
Theophylline		S	8–18 (20); baby 5–10	20–25; baby 15
Thiazinamium		S	0.05–0.15	0.3
Thiocyanate	62	S	1–12 (30)	35–50 (100)
Thiopental	63	S	1–5 (25–40)	10 (40–50)
pentobarbital	63		5–10	10–15
Thioridazine	64	S	0.2–1	2
mesoridazine			0.3	
sulforidazine		S	sum 0.75–1.5	sum 3
Timolol		S	T 0.005–0.05 P 0.02–0.1	–
Tiotixene		S	T 0.001–0.02 P 0.01–0.025	0.1
Tobramycine	65	S	T 0.5–1.5 P 5–10 (12)	T 2
Tocainide		S	4–10	25
Tofenacine		S	P 0.025–0.1	0.5–1
Tolbutamide		S	60–100	500
Trazodone		S	0.5–2.6 T 0.3–1.5 P 1.5–2.5	4
Trichloroethanol		S	5–15	40–70
Trichlorophenoxyacetic acid		S	–	200
Trifluoperazine		S	0.005–0.05	0.1–0.2
Triflupromazine		S	0.03–0.1	0.3–0.5
Trimethadione			20–40	–
dimethadione		S	500–1000	1000
Trimethoprim		S/U	1.5–2.5 (5–10)	15–20
Trimipramine		S	0.07–0.17 (0.3)	0.5

Table 1. (continued)

Parent Compound metabolite	Re-marks	Mat.	Ref. Concentration Therapeutic mg/l T = trough P = peak	Ref. Concentration Toxic mg/l T = trough P = peak
Valnoctamide		S	5	40
Valproic acid		S	50–100	150
Vancomycine		S	10–20 T 8–12 P 20–35	P 40
Verapamil	66	S	0.09–0.35	0.9
norverapamil		S	sum 0.15–0.6 T 0.05–0.4	sum 1
Vinblastine		S	P 0.25–0.4	–
Vincristine		P	P 0.3–0.4	–
Vinylbital		S	5–10	15
Warfarin		S	1–7 T 0.3–3 P 5–10	10–12
Zidovudine		S	T 0.1–0.3 P 1–1.5	2–3
Zopiclone		S	0.01–0.05 P 0.04–0.07	0.15

Table 2. Remarks to List of Therapeutic and Toxic Reference Values

1 Acetazolamide: for glaucoma 4–5 mg/l.

2 Acetone: Endogenic for diabetes: 5–20 mg/l blood; extreme diabetes: to 600 mg/l; chronic alcoholism: 40–150 mg/l blood; 100–700 mg/l urine; diabetic ketoacidosis: 325–450 mg/l blood; 450–900 mg/l urine; starving: 50 mg/l blood; 2200 mg/l urine; Toxic after acetone sniffing: 200–300 mg/l blood.

3 Acetylsalicylic acid: Directly metabolized to salicylic acid. Acetylsalicylic acid in serum is not stable in vitro.

4 Acyclovir: Orally: trough 0.2–0.5 mg/l; peak 0.5–1 mg/l; ED 50 for sensitive virus: 0.01–2 mg/l.

5 Aluminium: Patients on hemodialysis acceptable to 0.07 mg/l. Avoid contact with glass.

6 Penicillins and cephalosporins: Therapeutic concentrations depend on micro-organism and route of administration. (Instable: freeze at once).

7 Amsacrine: Very instable. Add 1 dr. lactic acid, then stable for 48 h at −20 °C. Therapeutic: 0.1–0.5 mg/l on continuous infusion. Half life length depends on liver function.

8 Arsenic: Toxicity of arsenic strongly depends on compound. As(I) much more toxic than As(V). Toxic urine levels: chronic 0.2–1 mg/l; acute 1 mg/l.

9 Azathioprine: Normally determined as 6-mercaptopurine (50 µl of 1,4-dithiothreitol 1 M in test tube). Mercaptopurine administration by infusion might give much higher peak levels.

10 Bromide: Normally up to 30 mg/L; Therapeutic 75–100 mg/l. With pyridostigmine therapy 50–150 mg/l. Bromism 200 mg/l.

11 Bromperidol: LLQ at lower therapeutic levels.

12 Bupivacaine: epidural anesthesia 0.25–0.75 mg/l.

13 Caffeine: Neonates with apnea 8–15 mg/l. 95% of the coffee drinkers 5 mg/l, and 5% 5–19 mg/l. During doping control acceptable up to 12 mg/l. Toxic 20–30 mg/l, some patients at levels > 50 mg/L.

14 Carbamazepin: Mood improvement: 7–12 mg/l. Coma: 24 mg/l. Carbamazepine levels above 25 mg/l with severe clinical picture: hemoperfusion. Normally carbamazepine-epoxide/carbamazepine ratio = 0.1. After intoxication the epoxide level can increase substantially.

15 Carbon monoxide: normal: 0–5% geriatric patients: 15%; smokers: 8–10%; toxic: 20–30%; lethal: 50% HbCO.

16 Clobazam: Metabolite after about one month on steady state. Only the therapeutic concentration of N-desmethylclobazam is important.

17 Clorazepate: Is only detectable after parenteral administration; changes at once into nordazepam in the acid of the stomach.

18 Clozapine: Adolescents with behavior disorders 0.125–0.4 mg/l. Therapeutically resistent 0.5–0.8 mg/l. At levels above 0.5 mg/l add valproic acid as prophylactic against convulsions. Alcohol potentiates the effect of clozapine.

19 Cocaine: In urine as benzoylecgonine. Detectable in urine after 2–3 days.

20 Co-trimoxazole: trimethoprim: therapeutic: 1.5–2.5 mg/l; toxic: 20 mg/L; sulfamethoxazole: therapeutic: 40–60 mg/L; toxic: 400 mg/l; acetylsulfamethoxazole: toxic: 100–200 mg/L; Pneumocystis carinii: sulfamethoxazole 100–200, trimethoprim 5–10 mg/L.

21 Cyclosporine A: Edetate whole blood

Therapeutic trough:	HPLC	TDx monoclonal
Liver transplantation:		
1–28 days	225–300 μg/L	250–315 μg/l
then	100–150 μg/l	125–200 μg/l
Heart transplantation:		
induction	250–325 μg/l	300–400 μg/l
maintenance	125–175 μg/l	150–250 μg/l
Lung transplantation:		
1– 5 days	375–425 μg/l	375–425 μg/l
6– 9 days	275–350 μg/l	275–350 μg/l
10–13 days	225–275 μg/l	250–300 μg/l
14–21 days	175–225 μg/l	200–275 μg/l
then	125–175 μg/l	150–200 μg/l
Kidney transplantation:		
1st month	175–200	175–200 μg/l
then	100–125	100–150 μg/l
For multiple sclerosis, rheumatism, asthma:		
1st week	150–200 μg/l	150–200 μg/l
after 1 month	100–150 μg/l	125–200 μg/l
Toxic trough: during several days		
	350–400 μg/l	400–500 μg/l

TDx total (with metabolites) conc. 2–2 times HPLC values
toxic: 1250 μg/l in blood.

22 Cytarabine: Add 500 μg tetrahydrouridine (THU) directly to test tube, otherwise cytarabine becomes Ara U. Ara U is the inactive metabolite.

23 Diazepam: Therapeutic status epilepticus: 0.25–0.5 mg/l; anxiolytic: 0.125–0.250 mg/l; eclampsi, tetanus: 1–1.5 mg/l; Metabolite nordiazepam (desmethyldiazepam) after 24 hrs of the same level of diazepam. Effect about the same as diazepam, probably less muscle relaxation.

24 Digoxin: Sample 5–6 hrs after intake. Level of 2–3 μg digoxin, with potassium level < 3.5 mmol/l could potentially be toxic.

25 Dinitro-ortho-cresol DNOC: unsafe: 40–60 mg/l; toxic: 60–80 mg/l; dangerous: 80–100 mg/l; hemoperfusion: 100 mg/l.

26 Dipyridamole: As vasodilator (peak 2.5 hrs after administration) 1–2 mg/l; antithrombolic 1.75 mg/l.

27 Disopyramid: Atrium fibrillation 2.5–3.5 mg/l; ventricular arrhythmia 3.5–7.0 mg/l; Nordisopyramide 20 % active; stronger anticholinergic.

28 Fluorine: Therapeutic concentrations for osteoporosis 0.08–0.15 mg/l after 24 hrs no medication. After 24 hrs the serum level is in equilibrium with the bone fluorine level. If no fluorine therapy levels up to 0.04 mg/l. In acute intoxication very high levels, with no relation to the severity of the case. Avoid any contact to glass.

29 5-Fluorouracil: 0.3–0.6 mg/l hematological (grade 0) and gastrointestinal side effects; 0.8 mg/l hematological (grade 1); 1 mg/l neurological toxicity.

30 Fluoxetine/Fluvoxamine: We have seen high concentration without any toxic effect. High concentration might give strong enzyme inhibition which could cause intoxication by the comedication.

31 Flupentixol: Therapeutic levels near the LLQ.

32 Fluphenazine: Depot injections 0.0025–0.005. This is under LLQ.

33 Flurazepam: sedation: 0.007 mg/l; sleep: 0.012–0.015 mg/l; deep sleep: 0.015 mg/l; metabolite desalkylflurazepam: therapeutic: 0.04–0.15 mg/l; toxic: 0.2 mg/l.

34 Gentamicin: With administration once daily higher peak levels are acceptable (up to 15 mg/l), but then the trough level must be < 0.5 mg/l.
35 Haloperidol: Extrapyrimidal side effects possible at therapeutic concentrations > 0.008 mg/l. Toxic: 0.01–0.05 mg/l. Sometimes haloperidol levels at > 0.05 mg/l might be necessary. LLQ = 0.005 mg/l.
36 Lead: (in heparinized whole blood) adults: up to 0.3 mg/l is acceptable; child: up to 0.25 mg/l is acceptable; acute toxic: 0.45 mg/l mental and physical disfunction; chronic: 0.4 mg/l. Treatment child: 0.5 mg/l; adult: 0.7–1 mg/l.
37 Lormetazepam: Therapeutic level under LLQ.
38 6-Mercaptopurine: With doses azathioprine 0.04–0.3 mg/l; very instable, add DTT.
39 Mercury: (inorganic mercury) Normal: up to 0.01 mg/l whole blood or urine. Without symptoms: 0.02–0.05 mg/l whole blood or urine. Toxicity of organic mercury depends on compound.
40 Methadone: Analgesic 0.1–0.3 mg/l. At methadone program for addicts 0.2–0.75 mg/l. Toxicity 1 mg/l; at first use toxic < 1 mg/l. Addicts could be resistent to levels much higher than 1 mg/l. In urine methadone is detectable for 2–4 days. Serum half life depends on pH urine.
41 Methotrexate: cytotoxic above 5–10 µg/l. Therapeutic: depends on kind of therapy, dose and time interval. Toxic: high dose MTX: 0.5 mg/l after 48 hrs, medium or low dose 0.2 mg/l after 48 hrs, 0.1 mg/l after 72 hrs after start 24 hrs infusion.
42 Midazolam: hypnotic at 0.08 mg/L. Postoperative awake at 0.04 mg/l. Patients with 0.11 mg/l midazolam and 0.03 mg/l hydroxymidazolam start to awaken. Metabolite hydroxymidazolam has short half life and has about 60–80% of the efficacy.
43 Morphine: Detectable in urine with FPIA as opiate (no distinction between morphine, heroin and codeine). Treshold 0.3 mg/l urine. Detectable 1–2 days, with urine at high pH for 4 days. Neonates under artificial breathing therapeutic 0.08–0.12 mg/l. Analgesic 0.04 mg/l. 6-Morphineglucuronide is the active metabolite (is still not measurable by us). For neonates toxic 0.15 mg/l; Adults: 0.15–0.5 mg/l; Addicts: toxic symptims might start at very high levels.
44 Naproxen: Nephrotic syndrome therapeutic: 250 mg/l.
45 Netilmicin: With daily administration high peak levels (up to 15–18 mg/l) acceptable, but then trough level < 1 mg/l. Cystic Fibrosis requires high peak levels up to 12–15 mg/l.
46 Nicotine/cotinine: Screening for smoking: FPIA (TDx) test in urine of saliva. Under 50 µg/l an HPLC or GLC method is available.
Heavy smokers: In the morning before the first cigarette nicotine 0.003–0.005 mg/l. At the end of the day 0.030–0.035 mg/l. In the morning before the first cigarette cotinine 0.175–0.275 mg/l. At the end of the day 0.25–0.34 mg/l.
Light smokers: In the morning before the first cigarette nicotine 0.001–0.002 mg/l. At the end of the day 0.001–0.005 mg/l. In the morning before the first cigarette cotinine 0.001–0.01 mg/l. At the end of the day 0.025–0.05 mg/l.
47 Nitrazepam: Can be reduced in the test tube by micro-organisms. Antiepileptic 0.05–0.12 mg/l; anxiolytic 0.03–0.05 mg/l.
48 Nomifensine: Therapeutic conjugated and unconjugated: Peak: 2–6 mg/l; Trough: 0.2–0.6 mg/l.
49 Orphenadrine: Metabolite nororfenadrine is tophenacine. Sum have the same activity.
50 Paracetamol: concentration calculated at 4 hrs after intake: non severe intoxication: 75–100 mg/l; rather severe intoxication: 100–300 mg/l; severe intoxication: 300 mg/l.
51 Paraldehyde: Sedative: 30–100 mg/l; Hypnotic: 100–300 mg/l; Anticonvulsive: 200–300 mg/l.
52 Parathion: Paraoxon is active metabolite which is very instable in blood; add Edetate to the sample and freeze at once. Determine pseudocholinesterase activity first. Workers, usually in contact with parathion: serum levels of 0.003–0.2 mg/l, without symptoms.
53 Pentobarbital: Decrease intracranial pressure (with artificial breathing) 25–40 mg/l. Also metabolite of thiopental.
54 Perphenazine: Sometimes effective at 0.0006–0.0011 mg/l. At 0.012 mg/l extrapyramidal symptoms. LLQ = 0.002 mg/l.
55 Phenazone: (Antipyrine) As liver function test: Serum half life < 10 hours, enzyme induction; > 10 hours, enzyme inhibition or disease.
56 Probenecid: Inhibition of penicillin excretion 40–60 mg/l; as uricosuric 100–200 mg/l.
57 Propafenone: Antiarrhythmic 0.2–1.1 mg/l; β-blockade > 0.8 mg/l. Dose 3 times higher: serum concentration 6–10 times higher. > 1.1 mg/l increase of neurologic side effects (particularly poor metabolizers).

58 Propranolol: Antihypertensive: 0.025–0.05 (max. 0.1 mg/l) β-adrenergic: 0.3 mg/l maximum reduction of the β-receptor.
59 Salicylic acid: Therapeutic: rheumatism adults: 200–300 mg/l; rheumatism children, trough level: 100–250 mg/l; anti-coagulant: 50–125 mg/l; analgesic, antipyretic: 30–100 mg/l; prostaglandinsynthetase inhibitor: 50–150 mg/l; Toxic: adults: 400–500 mg/l; children 300 mg/l.
60 Silver: Silver sulfadiazine ointment for burns: 0.06–0.6 mg/l.
61 Thallium: Toxic: 0.25–0.5 mg/24 h urine, 0.1–0.5 mg/l whole blood, Treatments up to 0.1 mg/24 h urine. Hemoperfusion: 1 mg/l whole blood (for 48 hrs after ingestion).
62 Thiocyanate: Acceptable: non smokers: 1–4 mg/l; smokers: 3–12 mg/l; nitroprussid infusion: 6–30 mg/l. Toxic: 35–50 mg/l; nitroprusside infusion: 60 mg/l; hemodialyses: 60–100 mg/l; lethal: 200 mg/l.

Indication determination:	Nitroprussidedose µg/kg/min	Number of days SCN > 60 mg/l
normal kidney function:	10	1–2 days
	5	3 days
	< 3	C < 60 mg/l
anuria:	10	1 day
	5	3 days
	3	6 days
	< 1.25	C < 60 mg/l

63 Thiopental: During artificial respiration empty EEG: 25–40 mg/l metabolite: pentobarbital therapeutic: 5–10 mg/l; toxic: 10–15 mg/l.
64 Thioridazine + mesoridazine + sulforidazine together:
Therapeutic: 0.75–1.5 mg/l; Toxic: 3 mg/l
Thioridazine-ringsulfoxide: Toxic: 0.9 mg/l.
65 Tobramycin: With administration once daily higher peak levels (up to 15 mg/l) are acceptable, if trough levels < 0.5 mg/l. Cystic fibrosis requires higher peak levels (up to 12 mg/l).
66 Verapamil: > 1 mg/l: low tension; 1st grade AV block + decompensation. norverapamil 20% activity of verapamil; (add up sum reference values).

References

1. Uges, D.R.A. (1987) Drug Analysis. Therapeutic Drug Monitoring and Clinical Toxicology by Dutch Hospital Pharmacists. Ziekenhuisfarmacie *3* suppl. 1, 73–77.
2. Merkus, F.W.H.M. (1980) The Serum Concentration of Drugs; Clinical relevance, theory and practice. Int. Congress Series 501. Excerpta Medica, Amsterdam.
3. Uges, D.R.A. (1982) Klinisch-farmaceutische en toxicologische bepalingen in Nederlandse Ziekenhuisapotheken. Pharm. Weekblad *117*: 697–710.
4. Uges, D.R.A. (1987) Referentiewarden van klinisch-farmaceutische en toxicologische bepalingen. Pharm. Weekblad *122*: 163–181.
5. Baselt, R.C., Wright, J.A., Cravery, R.H.(1975) Therapeutic and toxic concentrations of more than 100 toxicologically significant drugs in blood plasma or serum; a tabulation. Clin. Chem. *21*: 44–62.
6. Winek, C.L. (1976) Tabulation of therapeutic toxic and lethal concentrations of drugs and chemicals in blood. Clin. Chem. *22*: 832–836.
7. Stead, A.H., Moffal, A.C. (1983) A collection of therapeutic toxic and fatal blood drug concentrations in man. Hum. Toxicol.: 437–464.
8. Uges, D.R.A., Von Clarmann, M. (1990) Orientierende Angaben zu therapeutischen und toxischen Konzentrationen von Arzneimitteln und Giften in Blut, Serum oder Urin. Deutsche Forschungsgemeinschaft Mitteilung XV der Senatskommission für Klinisch-Toxikologische Analytik. VCH Verlagsgesellschaft mbH, Weinheim.

Index